Engineering Mathematics

Wind-induced collapse of oil storage tanks at Haydock, Lancashire, England, in 1967

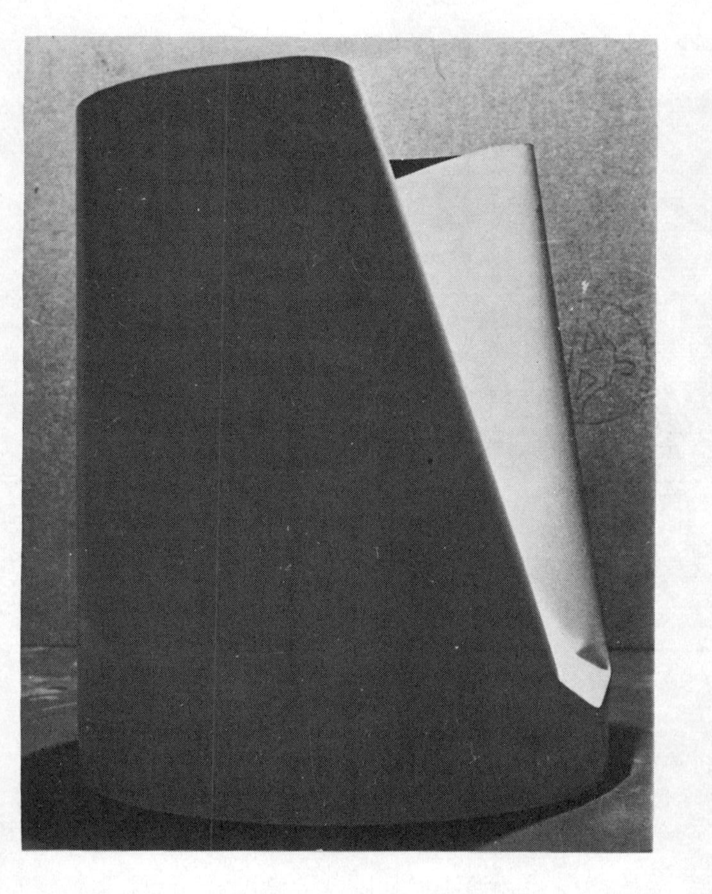

Photographs showing the wind tunnel simulation of the full-scale collapse of the shells in the previous photograph

Engineering Mathematics

A. C. BAJPAI

L. R. MUSTOE

D. WALKER

Loughborough University of Technology

In collaboration with

W. T. Martin

Massachusetts Institute of Technology

JOHN WILEY & SONS

Chichester · New York · Brisbane · Toronto

Library of Congress catalog card number
ISBN 73-21230

ISBN 0 471 04375 3 Cloth bound
ISBN 0 471 04376 1 Paper bound

Reprinted December 1975
Reprinted with corrections October 1977
Reprinted June 1978
Reprinted September 1979
Reprinted September 1980
Reprinted May 1982

PRINTED BY Unwin Brothers Limited,
The Gresham Press, Old Woking, Surrey, England
a member of the Staples Printing Group

Preface

It is well accepted that a good mathematical grounding is essential for all engineers and scientists. This book is designed to provide this grounding, starting from a fairly elementary level and is aimed at first year undergraduate science and engineering students in universities, polytechnics and colleges in all parts of the world. It would also be useful for students preparing for the Council of Engineering Institutions examinations in mathematics at Part 1 standard.

The basic concept of the book is that it should provide a motivation for the student. Thus, wherever possible, a topic is introduced by considering a real example and formulating the mathematical model for the problem; its solution is considered by both analytical and numerical techniques. In this way, it is hoped to integrate the two approaches, whereas most previous texts have regarded the analytical and numerical methods as separate entities. As a consequence, students have failed to realise the possibilities of the different methods or that on occasions a combination of both analytical and numerical techniques is needed. Indeed, in most practical cases met by the engineer and scientist the desired answer is a set of numbers; even if the solution can be obtained completely analytically, the final process is to obtain discrete values from the analytical expression.

The authors believe that some proofs are necessary where basic principles are involved. However, in other cases where it is thought that the proof is too difficult for students at this stage, it has been omitted or only outlined. For the numerical techniques, the approach has been to form a heuristically derived algorithm to illustrate how it is used and a formal justification is given only in the simpler cases.

Where a computer approach is possible, a flow diagram is provided; in the Appendix a summary is given of Fortran IV together with some specimen programs of problems or techniques discussed in the text. Many books are devoted solely to the teaching of computer programming; we therefore do not attempt to cover this topic in detail. It is hoped that students will either have prior knowledge of programming or will be studying this in a parallel course.

Throughout the text there is a generous supply of worked examples which illustrate both the theory and its application. Supplementary problems are provided at the end of each section.

Finally, we sincerely hope that both students and lecturers read the Open Letters which set the ethos and philosophy for the book.

A debt of gratitude to the following is acknowledged with pleasure:

Staff and students of Loughborough University of Technology and other institutions who have participated in the development of this text.

Mr. J. Mountfield of Warrington, Lancashire, for permission to use his photograph of the collapsed oil storage tanks.

Professor D. J. Johns of L.U.T. for providing the photographs of the wind tunnel tests on models of the oil tanks.

John Wiley and Sons for their help and co-operation.

The University of London and the Council of Engineering Institutions for permission to use questions from their past examination papers. (These are denoted by L.U. and C.E.I. respectively).

Mrs. J. Russell for typing the major part of the book.

Contents

Chapter Zero

Open Letters

OPEN LETTER TO STUDENTS

This is a book with a difference. It is different in that we do not seek simply to give you a grounding in those mathematical techniques you will need in your studies. Rather we hope that you will be encouraged to think mathematically. By this, we mean that after following through the text you will be able to look at some practical problem, think about forming the problem in mathematical terms, consider the possible ways of obtaining an answer to this mathematical problem and choose the most suitable way. Then you should be able to find an answer and furthermore interpret this answer in the context of the original problem. This is not to say that we do not place a great emphasis on building up your knowledge and skills in mathematical techniques. It is essential for you to be able to handle and manipulate mathematical formulae and equations: indeed, much of our book is devoted to the development of such mathematical ability. However, we regard this as only one part of the story and at the end of your study we hope that you will not have to ask the usual questions, *What use is all this mathematics?*, *Why do we have to study such abstract ideas?* and *How does this tie in with my other subjects?*

The whole scene is set in Chapter 1 and its sole purpose is to establish a way of thinking. For this reason there will appear to be little conventional mathematics in this chapter but we strongly plead with you to follow it closely. It will not be a waste of time we can assure you, and a careful study of the ideas involved will be repaid in that you will get a feeling for the relevance of the mathematical work and in that you will already be thinking along the right lines. Constant reference to this chapter will be made in the rest of the book and the whole development of the text hinges on a clear understanding of the principles expounded there.

We hope that you will enjoy using this book and wish you good luck in your studies.

OPEN LETTER TO TEACHERS

Perhaps you will have read the preface and the open letter to students. We should like to address a few remarks to you especially. This book is not for the lecturer who is content to approach engineering mathematics in the same way that he was taught. It tries to present an integrated study in two ways. Often the complaint has been made that mathematics is isolated from the engineering subjects and seems to bear little relevance; abstract ideas are studied with little attempt to link them to the engineering world. This book seeks, where possible, to introduce the techniques via practical examples. The second feature we want to emphasise is the welding together of numerical and analytical methods. We feel that the separation is artificial and we have tried to present a problem-oriented view in which the techniques that seem most suitable in a particular situation are employed; in any event, numerical methods, now well-established, often give a solution where the analytical techniques have failed and it is unrealistic to treat them as second-best. Indeed, some topics, for example, the behaviour of sequences, are better approached from a numerical view-point.

You will have noticed that we have asked the students to read carefully Chapter 1. We cannot emphasise the importance of this too strongly as, without so doing they will not really appreciate the flavour of the book. We are relying on your help in this matter. We hope that you will enjoy teaching your course along the lines we advocate; certainly we have found it a challenging and stimulating experience in the years we have taught via this approach.

Chapter One

Why Mathematics?

1.1 THE ROLE OF MATHEMATICS

In your previous mathematical work, the emphasis will have been placed most probably on the development of manipulative skills. The subject will have been broken down into fairly water-tight compartments and often it is not clear how one compartment impinges on the next. Some of the techniques studied may have appeared to be 'tricks' and this impression is encouraged by the fact that the numbers chosen are usually those that make the answers come out easily and exactly. This is not necessarily a criticism of the way you have learned mathematics up to this stage: in order to proceed it was necessary that you should have mastered many basic techniques and acquired much background information; this is the case in any other discipline, be it science, history, geography or languages.

However, the role of mathematics in the study of scientific or engineering problems goes deeper and wider than this. In such problems it may be that one specific answer is required or, more generally, the nature of the relationship between two or more of the variable quantities involved is sought in order that certain deductions may be drawn.

Some Simple Experiments

To clarify ideas, let us discuss four simple experiments.

(i) A uniform beam is simply supported near its ends and a load placed near its centre; the beam will be deflected. We may seek the maximum load that the beam can support before breaking or we may be interested in the profile that the beam adopts. More likely we shall be interested in predicting these features for beams at the design stage.

(ii) A liquid that has been heated and removed from the source of heat will cool; we shall be interested in the rate at which it cools, with a possible view to predicting the time that elapses before a specified temperature is reached.

(iii) If a loaded spring is set in motion, the position of the weight varies about its equilibrium position, sometimes below and sometimes above. We might wish to know the greatest depth the load reaches or how long it will be before the amplitude of the oscillations decreases below a certain amount.

(iv) A simple electrical circuit is set up in which the potential difference across the ends of a conductor is varied and the current passing through it is measured. The object might be to study the relationship between the potential difference and the current or to predict the current which flows through the conductor for a non-observed potential difference.

We can examine this last example more closely. Figure 1.1 below shows a typical set of results from the experiment.

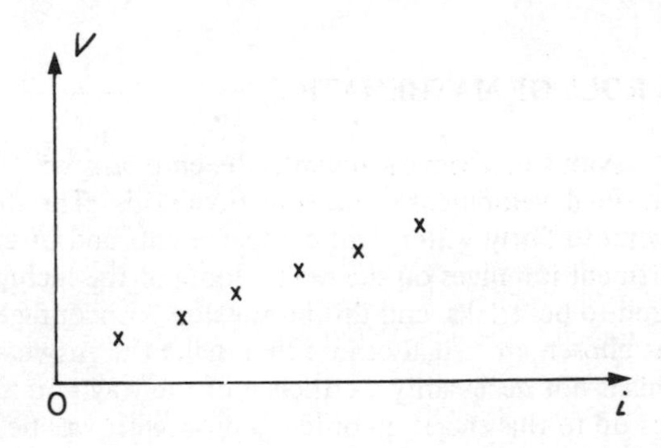

Figure 1.1

We call the potential difference V and the current i. The results suggest a straight line relationship of the form $V = V_0 + Ri$ where V_0 and R are constants. (The values of V_0 and R can be obtained graphically or by using a method employing directly the numerical values of the observations. The danger in using the graphical approach to fitting the straight line is that it is a subjective process and different people might obtain different results. However, the graphical approach does have one advantage sometimes in that a wrong observation can be eliminated from consideration.)

This relationship is the *mathematical model* for the physical situation and is an **empirical** formula, since it is based on experimental data and not on a background theory. However, this model can be used to predict values of i for non-observed values of V and vice-versa, and, provided the model gives reasonable predictions, it will suffice.

These predictions can be made graphically or by using a numerical formula, i.e. a formula which involves the specific numerical values of the observations.

There is little virtue in deducing a relationship $V = V_0 + Ri + 10^{-6} Ri^2$ if the extra term plays no part in the predictions, at least to the accuracy to which we are working. However, we must bear in mind that we can only safely predict *within* the range of observations – a process known as **interpolation**; the dangers attendant on predicting *outside* the range **(extraplation)** are shown in Figure 1.2.

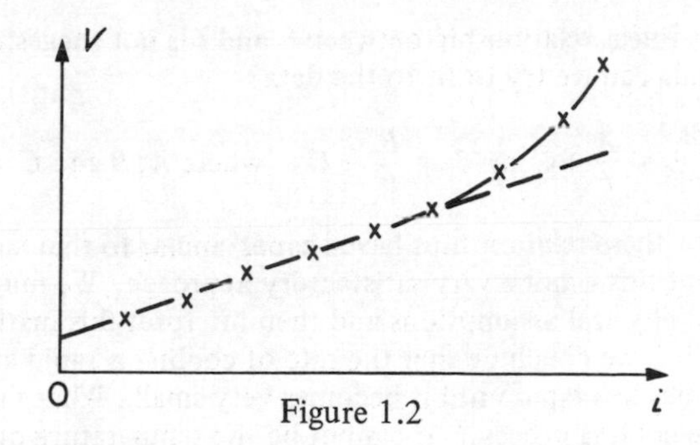

Figure 1.2

If we had an underlying theory based on certain assumptions, for example, concerning the forces on electrons, then we might know the limitations of the predictions from this **theoretical** model. One further point concerning observations is worth making and we illustrate this with the third experiment. The nature of the relationship between the position of the load on the spring, z, relative to its equilibrium position is shown. Had the observations been only those circled, Figure 1.3, a different impression could have emerged if there were no reference to the original problem.

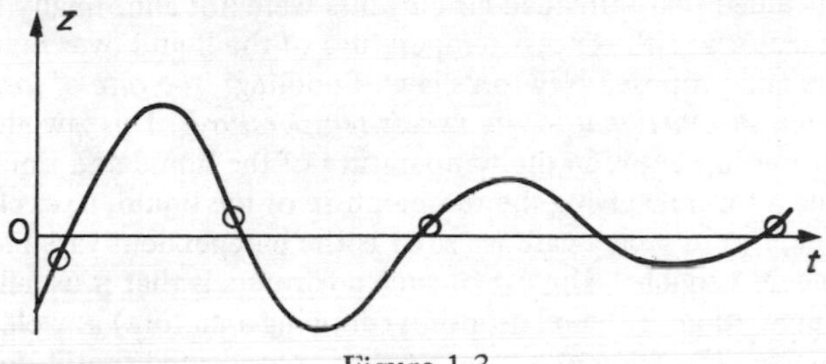

Figure 1.3

Models and Assumptions

We now turn our attention to the cooling liquid. A typical set of observations is shown in Figure 1.4, where θ is the temperature of the liquid at time t.

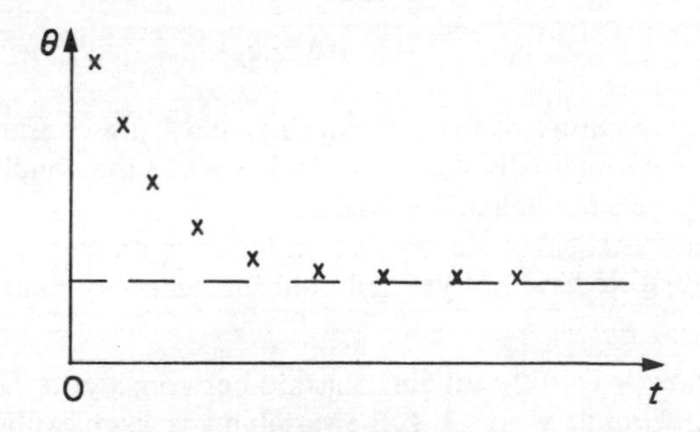

Figure 1.4

It is clear that a linear relationship between θ and t is not suggested. What empirical formula can we try to fit to the data?

How about $\theta = \dfrac{A}{t} + C$ or $\theta = \dfrac{B}{t^2} + C$ where A, B and C are constants?

The graphs of these relationships have shapes similar to that suggested by the observations, but this is not a very satisfactory approach. We must build a **model** based on physical **assumptions** and then interpret this mathematically. From observations we conclude that the rate of cooling is rapid at first and gradually becomes less rapid until it becomes very small. What factors do we think affect the cooling process? It cannot be the temperature of the liquid alone, since if a beaker of liquid helium were placed in a room at average temperature, the liquid would boil away. You will no doubt have remarked that this is hardly an example of a hot liquid; this example was used to emphasise that it is important to specify conditions correctly: by *hot* we mean *hot compared to the surrounding atmosphere.*

What would seem to be relevant is the excess temperature of the liquid over that of its surroundings. But is this all? What about air currents or the shape of the container or the material from which it is constructed? From experiments Newton concluded that, provided air currents were not abnormally large, the dominating factor was the excess temperature of the liquid over that of its surroundings and proposed Newton's law of cooling: *the rate of loss of heat from a liquid is proportional to the excess temperature.* This law gives an implicit relationship between the temperature of the liquid and time. We should like to obtain a formula giving the temperature of the liquid, θ, explicitly in terms of time, t†. In such a case we say t is the **independent variable** and θ is the **dependent variable.** The use of such a formula is that it usually allows a qualitative prediction of the relationship (drawing a picture) as well as a quantitative one. (Sometimes a formula is so complicated that it does not lend itself to easy interpretation either qualitatively or quantitatively and in such cases it is probably best to estimate particular values of the dependent variable at specified values of the independent variable by a method which sets out to do precisely this.)

The mathematical statement of Newton's Law is

$$\frac{d\theta}{dt} = -k(\theta - \theta_s)$$

where θ_s is the temperature of the surroundings and k is a constant for the liquid which has to be experimentally determined. If we add the condition that at $t = 0$, $\theta = \theta_0$, then we have a mathematical model.

From this mathematical model we can, in fact, obtain an expression for θ in terms of t, although we have not yet acquired the necessary mathematical know-

† In the formula $x^2 + y^3 = 2y$ the relationship between x and y is **implicit**, whereas in the formula $y = x^2 + 2$, the variable y is given **explicitly** in terms of x.

ledge. We shall study the techniques required later, and shall return to a discussion of this problem in Chapter 9.

It has been found that this model does compare reasonably well with observation, and gives a good representation of the temperature variation. As we have taken a very simple model, it is worth asking whether the addition of other factors might give a better representation. The difficulty is how to allow for these other factors in the formulation of a new model and we should bear in mind that the resulting model might be too complicated to yield a useful formula.

The **block diagram** below, Figure 1.5, shows the simplified process we have followed so far:

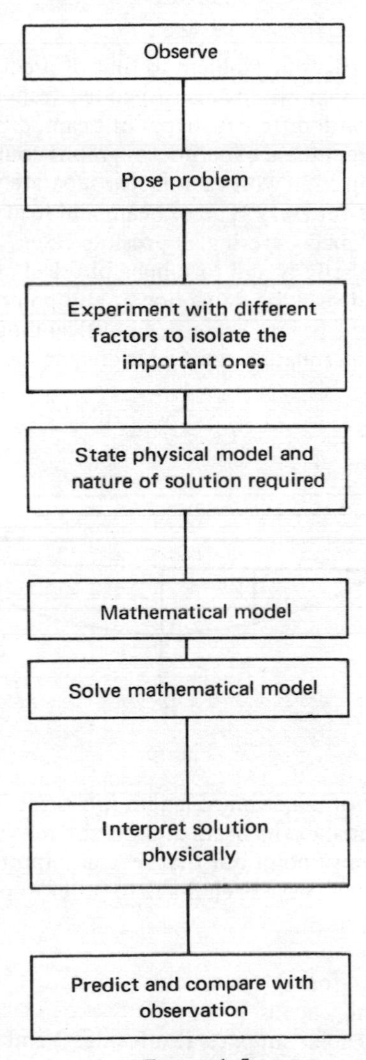

Figure 1.5

Let us now examine the problem of the loaded beam (Figure 1.6) with a view to studying more closely the assumptions made in building a model.

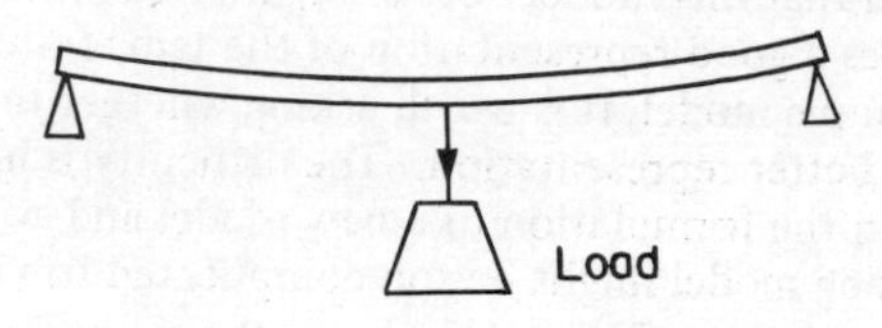

Figure 1.6

Suppose that we restrict our problem to that of predicting how far a given beam will deflect when a known load is suspended from its mid-point. We then carry out experiments with different shapes of beam, different beam materials and different loads. From these experiments graphs could be plotted showing the variation of the deflection with each factor separately. It is impossible in practice to draw graphs for every type of beam and load; therefore we try to find a mathematical formula covering all possible cases. This will involve going back to the theory of elasticity and forming a physical model which states an implicit relationship between the deflection at any point of the beam and certain physical properties of the beam. From this physical model represented in Figure 1.7, the following equation can be produced:

$$EI\frac{d^4y}{dx^4} = \frac{W}{l}$$

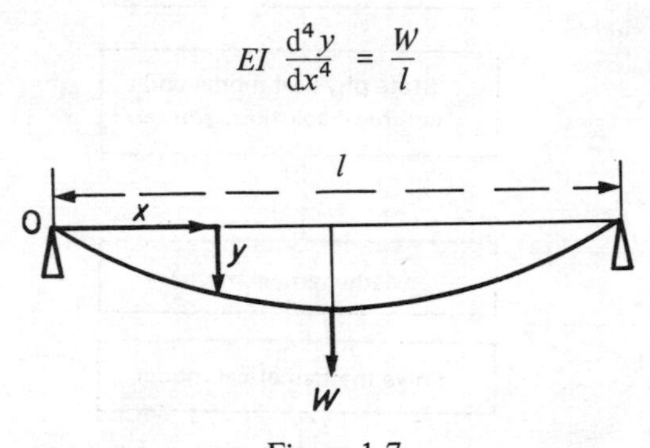

Figure 1.7

where W is the weight of the beam of length l, E is the Young's modulus of elasticity for the material of the beam and I is the relevant moment of inertia of its cross-section. We must point out that several important simplifying assumptions, given below, have been made in order to produce this model:

(i) the deflection is small
(ii) the beam is thin
(iii) the beam is of uniform cross-section
(iv) the beam is homogeneous
(v) the supports are point supports (knife edges) and are exactly at the ends of the beam.

The equation can be solved to give an explicit formula for y in terms of x (and E, I, W and l) so that we may predict the deflection at any point of a given beam with a given central load.

It has been found by engineers that, within the limits of the assumptions governing the model, the predictions are borne out fairly well in practice. But you should by now be asking some of the following questions:

(i) How small is a small deflection?
(ii) What is 'thin'? How can we cater for beams which are not 'thin'?
(iii) Can we take into account beams which are of non-uniform cross-section and which are non-homogeneous?
(iv) Does it matter if the supports are not knife edges and are not exactly at the ends of the beam?

There are two kinds of question here. The first kind involves a careful definition of the limitations of the model and the second is concerned with generalising the model to make it more widely applicable (e.g. non-uniform cross-section). One way of testing for the first is to check with experiment or subsequent observations: this approach could have embarrassing results. A second way is not to make the simplifying assumption of, for example, thinness, which is really in the realms of the second kind of question, i.e. generalising the model. Then one can look at the generalised solution and see how the thickness of the beam enters into it and under what circumstances its effects can be ignored.

All this assumes that a neat solution to the mathematical model exists. This is by no means always the case. It is time to consider an example which illustrates the need for a revision of the block diagram.

The Pitot Tube in an Air Speed Indicator: Sophistication of a Mathematical Model

We choose this example to show how a model can progressively be made more sophisticated to obtain better results, but where the question may be asked: *Is this extra sophistication a worthwhile exercise?*

You may not have met the *Pitot Tube* before – it is a commonly used device for measuring the speed of an aircraft or a body moving in a fluid. When the speed of the aircraft is constant, the problem of measuring its speed forward into still air is equivalent to measuring the speed of air past a stationary aircraft. That is, we imagine the air to be flowing past the body at rest rather than the body moving and the air at rest.

The pitot tube, as shown in Figure 1.8, consists of a slender tubular body aligned with the stream of air whose speed is to be measured. It is of such a shape that it does not interfere unduly with the airflow and therefore at some distance downstream from the nose of the tube the speed of the air will be almost the same as the speed upstream and the pressure will also be practically that of the free stream. At the nose, the stream meets the tube and is brought to rest so that there is a build-up of pressure and we get a pressure p_1 inside the tube at A, called the **stagnation pressure.** The pressure p just outside the tube at A will be equal to the pressure at B, which is measured at the static opening and is called the **static pressure.** The difference in pressures between A and B is

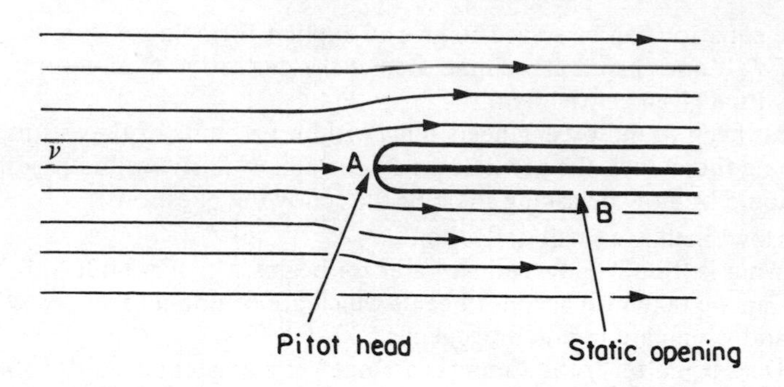

Figure 1.8

$(p_1 - p)$, which can be measured using a manometer. This difference in pressure is used to calculate the speed of the air stream which is usually read off automatically on an air speed indicator (differential pressure gauge).

The most simple model of airflow is one in which the air is assumed to be incompressible. This model gives rise to the energy equation

$$p + \frac{1}{2}\rho v^2 = p_1$$

where ρ is the density of the air and v its speed. Solving this equation, we obtain

$$v = \sqrt{\frac{2(p_1 - p)}{\rho}} \tag{1.1}$$

Air speed indicators for some low-speed commercial air planes are calibrated according to this equation using a value for ρ of 0.002378 slugs/ft^3.† (The density of air is taken to be that for a standard atmosphere at sea-level.)

Is this a good formula and how accurate is it? Well, first of all, since the density used is that of the density of air at sea-level we see that this formula will only be correct under sea-level conditions. It is well-known that the air gets *thinner* the higher we go above the earth, which means that ρ decreases as the altitude increases. Therefore the true air speed at altitude from this formula is

$$v = \sqrt{\frac{2(p_1 - p)}{\rho_{\text{altitude}}}} = \sqrt{\frac{2(p_1 - p)}{\rho_{\text{sea-level}}}} \cdot \sqrt{\frac{\rho_{\text{sea-level}}}{\rho_{\text{altitude}}}} \tag{1.2}$$

At 20 000ft the density is roughly 0.001266 slugs/ft^3.

† A slug is the unit of mass corresponding to a force of one pound weight in the same way that 1 kg corresponds to 1 Newton; it is equivalent to 14.5939 kg.

Hence the error factor at 20000 ft is

$$\sqrt{\frac{\rho_{\text{sea-level}}}{\rho_{\text{altitude}}}} = \sqrt{\frac{0.002378}{0.001266}}$$

$$= 1.37 \quad \text{(3 significant figures)}$$

An air speed indicator which is calibrated using (1.1) will always give an air speed which is too low.

A second consideration is that equation (1.1) assumes that the air is incompressible and this is not true. Compressibility does not have an appreciable effect below about 200 miles/hour, but at higher speeds we should use the *adiabatic* pressure-density law p/ρ^γ = constant, with $\gamma = 1.4$, which gives rise to the new model equation

$$\frac{p}{\rho} + \frac{(\gamma-1)}{\gamma}\frac{v^2}{2} = \frac{p_1}{\rho_1}$$

where ρ_1 is the density just inside the pitot tube at A.
Solving this for v, we get

$$v^2 = \frac{2\gamma}{\gamma-1}\left[\frac{p_1}{\rho_1} - \frac{p}{\rho}\right]$$

and since

$$\frac{p_1}{\rho_1{}^\gamma} = \frac{p}{\rho^\gamma}$$

then

$$\rho_1 = \rho\left(\frac{p_1}{p}\right)^{1/\gamma}$$

therefore

$$v^2 = \frac{2\gamma}{\gamma-1}\left[\frac{p_1}{\rho}\left(\frac{p}{p_1}\right)^{1/\gamma} - \frac{p}{\rho}\right]$$

$$= \frac{2\gamma}{\gamma-1}\,\frac{p^{(1/\gamma)}}{\rho}\left[p_1{}^{(\gamma-1)/\gamma} - p^{(\gamma-1)/\gamma}\right]$$

thus

$$v = \sqrt{\frac{2\gamma}{\gamma-1}\,\frac{p^{(1/\gamma)}}{\rho}\left[p_1{}^{(\gamma-1)/\gamma} - p^{(\gamma-1)/\gamma}\right]}$$

and with $\gamma = 1.4$

$$v = \sqrt{7 \frac{p^{5/7}}{\rho} \left[p_1{}^{2/7} - p^{2/7} \right]} \tag{1.3}$$

Compare equations (1.1) and (1.3) and we now see the difficulty introduced by making our model more sophisticated. We can easily measure $(p_1 - p)$ with a manometer, but to measure $(p_1{}^{2/7} - p^{2/7})$ is very much more difficult, and is the error introduced by using the simpler equation (1.1) appreciable at sea-level?

It can be shown that there is a 5% error at 480 mph which increases to 10% at about 675 mph.

Thus, the model has become very sophisticated in order to gain an accuracy of about 5% at 500 mph and it is a matter of judgement (and usually expense) as to whether the extra sophistication is worthwhile.

We have not told you the whole story, because in modern high speed aircraft the Mach number†, M, is measured rather than the air speed and it can be shown that

$$M^2 = \frac{2}{\gamma - 1}\left[\left(\frac{p_1 - p}{p} + 1\right)^{(\gamma - 1)/\gamma} + 1\right]$$

which is used for calibration of the Machmeter with the quantity $(p_1 - p)$ being measured directly by the instrument. Again, an error is introduced in this model by taking p in the denominator of the term $(p_1 - p)/p$ to have its value at sea-level.

We can expand these ideas in a block diagram, Figure 1.9, which illustrates the way in which the modelling process works in general.

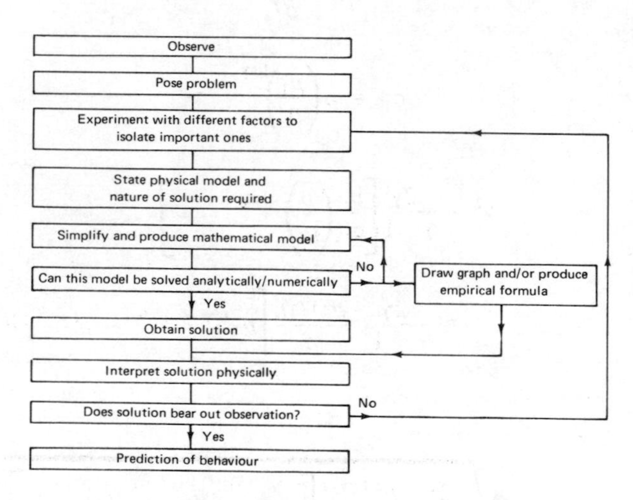

Figure 1.9

† The Mach number is the ratio of the speed of the aircraft to the local speed of sound.

1.2 SOLUTIONS TO MATHEMATICAL MODELS

Sometimes a mathematical model can be solved exactly in the sense that it yields either a set of values for the dependent variables involved or an accurate formula: by accurate, we mean as far as the model is concerned. Sometimes, however, the model does not yield this information because the necessary techniques of solution have not been developed. If simplification of the model yields results which are far from those found in practice, then we must look again at our model to see whether we can squeeze information out of it by different means.

In the first category, where we seek one answer or set of answers, sometimes an algebraic formula for solution exists as is the case with a quadratic equation: but such a formula is not always possible, e.g. $x^5 - 7x^2 + 3 = 0$; one method of solving this equation would be to draw a graph of $x^5 - 7x^2 + 3$ and find where this crosses the x-axis; alternatively, we can obtain successively better approximations to the correct answers by a so-called iterative technique which we shall discuss in Chapter 3.

In the second category, where the ideal would be a formula expressing the dependent variable explicitly in terms of the independent variables, the possibilities open are as shown in Figure 1.10.

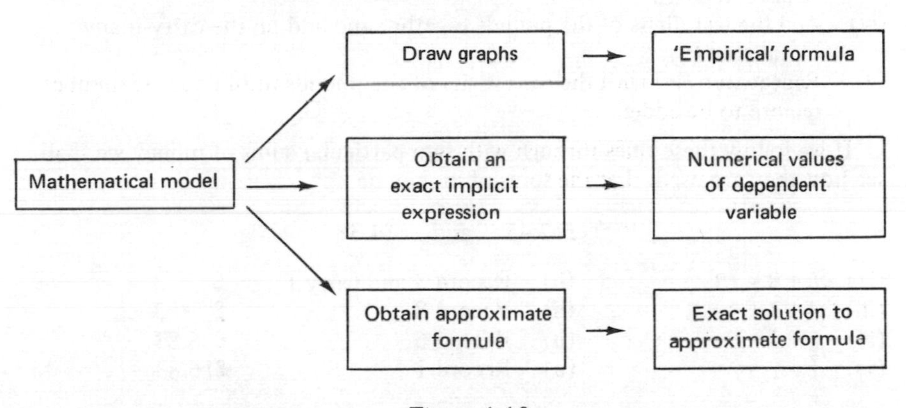

Figure 1.10

Any one of these approaches provides information about the dependent variable and is regarded by us as a solution to the model, albeit an *approximate solution.* In any event, an exact formula may still have to yield numerical values for purposes of prediction and it is not always possible to get an exact number from the formula. For example, if the formula for the beam problem turned out to involve $\sin x/l$, then the precision of our answer would be limited by, amongst other things, the accuracy of a table of sines. With approximate methods, we need some estimate of their accuracy, and, in any event, we need to discuss possible errors in the prediction of a solution to a problem.

1.3 ALGORITHMS AND FLOW CHARTS

We have already introduced some simple flow charts to enable us to discuss the procedure of model-making and we shall now deal with the concepts involved.

We first develop the idea of an **algorithm.** This is a systematic, unambiguous procedure for producing the answer to a problem (or indicating that no answer can be obtained) in a finite number of steps. It should be noted that this is an imprecise definition and much research has been done on making a more watertight definition; for our purposes, however, it will suffice and we shall merely illustrate our definition by examples.

Example 1

An example where an algorithm is used without most people realising that it underlies the set of rules they follow, is in the addition of two sums of money. We might express such rules as:

(i) Add the second digits of the pence together;
(ii) If this sum is less than 10, record it, and proceed to step (iii) or step (iv) as appropriate; if not, record the second digit and make a note to 'carry 1'.
(iii) Add the first digits of the pence together and add on the carry if required. Repeat step (ii).
(iv) Add the last digits of the pounds together and add on the carry if any; repeat step (ii).
(v) Repeat step (iv) with the next digits of the pounds until no more numbers require to be added.

If we follow these rules through with two particular sums of money we shall see how they are used. Let the sums of money be

£12.45 and £4.38

(i)	5 + 8 = 13	(ii)	Record 3 and carry 1	£ . 3
(iii)	4 + 3 + 1 = 8	(ii)	Record 8	£ .83
(iv)	2 + 4 = 6	(ii)	Record 6	£ 6.83
(v)	1 = 1	(ii)	Record 1	£16.83

STOP

Note that we have not covered the possibility of one or both of the sums containing ½p and you should endeavour to modify the set of rules to incorporate this feature. This set of rules involves conditional clauses of the form *if* ...... *then* ...... *otherwise* ...... and seems to allow flexibility.

Example 2

A second example is the sorting of a hand of 13 cards into the order clubs, diamonds, hearts, spades, where the order in each suit ascends 2 to ace. One possible method of sorting might be:

(i) Take the cards one by one until a club is encountered; if so, place it on one side, if not, then go to next suit in the sequence.
(ii) Repeat until all 13 cards have been examined.
(iii) Sort the clubs into ascending order.
(iv) Repeat steps (i) to (iii) for diamonds.
(v) Repeat steps (i) to (iii) for hearts.
(vi) Repeat steps (i) to (iii) for spades.

We have not really explained how to perform step (iii) and this could be incorporated into a full algorithm.

Problems I

1. Write an algorithm for multiplying together two decimal numbers.
2. Write an algorithm for sorting cards of one suit into ascending order.

We see that, even for simple problems, the written description of an algorithm can be complicated. A **flow chart** presents a semi-pictorial representation which acts as a more efficient means of communication. It plays a similar role to a blue-print. In constructing a flow chart we use several shapes of boxes connected by lines.

The *beginning* and *end* of the sequence of statements under consideration are indicated by an oval box:

Statements of intent, i.e. executable statements are indicated by a rectangular box:

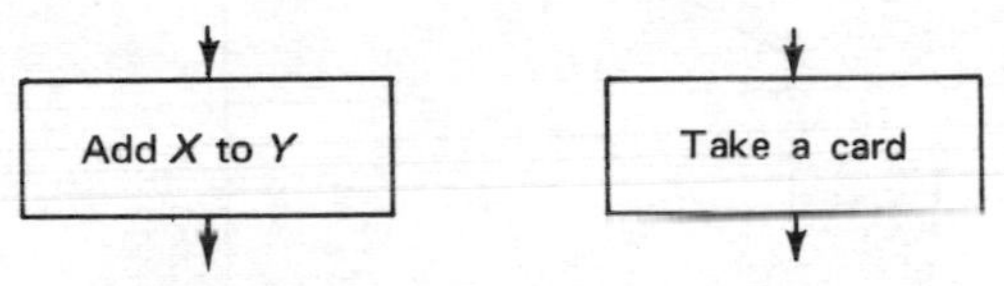

Decisions are indicated by a lozenge- or diamond-shaped box:

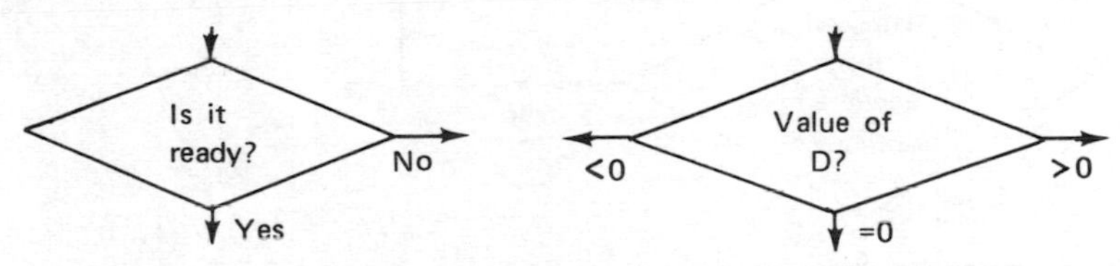

For their use in explaining computer programs, two other symbols are employed: the boxes for *input* and *output*

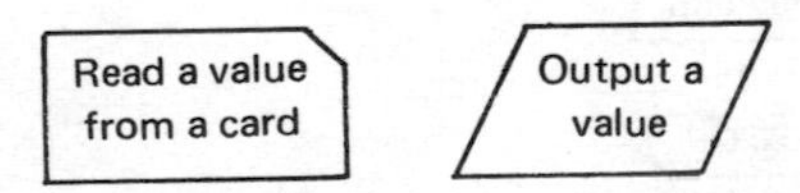

Conventions for these symbols differ, but we shall stick to the above. We now consider some examples of flow charts.

Example 3

Making a Telephone Call

Suppose we wish to call another person on a dialled telephone. The sequence of events would be as shown in Figure 1.11.

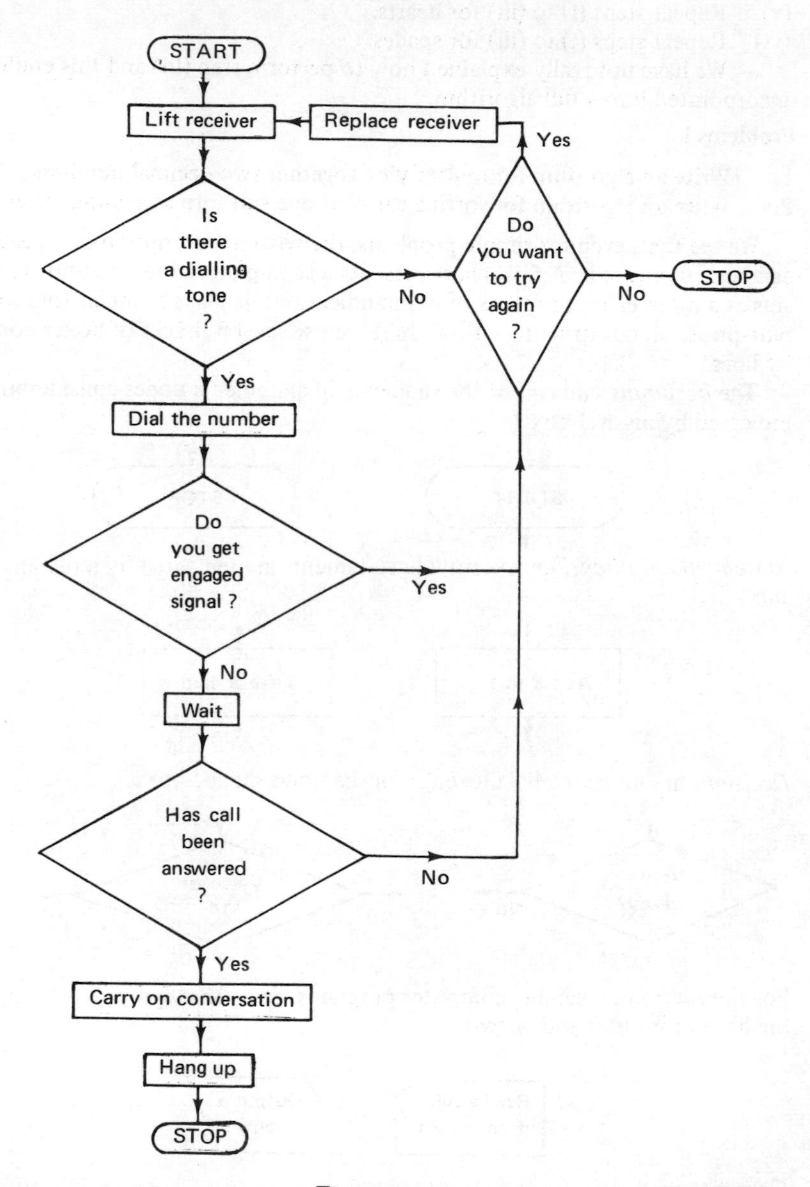

Figure 1.11

Example 4

Taking out a Library Book

A second every-day example is the taking out of a library book. A flow chart to cover the sequence of events is shown in Figure 1.12.

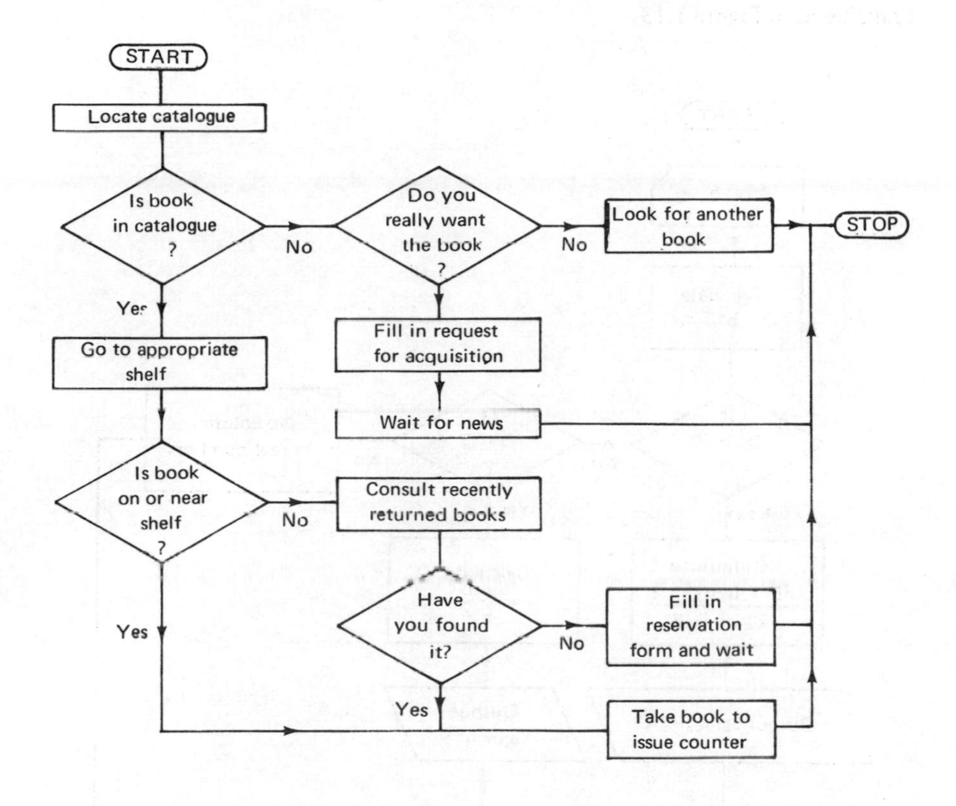

Figure 1.12

Problems II

Now try and produce a flow chart for entering a parked car and driving away.

Example 5

It is now time to consider a mathematical problem — you probably know how to solve the quadratic equation $x^2 + bx + c = 0$ using the formula

$$x = \frac{-b \pm \sqrt{b^2 - 4c}}{2}$$

We wish to write out a logical flow chart of operations to give the solution of this quadratic equation. The crux of the use of the formula is whether the discriminant $D = (b^2 - 4c)$ is positive, zero or negative and we shall have three branches as in Figure 1.13.

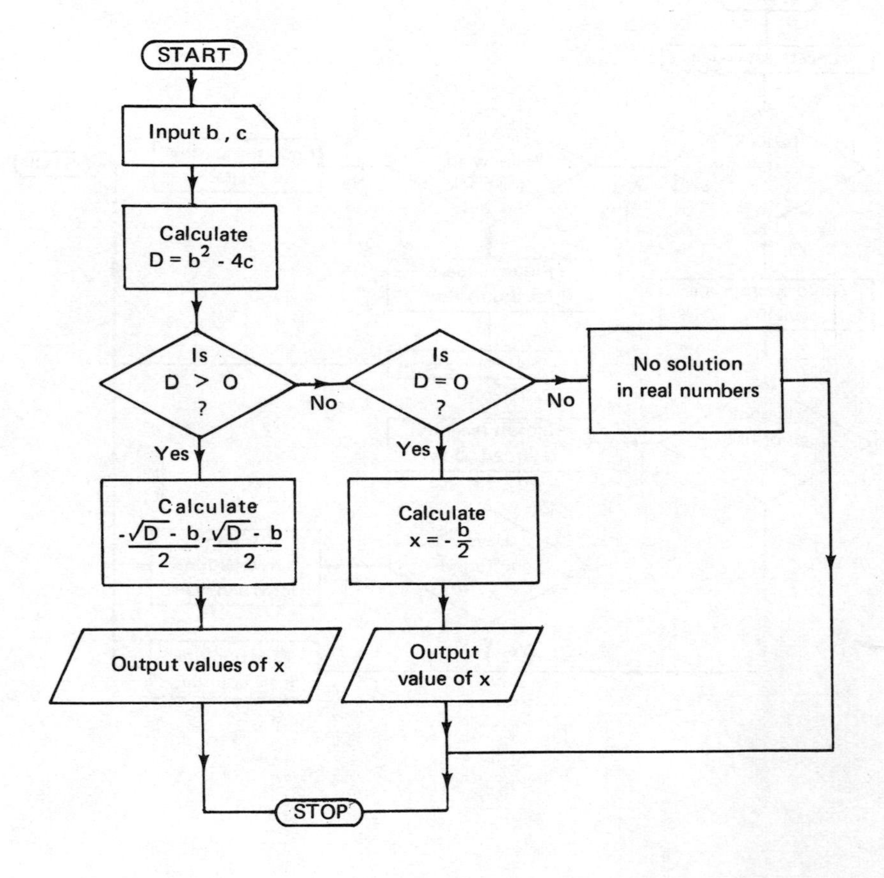

Figure 1.13

We could deal with the branching in one go and we reproduce in Figure 1.14 the relevant portion of flow chart.

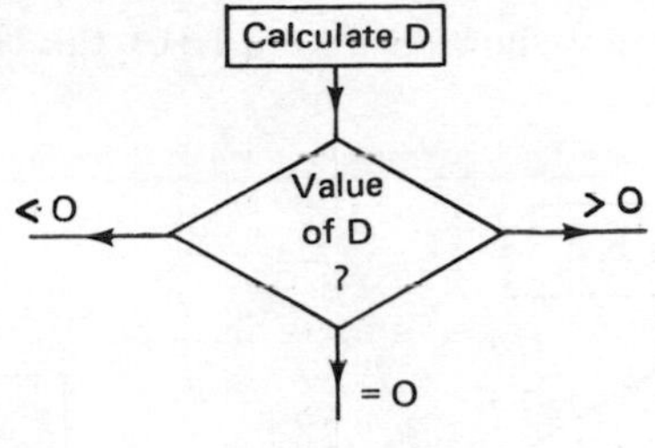

Figure 1.14

Example 6

A problem which appears simple, but which needs careful thought before programming on a computer is that of placing a set of three numbers in ascending order.

Follow through the chart, Figure 1.15, with several sets of three numbers, performing the calculations as you go; this kind of procedure is called a *dry run.*

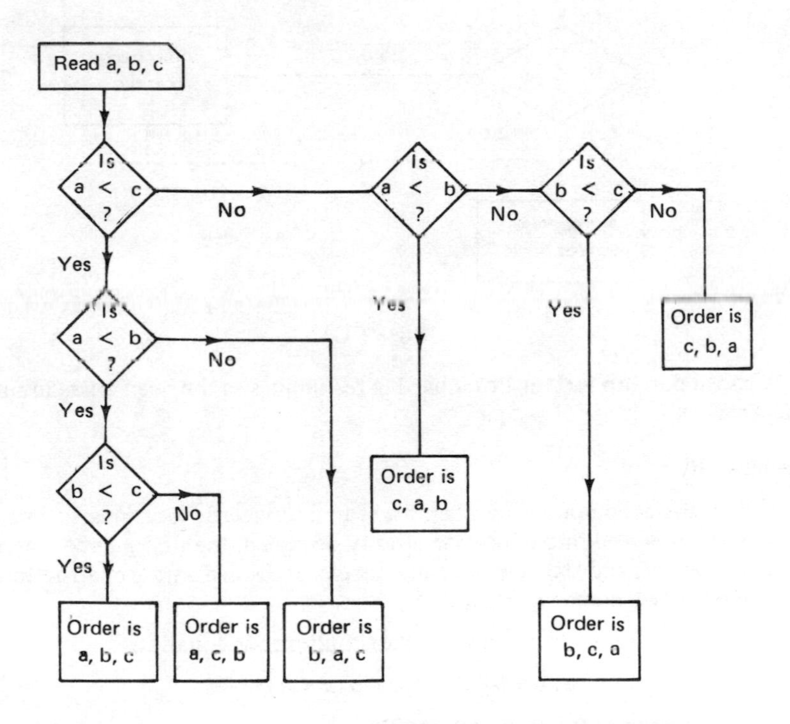

Figure 1.15

An alternative lay-out is now shown in Figure 1.16: this is particularly suitable for a computer program.

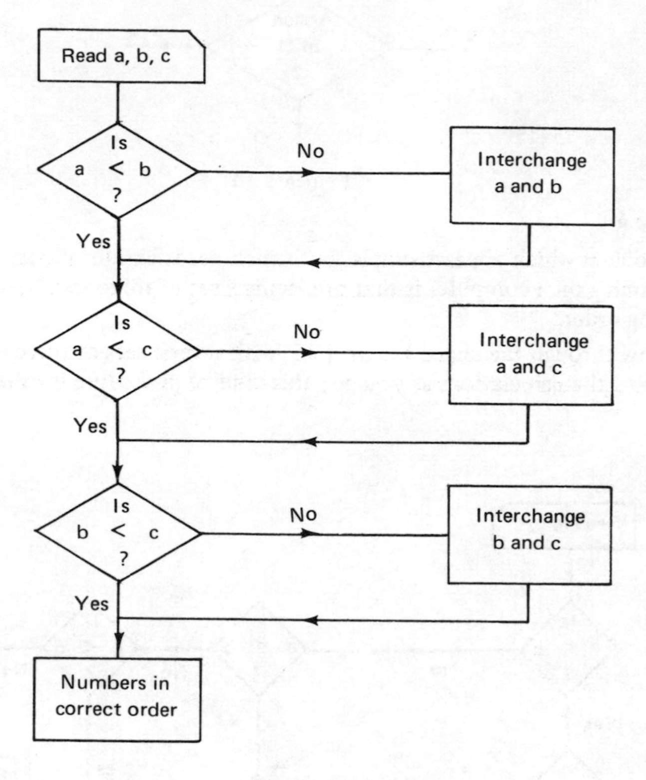

Figure 1.16

We shall develop further flow charting techniques as the need arises in further chapters.

Problems III

1. In a digital computer, numbers are stored in specific locations; if one number is read into a location already occupied, the first number there is obliterated and the new one takes its place. Devise a flow chart to interchange two numbers, a and b.

2. Construct a flow chart to solve the simultaneous equations

$$a_1x + b_1y = c_1, \quad a_2x + b_2y = c_2$$

taking special care with zero coefficients.

3. The polynomial $y = ax^3 + bx^2 + cx + d$ can be evaluated as

$$y = [(ax + b)x + c]x + d$$

Why is this a preferable method of evaluation on a computer? Produce a flow chart to read in the values of the variables and carry out the evaluation.

4. Develop a flow chart to read in two sets of numbers, each set in ascending order, and merge them into one set of numbers also in ascending order. Carry out a dry run on suitable sets of numbers.

5. Construct a flow chart to read in a set of numbers and find their average; since the number of members of the set may be a variable quantity this must be allowed for.

6. What could the flow chart in Figure 1.17 map out?

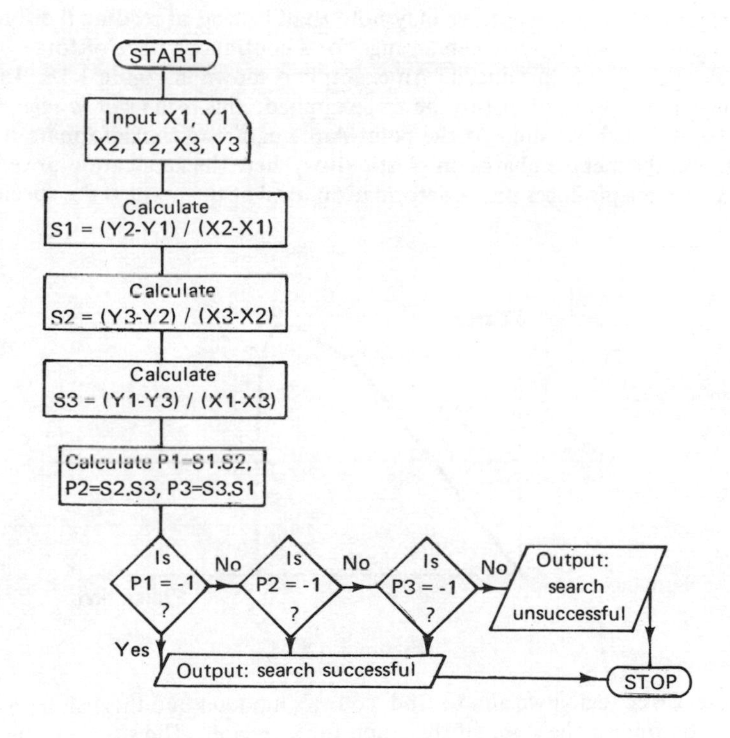

Figure 1.17

Hint: If (x_1, y_1) and (x_2, y_2) are the coordinates of two points in a plane,

$$\frac{y_2 - y_1}{x_2 - x_1}$$

is the slope of the straight line joining them.

7. Remembering that numbers are not stored exactly in a computer, make a suitable modification to the flow chart of Problem 6 to allow for this kind of error.

1.4 ACCURACY AND ERROR

We now turn our attention to the kinds of error which can arise in the determination of some physical quantity from an experiment.

In an experiment to measure Young's modulus of elasticity for a metal specimen, the metal is stretched by an applied stress and the modulus found from the formula

$$\text{Young's modulus, } E = \frac{\text{Force per unit area}}{\text{Elongation per unit length}}$$

As regards the experiment, we may note that, instead of reading the force and elongation from dials, we can arrange for a continuous trace of force versus elongation to be produced. An example is shown in Figure 1.18. Initially, the elongation is proportional to the force applied; this is the *elastic* region in which Hooke's Law is valid. At the point A the nature of the deformation changes and the metal embarks on plastic flow; here the application of very little extra force produces much deformation until at the point B the specimen fractures.

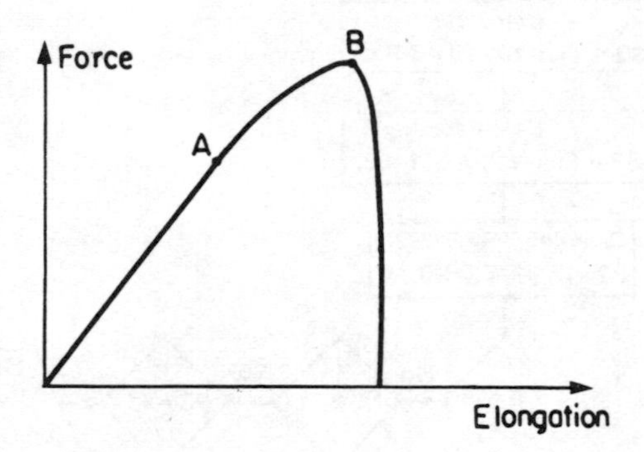

Figure 1.18

In the elastic region we aim to find Young's modulus and this can be partly achieved by finding the slope of the graph in the region. The slope can be found either by measurement from a trace produced by a sophisticated machine or by plotting observations manually on a graph and then making direct measurements from the graph by using the observations themselves or by using a numerical formula.

The cross-sectional area of the specimen which undergoes deformation has to be measured: in the case of a rectangular specimen this involves measuring average breadth and thickness using a micrometer screw gauge. As we shall see, one measurement may not suffice, even if the specimen is of constant area (to the accuracy of measurement possible). These quantities can be inserted into the formula

$$E = \frac{\text{STRESS}}{\text{STRAIN}}$$

$$= \frac{\text{LOAD APPLIED} \div \text{AREA}}{\text{EXTENSION} \div \text{ORIGINAL LENGTH}}$$

$$= 7.031 \times 10^9 \text{ kN/m}^2, \text{ say,}$$

and a result obtained. But just how accurate is the result, and how can we estimate the likely error? As quoted, we should expect to rely on the last figure given, i.e. we should have 4 significant figures accuracy.

(The number 3.102 is quoted to 4 significant figures, as are 3102.0, 0.003102 and 0.00310200; we count from the first non-zero digit and ignore any 'trailing' zeros if these occur after the decimal point. Note therefore that the first significant figure in the number 0.00123456 is 1 and, as it stands, it is quoted to 6 significant figures and 8 d.p. Also note that whereas 12340 has five significant figures, 0.12340 has only four.)

Where might errors have occurred? There are **three** main areas: experimental error, observational error and calculational error.

1. *Errors in Experimenting*

Suppose we determine the resistance of a wire by passing different measured currents through it and measuring the potential difference between its ends. The usual method would be to start with a small current and gradually increase it. But the wire heats proportionally to the square of the current and as it heats, its resistance changes with temperature. We may attempt to minimise this effect by decreasing the current again to average the readings but this may not be enough.

Again, if the temperature of the environment increases during the experimental checking of Boyle's Law, the results may be awry; increasing the pressure should be followed by decreasing the pressure.

This source of error is, however, not within our scope and we shall not discuss it further.

2. *Errors in Observation*

In taking any measurement an error is often made in the last figure read. One can aim for greater **precision**† by various devices, e.g. the vernier shown in Figure 1.19; 10 divisions on the vernier correspond to 9 on the main scale. To obtain the reading from Figure 1.20, we can first see that it lies between 13.0 and 13.1; since the vernier division 7 corresponds to a line on the main scale, there is a gap of (7 × 0.01) between 13.0 on the main scale and 0 on the vernier scale; hence the reading is nearer 13.07. More precise verniers can be (and have been) used.

The principle of the vernier is employed in the micrometer screw gauge,

---

† Note: For π, 3.1 is accurate to 1 d.p., but not very precise
3.14158 is more precise but not accurate to 5 d.p. (3.14159)

which is used to measure the thickness and breadth of the metal specimen. Even if these quantities are obtained by taking several readings of each, the discrepancies in the last figure could be attributable to actual random variations in the specimen or to error in observation. In some cases, the least step in the micrometer is only a factor of 10 smaller than the diameter to be measured.

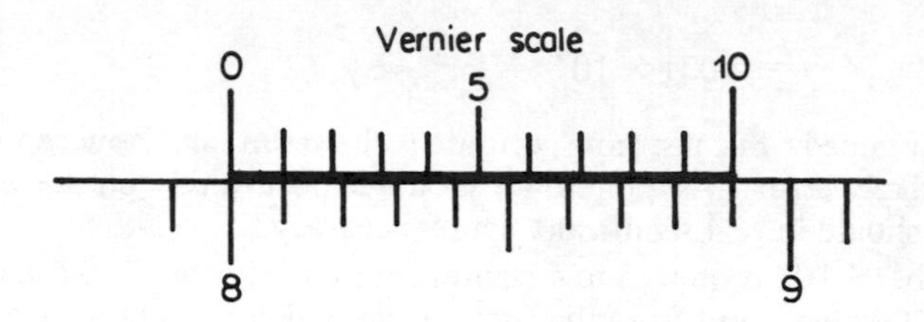

Figure 1.19

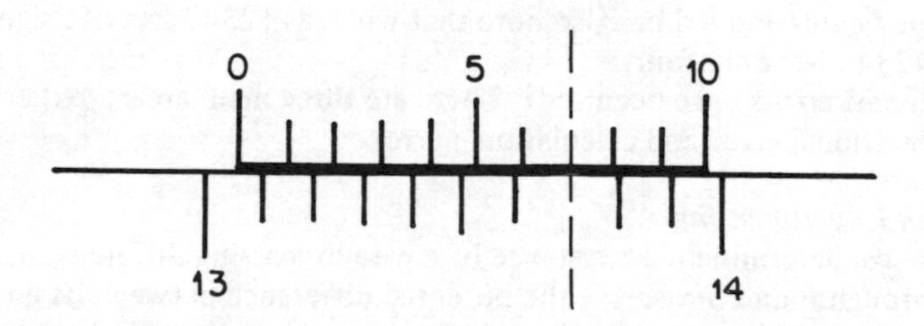

Figure 1.20

In order to investigate the error due to observation, suppose the distance between two parallel lines on a graph paper is found using a travelling microscope. If a large enough set of readings is taken, we should find that if we plotted frequency of occurrence against value of reading (as in Figure 1.21), the set of points approximates to a curve like that on Figure 1.22. The further we move from the 'average reading', the fewer times such a value has been observed and there are (about) as many readings above the 'average' as are below. The curve is in fact a theoretical curve — the *Gaussian curve* — which is a model for the statistical treatment of errors. A cruder method can be used: the readings are averaged and the deviations of each reading from this average noted; the result can be quoted as (average) ± (largest deviation) and this is unduly pessimistic, though possible.

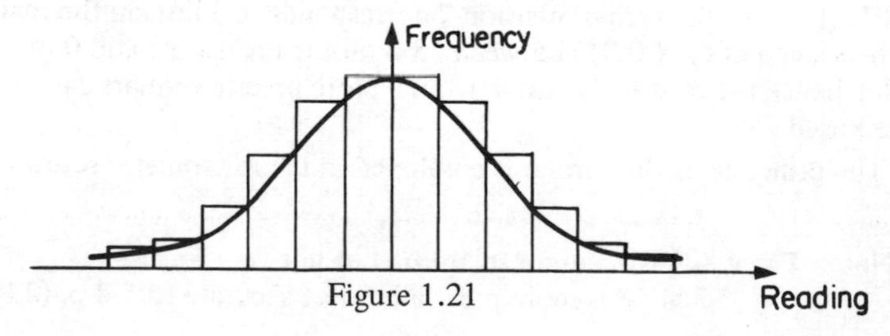

Figure 1.21

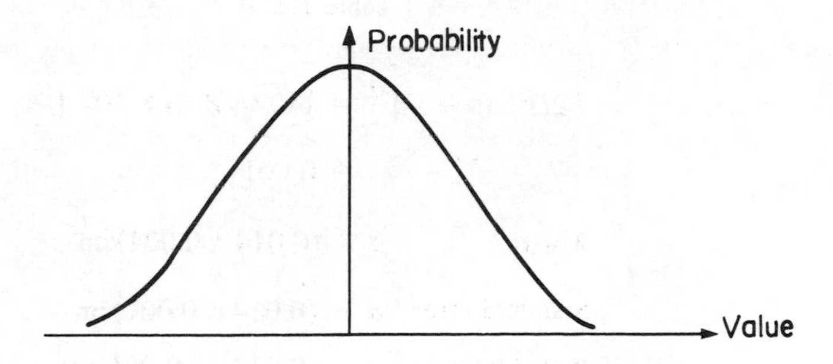

Figure 1.22

Alternatively, one can take the n deviations D_i and find ΣD^2_i and hence $[\Sigma D^2_i / n(n-1)]^{1/2}$ which is known as the standard error of the mean. Finally, one may quote the *Probable Error*, which is = 0.7 × the standard error.

As long as one is consistent, it is not vital as to which measure of error is used. See Table 1.1 for some observations and Table 1.2 for some measures of error.

A Gaussian curve is not the only distribution which arises: for example, if the diameters of spherical drops of liquid follow a Gaussian curve, their masses will not. A fuller treatment of all this will require statistical theory which will be developed in Chapters 11, 12 and 13.

Table 1.1

	Diameter (cm)	Deviation (cm)	
	d	D	D^2 (10^{-6})
	0.010	−0.004	16
	0.014	0	0
	0.016	+0.002	4
	0.012	−0.002	4
	0.018	+0.004	16
	0.014	0	0
TOTAL	0.084	0	40
MEAN	0.014	0	

Table 1.2

$$[\Sigma D^2/n(n-1)]^{1/2} = [40/(6 \times 5) \times 10^{-6}]^{1/2}$$

$$\simeq 0.001$$

Worst	$d = (0.014 \pm 0.004)$cm
Standard error	$d = (0.014 \pm 0.001)$cm
Probable error	$d = (0.014 \pm 0.001)$cm

All Results to 3 d.p.

3. *Errors in Calculation*

The use of a slide rule involves errors in observation, but we shall treat these as errors in calculation.

Let us consider the likely errors associated with multiplying two numbers by a slide rule: in Figure 1.23 we are trying to multiply 2.57 by 1.14 with a very poor slide rule. Errors will occur in placing the 1 of the first scale directly above 2.57 on the lower scale and in placing the wire marker exactly over 1.14 on the upper scale. The exact answer, 2.9298, cannot be obtained: the best we can do is estimate it to 2 d.p. and even then the last figure will be in doubt.

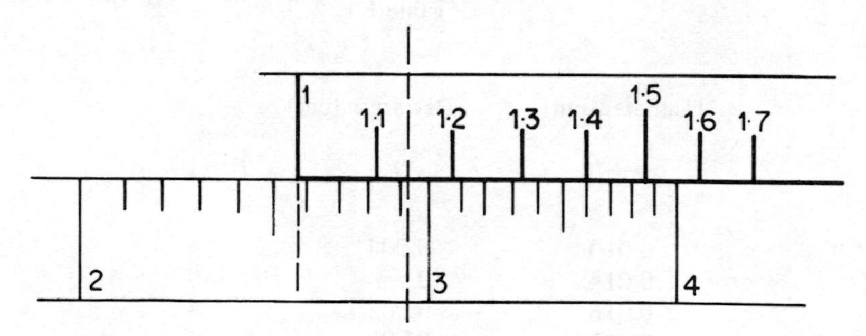

Figure 1.23

In the formula quoted for Young's modulus, assuming for the moment that the component values are accurate, could we obtain an accurate result? Clearly the answer is that we cannot, since even the simplest division, e.g. 2/3 produces an infinite decimal which we can either truncate or round-off.

Truncation means that to 4 stored places, 2/3 = 0.6666, i.e. we chop off all figures after the ones we store.

Round-off would produce 0.6667 (i.e. we quote the answer in the fourth decimal place to the nearest whole number, which is 7).

In either case we have not got the result exactly, and, if in the course of a lengthy computation there are many such arithmetic operations, the cumulative effects could be serious.

Examples

We consider some further examples:

3.323 is 3.32 rounded to 2 decimal places or truncated to 2 decimal places.

It is conventional to round up a 5, e.g.

2.105 is 2.11 and
−2.105 is − 2.11 both rounded to 2 decimal places.

It is always essential when a rounded answer is given to quote the decimal places of accuracy. If this is not done the quoted answer could be taken as exact.

We could get an answer more exactly using a desk calculator which can store 8 figures, but even this has its limitations, since

$$\frac{1}{3} \times 3 = 0.99999999 \neq 1$$

Even a digital computer has problems of how to store numbers which are both large and small. Let us consider a very simplified example as depicted below. Here we can store numbers to 2 places of decimals and two pre-decimal point digits, four spaces in all. The effect of trying to store four numbers is shown. The second is stored to within 1/2000 of its value, the third to within 1/350 of its value, but the fourth is lost altogether and the first has completely lost any meaning. Yet the computer wants to store very large and very small numbers to the same precision and the clue here is significant figures rather than decimal places.

Lost (left)	Stored	Lost (right)	
1 2 3 4	5 6 . 0 0		No significant figures kept
	1 2 . 3 4	5 6	3 s.f. accuracy
	0 . 1 2	3 4 5 6	2 s.f. accuracy
	0 . 0 0	1 2 3 4 5 6	None

The mass of an object is determined by different methods according to its magnitude. A small amount of a chemical may be weighed correct to 0.00001 gram on a good balance and a quoted mass of 0.86723 gram may be correct to 5 significant figures. On the other hand, the mass of a bag of potatoes may be accurate only to 0.02kg and it may be quoted as 2.54kg, where we have only two significant figures accuracy. Were we to quote the result as 2.543kg this would be implying greater accuracy than the problem demands. Again, if we multiply a mass correct to 0.1 gram by g (= 9.81) to obtain its weight, this is reasonable; there could be little justification in using a value for g correct to 6 d.p.

Floating-point Numbers

We alleviate the problem by the following artifice. In the following list of numbers, we have rewritten the four numbers by **floating** the decimal point and then by coding **times ten to the power of** as **E**; in this way we have achieved 6 significant figure accuracy by using only 10 spaces of storage, and we can handle numbers as small as 10^{-9} and as large as 10^{9}. Most reasonably-sized digital computers usually handle numbers in the range 10^{-76} to 10^{76}, can cope with 11 significant figures and can allow for negative numbers; they, in fact, work with binary arithmetic. Greater precision (almost double) can be obtained under special circumstances.

$$\begin{array}{l} 1.23456 \times 10^{5} \\ 1.23456 \times 10^{1} \\ 1.23456 \times 10^{-1} \\ 1.23456 \times 10^{-3} \\ \hline 1.23456\ \mathrm{E}+5 \\ 1.23456\ \mathrm{E}+1 \\ 1.23456\ \mathrm{E}-1 \\ 1.23456\ \mathrm{E}-3 \end{array}$$

Suppose we wish to add together two numbers in floating-point form:

$$1.23 \times 10^{-1} + 4.862 \times 10^{-2}$$

The numbers are 0.123 and 0.04862, and their sum is 0.17162 or 1.7162×10^{-1}; if we could not justify a fourth significant figure in the first number, we would have to write the sum as 1.72×10^{-1}. If we wish to multiply two such numbers, we proceed as follows:

$$(1.2 \times 10^{-1}) \times (1.3 \times 10^{4}) = (1.2 \times 1.3) \times 10^{-1+4} = 1.56 \times 10^{3}$$

Problems I

1. What is the error in rounding 2/7 to 3 d.p.; what is the error in truncating it to 2 d.p.?
2. Repeat Problem 1 for 3/8, 11/19.
3. Explain the idea of 'precision', 'accuracy' and 'significant figures' using as illustration 4/11, −13/17.
4. Carry out addition, subtraction, multiplication and division of the floating-point numbers 2.43×10^{-2} and -1.863×10^{-3}; give your answers in floating-point form.

Spread of Errors by Arithmetic Operations

In certain numerical methods which rely on many lengthy computations, rounding error can build up sufficiently to cast doubt on the eventual result, if not to destroy totally its validity. We must have some means of estimating these errors, and, whilst the error analysis of some numerical methods is a very involved (and unsatisfactory) process, we can at least attempt a brief study. Usually we adopt the most pessimistic approach, e.g. the possible error of the difference between two numbers is the sum of the possible errors in each number. We can investigate the effect of operating on two numbers.

Arithmetic of Errors

Let N be the exact value of some quantity and n an approximate value of the quantity such that

$$n = N + e$$

e	is the **Error**
$\lvert e \rvert$ or ABSF(e)	is the **Absolute error** (i.e. $= -e$ if $e < 0$) †
$\lvert e \rvert / \lvert N \rvert$	is the **Relative error**
$(\lvert e \rvert / \lvert N \rvert) \times 100$	is the **Percentage error**

Addition

Let n_1, n_2 be approximations to N_1, N_2 such that

$$n_1 = N_1 + e_1, \quad n_2 = N_2 + e_2$$

Hence, $n_1 + n_2 = (N_1 + N_2) + (e_1 + e_2)$

The absolute error is $\lvert e_1 + e_2 \rvert$ which can be shown to be $\leqslant \lvert e_1 \rvert + \lvert e_2 \rvert$

For example,
we let $N_1 = 2.6, N_2 = 3.1, n_1 = 2.55$ and $n_2 = 3.16$;
therefore $e_1 = -0.05, e_2 = +0.06$.
$n_1 + n_2 = 5.71, N_1 + N_2 = 5.7$.
Then $\lvert e_1 + e_2 \rvert = 0.01$, and $\lvert e_1 \rvert + \lvert e_2 \rvert = 0.05 + 0.06 = 0.11$

Subtraction

Using the above notation,

$$n_1 - n_2 = (N_1 - N_2) + (e_1 - e_2)$$

and the absolute error is $\lvert e_1 - e_2 \rvert \leqslant \lvert e_1 \rvert + \lvert e_2 \rvert$ (not proved here)

With the example, $n_1 - n_2 = -0.61$
$N_1 - N_2 = -0.5, \lvert e_1 - e_2 \rvert = 0.11$

We may summarise by saying that, in addition and subtraction, the absolute error in the result is, at most, the sum of the absolute errors.

---

† $\lvert -3.62 \rvert = 3.62, \quad \lvert 7.1 \rvert = 7.1$

Multiplication

$n_1 n_2 \doteqdot N_1 N_2 + (N_2 e_1 + N_1 e_2)$

We neglect the term $e_1 e_2$ since it is of the second order of smallness, assuming the individual errors are small.

$$\text{The relative error} \doteqdot \frac{|N_2 e_1 + N_1 e_2|}{|N_1 N_2|} = \left|\frac{e_1}{N_1} + \frac{e_2}{N_2}\right|$$

$$\leqslant \left|\frac{e_1}{N_1}\right| + \left|\frac{e_2}{N_2}\right|$$

Thus the relative error in the result is, at most, the sum of the individual relative errors.

In the example,

$$n_1 n_2 = 8.058$$

$$N_1 N_2 = 8.06, \text{ relative error} = \frac{0.002}{8.06} = 0.00025 \text{ (5 d.p.)};$$

$$\frac{|N_2 e_1 + N_1 e_2|}{|N_1 N_2|} = \frac{|3.1 \times -0.05 + 2.6 \times 0.06|}{2.6 \times 3.1} = \frac{0.001}{8.06} = 0.00012 \quad \text{(5 d.p.)}$$

$$\left|\frac{e_1}{N_1}\right| + \left|\frac{e_2}{N_2}\right| = \left|\frac{-0.05}{2.6}\right| + \left|\frac{0.06}{3.1}\right| = 0.0386 \quad \text{(4 d.p.)}$$

Division

A similar result applies here, i.e. the relative error is, at most, the sum of the individual relative errors.

The proof of this may be omitted at a first reading. We have

$$\frac{n_1}{n_2} = \frac{N_1 + e_1}{N_2 + e_2} = \left(N_1 + e_1\right)\frac{1}{N_2}\left(1 + \frac{e_2}{N_2}\right)^{-1}$$

$$\simeq \left(N_1 + e_1\right)\frac{1}{N_2}\left(1 - \frac{e_2}{N_2}\right) \quad \text{by the Binomial expansion}$$

$$= \left(\frac{N_1}{N_2} + \frac{e_1}{N_2}\right)\left(1 - \frac{e_2}{N_2}\right)$$

$$\simeq \frac{N_1}{N_2} + \frac{e_1}{N_2} - \frac{N_1}{N_2}\frac{e_2}{N_2} \quad \text{ignoring the term } e_1 e_2$$

$$\text{Therefore, relative error} = \left|\frac{e_1}{N_2} - \frac{N_1}{N_2}\frac{e_2}{N_2}\right| \Big/ \left|N_1/N_2\right|$$

$$= \left|\frac{e_1}{N_1} - \frac{e_2}{N_2}\right|$$

$$\leqslant \left|\frac{e_1}{N_1}\right| + \left|\frac{e_2}{N_2}\right| \qquad \text{as suggested †}$$

With the example,

$$\frac{n_1}{n_2} = 0.8070 \quad (4 \text{ d.p.}), \qquad \frac{N_1}{N_2} = 0.8387 \quad (4 \text{ d.p.});$$

$$\text{relative error} = \frac{0.0317}{0.8070} = 0.0393 \quad (4.\text{d.p.})$$

Let us consider now the error involved in finding the hypotenuse, a, of a right-angled triangle using Pythagoras' Theorem:

$$a^2 = b^2 + c^2$$

Suppose errors $\delta b, \delta c$ are made in b, c respectively, giving an error δa^2 in a^2 or an error Δa in a;

Then

$$a^2 + \delta a^2 = (b + \delta b)^2 + (c + \delta c)^2 \doteqdot b^2 + c^2 + 2b\delta b + 2c\delta c$$

i.e. $\qquad \delta a^2 = 2b\delta b + 2c\delta c$

Now

$$a + \Delta a \simeq (b^2 + c^2 + 2b\delta b + 2c\delta c)^{\frac{1}{2}}$$

$$= (a^2 + 2b\delta b + 2c\delta c)^{\frac{1}{2}}$$

$$= a\left(1 + \frac{2b}{a^2}\,\delta b + \frac{2c}{a^2}\,\delta c\right)^{\frac{1}{2}}$$

$$\simeq a\left(1 + \frac{b}{a^2}\,\delta b + \frac{c}{a^2}\,\delta c\right) \qquad \text{by the Binomial expansion}$$

$$= a + \frac{b}{a}\,\delta b + \frac{c}{a}\,\delta c$$

Thus

$$\Delta a = \frac{b}{a}\,\delta b + \frac{c}{a}\,\delta c$$

$$\left(\frac{\Delta a}{a} \times 100\right)\% \simeq \left(\frac{b}{a^2}\,\delta b + \frac{c}{a^2}\,\delta c\right) 100 \quad \% \qquad (1.4)$$

---

† Note that all these results apply approximately and only if e_1, e_2 are small

Example

If $b = 3$, and $c = 4$, then $a = 5$; but if b were measured as 3.01 and c as 3.98, a would be found as 4.990 (3 d.p.). This gives a percentage error of 0.2%.

From formula (1.4), the percentage error is

$$\left[\frac{3}{5^2} \times 0.01 + \frac{4}{5^2} \times (-0.02)\right] \times 100\,\%$$

$$= \left(\frac{-0.05}{25} \times 100\right)\% = 0.2\%$$

Problems II

1. Divide 4.2 by 2.3 assuming errors in each number of ±0.05. Consider the largest and the smallest values the fraction can take and check against the relative error as expressed above.
2. Likewise for +, × and − with these two numbers.
3. If, in the determination of the hypotenuse, a, of a right-angled triangle, b and c are read as 3.02 and 3.98 respectively, where each reading could be in error by ±0.03, calculate an approximate value for a and quote the actual and percentage errors.

Avoidance or Minimisation of Errors

We glanced at the problem of overcoming systematic experimental errors: we now briefly examine the other categories.

To minimise random error, a larger number of readings may be taken. Since the probable error $\propto 1/\sqrt{n}$, where n is the number of readings, one needs 100 times as many readings to give ten times the accuracy.

Often rounding errors can be minimised by low cunning; the troubles are at their worst when two approximately equal numbers are subtracted.

Consider

$$Z = \sqrt{X+1} - \sqrt{X} \qquad \text{where } X = 80$$

If we work to 6 s.f., we have $Z = 9 - 8.94427 = 0.05573$

If we now are only able to work to 3 s.f. we find $Z = 9 - 8.94 = 0.06$.

Using the identity $(\sqrt{X+1} - \sqrt{X})(\sqrt{X+1} + \sqrt{X}) \equiv 1$, we see that

$$Z = \frac{1}{\sqrt{X+1} + \sqrt{X}} = \frac{1}{17.9} = 0.0559,$$ which is more accurate, even though we are only working with 3 s.f.

Likewise we can overcome the problem of finding accurately the roots of a quadratic equation when one of those roots is very small.

Example

We work to 2 d.p. in the following example.

$$x^2 - 500x + 1 = 0$$

Solving by the formula method, we obtain

$$x = \frac{500 \pm \sqrt{250000 - 4.0}}{2}$$

$$\simeq \frac{500 \pm \sqrt{250000}}{2}$$

i.e. $x_1 = 500$, $x_2 = 0$; and these do not satisfy the equation, but x_1 is probably more accurate than x_2. However, since the product of the roots,

$$x_1 x_2 = 1 \dagger$$

$$x_2 \simeq \frac{1}{500} = 2.0 \times 10^{-3}$$

In fact, the values are $x_1 = 499.996$ (6 s.f.) and $x_2 = 2.000 \times 10^{-3}$ (4 s.f.) so we have clearly overcome our troubles.

A consideration of the formula used in experimental data may give a guide to the accuracy required in certain measurements. In the case of the simple pendulum the period of oscillation,

$$T = 2\pi\sqrt{\frac{l}{g}}$$

where l is the length of the pendulum; we assume that this is an exact model for the moment.

If this is used to estimate g, we note that

$$g = 4\pi^2 \frac{l}{T^2}$$

Now an error of $x\%$ in the reading of T will have the same effect on the estimate of g as an error of $2x\%$ in the reading of l. Usually, if we cannot measure l to less than 0.5% accuracy, there is little virtue in concentrating on achieving a reading of T to 0.005% accuracy.

(As an aside note one might mention as a general rule that it is wise to work to one more decimal place than the accuracy quoted.)

It is perhaps necessary to re-emphasise at this juncture that even an *exact* analytical formula resulting from solving an equation of a model is only as exact as the precision that can be obtained upon substituting numerical values into it. For example, if $y = \sin(\log x^2)$ then at each of the three operations: *square, take logs* and *find the sine*, accuracy may be lost.

Problems III

1. Find $Z = \sqrt{X + a} - \sqrt{X}$ to 3 d.p., where $X = 75$, $a = 0.21$.
   Work with 2 s.f., 3 s.f., then 4 s.f.
   Repeat, using an artifice similar to the one in the text.

---

† If x_1, x_2 are the roots of $ax^2 + bx + c = 0$, then $x_1 x_2 = c/a$

2. Find the roots of $x^2 + 48x + 1 = 0$ by the formula method working to 3 s.f. Compare your value for the smaller root with the value obtained by using the result that the product of the roots is 1.

Ill-Conditioning

Sometimes there are errors inherent in a system and no simple way of overcoming such trouble exists.

If we examine the two pairs of equations

$$\left.\begin{aligned} x + y &= 1 \\ x + 1.001y &= 0 \end{aligned}\right\} \quad (1.5) \qquad \text{and} \qquad \left.\begin{aligned} x + y &= 1 \\ x + 0.999y &= 0 \end{aligned}\right\} \quad (1.6)$$

we see that the solutions are respectively $x = +1001$, $y = -1000$ and $x = -999$, $y = +1000$. A change in one of the coefficients of about 0.2% has caused a radical change in the nature of the solution.

If we look at the geometrical aspect in Figure 1.24, we see that the two lines in each set are nearly parallel; we have exaggerated the situations to show that the two lines in either case do meet. The fact that their slopes are so nearly the same means that only a slight change in the slope of one causes it to overbalance in much the same way as the addition of a small mass to one pan of a chemical balance which is comparing almost equal masses could cause a reversal of the relative positions of the pans.

In our example the slopes are (to 3 d.p.) -1 and -0.999 for the first pair of equations (1.5) and -1 and -1.001 for the second pair (1.6). The difference in slopes for each pair is small.

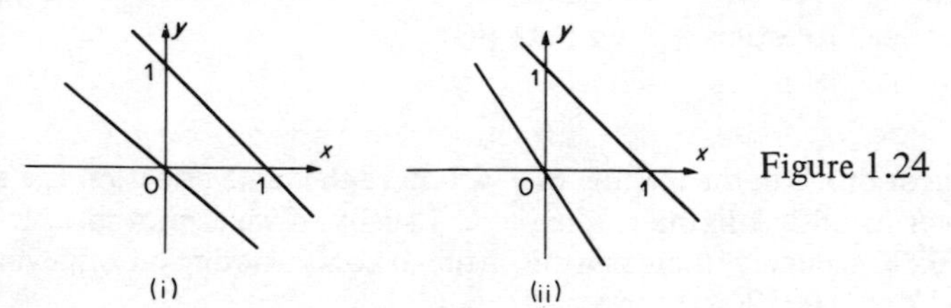

Figure 1.24

For the pair of equations

$$\left.\begin{aligned} a_1x + b_1y &= c_1 \\ a_2x + b_2y &= c_2 \end{aligned}\right\} \qquad (1.7)$$

where $a_1, b_1, a_2, b_2, c_1, c_2$ are constants,
the solution may be written

$$x = \frac{c_1b_2 - c_2b_1}{a_1b_2 - a_2b_1} \qquad y = \frac{c_2a_1 - c_1a_2}{a_1b_2 - a_2b_1}$$

and it is clear that if $a_1b_2 = a_2b_1$ then we have no solution. Ignoring special cases where b_2 or b_1 are zero, this condition becomes

$$\frac{a_1}{b_1} = \frac{a_2}{b_2}$$

which is equivalent to saying that the lines are parallel.

If $a_1 b_2 \doteq a_2 b_1$ then we might suppose that we are dealing with the *ill-conditioned* cases above.

However, consider the same two sets of equations (1.5) and (1.6), rewritten as follows:

$$\left.\begin{aligned} 1000x + 1000y &= 1000 \\ 1000x + 1001y &= 0 \end{aligned}\right\} \quad (1.8) \qquad \left.\begin{aligned} 1000x + 1000y &= 1000 \\ 1000x + 999y &= 0 \end{aligned}\right\} \quad (1.9)$$

The quantity $(a_1 b_2 - a_2 b_1)$ is 1000 for the first pair (1.8) and -1000 for the second (1.9) and can hardly be considered small, yet these sets of equations are precisely the same as the sets (1.5) and (1.6) considered earlier.

The clue here is that the value of $(a_1 b_2 - a_2 b_1)$ is small compared with either $a_1 b_2$ or $a_2 b_1$; note that we are not interested in the sign of these quantities, merely their size.

We shall need the symbol $\ll$, meaning **very much less than.**

We may interpret our condition as:

$$|a_1 b_2 - a_2 b_1| \ll |a_1 b_2|$$

$$\text{or} \quad |a_1 b_2 - a_2 b_1| \ll |a_2 b_1|$$

The danger with ill-conditioned systems of equations is that, due to their sensibility to changes in the coefficients, working with a slide rule may produce a quite different answer to that obtained on a desk calculator working to more significant figures.

Problems IV

1. Consider the special cases where b_2 or b_1 is zero
2. Show that the pair of equations

$$x - y = 2$$
$$30x - 30.01y = 60.02$$

is ill-conditioned. Solve working to 1 decimal place, 2 d.p. and 4 d.p. Examine separately the results of changing the -30.01 to -30 and the 60.02 to 60.

3. Repeat Problem 2 for the equations

$$x + 2y = 5$$
$$100x + 199.8y = 299.6$$

Checks

It is always sensible to carry out a check on numerical work. This can take the form of substituting supposed roots back into their generating equation, or it can be the use of the fact that the sum of the roots of a polynomial equation

$a_n x^n + a_{n-1}x^{n-1} + \ldots\ldots + a_0 = 0$ is $(-a_{n-1}/a_n)$. There are many such artifices.

Often when a fraction is being evaluated, it is wise to estimate the order of magnitude of the fraction. For example,

$$\frac{\pi^2 19.3}{4.8 \times 5.1} \simeq \frac{10 \times 20}{5 \times 5} = 8;$$

this compares with the value 7.75 obtained with a slide rule.

Problems V

1. Estimate the values of $\dfrac{4.1 \times 16.9}{2.3 \times 0.95}$ and $\dfrac{3.61}{1.52\pi}$
   Compare with values obtained by a slide rule or log tables.
2. Repeat Problem 1 for $\dfrac{17.2 \times 28.9}{-4.4 \times 81.8}$, $\dfrac{\pi^3 \times 41.3}{11.2 \times 7.63}$, $\dfrac{-\ 1.62}{1.85 \times 74.2}$, $\log_{10} 28$
3. Show that $x = 1, -2, 3$ are the roots of $x^3 - 2x^2 - 5x + 6 = 0$
4. By finding the sum of the supposed roots $x = 1.31, 4.8, -5.2$ of the equation

$$x^3 - 0.97x^2 + 25.43x + 33.7 = 0$$

check their validity. Check also the product of the roots which should be -33.7.

1.5 BUILD-UP OF KNOWLEDGE

From an engineering point of view the aims of our studies of mathematics are two-fold.

First of all we want to be able to solve physical problems. As pointed out earlier, a solution to a problem can take many forms and can be obtained by various means — either analytical or numerical with appropriate use of the computer. However, at the present time we do not have the necessary techniques and skills to solve many problems nor do we have sufficient knowledge of the functions and numbers involved in solutions. Therefore, before we can try to solve most problems we have got to build up our knowledge of mathematics with the associated skills and techniques available. In the following chapters we shall be studying various functions, number systems, analytical and numerical techniques and only when we have acquired a thorough working knowledge of these will it be possible to tackle the more demanding engineering problems. At all times we shall try, wherever possible, to link the mathematics to practical problems, but at intermediate stages we shall have to study topics which at first sight will appear to be abstract and of very little practical use. It is just like learning a language or, say, learning to play the guitar — we have got to build up our vocabulary and syntax. A question always asked by engineering students is: *Where shall I use this, and do I need to know such abstract mathematical ideas?* Well, we hope to show you, as the pages go by, just where you can use your mathematics and how useful it can be in your work.

Build-up of knowledge is also important in another way, in that the more we know the more sophisticated we can make our mathematical models for use in design work and the more we shall appreciate the limitations of our models. It was lack of knowledge which caused the well-known Tacoma Narrows Bridge failure where a wind speed of 42 mph caused the bridge to collapse, although it had been designed to resist a steady wind of at least 100 mph. It was certainly lack of knowledge which caused the wind-induced collapse of full-scale oil storage tanks in the dramatic first picture of our frontispiece. The expense of such a failure can be realised by comparing the size of the tanks with the size of the workmen's hut in the foreground. The collapse was subsequently modelled successfully in a series of wind-tunnel studies; see the second page of the frontispiece. The aim of the tests was to understand and measure the form of the collapse mechanism and to produce an empirical criterion against which a detailed analytical treatment could be compared. Had these tests being performed and a mathematical model been available, the design of the tanks would have been improved.

Although you, as an engineer, will perhaps not be working on the mathematical solutions and mathematics involved in design, you will probably be working in a team and it will be necessary for you to be able to converse with those who are formulating mathematical models and providing solutions. You will also be able to understand the assumptions made in forming these models and thence the limitations of their use. We think, therefore, that you should build up your knowledge of mathematics and become, if not an expert in the language, certainly competent enough to be able to read on to more advanced work.

SUMMARY

After reading through this chapter you should be aware of the role played by mathematics in your engineering subjects. The concepts of a mathematical model and a solution to such a model should be understood; in particular you must realise that a mathematical model is built on a number of assumptions and, therefore, any solution is valid only as long as these assumptions hold.

The value of flowcharting a problem should be recognised; it is a means of analysing the nature of the problem. Then you ought to appreciate the kinds of error that arise in numerical calculations and the ways in which we can estimate the error in a result; you should, further, have an idea of the methods for cutting down errors, where possible.

Unlike other chapters in this book, this one has emphasised the idea of mathematical modelling rather than dwelling on techniques.

Chapter Two

Elementary Ideas I–Functions and Sets

2.0 INTRODUCTION

In this chapter we intend to cover the basic mathematics required for our study of more advanced topics. Additionally we shall introduce the notations which will be used throughout the remainder of the book. This chapter is furthermore intended as a bringing together and bringing to the same level students who may have followed different courses up to the present time. You may have followed a *modern* syllabus, a *traditional* syllabus, an *engineering science* based or *engineering* course of study. Some of the material covered will be familiar but it will be worthwhile to learn the notations used and to see the particular ways of looking at the fundamental material covered.

We start by defining real numbers which you will have met before.

2.1 REAL NUMBERS

The real numbers consist of several types as follows:

(i) *Cardinal numbers*
These are the numbers used in counting, otherwise known as the **natural** or **whole** numbers. They are 1, 2, 3, 4, 5, ...

(ii) *Integers*
This is the name given to the collection of numbers which includes the cardinals, zero and the negative whole numbers. For example,
........., −2000, ..., −56, ..., −2, −1, 0, 1, ..., 21, ..., 7562, ...

(iii) *Rational numbers*
These are the numbers of the form p/q where p and q are integers, except that $q \neq 0$ ($\neq$ means *cannot equal*). For example, 2/3, 3/4, 4/7, 15/8 are rational numbers. We can, of course, express these in decimal form and the characteristic property of a rational number is that it either *terminates* or *recurs.*
Take the examples quoted

$$\frac{2}{3} = 0.66666..............$$

$$\frac{3}{4} = 0.75$$

$$\frac{4}{7} = 0.571428571428571428\ldots\ldots\ldots\ldots$$

$$\frac{15}{8} = 1.875$$

(iv) *Irrational numbers*
A number which does not recur or terminate is said to be **irrational**. You will already have met $\pi = 3.14159265\ldots\ldots$ and we can prove that $\sqrt{2} = 1.414213563\ldots\ldots$ is an irrational number. Of course we can make up an irrational number easily as most of the decimals we write down are in fact irrational; for example, $-11.2233445566778899\,111\,222\,333\ldots\ldots\ldots$ Notice the pattern of two 1's, two 2's, ..., and then three 1's, three 2's, etc.

Notice that the integers and cardinal numbers are all rational numbers so that we can say that the collection of **real numbers** consists of the collection of both the rationals and irrationals.

Before looking further into the properties of the real numbers we shall now consider the fundamentals of the theory of the collections of things. This is called the *theory of sets.*

2.2 SETS

At the outset the theory of sets may seem to have very little practical significance. We are going to study this theory for two reasons. Firstly it is basic to many branches of mathematics and is particularly useful in the initial ideas of functions. Secondly the ideas of set algebra are fundamental in probability theory which we shall meet in Chapter 11 when we start our study of statistics. The theory has practical applications in the theory of switching circuits and Boolean Algebra.

Definition of a Set

A set comprises a collection of objects or elements, each of these objects possessing some property which is chosen to define membership of the set. The definition of membership can either be by a list of members or by a rule which unambiguously determines membership. Here are some examples:

(i) the numbers 1, 2, 6, 9
(ii) the consonants of the English alphabet
(iii) all integers
(iv) all living people
(v) all Shakespeare plays
(vi) all people attending classes at a Technical College in the year 1972
(vii) 1, 3, 5, 7, 9, ...
(viii) the solutions of the equation $x^2 - 3x + 2 = 0$
(ix) 6
(x) all right-angled triangles.

Several points arise from these examples. We see that (i), (ix) and, to a certain extent, (vii) define membership by a list of members whereas the others prescribe rules for membership. (vii) could have been defined alternatively by the rule *all odd positive integers.*

Examples (i), (ii), (iv), (v), (vi), (viii) and (ix) are all **finite sets** having a finite number of members, whereas (iii), (vii) and (x) are **infinite sets** containing an infinite number of members.

Notation

Members of a set are denoted by lower case letters *a, b, c,* ......, *x,* ...... and the sets themselves by capital letters *A, B, C,* ....

We write $A = \{1, 3, 5, ...\}$ which says *the set A consists of the odd positive integers.* Any odd positive integer belongs to the set A which we write as $x \in A$. If an object does not belong to a set A we write $x \notin A$.

Example 1

If we take I as the set of all integers, then $3 \in I$ but $\pi \notin I$.

It is often useful to have a notation indicating the rule by which the set is defined. Suppose we wish to say: If I is the set of positive integers then the set A consists of those numbers x belonging to I such that $x^2 + 1$ is less than or equal to 17. We would write:

$$\text{If } I = \{1, 2, 3, ...\}, \qquad A = \{x \in I : x^2 + 1 \leqslant 17\}$$

(The *colon* : stands for *such that.*)

Of course in this case we can list the members of the set and write $A = \{1, 2, 3, 4\}$ Notice $5 \notin A$ since $5^2 + 1 > 17$, and similarly for any larger integer.

Example 2

(viii) could be written as

$$R = \{\text{Real numbers}\}, \qquad A = \{x \in R : x^2 - 3x + 2 = 0\}$$

We don't need always to have the first bracket, and indeed x does not have to be a number. For example (v) could be written

$$A = \{x : x \text{ is a Shakespeare play}\}$$

Try writing the other examples in one of these forms.

The empty or null set

It is possible for a set to contain no members. It is then called the **empty** or **null set** and is denoted by $\emptyset$. For example, if $R_+ = \{\text{Positive real numbers}\}$, $A = \{x \in R_+ : x^2 + 3x + 2 = 0\}$, then $A = \emptyset$ since the roots of $x^2 + 3x + 2 = 0$ are -1 and -2 and no x satisfies both requirements.

Equality of two sets

Two sets A and B are said to be equal if A and B contain precisely the same elements. We then write $A = B$. The following conditions must both be satisfied:

(i) if $x \in A$, then $x \in B$ $(x \in A \Rightarrow x \in B)$ ($\Rightarrow$ is read *implies that*)
(ii) if $x \in B$, then $x \in A$ $(x \in B \Rightarrow x \in A)$

Example 3

$I = \{1, 2, 3, ...\}$, $A = \{x \in I : x^2 + 1 \leqslant 17\}$

$B = \{x \in I : 1 \leqslant x \leqslant 4\}$

Both A and B have only the members $1, 2, 3, 4\}$
Therefore $A = B$

Subsets

If we have two sets A and B which are such that all the elements of A are also members of B (but not necessarily vice-versa) then we say A is a subset of B and write $A \subset B$.

For example: If A = {cardinals} and B = {integers}, then $A \subset B$.

If C = {rationals} and R = {real numbers}, then $C \subset R$

Note that if $A \subset B$ and $B \subset C$ then $A \subset C$. You should convince yourself of this by considering a few examples. If equality of two such sets is possible we write $A \subseteq B$. Examine examples (i) – (x) and write a list showing which set is a subset of one of the others.

Disjoint sets

If two sets have no elements in common we say that they are **disjoint**; examples (ii) and (iii) are disjoint as are (v) and (vi). This leads us to the idea of the common membership of two sets called the intersection of A and B.

Intersection

The intersection of two sets A and B is defined as the set of elements which belong to both A and B. This is written $A \cap B$ (read as *A intersection B*). We could write $A \cap B = \{x : x \in A \text{ and } x \in B\}$.

Thus, for example, if $A = \{1, 2, 3, 4, 5\}$, $B = \{2, 3, 4, 6, 8, 9$ then $A \cap B = \{2, 3, 4\}$. For this example show, by considering each side separately, that

$$A \cap (A \cap B) = A \cap B \quad \text{and} \quad B \cap (A \cap B) = A \cap B$$

Union of two sets

The union of two sets, A and B, is defined as the set consisting of members belonging to either A or B or to both A and B. This is written $A \cup B$ (read as *A union B*). Again, if $A = \{1, 2, 3, 4, 5\}$, $B = \{2, 3, 4, 6, 8, 9\}$

then $A \cup B = \{1, 2, 3, 4, 5, 6, 8, 9\}$

Show for these sets that $A \cap (A \cup B) = A$ $\quad A \cup (A \cap B) = A$

$B \cap (A \cup B) = B$ $\quad B \cup (A \cap B) = B$

Venn diagrams

The ideas of unions, intersection and other combinations of sets may be illustrated by means of **Venn diagrams**. The members of a set are represented by the points within a closed curve, the shape of the curve being immaterial, but usually drawn as an oval, as shown in Figure 2.1.

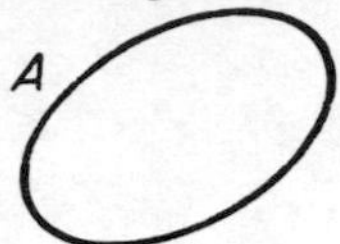

Figure 2.1

The relationship between the curves then gives the relationship between the sets. In Figure 2.2 we represent, using Venn diagrams

(i) disjoint sets
(ii) the intersection of two sets
(iii) union
(iv) subset

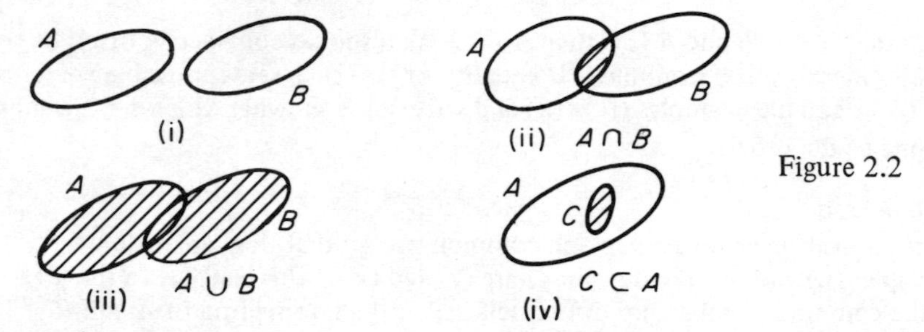

Figure 2.2

Notice that the shaded portions represent the appropriate set. Let us try to construct $A \cap (B \cup C)$ in two stages (see Figure 2.3).

First of all $B \cup C$ is as shown

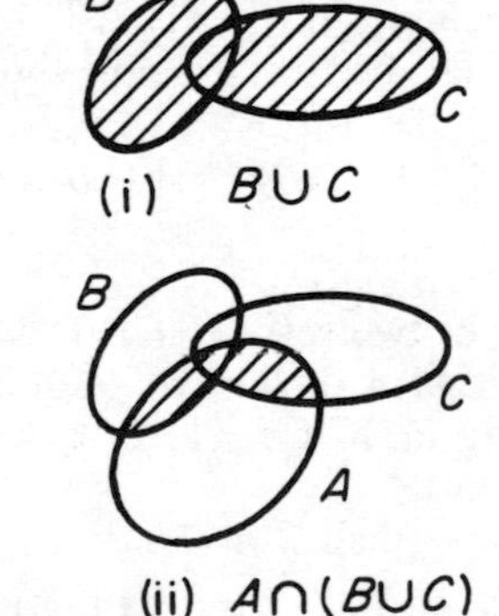

Now $A \cap (B \cup C)$ means the set of elements which belong to both $B \cup C$ and A and we get

Figure 2.3

Using Venn diagrams we can also show the equality of two expressions. For example, $(A \cup B) \cap (A \cup C) = A \cup (B \cap C)$. The left-hand side is made up of the intersection of $A \cup B$ and $A \cup C$ which is built up as follows in Figure 2.4.

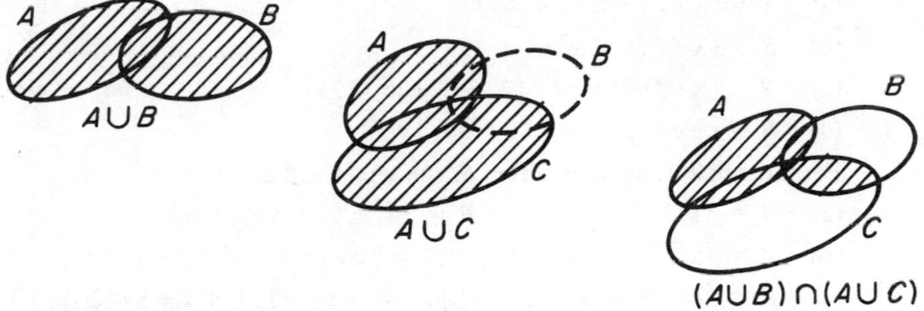

Figure 2.4

The right hand side is the union of A and $B \cap C$ which is built up as follows in Figure 2.5.

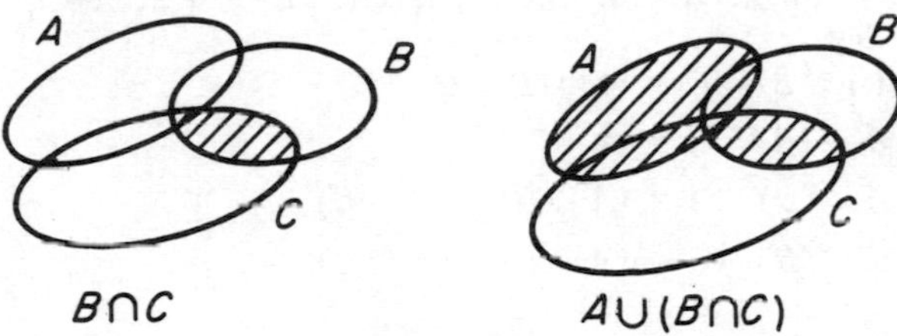

Figure 2.5

It is seen that the two expressions represent the same region and are therefore equal.

By considering Venn diagrams it is easy to show the following **Basic Set Operations** and you should try to prove each one.

1. $A \cup A = A \cap A = A$
2. $A \cap B = B \cap A$ (order does not matter)
3. $A \cup B = B \cup A$ (order does not matter)
4. $(A \cup B) \cup C = A \cup (B \cup C)$ (brackets can be moved and there will be no confusion if we write $A \cup B \cup C$)
5. $(A \cap B) \cap C = A \cap (B \cap C)$ (we can write this as $A \cap B \cap C$)
6. $A \cup (B \cap C) = (A \cup B) \cap (A \cup C)$ (notice brackets cannot be moved here)
7. $A \cap (B \cup C) = (A \cap B) \cup (A \cap C)$ (notice brackets cannot be moved here)
8. $A \cup (A \cap B) = A$
9. $A \cap (A \cup B) = A$

Problems

1. Enumerate the elements in the following sets, where I is the set of all integers both positive and negative
   (a) $S = \{x \in I : 5 < x^2 < 46\}$
   (b) $S = \{x \in I : 4 < x^3 + 1 \leqslant 102\}$
   (c) $S = \{x \in I \text{ and } y \in I : 12 < x^2 + y^2 \leqslant 25\}$
   (d) $S = \{x \in I : x^2 - 4x + 3 = 0\}$
2. Form the union and intersection of the sets A and B if
   (a) $A = \{1, 3, 5, 7, 9, 11\}, \quad B = \{1, 2, 3, 4, 5, 6, 7, 8\}$
   (b) $A = \{x \in I : 3 < x < 10\}, \quad B = \{x \in I : 0 < x^2 + x + 1 < 42\}$
   (c) $A = \{0, \sqrt{2}, 6, \pi, 5, 8, 9, 10\}, \quad B = \{x \in I : 0 < x + 5 < 20\}$
3. Verify the basic set operations (8) and (9) for the sets A and B in problem 2.
4. If $A = \{1, 2, 4, 5\}, \quad B = \{2, 3, 5, 6\}, \quad C = \{4, 5, 6, 7\}$, show that
   (a) $(A \cup B) \cup C = A \cup (B \cup C)$
   (b) $A \cup (A \cap B) = A$
   (c) $A \cup (B \cap C) = (A \cup B) \cap (A \cup C)$
   (d) $A \cap (B \cup C) = (A \cap B) \cup (A \cap C)$

Universal set and Complementation

Defining a set A not only determines those elements which belong to A, but also determines those which do not belong to A. For example, if I is the set of integers, then I also defines another set of elements which are not integers. The second set is called the **Complement** of the first and is written I'.

I', of course, contains every element which is not an integer. For example, it contains all the mountains of the world, all of Shakespeare's plays, and every object which is not an integer. Now it is unlikely if we are considering I as the set of integers that we require I' to contain all such non-integers. We are more likely, especially in mathematics, to be restricting the objects under discussion to the set of real numbers or perhaps the set of rational numbers. We therefore define a **Universal Set** which we define to be the set of all elements under consideration. Having determined what elements of this universal set belong to a particular set, the elements remaining make up the complement. For example, if R is the set of rational numbers and the universal set is the set of real numbers, then R' is the set of real numbers which are not rational. In other words, R' is the set of irrational numbers.

The following can now be added to the laws of set algebra.

(i) $A \cup A' = I$ (I is here the universal set)
$A \cap A' = \emptyset$ (the null set)

(ii) $I' = \emptyset, \quad \emptyset' = I$

(iii) De Morgan's Laws

$$(A \cup B)' = A' \cap B'$$
$$(A \cap B)' = A' \cup B'$$

We usually represent the universal set on a Venn diagram as a rectangle, as shown in Figure 2.6. Thus for R = rationals and R' = the irrationals we have:

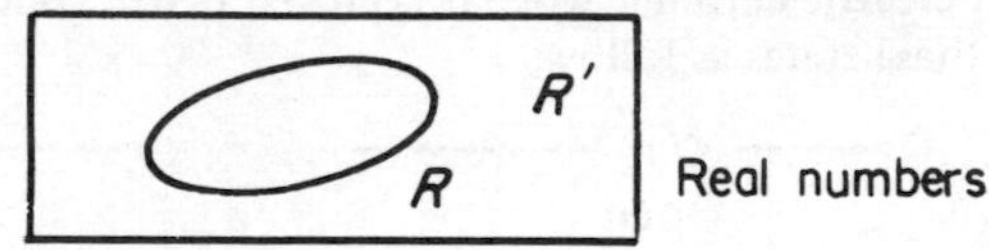

Figure 2.6

We shall show, using Venn diagrams, the validity of the first of De Morgan's laws, the other one being left as an exercise for the reader. Consider the left hand side:

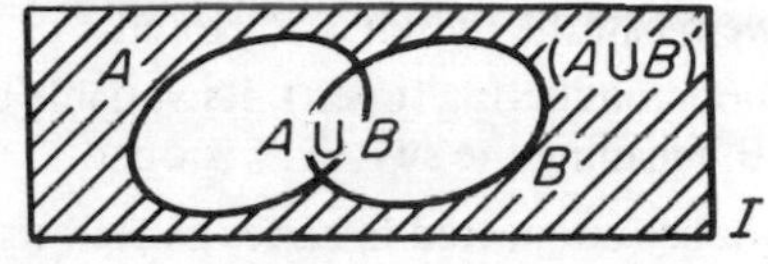

The right hand side means all the elements common to both A' and B'.

Now A' is represented by

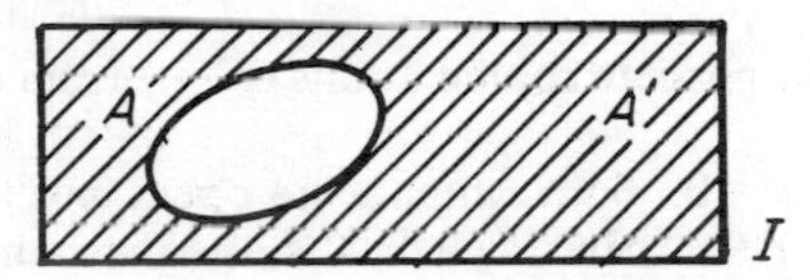

and similarly for B'.

Thus $A' \cap B'$, is represented by

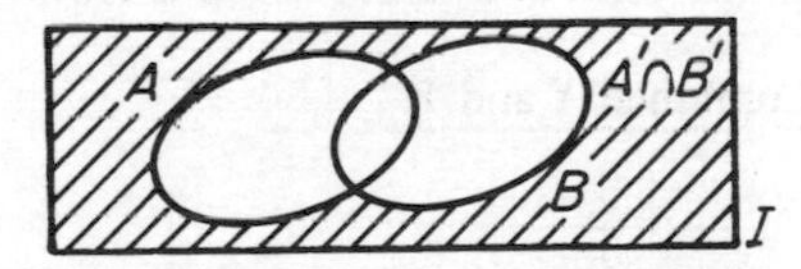

and clearly both sides of the equation are represented by the same region.

2.3 AN INTRODUCTION TO SWITCHING CIRCUITS

With the advent of the telephone and more recently of the electronic computer, it became necessary to study the properties of switching circuits used as routing devices. The telephone system, for example, uses switches to route a particular call through a network of telephone lines. We shall only study a small part of the subject of switching circuits but include it for interest and to show one application of set theory. In effect we shall set up mathematical models of simple electric switching situations. We assume direct electric current (d.c.) and that all the switches are *idealised*, by which we mean that they either conduct all the current available or none at all. Resistance and capacitance are ignored in the theory.

We consider *two-state* swtiches – that is, devices which are always in one of the two possible positions, *open* or *closed*. When the switch is open, it will not conduct electric current; when it is closed it will conduct electric current. We denote these states as follows

We shall denote switches by variables X, Y, Z, ... etc. These variables can only take two values since the switch is either open or closed and we use the convention:

(a) If a switch is conducting (closed), its variable assumes the value 1 and if we write $X = 1$, we mean the switch X is closed.

(b) If a switch is non-conducting (open), its variable assumes the value 0 and we write $X = 0$, meaning the switch X is open.

Consider two switches connected in *series* as follows

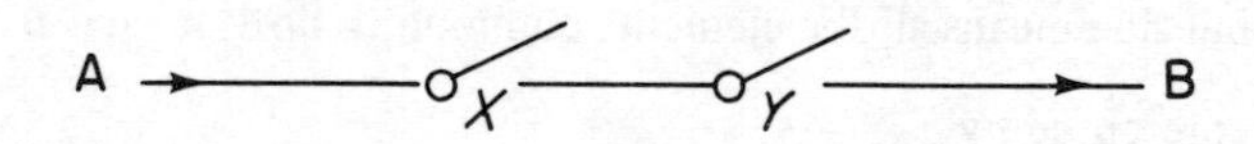

Let us consider the possible combinations of open and closed for this simple circuit

(i) $X = 0$ and $Y = 0$. Both switches are open; no current can pass.
(ii) $X = 0$ and $Y = 1$. Since switch X is open, no current can get from A to B.
(iii) $X = 1$ and $Y = 0$. Again no current can get from A to B.
(iv) $X = 1$ and $Y = 1$. Both switches closed, therefore current can pass from A to B.

We can write this in the form of a table, called a **Truth Table.**

X	Y	Combined X and Y
0	0	0
0	1	0
1	0	0
1	1	1

Consider two switches connected in *parallel* as follows

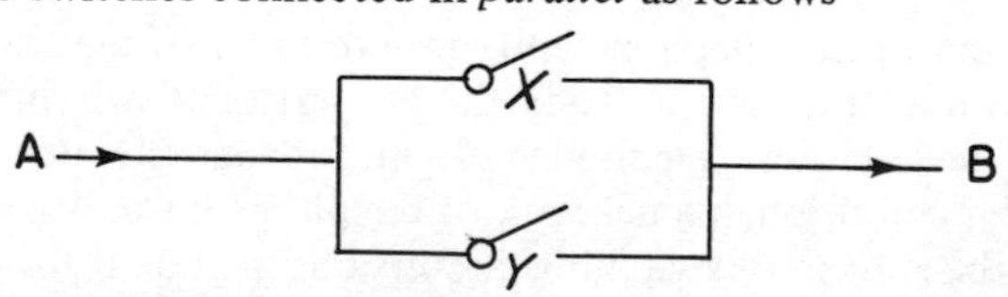

If you now go through the possible combinations of open and closed you will construct the following truth table

X	Y	Combined X and Y
0	0	0
0	1	1
1	0	1
1	1	1

What has all this to do with sets? Well, consider the following:

Suppose we are given an object a and a given set X. If $a \in X$, denote the situation by 1, and if $a \notin X$, denote the situation by 0. Given two sets X and Y, consider the intersection of X and Y, that is $X \cap Y$. You will remember that intersection refers to the set of objects belonging to *both* X and Y. Thus we would have the *truth table*

X	Y	$X \cap Y$	
0	0	0	Object a does not belong to X or Y and therefore not to $X \cap Y$
0	1	0	Object a does not belong to *both* X and Y
1	0	0	
1	1	1	Object a belongs to both X and Y

Compare this table with the table for switches connected in series. We see that the combined switch behaves just like the intersection operation and this could therefore be used to represent it.

Now form the truth table of $X \cup Y$ and compare with the parallel switches circuit. You should find that union could be used to represent the combined switch.

The use of the above representations can be extended to more complicated circuits. Consider the circuit in Figure 2.7.

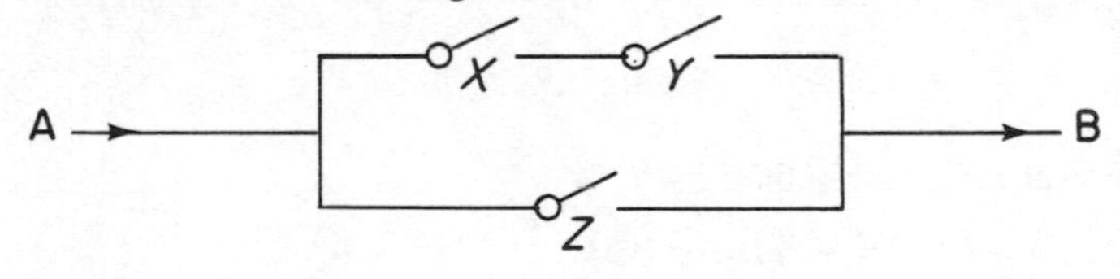

Figure 2.7

The circuit consists of two parallel branches, one of which has X and Y in series. The combined switch can be represented by

$$T = \underbrace{(X \cap Y)}_{X \text{ and } Y \text{ in series}} \quad \underbrace{\cup}_{\text{in parallel with}} \quad (Z)$$

We can use set algebra to transform this as follows

$$T = Z \cup (X \cap Y) = (Z \cup X) \cap (Z \cup Y)$$

This last expression represents the combined switch

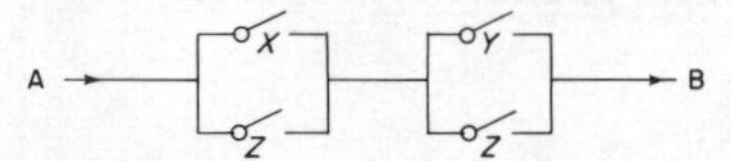

which is equivalent to the first one. We can always check that this is so by forming a *truth table* for each of them as follows.

X	Y	Z	$(X \cap Y) \cup Z$	$(Z \cup X) \cap (Z \cup Y)$
0	0	0	$0 \cup 0 = 0$	$0 \cap 0 = 0$
1	0	0	$0 \cup 0 = 0$	$1 \cap 0 = 0$
1	1	0	$1 \cup 0 = 1$	$1 \cap 1 = 1$
1	0	1	$0 \cup 1 = 1$	$1 \cap 1 = 1$
0	1	1	$0 \cup 1 = 1$	$1 \cap 1 = 1$
1	1	1	$1 \cup 1 = 1$	$1 \cap 1 = 1$

The same truth table is obtained for each combined switch and they are thus equivalent. The inference of course in the second is that both Z switches must be open or closed at the same time.

Problems I

1. Write down set representations of the circuits in Figure 2.8.

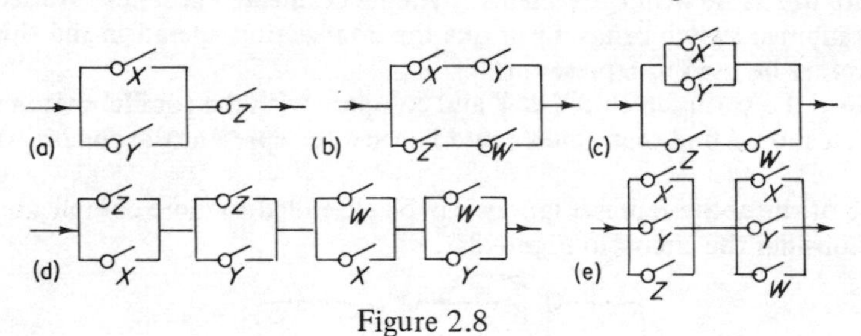

Figure 2.8

2. Draw circuits represented by
   (a) $X \cup (Y \cap Z \cup W)$ (b) $X \cap (Y \cup Z)$
   (c) $X \cup (X \cap Y \cap Z)$ (d) $(X \cup Y \cup Z) \cap (X \cup Y)$
3. Show that (1c) and (1e) are equivalent.
4. Show that (lb) and (1d) are equivalent.
5. Find equivalent circuits to (1b) and (1c) by using set algebra.

Multiple Contacts, Complementation

We saw in a previous example that two switches marked Z appeared. We regard these as two electrically independent switches which are always operating in the same state, i.e. they open together and close together. (Such a situation is achieved physically by the use of *relays* which are electromagnetic devices.)

Indeed we can have more than two switches acting in this way. They are called **Multiple Contacts.**

It is often necessary in a circuit to have two switches which are always in an opposite state. That is, when one is open, the other is closed and vice-versa. The second switch is called the **Complement** of the first. If X denotes the first switch then X' will denote the second. The operation of such switches is described in the following truth table.

X	X'
0	1
1	0

Notice that the use of complement here is in exactly the same sense as that used in set-algebra, the universal set being $\{0,1\}$.

Problems II

1. Show by means of a truth table that the complement of $X \cup Y$ is $X' \cap Y'$
2. Show the following pairs are complements by using both set algebra and a truth table.
   (a) $X \cap Y', \quad X' \cup Y$
   (b) $X \cup Y' \cup Z', \quad X' \cap Y \cap Z$
   (c) $X' \cap (Y \cup Z), \quad X \cup (Y' \cap Z')$

Design of Simple Circuits

In the practical design of simple circuits we often have to approach the problem first of all from a truth table because we know what the desired circuit should do. The flow chart (Figure 2.9) for the design is

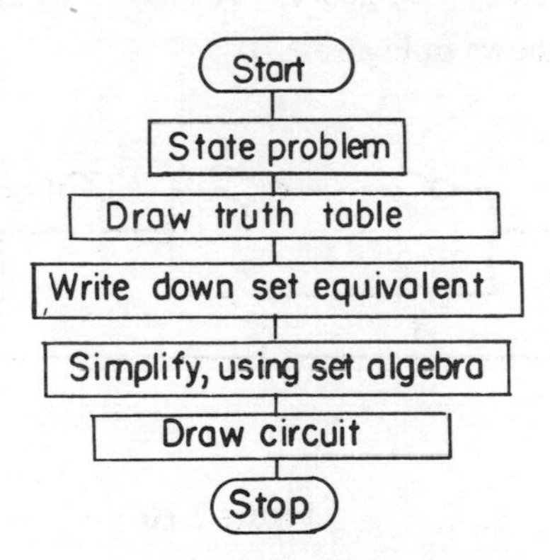

Figure 2.9

Example 1

Let us now take as an example the problem of lighting a staircase. There is a light at the top of the stairs and there are switches operating the light at both the bottom and top of the stairs. Can we design a circuit so that operating either switch changes the state of the light?

Assume the switches are X and Y and let us suppose that initially X is on and Y is on, when the light is on.

i.e.

X	Y	L
1	1	1

($L = 1$ since the light is on)

If we change X to 0, the light goes off and $L = 0$.
If we now change Y to 0, the light should come on giving $L = 1$, and so on.
We can thus build up the truth table of the circuit as

X	Y	L	
1	1	1	(a)
0	1	0	(b)
0	0	1	(c)
1	0	0	(d)

Concentrate your attention on the rows where there is a flow of current, i.e. row (a) and row (c). We see that (a) could be realised by $X \cap Y$, and row (c) could be realised by $X' \cap Y'$.

Thus we try for the circuit function

$$L = (X \cap Y) \cup (X' \cap Y')$$

If it is satisfied by (b) and (d) then it is the circuit required.

Testing for these, we find this is so, for

$$\text{when } X = 0 \text{ and } Y = 1, \text{ then } L = (0 \cap 1) \cup (1 \cap 0) = 0 \cup 0 = 0$$
$$\text{and when } X = 1 \text{ and } Y = 0, \text{ then } L = (1 \cap 0) \cup (0 \cap 1) = 0 \cup 0 = 0$$

The circuit is thus as shown in Figure 2.10

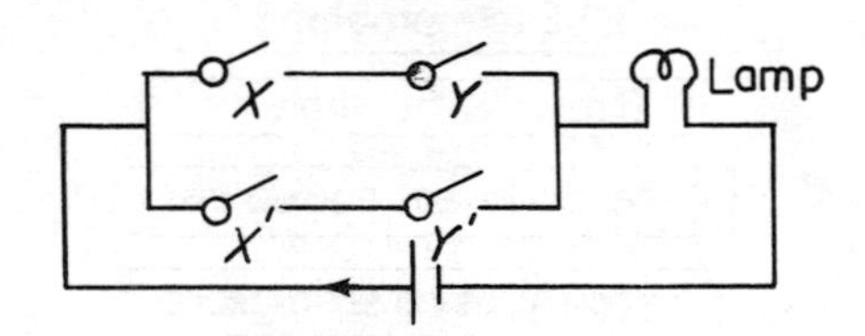

Figure 2.10

Example 2

Let us take one more example. Suppose we are given the truth table

X	Y	Z	L	
1	1	1	1	(a)
1	1	0	1	(b)
1	0	1	0	(c)
1	0	0	0	(d)
0	1	1	1	(e)
0	1	0	0	(f)
0	0	1	1	(g)
0	0	0	0	(h)

Concentrating again on the rows where L is 1, we see that (a) suggests $X \cap Y \cap Z$, (b) suggests $X \cap Y \cap Z'$, (e) suggests $X' \cap Y \cap Z$ and (g) suggests $X' \cap Y' \cap Z$.

Hence we take for L

$$L = (X \cap Y \cap Z) \cup (X \cap Y \cap Z') \cup (X' \cap Y \cap Z) \cup (X' \cap Y' \cap Z)$$

The circuit corresponding to this is given in Figure 2.11.

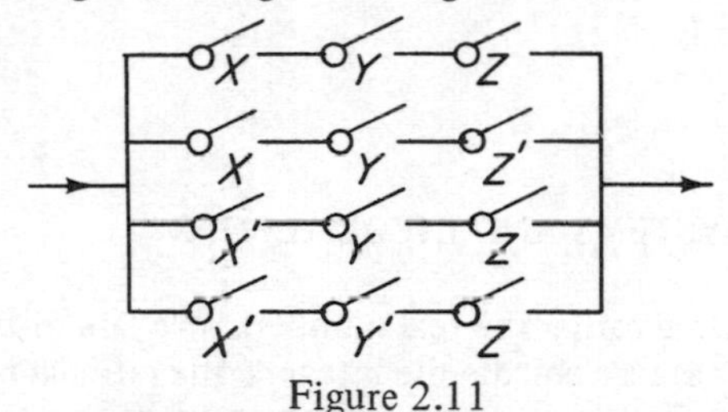

Figure 2.11

Simplifications of circuits such as this can be achieved by using set algebra, but we do not wish to deal with it here.

Problems III

1. Given the following truth tables, find the corresponding circuits

(a)

X	Y	L
1	1	0
1	0	1
0	1	1
0	0	0

(b)

X	Y	Z	L
1	0	0	0
1	1	0	1
1	0	1	1
0	1	0	0
0	0	1	1
0	1	1	1
1	1	1	1

(c)

X	Y	Z	L
1	0	0	0
1	1	0	0
1	0	1	1
0	1	0	0
0	1	1	1
0	0	1	0
1	1	1	1

2. Simplify the circuits in Problem 1, if possible.
3. Show that 1(b) and 1(c) are equivalent to

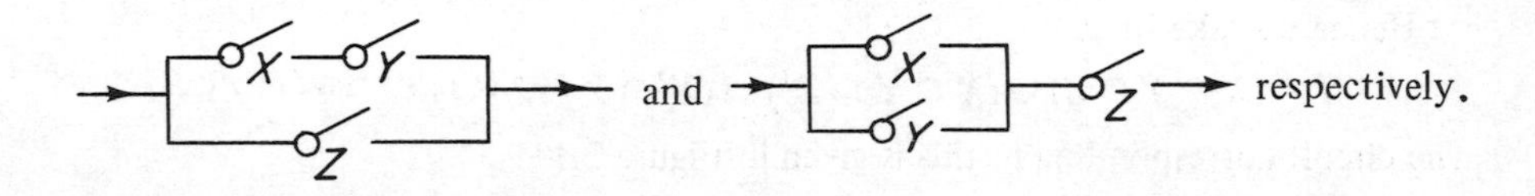

2.4 NUMBER SYSTEMS AND INEQUALITIES

We have considered earlier the real numbers. We talk of the set of real numbers which has as its elements the integers, the rational numbers and the irrational numbers. On a Venn diagram they could be represented as follows

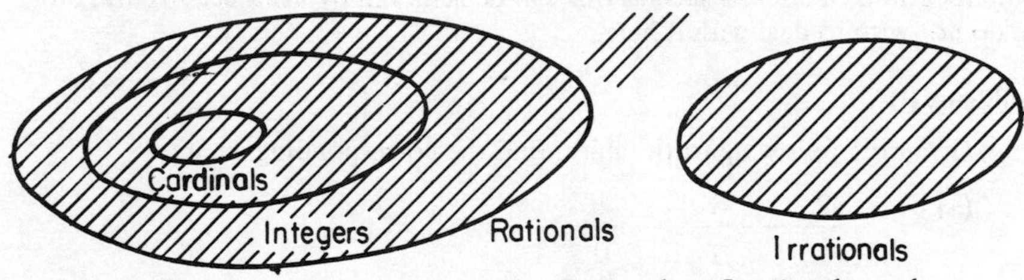

Notice that Real numbers R = Rationals $\cup$ Irrationals, and
Cardinals $\subset$ Integers $\subset$ Rationals

Contained in the set of integers is the subset of PRIME NUMBERS which are defined as being only divisible by 1 and the number itself. The set of positive prime numbers is $\{1, 2, 3, 5, 7, 11, 13, 17, 19, \ldots\ldots\}$.

We can represent all the real numbers by points on a line: the **Real Line**.

Considering first the subset of the integers, we can plot these on the real line as follows:

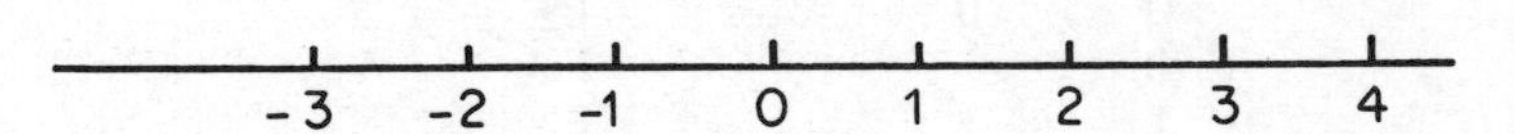

We say that there is a **one-to-one correspondence** between the integers and these points on the real line. By this we mean that

(i) to every integer there corresponds one, and only one, point on the real line, and

(ii) every one of the plotted points represents one, and only one, of the integers.

We can now turn to the rational numbers and mark off points on the real line corresponding to all the elements of this set besides the integers. This is much more difficult because between any two integers there is an infinite number of rationals. Even though there are an infinite number of rationals there will still be gaps on the line where no rational exists. It is shown in higher mathematical texts that **all** these remaining points are taken up by the irrational numbers. That is, there is a one-to-one correspondence between the set of real numbers of the set of points on the real line. This means that

(i) to **every** real number, there corresponds one, and only one, point on the real line, and

(ii) **every** point on the real line represents one, and only one, real number.

Inequalities

If we choose any two numbers on the real line then you will see that the right hand one is the greater.

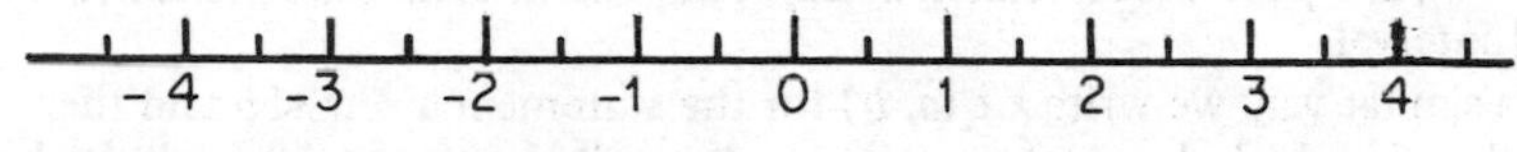

For example:

2 is greater than 1
3 is greater than -1
1.2 is greater than 0.9
-1.2 is greater than -2
1.1 is greater than -1.2

If one number, x, is greater than another number, y, we write $x > y$, so that in the above examples

$$2 > 1, \quad 3 > -1, \quad 1.2 > 0.9, \quad -1.2 > -2, \quad -1.1 > -1.2$$

Of course, we can say 1 is less than 2, -2 is less than -1.2 and so on, and we can write $1 < 2$, $-1 < 3$, $-2 < -1.2$ etc.

We can extend these ideas and write statements such as $x \geqslant 3$ (x is greater than or equal to 3) or $x < -1.5$ (x is less than -1.5). On the real line, the first of these would be represented by the set of points to the right of and including 3, and the second would be represented by the set of points to the left of -1.5 and not including 1.5

-4 -3 -2 -1 0 1 2 3 4

In each case the sets are said to represent semi-infinite intervals of the real line since they each extend to infinity in one direction only.

Again we can write statements such as $-3 \leqslant x \leqslant -2$ or $-1 < x < 1$ or perhaps $3 \leqslant x < 5.5$. These sets of points would be represented on the real line by **finite intervals**, the • indicating that the point is included, ∘ indicating that the point is not included in the set.

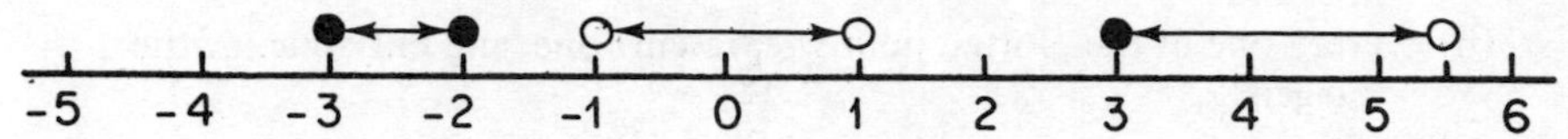

Notice that the interval for $-1 < x < 1$ is symmetrical about the origin and represents points which are at a distance less than one from the origin. We write this as $|x| < 1$ and say the modulus of x (or mod x) is less than 1. Alternatively it can be read as the absolute value of x is less than 1 – this means more because it says the numerical value of x, whatever its sign, is less than 1. We can generalise to statements like $|x - 3| \leqslant 5$, which means that the distance of x from the number 3 is less than or equal to 5, and on the real line this is represented by the set of points in the interval $-2 \leqslant x \leqslant 8$.

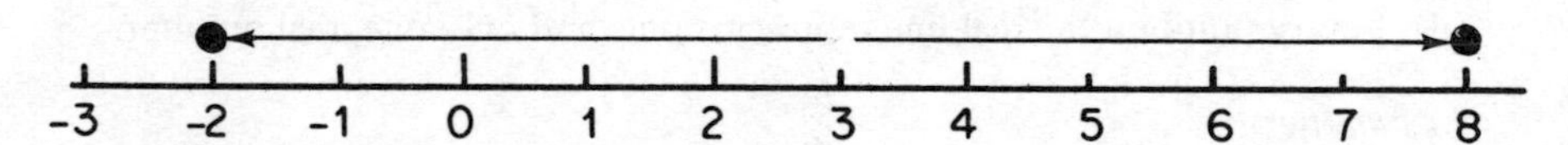

We introduce now a notation which is often used to denote intervals. We write $x \in [a, b]$ for the statement $a \leqslant x \leqslant b$. The interval $[a, b]$ is called a **closed** interval.

In a similar way we write $x \in (a, b)$ for the statement $a < x < b$ and the interval (a, b) which does not contain a or b is called an **open** interval. It should be perfectly clear what we mean by the intervals $(a, b]$ and $[a, b)$. The first of these does *not* contain a while the second does *not* contain b.

Sometimes it is not easy, in the first instance, to see what an inequality statement means and we have to *unravel* the statement to understand its implication; this necessitates rules for handling inequalities. We summarise these rules below.

Inequality Rules

(i) $a > b$ if and only if $a - b$ is some positive number.
For example, $-1 > -2$ since $-1 - (-2) = +1$, a positive number.

(ii) If $a > b$, then $a + c > b + c$. This says we can add a real number c (positive or negative) to each side of the inequality and not change the sense of the inequality.
For example, if we are given $2x - 7 > x + 2$
then $2x - 7 + 7 > x + 2 + 7$
or $2x > x + 9$
so that $2x - x > x + 9 - x$
or $x > 9$

(iii) If $a > b$ and $\lambda > 0$, then $a\lambda > b\lambda$. That is, we can multiply by a positive number without changing the sense of the inequality.
For example, $3x - 4 < x - 8 \Rightarrow 2x < -4 \Rightarrow x < -2$

(iv) If $a > b$ and $\lambda < 0$, then $a\lambda < b\lambda$. That is, multiplying by a negative number *changes* the sense of the inequality.
For example, $x - 5 < 5x - 2 \Rightarrow -4x < 3 \Rightarrow x > -3/4$

(v) If $a > b$ and $c > d$, then $a + c > b + d$.

(vi) If $ab > 0$, then *either* $a > 0$ and $b > 0$
or $a < 0$ and $b < 0$

(vii) If $ab < 0$, then *either* $a > 0$ and $b < 0$
or $a < 0$ and $b > 0$

All these rules are still valid if $>$ is replaced by $\geqslant$ and $<$ by $\leqslant$.

As an example of the use of the last two rules consider the statement $x^2 - 3x + 2 > 0$.

We can factorise the left hand side and obtain $(x - 2)(x - 1) > 0$.
Hence *either* (a) $x - 2 > 0$ and $x - 1 > 0$
or (b) $x - 2 < 0$ and $x - 1 < 0$

On the real line we have for (a):

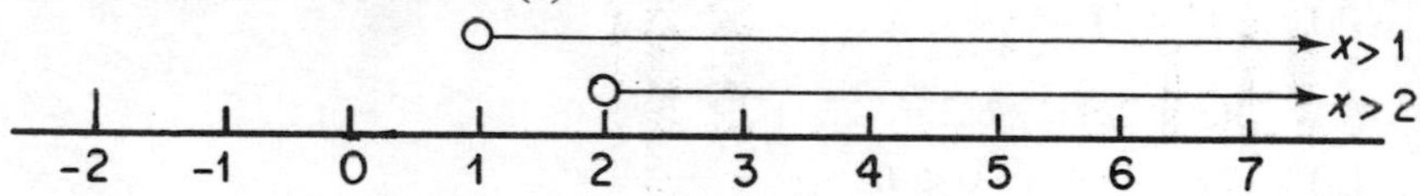

To satisfy both statements that $x > 2$ *and* $x > 1$ we must have $x > 2$.

On the real line we have for (b):

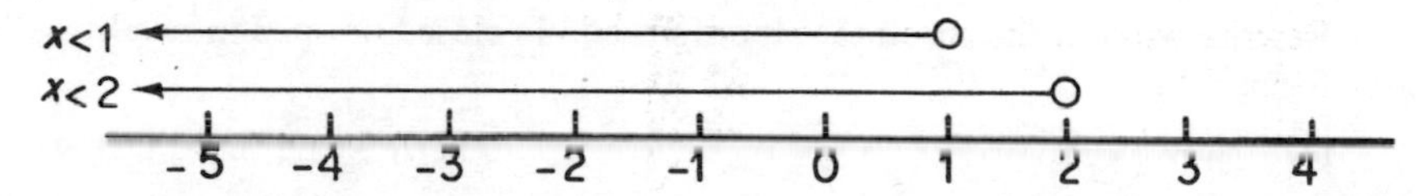

To satisfy both these statements, $x < 1$.
Hence we have $x > 2$ or $x < 1$ if $x^2 - 3x + 2 > 0$.

(Try $x = 1.5$ in the l.h.s.)

Suppose we have been given $x^2 - 3x + 2 < 0$.

Then $(x - 1)(x - 2) < 0 \Rightarrow$ either (i) $(x - 1) > 0$ and $(x - 2) < 0$
or (ii) $(x - 1) < 0$ and $(x - 2) > 0$

(i)
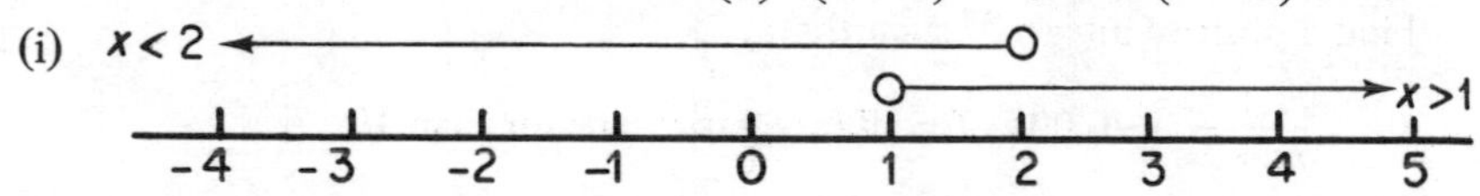

We must have $1 < x < 2$ to satisfy both these.

(ii)
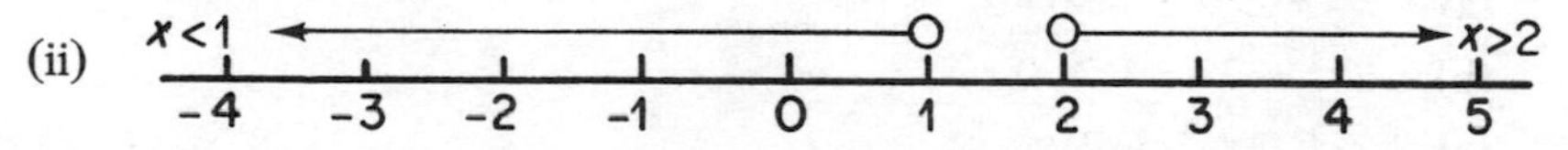

The solution set here is empty – there are no points such that $x < 1$ and $x > 2$.

Thus for $x^2 - 3x + 2 < 0$, $1 < x < 2$. Alternatively $x \in (1, 2)$.

Problems

1. Represent on the real line the sets
   (a) $\{-3, 2, 6, -2\}$
   (b) $\{x : x < 2\}$
   (c) $\{x : -1 < x \leqslant 4\}$

2. Solve and represent the solutions to
   (a) $6 - 3a > 2a + 1$
   (b) $\frac{1}{3}(b + 3) \geqslant \frac{3}{4}$
   (c) $5(3 - 2p) < -6$
   (d) $(2x - 1)(4 - x) < 0$
   (e) $(y + 2)(y - 5) < 0$
   (f) $|x| \leqslant 0$

3. Rewrite without the absolute value signs and in closed or open interval forms
   (a) $|x| < 4$
   (b) $|x - 3| < 2$
   (c) $|3x + 2| < 9$

4. Rewrite with absolute value signs
   (a) $-3 < x < 7$
   (b) $3 < x < 9$

5. Find a positive integer M such that

   $$\frac{1}{m + 2} < 0.001 \quad \text{for all integers } m \text{ greater than } M.$$

6. Show that for all values of x
   (a) $x^2 + 2x + 8 > 0$
   (b) $2x^2 - 3x + 4 > 0$
   (c) $2x^2 - x > -2$
   (d) $(x - 1)^2 > x - 4$

7. Determine the set of numbers x for which the following conditions are satisfied

(a) $|2x^2 + 3x - 2| < 1$

(b) $|6x - 3x^2| \leqslant 4$

(c) $|x^3 + x| \geqslant 2$

(d) $\dfrac{1}{x} < \dfrac{1}{4}$

(e) $100 < \dfrac{1}{x} < 200$

2.5 RELATIONS AND FUNCTIONS

One of the most important, if not the most important, concepts occurring in calculus is that of a function. The idea of a function is paramount in our formulation of the interaction between natural phenomena. For example, we talk loosely of the profit of a company as being a function of raw materials, plant, labour employed, taxation, and so on. In mechanics we speak of the position of a body, its velocity and acceleration as being all functions of time. The question is: *What do we mean by saying some quantity is a function of another?* We need to have a strict mathematical definition of what *is* a function and what *is not* a function before we can proceed with the ideas of derivative, integral and further work in the calculus.

We start by considering **Relations.**

Definition: A relation is a set of ordered pairs.

Example 1

The speed of a car (v m.p.h.) was measured at various times (t minutes) after starting from rest and gave

t (minutes)	0	1	2	3	4	5	6
v (m.p.h.)	0	20	36	50	60	60	60

We can write this relation as a set of ordered pairs

$$R_1 = \{(0,0) \quad (1,20) \quad (2,36) \quad (3,50) \quad (4,60) \quad (5,60) \quad (6,60)\}$$

Example 2

In a local football league, 5 teams played each other twice, once at home and once away. The scores are shown in the table.

		Away Team				
		A	B	C	D	E
Home Team	A	–	1–0	0–0	2–4	3–0
	B	2–0	–	1–1	4–2	1–0
	C	1–3	1–0	–	0–0	2–2
	D	4–1	0–3	1–2	–	2–1
	E	4–5	0–1	1–0	0–1	–

For example, A beat E 3–0 at home and also beat E 5–4 away from home. From the table we can construct the following information where (A, B) means A playing B, A at home.

Home wins: (A,B) (A,E) (B,A) (B,D) (B,E) (C,B) (D,A) (D,E) (E,C)

Draws: (A,C) (B,C) (C,D) (C,E)

Aways: (A,D) (C,A) (D,B) (D,C) (E,A) (E,B) (E,D)

We have now got three relations, for example, the relation connecting the teams that drew is the set of ordered pairs

$$R_2 = \{(A,C) \ (B,C) \ (C,D) \ (C,E)\}$$

Example 3

Let Z^+ be the set of positive integers and consider the set of ordered pairs $\{(m, n)\}$ such that $m \in Z^+, n \in Z^+$ and $m + n \leqslant 5$.

We get the relation as the set of ordered pairs

$$R_3 = \{(1,1) \ (1,2) \ (1,3) \ (1,4) \ (2,1) \ (2,2) \ (2,3) \ (3,1) \ (3,2) \ (4,1)\}$$

Example 4

A car factory produces 4 different models and the numbers of each model produced on a certain day is given by

Model 1	Model 2	Model 3	Model 4
200	500	150	400

This gives us the relation

$$R_4 = \{(1,200) \ (2,500) \ (3,150) \ (4,400)\}$$

Now we can represent relations graphically by using coordinate axes and plotting the ordered pair (x, y) as point P, as in Figure 2.12 (O is called the origin, x the

abscissa and y the ordinate)

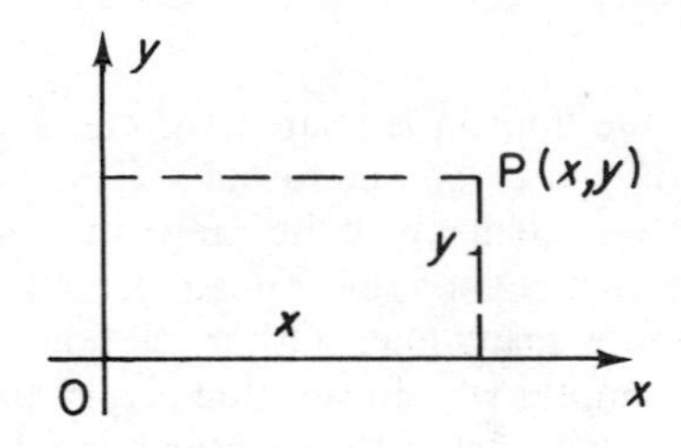

Figure 2.12

The relations R_1, R_2, R_3, R_4 given in the examples will then be represented by the following diagrams, Figure 2.13 (i)–(iv)

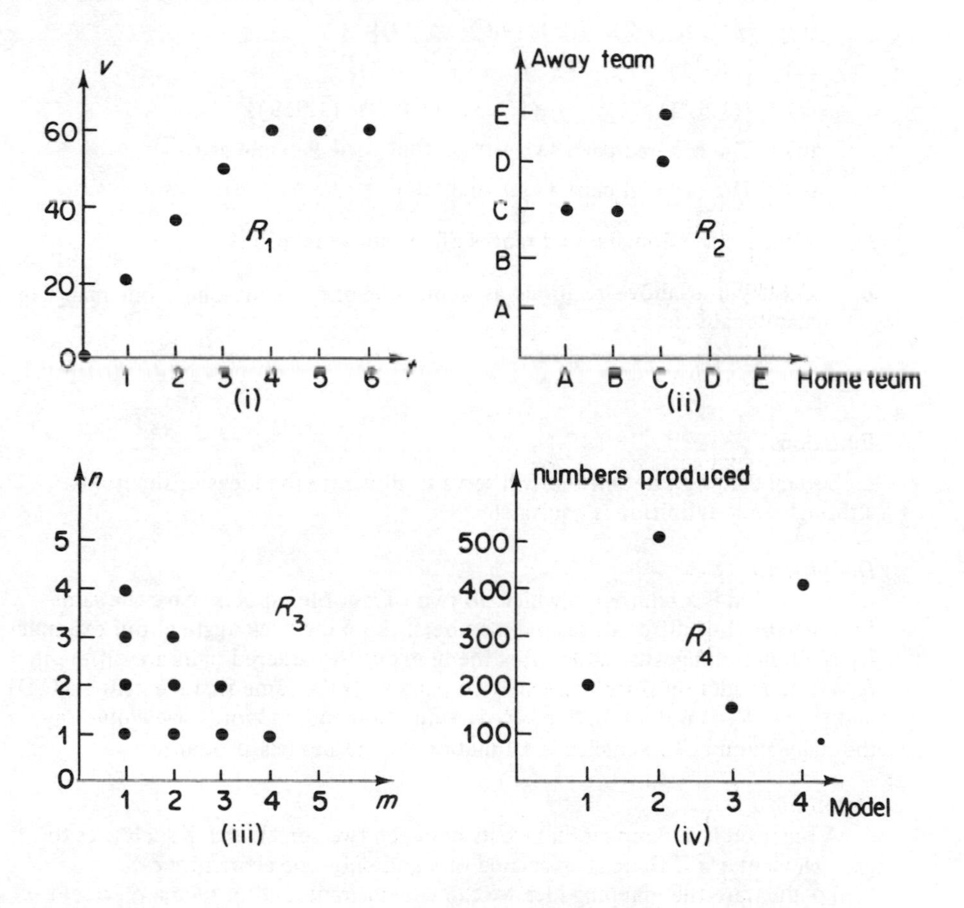

Figure 2.13

These are all examples of **Graphs**. The x values (on the horizontal line) constitute a set X called the **Domain** and the set of y values constitute a set Y called the **Range** or **Co-domain**.

If each element of the domain is related to one and only one element of the range, the relation is said to be **one-to-one**. If one element of the domain is related to more than one element of the range the relation is **one-many**. Finally, if two or more elements of the domain are related to the same element of the range, the relation is **many-one**. Other relations are called **many-many**. Looking back at our examples we can see that R_1 is many-one since members 4, 5, 6 of the domain are all related to the member 60 of the range. R_2 is many-many, R_3 is many-many and R_4 is one-to-one.

Problems I

1. Draw graphs of the following relations (Z is set of integers)
   (a) $\{(1,1)\ (1,2)\ (3,5)\ (4,6)\ (5,20)\}$
   (b) $\{(1,-1)\ (1,8)\ (2,6)\ (3,6)\ (3,-2)\}$
   (c) $\{(1.5, 2)\ (1.6, 3)\ (1.7, 5)\ (1.8, 7)\ (1.9, 9)\}$
   (d) The ordered pairs (x,y) such that $x < y$ $(x,y \in Z)$
   (e) The ordered pairs (x,y) such that $|x - y| < 6$ $(x,y \in Z)$

2. What are the domains and ranges of the above relations?

3. Classify the above relations as being one-one, many-one , one-many or many-many.

Functions

Several different definitions will serve to illustrate the ideas of functions although each definition is equivalent.

Definition 1

A function is a relation in which no two of its ordered pairs have the same first member but different second members. Looking back again at our examples, R_1 is a function because all the first members of the ordered pairs are different. R_2 is not a function since there are two pairs with the same first element – (C,D) and (C,E). R_3 is not a function. R_4 is a function and, in words, we would say the daily number of vehicles is a function of the models produced.

Definition 2

A function is a mapping that exists between two sets X and Y such that to each element $x \in X$ there is associated one and only one element $y \in Y$.

To illustrate the mapping idea we can use diagrams and in such a representation (Figure 2.14), relation R_1, which is a function, would be

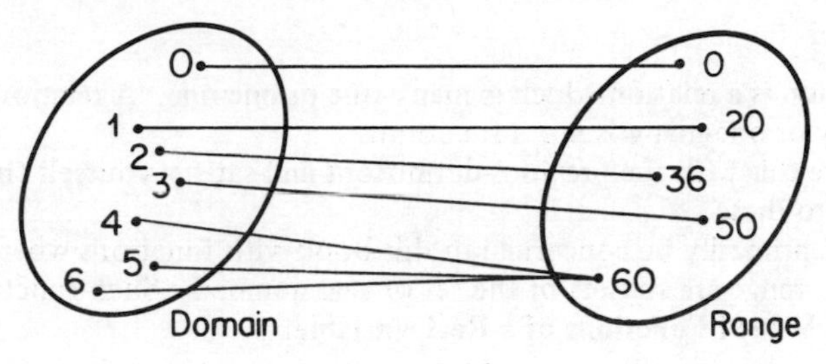

Figure 2.14

Notice elements 4, 5, 6 all related to 60, but this does not violate the definition.

Relation R_2 would be as represented in Figure 2.15

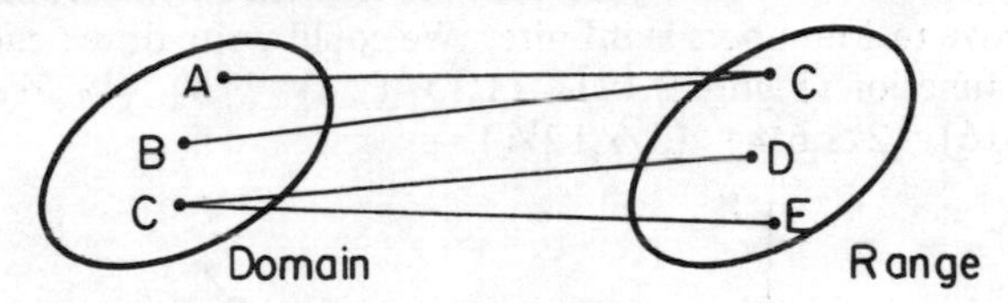

Figure 2.15

Notice this relation violates the definition of a function since C is related to two points, D and E, of the range.

We can draw similar diagrams (Figure 2.16) for R_3 and R_4 to get

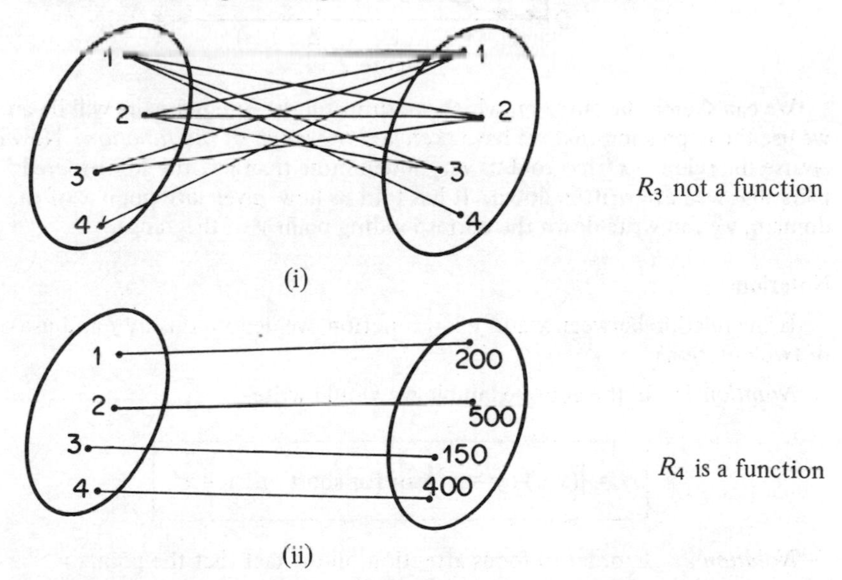

Figure 2.16

Definition 3

A function is a relation which is many-one or one-one. A relation which is many-many or one-many is *not* a function.

Compare this with the previous definitions and satisfy yourself that it is equivalent to these.

We shall primarily be concerned in this book with functions where the domain and range are subsets of the set of real numbers. Such functions are called **Real Valued Functions of a Real Variable.**

Example 5

Very often we are given some rule by which we can obtain the ordered pairs of the function. For example, we might be given that the function is the set of ordered pairs $\{(x, y)\}$ such that $x \in R^+, y \in R^+$ (+ve reals) and $y = x^2$. We could never write down all the ordered pairs that are elements of this function since the set of positive real numbers is infinite. We could write down enough to draw a **graph** of the function (Figure 2.17) – (1,1) (2,4) (3,9) $(\frac{1}{4},\frac{1}{16})$ $(\frac{1}{2},\frac{1}{4})$ $(1\frac{1}{2},2\frac{1}{4})$ (4,16) $(2\frac{1}{2},6\frac{1}{4})$ $(3\frac{1}{2},12\frac{1}{4})$.

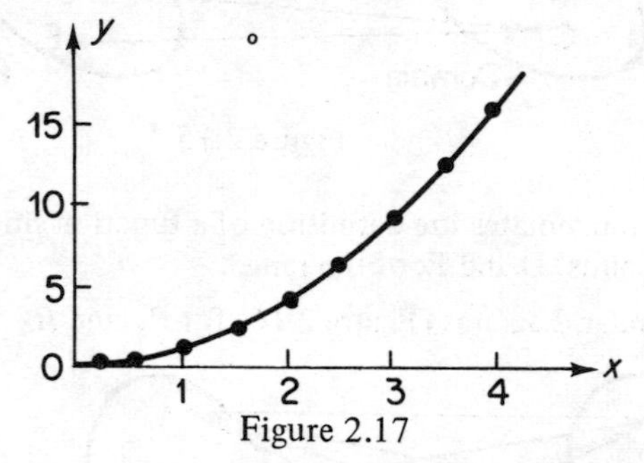

Figure 2.17

We can sketch the curve on which we think all the ordered pairs will lie and we use the expression that we have *sketched the graph of the function*. Now of course the rule $y = x^2$ has told us very much more than just the few ordered pairs that we have written down. It has told us how, given *any* point x of the domain, we can write down the corresponding point y of the range.

Notation

If the relation between x and y is a function, we denote this by f and use one of two notations

Notation 1: In the above example we would write

$$f = \left\{(x,y) : y = x^2\right\} \text{ or for short } f : y = x^2$$

Notation 2: In order to focus attention on the fact that the points y arise or are *mapped* from the points x of the domain by using a function rule f, we write

$$y = f(x) = x^2 \qquad \text{[read "}y\text{ equals }f\text{ of }x\text{"].}$$

The first notation will have been met by students who have followed a "modern maths" syllabus. However, from now on we shall use the second notation, unless further clarity is desired at some point.

As was mentioned in Chapter 1, when considering certain relations (these were in fact functions) the symbol denoting an arbitrary point in the domain is called the **independent variable** of f while a symbol standing for a point in the range is called the **dependent variable.** Here x is the independent, y the dependent variable.

Further Examples of Functions and their Graphs

Example 6

$y = f(x) = 2x + 1, \quad x,y \in R$

We can work out a few of the ordered pairs giving, for example, (0,1) (1,3) (2,5) (3,7) (4,9) and we sketch the graph of the function as follows – in fact, a straight line as in Figure 2.18.

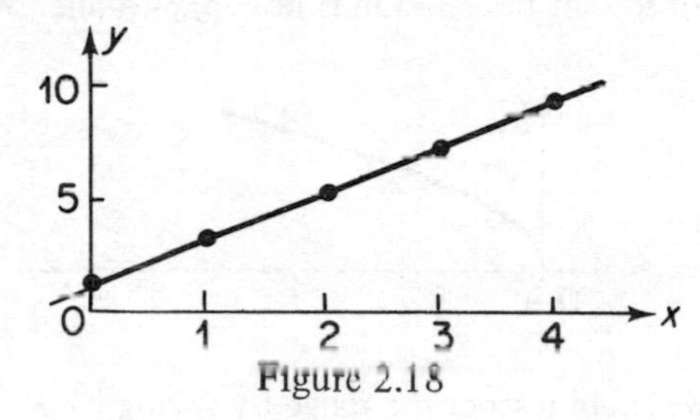

Figure 2.18

Example 7

$y = f(x) = |x|$

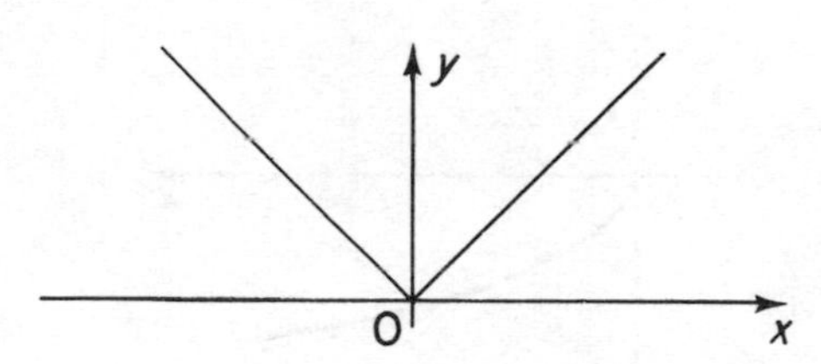

Example 8

$y = f(x) = \dfrac{|x|}{x}, \quad x \neq 0$

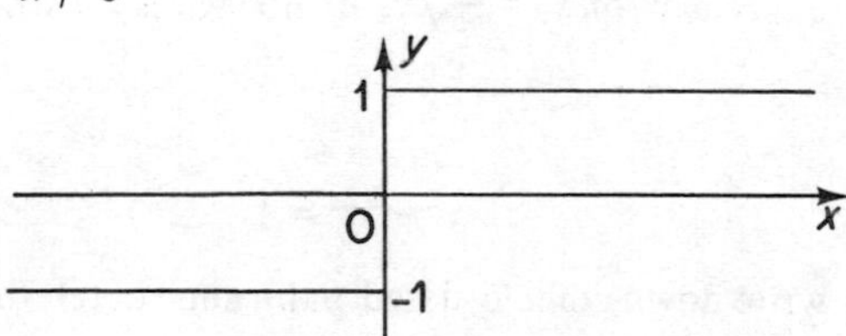

Examples of Relations that are not Functions

We may be given a formula defining a relation for which we can draw a graph but which is not a function.

Example 9

$$y = \pm\sqrt{x}$$

Working out a few ordered pairs we shall get (0,0) (1,1) (1,–1) (4,2) (4,–2) and we can see straight away that this is not a function since the relation is one-many. The square root of x always gives + or – the absolute value of $\sqrt{x}$. The graph is as follows

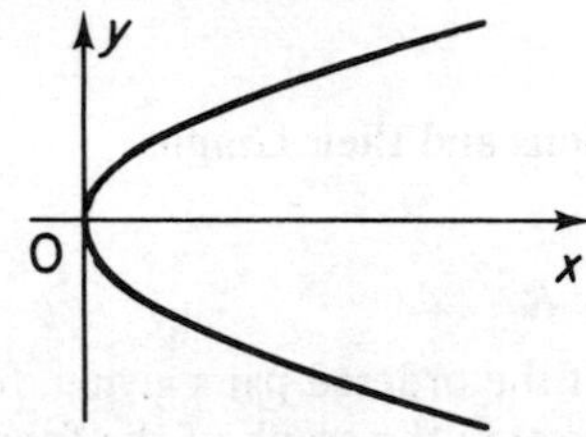

If we restrict the range to say $y \in R^+$ then this is a function and we only get the positive square root so that the relation is now one-to-one. We could then write $y = f(x) = +\sqrt{x}$

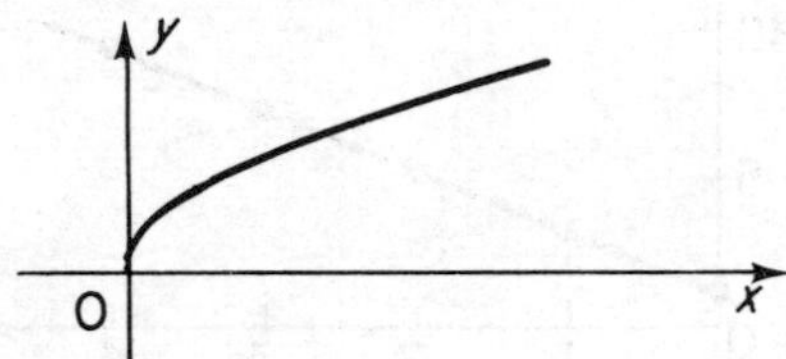

In the same way we could restrict the range by saying $y \in R^-$ so that we only consider the negative square root which then has the graph

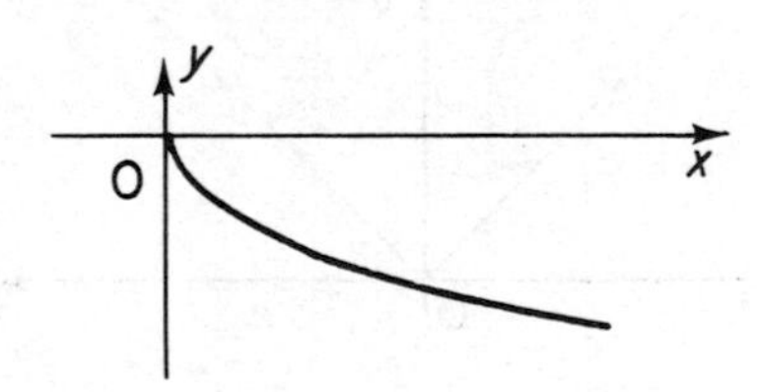

We could write this as $y = f(x) = -\sqrt{x}$. If no sign is shown, it is assumed that $y = +\sqrt{x}$.

Example 10

$$x^2 + y^2 = 1$$

Again we can write down some ordered pairs and sketch the graph as follows

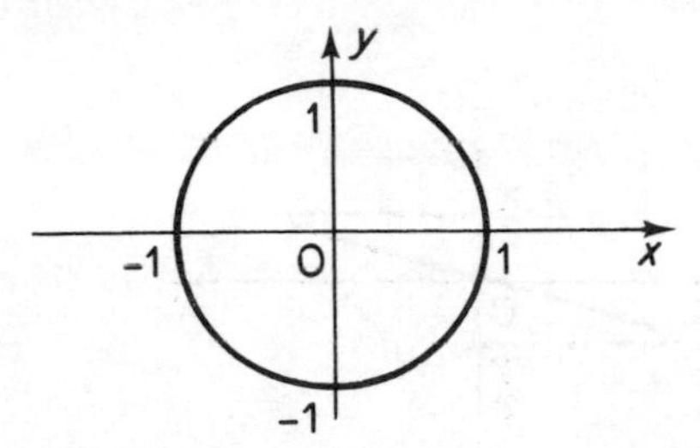

This is a circle of radius 1 and does not represent a function since more than one value of y arises from each value of x between -1 and $+1$.

Again we can make this into a functional relationship if we restrict our attention to the part of the graph where $y \geqslant 0$

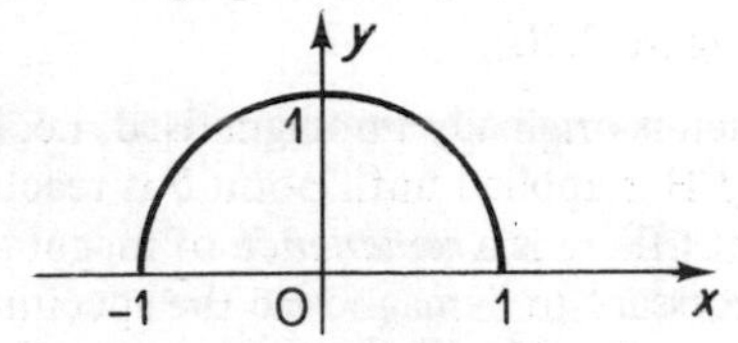

Example 11

Later on in this chapter we shall be considering the function $y = f(x) = \sin x$. Now the inverse of $y = \sin x$ is $x = \sin^{-1} y$, which means x is the angle whose sine is y.

We shall now look at the formula $y = \sin^{-1} x$. If you look in tables and construct some number pairs you will find, remembering that $\sin(x \pm 2n\pi) = \sin x$ for n integer, that the graph can be sketched as in Figure 2.19.

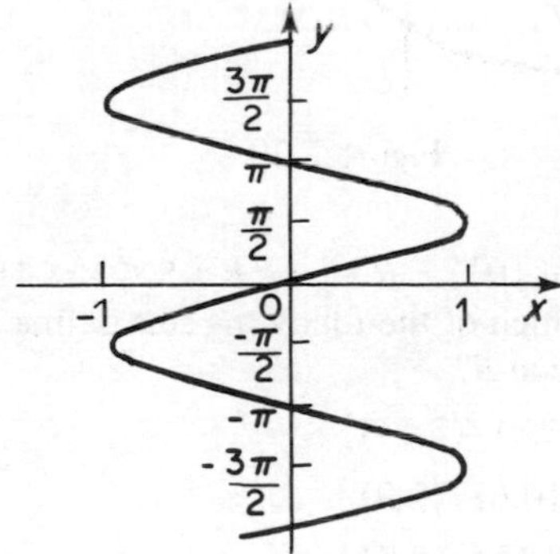

Figure 2.19

This is not a function since for each x between $+1$ and -1 there is an infinite number of values of the range.

It will be necessary in our further work on the calculus to restrict the range for $y = \sin^{-1} x$ so that it is a function and although the choice of range is ours, it is usual to take $-\pi/2 \leqslant y \leqslant \pi/2$ so that we get

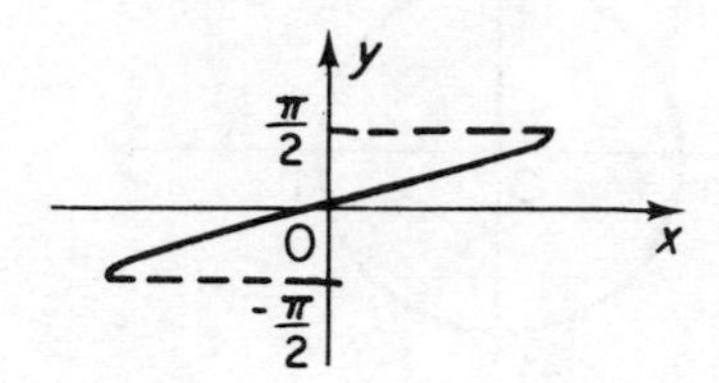

This is a function since one value of x gives rise to only one value of y.

Example 12

Hysteresis Loop (refer to Figure 2.20).

A ferromagnetic specimen is originally unmagnetised, i.e. $\mathbf{B} = \mathbf{0}$, at point a; an increasing magnetic field **H** is applied until point b is reached, when the field is then decreased to zero, but there is a *remanance* of magnetisation at c; a further *coercive force* is necessary to demagnetise the specimen. Continuation to point e, and a further reversal of **H**, allows us to trace out a closed loop: the *hysteresis loop*. The relation between **B** and **H** is many-many.

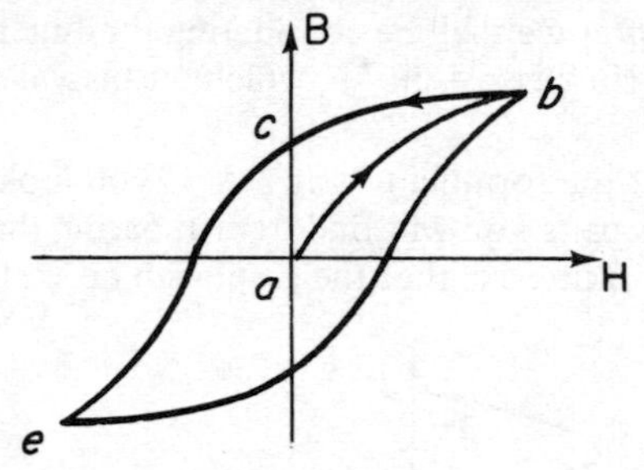

Figure 2.20

Problems II

1. If $A = \{x \in R : 0 \leqslant x \leqslant 10\}$, $B = \{x \in R : 5 \leqslant x \leqslant 15\}$, where R is the set of real numbers, which of the following sets define a function with domain in A and range in B?

   (a) $\{(0,6)\ (2,6)\ (3,8)\ (5,14)\}$

   (b) $\{(2,5)\ (3,9)\ (10,6)\ (2,9)\}$

   (c) $\{(0,6)\ (4,3)\ (7,11)\ (8,4)\}$

   (d) $\{(2,10)\ (5,11)\ (6,13)\ (2,12)\ (7,20)\}$

2. Write down the largest domain for which each of the following functions is defined

   (a) $f(x) = x^3 + 3$

   (b) $f(x) = x^2 + \sqrt{x - 1}$

(c) $f(x) = x^2 + \sqrt{1 - x^2}$

(d) $f(x) = \dfrac{1}{x} + x$

(e) $f(x) = \sqrt{x^2 - 1}$

3. State whether the mappings defined by the functions in Problem 2 are one-one or many-one.

4. Sketch the graphs of the following functions in the stated domains of definition

(a) $f(x) = x^3$ for $x \in [-3,3]$

(b) $f(x) = x^2 - 3x + 2$ for $x \in [1,2]$

(c) $f(x) = |x| - x$ for $x \in [-3,3]$

(d) $f(x) = x + \dfrac{1}{x}$ for $x \in [\tfrac{1}{2},5]$

5. Which of the following in Figure 2.21 are the graphs of functions?

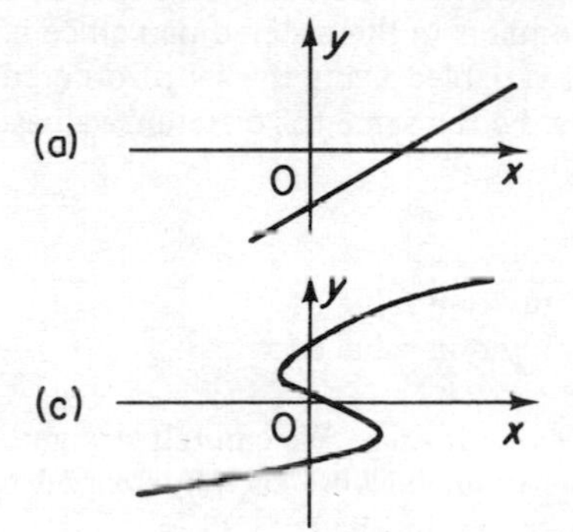

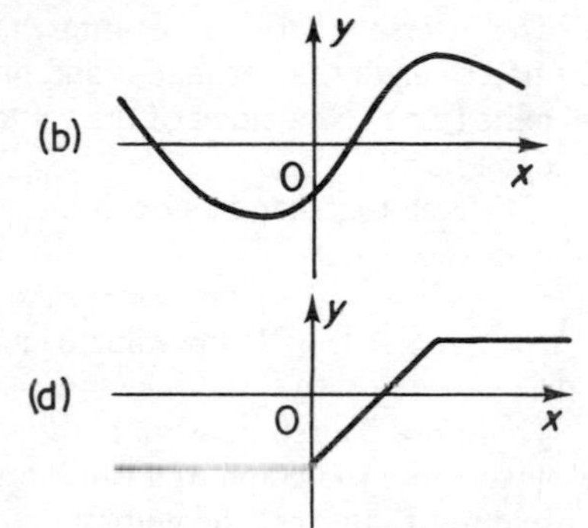

Figure 2.21

Inverse Functions

You have probably already met in physics or thermodynamics the following law, called Boyle's Law. This is an idealised law which says "the pressure and volume of a gas held at constant temperature obey the law pV = constant". In other words, if we were doing an experiment in which we measured the volume of a gas for different pressures we would find that V is a function of p, namely

$$V = f(p) = \frac{\text{constant}}{p}$$

Written in this way, p is the independent variable and V is the dependent variable. However, we could always have done the experiment in reverse: that is, we could measure the different pressures arising from different volumes and we would then obtain

$$p = g(V) = \frac{\text{constant}}{V}$$ (g used here for function to distinguish between the two functions)

This time V is the independent variable and p is the dependent variable. Of course the nature of the function g is dependent on the function f.

The function g is called the **Inverse Function** of f and must satisfy the identity $f[g(V)] \equiv V$ for all volumes V in the domain of g.

We now consider the general question: given a function f, can we find a function g that reverses the action of f? If it does exist we call it the inverse function and often write it as f^{-1}.

We know that if f is a function then one value of the domain produces only one value of the range. That is, f is many-one and cannot be one-many. If the function g exists, it must also be many-one and cannot be one-many. Therefore it follows that a function f will only possess an inverse function if it is **one-to-one.**

Put in another way, we know that the function f is the set of ordered pairs (x,y) and none of the values of x must be the same for different values of y. The inverse function must interchange the numbers in the ordered pair since it interchanges the dependent and independent variables giving the set of ordered pairs (y,x). Now none of the values of y must be the same for different values of x.

Taken together we now have

$$\begin{cases} \text{one value of } x \text{ produces only one value of } y \\ \text{one value of } y \text{ produces only one value of } x \end{cases}$$

Therefore the relationship between y and x is one-to-one. We can tell straight away from the graph of a function if it is one-to-one because each horizontal line must intersect the graph at most once as shown in Figure 2.22

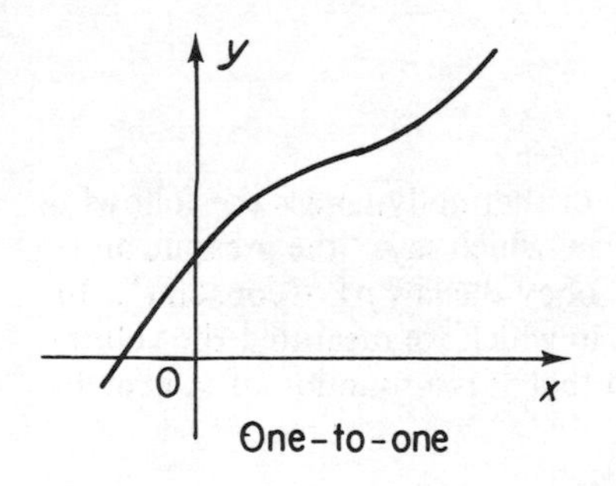

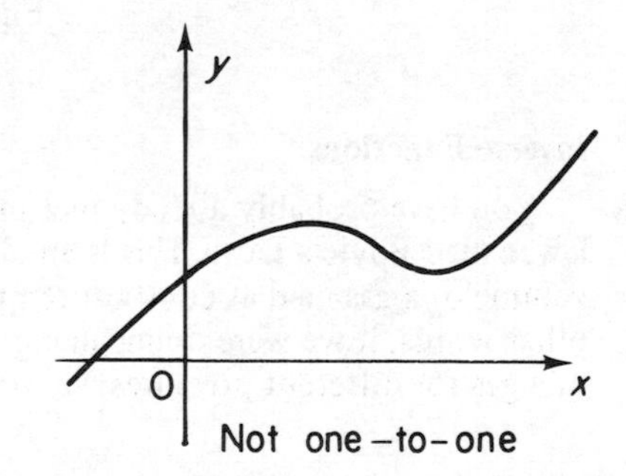

Figure 2.22

We have already met in Examples 9 and 11, functions which do not have an inverse which is also a function unless we restrict the range in some way.

In Example 9 we considered $y = \sqrt{x}$ which is the inverse of the function $y = x^2$. The graphs of these are shown in Figure 2.23

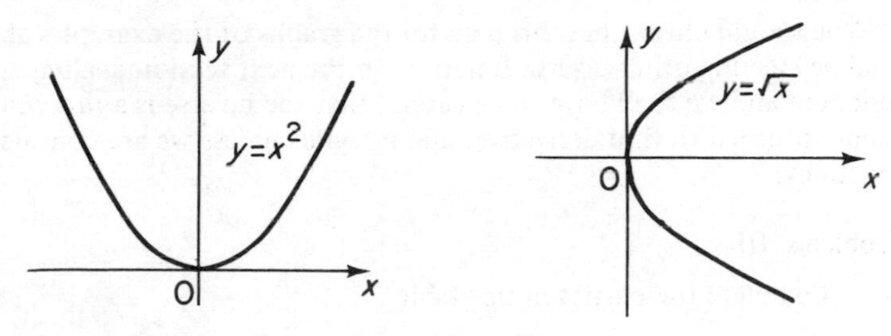

Figure 2.23

The relationships are certainly not one-to-one and we can only call $y = \sqrt{x}$ the inverse *function* if we restrict the range to $y \geqslant 0$ and write $y = +\sqrt{x}$, meaning the positive square root.

In Example 11 we considered $y = \sin^{-1} x$ which is the inverse of the function $y = \sin x$. The graphs of these are shown in Figure 2.24

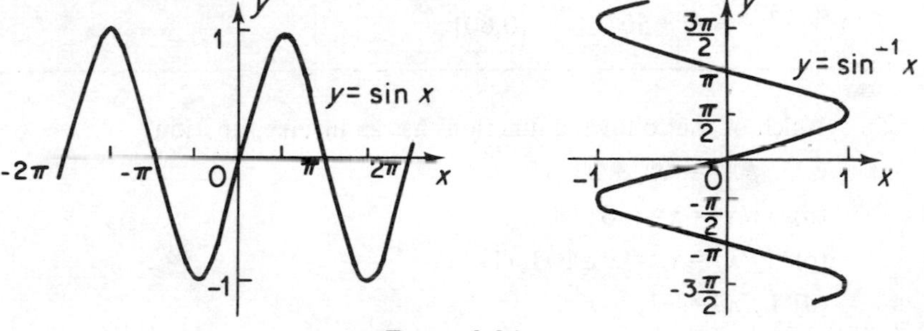

Figure 2.24

Again the inverse is not a *function* unless we restrict the range of $y = \sin^{-1} x$ and as before we usually take $-\pi/2 \leqslant y \leqslant \pi/2$.

Geometrically the graph of the inverse function f^{-1} is drawn by reflecting the graph of $y = f(x)$ in the line $y = x$, as illustrated in Figure 2.25

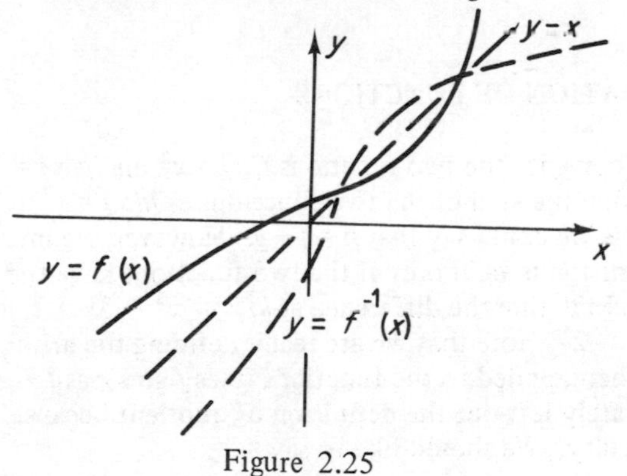

Figure 2.25

You should check that this is so for the graphs of the examples above. We shall be meeting other inverse functions in the next section dealing with standard functions and we shall have to be careful that the inverse is a *function*, since we cannot proceed to find derivatives and integrals unless we are considering functions.

Problems III

1. Complete the entries in this table

f	Domain	Is mapping one-one?	f^{-1} when it exists
x	$[-5,2]$		
$\dfrac{1}{2+x}$	$[1,4]$		
$\sin x$	$[-\pi/4, \pi/2]$		
$x^2 - 60x + 50$	$[0,60]$		

2. Which of the following functions has an inverse function?
   (a) $f(x) = 6x^2 + 2$
   (b) $f(x) = x^3 - 3$
   (c) $f(x) = x, \quad x \in [-1,+1]$
   (d) $\begin{cases} f(x) = -1, & x < -1 \\ f(x) = +1, & x \geqslant +1 \end{cases}$
   (e) $f(x) = \sqrt{1 - x^2}$

3. Find, if it exists, the inverse function of each of the functions defined in Problem 2.

2.6 COMBINATION OF FUNCTIONS

Suppose we consider the two functions $f(x) = x^2$ and $g(x) = 3x - 2$. Then it is natural to define the **sum** of the two functions as $h(x) = x^2 + 3x - 2$. In function symbols we could say that $h = f + g$. However, we must bear in mind that such a definition is valid only if the two functions have the same domain. Likewise we could define the **difference** as $k(x) = x^2 - 3x + 2$, and the **product** as $p(x) = x^2(3x - 2)$; note that we are really defining the arithmetic operations of $+$, $-$ and $\cdot$ when applied to the functions rules f and g as $f + g$, $f - g$ and $f.g$. We have deliberately left out the definition of **quotient** because there is an additional difficulty. We should like to say

$$q(x) \equiv \frac{f(x)}{g(x)} = x^2/(3x-2)$$

but the denominator vanishes at $x = 2/3$ and so we must define $q(x)$ on the domain which is the set of real numbers with the element 2/3 removed.

These ideas clearly generalise to more complicated function combinations. For example,

$$f(x) = \frac{2x^3 + \sqrt{x+2}}{x+1}$$

Here it is not quite so easy to specify its domain (the importance of such specification will become apparent in later chapters). We could break the function down into smaller components in many ways. We could write f as $(g+h) \div k$ where $g(x) = 2x^3$, $h(x) = \sqrt{x+2}$ and $k(x) = x+1$. Notice that for h to be defined, x must be $\geqslant -2$ and in order that the division can occur, x must not equal -1. The domain is then $\{x : x \neq -1 \text{ and } x \geqslant -2\}$.

Another way of combining functions is to apply them successively. Consider the two functions $f(x) = \frac{1}{2}x$ (halve) and $g(x) = x + 2$ (add two). Suppose we first form $f(x)$ and then apply the rule g. In words this says that we halve the given value of x and then add 2 to the result, i.e. $g[f(x)] = \frac{1}{2}x + 2$†. However, if we reverse the order of application, i.e. add two then halve the result we get $f[g(x)] = \frac{1}{2}(x+2) = \frac{1}{2}x + 1$†. Clearly the order of application matters. And there is a further complication: having first formed $f(x)$ we can only form $g[f(x)]$ if $f(x)$ lies in the domain of g. Consider $f(x) = \sqrt{x}$ (positive square root) and $g(x) = 4x$; f has as domain the set of positive real numbers. Now the co-domain of f is the set of positive real numbers and g can certainly be applied to this set to give $g[f(x)] = 4\sqrt{x}$, where x belongs to the set of positive real numbers. However, g has all real numbers in the domain, and it is possible for $g(x)$ to be negative. Now $f[g(-4)]$ cannot be defined and so care is needed; even if we restrict x to the positive real numbers, the domain of $g[f(x)]$ is not necessarily the same as that of $f[g(x)]$.

Consider now $f(x) = \sqrt{x}$, $g(x) = x - 2$. Here $f[g(x)] = \sqrt{x-2}$ while $g[f(x)] = \sqrt{x} - 2$. Considering $f[g(x)]$, we see that x must be real and $\geqslant 2$, while for $g[f(x)]$ we must have x as real and $\geqslant 0$. Thus in this case the domains of $f[g(x)]$ and $g[f(x)]$ are not the same.

We have already met the idea of *undoing* a function to see whether the inverse rule is itself a function. With a composite function, we must undo the functions in reverse order, as one would take off first the outermost of several wrappings comprising a parcel.

If $f(x) = 2x$, $g(x) = x + 1$ then $f[g(x)] = 2(x+1)$ and $g[f(x)] = 2x + 1$. To undo $f[g(x)]$* we note that it was formed by *adding one* then *doubling;* the inverse operations are *subtracting one* and *halving* and if these are applied in the reverse order, x becomes first $\frac{1}{2}x$ and then $\frac{1}{2}x - 1$. Thus $f[g(x)]^{-1} = \frac{1}{2}x - 1$.

---

† Sometimes $g[f(x)]$ is written $(f{\circ}g)(x)$ and $f[g(x)]$ is written $(g{\circ}f)(x)$.

* The inverse mapping is more easily seen as $(g{\circ}f)^{-1}$ and hence as $f^{-1}{\circ}g^{-1}$

This is written $g^{-1}[f^{-1}(x)] = \frac{1}{2}x - 1$. In this case the inverse mapping is a function. Similarly the inverse of $g[f(x)]$ is written $f^{-1}[g^{-1}(x)]$; again, the inverse mapping is a function.

Troubles with attempted inversion arise in a similar way to the inversion of simple functions.

Problems

1. Give formulae for the functions $f + g$, $f - g$, $f.g$, $f \div g$ and $g \div f$ with the appropriate domains when $f(x) = x - 3$, $g(x) = x^2 - 1$.
2. Repeat Problem 1 for $f(x) = x^2 - 2x + 1$, $g(x) = x - 1$
3. If $f(x)$, $g(x)$ are as in Problems 1 and 2, give formulae for $f[g(x)]$ and $g[f(x)]$ stating the domains in each case.
4. If x belongs to the domain of f and $f(x) = x$, it is called a *fixed point* of the function. What other requirement is there on x? Find the fixed points of $f(x) = x$, x^2, $x^2 - x + 1$, $x^2 - 3x + 3$.
5. Give an example of $f(x)$ such that $f[f(x)] = x$ and an example of $g(x)$ such that $g[g(x)] = g(x)$; give the domain as well as the formula.

2.7 PARTIAL FRACTIONS

We know how to put expressions such as

$$\frac{1}{x-3} - \frac{2x-3}{x^2+x+1} + \frac{2}{x+3}$$

on to a common denominator and obtain the expression

$$\frac{(x+3)(x^2+x+1) - (2x-3)(x-3)(x+3) + 2(x-3)(x^2+x+1)}{(x-3)(x+3)(x^2+x+1)}$$

This simplifies to

$$\frac{x^3+3x^2+18x-30}{x^4+x^3-8x^2-9x-9}$$

It is often useful to be able to perform the reverse process. That is, if we are given the expression

$$\frac{x^3+3x^2+18x-30}{x^4+x^3-8x^2-9x-9}$$

to split it into its component parts

$$\frac{1}{x-3} - \frac{2x-3}{x^2+x+1} + \frac{2}{x+3}$$

The parts $1/(x-3)$, $-(2x-3)/(x^2+x+1)$ and $2/(x+3)$ are called **Partial Fractions.**

The rules for splitting an expression into its partial fractions are:

1. The given expression must be a **Proper Fraction** – by this we mean that the degree† of the numerator must be less than the degree of the denominator. If it is not a proper fraction, then we must divide out to produce a proper fraction,

e.g. $$\frac{3x^3 + x^2 - 2x + 1}{x^2 - 3x + 2}$$

is not a proper fraction (degree of the numerator is 3, degree of the denominator is 2).

Hence we divide out

$$\begin{array}{r} 3x + 10 \\ x^2 - 3x + 2 \,\big)\, \overline{3x^3 + x^2 - 2x + 1} \\ 3x^3 - 9x^2 + 6x \\ \hline 10x^2 - 8x + 1 \\ 10x^2 - 30x + 20 \\ \hline 22x - 19 \end{array}$$

The expression is $3x + 10 + \dfrac{22x - 19}{x^2 - 3x + 2}$

and we would split the second part of the expression, which is a proper fraction, into partial fractions.

2. The denominator is factorised as far as is possible.

3. Suppose the expression is

$$\frac{f(x)}{(x - \alpha)(x - \beta)^3 (x^2 + \gamma x + \epsilon)(x^2 + \nu x + \eta)^2}$$

factorised as far as possible. This expression will have partial fractions of the form

$$\frac{A}{x-\alpha} + \underbrace{\frac{B}{x-\beta} + \frac{C}{(x-\beta)^2} + \frac{D}{(x-\beta)^3}}_{\text{all arising from } (x-\beta)^3 \text{ factor}} + \frac{Ex+F}{x^2+\gamma x+\epsilon} + \underbrace{\frac{Gx+H}{(x^2+\nu x+\eta)} + \frac{Ix+J}{(x^2+\nu x+\eta)^2}}_{\text{both arising from } (x^2+\nu x+\eta)^2 \text{ factor}}$$

The constants $A, B, C, \ldots\ldots\ldots, J$ are as yet unknown. Notice that in a partial fraction for a non-repeated factor, the numerator is degree one less than the denominator. A repeated factor must have all the possible lower degree fractions.

† degree of a polynomial expression is the number of the highest power of x occurring in the expression

4. The constants can be found in several ways as will be illustrated in the following examples.

Example 1

$$\frac{3x-2}{(x^2-3x+2)}$$

This is a proper fraction and can be written in factorised form as

$$\frac{3x-2}{(x-1)(x-2)}$$

It will have partial fractions

$$\frac{A}{x-1}+\frac{B}{x-2}$$

To find A and B we put this back on a common denominator and compare with the given expression.

Thus $$\frac{3x-2}{(x-1)(x-2)} \equiv \frac{A(x-2)+B(x-1)}{(x-1)(x-2)}$$

We put *identically equal* because it must be true for all values of x. Hence

$$3x-2 \equiv A(x-2)+B(x-1) \qquad (2.1)$$

Method 1:
Compare coefficients in the numerator [use identity (2.1)]

$$(A+B) = 3 \quad \text{(coefficients of } x\text{)}$$
$$-2A-B = -2 \quad \text{(constant term)}$$

adding

$$A = -1$$

therefore

$$B = 4.$$

Method 2:
Put x equal to any convenient number in equation (2.1).
Put $x = 2$, L.H.S = 4, R.H.S. = B therefore $B = 4$
Put $x = 1$, L.H.S. = 1, R.H.S. = $-A$ therefore $A = -1$

Thus

$$\frac{3x-2}{x^2-3x+2}=\frac{-1}{x-1}+\frac{4}{x-2}$$

An equivalent process to Method 2 is called the **cover-up** rule.
We write

$$\frac{3x-2}{(x-1)(x-2)}=\frac{?}{x-1}+\frac{?}{x-2}$$

To find the numerator of the first partial fraction, we put $x = 1$ (from the factor $x - 1$) in the L.H.S. after covering up the $(x - 1)$ factor giving $1/-1 = -1$. For the second partial fraction $?/(x - 2)$ we put $x = 2$ in the L.H.S. after covering up the factor $x - 2$ giving $4/1 = 4$.

Thus

$$\frac{3x-2}{(x-1)(x-2)} = \frac{-1}{x-1} + \frac{4}{x-2}$$

Example 2

$$\frac{2x^2 - x}{(x-1)(2x+1)(x+2)}$$

Let

$$\frac{2x^2 - x}{(x-1)(2x+1)(x+2)} = \frac{A}{x-1} + \frac{B}{2x+1} + \frac{C}{x+2}$$

therefore

$$\frac{2x^2 - x}{(x-1)(2x+1)(x+2)} \equiv \frac{A(2x+1)(x+2) + B(x-1)(x+2) + C(x-1)(2x+1)}{(x-1)(2x+1)(x+2)}$$

therefore

$$2x^2 - x \equiv A(2x+1)(x+2) + B(x-1)(x+2) + C(x-1)(2x+1) \quad (2.2)$$

Method 1:

Compare coefficients in (2.2)

x^2: $\quad 2 = 2A + B + 2C$

x : $\quad -1 = 5A + B - C$

constant: $\quad 0 = 2A - 2B - C$

(The first equation) + 2(second equation) gives $0 = 12A + 3B$

(The second equation) – (third equation) yields $-1 = 3A + 3B$

Hence

$$A = \frac{1}{9} \quad B = \frac{-4}{9} \quad C = \frac{10}{9}$$

Method 2:

Put x equal to any convenient number in (2.2)

Put $x = 1$, $\quad 1 = 9A \quad$ therefore $\quad A = 1/9$

Put $x = -1/2$ (makes $2x + 1 = 0$), $1 = (-9/4)B \quad$ therefore $B = -4/9$

Put $x = -2$, $\quad 10 = 9C$, $\quad C = 10/9$

Thus

$$\frac{2x^2 - x}{(x-1)(2x+1)(x+2)} = \frac{1/9}{x-1} - \frac{4/9}{2x+1} + \frac{10/9}{x+2}$$

Alternatively, by cover-up rule,

$$\frac{2x^2 - x}{(x-1)(2x+1)(x+2)} = \frac{?}{x-1} + \frac{?}{2x+1} + \frac{?}{x+2}$$

Putting $x = 1$ in L.H.S. after covering up $(x - 1)$ factor gives
first coefficient $= 1/9$

Putting $x = -1/2$ in L.H.S. after covering up $(2x + 1)$ factor gives
second coefficient $= -1/(9/4) = -4/9$

Putting $x = -2$ in L.H.S. after covering up $(x + 2)$ factor gives
third coefficient $= 10/9$

Example 3

$$\frac{x^3 - x + 1}{(x-1)(x^2 + 2x + 1)}$$

This is an improper fraction since degree of the numerator = degree of denominator = 3.

Thus we must divide first

$$\begin{array}{r} 1 \\ x^3 + x^2 - x - 1 \,\overline{)\, x^3 + 0 \;\; - x + 1} \\ x^3 + x^2 - x - 1 \\ \hline -x^2 \qquad +2 \end{array}$$

therefore

$$\frac{x^2 - x + 1}{(x-1)(x^2 + 2x + 1)} = 1 - \frac{x^2 - 2}{(x-1)(x^2 + 2x + 1)} = 1 - \frac{x^2 - 2}{(x-1)(x+1)^2}$$

Let

$$\frac{x^2 - 2}{(x-1)(x+1)^2} \equiv \frac{A}{x-1} + \frac{B}{x+1} + \frac{C}{(x+1)^2}$$

Putting the R.H.S. back on the *same* common denominator

$$x^2 - 2 \equiv A(x+1)^2 + B(x-1)(x+1) + C(x-1) \qquad (2.3)$$

We shall use a combination of methods 1 and 2 here.

In equation (2.3),

put $x = 1$ giving $-1 = 4A$ therefore $A = -1/4$

put $x = -1$ giving $-1 = -2C$ therefore $C = 1/2$

Compare the coefficient of x^2: $1 = A + B$ therefore $B = 5/4$

Hence

$$\frac{x^3 - x + 1}{(x-1)(x^2 + 2x + 1)} = 1 + \frac{1/4}{x-1} - \frac{5/4}{x+1} - \frac{1/2}{(x+1)^2}$$

We stress here that the cover-up rule will only work for *linear factors* and for the *highest power of a repeated linear factor.* In other words, we can find A and C using the rule but not B. Using the cover-up rule we would proceed as follows:

$$\frac{x^3 - x + 1}{(x-1)(x^2+2x+1)} = 1 - \frac{x^2 - 2}{(x-1)(x^2+2x+1)}$$

and

$$\frac{x^2 - 2}{(x-1)(x+1)^2} = \frac{?}{x-1} + \frac{B}{x+1} + \frac{?}{(x+1)^2}$$

$$\equiv \frac{-1/4}{x-1} + \frac{B}{x+1} + \frac{1/2}{(x+1)^2}$$

To find B, we must put the expression back on a common denominator giving

$$x^2 - 2 \equiv -\tfrac{1}{4}(x+1)^2 + B(x^2-1) + \tfrac{1}{2}(x-1)$$

Comparing the coefficient of x^2 in this identity gives B as before.

Example 4

$$\frac{2x-10}{(x^2+2x+5)(x^2-2x+1)(x+3)} = \frac{2x-10}{(x^2+2x+5)(x-1)^2(x+3)}$$

We shall use the cover-up rule first of all.

We get

$$\frac{2x-10}{(x^2+2x+5)(x-1)^2(x+3)} \equiv \frac{Ax+B}{x^2+2x+5} + \frac{C}{x-1} + \frac{?}{(x-1)^2} + \frac{?}{x+3}$$

$$\equiv \frac{Ax+B}{x^2+2x+5} + \frac{C}{x-1} - \frac{1/4}{(x-1)^2} - \frac{1/8}{x+3}$$

To find A, B, C we put the expression back on the *same* common denominator, giving

$$2x - 10 \equiv (Ax+B)(x-1)^2(x+3) + C(x^2+2x+5)(x-1)(x+3) - \frac{1}{4}(x^2+2x+5)(x+3) - \frac{1}{8}(x^2+2x+5)(x-1)^2 \qquad (2.4)$$

In equation (2.4),

compare coefficients of x^4: $\quad 0 = A + C - 1/8$

constant: $\quad -10 = 3B - 15C - 15/4 - 5/8$

Put $x = -1$ $\quad -12 = (-A+B)8 - 16C - 2 - 2$

Hence, on rearranging,

$$\begin{aligned} A + C &= 1/8 \\ B - 5C &= -15/8 \\ A - B + 2C &= 1 \end{aligned}$$

These give

$$\begin{aligned} B - 5C &= -15/8 \\ -B + C &= 7/8 \end{aligned}$$

Solving these, we have

$$C = 1/4, \quad B = -5/8, \quad A = -1/8$$

Hence

$$\frac{2x - 10}{(x^2 + 2x + 5)(x - 1)^2(x + 3)} \equiv \frac{-(x + 5)}{8(x^2 + 2x + 5)} + \frac{1/4}{x - 1} - \frac{1/4}{(x - 1)^2} - \frac{1/8}{x + 3}$$

This was a difficult example and you are unlikely to meet such complicated expressions in practice, but notice how we used a combination of the different methods.

Problems

1. Express in Partial Fractions the following expressions using both comparison of coefficients, substitution and the cover-up rule methods.

   (a) $\dfrac{3x - 1}{2x^2 - x - 1}$

   (b) $\dfrac{2x^3 + 3x^2 - 3}{2x^2 - x - 1}$

   (c) $\dfrac{8 - x^2}{4 - x^2}$

   (d) $\dfrac{3x^2 + 2x + 1}{(x - 1)(x + 2)(x + 3)}$

2. Express in Partial Fractions using any convenient method

   (a) $\dfrac{1}{(x + 1)(x^2 + 9)}$

   (b) $\dfrac{3x - 2}{(x - 1)^2(x^2 + 3x + 1)}$

   (c) $\dfrac{4x^2 + 2x - 1}{x^2(2x + 1)}$

   (d) $\dfrac{x}{x^3 + 1}$

2.8 STANDARD FUNCTIONS OF THE CALCULUS

The Trigonometric Functions, their Graphs and Inverses

You will already have met in trigonometry, expressions for the sine, cosine and tangent of an angle. We now remind you of the relevant properties and then consider the functions $f(x) = \sin x$, $f(x) = \cos x$ and $f(x) = \tan x$. Later we shall consider the inverses of these functions.

Given the right-angled triangle ABC

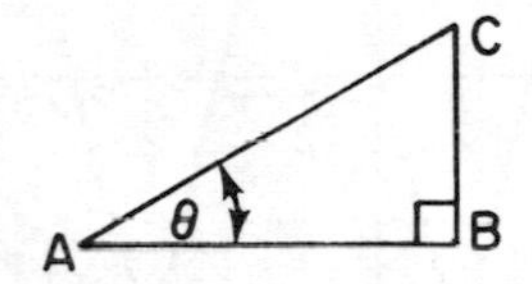

we define $\sin\theta = BC/AC$, $\cos\theta = AB/AC$, $\tan\theta = CB/AB$.
For angles greater than 90° we use the well-known mnemonic

sin	all
tan	cos

This summarises the following rules: that the sine, cosine and tangent of an angle between 0° and 90° are all positive. For angles between 90° and 180°, only the sine is positive (cosine and tangent are negative) and similarly for angles 180° to 270° and 270° to 360°. This, together with the properties $\sin(\theta + n.360^\circ) = \sin\theta$, $\cos(\theta + n.360^\circ) = \cos\theta$, $\tan(\theta + n.360^\circ) = \tan\theta$ for all $n \in Z$ (the set of all positive and negative integers), enables us to find the sine, cosine and tangent of any angle.

The following identities should be familiar

$$\sin^2 A + \cos^2 A = 1$$

$$\sin(A \pm B) = \sin A \cos B \pm \cos A \sin B$$

$$\cos(A \pm B) = \cos A \cos B \mp \sin A \sin B$$

$$\tan(A \pm B) = \frac{\tan A \pm \tan B}{1 \mp \tan A \tan B}$$

$$\sin 2A = 2 \sin A \cos A$$

$$\cos 2A = \cos^2 A - \sin^2 A = 2\cos^2 A - 1 = 1 - 2\sin^2 A$$

$$\tan 2A = \frac{2\tan A}{1 - \tan^2 A}$$

The following are useful results to remember

$\sin 0 = 0$, $\sin 30^\circ = 1/2$, $\sin 45^\circ = 1/\sqrt{2}$, $\sin 60^\circ = \sqrt{3}/2$, $\sin 90^\circ = 1$

$\cos 0 = 1$, $\cos 30^\circ = \sqrt{3}/2$, $\cos 45^\circ = 1/\sqrt{2}$, $\cos 60^\circ = 1/2$, $\cos 90^\circ = 0$

$\tan 0 = 0$, $\tan 30^\circ = 1/\sqrt{3}$, $\tan 45^\circ = 1$, $\tan 60^\circ = \sqrt{3}$, $\tan 90^\circ = \infty$

We now consider the functions based on sine, cosine and tangent.

(i) We define the **Sine Function** as $y = f(x) = \sin x$ where the domain of x is $\{x : x \text{ is real}\}$. Further, x is always measured in radians. Using tables, the above mnemonic, and the fact that $\sin(x + n.2\pi) = \sin x$ for all $n \in Z$, we can draw the graph (Figure 2.26) of $y = f(x) = \sin x$ and obtain

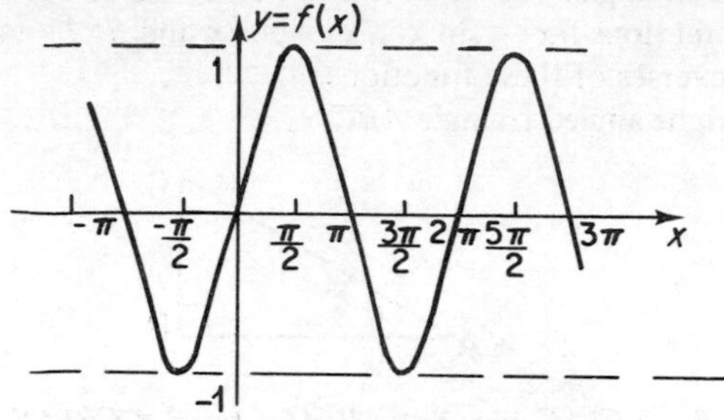

Figure 2.26

This is a **Periodic Function**. That is, the basic shape is repeated over and over again. Its **Period** is 2π – that is, it repeats itself every 2π. It is an **Odd** function. By this we mean that if the y-axis is regarded as a mirror, then the image in the mirror of the graph for $x > 0$ is exactly the same shape but is multiplied by (-1). For example the following is an odd function

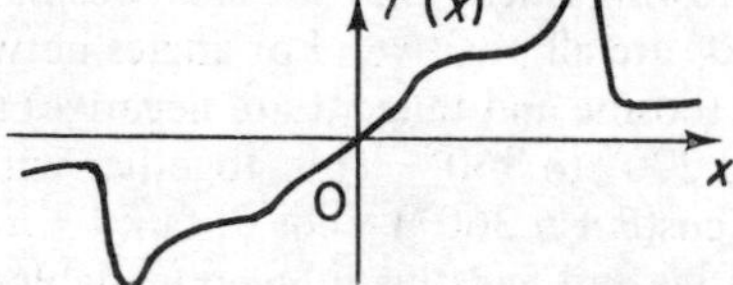

Note also that the range of the sine function is $-1 \leqslant f(x) \leqslant 1$.

(ii) $y = f(x) = \cos x$ for $x \in R$ (Real Numbers) is defined as the **Cosine Function.** The graph of this function (Figure 2.27) is

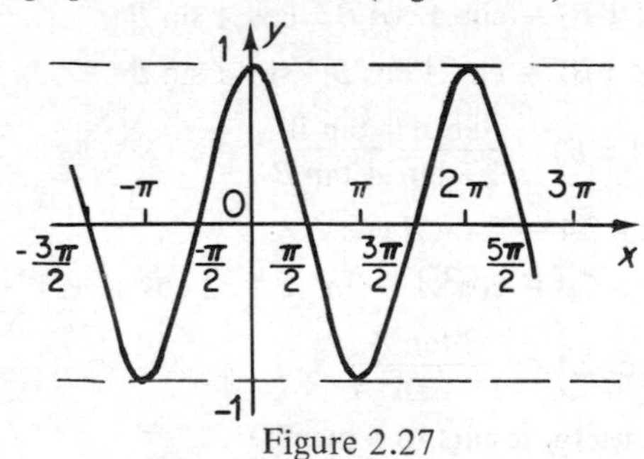

Figure 2.27

It is a *periodic function* of period 2π and $\cos(x + 2n\pi) = \cos x$ for $n \in Z$. It is an **even** function. This time the image of the graph for $x > 0$ in the y-axis is that for $x < 0$. We express this fact by saying $\cos(-x) = \cos x$.

(iii) $y = f(x) = \tan x = \left\{\dfrac{\sin x}{\cos x}\right\}$ for $x \in R$ is defined as the **Tangent Function.**
The graph of this function (Figure 2.28) is as follows

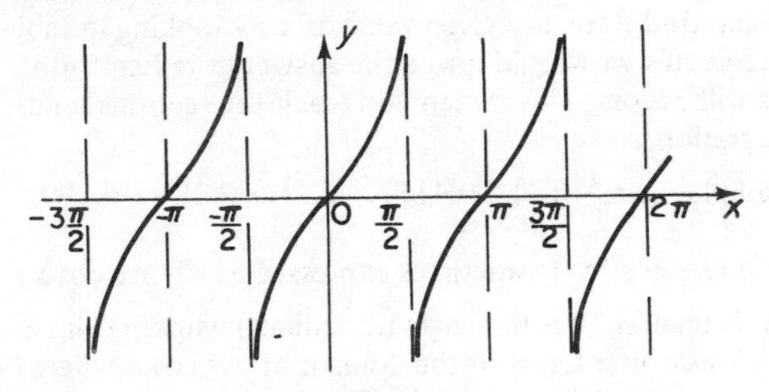

Figure 2.28

It is *periodic* of period π so that we write $\tan(x + n\pi) = \tan x$ for $n \in Z$.
It is an **odd** function. That is, $\tan(-x) = -\tan x$.

As an exercise, you should draw the graphs of $y = f(x) = 1/(\sin x) = \operatorname{cosec} x$, $y = f(x) = 1/(\cos x) = \sec x$ and $y = f(x) = 1/(\tan x) = \cot x$. Also write down the identities satisfied for compound angles by these functions and show that

$$\sec^2 A = 1 + \tan^2 A$$
$$\operatorname{cosec}^2 A = 1 + \cot^2 A$$

You will find the graphs of these functions in the Appendix at the end of the book.

The inverse trigonometric functions

(i) $y = f(x) = \sin^{-1} x$ (sometimes expressed as $\arcsin x$)

This is read as "y is the angle (in radians) whose sine is x."
Now the function $y = \sin x$ is not one-to-one and so we can only talk of its inverse relation as being a function if we restrict the range. The customary choice of range for the inverse function is $[-\pi/2 \text{ to } \pi/2]$. The graph of the inverse function is sketched in Figure 2.29, the dotted parts being inadmissible for our definition of the inverse function. It can easily be obtained by drawing the image in the line $y = x$ of the graph of $y = \sin x$.

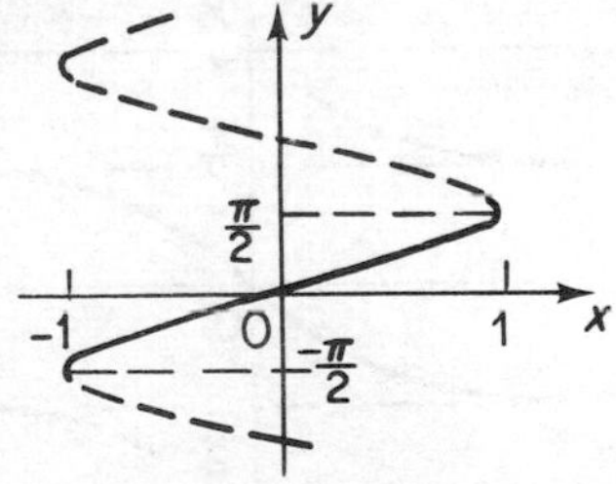

Figure 2.29

Notice that the domain of the function is the set of numbers x such that $-1 \leqslant x \leqslant 1$. For $|x| > 1$, the function is not defined. It is **not** *periodic*. It is an *odd* function, that is $\sin^{-1}(-x) = -\sin^{-1} x$.
We can find y for any given value of x by looking in tables but for work in the calculus we should quote the answer in radian form. The reason for this will become clear when you reach the sections on differentiation and integration.

Examples $\sin^{-1}(0.4) = 0.4117$, $\sin^{-1}(-0.2) = -0.1993$

(ii) $y = f(x) = \cos^{-1} x$ (sometimes expressed as $y = \text{arc}\cos x$)

This is read as "y is the angle (in radians) whose cosine is x".
We choose an interval of the domain of $y = \cos x$ where $\cos x$ is one-to-one. Usually this is taken as $[0,\pi]$. Then the inverse function has a graph as in Figure 2.30 with range $[0,\pi]$ (the dotted parts only being shown to show the relationship to $y = \cos x$)

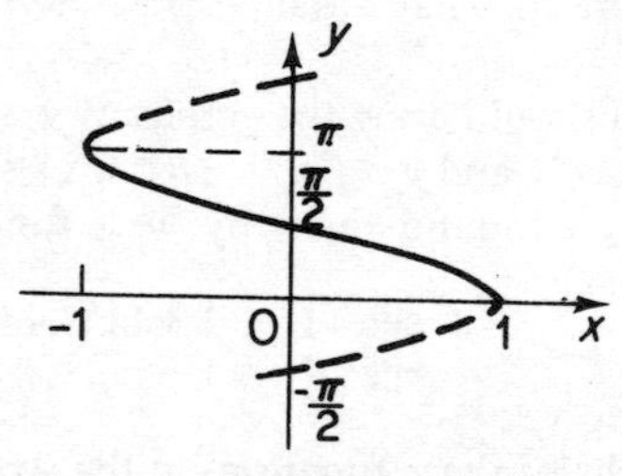

Figure 2.30

Notice that $y = \cos^{-1} x$ is defined only for x such that $-1 \leqslant x \leqslant 1$. It is **not** *periodic*.
It is *neither even nor odd*.

Examples $\cos^{-1}(0.5) = 1.0472$, $\cos^{-1}(-0.3) = 2.8394$ from tables.

(iii) $y = f(x) = \tan^{-1} x$ (or $y = \text{arc}\tan x$)

If we restrict the domain of the function $y = \tan x$ to the interval $[-\pi/2, \pi/2]$ then it is one-to-one and the inverse function can be defined with range $[-\pi/2, \pi/2]$. The sketch of this inverse function is as in Figure 2.31.

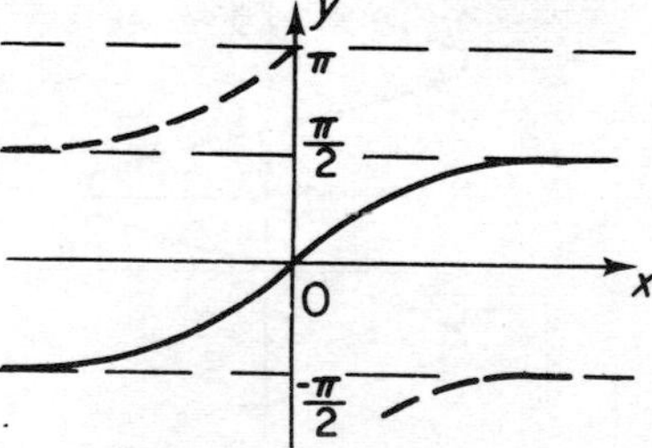

Figure 2.31

Notice that x may take any value from $-\infty$ to $+\infty$. It is **not** *periodic.* It is an *odd* function, that is, $\tan^{-1}(-x) = -\tan^{-1}x$.

Numerical examples are $\tan^{-1}(-191) = -1.5656$, $\tan^{-1}(2) = 1.1071$. Note again the use of radians !

You should now consider the inverse functions $y = f(x) = \text{cosec}^{-1}x$, $y = f(x) = \sec^{-1}(x)$, $y = f(x) = \cot^{-1}(x)$. Define a suitable interval and state whether they are periodic functions and whether they are even or odd or neither. With your definition then find $\text{cosec}^{-1}(0.3)$, $\sec^{-1}(\pm 0.4)$ and $\cot^{-1}(\pm 0.7)$.

You will find the graphs of these inverse functions in the Appendix.

The logarithmic function

You have already met the use in calculations of logarithms to the base ten. They have the properties

(i) $\log_{10}(a \times b) = \log_{10}a + \log_{10}b$
(ii) $\log_{10}(a^n) = n \log_{10}a$
(iii) $\log_{10}(1) = 0$
(iv) $\log_{10}(10) = 1$

It can be shown that we can replace the base 10 in the above by any real base we care to choose. For example, with a base α we get

$$\begin{aligned} \log_\alpha (a \times b) &= \log_\alpha a + \log_\alpha b \\ \log_\alpha (a^n) &= n \log_\alpha a \\ \log_\alpha (1) &= 0 \\ \log_\alpha (\alpha) &= 1 \end{aligned}$$

The whole theory of logarithms hinges on the fact that

$$\text{If } x = \log_\alpha a \text{ then } a = \alpha^x$$

A special case is where the base is e, which in the section on sequences is defined by

$$e = \lim_{n \to \infty} \left(1 + \frac{1}{n}\right)^n$$

The value of e is 2.718......... Logarithms to the base e arise in many physical situations and are called **Natural Logarithms.**

Thus we define the natural logarithmic function as

$$y = \log_e x \quad \text{with } e = 2.718......$$

(In science and engineering the notation $y = \ln x$ is often used.)

It satisfies, of course, the rules

$$\begin{aligned} \log_e x_1 + \log_e x_2 &= \log_e (x_1 x_2) \\ \log_e x^n &= n \log_e x \end{aligned}$$

$$\log_e(1) = 0, \quad \log_e(e) = 1$$

$$\text{and if } y = \log_e x, \text{ then } x = e^y.$$

Its graph can be drawn by referring to tables and is sketched in Figure 2.32.

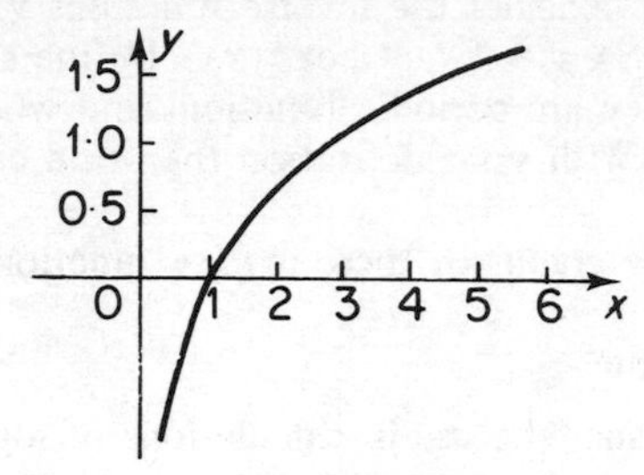

Figure 2.32

Notice that $y = \log_e x$ is only defined for $x \geqslant 0$ and that as x approaches 0 through positive values of x (usually written $x \downarrow 0$) then $\log_e$ tends to $-\infty$. Notice that it has a one-to-one relation and as such must possess an inverse which will be the inverse function. This inverse function is called the **Exponential Function** and details are given in the next sub-section.

The exponential function

Remembering the result that if

$$y = \log_e x \text{ then } x = e^y$$

we conclude that the function which will *undo* the function $y = \log_e x$ is the function $y = e^x$. This is the **exponential function**. (For convenience in printing and for large expressions we often write $y = \exp(x)$ instead of $y = e^x$.)

A sketch of its graph is obtained either by the use of tables or by reflecting the graph of $y = \log_e x$ in the line $y = x$. It is shown in Figure 2.33.

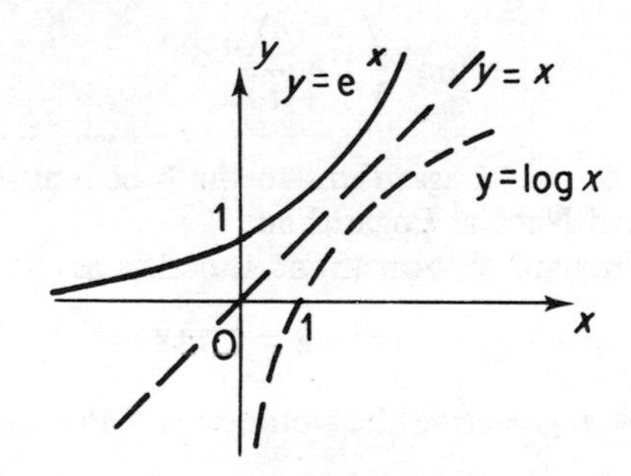

Figure 2.33

Notice that $y = e^x$ is defined for all values of x and that as x tends to a large negative value, e^x tends to zero.

Besides the usual indicial law that $e^{x_1}.e^{x_2} = e^{x_1+x_2}$ we have the very important result that

$$e^{\log_e x} = x$$

We can prove this easily; by putting

$$z = e^{\log_e x}$$

and taking logarithms to the base e we get

$$\log_e z = \log_e (e^{\log_e x}) = (\log_e x)(\log_e e) = (\log_e x).1 = \log_e x$$

Hence $z = x$.

We mention also the associated function $y = e^{-x} = 1/e^x = (1/e)^x$, very often called the *curve of exponential decay,* which has a graph as shown in Figure 2.34.

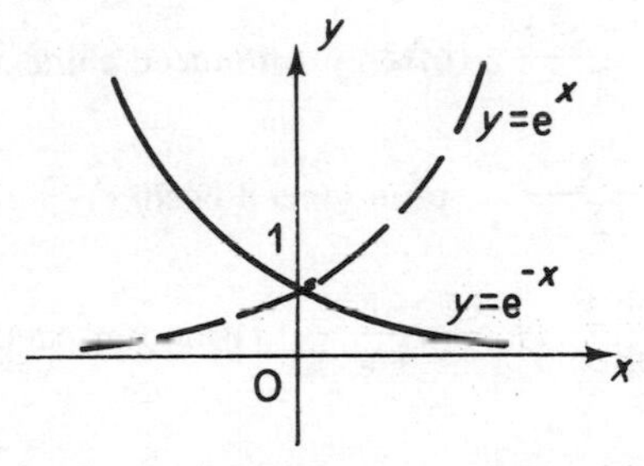

Figure 2.34

Exponential decays occur widely in physical situations – indeed we have already met this shape in Chapter 1, when looking at the cooling of a liquid.

One other associated curve is that of the function $y = a^x$ for some number $a > 1$. Its graph, Figure 2.35, is of the same general shape as $y = e^x$.

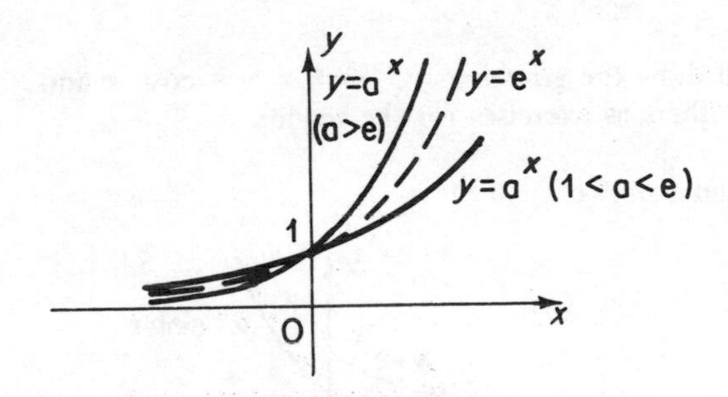

Figure 2.35

We note that $y = a^x = e^{x \log_e a}$.

It is worth mentioning here that e^x increases more rapidly than any power of x whilst $\log_e x$ increases more slowly than any power of x.

Since the $\log_e$ function is used so much in calculus and in science and engineering, in the rest of this *book we shall drop the* e *and write* $\log x$. It will only be when bases other than e are used that we shall show the base.

Arising from the exponential function we can introduce some new functions, called the **Hyperbolic Functions.** These are now examined, together with their inverses.

The hyperbolic functions

These functions which are combinations of the exponential function occur widely in applications, especially engineering, and have properties reminiscent of the trigonometric functions. As the name suggests, they were introduced originally by mathematicians to deal with work on a curve called the hyperbola. The hyperbola is discussed later in Chapter 4.

The hyperbolic functions are defined as follows:

$$\sinh x = \frac{e^x - e^{-x}}{2} \quad \text{(often pronounced shine } x \text{ or sinch } x\text{)}$$

$$\cosh x = \frac{e^x + e^{-x}}{2} \quad \text{(pronounced cosh } x\text{)}$$

$$\tanh x = \frac{\sinh x}{\cosh x} = \frac{e^x - e^{-x}}{e^x + e^{-x}} \quad \text{(pronounced than } x \text{ or tanch } x\text{)}$$

$$\operatorname{cosech} x = \frac{1}{\sinh x}$$

$$\operatorname{sech} x = \frac{1}{\cosh x}$$

$$\coth x = \frac{1}{\tanh x}$$

We shall draw the graphs of $y = \sinh x$, $y = \cosh x$ and $y = \tanh x$ only, leaving the others as exercises for the reader.

(i) $y = \sinh x = \frac{1}{2}(e^x - e^{-x})$

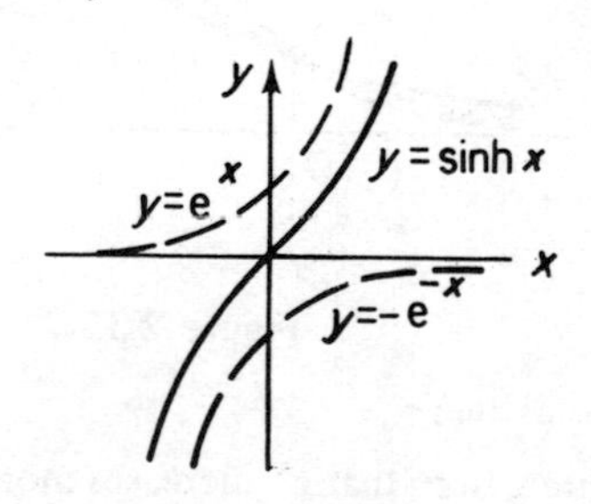

Figure 2.36

We can draw the graph as shown in Figure 2.36 by drawing the curve half-way between the graphs of e^x and $-e^{-x}$ since $\sinh x = \frac{1}{2}[e^x + (-e^{-x})]$.
Notice that

1. $y = \sinh x$ is an odd function, $\sinh(-x) = -\sinh x$
2. $\sinh(0) = 0$
3. $\sinh(A \pm B) = \sinh A \cosh B \pm \cosh A \sinh B$.

$$\left[\sinh(A + B) = \frac{e^{A+B} - e^{-A-B}}{2}\right.$$

$$= \frac{e^A . e^B - e^{-A} . e^{-B}}{2}$$

$$= \frac{(e^A - e^{-A})}{2}\frac{(e^B + e^{-B})}{2} + \frac{(e^A + e^{-A})}{2} . \frac{(e^B - e^{-B})}{2}$$

$$\left. = \sinh A \cosh B + \cosh A \sinh B\right]$$

There are other similarities to the sine function which we shall meet later. However, two major differences are that the function $y = \sinh x$ is **not periodic** and also that it is a one to one relation.

(ii) $y = \cosh x = \dfrac{e^x + e^{-x}}{2}$

The sketch of the graph is thus obtained, as in Figure 2.37, by taking the average of e^x and e^{-x}.

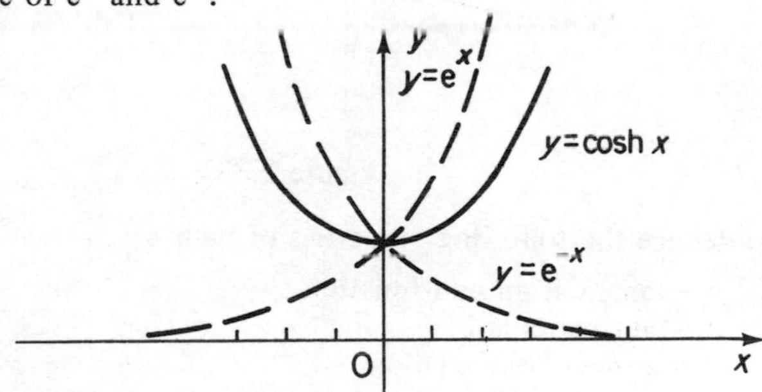

Figure 2.37

Notice that

1. $y = \cosh x$ is **even**, $\cosh(-x) = \cosh x$
2. $\cosh(0) = 1$
3. $\cosh(A \pm B) = \cosh A \cosh B \pm \sinh A \sinh B$

$$\left[\text{R.H.S.} = \frac{e^A + e^{-A}}{2} . \frac{e^B + e^{-B}}{2} \pm \frac{e^A - e^{-A}}{2} . \frac{e^B - e^{-B}}{2}\right.$$

$$= \frac{e^{A \pm B} + e^{-(A \pm B)}}{2}$$

$$= \text{L.H.S.} \,]$$

Compare this with $\cos(A + B)$ but note the difference in sign. The major difference again is that the function $y = \cosh x$ is **not periodic.** We shall have to be careful when defining its inverse since the relation is not one-to-one.

(iii) $y = \tanh x = \dfrac{e^x - e^{-x}}{e^x + e^{-x}}$

By writing this in the form $y = (e^{2x} - 1)/(e^{2x} + 1)$ or in the alternative form $y = (1 - e^{-2x})/(1 + e^{-2x})$, you should show that

1. when $x = 0$, $y = 0$
2. when x tends to a large negative value, y tends towards -1
3. when x tends to a large positive value, y tends to $+1$

and that the graph of $y = \tanh x$ is as shown in Figure 2.38.

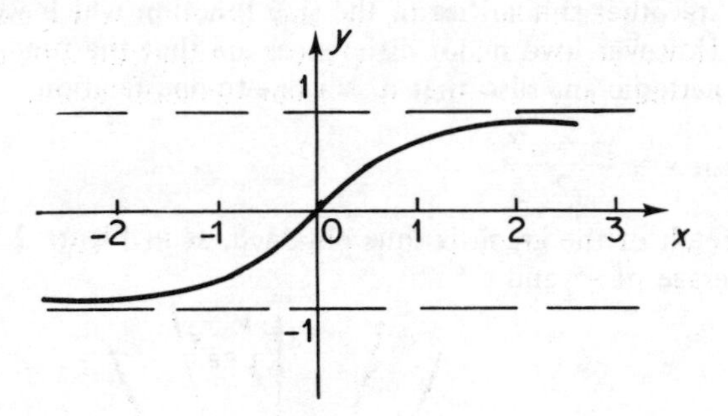

Figure 2.38

Also deduce the following properties of $\tanh x$

(a) $y = \tanh x$ is an **odd** function
(b) it is **not** periodic
(c) it is a one-to-one relation
(d) $\tanh(A \pm B) = \dfrac{\tanh A \pm \tanh B}{1 \pm \tanh A \tanh B}$

Perhaps the most important results are the connections between $\sinh x$, $\cosh x$ and $\tanh x$ which are stated without proof.

1. $\cosh^2 A - \sinh^2 A = 1 \Leftrightarrow \cosh^2 A = 1 + \sinh^2 A$
   $\Leftrightarrow \sinh^2 A = \cosh^2 A - 1$ ($a \Leftrightarrow b$ means $a \Rightarrow b$ **and** $b \Rightarrow a$)
2. $\sinh 2A = 2 \sinh A \cosh A$
3. $\cosh 2A = 2\cosh^2 A - 1 = 1 + 2\sinh^2 A$

4. $\cosh A + \sinh A = e^{A}$
5. $\cosh A - \sinh A = e^{-A}$
6. $(\cosh A \pm \sinh A)^n = \cosh nA \pm \sinh nA$
7. $\text{sech}^2 A = 1 - \tanh^2 A$
8. $\text{cosech}^2 A = \coth^2 A - 1$

Again, note the similarity with the trigonometric results apart from an occasional difference in sign. In fact we can write down any hyperbolic identity from the corresponding trigonometric identity by using the following rule (we cannot prove this yet). The rule says: "Cosine is replaced by cosh, sine is replaced by sinh and tangent is replaced by tanh, but if there is an explicit or implicit sine × sine, this must be replaced by −sinh × sinh".
Let us take an example.

$$\cos 3A = 4 \cos^3 A - 3 \cos A$$

$$\sin 3A = 3 \sin A - 4 \sin^3 A$$

The corresponding hyperbolic identities are

$$\cosh 3A = 4 \cosh^3 A - 3 \cosh A$$

$$\sinh 3A = 3 \sinh A + 4 \sinh^3 A$$

Note $\sin^3 A = (\sin A \times \sin A) \times \sin A$ which has an implied sine × sine therefore gives $(-\sinh A \times \sinh A) \times \sinh A$.

Now write down the hyperbolic equivalents of the following

(a) $\tan 2A = 2 \tan A/(1 - \tan^2 A)$

(b) $\sin 4A = 4 \sin A \cos A - 8 \sin^3 A \cos A$

The inverse hyperbolic functions

(i) $y = \sinh^{-1} x$

y is the number whose sinh is x and an equivalent expression is $x = \sinh y$. Looking back at the graph of $y = \sinh x$ we see that $y = \sinh x$ is a one-to-one relationship. Thus the inverse function exists without any restriction on the range. A sketch of its graph is obtained by a mirror image of $y = \sinh x$ in the line $y = x$ and is shown in Figure 2.39.

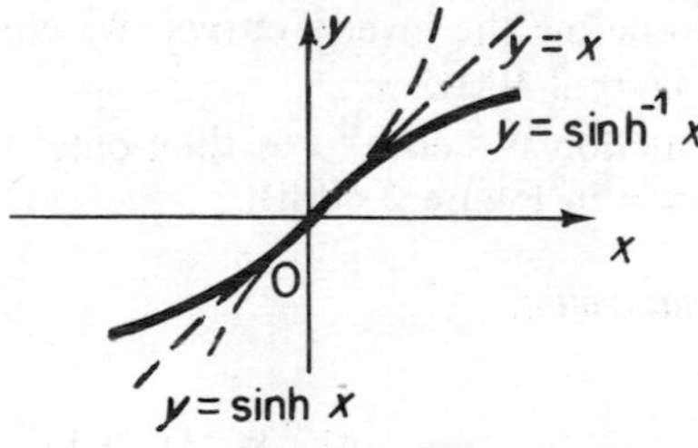

Figure 2.39

Logarithmic equivalent

If

$$y = \sinh^{-1} x$$

then

$$\sinh y = x$$

that is,

$$\frac{e^y - e^{-y}}{2} = x$$

Multiplying by $2e^y$ and rearranging we get

$$(e^y)^2 - 2x\, e^y - 1 = 0$$

Solving this quadratic in e^y yields

$$e^y = x \pm \sqrt{x^2 + 1}$$

Now $\sqrt{x^2 + 1} > x$ and the negative sign would give a negative value for e^y which is impossible. Thus only the positive sign is admissible and

$$e^y = x + \sqrt{x^2 + 1}$$

Thus

$$y = \log(x + \sqrt{x^2 + 1})$$

That is,

$$\sinh^{-1} x = \log(x + \sqrt{x^2 + 1})$$

Example

$$\sinh^{-1}(½) = \log(½ + \sqrt{5/4}) = \log \frac{1 + \sqrt{5}}{2}$$

$$= 0.4810 \text{ (4 d.p.)}$$

Thus the logarithmic equivalent is a much more convenient method of evaluating the inverse function.

(ii) $y = \cosh^{-1} x$

The relation $y = \cosh x$ is not one-to-one and we shall have to restrict the domain to define the inverse curve. We choose only the positive x-axis as in Figure 2.40(i).

The inverse function $y = \cosh^{-1} x$ is then only defined for $y \geqslant 0$ and has a graph as shown in Figure 2.40(ii).

Logarithmic equivalent

If

$$y = \cosh^{-1} x \quad (x \geqslant 1)$$

then

$$\cosh y = x$$

or

$$\frac{e^y + e^{-y}}{2} = x$$

Multiplying by $2e^y$ and rearranging

$$(e^y)^2 - 2x\, e^y + 1 = 0$$

Solving we get

$$e^y = x \pm \sqrt{x^2 - 1}$$

The positive sign must be taken to obtain a point on the inverse function curve. Therefore

$$e^y = x + \sqrt{x^2 - 1}$$

or

$$y = \log(x + \sqrt{x^2 - 1})$$

Hence

$$\cosh^{-1} x = \log(x + \sqrt{x^2 - 1})$$

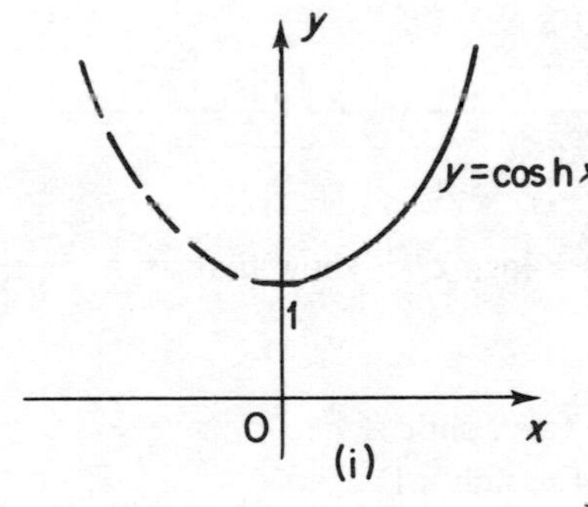

Figure 2.40

(iii) $y = \tanh^{-1} x$

You should now proceed in a similar way to the above to consider this function and establish the following facts.

1. the inverse function exists without modification of the range
2. its graph is of the form (see Figure 2.41)
3. the function only exists for $|x| < 1$
4. $\tanh^{-1} x = \frac{1}{2} \log [(1 + x)/(1 - x)]$

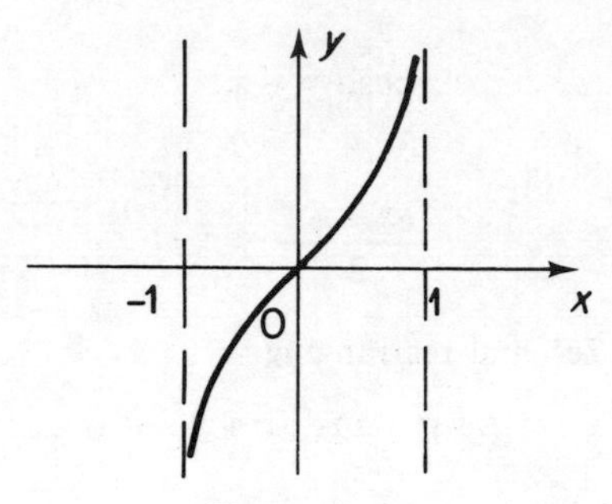

Figure 2.41

Problems

1. Find $\sin^{-1}(0.65)$, $\sin^{-1}(-0.1)$
   $\cos^{-1}(0.732)$, $\cos^{-1}(-0.836)$
   $\tan^{-1}(3.3)$, $\tan^{-1}(-15)$

2. Show that $\cos^{-1}x = \dfrac{\pi}{2} - \sin^{-1}x$

3. Without using tables, show that

   (a) $2\sin^{-1}\dfrac{3}{5} = \sin^{-1}\dfrac{24}{25}$

   (b) $\tan^{-1}\dfrac{63}{16} - \cos^{-1}\dfrac{12}{13} = \sin^{-1}\dfrac{4}{5}$

4. Show that $\log_9 27 = \dfrac{3}{2}$

5. If $\log_e y = 3\log_e x - \frac{1}{2}\log_e(x+2) + \log_e x^{3/2}$, show that $y^2 = \dfrac{x^9}{x+2}$

6. Prove, using basic definitions, that

   (a) $\cosh 2A = 2\cosh^2 A - 1 = 1 + 2\sinh^2 A$

   (b) $(\cosh A + \sinh A)^n = \cosh nA + \sinh nA$

   (c) $\text{sech}^2 A = 1 - \tanh^2 A$

   (d) $2\sinh A \sinh B = \cosh(A+B) - \cosh(A-B)$

7. Evaluate $\cosh(2)$, $\sinh(0.5)$, $\tanh(-0.3)$
   $\sinh^{-1}(1.5)$, $\cosh^{-1}(3)$, $\tanh^{-1}(1/4)$
   $\tanh^{-1}(-0.25)$, $\cosh^{-1}(1.4)$, $\sinh^{-1}(-2)$
   (Hint:for inverse functions – use log equivalents)

8. If $\text{cosech}\, x = \dfrac{-9}{40}$, find $\cosh x$ and $\tanh x$

9. Prove that

(a) $\text{sech}^{-1} x = \log \dfrac{1 + \sqrt{1 - x^2}}{x}$

(b) $\text{cosech}^{-1} x = \log \dfrac{1 + \sqrt{1 + x^2}}{x}$

10. Prove that $\tanh^{-1} \dfrac{x^2 - 1}{x^2 + 1} = \log x$

2.9 TABULATION OF FUNCTIONS

Engineers work frequently with tables. These can be of a mathematical nature (e.g. tables of logarithms, or sines) or can be tables obtained by experiment for design purposes. The extract below, Table 2.1, is taken from a table of values for the pressure of saturated water in bars, P, as a function of temperature in degrees centigrade, T.

Table 2.1

T	10	11	12	13	14	15
P	0.01227	0.01312	0.01401	0.01497	0.01597	0.01704

Often we wish to estimate a value that has not been tabulated. Let us first study how we accomplish this using logarithmic tables (to the base 10) where four figures are quoted. Here is a typical row from such a table (see Table 2.2).

Table 2.2

	0	1	2	3	4	5	6	7	8	9	1	2	3	4	5	6	7	8	9
20	3010	3032	3054	3075	3096	3118	3139	3160	3181	3201	2	4	6	8	11	13	15	17	19

Thus the logarithm of 2.06 is read as 0.3139. To find $\log_{10} 2.064$ we add on the *proportional part* under the '4' column, i.e. 8, to obtain 0.3147. If we wanted $\log_{10} 2.084$ we now add 8 to 0.3181 to obtain 0.3189. It is not coincidence that the proportional part to be added on is 0.0008 in both cases. The amount to be added on to the tabulated values $\log_{10} 2.00$, $\log_{10} 2.01$, ... $\log_{10} 2.09$ for obtaining $\log_{10} 2.004$, $\log_{10} 2.014$, ..., $\log_{10} 2.094$ respectively will vary slightly and 0.0008 is the average amount. Later in the table of logarithms you will see that the nine proportional parts for each row are not all distinct since, to 4 figures, they cannot be distinguished, yet earlier in the table, the rows are split into two, each having its own proportional parts. We have already drawn the graph of $y = \log_e x$ (see Figure 2.32) and the graph of $y = \log_{10} x$ is similar in shape. A glance at the $\log_{10} x$ graph in the region of 1 will show a decreasing slope as x increases and thus values tend to differ less.

Suppose we wanted a more accurate value of $\log_{10} 2.064$ than the proportional parts allow; we are limited to 4 d.p. accuracy by the tabulated values but we might be able to improve on our previous estimate. Again, suppose we wanted the pressure of saturated water at 12.5°C. We are then seeking a means of interpolation.

Let us focus our attention on three extracts from a table of tangents (see Table 2.3)

Table 2.3

x°	$\tan x$	x°	$\tan x$	x°	$\tan x$
0	.0000	40	.8391	80	5.6713
1	.0175	41	.8693	81	6.3138
2	.0349	42	.9004	82	7.1154
3	.0524	43	.9325	83	8.1443
4	.0699	44	.9657	84	9.5144
5	.0875	45	1.0000		

Suppose we wish to estimate $\tan 1^\circ 30'$. We may say that since $1^\circ 30'$ lies halfway between 1° and 2° then the tangent value lies approximately halfway between 0.0175 and 0.0349, i.e. 0.0262. In fact, this is the value of $\tan 1^\circ 30'$ to 4 d.p. Similar reasoning would suggest that the value of $\tan 1^\circ 20'$ is one third of the way between 0.0175 and 0.0349, i.e. $0.0175 + (1/3)(0.0349 - 0.0175) = 0.0233$ which again agrees with the tables. We are effectively assuming a linear relationship for $\tan x$ over the range $x = 1^\circ$ to $x = 2^\circ$. In our case the assumption is a fair one and we are using the process called **Linear Interpolation.**

For linear interpolation, if $x = x_0 + ph$, where h is the (constant) table spacing in x, and p is the fraction of the interval under consideration (see Figure 2.42), then

$$f(x_0 + ph) \simeq f(x_0) + p\,[f(x_0 + h) - f(x_0)]$$

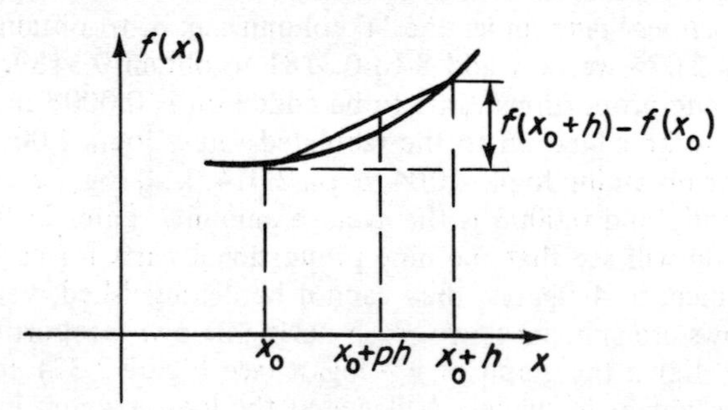

Figure 2.42

Will this process always produce such good results? Let us now estimate tan 40°30′ by the same method. We obtain

$$0.8391 + \frac{1}{2}\left(0.8693 - 0.8391\right) = 0.8542$$

against the tabulated value of 0.8451. Estimating tan 81°30′ produces 6.7146 against the tabulated value of 6.6912. It is clear that the assumption of linear behaviour is not justified in these cases and a more accurate method must be found.

If we return to the logarithm table extract, Table 2.2, let us estimate by linear interpolation the values of $\log_{10} 2.064$ and $\log_{10} 2.004$.

$$\begin{aligned}\log_{10} 2.064 &= 0.3139 + \frac{4}{10}\left(0.3160 - 0.3139\right)\\ &= 0.3139 + 0.4 \times 0.0021\\ &= 0.3147 \quad \text{(4 d.p.)}\end{aligned}$$

$$\begin{aligned}\log_{10} 2.004 &= 0.3010 + \frac{4}{10}\left(0.3032 - 0.3010\right)\\ &= 0.3010 + 0.4 \times 0.0022\\ &= 0.3019 \quad \text{(4 d.p.)}\end{aligned}$$

For $\log_{10} 2.004$ we add on 0.0009 instead of the mean proportional part of 0.0008. Accuracy in the tables has been sacrificed for convenience and size of tables.

Since we are dogged by round-off in most practical tables, the main features tend to be somewhat obscured. For the moment we turn our attention to tables for simple mathematical expressions where the tabulated values are exact. An example is a table of squares of integers (Table 2.4)

Table 2.4

x	x^2	Differences
11	121	
		23
12	144	
		25
13	169	
		27
14	196	
		29
15	225	

Here we have taken the differences of the tabulated values and we can see a pattern emerging. If we wished to continue this pattern we might expect the next difference to be 31 yielding a value of 256 for $(16)^2$ which is correct. We have extrapolated the table successfully but this is a more dangerous process than interpolation (see Section 1.1.)

Consider now the table of values for $f(x) = 2x - 3$ in Table 2.5.

Table 2.5

x	$f(x)$	Differences
4	5	
		2
5	7	
		2
6	9	
		2
7	11	

Here the differences are **constant**. The values in the column of differences as obtained in these two examples are called **First Differences**. We can extend such tables to obtain differences of differences (*second differences*). For $f(x) = x^2$, we obtain Table 2.6.

Table 2.6

x	x^2	First Differences	Second Differences
11	121		
		23	
12	144		2
		25	
13	169		2
		27	
14	196		2
		29	
15	225		

This time the **second** differences are constant.

Consider $f(x) = 2x - x^2$

Table 2.7

x	$f(x)$	1st	2nd
0	0		
		1	
1	1		−2
		−1	
2	0		−2
		−3	
3	−3		−2
		−5	
4	−8		−2
		−7	
5	−15		

The second differences are again constant. Is this a general result? Putting the process on a formal basis; if the difference between $f(x_0)$ and $f(x_0 + h)$ be denoted $\Delta f(x_0)$, that is, $\Delta f(x_0) = f(x_0 + h) - f(x_0)$, then suppose $f(x) = a + bx$, where a and b are constants.

$$f(x_0) = a + bx_0, \qquad f(x_0 + h) = a + b(x_0 + h) = a + bx_0 + bh$$

therefore $f(x_0 + h) - f(x_0) \equiv \Delta f(x_0) = bh$ which is constant if the spacing of values, h, is constant.

Similarly, let us take $f(x) = a + bx + cx^2$; a, b, c constants.

Then

$$f(x_0) = a + bx_0 + cx_0^2$$

$$f(x_0 + h) = a + b(x_0 + h) + c(x_0 + h)^2$$

and

$$\Delta f(x_0) \equiv f(x_0 + h) - f(x_0) \equiv 2cx_0h + ch^2 + bh$$

Likewise

$$f(x_0 + 2h) = a + b(x_0 + 2h) + c(x_0 + 2h)^2$$

and

$$\Delta f(x_0 + h) \equiv f(x_0 + 2h) - f(x_0 + h) = 2cx_0h + 3ch^2 + bh$$

The second difference, $\Delta f(x_0 + h) - \Delta f(x_0)$, denoted by $\Delta^2 f(x_0)$, is given by $\Delta^2 f(x_0) \equiv 2ch^2$ and is constant for constant spacing.

Thus the second differences of a second degree polynomial are constant. It would be tempting to conclude that the third differences of a cubic are constant. (Prove this algebraically and from a suitable table of values; you should consider only $y = dx^3$ for d constant – why can we ignore the terms in x^2, x and the constant term?)

Consider now the tables for $f(x) = 2x - x^2$ below for non-integral values of x (Table 2.8 and Table 2.9).

Table 2.8

x	$f(x)$	First differences	Second differences
1.0	1.00		
		−0.01	
1.1	0.99		−0.02
		−0.03	
1.2	0.96		−0.02
		−0.05	
1.3	0.91		−0.02
		−0.07	
1.4	0.84		−0.02
		−0.09	
1.5	0.75		

Table 2.9

x	$f(x)$	First differences	Second differences
1.0	1.00		
		−1	
1.1	0.99		−2
		−3	
1.2	0.96		−2
		−5	
1.3	0.91		−2
		−7	
1.4	0.84		−2
		−9	
1.5	0.75		

In Table 2.9 the differences are recorded as integers for ease; when used in calculations they must be reinstated to their true values. (Some tables even omit the signs, but this is not to be recommended.)

Example 1

Let us now examine the dangers of round-off. We have produced below two tables (Table 2.10) for $f(x) = x^3 - 2x^2 + 1$, the first keeping exact values and the second using values of $f(x)$ correct to 2 d.p.

Table 2.10

x	$f(x)$	1st diff	2nd diff	3rd diff	x	$f(x)$	1st	2nd	3rd	4th	5th	6th
0	1.000				0	1.00						
		−19					−2					
0.1	0.981		−34		0.1	0.98		−3				
		−53		6			−5		0			
0.2	0.928		−28		0.2	0.93		−3		0		
		−81		6			−8		0		3	
0.3	0.847		−22		0.3	0.85		−3		3		−11
		−103		6			−11		3		−8	
0.4	0.744		−16		0.4	0.74		0		−5		16
		−119		6			−11		−2		8	
0.5	0.625		−10		0.5	0.63		−2		3		−10
		−129		6			−13		1		−2	
0.6	0.496		−4		0.6	0.50		−1		1		−1
		−133		6			−14		2		−3	
0.7	0.363		2		0.7	0.36		1		−2		5
		−131		6			−13		0		2	
0.8	0.232		8		0.8	0.23		1		0		
		−123		6			−12		0			
0.9	0.109		14		0.9	0.11		1				
		−109					−11					
1.0	0.000				1.0	0.00						

Note:

(i) In the second table the third differences are not constant, and hence the fourth and higher differences are not zero.

(ii) The values of the fourth and higher differences are due to round-off errors.

(iii) The magnitudes of third differences are smaller than those to the left and to the right.

(iv) The magnitude of differences higher than the fourth tend to increase and signs follow no discernible pattern.

The general rule is that differences which are larger in magnitude than the predecessors should be ignored as should all higher differences.

We can analyse the situation as follows: rounding errors have their greatest effect when entries in a table have errors which are alternately ½ and −½ in the last decimal place shown.

Errors	1st diff	2nd diff	3rd diff	4th diff	5th diff
½					
	−1				
−½		2			
	1		−4		
½		−2		8	
	−1		4		−16
−½		2		−8	
	1		−4		
½		−2			
	−1				
−½					

In general, the magnitude of the maximum error in the nth difference is 2^{n-1}. Thus, in our table of rounded-off values, the third differences have an absolute bound due to errors of 2^2, i.e. 4, in the last place shown. This means that if they have values less than 4 they could be due entirely to round-off and so might not be worth including in calculations. We are really looking for differences which could have arisen by errors alone; note that if the second differences were below 2 this could also be due to round-off. However, the actual bounds are seldom reached in practice and we are being over-generous in our criteria.

Example 2

Let us pursue a second example. In Table 2.11,values of tan x have been taken from 4 figure tables and rounded to 3 d.p.

Table 2.11

x^o	tan x	1st	2nd	3rd	4th	5th
0	0.000					
		18				
1	0.018		−1			
		17		1		
2	0.035		0		0	
		17		1		−2
3	0.052		1		−2	
		18		−1		
4	0.070		0			
		18				
5	0.088					
		(1)	(2)	(4)	(8)	(16)

Absolute error bound is shown in parentheses, thus: ( )

We conclude here that second and higher differences should be neglected.

Extension of a Table using Differences

If we have a table where a constant set of differences has been reached then we may extend the table in either direction. Table 2.12 is exact. We wish to estimate $f(1.0)$.

Table 2.12

x	$f(x)$			
0	0			
		24		
0.2	0.024		144	
		168		144
0.4	0.192		288	
		456		144
0.6	0.648		432	
		888		144
0.8	1.536		576	
		1464		
1.0	3.000			

The numbers below the bold line are added from right to left.

Now try and show that $f(-0.2) = -0.024$

If the table is inexact due to round-off, the answers will be approximate.

If the table has not got almost constant differences, the extrapolation process will be poor as we shall see in Section 6.6.

Errors in Tabulated Values

There is another source of error in a difference table: due to a mistake in a value for $f(x)$. We shall not dwell too deeply on this, but merely indicate the approach. In the exact table for $f(x) = x^3 - 2x^2 + 1$, Table 2.13, the value of $f(0.5)$ has been entered as 0.635 instead of 0.625.

Table 2.13

x	$f(x)$	1st	2nd	3rd	4th
0	1.000				
		−19			
0.1	0.981		−34		
		−53		6	
0.2	0.928		−28		0
		−89		6	
0.3	0.847		−22		10
		−103		16	
0.4	0.744		−6		40
		−109		−24	
0.5	0.635		−30		60
		−139		36	
0.6	0.496		6		40
		−133		−4	
0.7	0.363		2		10
		−131		6	
0.8	0.232		8		0
		−123		6	
0.9	0.109		14		
		−109			
1.0	0.000				

We see that the fourth differences, instead of being zero all the way down have a perturbation. The error has spread its effects as we move into the higher differences. To consider the full effects, we take a column of zeros with a simple entry of ϵ and see in Table 2.14 how the effect spreads.

Table 2.14

0				
	0			
0		0		
	0		0	
0		0		ϵ
	0		ϵ	
0		ϵ		-4ϵ
	ϵ		-3ϵ	
ϵ		-2ϵ		6ϵ
	$-\epsilon$		3ϵ	
0		ϵ		-4ϵ
	0		$-\epsilon$	
0		0		ϵ
	0		0	
0		0		
	0			
0				

Note that the errors follow the pattern of binomial coefficients, i.e. the third differences are multiplied by the coefficients of $(1-x)^3$, the fourth differences are multiplied by the coefficients of $(1-x)^4$, etc.

Hence in a table of exact values of a polynomial, at some stage the differences should all be zero and the presence of such a pattern as above will become evident. We can then trace back to the source.

Consider the table below, Table 2.15, of $f(x) = 2x - x^2$

Table 2.15

1.0	1.00			
		−1		
1.1	0.99		−2	
		−3		−3
1.2	0.96		−5	
		−8		9
1.3	0.88		4	
		−4		−9
1.4	0.84		−5	
		−9		3
1.5	0.75		−2	
		−11		
1.6	0.64			

This suggests an error of −0.03 in $f(1.3)$ and it should be 0.91.

If the table has rounded-off entries or if the function is not a polynomial we can make only an estimate of the error. It is to be hoped that the effects of the error outweigh that due to round-off. The following values in Table 2.16 are taken from four-figure tangents with the value for tan 25° 'doctored'.

Table 2.16

$x°$	tan x	1st	2nd	3rd	4th
18°	0.3249				
		194			
19°	0.3443		3		
		197		−1	
20°	0.3640		2		1
		199		0	
21°	0.3839		2		2
		201		2	
22°	0.4040		4		−4
		205		−2	
23°	0.4245		2		−23
		207		−25	
24°	0.4452		−23		105
		184		80	
25°	0.4636		57		−160
		241		−80	
26°	0.4877		−23		107
		218		27	
27°	0.5095		4		−27
		222		0	
28°	0.5317		4		1
		226		1	
29°	0.5543		5		−2
		231		−1	
30°	0.5774		4		2
		235		1	
31°	0.6009		5		
		240			
32°	0.6249				

After the third differences, those outside the error range begin to increase and so the column of the ratio of third differences is the highest one which is worthwhile studying.

If there is an error of ϵ in the tabulated value of tan 25°, then we know that the resulting errors in the third differences are $\epsilon, -3\epsilon, 3\epsilon, -\epsilon$. The relevant third differences in the table are −25, 80, −80, 27 and looking at the values of the third differences outside the error range we would expect the true values to be at most 0, 1 or 2 in magnitude. This suggests that the error in tan 25° is of order −26. Let us try $\epsilon = -25$, $\epsilon = -26$ and $\epsilon = -27$.

Now an error of −25 gives resulting errors of −25, 75, −75, 25
an error of −26 gives resulting errors of −26, 78, −78, 26
an error of −27 gives resulting errors of −27, 81, −81, 27

The deviations of these from the values of the third differences are respectively 0, 5, 5, 2; 1, 2, 2, 1 and 2, 1, 1, 0 with totals 12, 6 and 4 respectively.

If we choose the error which makes the total deviation as small as possible we choose $\epsilon = -27$, making tan 25° = 0.4636 + 0.0027 = 0.4663. In fact this was the error in the tabulated value of tan 25° and the amended difference table inside the error range is as shown in Table 2.17.

Table 2.17

				2
			4	
		211		−1
tan 25°	0.4663		3	
		214		1
			4	
				0

Problems

1. Tabulate the values of cos x for $x = \pi/2, 3\pi/2, 5\pi/2$ radians and calculate cos $11\pi/12$ by linear interpolation. Comment.
   Now tabulate the values of cos x for $x = \pi/2, 2\pi/3, 5\pi/6, \pi$, and repeat likewise for $x = 160°, 170°$. Conclusions?

2. From a four figure table of reciprocals, obtain 1/1.817 by linear interpolation and compare with the value calculated using mean differences. Why the discrepancy?

3. Form a difference table from the pressure of saturated water P and temperature T, as far as the differences seem useful. (Table 2.1)

4. Evaluate S – the sum of the squares of the first n natural numbers (hint: take $n = 0,1,......,6$ for the table); form a difference table and deduce the degree of the polynomial. Fit suitable data to show that it is $[1/6]\,[n\,(n+1)(2n+1)]$.

5. Tabulate $f(x) = 3x^3 - 2x^2 + x$ for $x = 0(2)10$† until constant differences are reached.

6. Tabulate $4.2x^3 - 0.3x^2 + 2.7x - 0.8$ exactly for $x = 0(0.1)1.0$; extend the table to $x = -0.2$ and $x = -0.1$. Now write the table correct to 2 d.p. and form a difference table. What are the highest differences that can be used?

7. Locate and correct the mistake in the table below
   (a) take values stated
   (b) round-off to 1 d.p. and repeat

x	0.2	0.3	0.4	0.5	0.6	0.7	0.8	0.9	1.0	1.1
$f(x)$	10.08	12.75	16.00	19.89	24.84	29.83	36.00	43.05	51.04	60.03

† $x = 0$ to 10 in steps of 2

8. The results from a tensile test are shown: estimate the stress at a strain of 2500×10^{-6} mm/mm using linear interpolation. Why can we use linear interpolation here?

Stress MN/m^2	Strain mm/mm
0	0
30	1003×10^{-6}
60	2007×10^{-6}
90	3009×10^{-6}
120	4013×10^{-6}
150	5017×10^{-6}

9. The absorption coefficients for a carpet with felt underlay are

125 Hz	250 Hz	500 Hz	1 KHz	2 KHz
0.1	0.25	0.5	0.5	0.6

["Noise" – Rupert Taylor (Penguin)]

What would be your estimate of the absorption coefficient at 1/3 KHz, 3 KHz?

10. Produce a difference table from data for air density ρ and estimate density at altitude 5,500 and 20,000m.

Altitude $m\,(10^3)$	ρ kg/mm^3
0	1.2256
1	1.1121
2	1.0069
3	0.9095
4	0.8194
5	0.7364
6	0.6599
7	0.5897
8	0.5253
9	0.4665
10	0.4128

SUMMARY

In this chapter you have met a wide range of ideas. Many fundamental concepts have been introduced as well as a number of manipulative techniques.

The first concept is that of sets. You should recognise their basic role in mathematics and be able to perform simple set algebra.

The second concept is that of functions. So much of the mathematics you will study involves functions that we cannot over-emphasise their importance. It is absolutely crucial that you feel confident in handling functions and understand precisely what a function is – and what a function is not. If you are applying general results for functions to a particular problem, then you must be certain that all the relationships between the problem variables are functional relationships; if some are not then great care is needed in the interpretation of the results. The standard functions of the calculus and their properties must become completely familiar to you and you need to be able to manipulate them (and functions in general) without difficulty.

The handling of inequalities should present no problems and you should be able to split a ratio of polynomials into its partial fractions. Finally, the formation of a table of differences from a tabulated function and the location of any errors the table may contain should be added to your mathematical repertoire.

Chapter Three

Elementary Ideas II–Limits and Continuity

3.1 SEQUENCES AND LIMITS

In the study of a slender column of length l built in at its lower end A, pinned and laterally supported at its upper end B, as shown in Figure 3.1, it is necessary to solve the equation

$$\tan u = u \tag{3.1}$$

$$\text{where } u^2 = \frac{Pl^2}{EI}$$

for P the *buckling load;* EI is the flexural rigidity of the column.

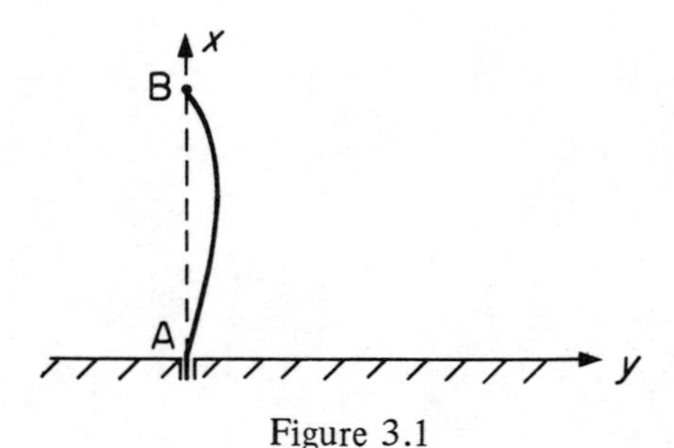

Figure 3.1

How can we obtain the root of equation (3.1), if no formula exists? It may seem naïve to suggest that we make a guess or even a succession of guesses; however, as long as we do not guess wildly but use the information from previous guesses in an organised way we can tie down the value of the root quite accurately. In order to see clearly the way that the mechanism for guesswork behaves, we shall take some simple abstract problems to which the answers are already known; this has two advantages: firstly, it prevents the physical aspects of a problem obscuring the basic mathematical processes, secondly, it allows us to see how the guesses compare with the known answer. This latter advantage has a wider importance for, if we can test a method against a known result, we can evaluate its strengths and weaknesses and hence apply it confidently in situations where the result is not known by other means.

Suppose we attempt to find $\sqrt{6.7}$ to 2 d.p. We could proceed as follows. Since $2^2 = 4$ and $3^2 = 9$, the value we seek is between 2.00 and 3.00. Now $(2.5)^2 = 6.25$ and $(2.6)^2 = 6.76$ and so the value is between 2.50 and 2.60. It would seem reasonable to try 2.58 as a next guess; $(2.58)^2 = 6.6564$, and $(2.59)^2 = 6.7081$, and we can choose 2.59 as the nearest estimate of $\sqrt{6.7}$ to 2 d.p. We have our answer, but it required a judgement to guess 2.5 and 2.58 as likely candidates; in other words, we do not have an efficient strategy for choosing the next estimate.

We turn our attention to the finding of $\sqrt{4}$ by a method of successive guesses. It is helpful to consider the geometrical equivalent: the determination of the length of the side of a square of area 4 units [Figure 3.2(i)]. (In algebraic terms, we seek the positive root of $x^2 = 4$)

Suppose we guess that $x = 1$. Instead of a square we have a rectangle of area 4 with sides of lengths 1 and 4 [Figure 3.2(ii)]. Clearly the sides of length 4 are too long, those of length 1 too short. The length we require is somewhere in between and we choose to guess at the arithmetic mean†, 2.5. To test the accuracy of this guess, we divide the area 4 by this length and obtain 1.6 [Figure 3.2(iii)]. Clearly the number we seek lies between 2.5 and 1.6 and we choose a next estimate as $\frac{1}{2}(2.5 + 1.6) = 2.05$. We seem to be approaching the true answer of 2 but we cannot be sure.

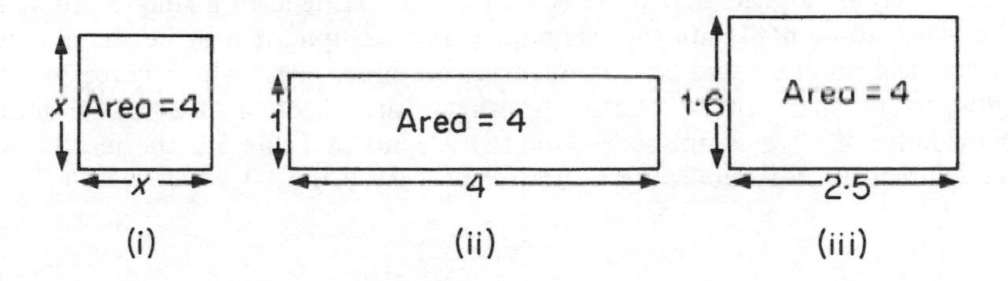

Figure 3.2

To investigate further, we must put the method on an algebraic footing. Let x_0 be the initial guess of 1. Then the next guess, $x_1 = \frac{1}{2}(x_0 + 4/x_0)$ and the guess of 2.05 is $x_2 = \frac{1}{2}(x_1 + 4/x_1)$. We can thus define the method as $x_0 = 1$, $x_{n+1} = \frac{1}{2}(x_n + 4/x_n)$; we have given an initial guess and a rule for generating successive guesses. In Table 3.1 we show the results to 9 d.p. in the left-hand column (using a desk calculator) and rounded to 2 d.p. in the right-hand column. Notice that rounding helps in getting the right answer.

† The arithmetic mean of two numbers a and b is $\frac{1}{2}(a + b)$

Table 3.1

n	x_n	
0	1	1
1	2.5	2.5
2	2.05	2.05
3	2.000609756	2.00
4	2.000000092	2.00

NOTE: $(2.000000092)^2 = 4.000000368$

After 4 applications of the rule we have got very close to the exact value of the root, but would we be able to get it exactly if we applied the rule more times? First, we must check that the formula is consistent: if we substitute $x_n = 2$ then you can see that $x_{n+1} = 2$; however if we substitute $x_{n+1} = 2$ we have that $2 = \frac{1}{2}(x_n + 4/x_n)$ and a little algebra will show that $x_n = 2$ is the only result. What does this tell us? Only if we guess initially the exact answer will the sequence of guesses provide the exact answer; otherwise we shall never obtain it. Of course, this is a theoretical statement and in practice we can get the value correct to a number of decimal places which depends on the precision of the calculating aid we employ and the build-up of round-off error. (It may be that round-off prevents us getting the value we seek accurate to the full number of decimal places the aid can provide.)

It may be argued that there is a danger in taking such a simple problem on which to demonstrate the technique; for example, it may be the case with a practical problem that we cannot make an initial guess which is reasonably close to the value sought. Let us investigate the effect of other initial guesses. Obviously, $x_0 = 0$ is not acceptable (why?) but in Table 3.2 the results using initial guesses of 10 and 64 are shown; you try using -3.

Table 3.2

n	x_n (to 4 d.p.)	
0	10	64
1	5.2	32.0625
2	2.9846	16.0936
3	2.1624	8.1710
4	2.0060	4.3302
5	2.0000	2.6269
6	2.0000	2.0748

Why the almost linear decrease early on?

It would seem that the initial guess merely delays the achievement of a particular accuracy. This can be shown theoretically, but we must always bear in mind that in practice round-off may spoil matters.

Let us now investigate the application of this approach to finding $\sqrt{2}$. You may know that $\sqrt{2}$ is irrational and hence cannot be exactly represented by a finite decimal†; since the approximation method will not give it exactly anyway, this does not matter. However, we could stop the sequence of approximations to $\sqrt{4}$ when the last approximation was as close to the known result of 2 as was required; we can here assume that we have only a vague idea of the value of $\sqrt{2}$ and consequently cannot be sure when we have the value correct to, say, 4 d.p. If we examine the tables again we see that the approximations, as well as approaching 2, are getting closer together. This will be our criterion for stopping the sequence of guesses: if we require an accuracy of 4 d.p. we shall stop when two consecutive approximations differ by less than .0001. We start with $x_0 = 1$ and the **recurrence formula** $x_{n+1} = ½(x_n + 2/x_n)$. Table 3.3 shows the sequence of approximations as calculated on a machine.

Table 3.3

n	x_n
0	1
1	1.5
2	1.416666666
3	1.414215700
4	1.414213562
5	1.414213562

We see that the last two approximations agree to 9 d.p. and we could quote this value as the result required.

Suppose we generalise this method to one which finds the positive square root of any non-negative number, A; the *iterative formula* becomes

$$x_{n+1} = \frac{1}{2}\left(x_n + \frac{A}{x_n}\right) \tag{3.2}$$

It is valuable at this stage to flow chart the process (Figure 3.3). We let XOLD and XNEW take the role of x_n and x_{n+1}, TOL stand for the accuracy (tolerance) required.

You should modify the flow chart to incorporate a check that the number of times the formula is used does not exceed a pre-specified number.

---

† one which has a finite number of decimal places

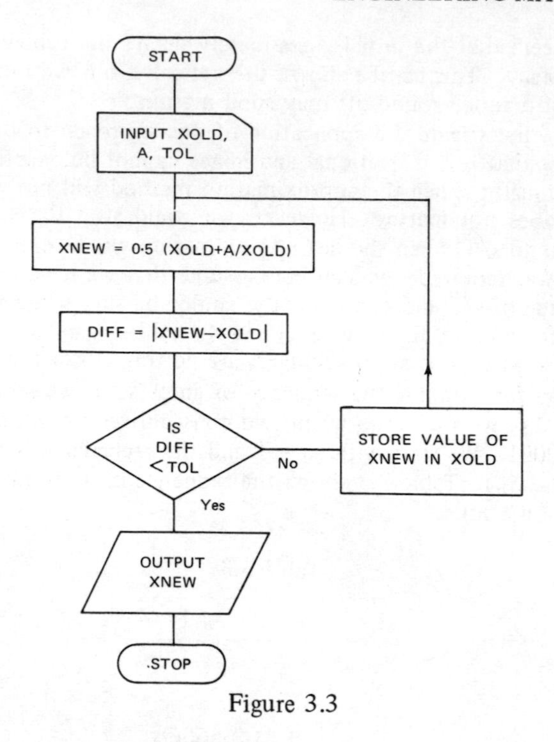

Figure 3.3

Problems I

1. The *geometric mean* of two positive numbers, a, b, is defined to be $\sqrt{ab}$. Show that the iterative formula (3.2) for square roots approximates at each step the geometric mean by the arithmetic mean. Show that, in general, the geometric mean of two numbers is less than or equal to their arithmetic mean; under what circumstances is equality achieved? Comment on the inequality by reference to the successive approximations in Tables 3.1 and 3.3.

2. Verify that the formula $x_{n+1} = \frac{1}{2}[x_n + (2/x_n)]$ yields $\sqrt{2}$ as an approximation only if $\sqrt{2}$ is fed in as x_n. In the ideal state, $x_{n+1} = x_n$; show that by writing $x_{n+1} = x_n = x$ the resulting equation has $\sqrt{2}$ as its positive root. Where does the other root come from?

3. Starting from an equivalent geometrical problem, develop two possible *iterative formulae* for finding $\sqrt[3]{A}$ where A is a positive number. Test both on the case $A = 8$ and then use the preferred one to calculate $\sqrt[3]{10}$ to 2 d.p. Write and run a computer program to calculate $\sqrt[3]{10}$ to 5 d.p. and output the result. Arrange for the result to be cubed as a check on its accuracy. Consider the iteration formula $x_{n+1} = 8/x_n^2$; start with guesses 1.8 and 2.2 and apply the formula three times in each case. What conclusion can you draw?

We have obviously touched on a wealth of problems that need answering. First we make a definition: a set of objects (these may not necessarily be all distinct) which are arranged in a definite order is called a **sequence**. Examples are

(i) the days in a week
(ii) the letters and digits forming a postal code
(iii) the tasks performed to start a car
(iv) the integers 1 to 5 arranged in ascending order
(v) the numbers 1, 4, 7, 10, 13, ...†

We shall confine ourselves to sequences of numbers and we distinguish between *finite* sequences and *infinite* sequences depending on the number of elements of the sequence. An infinite sequence has an infinite number of elements and can be specified by an incomplete list as in Example (v) above, or by a rule which gives the nth term in the sequence [e.g. $u_n = 3n - 2$ for (v)] or by a *recurrence relation* which tells us how to generate one term from others [e.g. $u_{n+1} = u_n + 3$ with $u_1 = 1$ for (v)].

Let us consider some more examples.

(vi) the sequences 3, 8, 15, 24, ... can be specified by $u_n = n(n+2)$; from this formula we can say that $u_{n+1} = (n+1)(n+3)$, replacing n by $(n+1)$.

(vii) the sequence specified by $u_{n+1} = 2u_n$, $u_1 = 2$ can also be written as

$$u_{n+1} = 2^2 u_{n-1} = 2^3 u_{n-2} = \ldots\ldots = 2^n u_1 = 2^{n+1} \quad \text{or} \quad u_n = 2^n;$$

further, it can be specified as 2, 4, 8, 16, ... This sequence is an example of a *geometric progression* a, ar, ar^2, ... where one term is generated from its predecessor by multiplication by a common ratio r.

We saw earlier that each sequence of approximations to $\sqrt{4}$ approached the value of 2 and we may refer to the number 2 as the **LIMIT** of that (infinite) sequence and we say that the sequence **converges** to the limit 2. We shall need some general definitions which will allow us to describe the behaviour of infinite sequences. To do this we choose some simple, abstract examples.

(viii) the sequence $u_n = (1/2)^n$, i.e. 1/2, 1/4, 1/8, 1/16, ...
(ix) the sequence $u_n = 1 - (1/2)^n$ = 1/2, 3/4, 7/8, 15/16, 31/32, 63/64 ......
(x) the sequence 1, 0, 1, 0, 1, ...
(xi) the sequence 1, −2, 3, −4, 5, ...
(xii) the sequence 1/2, 1/2, 3/4, 3/4, 3/4, 7/8, 7/8, 7/8, 7/8, 15/16, ...
(xiii) the sequence −1, −2, −4, −8, ...

The first concept we must define is that of an **increasing** sequence. If $u_{n+1} \geqslant u_n$ all through the sequence, it is said to be *monotonically* increasing. If $u_{n+1} > u_n$ all through the sequence, it is said to be *strictly* increasing.

Hence sequences (ix) and (xii) are both monotonically increasing and, in addition, sequence (ix) is strictly increasing. Similarly, we say that sequence (viii) is strictly decreasing. Sequences (x) and (xi) are said to *oscillate*.

† An example of an *arithmetic progression* which is in general a, $a+d$, $a+2d$, ...

The second concept is that of **boundedness**. A sequence is said to have an *upper bound M* if no term $u_n > M$. For sequence (viii), the number 47 is clearly an upper bound and so are 7, 2, 2/3, etc. We shall define a *least upper bound* (l.u.b.) for a sequence to be that number M^* such that

(a) $u_n \leqslant M^*$ for all terms u_n of the sequence, and
(b) if we take any number $M^+ < M^*$ we can find at least one term of the sequence which exceeds M^+ (in other words, nothing smaller will do).

For sequence (viii) the l.u.b. is 1/2 and equality is achieved with the initial term; for sequence (ix) the l.u.b. is 1 (convince yourself by taking a number just less than 1 and find a value for n such that u_n, u_{n+1}, u_{n+2} ..... are all greater than your number).

The l.u.b.'s for sequences (x) and (xii) are both 1, that for (xiii) is -1.

In a similar fashion we can define a *lower bound m* and a *greatest lower bound* (g.l.b.), m^*, and you should verify that the g.l.b.'s for sequences (viii), (ix), (x) and (xii) are respectively 0, 1/2, 0, 1/2. Sequences (vi) and (vii) both have g.l.b.'s but neither has an upper bound. Such sequences are said to **diverge**. Sequence (xi) has neither an upper nor a lower bound and is said to oscillate infinitely in contrast to (x) which oscillates finitely. Sequence (xiii) also diverges (to what?).

Let us now put on diagrams (Figure 3.4) the schematic behaviour of some of these sequences, identified by the sequence number shown on the graph. The terms are shown by bold dots and are joined by broken lines for convenience in discerning any trends. The diagrams are **not** to scale.

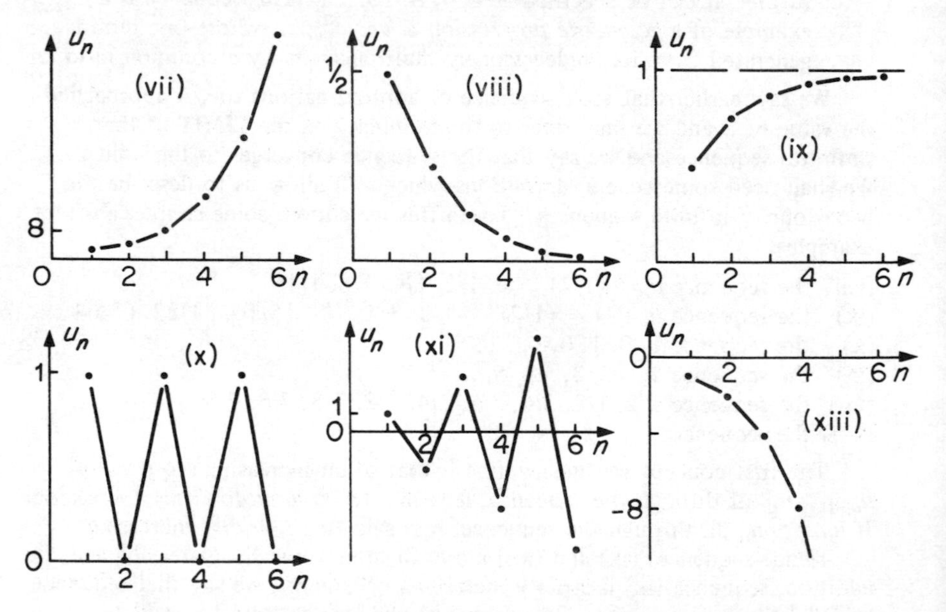

Figure 3.4

It is now time to tie down the concept of the **limit of a sequence**. The sequence whose typical term is u_n (often written $\{u_n\}$) is said to have a limit L (or to converge to the limit L) if for every positive number ϵ there is a term u_N in the sequence after which every term is within the range of values $(L - \epsilon, L + \epsilon)$. Notice three points: first, that the criterion must hold for *every* positive number ϵ no matter how small; second, that the critical term of the sequence, u_N, will not necessarily be the same for all possible ϵ; third, we are only interested in the long-term behaviour of the sequence since, with a an infinite number of terms, the earlier ones are but a minute fraction of the total. We illustrate these *three* points by examples.

The **probability** of an event may be regarded as a pre-assigned number, based on theoretical grounds or as long-range relative frequency. The results of tossing a presumed unbiased coin are tabulated below in Table 3.4. You can see that after a 'jerky' start, things settle down to a frequency close to 0.5. Whether the figure of 0.5 is justified is discussed in Chapter 11.

Table 3.4

Number of *tosses* to date	1	2	3	4	5	10	100	1000	10000
Number of *heads*	0	1	2	3	3	6	52	511	5054
Relative frequency	0	0.5	0.67	0.75	0.6	0.6	0.52	0.511	0.5054

It is important to consider the *tail* of a sequence since the first few terms may be atypical and, as in this example, more subject to influence by single events.

Let us look again at sequence (x), i.e. 1, 0, 1, 0, 1, ... Certainly if $\epsilon = 2$ we might try and claim a limit, L, of 0.5, since all terms of the sequence lie between $0.5 - 2$ and $0.5 + 2$, i.e. in the range (-1.5 to 2.5). Indeed we can take $\epsilon = 1$ to obtain the range (-0.5 to 1.5); but consider what happens if $\epsilon = 1/4$: all terms lie outside the range (0.25, 0.75) and the criterion must hold for **all** $\epsilon > 0$. Hence the sequence does not converge.

If we turn again to sequence (viii), $u_n = (1/2)^n$, we have said that the limit L is 0. We attempt to justify this. Suppose we choose $\epsilon = 1/64$; we look for a critical term such that thereafter, terms of the sequence lie in the interval $[-1/64, 1/64]$. But the sixth term is 1/64 and clearly, all succeeding terms lie in the required interval. Similarly, if we take $\epsilon = 1/256$ we note that all terms after the eighth are in the range of values $[-1/256, 1/256]$.

To generalise this argument, first suggest how you would choose N if ϵ was of the form $1/2^r$ where r is a positive integer. Then decide what you would do for a value of ϵ in the interval $(1/2^{r+1}, 1/2^r)$; for example, if $\epsilon = 1/230$ then $1/256 < \epsilon < 1/128$ and $r = 7$.

Sequence (ix) needs a little more care. We claim that the limit is 1. Now take $\epsilon = 1/16$; we seek that term after which all terms lie in the range of values (15/16, 17/16) and we see that the critical term, u_N, is the fourth. If we take $\epsilon = 1/64$ we must advance past the sixth term. You should try and show that if $\epsilon = 1/2^r$ then $N = r$, i.e. $N = \log \epsilon / \log \frac{1}{2}$. This is a complicated formula for N but the fact that as ϵ decreases N increases is clear as it was for sequence (viii).

Alternative ways of writing the statement that the limit of $\{u_n\}$ is L are

$$\lim_{n\to\infty} u_n = L$$

and

given any $\epsilon > 0$, *there exists an integer* N *such that* $|u_n - L| < \epsilon$ *when* $n > N$.

Obviously, from a practical viewpoint, the rate at which convergence takes place is important and will be studied later.

It would be tedious if the limit of every sequence had to be verified by resort to the definition. To aid us in this task we quote some rules for sequences in general:

Rules

Let $\{u_n\}$ and $\{v_n\}$ be sequences such that $\lim_{n\to\infty} u_n = U$, $\lim_{n\to\infty} v_n = V$.

1. The sequence $\{u_n + v_n\}$ which has terms $u_1 + v_1$, $u_2 + v_2$ etc. has limit $U + V$.

Examples

(i) The sequence $2\frac{1}{2}, 2\frac{1}{4}, 2\frac{1}{8}, 2\frac{1}{16}, \ldots\ldots$ can be regarded as the sum of the sequences 2, 2, 2, 2, ... and $\frac{1}{2}, \frac{1}{4}, \frac{1}{8}, \frac{1}{16}, \ldots\ldots$ Since these have limit 2 and 0 respectively, the sum sequence converges and has limit 2.

(ii) The sequences $2\frac{1}{2}, 2\frac{1}{4}, 2\frac{1}{8}, 2\frac{1}{16}, \ldots\ldots$ and $\frac{1}{2}, \frac{3}{4}, \frac{7}{8}, \frac{15}{16}, \ldots\ldots$ have limits 2 and 1 respectively. The difference sequence 2, $1\frac{1}{2}$, $1\frac{1}{4}$, $1\frac{1}{8}$, ... has limit $2 - 1 = 1$.

(iii) The sequence $2\frac{1}{2}, 3\frac{1}{3}, 4\frac{1}{4}, 5\frac{1}{5}, \ldots\ldots$ is the sum of the sequences 2, 3, 4, 5, ... and $\frac{1}{2}, \frac{1}{3}, \frac{1}{4}, \frac{1}{5}, \ldots\ldots$. Since the first of these has no limit, the sum sequence has no limit.

2. The sequence $\{\alpha u_n\}$ which has terms αu_1, αu_2, etc. has limit αU.

Example

With $\alpha = 3$, we see that the sequence $\frac{3}{2}, \frac{3}{4}, \frac{3}{8}, \frac{3}{16}, \ldots\ldots$ converges and has limit $3 \lim_{n\to\infty} \left\{\frac{1}{2}, \frac{1}{4}, \frac{1}{8}, \frac{1}{16}, \ldots\ldots\right\} = 3 \times 0 = 0$.

3. The sequence $\{u_n v_n\}$ has limit UV.

Example

The sequences $1\frac{1}{2}, 1\frac{3}{4}, 1\frac{7}{8}, 1\frac{15}{16}$ and $2\frac{1}{2}, 2\frac{1}{4}, 2\frac{1}{8}, 2\frac{1}{16}, \ldots\ldots$ both have limit 2. The product sequence 15/4, 63/16, 255/64, 1023/256, ... has limit 4.

[Note that (2) follows from (3) by taking $\{v_n\}$ to be the constant sequence $\alpha, \alpha, \alpha, \ldots\ldots$]

4. The sequence $\{u_n/v_n\}$ has limit U/V PROVIDED $V \neq 0$.

Example

With $\{u_n\} = 1, 1, 1, 1, ...$ and $\{v_n\} = 1/2, 3/4, 7/8, 15/16$. We find $\{u_n/v_n\} = 2/1, 4/3, 8/7, 16/15$ has limit 1.

Note that we can say nothing about the limit of the sequence 2/1, 4/1, 8/1, 16/1 from these rules, where $\{v_n\} = 1, 1, 1, ...$ though common sense suggests it diverges. (We use the symbolism $u_n \to \infty$ to indicate that the terms in the sequence increase without limit.)

We now examine these rules applied to some more examples.

Examples

(i) $\lim\limits_{n\to\infty} \dfrac{2n - 1}{3n + 4}$

Here it is useless to write the nth term as the quotient of $2n - 1$ and $3n + 4$ since neither $\{2n - 1\}$ nor $\{3n + 4\}$ converges. However, if we write $(2n - 1)/(3n + 4)$ as $(2 - 1/n)/(3 + 4/n)$ we may use Rule 1 to say that the limit of the numerator $2 - 1/n$ is 2 and that of the denominator is 3, then by Rule 4, the limit we seek is 2/3.

(ii) $\lim\limits_{n\to\infty} \dfrac{n^2 + 3n + 1}{2n^2 - 4n - 2}$

Division by n^2 in both numerator and denominator gives the nth term as $(1 + 3/n + 1/n^2)/(2 - 4/n - 2/n^2)$ and, by extension of the above argument, the limit of the sequence is 1/2.

(iii) $\lim\limits_{n\to\infty} \dfrac{2n^{-1} + 3n^{-4}}{5n^{-1} - 2n^{-2}}$

Multiplication of top and bottom by n gives the nth term as $(2 + 3n^{-3})/(5 - 2n^{-1})$ and the limit as 2/5.

(iv) $\lim\limits_{n\to\infty} \dfrac{1 + 2 + 3 + ... + n}{n} = \lim\limits_{n\to\infty} \dfrac{\frac{1}{2}n(n + 1)}{n} = \lim\limits_{n\to\infty} \dfrac{1}{2}(n + 1)$

Whilst the limit does not exist (as a finite quantity), since $\frac{1}{2}(n + 1) \to \infty$ as $n \to \infty$ it is customary to write $\lim\limits_{n\to\infty} \frac{1}{2}(n + 1) = \infty$.

Notice that we use the symbols $\to \infty$ since infinity is a state to be approached rather than achieved, and this is emphasised by equating a limit to ∞.

We have *two* statements which are of *importance:*

(a) an increasing sequence bounded above tends to a limit.
(b) a decreasing sequence bounded below tends to a limit.

Examples

(a) The sequence $1, 1\frac{1}{2}, 1\frac{3}{4}, 1\frac{7}{8}, 1\frac{15}{16}, ...$ is increasing and bounded above by 2; its limit is 2.

(b) The sequence 1, ½, ¼, ⅛, 1/16, ... is decreasing and bounded below by 0; its limit is 0.

Finally, let us return to the formula for approximations to $\sqrt{2}$:

$$x_{n+1} = \frac{1}{2}\left(x_n + \frac{2}{x_n}\right)$$

Let us suppose that the limit of the sequence $\{x_n\}$ is L; then, clearly, the limit of the sequence $\{x_{n+1}\}$ is L. By applying the rules for combinations of sequences we see that

$$\lim_{n\to\infty} \frac{1}{2}\left(x_n + \frac{2}{x_n}\right) = \frac{1}{2}\lim_{n\to\infty}\left(x_n + \frac{2}{x_n}\right)$$

$$= \frac{1}{2}\lim_{n\to\infty} x_n + \frac{1}{2}\lim_{n\to\infty}\frac{2}{x_n}$$

$$= \frac{1}{2}L + \frac{1}{2}\cdot\frac{2}{L}$$

But this limit must be also $\lim_{n\to\infty} x_{n+1}$ which is L. Then $L = ½L + 1/L$, i.e. $½L = 1/L$ or $L^2 = 2$, giving $L = \pm\sqrt{2}$. If x_0 is positive we get a sequence of positive numbers, and $L = +\sqrt{2}$, whereas, if x_0 is negative, $L = -\sqrt{2}$.

You should follow this argument replacing the iteration formula by the general one for finding $\sqrt{A}$.

The number e

As we shall see in many examples later on, the following limit is of importance:

$$\lim_{n\to\infty}\left[\left(1 + \frac{1}{n}\right)^n\right]$$

Write a computer program to evaluate the first 20 terms of this sequence; the limit value is 2.7183 to 4 d.p. You will see that convergence is slow. Convergence to a limit may be shown by demonstrating that the sequence is increasing and bounded above by 3.

Problems II

1. Write, in each case, as a list, the sequence specified by

   (i) $u_n = 2n(2n - 1)$

   (ii) $u_{n+1} = 3u_n - u_{n-1}$ given $u_1 = 0, u_2 = 2$

2. List the first 4 elements of the sequence $u_n = (-1)^n + 3^n$

3. Graph the first 6 terms of the sequences

   (i) $u_n = 3$; (ii) $u_n = \dfrac{2n - 1}{n}$

(iii) $u_n = n(-1)^n$ (iv) $u_n = 2 + \frac{(-1)^n}{n}$

State whether the sequences are bounded above or below, whether they are increasing, decreasing or oscillatory. If any is convergent, state its limit.

4. Repeat Problem 3 with the sequences:

(i) $u_n = 2n$ (ii) $u_n = \frac{2}{n}$

(iii) $u_n = 1 - \frac{1}{n}$ (iv) $u_n = 10^{-4} + \frac{1}{n}$

(v) $u_n = \frac{2n + 2}{n - 1}$ (vi) $u_n = \left(n + \frac{1}{n}\right)^2$

(vii) $u_n = (-n)^n$

5. Find an interval which contains all but the first 10 terms of the sequence $\{1 + (-2)^n/n\}$. Find a similar interval which excludes the first 1 000 terms.

6. Let $\{u_n\} = 1\frac{1}{2}, 1\frac{1}{3}, 1\frac{1}{4}, 1\frac{1}{5}, \ldots\ldots$ and $\{v_n\} = 1, 1, 1, 1, \ldots\ldots$ What is $\{u_v + v_n\}$ and what are the limits of the three sequences? Find the limits of $\{u_n.v_n\}$ and $\{u_n/v_n\}$.

Repeat for $\{v_n\} = 1, 3, 5, 7, \ldots\ldots$
and for $\{v_n\} = 3, 2\frac{1}{3}, 2\frac{1}{5}, 2\frac{1}{7}, \ldots\ldots$

3.2 CONTINUED FRACTIONS

It is often the case that in the manufacture of spiral gears, cams and screws a velocity ratio is required which is not exactly obtainable with the gears available or which has to be achieved by a combination of gears. Continued fractions are useful in this context. We shall consider only ratios less than 1 in the worked examples.

Example 1

We start with the ratio 43/143; we can divide 43 into 143 to obtain

$$\frac{43}{143} = \frac{1}{3 + \frac{14}{43}}$$

It is clear that $C_1 = 1/3$ is a first approximation to 43/143. But 14/43 may be written as $1/(3 + 1/14)$ and so

$$\frac{43}{143} = \cfrac{1}{3 + \cfrac{1}{3 + \cfrac{1}{14}}}$$

and a second approximation to 43/143 is

$$C_2 = \cfrac{1}{3 + \cfrac{1}{3}} = \frac{1}{\frac{10}{3}} = \frac{3}{10}$$

Since 1/14 has numerator 1, the process stops. The value of 43/143 to 3d.p. is 0.301.

Example 2

A second example is 226/361. This becomes successively

$$\cfrac{1}{1 + \cfrac{135}{226}} \qquad \cfrac{1}{1 + \cfrac{1}{1 + \cfrac{91}{135}}} \qquad \cfrac{1}{1 + \cfrac{1}{1 + \cfrac{1}{1 + \cfrac{44}{91}}}} \qquad \cfrac{1}{1 + \cfrac{1}{1 + \cfrac{1}{1 + \cfrac{1}{2 + \cfrac{3}{44}}}}}$$

$$\cfrac{1}{1 + \cfrac{1}{1 + \cfrac{1}{1 + \cfrac{1}{2 + \cfrac{1}{14 + \cfrac{2}{3}}}}}} \qquad \cfrac{1}{1 + \cfrac{1}{1 + \cfrac{1}{1 + \cfrac{1}{2 + \cfrac{1}{14 + \cfrac{1}{1 + \cfrac{1}{2}}}}}}}$$

The successive approximations (which are called **Convergents**) are:

$$C_1 = 1, \qquad C_2 = \frac{1}{1 + 1},$$

$$C_3 = \cfrac{1}{1 + \cfrac{1}{1 + \cfrac{1}{1}}}, \qquad C_4 = \cfrac{1}{1 + \cfrac{1}{1 + \cfrac{1}{1 + \cfrac{1}{2}}}},$$

$$C_5 = \cfrac{1}{1+\cfrac{1}{1+\cfrac{1}{1+\cfrac{1}{2+\cfrac{1}{14}}}}}, \quad \text{and} \quad C_6 = \cfrac{1}{1+\cfrac{1}{1+\cfrac{1}{1+\cfrac{1}{2+\cfrac{1}{14+\cfrac{1}{1}}}}}}$$

$$\text{i.e. } C_1 = 1, \quad C_2 = 0.5, \quad C_3 = \cfrac{1}{1+\cfrac{1}{2}} = \cfrac{1}{\frac{3}{2}} = 0.\dot{6}$$

$$C_4 = \cfrac{1}{1+\cfrac{1}{1+\cfrac{2}{3}}} = \cfrac{1}{1+\frac{3}{5}} = \cfrac{1}{\frac{8}{5}} = 0.625$$

$$C_5 = \frac{72}{115} = 0.6261 \text{ (4 d.p.) and } C_6 = \frac{77}{123} = 0.6260 \text{ (4 d.p.)}$$

Check that you can obtain C_5 and C_6. The value of 226/361 to 4 d.p. is 0.6260.

Notice two things about these examples. First, the successive approximations are alternately greater and less than the actual value. Second, that we fairly readily get a close approximation to the actual value (see Figure 3.5).

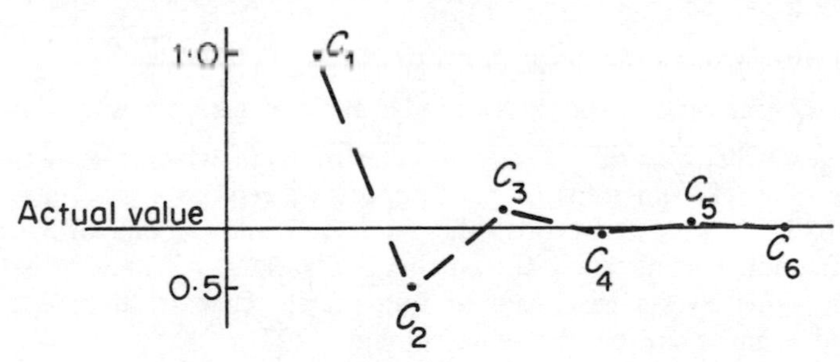

Figure 3.5

There is a space-saving way of writing continued fractions; for example, C_5 for 226/361 could be written

$$\frac{1}{1+}\,\frac{1}{1+}\,\frac{1}{1+}\,\frac{1}{2+}\,\frac{1}{14}$$

There is a tabular method of breaking down a ratio into a continued fraction. We develop the method stage by stage with 43/143 .

43	143	3

43	143	3
	129	

43	143	3
	129	
	14	

43	143	3
	129	3
	14	

43	143	3
42	129	3
	14	

43	143	3
42	129	3
1	14	

43	143	3 ←
42	129	3 ←
1	14	14 ←
	14	
	0	

Look back at the expansion and note where these numbers appear.

We have drawn the table (Table 3.5) for 226/361. Try and follow it through. You should compare it with the expansion method on page 118.

Table 3.5

226	361	1	←
135	226	1	←
91	135	1	←
88	91	2	←
3	44	14	←
2	42	1	←
1	2	2	←
	2		
	0		

Problems

1. In cutting a screw we have the formula

   gear ratio = (pitch required)/(pitch of lead screw)

   The gear wheels usually have a number of teeth which are in the range 20 to 120 in steps of 5. Suppose we require a gear ratio of 159/635. If we can factorise the numerator and denominator so that the fraction becomes $(x/y).(z/w)$ then we seek x, y, z, w in the stated range. This gives a gear train of four gears. Then x drives y, z drives w and y and z are on the same spindle.

   Show that the ratio can be factorised into $\dfrac{30}{50} \times \dfrac{53}{127}$.

   Express 53/127 as a continued fraction and show that the fourth convergent can be expressed as 50/120. Compare the approximate gear ratio with the required ratio.

2. Express as a continued fraction 26/139, 157/373, 146/2371, 59/43.

3. Calculate all the convergents in each case of Problem 2.

4. Find a suitable gear train for 877/647, 102/397, given that the wheels available have 20(5)120 teeth.

5. If $p/q = r/s$, for what values α, β does $(\alpha p + \beta r)/(\alpha q + \beta s)$ share their common value?

6. Show that $\frac{p}{q} < \frac{p + r}{q + s} < \frac{r}{s}$. (Note that it is implicit here that $\frac{p}{q} < \frac{r}{s}$ and p, q, r, s are all positive.)
Verify this result with the fractions 2/7 and 5/12.

3.3 FUNCTIONS OF A DISCRETE VARIABLE - INDUCTION

Sometimes the variables being studied are **continuous**, e.g. time, temperature, angle of twist, but sometimes they are **discrete**, e.g. if the total voltage in a circuit is provided by a number of batteries each of a given voltage V then the total voltage can only be an integral multiple of V (V, $2V$, $3V$, ...). Another important class of examples arises in the consideration of probabilities of discrete events, e.g. the probabilities of the various possible scores with two dice; the domain is the set $\{2, 3, 4, ..., 12\}$ A special case is when the domain of the function is the set of natural numbers: $\{1, 2, 3, 4, ...\}$ We can regard a sequence as a function of this type so that $\{u_n\}$ where $u_n = 2n - 1$ may be written $f(n) = 2n - 1$.

We consider a class of situations where the problem is expressed in terms of integers. It often happens that we need to verify that a conjecture holds for all values of a set; for example, it can be claimed that the formula $S_n \equiv 1 + 2 + 3 + ... + n = \frac{1}{2}n(n + 1)$ is valid for all positive integral values of n. It would be impossible to prove the conjecture true for each value of n in turn and it would be dangerous to prove it to be true for a few values of n and hope it is true for other values. The proof by mathematical induction is in two parts: first we show the conjecture to be true for the smallest value of n possible (usually $n = 1$), then we show that if the conjecture holds for a particular, but unspecified, value of n, say p, then it holds for the next value of n, i.e. $p + 1$. Two examples should help fix ideas.

Example 1

We start with the conjecture that

$$S_n = 1 + 2 + 3 + ... + n = \frac{1}{2}n(n + 1) \tag{3.3}$$

When $n = 1$ the sum consists of the term 1 only and the formula yields $\frac{1}{2}.1(1 + 1) = 1$, hence the conjecture is true in this case. Now we assume it to be true for $n = p$, i.e. $1 + 2 + ... + p = \frac{1}{2}p(p + 1)$. To investigate whether it holds true for $n = p + 1$, add the number $(p + 1)$ to both sides of this equation: $1 + 2 + ... + p + (p + 1) = \frac{1}{2}p(p + 1) + (p + 1)$. Now the right-hand side may be simplified to $[(p + 1)/2]\,[p + 2]$ and this is just the result we would have got by substituting $(p + 1)$ for n in formula (3.3). Hence, if the conjecture is true when $n = p$, it is true when $n = p + 1$. But we have shown directly that the conjecture is true when $n = 1$ and this second part has shown that the conjecture is true when $n = 2$ (put unspecified p equal to 1). Since this result is true when $n = 2$, a further application of the second part shows the conjecture to be true when $n = 3$ (put $p = 2$). In this way we prove the conjecture true for $n = 4, 5, 6, \ldots$. We have thus moved from the particular to the general – as opposed to the usual mathematical procedure.

Example 2

As a second example, consider the possibility that $n! > 2^n$. It is certainly not true when $n = 1$ or when $n = 2$ or 3. But when $n = 4$, $4! = 24 > 16 = 2^4$. This is the smallest value of n for which the inequality holds. Now let us suppose that $p! > 2^p$ for some unspecified $p \geqslant 4$.

Multiply both sides of the inequality by $(p + 1)$; then $(p + 1)p! > (p + 1)2^p$, i.e. $(p + 1)! > (p + 1)2^p$ and since $p + 1 > 2$, $(p + 1)2^p > 2^{p+1}$; hence $(p + 1)! > (p + 1)2^p > 2^{p+1}$ and so $(p + 1)! > 2^{p+1}$, proving the conjecture true when $n = p + 1$.

We conclude that $n! > 2^n$ for $n = 4, 5, 6, \ldots$

Notice (i) that such conjectures may be proved by other means, and
(ii) that we must have a conjecture in terms of n before the method of mathematical induction is applicable. A problem such as the determination of $1^2 + 2^2 + 3^2 + \ldots + n^2$ may best be tackled by a different approach.

Problems

1. Consider $S_n = a + (a + d) + \ldots + a + (n - 1)d$: the sum of the first n terms of an arithmetic progression. Write down the terms of the right-hand side in reverse order and obtain an expression for $2S_n$ which can be simplified by taking a common factor from each term. Hence deduce $S_n = \frac{1}{2}n[2a + (n - 1)d]$. Verify this formula by induction.

2. Consider $S_n = a + ar + ar^2 + \ldots + ar^{n-1}$. Write down expressions for rS_n and $(1 - r)S_n$ and hence show $S_n = a(1 - r^n)/(1 - r)$. Verify the formula by induction.

3. Show by induction

(i) $1^2 + 2^2 + 3^2 + \ldots + n^2 = \dfrac{1}{6}n(n + 1)(2n + 1)$

(ii) $1^3 + 2^3 + \ldots + n^3 = \dfrac{1}{4}n^2(n + 1)^2$

(iii) $n^{n-1} \geqslant n!$

(iv) $(a + b)^n = \displaystyle\sum_{r=0}^{n} \binom{n}{r} a^{n-r}b^r$

$$\equiv a^n + na^{n-1}b + \frac{n(n-1)}{2!}a^{n-2}b^2 + \ldots + \frac{n(n-1)\ldots(n-r+1)}{r!}a^{n-r}b^r + \ldots + nab^{n-1} + b^n$$

3.4 LIMITS OF FUNCTIONS AND CONTINUITY

We have said that a sequence could be regarded as a function defined on the natural numbers and we have discussed the limits of sequences. We might ask whether any useful information could be obtained from studying the application of the limiting process to other functions: this is the analytical approach to the study of functions.

A keystone in this study are the fundamental ideas of **continuity** and **discontinuity**. These concepts occur in physical situations: across a shock front in a gas the velocity and pressure of the gas change abruptly; the depth of water in a channel changes abruptly at a sluice gate, see Figure 3.6. We say that discontinuities occur in such cases.

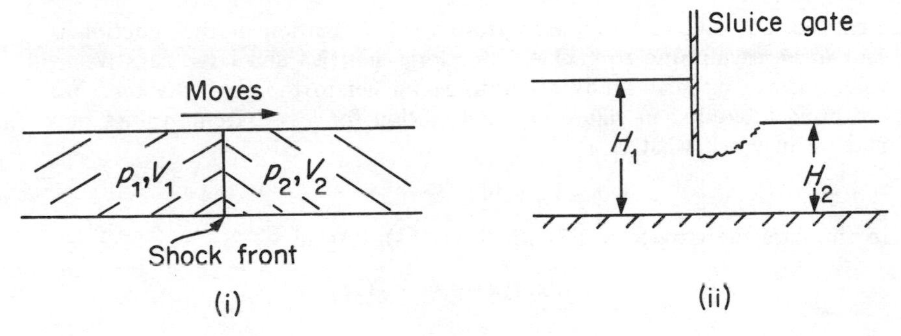

Figure 3.6

The idea of continuity can be described in terms of limits and we start with some basic idea on limits of functions.

We can see that

$$\lim_{n\to\infty} \left(1 - \frac{1}{n}\right) = 1 \dagger$$

and it would seem natural to suppose that

$$\lim_{x\to\infty} (1 - 1/x) = 1;$$

we note that whereas n approached infinity via a discrete set of values, x moves through a continuous range of values. Notice that $1 - (1/x)$ is not defined at $x = 0$ (see Figure 3.7).

We may liken the special case of finding the limit of the sequence $\{1 - (1/n)\}$ to walking on stepping stones placed on those values of x which are integral, instead of continuously proceeding to the limit as we do when considering the function $f(x) = 1 - 1/x$, defined for all $x \neq 0$. But other possibilities open up for the function $1 - 1/x$; what happens to $1 - 1/x$ as x becomes more and more negative? Since $-1/x$ is positive and becomes smaller, $1 - 1/x$ approaches the value 1 from above. In this

† The sequence is bounded above by 1 and is strictly increasing.

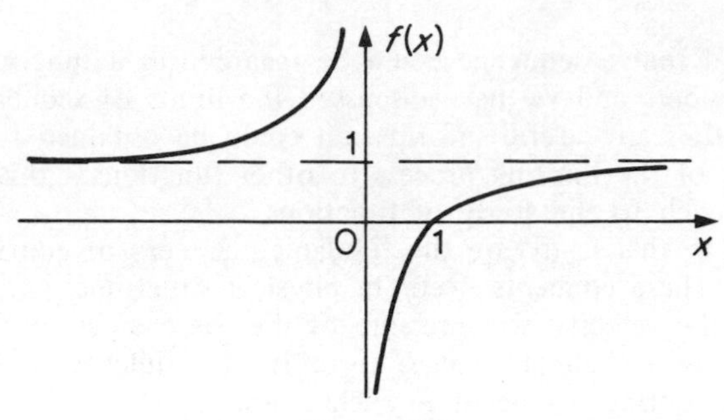

Figure 3.7

example, 1 is said to be an **asymptote** to the function or the function is said to be **asymptotic** to $f(x) = 1$ for large positive and large negative values of x. We shall study asymptotes further in the next chapter. We are often interested in values of the function for less extreme values of x and we may ask what is

$$\lim_{x \to 3} f(x);$$

in this case the answer is 2/3, which is $f(3)$. Again

$$\lim_{x \to 1} f(x) = 0 = f(1)$$

but this pattern breaks down for $\lim_{x \to 0} f(x)$

since $f(0)$ is not defined.

Some further examples should help clarify ideas.

1. If $f(x) = x^2 - 2$, then we would expect $\lim_{x \to 3} f(x) = 7 = f(3)$.
2. If $f(x) = \dfrac{x^2 - 1}{x - 1}$ then the function is not defined at $x = 1$.

In order to evaluate $\lim_{x \to 1} f(x)$ we must first simplify $f(x)$ to $(x + 1)$, which we can do as long as $x \neq 1$, and then take the limit to obtain the result 2

However, we could expect $\lim_{x \to 3} f(x)$ to be $(3^2 - 1)/(3 - 1)$, i.e. 4, which is equal to $f(3)$.

What we have been doing is applying the rules for sequences in a slightly different form; we may state these modified rules as follows. Assuming that all limits quoted exist, if

$$\lim_{x \to a} f(x) = A, \qquad \lim_{x \to a} g(x) = B,$$

(i) $\lim_{x \to a} [f(x) + g(x)] = A + B$

(ii) $\lim_{x \to a} [f(x).g(x)] = A.B$ [note the special case $f(x)$ is a constant C, then $\lim_{x \to a} f(x) = C$]

(iii) $\lim_{x \to a} [f(x)/g(x)] = A/B$, *provided* $B \neq 0$.

We have to some extent jumped the gun, and what we really need is a definition of what is meant by $\lim_{x \to a} f(x) = A$.

We effectively want to say that the closer x approaches the value a then the closer $f(x)$ is to A. We measure closeness of x to a by means of the gap between them: $|x - a|$; similarly, a measure of the closeness of $f(x)$ to A is $|f(x) - A|$. We want to say that given any error $\epsilon > 0$, however small, we can confine $f(x)$ within the range $(A-\epsilon, A+\epsilon)$ by confining x to some interval $(a-\delta, a+\delta)$; the less freedom we allow $f(x)$, the narrower this last interval must be. The important point is that we can always find a value for δ. As with sequences, we must guess A first.

Consider $f(x) = 1 + 4x^2$. We shall assert that

$$\lim_{x \to 0} f(x) = 1,$$

and try to prove this. From the definition we must take $A = 1$ and consider $|f(x) - 1| = |4x^2|$. Suppose we allow an error of 0.01, then we must confine x to some range about 0 so that $|4x^2| < 0.01$, i.e. $4x^2 < 0.01$, i.e. $x^2 < 0.0025$ and hence x lies in the interval $(-0.05, 0.05)$. Again, were we to allow an error of 0.0001 we should be forced to restrict x to the interval $(-0.005, 0.005)$ and if we take a general error ϵ we must restrict x to the interval $(-\frac{1}{2}\sqrt{\epsilon}, \frac{1}{2}\sqrt{\epsilon})$. It must be emphasised that, in general, the application of the formal definition is a process involving much algebraic manipulation which we shall not pursue here.

Returning to the graph of $f(x) = 1 - 1/x$ we see that the definition of $\lim_{x \to 0} f(x)$ has said nothing about the direction from which we approach 0. Someone journeying to $x = 0$ along the left-hand portion of the graph would obtain a very different impression of the behaviour of $f(x)$ near $x = 0$ from a traveller approaching that value of x from the right. Clearly the graph is discontinuous and we say that $f(x)$ has an infinite discontinuity at $x = 0$. A less dramatic example is the graph of the current in a simple circuit before and after its direction has been instantaneously reversed.

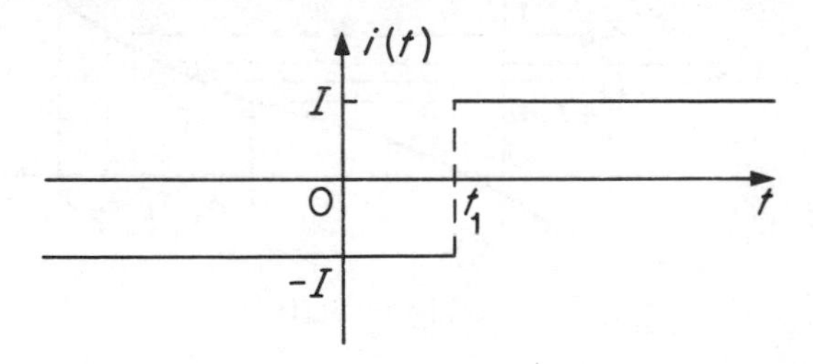

Figure 3.8

In order to complete the definition of the function, let us suppose

$$i(t)\begin{cases} = -I & \text{for } t \leqslant t_1 \\ = I & \text{for } t > t_1 \end{cases}$$

(see Figure 3.8)

This function has a finite discontinuity at $t = t_1$.

We can see that the limit of $i(t)$ as t approaches t_1 from below is $-I$; this is written

$$\lim_{t \uparrow t_1} i(t) = -I$$

and the limit of $i(t)$ as t approaches t_1 from above is I, written as

$$\lim_{t \downarrow t_1} i(t) = I$$

and only the former limit is actually equal to $i(t_1)$. Now we give some graphs in Figure 3.9 of 'invented' functions to illustrate the possibilities of discontinuity.

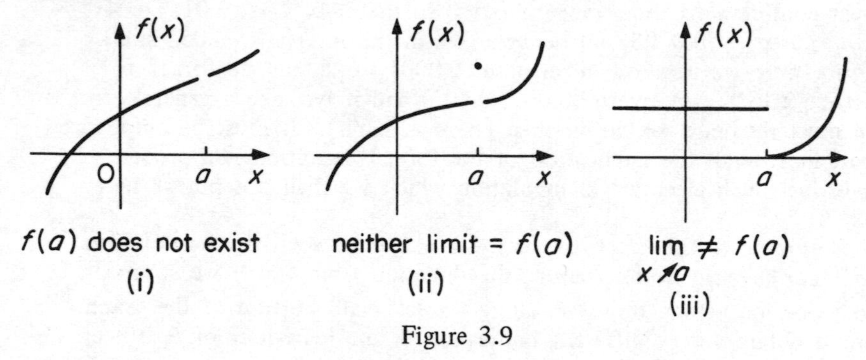

Figure 3.9

We may represent continuity pictorially in Figure 3.10 below.

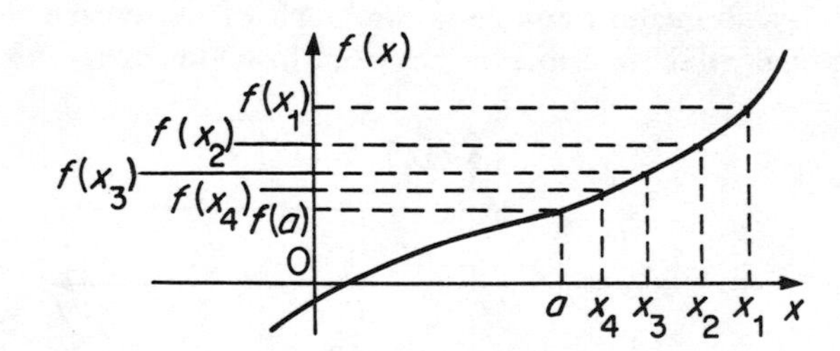

Figure 3.10

Consider a sample of x-values, x_1, x_2, x_3, x_4 which are getting closer to a. Then if $f(x)$ is continuous at $x = a$, the values $f(x_1), f(x_2), f(x_3), f(x_4)$ get closer to $f(a)$; by taking x close enough to a we can ensure that $f(x)$ is as close to $f(a)$ as required. Note that we should need to be able to repeat the argument with values of x to the left of a, sometimes written as $a - 0$ or $a-$.†

In fact we say that $f(x)$ is **continuous** at $x = a$ if $f(x)$ is defined for all values of x in some interval containing a and if $\lim_{x \to a} f(x) = f(a)$. This means, as for sequences, that we are only concerned with the tail of function values, i.e. we need concern ourselves only with the immediate vicinity of a. Hence the function $i(t)$ is continuous at all points save $t = t_1$. A function $f(x)$ which is continuous at all points on which it is defined is called a **continuous function**. As is the case with all analytical methods, the basic results apply at a point.

As the idea of continuity is linked to that of sequences there are analogous rules for combinations of functions. If $f(x)$ and $g(x)$ are both continuous at $x = a$, then so are $f(x) + g(x)$, $f(x).g(x)$ and, *provided* $g(a) \neq 0$, $f(x)/g(x)$. Also, if $f(x)$ is continuous at $x = a$ and $g(y)$ is continuous at $y = f(a)$ then $g[f(x)]$ is continuous at $x = a$.

We can establish the continuity of some functions straight from the definition and by use of these rules extend the number of functions which can be shown to be continuous. For example, we can show directly that $f(x) = C$, a constant, and that $g(x) = x$ are continuous and by repeated application of the rules any polynomial $a_n x^n + \ldots\ldots + a_1 x + a_0$, where a_i are constants, is continuous.

We state some useful results.

1. $\sin x$, $\cos x$ are continuous everywhere
2. $\tan x$ and $\sec x$ are continuous everywhere except at odd multiples of $\pi/2$
3. $\cot x$ and $\operatorname{cosec} x$ are continuous except at multiples of π
4. $1/(x - a)$ is continuous everywhere except at $x = a$

You should remember that the idea of continuity is a simple one but that the mathematician has to formalise its definition to be rigorous. For our purposes, the simple-minded approach will suffice.

We conclude by examining an important limit:

$$\lim_{x \to 0} \frac{\sin x}{x}$$

where x is measured in radians.

First we restrict ourselves to values of x in $0 < x < \pi/2$. Let O be the centre of a circle radius OP = OR and PQ be perpendicular to OP. From Figure 3.11 we can see that ΔOPR < sector OPR < ΔOPQ, i.e.
$\frac{1}{2}\text{OP} \,.\, \text{OR} \sin x < \frac{1}{2}(\text{OP})^2 \,.\, x < \frac{1}{2}\text{OP} \,.\, \text{PQ}$. But PQ = OP $\tan x$, and on dividing by $\frac{1}{2}(\text{OP})^2$,

$$\sin x < x < \frac{\sin x}{\cos x}$$

---

† Similarly, approaching a from the right will be written $a + 0$ or $a+$.

i.e. if we restrict ourselves to acute angles

$$1 < \frac{x}{\sin x} < \frac{1}{\cos x} \tag{3.4}$$

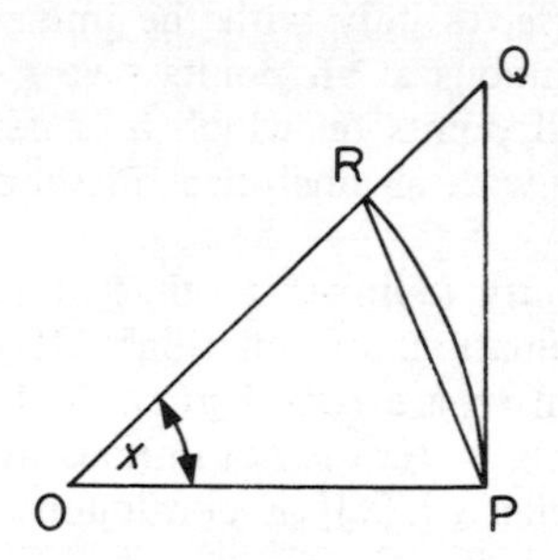

Figure 3.11

Now $\cos x$ is a continuous function and as $x \searrow 0$, $\cos x \to \cos 0 = 1$ and so $1/\cos x \to 1$. If in equation (3.4) we let $x \searrow 0$, $x/\sin x$ is squeezed between 1 and something which is approaching 1; therefore its limiting value must also be 1. Hence $\lim\limits_{x \searrow 0} x/\sin x = 1$ and by the rule for quotients $\lim\limits_{x \searrow 0} (\sin x)/x = 1$. Since $[\sin(-x)]/-x = (\sin x)/x$ we have $\lim\limits_{x \nearrow 0} (\sin x)/x = 1$. It follows that $\lim\limits_{x \to 0} (\sin x)/x = 1$.

It would be interesting for you to support these ideas by writing a computer program to print out values of $(\sin x)/x$ for $x = 1, 0.99, 0.98, \ldots.. 0.01$.

Problems

1. Determine $\lim\limits_{x \to 2} (x^2 + 2)$; $\lim\limits_{x \to -1} \dfrac{x^2 + 4x + 3}{x + 1}$; $\lim\limits_{x \to -1} \dfrac{x^2 + 4x + 4}{x + 1}$

2. Examine for discontinuities $\dfrac{1}{x}$, $\dfrac{2x - 4}{x^2 - 3x + 2}$

$$f(x) = |x| \qquad g(x) = \begin{cases} x, & x \geqslant 0 \\ 0, & x < 0 \end{cases} \qquad h(x) = \begin{cases} x + 1, & x > 1 \\ 2 - x, & x \leqslant 1 \end{cases}$$

3. The function $\text{sign}(x)$ takes the value 1 if $x > 0$, -1 if $x < 0$ and $\text{sign}(0) = 0$. Draw its graph.

4. Outline the steps by which you might show that $x^2 \sin x + 2 \cos x$ is a continuous function.

5. Show that $\lim\limits_{x \to 0} \dfrac{\sin \lambda x}{x} = \lambda$, and by noting that $\cos \lambda x = 1 - 2 \sin^2 \tfrac{1}{2}\lambda x$

   deduce that $\lim\limits_{x \to 0} \dfrac{1 - \cos \lambda x}{x^2} = \dfrac{\lambda^2}{2}$

6. $f(x)$ is **of the order of** $g(x)$, written $f(x)$ is $O[g(x)]$, in the region of $x = a$ if $f(x)/g(x)$ remains finite as $x \to a$. Thus since

$$\frac{x^3 + x - 1}{x^3 + 2} = \frac{1 + (1/x^2) - (1/x^3)}{1 + (2/x^3)} \to 1 \text{ as } x \to \infty$$

we say $x^3 + x - 1$ is $O(x^3 + 2)$ as $x \to \infty$.
More usually we compare with powers of x and we could say $x^3 + x - 1$ is $O(x^3)$ as $x \to \infty$. (Note that we choose the smallest power of x for which the statement is true.) Let $O(1)$ denote a constant.
Show $3 \cos x$ is $O(1)$ as $x \to 0$, $x^2 + 0.001x^4$ is $O(x^4)$ as $x \to \infty$, that

$(x - 1)(2x + 3)$ is $O(x^2)$ and hence $\dfrac{2x^2 + x - 3}{x^2 - 4} \to 2$ as $x \to \infty$.

7. Show $2 + (1/x^2)$ is $O(1)$ for large x and $(6x^3 + 2)/(x^2 - 1)$ is $O(x)$ for large x: investigate the behaviour for large x of

$$\frac{x}{2\sqrt{x^3 + 2}}, \quad \frac{2x^2 \cos x}{x^2 + 1}$$

8. $f(x)$ is of **smaller order** than $g(x)$, written $f(x)$ *is* $o[g(x)]$ *as* $x \to a$, if in the region of $x = a$, $f(x)/g(x) \to 0$ as $x \to a$.
Usually we take a to be zero. For example, $x^4 + 4$ is $o(x^3)$ as $x \to 0$. Investigate the behaviour of the functions in Problem 7 as $x \to 0$.

3.5 RATES OF CHANGE – DIFFERENTIATION

When one variable is related to another we are interested in studying the change in one variable induced by a change in the other. An important feature of the relationship is the *rate* at which the dependent variable changes with the independent variable. A particular class of problems is the study of the rate of change of quantities which vary with time: e.g. distance (whose time rate of change is speed), or work (whose time rate of change is power), or angle turned through (whose time rate of change is angular velocity). The rate at which water is being taken from a reservoir may be crucial as is the speed at which a space probe re-enters the earth's atmosphere. Other kinds of rates of change are important: for example, the rate of change of strain with applied stress for a metal specimen in a tensile test or the rate of change of temperature along a rod. Often we need to know as far as possible the instantaneous rate of change of a variable: certain materials may fracture if the rate of change of stress even momentarily exceeds a critical figure. Sometimes we are forced to consider *AVERAGE RATES OF CHANGE*, and this we do first. We use the data tabulated in Problem 8 on page 104. The results are plotted on Figure

3.12(i). The graph is effectively a straight line and if we seek the average rate of change of strain with stress we note that this is almost constant over the range of observations. We may read the slope of the graph anywhere, but we note that the strain rises from 0 to 5017×10^{-6} while the stress increases from 0 to 150; the average rate of change is then $(5017 \times 10^{-6})/150$. Such straight line relationships are rare and the relationship between the current passing through a varistor and the potential difference across its ends is an example of a non-linear equation. Experimental data is shown below and graphed in Figure 3.12(ii).

Table 3.6

E (volts)	0	5	10	15	20
I (amps)	0	0.26	0.41	0.56	0.68

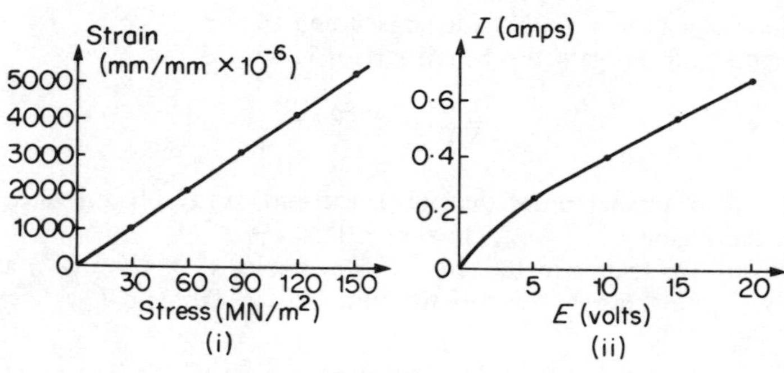

Figure 3.12

The average rates of change of I with E are 0.052 in the range 0 to 5 volts, 0.03 in the range 5 to 10 volts, 0.03 in the range 10 to 15 volts and 0.024 in the range 15 to 20 volts.

If we calculate the average rates of change in the intervals 0 to 10 volts, and 10 to 20 volts, the results are 0.041 and 0.027 respectively. The average rate of change over the whole interval is 0.034. Therefore in the region of 8 volts, we get different estimates depending on the interval chosen. It is clear that the smaller the interval, the more accurate the estimate, but the ideal is the (unattainable) local rate of change.

We now consider a second example. The speed of water in a channel was measured at various depths and tabulated as in Table 3.7. †

Table 3.7

Depth, cm	0	25	75	150	200
Speed, cm/sec	30	33	37	29	11

† One would presume physical reasons for such irregularly spaced data.

The shear stress in the water requires the local rate of change of speed with depth. To estimate this at 75cm we could take the average rate of change over the interval 0 to 150cm, i.e. −0.007 (3 d.p.). But the average from the interval 0 to 75 is 0.093 (3 d.p.) and the average from the interval 75 to 150 is −0.107 (3 d.p.). Which of these figures, if any, is a realistic estimate of the rate of change we seek? We might suppose that the average must be taken over an interval which has 75 near its centre and that the smaller the interval the better. However, if we have to estimate the rate of change at 200cm we are forced to take an interval having 200 as its right-hand end, but we would still suppose that the smaller the interval the better.

To test the wisdom of these suppositions we can try out our ideas on a known function, $f(x) = x^2$ and see how they are borne out in practice. The graph of part of the function is shown in Figure 3.13 and a table of certain values of the function is shown as Table 3.8.

Table 3.8

x	0	1/4	1/2	3/4	1	5/4	3/2	7/4	2
f	0	1/16	4/16	9/16	16/16	25/16	36/16	49/16	64/16

Let us endeavour to estimate the *local* rate of change of the function at $x = 1$ from the tabulated values. First we consider the estimates from the intervals with $x = 1$ as left-hand end.

[1, 1¼] gives an estimate of $\left(\frac{25}{16} - \frac{16}{16}\right)\Big/\frac{1}{4} = 2.25$

[1, 1½] gives an estimate of $\left(\frac{36}{16} - \frac{16}{16}\right)\Big/\frac{1}{2} = 2.5$

[1, 1¾] gives an estimate of $\frac{33}{16}\Big/\frac{3}{4} = 2.75$, and

[1, 2] gives an estimate of 3

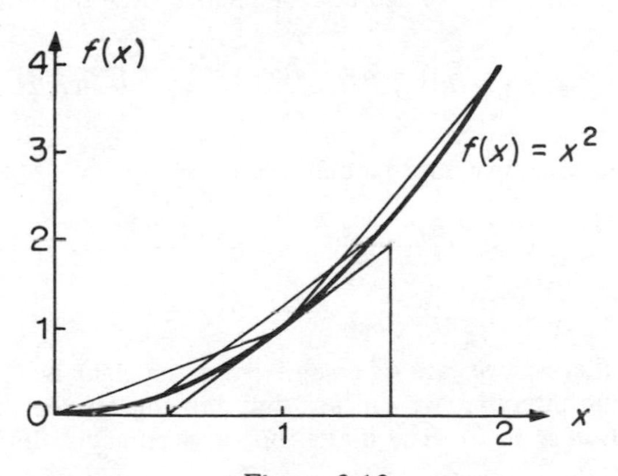

Figure 3.13

With intervals having $x = 1$ as right-hand end we obtain the following

$[\frac{3}{4}, 1]$ gives 1.75 $[\frac{1}{2}, 1]$ gives 1.5
$[\frac{1}{4}, 1]$ gives 1.25 $[0, 1]$ gives 1

With intervals symmetrically straddling $x = 1$, the results are:

$[\frac{3}{4}, 1\frac{1}{4}]$ gives 2 $[\frac{1}{2}, 1\frac{1}{2}]$ gives 2
$[\frac{1}{4}, 1\frac{3}{4}]$ gives 2 $[0, 2]$ gives 2

This consistency seems convincing and a glance at Figure 3.13 where the estimates of 3, 1 and 2 are shown demonstrates this. The local rate of change is influenced by both changes just before and just after, not one or other of these. The **tangent** at $x = 1$ to the curve representing $f(x)$ is drawn and this is found to have slope 2. We note that the left-hand or right-hand estimates improve as the interval becomes smaller. It would seem that the local rate of change is suitably measured by the tangent to the curve at the appropriate point. Let us try and investigate the situation algebraically. We refer to Figure 3.14.

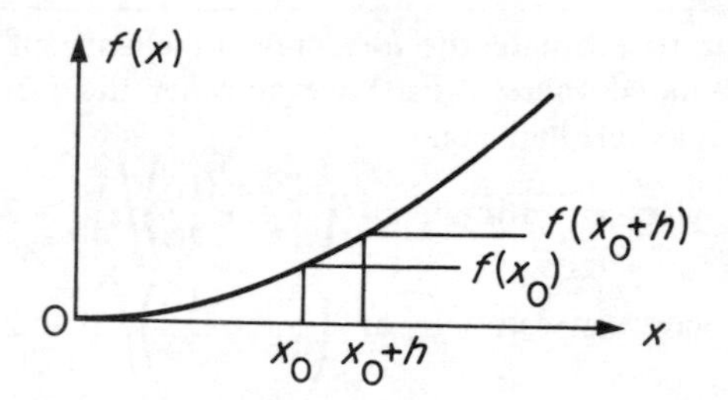

Figure 3.14

We have measured the average rate of change over the interval

$$[x_0, x_0 + h] \text{ by } \frac{f(x_0 + h) - f(x_0)}{x_0 + h - x_0} \quad \text{i.e.} \quad \frac{f(x_0 + h) - f(x_0)}{h}$$

In this particular example the fraction becomes

$$\frac{(x_0 + h)^2 - x_0^2}{h} = \frac{2x_0 h + h^2}{h}$$
$$= 2x_0 + h$$

Since $x_0 = 1$, the average rate of change from the right is $2 + h$, where h is the length of the interval; we can see that the smaller the interval, the closer the fraction is to 2. The limitation on accuracy is imposed by the tabular spacing, h.

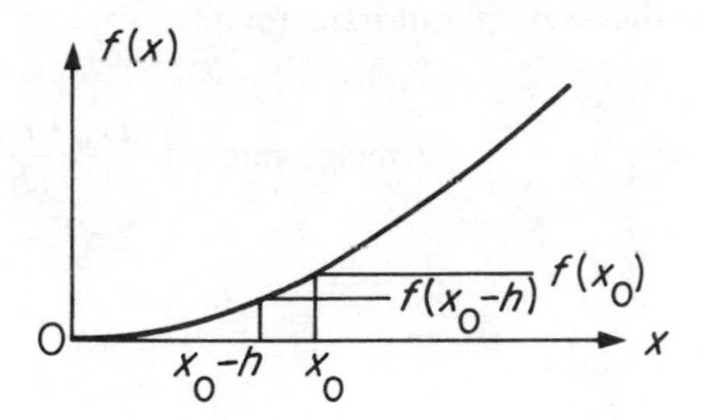

Figure 3.15

Refer now to Figure 3.15 above. Repeating the exercise for the average rate over the interval $[x_0 - h, x_0]$ we obtain the fraction

$$\frac{f(x_0) - f(x_0 - h)}{h} = \frac{2x_0h - h^2}{h} = 2x_0 - h$$

With $x_0 = 1$, we obtain $2 - h$ and, this time, as h gets smaller, the average rate approaches 2 from below.

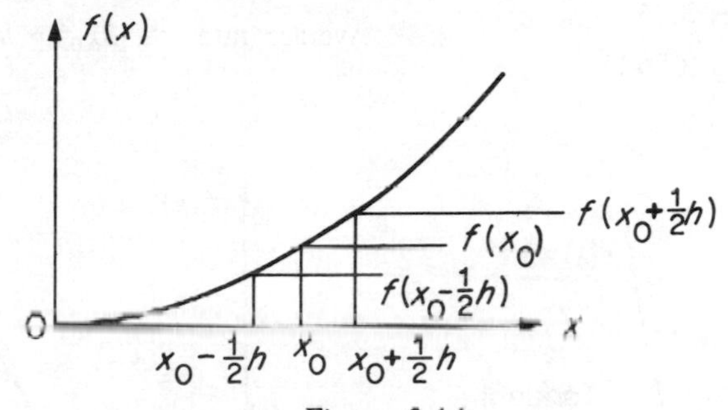

Figure 3.16

Finally we consider Figure 3.16 which represents the case of an interval with x_0 as its mid-point. The average rate is

$$\frac{f(x_0 + \frac{1}{2}h) - f(x_0 - \frac{1}{2}h)}{h} = \frac{2x_0h}{h} = 2x_0$$

Hence, at $x_0 = 1$, this rate is 2 irrespective of the length of interval. This seems a speciality of $f(x) = x^2$ so let us consider $f(x) = x^3$ and repeat the whole procedure; the graph is shown in Figure 3.17(i), and the values given in Table 3.9.

Table 3.9

x	0	1/4	1/2	3/4	1	5/4	3/2	7/4	2
$f(x)$	0	1/64	8/64	27/64	64/64	125/64	216/64	343/64	512/64

From the right-hand side the average rates are found.

[1, 1¼] gives 61/16 = 3.8125
[1, 1½] gives 4.75
[1, 1¾] gives 5.8125
[1, 2] gives 7

$$\text{Average rate} = \frac{(x_0+h)^3 - x_0^3}{h} = 3x_0^2 + 3x_0h + h^2$$

From the left-hand side we have

[¾, 1] gives 2.3125
[½, 1] gives 1.75
[¼, 1] gives 1.3125
[0, 1] gives 1

$$\text{Average rate} = \frac{1}{h}[x_0^3 - (x_0-h)^3] = 3x_0^2 - 3x_0h + h^2$$

And finally, the straddling intervals:

[¾, 1¼] gives 3.0625
[½, 1½] gives 3.25
[¼, 1¾] gives 3.5625
[0, 2] gives 4

$$\text{Average rate} = 3x_0^2 + h^2$$

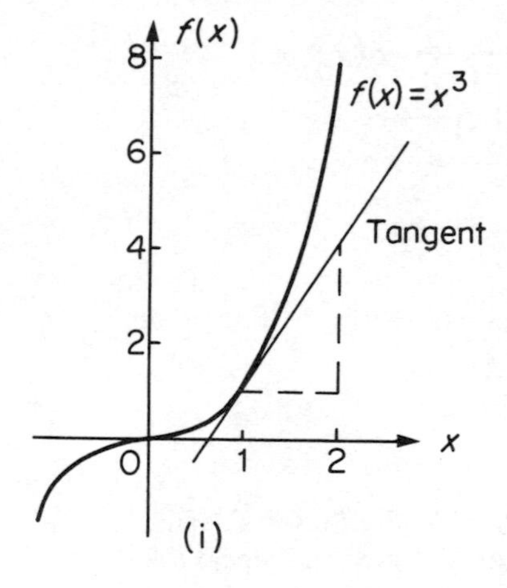

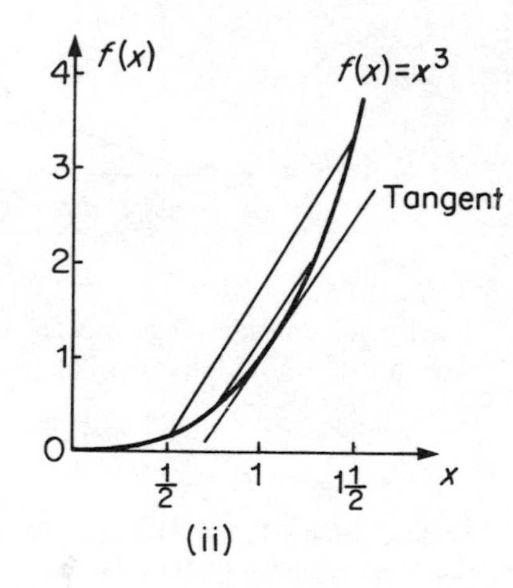

Figure 3.17

The tangent drawn at $x_0 = 1$ has a slope of 3 and this can be imagined as the limit of the slopes of the lines joining end points of the straddling intervals, see Figure 3.17(ii). We must note that the last set of estimates are all over-estimates and this is because the values of $f(x)$ for $x > 1$ are larger than the values for $x < 1$ and tend to dominate them in the calculation of the average rate. Ideally then we should estimate the average rate by the smallest straddling value that the table of values will permit. Note further that if the tabular spacing is h the function values to hand will be $f(x_0 - h)$, $f(x_0)$ and $f(x_0 + h)$ and if we require an estimate of the local rate at x_0 we must use

$$\frac{f(x_0 + h) - f(x_0 - h)}{2h}$$

We shall refer to this as the *central formula.* If we require the estimate half-way between two tabulated points $f(x_0)$ and $f(x_0 + h)$, the formula becomes

$$\frac{f(x_0 + h) - f(x_0)}{h}$$

Further notice that if we want an estimate at either end of the tabulated values, then the central formula is not applicable. If the local rate is required at a tabulated point where the tabular spacing is h, it may be argued that the straddling formula would require an interval of length $2h$, whereas the best left-hand or right-hand estimates would require an interval of length, h, only. However, for $f(x) = x^2$, the best central formula gives 2 as opposed to 2.25 or 1.75 and for $f(x) = x^3$, it gives 3.065 as opposed to 3.8125 or 2.3125.

Problems I

1. As accurately as possible, estimate the rate of change of strain with stress at stresses of 0, 50, 150 MN/m^2 from the values in Problem 8 on page 104.

2. Estimate as accurately as possible the rate of change of density of air with altitude from the table in Problem 10 on page 104. Make estimations at sea-level, 5000 metres and 10 000 metres.

3. In the study of consolidation of soil the following readings obtained for the *time factor* T_v against the mean degree of consolidation $\overline{U}$. Establish the rates of change of T_v with $\overline{U}$ for $\overline{U} = 0, 0.3, 0.8$.

$\overline{U}$	0	0.1	0.2	0.3	0.4	0.5	0.6	0.7	0.8	0.9	1.0
T_v	0	0.009	0.032	0.072	0.130	0.201	0.294	0.413	0.581	0.865	∞

4. The angle turned through by a shaft was measured and the results tabulated as follows:

t (seconds)	0.0	0.2	0.4	0.6	0.8	1.0
θ (radians)	−0.003	0.057	0.148	0.282	0.470	0.706

Estimate the angular velocity $\dot{\theta}$ when $t = 0, 0.4, 0.8, 1.0$.

Rate of change from a formula

We have seen that a table of values places a restriction on the accuracy of estimation of the local rate of change. If we possess a formula for the functional relationship being studied then we have no restriction on the value of h. Since, in general, the smaller h the more accurate the estimate and since we have intuitively seen that the tangent which measures the local rate of change is the limit of the approximating slopes, we investigate the possibility of the limit of the sequence of approximations as $h \to 0$. We can turn the approximations for the case $f(x) = x^2$ into terms of sequences by assuming $h = 1/n$; then as $h \to 0$, $n \to \infty$. The left-hand approximation becomes $2x_0 - 1/n$ and the right-hand approximation becomes $2x_0 + 1/n$; it is clear that both expressions have limit $2x_0$ as $n \to \infty$. For the approximations for the example $f(x) = x^3$ the expressions become $3x_0^2 + 3x_0 . 1/n + 1/n^2$ and $3x_0^2 - 3x_0 . 1/n + 1/n^2$: both have $3x_0^2$ as limit. The central formula yields $3x_0^2 + 1/n^2$ and this too has $3x_0^2$ as limit which we might expect to be approached more rapidly. Before we examine the snags let us work through two more examples.

Consider $f(x) = \sin x$. The central formula for average rate of change becomes

$$\frac{\sin(x_0 + h) - \sin(x_0 - h)}{2h}$$

$$= \frac{\sin x_0 \cos h + \cos x_0 \sin h - (\sin x_0 \cos h - \cos x_0 \sin h)}{2h}$$

$$= \frac{2 \cos x_0 \sin h}{2h}$$

$$= \cos x_0 . \frac{\sin h}{h}$$

Then, as $h \to 0$, $\dfrac{\sin h}{h} \to 1$ and so the rate of change becomes $\cos x_0$.†

Table 3.10 is an extract from a table of sines:

Table 3.10

x (radians)	0.9	0.96	0.98	1.0	1.02	1.04	1.1
$\sin x$	0.7833	0.8192	0.8305	0.8415	0.8521	0.8624	0.8912

For the interval [0.9, 1.1] the average rate of change of $\sin x$ is 0.5395; from [0.96, 1.04] it is 0.54 and from [0.98, 1.02] it is 0.54. These compare with the local rate of change, $\cos 1.0 = 0.5403$. The close agreement suggests that in this range $\sin x$ is well approximated by a straight line. The precision of the tables presents a better approximation.

† This result applied when h was measured in radians, consequently the same restriction holds for the whole derivation.

We note that we have proceeded to the limit at a point x_0. As with continuity of a function we define the property at a point and hope it generalises to all points. We assume that a limit exists from either side and that the values are the same. We have a definition: the function $f(x)$ is said to be **differentiable** at $x = x_0$ if

$$\lim_{x \to x_0} \frac{f(x) - f(x_0)}{x - x_0}$$

exists; the value of the limit is called the **derivative** of $f(x)$ at $x = x_0$, and the process of finding it is called **differentiation**. Observe that we have used a notation for limit which allows it to be approached from either side and which assumes the same value for the two approaches.

Further, if the function is differentiable at every point in its domain it is called a **differentiable function**. In such a case, to each x_0 in the domain we may associate a number: the value of the derivative at x_0. This will produce a new function which is called the **derived function**. For $f(x) = x^2$ we have seen that its derivative at x_0 is $2x_0$ and so the derived function, written $f'(x)$, is $2x$. For $f(x) = x^3$, $f'(x) = 3x^2$ and for $f(x) = \sin x$, $f'(x) = \cos x$. It would be tedious if the derived function had to be developed from the definition directly but as the underlying concept is that of a limit the usual rules apply, e.g. *the sum of two differentiable functions is differentiable and the value of the derivative of the sum at any point is the sum of the separate derivatives*, that is, $\{[f(x) + g(x)]' = f'(x) + g'(x)\}$. We take up these rules further in Chapter 6.

If the function is not differentiable at just one point it ceases to be called differentiable (c.f. continuity); the breakdown can occur if the limit does not exist at all, or if different values are obtained from the two directions.

It is important to note that the derivative $f'(x_0)$ at each point x_0 at which it exists gives the value of the slope of the graph of $f(x)$ at that point.

Snags

We first examine

$$f(x) = \begin{cases} 1, & x \geqslant 0 \\ -1, & x < 0 \end{cases}$$

which is, by our criterion, a function. Its graph is shown in Figure 3.18.

The local rate of change at all points $x_0 > 0$ is 0 as it is for $x_0 < 0$. The trouble occurs at $x_0 = 0$ because a *finite* change (2 units) takes place at a *point*, i.e. over an interval of zero length.

The right-hand approximation,

$$\frac{f(x_0) - f(0)}{x_0 - 0} = \frac{1 - 1}{x_0} = 0$$

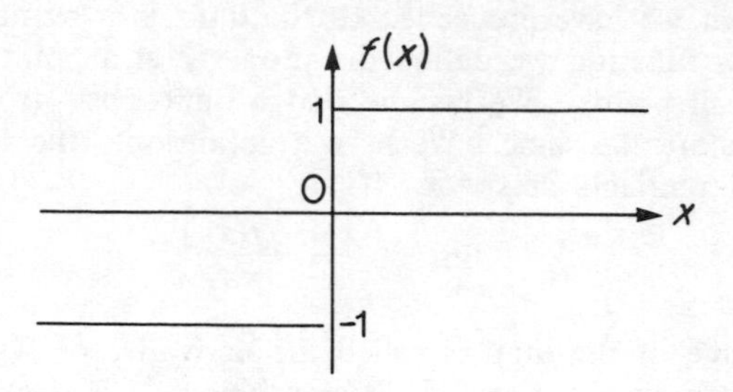

Figure 3.18

and, as $x_0 \searrow 0$, the fraction stays zero. The left-hand approximation,

$$\frac{f(0) - f(x_0)}{0 - x_0} = \frac{1 - (-1)}{-x_0} = \frac{2}{x_0}$$

has no limit as $x_0 \nearrow 0$.

This is an illustration of a general result that *if a function is not continuous at a point it is not differentiable there.* Danger will also exist in practice if a relationship is almost discontinuous.

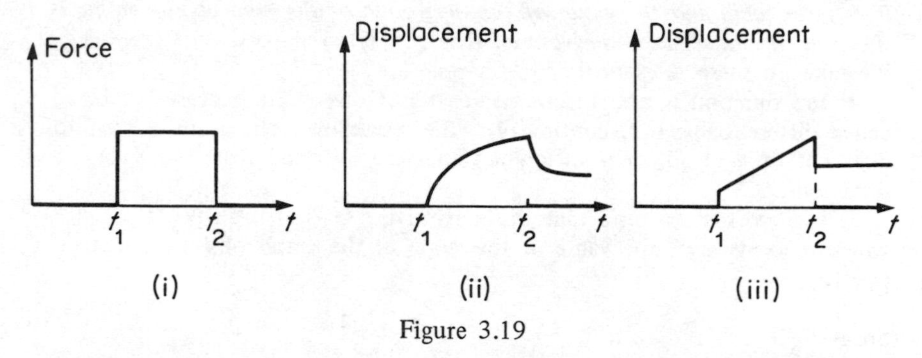

Figure 3.19

Figure 3.19(ii) shows the relationship between displacements of a specimen of asphalt subjected to a compressive force behaving as in Figure 3.19(i). The rapid changes at t_1 and t_2 cause very high rates of change. The force is suddenly applied at $t = t_1$ and removed at $t = t_2$.

An idealised model of the situation would behave as in Figure 3.19(iii). This idealised relationship is discontinuous at $t = t_1$ and $t = t_2$. However, we must not rush into the trap that all functions which are continuous are differentiable. Reflect on $f(x) = |x|$ which is shown in Figure 3.20.

This can also be written

$$f(x) = \begin{cases} x, & x \geqslant 0 \\ -x, & x \leqslant 0 \end{cases}$$

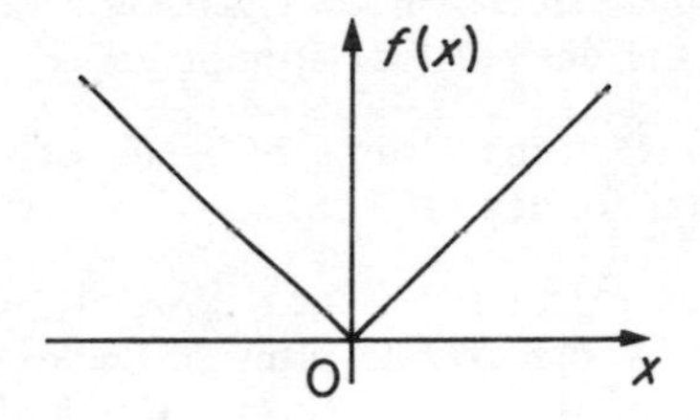

Figure 3.20

It is clear that the derivative of $f(x)$ is 1 for $x > 0$ and -1 for $x < 0$. Trouble is present at $x = 0$; if $x_0 > 0$

$$\frac{f(x_0) - f(0)}{x_0 - 0} = \frac{x_0}{x_0} = 1$$

whilst if $x_0 < 0$,

$$\frac{f(0) - f(x_0)}{0 - x_0} = \frac{-(-x_0)}{-x_0} = -1$$

Conflict has occurred and so we conclude that $f(x)$ is not differentiable at $x = 0$ despite being continuous there. (It is true, in general, that a differentiable function is continuous.)

It is worth mentioning that the derived function (excluding $x = 0$ from consideration) of $f(x) = |x|$ is the function

$$f(x) = \begin{cases} 1, & x > 0 \\ -1, & x < 0 \end{cases}$$

The value of $f(0)$ does not exist.

It is a general principle that the process of differentiation tends to produce a more badly behaved function than the original and numerically is a process to be avoided where possible. We have not yet mentioned how to estimate the error involved; this we do in Chapter 6.

Problems II

1. Consider Table 3.11 which is an extract from a table of Naperian logarithms.

Table 3.11

x	3.6	3.8	3.9	4.0	4.1	4.2	4.4
$\log_e x$	1.2809	1.3350	1.3610	1.3863	1.4110	1.4351	1.4816

From the interval [3.6, 4.4] the average rate of change is 0.251 (3 d.p.)
From the interval [3.8, 4.2] the average rate of change is 0.250 (3 d.p.)
From the interval [3.9, 4.1] the average rate of change is 0.250 (3 d.p.)

Repeat for similar entries around 2.0, 3.0, 5.0. What do you conjecture about the derived function of $\log_e x$?

2. Repeat Problem 1 using a table of reciprocals and suggest a suitable derived function for $f(x) = 1/x$.

3. Tabulate $f(x) = x^3$ for $x = -\frac{1}{4}, -\frac{1}{2}, -\frac{3}{4}, -1, -1\frac{1}{4}, -1\frac{1}{2}, -1\frac{3}{4}, -2$. Estimate $f'(0)$ by the average rates of change over the intervals $[-\frac{1}{4}, \frac{1}{4}], [-\frac{1}{2}, \frac{1}{2}], \ldots\ldots, [-2, 2]$. Compare with the values from intervals $[0, \frac{1}{4}], [0, \frac{1}{2}], \ldots\ldots, [0, 2]$.
Comment.

4. Tabulate $f(x) = \sinh x$ for $x = 1.0(0.1)2.0$†. Produce successive approximations to $f'(1.5)$ and compare with $\cosh 1.5$. Repeat for $f(x) = \cosh x$ and compare with $\sinh 1.5$.

5. A number a is part of data for an experiment. In the course of some calculations, it is necessary to form a^2. If the original number was subject to a small data error ϵ the value of a^2 is in error by $2a\epsilon$ (to first order). The function of squaring has thus multiplied the error by a **scale factor** of $2a$.
Suppose we wish to solve $x^3 = 9$. We know that the true value is nearer to 2 than 3. Calculate s, the *scale factor* for the function $f(x) = x^3$ at $x = 2$, hoping that this is a good approximation to the average scale factor over the interval $[2, \sqrt[3]{9}]$. {Note that $f(x) = x^3$ maps $[2, \sqrt[3]{9}]$ into $[8, 9]$.}
If the true value of $\sqrt[3]{9}$ is $2 + \epsilon$, then the 'error' ϵ is scaled into an error of $s\epsilon$; but $s\epsilon = 9 - 8 = 1$, hence we can find ϵ approximately and obtain a better estimate of $\sqrt[3]{9}$.
Starting with this new value, repeat the process to obtain a second revised estimate of $\sqrt[3]{9}$. Repeat, if necessary, until $\sqrt[3]{9}$ is obtained correct to 2 d.p. (compare with tables).

6. Repeat Problem 5 to find $\sqrt{17}$ to 2 d.p.

7. Obtain from the definition, the derived function of $f(x) = ax^3 + bx^2 + cx + d$, where a, b, c, d are constants.

8. Obtain the derived function of each of the following:
$\cos x$, $\sin 2x$, $\cos 3x$.

9. We seek the derived function of $1/x$. Obtain the expression $-(1/a)(a+h)^{-1}$ as an approximation to the derivative at $x = a$. Write $(a+h)^{-1}$ as

$$\frac{1}{a}\left(1 + \frac{h}{a}\right)^{-1}$$

....

† 1.0, 1.1, 1.2, ..., 1.9, 2.0.

and expand by the binomial theorem. Hence deduce the derived function as $-1/x^2$. By a similar procedure deduce that the derived function of $1/x^2$ is $-2/x^3$.

10. Draw a flow chart and write a computer program to calculate the approximation $(-1/a)(a+h)^{-1}$ for $h = 1.0(0.05)0.5$, and $a = 2.0$. Compare with the known derivative.

11. For the extract from the table of sines on page 136 , compute the best estimate of the derivative at the points 0.96(0.02)1.04. These should be close to values of the cosines of these angles. If we start with these values and repeat the process we shall obtain 3 values which are approximations to $-\sin 0.98$, $-\sin 1.0$ and $-\sin 1.02$ (the derivative of $\cos x$ is $-\sin x$). Repeating the process twice more should leave us with one value, approximately $\sin 1.0$ (the derivative of $-\sin x$ is $-\cos x$ and the derivative of $-\cos x$ is $\sin x$). What do you observe about the accuracy of the two estimates of $\cos 1.0$ and the two estimates of $\sin 1.0$?

12. It is clear that the methods for estimating $f'(a)$ replace the tangent by straight line chords. It is claimed that the central formula gives the same estimate as would a parabola through the points D, E, F (see Figure 3.21). To investigate this claim, let the parabola be of the form $g(x) = \alpha(x - x_2)(x - x_1) + \beta(x - x_1) + \gamma$.

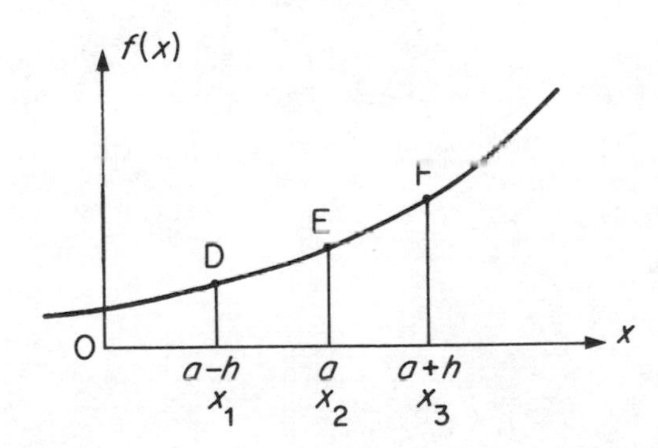

Figure 3.21

Find $g'(x)$ and hence $g'(x_2)$. Since $g(x) = f(x)$ at x_1, x_2, x_3, determine α, β and γ, and hence show that $g'(x_2)$ is the central formula. Why should that make this formula more accurate than the others?

SUMMARY

This chapter is mainly concerned with the process of moving to a limit, which is the basic idea behind the calculus. We want you to realise that the limiting process is fundamental to the concept of differentiation and, since limits do not always exist, then not all functions can be differentiated.

The concept of limit of a sequence was introduced first since we expect you to be able to appreciate the limiting behaviour of sequences of numbers. You are not expected to be able to find the limit of *any* sequence if this would involve very difficult algebra, but we do want you to have acquired the skill to find the limits of certain types of sequence.

The extension of the idea of limit to the case of functions is important because you should appreciate that the calculus is concerned with the behaviour of a function at and near a point, hence the alternative name – analysis. When we differentiate a function by calculus we usually apply general rules to the functional expression to obtain an expression for the derived function. However, we want you to realise that the differentiation process is applied strictly at a point and it is important to know if the function misbehaves at any point in its domain. We have introduced you to the differentiation of certain simple functions so that you can see the process in action.

It is expected that you understand the concept of continuity of a function. We have included the section on Mathematical Induction so that you may add it to your stock of techniques.

Chapter Four

Geometry, Vectors and Complex Numbers

4.1 ELEMENTARY COORDINATE GEOMETRY

It is assumed that the reader has already met some of the standard results in elementary coordinate geometry. We collect a few of them here for revision purposes.

Distance between two points

By Pythagoras' Theorem, the distance between the two points A and B in Figure 4.1 is

$$\sqrt{(x_1 - x_2)^2 + (y_1 - y_2)^2} \tag{4.1}$$

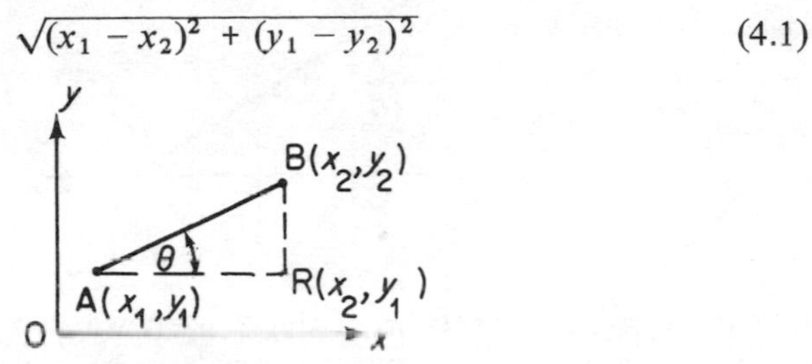

Figure 4.1

Gradient of a line

The **gradient** or slope of the line shown in Figure 4.1 is defined as tan θ (usually denoted by m). From this figure

$$m = \tan\theta = (y_2 - y_1)/(x_2 - x_1) \tag{4.2}$$

The straight line

The equation of a straight line can take one of many forms, depending on the information given.

(i) Line through two points, (x_1, y_1) and (x_2, y_2)

$$y - y_1 = \left(\frac{y_2 - y_1}{x_2 - x_1}\right)(x - x_1) \tag{4.3}$$

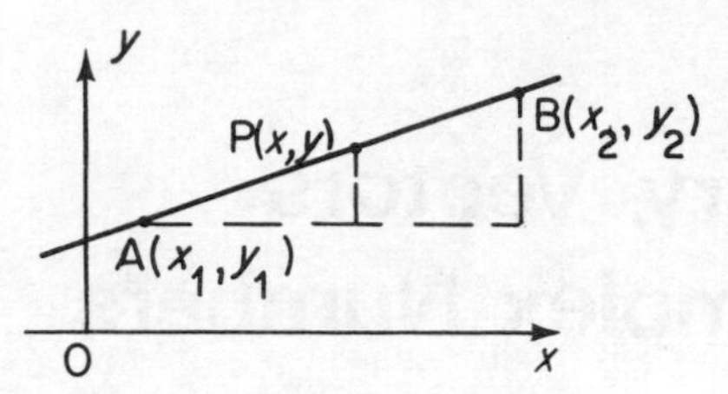

Figure 4.2

Equation (4.3) gives the coordinates (x, y) of any point P on the line. The result follows from the similar triangles shown in Figure 4.2.

(ii) Line through a given point (x_1, y_1) with a given slope m

$$y - y_1 = m(x - x_1) \tag{4.4}$$

(iii) Line making known intercepts on the axes

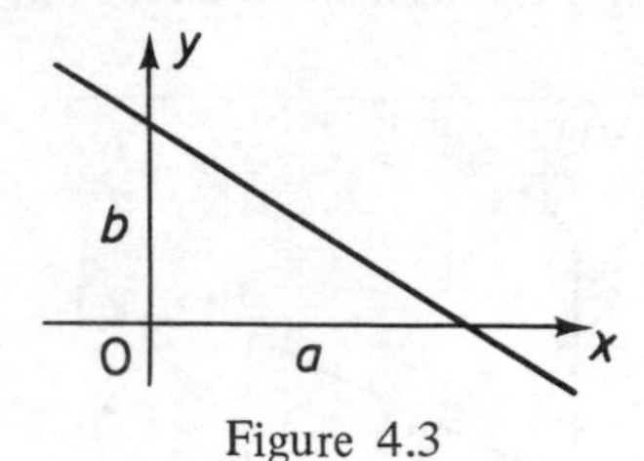

Figure 4.3

Suppose the intercepts are a and b as shown in Figure 4.3. Then the line passes through the points $(a, 0)$ and $(0, b)$. Using (4.2) the equation of the line is

$$y - 0 = \left(\frac{b - 0}{0 - a}\right)(x - a)$$

which simplifies to

$$\frac{x}{a} + \frac{y}{b} = 1 \tag{4.5}$$

(iv) Line with slope m and intercept c on the y-axis

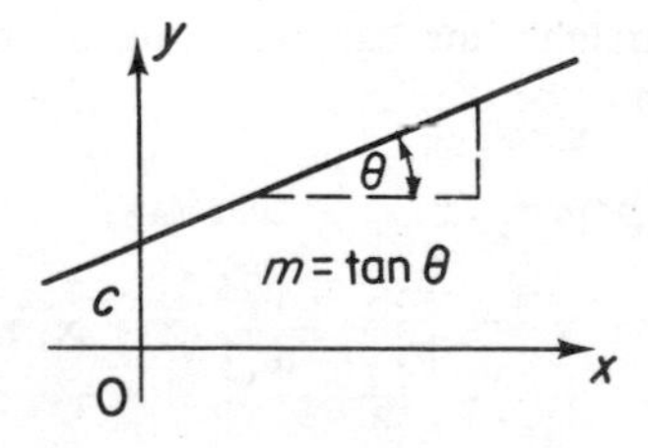

Figure 4.4

Here the slope is given and a point on the line is $(0, c)$. Hence the equation of the line, using (4.4) is

$$y = mx + c \tag{4.6}$$

(v) General form of the equation

Other forms exist, but they are all rearrangements of the general equation

$$ax + by + c = 0 \tag{4.7}$$

Writing this as $y = -ax/b - c/b$ and comparing with (4.6) we see that this has slope $-a/b$ and intercept on the y-axis $-c/b$.

Angle between two lines

For the lines shown in Figure 4.5,

$$m_1 = \tan\theta_1 \quad \text{and} \quad m_2 = \tan\theta_2$$

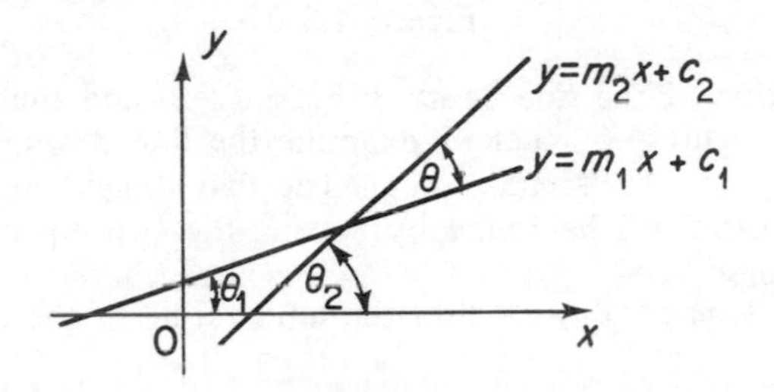

Figure 4.5

Let $\theta = \theta_2 - \theta_1$ be the angle between the lines, then

$$\tan\theta = \tan(\theta_2 - \theta_1) \equiv \frac{\tan\theta_2 - \tan\theta_1}{1 + \tan\theta_2 \tan\theta_1}$$

$$= \frac{m_2 - m_1}{1 + m_2 m_1} \tag{4.8}$$

If $m_1 = m_2$ then $\tan\theta = 0$ and the lines are parallel.
If $m_1 m_2 \to -1$ $\tan\theta \to \infty$ and $\theta \to \pi/2$; i.e. the condition for two lines to be perpendicular is that the product of their gradients is equal to -1.

Example

If two lines have slopes 1.0 and 2.0 then $\tan\theta = (2 - 1)/(1 + 2) = 1/3$, but if we took the slopes as 2.0 and 1.0, $\tan\phi = (1 - 2)/(1 + 2) = -1/3$; all this means is that we get the obtuse angle between the lines instead of the acute angle.

Intersection of two lines

If the lines be $y = m_1x + c_1$ and $y = m_2x + c_2$ they intersect when x and y satisfy both equations simultaneously, i.e. when

$$x = \frac{c_2 - c_1}{m_2 - m_1}, \qquad y = \frac{m_1c_2 - c_1m_2}{m_2 - m_1} \tag{4.9}$$

Note that this result breaks down if $m_1 = m_2$ as you might expect, since the lines are then parallel.

Shortest distance of a point from a line

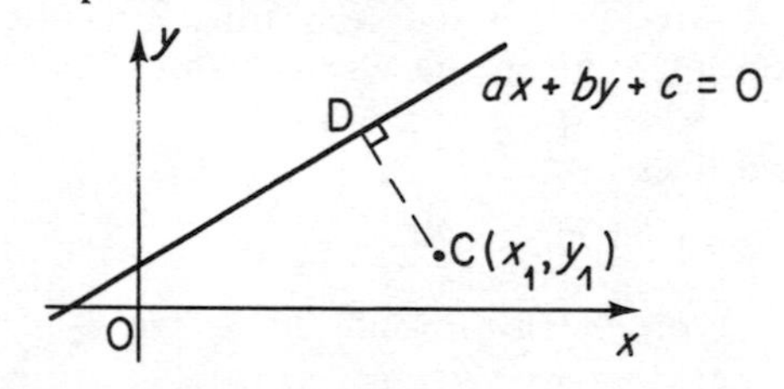

Figure 4.6

Let the equation of the line be $ax + by + c = 0$ and the point be C (x_1, y_1); see Figure 4.6. Let us examine the line through C with slope m: namely, $y - y_1 = m(x - x_1)$. The two straight lines intersect at D, whose coordinates will be found by solving the two equations simultaneously for x.

Since $y = y_1 + m(x - x_1)$ we find on substituting in the equation of the given line that

$$ax + by_1 + bm(x - x_1) + c = 0$$

i.e.
$$x = \frac{-by_1 + bmx_1 - c}{a + bm} = x_1 - \frac{by_1 + ax_1 + c}{a + bm}$$

Substituting back for x we obtain

$$y = y_1 - \frac{m(by_1 + ax_1 + c)}{a + bm}$$

The square of the distance we seek is

$$d^2 \equiv (x - x_1)^2 + (y - y_1)^2 = \frac{(by_1 + ax_1 + c)^2}{(a + bm)^2}(1 + m^2)$$

By intuition we expect that the shortest distance occurs where the slope of CD is perpendicular to the line $ax + by + c = 0$. That occurs when $m = b/a$. Then the distance

$$d = \left| \frac{ax_1 + by_1 + c}{\sqrt{a^2 + b^2}} \right| \tag{4.10}$$

Example

Consider Figure 4.7.

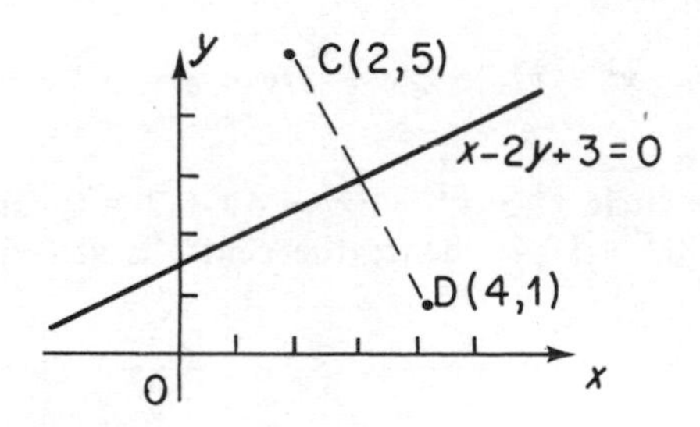

Figure 4.7

The distance of the point C from the given line is given by

$$\left|\frac{1.2 - 2.5 + 3}{\sqrt{1^2 + (-2)^2}}\right| = \left|\frac{-5}{\sqrt{5}}\right| = \sqrt{5}$$

The distance of the point D from the line is

$$\left|\frac{1.4 - 2.1 + 3}{\sqrt{1^2 + (-2)^2}}\right| = \left|\frac{5}{\sqrt{5}}\right| = \sqrt{5}$$

Notice that the points are equi-distant from the line but the fact that the quantity inside the modulus signs was positive in one case and negative in the other suggests that the points are on opposite sides of the line.

The Circle

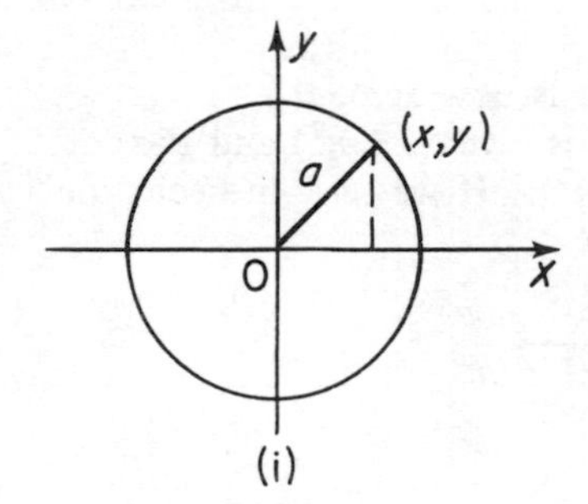

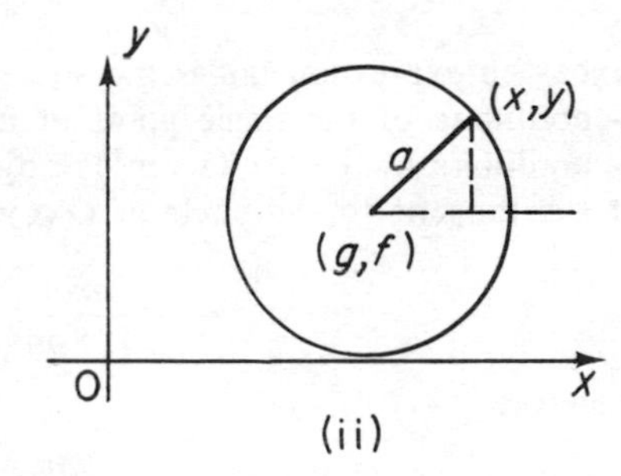

Figure 4.8

The circle of radius a, centre the origin, has (by Pythagoras' Theorem) the equation

$$x^2 + y^2 = a^2 \tag{4.11a}$$

The circle of the same radius but with centre (g, f) has the equation

$$(x-g)^2+(y-f)^2=a^2 \tag{4.11b}$$

This is often written as

$$x^2+y^2-2gx-2fy+c=0 \tag{4.11c}$$

where $c=g^2+f^2-a^2$.

For example, the circle $x^2+y^2+3x-4y+2=0$ can be written as $(x+3/2)^2+(y-2)^2=17/4$; hence the centre is at $(-3/2, 2)$ and the radius is $\sqrt{17}/2$.

Tangent at a point

Consider the intersection of the general line $y=mx+c$ with the circle $x^2+y^2=a^2$. These equations can be solved simultaneously to yield

$$x^2+(mx+c)^2=a^2$$

i.e.

$$x^2(1+m^2)+2mcx+c^2-a^2=0$$

This equation has roots

$$x=\frac{-2mc\pm\sqrt{4m^2c^2-4(1+m^2)(c^2-a^2)}}{2(1+m^2)}$$

giving the x-coordinates of the points of intersection with the general line. For the special case where the line is a tangent, the two points of intersection will coincide and the equation will have equal roots. The condition for equal roots is

$$4m^2c^2=4(1+m^2)(c^2-a^2)$$

i.e.

$$c^2=a^2(1+m^2)$$

Since the expression under the square root sign is now zero, the x-coordinate of the single point of intersection is $-mc/(1+m^2)$ and the y-coordinate is $[-m^2c/(1+m^2)+c]=c/(1+m^2)$. If we seek the equation of the tangent to the circle at (x_1, y_1) we have

$$x_1=-\frac{mc}{1+m^2},\qquad y_1=\frac{c}{1+m^2}$$

therefore

$$m=-\frac{x_1}{y_1}$$

and so

$$y_1=\frac{c}{1+x_1^2/y_1^2}=\frac{y_1^2c}{x_1^2+y_1^2}=\frac{y_1^2c}{a^2}$$

Hence

$$c=\frac{a^2}{y_1}$$

and the required equation is

$$y = \left(-\frac{x_1}{y_1}\right) x + \frac{a^2}{y_1}$$

or

$$xx_1 + yy_1 = a^2 \tag{4.12a}$$

Note that the **normal** to the circle at (x_1, y_1) must have slope y_1/x_1 and so its equation is

$$y - y_1 = \frac{y_1}{x_1}(x - x_1)$$

or

$$y = \frac{y_1}{x_1} x \tag{4.12b}$$

Problems I

1. Find the distance between the following pairs of points: $(6,-4), (2,-2)$; $(2,1), (3,-4)$; $(-1,-2), (-3,-4)$.

2. Write down the equation of each of the following lines:
   (i) which passes through $(2,1)$ and $(-7,-3)$
   (ii) which has gradient -2 and passes through $(6,1)$
   (iii) which cuts the axes at $(0,3)$ and $(-2,0)$

3. Find the distance of the given line from the given points
   (i) $(0,0)$ and $x + y - 2 = 0$
   (ii) $(-1,-1)$ and $2x + 3y - 6 = 0$
   (iii) $(2,3)$ and $y = 2x - 1$

4. Find the point of intersection of, and the acute angle between the lines
   (i) $2x + 3y = 7$ and $x - y = 1$
   (ii) $2x + 4y = 1$ and $4x + 8y = 2$
   (iii) $x - 3y = 1$ and $y = 2 - 3x$
   (iv) $3x + 2y + 3 = 0$ and $y = -(3/2)x - 1$

5. Show that the points $(2,-3)$, $(4,8)$, $(-2,5)$ are the vertices of a right-angled triangle; find the equation of the line joining the mid-points of the two shorter sides and find the equation of the perpendicular bisector of the hypotenuse.

6. The points $(-2, -1)$ and $(-1, 6)$ are adjacent vertices of a square. Find the other vertices and area of the square.

7. Find the area of a triangle with vertices (x_1, y_1), (x_2, y_2), (x_3, y_3)

8. Find the equations of the circles (i) centre $(-1,6)$, radius 4
   (ii) centre $(3,4)$, radius 3.

9. What are the centre and radius of the circles
   (i) $x^2 + y^2 + x - 3y + 8 = 0$
   (ii) $x^2 + y^2 + x - 3y - 8 = 0$
   (iii) $x^2 + y^2 - 2y = 0$

10. Find an equation for the tangent to the circle (4.11c) at the point (x_1, y_1); obtain this equation for the circle $x^2 + y^2 - 3x + y - 6 = 0$ at $(0, -3)$. Find the equation of the normal at (x_1, y_1) to the circle (4.11c).

11. Find the length of a tangent from the point (x_1, y_1) to the circle (4.11c) and hence find this length for the circle $x^2 + y^2 + x - 2y - 2 = 0$, given the points $(1, 3)$, $(2, 5)$, $(1, 2)$ and $(1, 1)$ respectively. What do you conclude?

12. Find the condition that two circles intersect. Two circles are said to be **orthogonal** if their tangents at the point of intersection are perpendicular. Sketch two orthogonal circles and derive the condition for orthogonality. Are the following pairs orthogonal?
   (i) $x^2 + y^2 + 6x - 4y + 5 = 0$ and $x^2 + y^2 + 2x + 5y - 9 = 0$
   (ii) $x^2 + y^2 + 6x - 4y + 5 = 0$ and $x^2 + y^2 - 4x + 7y - 3 = 0$
   (iii) $x^2 + y^2 + 6x - 4y + 5 = 0$ and $x^2 + y^2 + 3x + 5y - 19 = 0$

   The equation of the common chord, if intersection occurs, is found by subtracting the equations of the two circles. Do so for each case.

13. A circle is drawn with the points $A = (2, 1)$ and $B = (7, 4)$ as ends of a diameter. Take a point C on the circumference of the circle and show that $A\hat{C}B = 90°$.

Conic sections

Three curves which can be obtained by intersecting a double cone with a plane are of importance theoretically and sometimes practically. Refer to Figure 4.9(i). These are called **conic sections.** A cut such as AA produces a circle; an oblique cut AC produces an **ellipse.** A cut ED parallel to a generator of the cone yields a **parabola** whilst a cut AB parallel to the central axis of the cone gives rise to a **hyperbola** (which has two branches).

We look at these last three curves in turn, but first we note an alternative definition.

In Figure 4.9(ii) the point P is constrained to move so that the ratio $e = PS/PM$ is constant. BC is a *fixed line* called the **directrix** and S is a *fixed point* called the **focus.** The curve will be *symmetrical* about the axis CQS and Q is a **vertex** of the curve.

If the **eccentricity** $e < 1$ the curve traced out is an *ellipse*
if $e = 1$ the curve traced out is a *parabola*
if $e > 1$ the curve traced out is a *hyperbola*

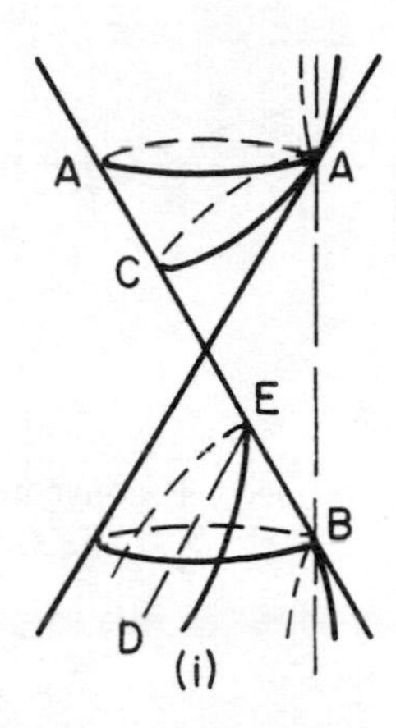

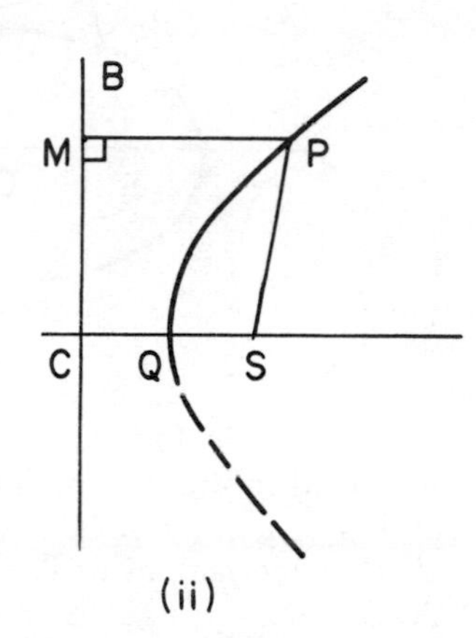

Figure 4.9

The Ellipse

This *oval* curve has two foci and two directrices. The equation in its simplest form is

$$\frac{x^2}{a^2} + \frac{y^2}{b^2} = 1 \tag{4.13}$$

when the curve is symmetrical about *both* the x- and the y-axes, as shown in Figure 4.10. QQ′ is called the **major axis** and RR′ the **minor axis**. S and S′ are the foci, BC and BC′ are the directrices. Note that if $a = b$ we recover the equation for a circle.

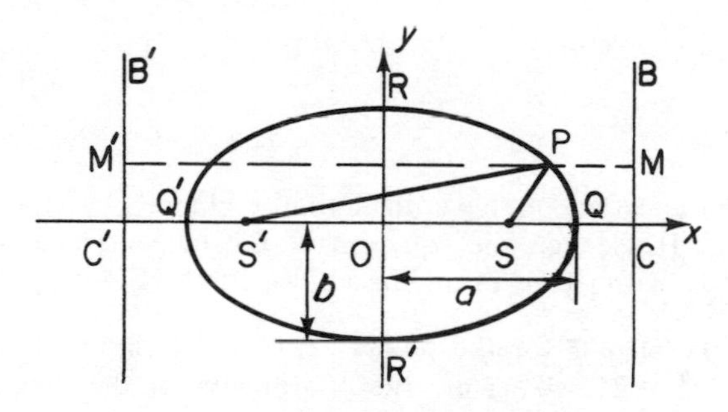

Figure 4.10

As an example of an ellipse occurring in practice you may know that planets move round the sun in elliptical orbits with the sun as one focus.

We first state some properties of the ellipse. Reference to Figure 4.11 is also necessary.

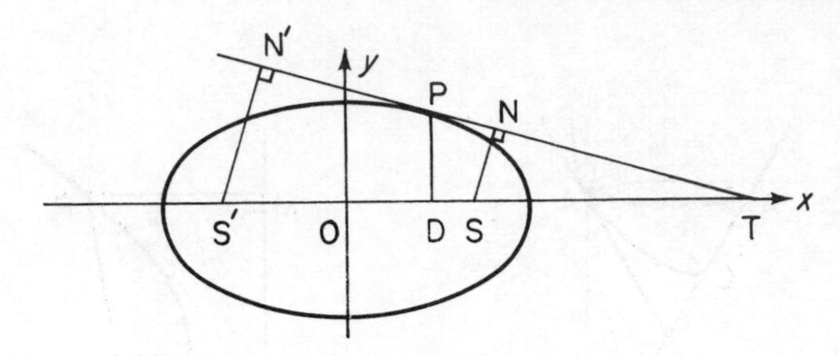

Figure 4.11

(i) S′P + PS = $2a$ for all P on the ellipse (this property is convenient to use when drawing an ellipse)
(ii) $e^2 = 1 - b^2/a^2$ (implies $b < a$)
(iii) if PT is the tangent at P, OD.OT = a^2
(iv) PD.OT = b^2
(v) S′N′.SN = b^2
(vi) S is the point $(ae, 0)$, S′ is $(-ae, 0)$, C the point $(a/e, 0)$.

The line $y = mx + c$ intersects the ellipse where $x^2(b^2 + a^2m^2) + 2a^2cmx + (a^2c^2 - a^2b^2) = 0$. The line will be tangential to the ellipse if this equation has a repeated root; the condition is $c = \pm\sqrt{a^2m^2 + b^2}$ and the value of x at the intersection is then $x_1 = (-a^2cm)/(b^2 + a^2m^2)$. Hence the corresponding value of y is $y_1 = cb^2/(b^2 + a^2m^2)$. We can see that $m = -(x_1b^2)/(y_1a^2)$ and hence $c = (a^2y_1^2 + b^2x_1^2)/a^2y_1$; since (x_1, y_1) lies on the ellipse, $(x_1^2/a^2) + (y_1^2/b^2) = 1$, therefore $c = a^2b^2/a^2y_1$ and the equation of the tangent is

$$y = -\frac{x_1 b^2 x}{y_1 a^2} + \frac{b^2}{y_1}$$

or

$$\frac{xx_1}{a^2} + \frac{yy_1}{b^2} = 1 \tag{4.14}$$

Problems II

1. Find the equation of the normal to the ellipse (4.13) at the point (x_1, y_1). Hence find the equation of the tangents and normals at the points $(1, 1)$ and $(-1, 1)$ on the ellipse $25x^2 + 144y^2 = 169$.

2. Repeat Problem 1 for the points $(2, 1)$ and $(-2, 1)$ on the ellipse $4x^2 + 9y^2 = 25$. What are the coordinates of the foci and the equations of the directrices? What are the lengths of the major and minor axes?

3. Given that a tangent to the ellipse (4.13) is $y = m_1x + \sqrt{a^2m_1^2 + b^2}$, and that (X, Y) is a point on this tangent, find the equation of a second tangent to the ellipse which is perpendicular to the first and

which also passes through (X, Y). Hence show that the locus of (X, Y) is a circle. This circle is called the **director circle** of the ellipse.

4. Find the equation of the **auxiliary circle** – the circle which circumscribes the ellipse.

5. The line $y = 4x$ cuts the ellipse $x^2/9 + y^2/25 = 1$. Find the coordinates of the points of intersection and the length of the intersecting chord.

6. Find the equation of the tangents to the ellipse $x^2 + 4y^2 = 4$, from the points $(2, 2)$, $(\sqrt{3}, \frac{1}{2})$, $(1, \frac{1}{2})$.

The Parabola

In its simplest form the parabola has equation

$$y^2 = 4ax \tag{4.15}$$

where the axes are as shown in Figure 4.12(i).

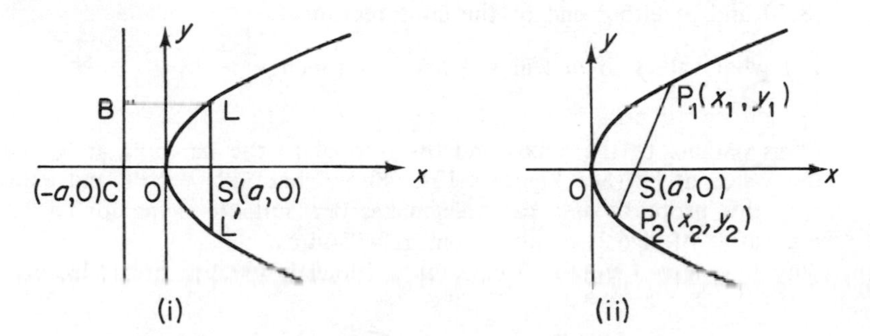

Figure 4.12

Note that $e = 1$ and so the coordinates of C and S are consistent with those for the ellipse.

The chord L′SL is referred to as the **latus rectum.**

It is part of the line $x = a$ which intersects the parabola at $y^2 = 4a^2$, i.e. at $y = \pm 2a$; hence the length of the latus rectum is $4a$. We could have deduced this from the fact that L lies on the parabola and so L′S = LS = BL = CO + OS = $2a$.

The intersection of the line $y = mx + c$ with the parabola (4.15) is given by $m^2x^2 + (2mc - 4a)x + c^2 = 0$.

For a repeated root, $c = a/m$ and the equation of the tangent is

$$y = mx + \frac{a}{m} \tag{4.16a}$$

The intersection point is given by $x_1 = a/m^2$ and $y_1 = 2a/m$.
Then $m = 2a/y_1$ and the equation of the tangent may be written

$$yy_1 = 2a(x + x_1) \tag{4.16b}$$

Note that its slope is $2a/y_1$.

A chord which passes through the focus of the parabola is called a **focal chord.** Let the chord be

$$y = mx + c$$

then since it passes through S, $0 = ma + c$, i.e. $c = -ma$; hence the equation of the focal chord is $y = m(x + a)$.

Problems III

1. Show that the equation of the normal to the parabola $y^2 = 4ax$ at (x_1, y_1) is

$$y + \frac{y_1}{2a}x = \left(\frac{x_1}{2a} + 1\right)y_1$$

2. Find the equations for the tangents and the normals to $y^2 = 8x$ at $(8, 8)$ and at either end of the latus rectum.

3. For what values of m will $y = mx + 3$ touch $y^2 = x$, $y^2 = 4x$, $y^2 = 9x$?

4. PF is parallel to the x-axis and the normal to the parabola at P meets the x-axis at N. (See Figure 4.13). Show that $S\hat{P}N = N\hat{P}F$ and explain why this property of a parabola makes it a suitable shape for an intensifier of signals coming from a far source.
Why is it also a suitable shape for a domestic electric fire reflector?

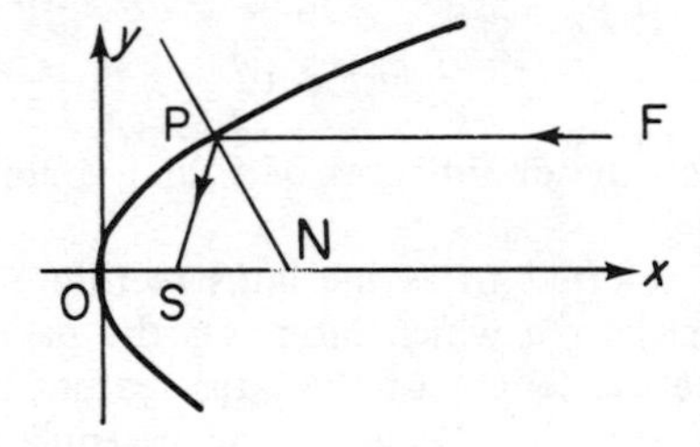

Figure 4.13

The Hyperbola

In its simplest form the equation of the hyperbola is

$$\frac{x^2}{a^2} - \frac{y^2}{b^2} = 1 \tag{4.17}$$

where the axes are chosen as shown in Figure 4.14.

The coordinates of S are $(ae, 0)$ and those of C are $(a/e, 0)$. The x-axis is called the **transverse axis**; BC and B′C′ are the *directrices.*

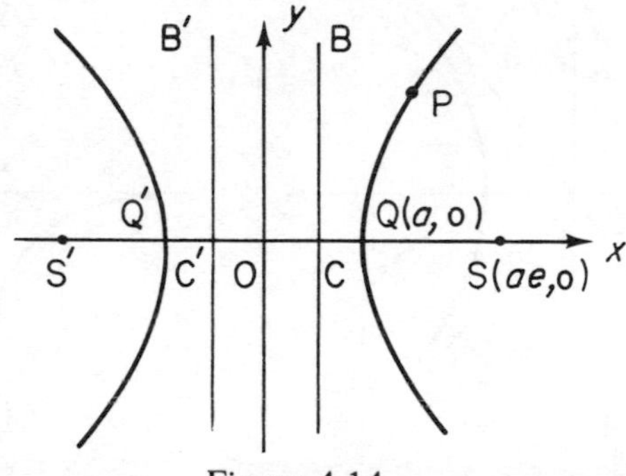

Figure 4.14

We collect some results which we leave for you to prove:

(i) the equation of the tangents of slope m is $y = mx \pm \sqrt{a^2m^2 - b^2}$
(ii) the equation of the tangent at (x_1, y_1) is $xx_1/a^2 - yy_1/b^2 = 1$ (why might you have expected this?)
(iii) the equation of the director circle is $x^2 + y^2 = a^2 - b^2$
(iv) $|SP - S'P| = 2a$

(Can you detect a link between these results and those for the ellipse?)

We now introduce the concept of an **asymptote.**

This is a straight line which is a tangent to the curve at infinity. The curve is then said to be *asymptotic* to the straight line. The phrase *at infinity* is a dangerous one to use, but its meaning should be clear. To investigate the possibility of asymptotes for the hyperbola, let us consider its intersections with the line $y = mx + c$. These are given by

$$(b^2 - a^2m^2)x^2 - 2a^2mcx - a^2(c^2 + b^2) = 0$$

Our requirement is that both roots occur at infinity; we can find the criterion more effectively by writing $u = 1/x$ so that as $x \to \infty$, $u \to 0$. The points of intersection are then given by the equation

$$a^2(c^2 + b^2)u^2 + 2a^2mcu - (b^2 - a^2m^2) = 0$$

We require this equation to have $u = 0$ as its repeated root and it must therefore take the form $u^2 = 0$. Hence

$$2a^2mc = 0 \qquad \text{and} \qquad b^2 - a^2m^2 = 0$$

That is, $m = \pm b/a$ and $c = 0$. The asymptotes exist and are $y = (b/a)x$ and $y = -(b/a)x$. This means that for large positive or negative x the values of y on the curve can be approximated by $\pm(b/a)x$.

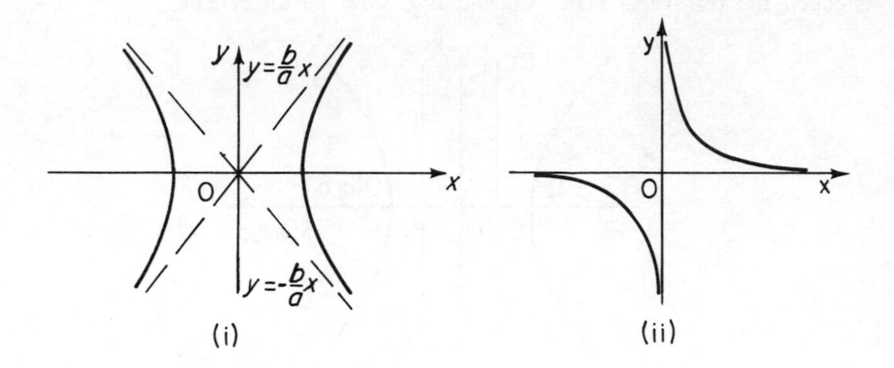

Figure 4.15

Two special cases are of interest (refer to Figure 4.15 above).

(i) The asymptotes are at right angles, the curve is called a **rectangular hyperbola**, and then $b^2/a^2 = +1$. The hyperbola is

$$x^2 - y^2 = a^2 \tag{4.18}$$

(ii) The asymptotes are taken as the coordinate axes. We shall see later when we study rotation of axes that the equation of the rectangular hyperbola becomes

$$xy = c^2 \tag{4.19}$$

This can be rearranged to $y = c^2/x$, $x \neq 0$ and is therefore a function of x with domain = $\{x : x \text{ is real and } x \neq 0\}$.

Problems IV

1. Sketch the hyperbola $x^2/4 - y^2/9 = 1$ and the **conjugate hyperbola** $x^2/4 - y^2/9 = -1$. Sketch the asymptotes.

2. Find the equation of the tangent to the hyperbola (4.19).

3. Find the equation for the normals at (x_1, y_1) to each of the hyperbolae (4.17), (4,18), (4,19).

4. Find the equation of the tangent and the normal to the hyperbola $4x^2 - 9y^2 = 36$, at the points $(3, 2)$, $(-3, -2)$. Determine the asymptotes.

5. Find a suitable meaning for the *latus rectum* of the hyperbola and determine its length. Guess the length of the latus rectum for an ellipse and verify your guess directly.

6. Find l, m and n for a tappet cam circle radius 3 cm (Figure 4.16).

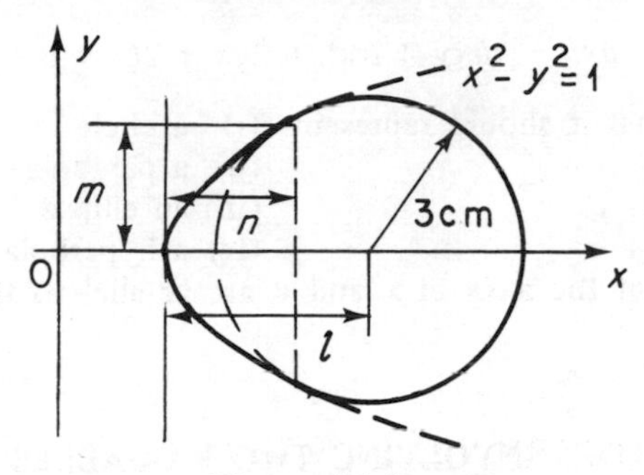

Figure 4.16

Translation of axes

We now investigate the taking of new axes $YO'X$ parallel to the axes yOx.

Let P have coordinates (x, y) in the axes yOx and coordinates (X, Y) relative to the new axes $YO'X$. See Figure 4.17.

Then

$$x = X + p$$
$$y = Y + q \tag{4.20}$$

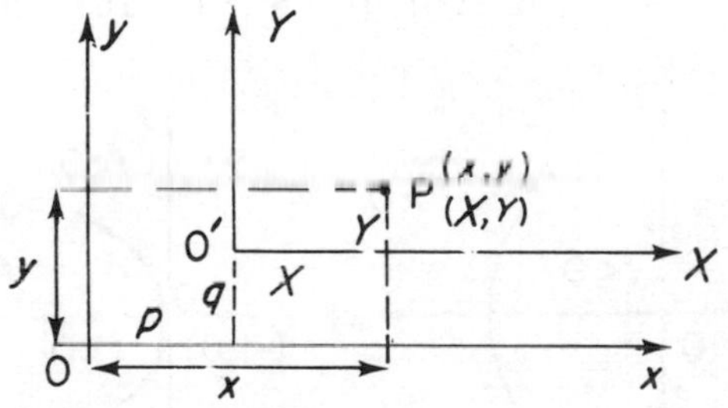

Figure 4.17

Let us apply this to a circle with centre O' at (x_0, y_0) and radius a. Take new axes $XO'Y$ through O' parallel to the axes xOy. Relative to O', the equation of the circle must be $X^2 + Y^2 = a^2$. But $x = X + x_0$ and $y = Y + y_0$, so that $X = x - x_0$ and $Y = y - y_0$. Hence relative to the original axes xOy the equation becomes $(x - x_0)^2 + (y - y_0)^2 = a^2$.

Problems V

1. Find the equation of a parabola with focus at $(3, 2)$ and directrix $x = -3$.

2. Transform the standard equations of a parabola, an ellipse and a hyperbola to the translated set of axes with origin at the point (p, q). All the equations are of second order in x and y. Ignoring the

degenerate cases where $a = h = b = 0$ state what are the conditions on the coefficients in the general equation of second degree

$$ax^2 + 2hxy + by^2 + 2gx + 2fy + c = 0$$

in order that it should represent (i) a circle
(ii) a parabola
(iii) an ellipse
(iv) a hyperbola

(assume that the axes of x and y are parallel to the axes of symmetry).

4.2 INEQUALITIES INVOLVING TWO VARIABLES

Consider the equation $x^2 + y^2 = a^2$; we may write it as $f(x, y) \equiv x^2 + y^2 - a^2 = 0$. All points on the circumference satisfy $f(x, y) = 0$. It can be shown that points inside the circle satisfy $f(x, y) < 0$ and points outside the circle satisfy $f(x, y) > 0$. See Figure 4.18(i). This is an example of a general principle that the plane is divided into three regions by a simple curve (i.e. one which contains no loops). For example, the parabola $y^2 = 4x$ is rewritten $f(x, y) \equiv y^2 - 4x = 0$. To find which region corresponds to $f(x, y) < 0$ and which to $f(x, y) > 0$ all we need do is take a point in either region and evaluate $f(x, y)$ there. Then, for example, $f(1, 0) = -4 < 0$ or $f(-1, 0) = 4 > 0$. The result is shown in Figure 4.18(ii).

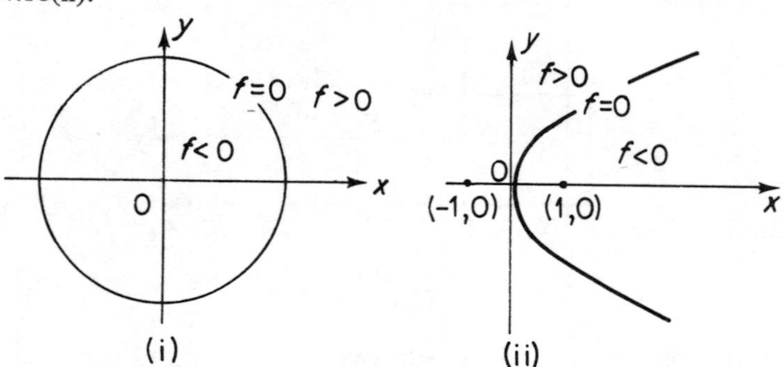

Figure 4.18

The inequalities produced by straight lines are of importance. Consider the line $x + y - 3 = 0$, then $f(x, y) = x + y - 3$ and $f(0, 0) < 0$. Hence the plane is divided into three distinct regions as shown in Figure 4.19(i).

Special cases relate to the coordinate axes. The region $x \geqslant 0$ lies on and to the right of the y-axis, the region $y \geqslant 0$ lies on and above the x-axis and so the shaded region in Figure 4.19(i) satisfies the simultaneous inequalities, $x \geqslant 0$, $y \geqslant 0$, $x + y - 3 \leqslant 0$; such a region is known as the **region of feasible solutions** to this set of inequalities.

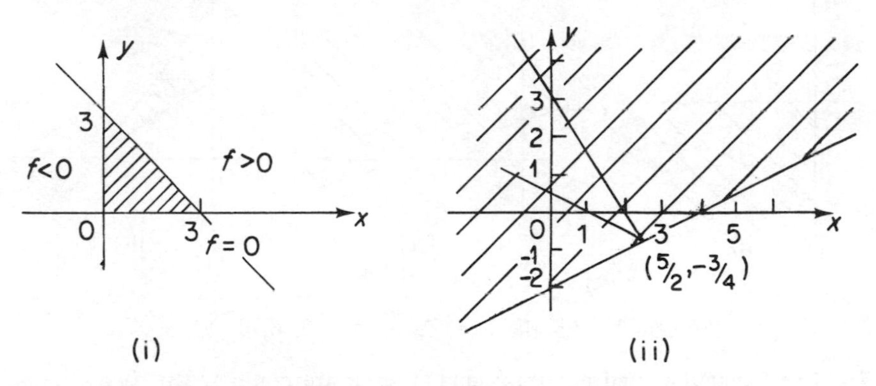

Figure 4.19

Consider the statement $3x + 2y - 6 < 2x + 4y - 2$. We can rearrange this to $3x + 2y - 6 - (2x + 4y - 2) < 0$, i.e. $x - 2y - 4 < 0$. This region is shaded in Figure 4.19(ii) and the lines $3x + 2y - 6 = 0$ and $2x + 4y - 2 = 0$ are sketched.

Note that the line $x - 2y - 4 = 0$ passes through the point of intersection of the original two lines, viz $(5/2, -3/4)$. The values of x and y at all points in the shaded region satisfy the given inequality.

Problems

1. Sketch the set of solutions of

   (i) $x^2 + 2y^2 > 16$
   (ii) $3x^2 + 4y^2 < 25$
   (iii) $(x + y)^2 > 0$
   (iv) $x + 2y < 4$
   (v) $2x - y > 3$

2. Shade the region where the three inequalities $x + y < 1$, $x + 2y < 4$, $x - 3y < 5$ are satisfied simultaneously and locate the points of intersection of pairs of lines. What happens to the region if the condition $x + 2y < 6$ is added? What happens if the condition $x + 2y < 3$ is added?

3. Sketch the region where all the inequalities $x \geqslant 0$, $y \geqslant 0$, $x + y < 3$, $x + 2y < 4$ are satisfied. Locate the vertices of this *convex region.* Suppose we add the requirement $3x + 4y > 10.1$, what do you conclude graphically? Verify this algebraically.

4. Describe the regions in Figure 4.20 by a set of simultaneous inequalities.

5. On the diagram to Figure 4.20(iii) superimpose the lines $2x + y = 0$, $2x + y = ½$, $2x + y = 1$, $2x + y = 8$. What is the greatest value of c for which $2x + y = c$ has a point in common with the shaded region. What is the least value?

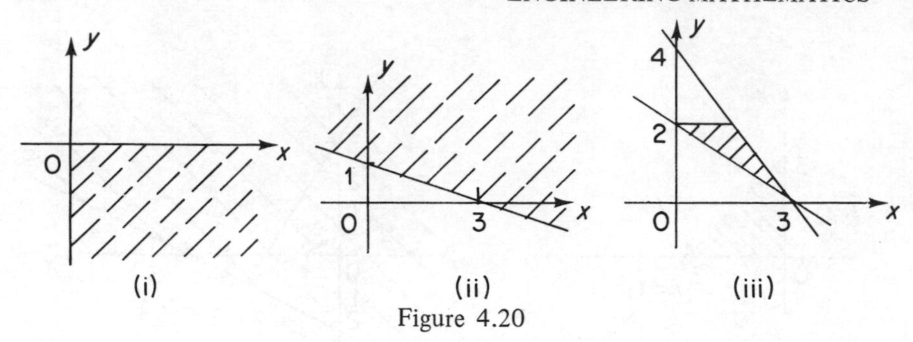

Figure 4.20

6. Repeat Problem 5 with the lines $2x + 3y = 0, 6, 9, 12$.

7. Two manufactured articles X and Y each are composed of two chemicals A and B. X comprises 2 g of A and 4 g of B, and Y comprises 3 g of A and 3 g of B. The profit on one article of X is 3 pence and on one of Y is 4 pence. If there is at most 100 g of A available and 200 g of B available, and if there is a demand for at least 20 articles of X and 15 articles of Y, how many of each should be manufactured if the profit is to be maximised? (Hint: let x of X be manufactured and y of Y).

   The cost to a retailer of an article of X is 10 p and one of Y is 20 p. Assuming he must satisfy the demand, what number of each should he request to minimise his costs?

4.3 SIMPLE IDEAS ON CURVE SKETCHING

We speak here of *sketching* a curve as distinct from *plotting*. By this we mean that we examine the salient features of the curve and produce a sketch of its general shape. (Each point of the curve is *not* plotted accurately.)

We have enough results already to make a reasonable attempt at sketching the main features of some curves. As we develop further techniques we shall be able to extend the range of functions we can sketch. The idea will be to deduce the maximum amount of information with the minimum of calculation.

First we collect some results from Section 3.5 about derivatives.

$f(x)$	$f'(x)$	$f(x)$	$f'(x)$
a, constant	0	a/x	$-a/x^2$ $[x \neq 0]$
cx, c constant	c	b/x^2	$-2b/x^3$ $[x \neq 0]$
bx^2, b constant	$2bx$	$g(x) + h(x)$	$g'(x) + h'(x)$
ax^3, a constant	$3ax^2$		

If $f'(x) > 0$ in some interval, then $f(x)$ is said to be **increasing** there; if $f'(x) < 0$ in some interval, then $f(x)$ is said to be **decreasing** there. For example, let $f(x) = 3x^2 - 2x - 1 = (3x + 1)(x - 1)$ so that $f'(x) = 6x - 2$.

Then $f'(x) > 0$ if $6x - 2 > 0$, i.e. if $x > 1/3$, and $f'(x) < 0$ if $x < 1/3$. A sketch of $f(x) = 3x^2 - 2x - 1$ is given in Figure 4.21.

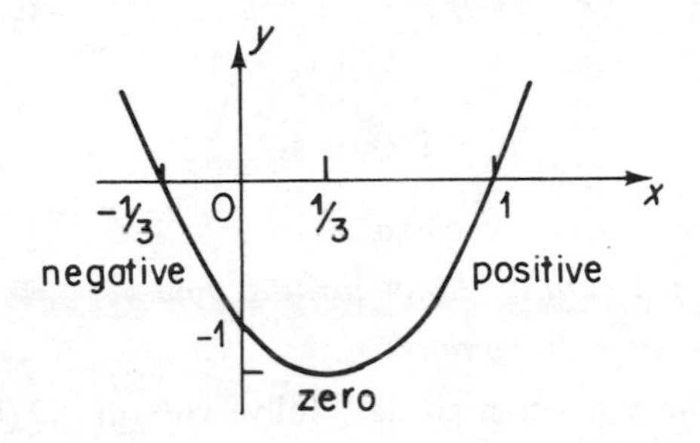

Figure 4.21

Note that $f(x) = 0$ when $x = 1$ or $-1/3$, also $f(0) = -1$; and that $f(x)$ is decreasing in the interval $(-\infty, 1/3)$ and increasing in the interval $(1/3, +\infty)$.

It is clear that where the function increases, its slope is positive and where it decreases its slope is negative. Where $f'(x) = 0$ we have a **turning point**. In the example shown $x = 1/3$ is a turning point and is a **local minimum**. We use *local* because it may not be the *absolute* minimum of the function. [If the equation is of the form $y = f(x)$, we often write dy/dx for $f'(x)$.]

We list the features to be examined when sketching a curve.

(a) *Symmetry about either axis.*
If in the equation governing the curve x can be replaced by $-x$ without altering the equation, the curve is *symmetrical about the y-axis*. You will remember that functions with this property are called **even functions**, for example Figure 4.22(i); examples of such functions are those which have an equation where x appears only as x^2, x^4 etc. If the effect of replacing x by $-x$ in a functional equation is to reverse the signs of all the terms involving x the function is called **odd**, for example Figure 4.22(ii).
We may similarly define symmetry about the x-axis: Figure 4.22(iii).

(b) *Any restrictions on the ranges of values of x and y*
For example, if $y^2 = x$ we need not consider $x < 0$.

(c) *Intersections of the curve with the axes*
These are found by putting x or y equal to zero.

(d) *The behaviour of the function for large values of x, both positive and negative.*

(e) *The behaviour near the origin if this lies on the curve*

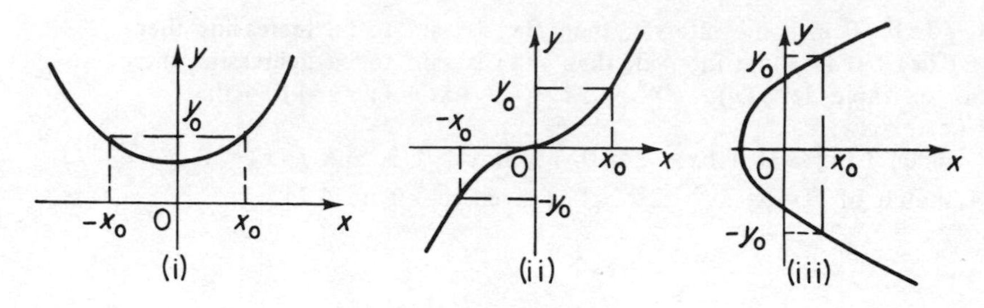

Figure 4.22

(f) *Any discontinuities in the curve*

(g) *The location and nature of any turning points*

(h) *The behaviour near discontinuities.*

We are not yet in a position to deal fully with (g) and (h); however, we consider a few examples.

Example 1: $y = 1 - x^2$ (Figure 4.23)

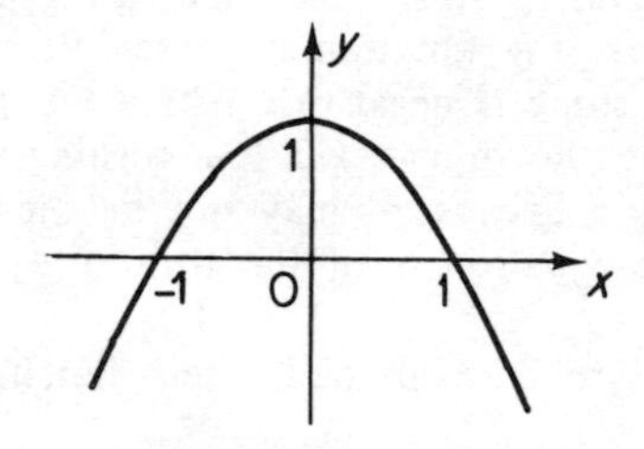

Figure 4.23

(i) the curve is even and has symmetry about the y-axis,
(ii) y must be $\leqslant 1$,
(iii) $x = 0 \Rightarrow y = 1$; $y = 0 \Rightarrow x = \pm 1$,
(iv) $|x|$ large $\Rightarrow y$ large and negative,
(v) $dy/dx = -2x$ and there is a turning point at $x = 0$ which is clearly a local maximum.

Example 2: $y^2 = 1 - x^2$ (Figure 4.24)
(i) the curve has symmetry about both axes,
(ii) $y^2 \geqslant 0 \Rightarrow -1 \leqslant x \leqslant 1$, similarly $-1 \leqslant y \leqslant 1$,
(iii) when $x = 0$, $y = \pm 1$ and $y = 0 \Rightarrow x = \pm 1$.

In fact, we know that this equation represents a circle, centre the origin and radius 1.

(We see from the sketch that $dy/dx = 0$ at $x = 0$ and is infinite at $x = \pm 1$.)

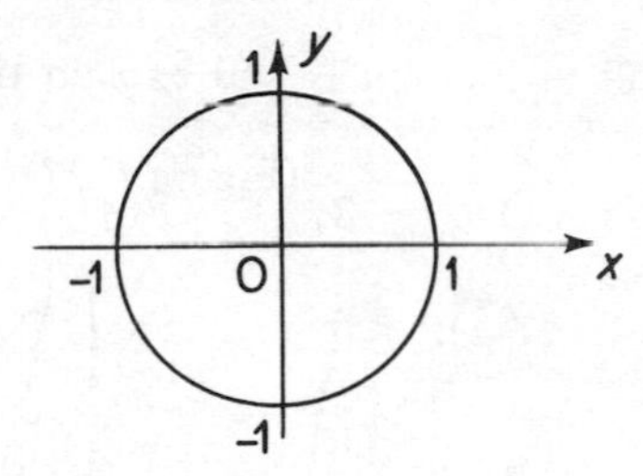

Figure 4.24

Example 3: $y = x(x^2 - 4) = x^3 - 4x$ (Figure 4.25)

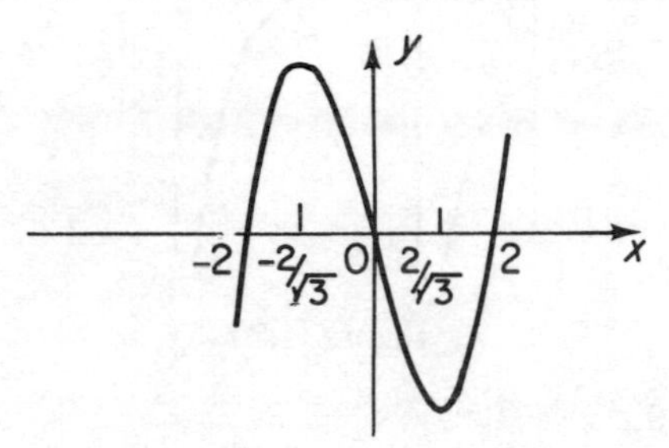

Figure 4.25

(i) no symmetry,
(ii) $x = 0 \Rightarrow y = 0$,
$y = 0 \Rightarrow x = 0$ or $x = \pm 2$,
(iii) x large +ve $\Rightarrow y$ large +ve,
(iv) x large −ve $\Rightarrow y$ large −ve.

$dy/dx = 3x^2 - 4$ and this is > 0 for $x \in (-\infty, -2/\sqrt{3})$ or $(2/\sqrt{3}, \infty)$
< 0 for $x \in (-2/\sqrt{3}, 2/\sqrt{3})$.

Example 4: $y = (x - 1)^2(x - 2)$ (Figure 4.26)

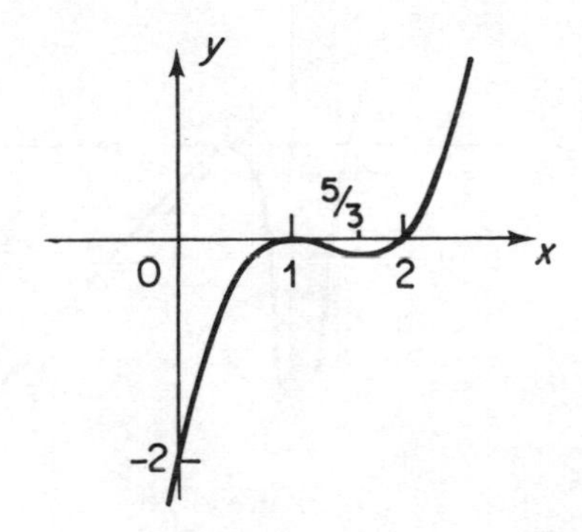

Figure 4.26

We leave you to fill in the details and explain the behaviour at $x = 1$.

Example 5: $y = \dfrac{x-2}{(x-1)(x-3)}$ (Figure 4.27)

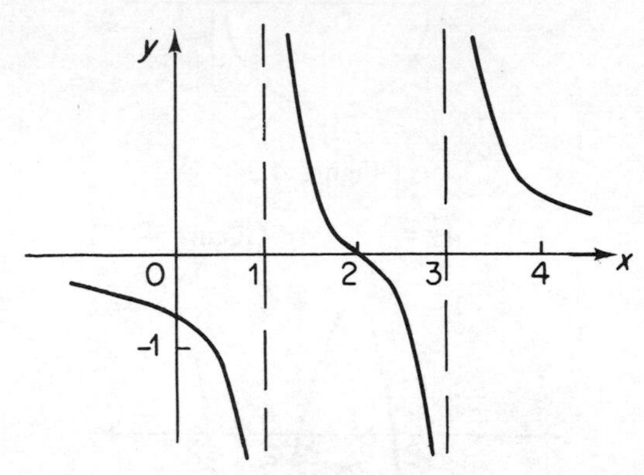

Figure 4.27

(i) $x = 2 \Rightarrow y = 0$; $x = 0 \Rightarrow y = -2/3$,

(ii) for large $|x|$, y can be approximated by $x/x.x$, i.e. $1/x$. Hence as $x \to \infty$, $y \searrow 0$, and as $x \to -\infty$, $y \nearrow 0$,

(iii) as $x \nearrow 1$, $y \to -\infty$ and as $x \searrow 1$, $y \to \infty$.
Further, as $x \nearrow 3$, $y \to -\infty$ and as $x \searrow 3$, $y \to \infty$.

We have not yet the skills to determine dy/dx at $x = 2$, but it is in fact -1.

Example 6: $y = -x + 1 + \dfrac{1}{x} - \dfrac{1}{x^2}$ (Figure 4.28)

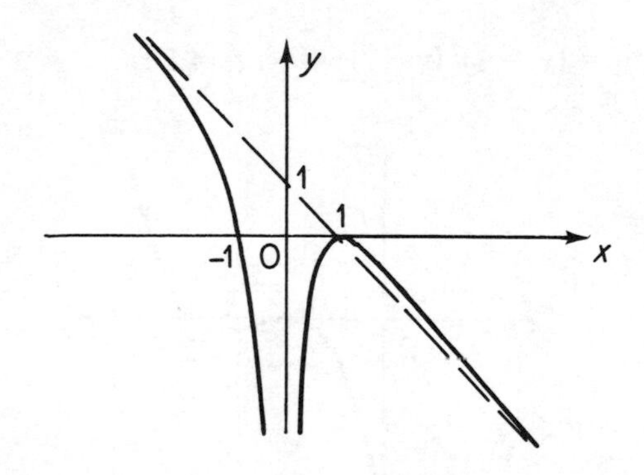

Figure 4.28

Here the important features are of an *asymptotic* nature.

For large positive x, y is large and negative; but we can go further and ignore the terms $1/x$ and $1/x^2$ to say that y is asymptotic to $(-x+1)$ for large positive x (and also for large negative x).

Since, for large values of x the term $1/x$ dominates $-1/x^2$, we see that the curve is *above* the asymptote for large positive x and *below* it for large negative x. As x approaches zero, the dominant term is $-1/x^2$.

$$\frac{dy}{dx} = -1 - \frac{1}{x^2} + \frac{2}{x^3} = \frac{-(x^3 + x - 2)}{x^3} = \frac{-(x-1)(x^2 + x + 2)}{x^3}$$

Note (i) dy/dx is infinite at $x = 0$ and zero at $x = 1$
(ii) $y = 0$ when $x = 1$.

Problems

Sketch the following curves:

1. $y = x^2 + x + 1$
2. $y^2 = 4 + 2x^2$
3. $y^2 = 4 - 2x^2$
4. $y = x^2(x - 4)$
5. $y = (x - 1)(x - 3)^2$
6. $y = (x - 1)/(x - 2)(x - 3)$
7. $y = (x - 3)/(x - 1)(x - 2)$
8. $y = (2 - x^2)/(2 + x^2)$
9. $y = x/(1 + x^2)$ [you may assume $dy/dx = (1 - x^2)/(1 + x^2)^2$]
10. $y = -x + 1 - 1/x$
11. $y = -x + 1 + 1/x^2$

4.4 POLAR COORDINATES

In the *Cartesian* system, we locate a point by its distance from the two axes. In the *polar* system we locate a point by its distance r from the origin and by the angle θ between the x-axis and the line joining the point to the origin as shown in Figure 4.29(i).

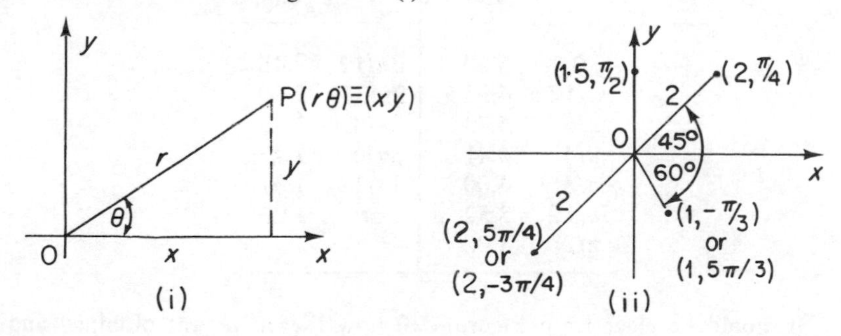

Figure 4.29

It can be seen that the equations of transfer from one coordinate system to the other are

$$x = r\cos\theta, \quad y = r\sin\theta \tag{4.21a}$$

or

$$r = \sqrt{x^2 + y^2}, \quad \tan\theta = y/x \tag{4.21b}$$

It is important to know the values of y and x separately; for instance, the information $\tan\theta = -1$ does not allow us to distinguish between $\theta = 135°$ and $\theta = 315°$. (It is always advisable to plot the point on a diagram.) Some points are marked on Figure 4.29(ii). We shall assume that r is always positive. Sometimes angles in the third and fourth quadrants are taken negative for ease in calculation.

It is customary to describe curves in polar coordinates in the form $r = f(\theta)$. If $f(\theta) = f(-\theta)$ the curve is symmetrical about the x-axis; in particular, this will be so if $r = f(\cos\theta)$ since $\cos\theta$ is even. However, if $r = f(\sin\theta)$ there will be symmetry about the y-axis ($\theta = \pi/2$).

The equation $r = 3\sin\theta$ is defined for $0 \leqslant \theta \leqslant \pi$ only (since r is positive). If we attempt to convert this to cartesian coordinates we obtain

$$r^2 = 9\sin^2\theta \quad \text{or} \quad r^4 = 9r^2\sin^2\theta$$

and since $r^2 = x^2 + y^2$ and $y = r\sin\theta$ we get

$$(x^2 + y^2)^2 = 9y^2$$

That is

$$x^2 + y^2 = 3y$$

This last equation is valid for $y \geqslant 0$ which corresponds to $0 \leqslant \theta \leqslant \pi$.

We can similarly convert from an x-y equation to its polar form by putting $x = r\cos\theta$, $y = r\sin\theta$.

Example 1 $r = 3 + 2\cos\theta$

Since $r = f(\cos\theta)$ we can assume symmetry about the x-axis. We tabulate r against θ for $\theta = 0(\pi/12)\pi$.

θ	r (2 d.p.)	θ	r (2 d.p.)
0	5.00	$7\pi/12$	2.48
$\pi/12$	4.93	$2\pi/3$	2.00
$\pi/6$	4.73	$3\pi/4$	1.59
$\pi/4$	4.41	$5\pi/6$	1.27
$\pi/3$	4.00	$11\pi/12$	1.07
$5\pi/12$	3.52	π	1.00
$\pi/2$	3.00		

It should be clear from Figure 4.30 how the curve was obtained and you should extend the plotting of points for θ between π and 2π to obtain the dashed curve.

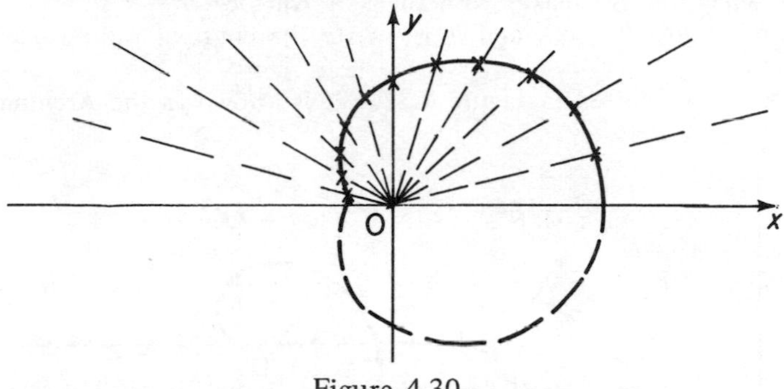

Figure 4.30

Example 2 $r = \sin \theta$

Since $r = \sin \theta$ we can assume symmetry about the y-axis and since $r \geqslant 0$, we have only to consider $0 < \theta < \pi$. A table of values is provided. Extend this to obtain the dashed part of the curve shown in Figure 4.31.

θ	r (2 d.p.)
0	0.00
$\pi/12$	0.26
$\pi/6$	0.50
$\pi/4$	0.71
$\pi/3$	0.87
$5\pi/12$	0.97
$\pi/2$	1.00

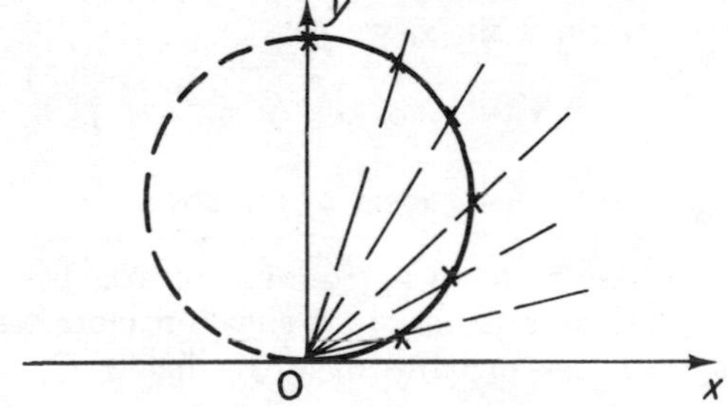

Figure 4.31

Example 3 $r = a\theta$

This arises when a rod rotates with constant angular velocity, ω, about one end and a particle P moves along the rod with a constant velocity, u, relative to the rod. After time t, the point P will be a distance $a\theta$ from A and the angle that AP makes with the axis will be $\omega t = \theta$.

Hence $r = ut$, $\theta = \omega t$, and if we write the constant ratio u/ω as a then $r = a\theta$.

The curve is plotted in Figure 4.32. It is known as the **Archimedean spiral**.

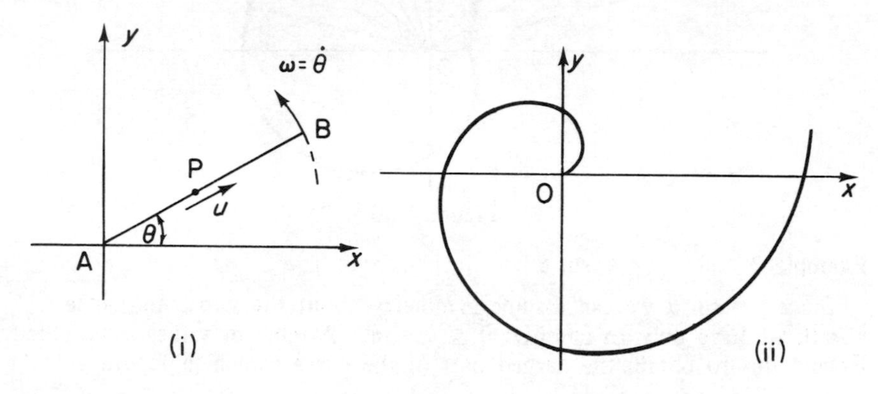

Figure 4.32

Rotation of Axes

We now turn to the problem of rotating the coordinate axes [refer to Figure 4.33(i)]. Consider a point P which has polar coordinates (r, θ) in the old axes yOx. After rotation through an angle α anti-clockwise the axes become $y'Ox'$ and the coordinates of P in this new set of axes are $(r, \theta - \alpha)$. To obtain the cartesian equivalent (x', y') we write $x' = r\cos(\theta - \alpha)$, $y' = r\sin(\theta - \alpha)$.

Now in the old axes we have $x = r\cos\theta$, $y = r\sin\theta$. Hence, since $x' = r\cos\theta\cos\alpha + r\sin\theta\sin\alpha$, we get

Similarly,

$$\left.\begin{aligned} x' &= x\cos\alpha + y\sin\alpha \\ y' &= -x\sin\alpha + y\cos\alpha \end{aligned}\right\} \tag{4.22a}$$

(If α is negative we have a clockwise rotation of axes.)

We can often rotate axes to make an equation more easy to understand. In such cases we need x, y in terms of x', y', that is

$$\left.\begin{aligned} x &= x'\cos\alpha - y'\sin\alpha \\ y &= x'\sin\alpha + y'\cos\alpha \end{aligned}\right\} \tag{4.22b}$$

Why would you have expected this result?

Example

If we take the equation $\frac{x^2}{2} + \frac{y^2}{2} - xy = 2\sqrt{2}\,a(x + y)$ we can write it in the form

$$\left(\frac{x}{\sqrt{2}} - \frac{y}{\sqrt{2}}\right)^2 = 4a\left(\frac{x}{\sqrt{2}} + \frac{y}{\sqrt{2}}\right)$$

If we rotate the coordinate axes anti-clockwise through 45°, we can use equations (4.22b) with $\alpha = 45°$ to obtain $x = (x' - y')/\sqrt{2}$, $y = (x' + y')/\sqrt{2}$ and substitution gives $(y')^2 = 4ax'$, which is the equation of a parabola with axis 45° to the horizontal as shown in Figure 4.33(ii).

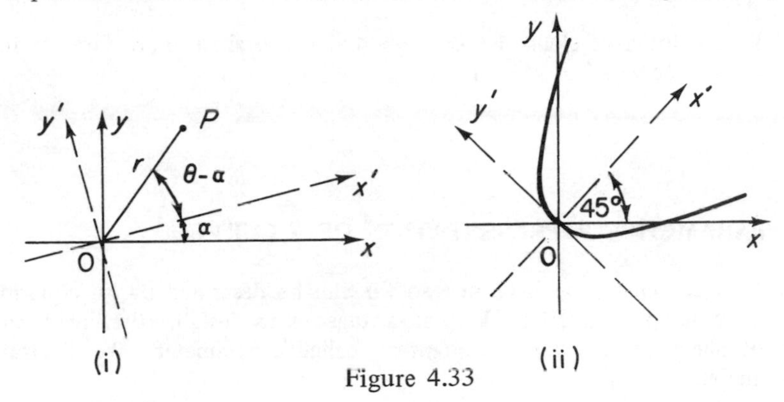

Figure 4.33

More complicated cases can arise if the rotation of axes is about a point other than the origin.

Problems

1. Find the polar form of (i) $y^2 = x^2(1 - x^2)$
   (ii) $y^2 = x^2(y^2 + x^2)$
   (iii) $(x^2 + y^2)(x - y)^2 = 4$

2. Find the Cartesian form of (i) $r = 3\cos\theta + 4\sin\theta$
   (ii) $r^2 = 2\sec 4\theta$
   (iii) $r = 1 - \tan^2\theta$

3. Sketch the curves $r = 2\sin^2\theta$, $r = 3\cos\theta$, $r = -2\cos\theta$, $r = \cos^2\theta$, $r = \sin 2\theta$, $r = \cos 2\theta$, $r = \sin 3\theta$.

4. Sketch the curves $r = 1 + \cos\theta$, $r = 1 + 2\cos\theta$, $r = 2 + \cos\theta$. Try and sketch $r = a + b\cos\theta$.

5. Show that (i) $r = 2a\cos\theta$ represents a circle centre, in polar coordinates, $(a, 0)$
   (ii) $r = 2a\cos(\theta - \alpha)$ represents a circle, centre (a, α).
   Deduce a special case by putting $\alpha = \pi/2$.
   Sketch the curves on common axes.

6. Calculate algebraically the points of intersection of $r^2 = 9 \cos \theta$ and $r = (1 + \cos \theta)$. Sketch the curves and locate the points of intersection graphically. What do you conclude?

7. The lemniscate $r^2 = a^2 \sin 2\theta$ is used in constructing transition curves in roads. For small values of r and θ the curve is approximated by $y = Kx^3$ for a suitable value of K. Take $a = 1$ and plot the two curves on common axes.

8. Write down the new coordinates (x', y') of a point when the axes are rotated 60° clockwise, 90° anti-clockwise.

9. Rotate the axes clockwise through 45° to obtain a more familiar form for the curve $x^2 - y^2 = 1$.

4.5 PARAMETRIC REPRESENTATION OF A CURVE

Each of the curves we have met so far can be described by an equation involving x and y; sometimes it is advantageous to describe the curves in terms of one subsidiary variable quantity, called a **parameter**. We illustrate by examples.

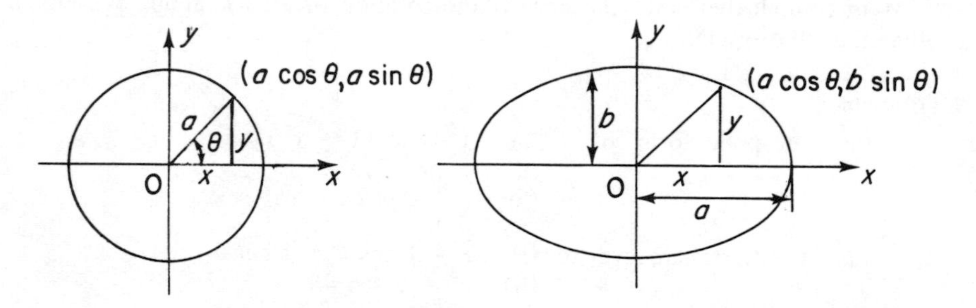

Figure 4.34

Circle and Ellipse

We can see from Figure 4.34 (i) that if we make the substitutions $x = a \cos \theta$, $y = a \sin \theta$ then $x^2 + y^2 = a^2 \cos^2\theta + a^2 \sin^2\theta = a^2$, and so a point on this circle can be represented as $(a \cos \theta, a \sin \theta)$.

As θ varies from 0 to 2π, so we describe in turn all the points on the circle's circumference. Instead of the two variables x and y we now have one variable θ. An extension to the ellipse is fairly evident and is shown in Figure 4.34(ii). We write $x = a \cos \theta$ and $y = b \sin \theta$, so that $x^2/a^2 + y^2/b^2 = \cos^2\theta + \sin^2\theta = 1$. ($\theta$ here is called the *eccentric angle.*)

Tangent to the circle

Suppose we now seek an equation for the tangent to the circle at any point $P(x_1,y_1)$. We saw previously that this could be written $xx_1 + yy_1 = a^2$; substituting in terms of the parameter θ_1 at P we get the equation of the tangent in the form $xa \cos \theta_1 + ya \sin \theta_1 = a^2$; which reduces to $x \cos \theta_1 + y \sin \theta_1 = a$.

Tangent to the ellipse

From the previous result for a tangent to an ellipse at the point (x_1,y_1), namely $xx_1/a^2 + yy_1/b^2 = 1$, we obtain the tangent at $(a \cos \theta_1, b \sin \theta_1)$ in the form

$$\frac{x \cos \theta_1}{a} + \frac{y \sin \theta_1}{b} = 1$$

Parabola

If we now turn to the parabola, $y^2 = 4ax$, we see that a possible parametric representation is $x = at^2$, $y = 2at$. Values of t at various points on the parabola are shown in Figure 4.35(i). Each point on the parabola is associated with a unique value of t and vice-versa.

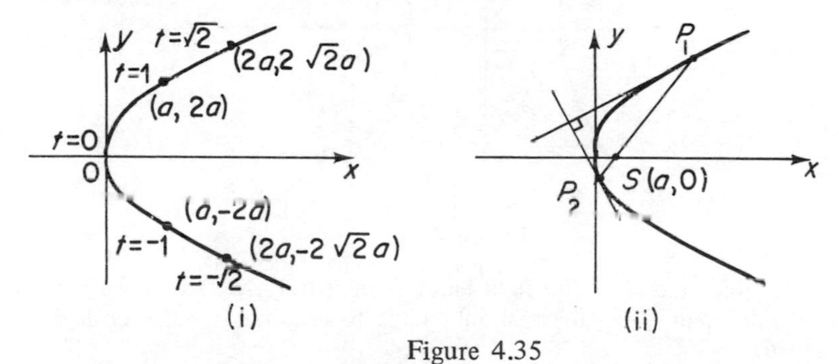

Figure 4.35

Note that x and y are each functions of t and t is a function of y, but t is not a function of x in the strict sense.

The tangent at any point (x_1,y_1) was found to be $yy_1 = 2a(x + x_1)$ and this becomes

$$y.2at_1 = 2a(x + at_1^2)$$

that is,

$$t_1y = x + at_1^2$$

We can see that the tangent has slope $1/t_1$ and this is consistent with the graph. (For example, the slope decreases as t increases.) It should be clear that the gradient of the normal at t_1 is $-t_1$ and the equation of the normal at t_1 is $y - 2at_1 = -t_1(x - at_1^2)$ or $y + t_1x = 2at_1 + at_1^3$.

Returning to the focal chord defined earlier, we may characterise the end points by t_1 and t_2. The equation of the chord joining points t_1 and t_2 is

$$y - 2at_1 = \frac{2at_2 - 2at_1}{at_2^2 - at_1^2}(x - at_1^2)$$

and since this is satisfied by the point $(a, 0)$ we have

$$-2at_1 = \frac{2a}{t_2 + t_1}(1 - t_1^2)$$

which reduces to $t_2 t_1 = -1$.

Notice that $1/t_1 \cdot 1/t_2 = -1$ and so the tangents at the end-points of the focal chord are perpendicular [see Figure 4.35(ii)].

Hyperbola

The equation $x^2/a^2 - y^2/b^2 = 1$ is satisfied by $x = a \cosh \theta$, $y = b \sinh \theta$ (see Figure 4.36).

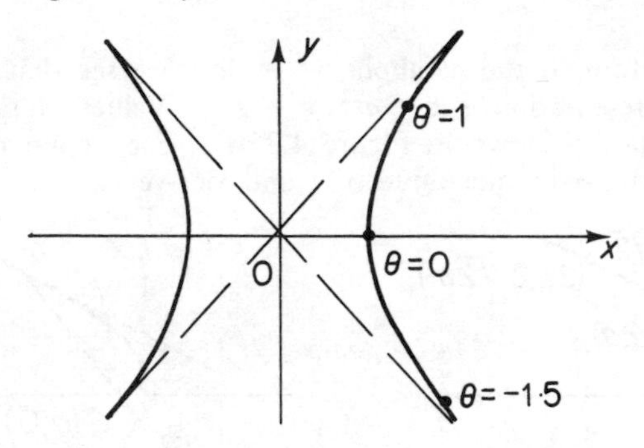

Figure 4.36

Note that this caters for the right-hand branch only so that if we wanted to cover the left-hand branch we should have to consider $x = -a \cosh \theta$, $y = b \sinh \theta$.

The tangent at θ_1 is $(x/a) \cosh \theta_1 - (y/b) \sinh \theta_1 = 1$, or $x/a - (y/b) \tanh \theta_1 = 1/\cosh \theta_1$.

Let $\theta_1 \to \infty$ then $\tanh \theta_1 \to 1$ and $\cosh \theta_1 \to \infty$, hence the tangent becomes $x/a - y/b = 0$ which is the equation for the asymptote, as found previously.

Cycloid

The path traced out by a point on the circumference of a circle which rolls without slipping on a straight line is called a cycloid [see Figure 4.37(i)].

It has parametric equations

$$x = a(\theta - \sin \theta)$$
$$y = a(1 - \cos \theta)$$

Here it is difficult to eliminate θ and find an equation connecting y and x. Indeed, even if we do find the equation, it is of very little use to us. By taking different values of θ we can easily sketch the curve as shown in Figure 4.37(ii). The values of θ taken are marked.

Try and see how it was obtained. Why not convince yourself that this is in the correct shape by marking a point on the rim of a circular object and seeing how it moves.

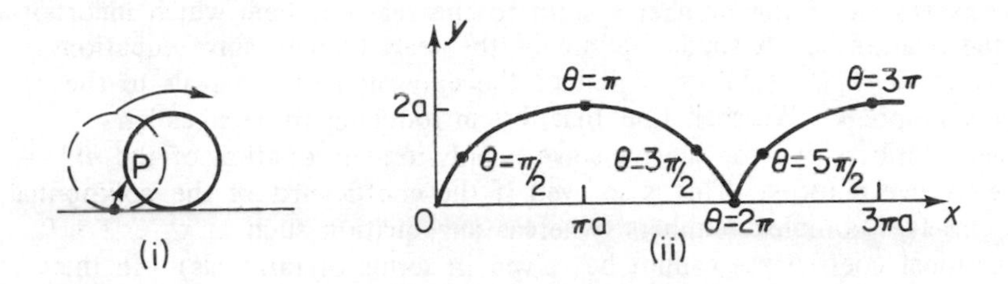

Figure 4.37

Problems

1. Show that the rectangular hyperbola can be parametrised into $x = ct$, $y = c/t$. Sketch the curve and mark on it suitable values of t. Show that at the point t_1 the equation of the tangent is $t_1^2 y + x = 2ct_1$.

2. Find where the tangents to $xy = c^2$ at t_1 and t_2 intersect.

3. Trace the curve whose parametric representation is $x = a(1 + \cos\theta)$, $y = a\sin\theta$. Interpret it algebraically by first eliminating θ from the equations for x and y.

4. A rod OA of length 50 cm rotates at 7 rad/sec and a rod AB of equal length, hinged at A to the rod OA, rotates about A in the opposite sense at 1 rad/sec. B is the point (x, y). In terms of time t, obtain expressions for x and y, assuming that OAB lay originally along the x-axis.

5. Eliminate θ from the equations $x = \cos^3\theta$, $y = \sin^3\theta$. Try and sketch the curve from this new equation and from the parametric equations.

6. Let P_1 and P_2 be (x_1, y_1) and (x_2, y_2) respectively. Let $P(x, y)$ be a point on this line such that $PP_1/P_2P_1 = \mu$ (note the meaning of $\mu < 0$). Find expressions for x and y in terms of μ. Show on a diagram the points $\mu = 0$, $\mu = 1$, $\mu = -3$, $\mu = 1/2$, $\mu = 2$.

4.6 COMPLEX NUMBERS

Introduction

When mathematicians encounter equations whose solutions cannot be expressed in terms of known functions they often define the solution to be a new function and set about determining its main features. In the past the failure of the existing rational numbers to provide a solution to $x^2 = 2$ led to the extension of the number system to the real numbers which incorporated the irrationals. A similar failure of the reals to help solve equations such as $x^2 = -1$, $(x - 3)^4 = -1$ led to the extension of the reals to the **complex numbers.** We shall find that, by introducing these so-called complex numbers, we can always solve a polynomial equation of the nth degree to give n roots. This is so even if the coefficients of the polynomial are themselves complex numbers (whereas an equation such as $x^2 - 2 = 0$ with rational coefficients cannot be solved in terms of rationals). In this sense the complex numbers are *complete.*

Since we wish to be able to solve an equation such as $x^2 = -1$ which gives $x = \pm\sqrt{-1}$, we introduce a quantity i which obeys the rule $\text{i}^2 = -1$. Then for this equation $x = \pm\text{i}$.

We now define a complex number to be of the form $a + \text{i}b$ where a and b are real numbers. For example, $3 + \text{i}2$, $\sqrt{3} + \text{i}\pi$, $0 - \text{i}\pi$, $\sqrt{2} - \text{i}2.53$ are all complex numbers; the real numbers themselves can be regarded as special cases of the complex numbers for we can say for example $2 = 2 + \text{i}0$, $\sqrt{2} = \sqrt{2} + \text{i}0$. (Many engineers and books use j instead of i.)

You may rightly point out that all this is rather arbitrary and remote from your engineering subjects. However, for reasons which will become clearer as you read through this book, complex numbers and associated quantities act as a useful vehicle in the course of many calculations and save a great deal of work.

Having introduced these numbers we must develop rules for the arithmetic operations involving them, but we must do so in a way which is consistent with the special case of real numbers $[b = 0]$.

We usually write $z \equiv x + \text{i}y$ for a general complex number and call x the **real part** of z, written $\mathcal{R}(z)$, and y the **imaginary part** of z, written $\mathcal{I}(z)$. Numbers of the form $\text{i}b$ are called purely imaginary. Hence the real part of $-7 + 3\text{i}$ is -7 and the imaginary part is 3; $\text{i}\pi$ is a *purely imaginary* number.

We define addition as one might expect: an example first

$$(3 - 2\text{i}) + (1 + 4\text{i}) = (3 + 1) + (-2 + 4)\text{i} = 4 + 2\text{i}$$

Note that we are defining the + between the parentheses on the left-hand side of the equation. Subtraction follows similarly:

$$(3 - 2\text{i}) - (1 + 4\text{i}) = (3 - 1) + (-2 - 4)\text{i} = 2 - 6\text{i}$$

In the general case for two complex numbers $(a + b\text{i})$ and $(c + d\text{i})$

$$(a + bi) + (c + di) = (a + c) + (b + d)\mathrm{i} \tag{4.23a}$$

$$(a + bi) - (c + di) = (a - c) + (b - d)\mathrm{i} \tag{4.23b}$$

Multiplication is carried out as in ordinary algebra except that wherever i^2 occurs, it is replaced by -1. For example,

$$(3 - 2\mathrm{i}).(1 + 4\mathrm{i}) = 3 - 2\mathrm{i} + 12\mathrm{i} - 8\mathrm{i}^2 = 3 - 2\mathrm{i} + 12\mathrm{i} + 8 = 11 + 10\mathrm{i}$$

More generally,

$$(a + bi).(c + di) = (ac - bd) + (bc + ad)\mathrm{i} \tag{4.24}$$

Before turning to division we need two further concepts. Two complex numbers are **equal** *if their real parts are equal and their imaginary parts are equal.* To see this, suppose $a + bi = c + di$; then $(a - c) = (d - b)\mathrm{i}$, and on squaring $(a - c)^2 = (d - b)^2.-1$, i.e. $(a - c)^2 + (d - b)^2 = 0$. This is a relationship involving real numbers and can *only* be satisfied if $a = c$ and $b = d$, thus showing that both the real and imaginary parts must be equal.

To each complex number $a + bi$ there corresponds a unique complex number $a - bi$ called the **conjugate.** For example, the conjugate of $3 - 2\mathrm{i}$ is $3 + 2\mathrm{i}$.

The conjugate of z is denoted by $\bar{z}$.

Observe that if z is real, $\bar{z} = z$ and if z is *purely imaginary,* $\bar{z} = -z$. Now it can be shown that $z.\bar{z}$ is *always* real; for if $z = x + \mathrm{i}y$, then $\bar{z} = x - \mathrm{i}y$ and hence $z\bar{z} = (x + \mathrm{i}y)(x - \mathrm{i}y) = x^2 - \mathrm{i}^2y^2 = x^2 + y^2$. This property allows us to carry out division of two complex numbers – we *multiply top and bottom by the conjugate of the denominator,* e.g.

$$\frac{3 - 2\mathrm{i}}{1 + 4\mathrm{i}} = \frac{3 - 2\mathrm{i}}{1 + 4\mathrm{i}} \cdot \frac{1 - 4\mathrm{i}}{1 - 4\mathrm{i}} = \frac{3 - 2\mathrm{i} - 12\mathrm{i} + 8\mathrm{i}^2}{1 - 16\mathrm{i}^2} = \frac{-5 - 14\mathrm{i}}{17} = \frac{-5}{17} - \frac{14}{17}\mathrm{i}$$

In algebraic terms,

$$\frac{z_1}{z_2} = \frac{z_1}{z_2} \cdot \frac{\bar{z}_2}{\bar{z}_2} = \frac{z_1\bar{z}_2}{z_2\bar{z}_2} \tag{4.25}$$

and the denominator is real.

Note, in particular, that $1/\mathrm{i} = -\mathrm{i}$.

We have said that the complex numbers include the reals as a special case; you should verify that the arithmetic operations defined above are valid for real numbers.

Since a complex number comprises two real numbers which are independent, we may use the notation $z \equiv (x, y)$ and this suggests an analogy to coordinate geometry. We visualise a complex number as a point on a plane or as the line joining this end-point to the coordinate origin. The points $z_1 = 3 + \mathrm{i} \equiv (3, 1)$, $z_2 = 1 - \mathrm{i} \equiv (1, -1)$, $z_3 = -2 + 4\mathrm{i} \equiv (-2, 4)$, $z_4 = -1 - 2\mathrm{i} \equiv (-1, -2)$, $z_5 = -3$, $z_6 = -2\mathrm{i}$, are shown in Figure 4.38 which is an example of an **Argand diagram.** Note that there is a *one-to-one correspondence* between the complex numbers and points in the plane.

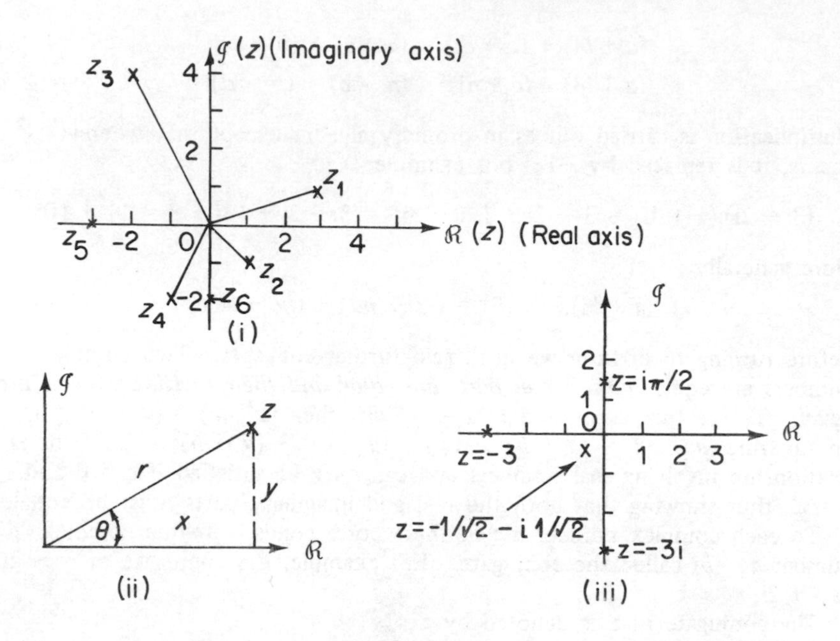

Figure 4.38

This diagrammatic representation suggests the use of polar coordinates (r, θ) as an alternative to (x, y) coordinates. r is called the **modulus** of the complex number and is sometimes written $|z|$; it corresponds to the distance of the point z from the origin and this is really what is meant by the real equivalent $|x|$. θ is called the **argument** or amplitude of a complex number and written arg(z). Notice that for real positive numbers, $\theta = 0$, for real negative numbers $\theta = \pi$ and a purely imaginary number will have an argument $\theta = \pi/2$ or $\theta = 3\pi/2$. An alternative notation to (r, θ) is $r\underline{/\theta}$.

We use the relations $x = r \cos \theta$, $y = r \sin \theta$ to convert to polars and $r = \sqrt{x^2 + y^2}$, $\tan \theta = y/x$ to convert from polars. One danger is that in the range $[0, 360^\circ]$ there are two values of θ for each value of $\tan \theta$ and to avoid ambiguity *it is safer to plot the desired point on an Argand diagram.* (The convention is usually used that θ is taken to be in the range $-\pi < \theta \leqslant \pi$)

Let us take a few examples (i) $z = -3$, (ii) $z = \mathrm{i}\,\pi/2$, (iii) $z = -3\mathrm{i}$, (iv) $z = -1/\sqrt{2} - \mathrm{i}\,1/\sqrt{2}$. We plot these first on an Argand diagram [Figure 4.38(iii)]. The equivalent polar forms are then $(3, 180^\circ)$, $(\pi/2, 90^\circ)$, $(3, -90^\circ)$, $(1, -135^\circ)$.

Example 1

Plot $2\underline{/45^\circ}$ on an Argand diagram and find the equivalent cartesian form. The diagram is Figure 4.39(i). $x = 2 \cos 45^\circ = \sqrt{2}$, $y = 2 \sin 45^\circ = \sqrt{2}$, and the equivalent form is $\sqrt{2} + \sqrt{2}\mathrm{i}$.

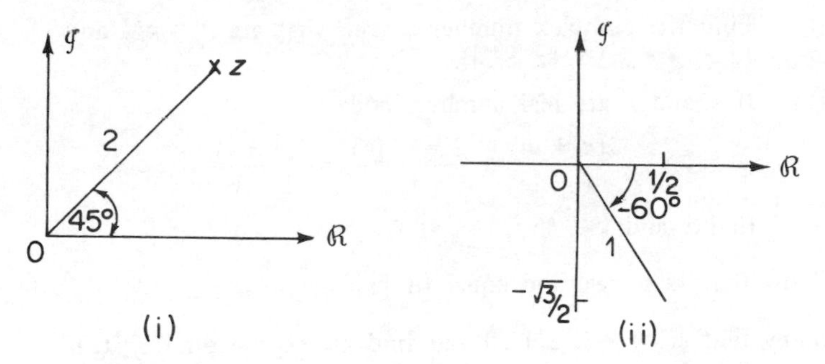

Figure 4.39

Example 2

Determine the polar form of $1/2 - (\sqrt{3}/2)i$ and plot on an Argand diagram.

$$r = \sqrt{\left(\frac{1}{2}\right)^2 + \left(\frac{-\sqrt{3}}{2}\right)^2} = \sqrt{\frac{1}{4} + \frac{3}{4}} = 1; \qquad \tan\theta = -\frac{\sqrt{3}}{2}\bigg/\frac{1}{2} = -\sqrt{3}$$

Now there are two angles, $120°$ and $-60°$ which have tangent of $-\sqrt{3}$ and if we note that $r\cos\theta = 1/2 > 0$ and $r\sin\theta = -\sqrt{3}/2 < 0$ we can see that θ lies in the fourth quadrant, selecting $-60°$ as the required angle. A plot on an Argand diagram would have established this straight away and it is wiser to do this as a first step. Then the polar form required is $1\underline{/-60°}$. See Figure 4.39(ii).

Problems I

1. Mark on an Argand diagram the points representing the following numbers: $3 + 4i$, $-2 + 3i$, $-3 - 2i$, $2 - 4i$.

2. Express the following complex numbers in polar form:
   3, $2i$, -1, $-2i$, $5 + 12i$, $-\sqrt{3} + i$, $-6 - 8i$, $\sqrt{2} - i\sqrt{2}$, $2 + 2i\sqrt{3}$, $-3 + 2i$, $-1 - i$, $1 - 2i$.

3. Express the following complex numbers in cartesian form:
   $\sqrt{2}\underline{/\pi/4}$ $2\underline{/140°}$, $2\underline{/-5\pi/6}$, $5\underline{/-55°}$

4. Simplify, giving answer in cartesian form: $(5 - 3i)(2 + i) - \dfrac{4(3 - i)}{1 - i}$

5. (i) Find the roots of the equation $x^2 + 4x + 13 = 0$ and express them in the polar form.

   (ii) If $b^2 < ac$, show that the roots of the equation $ax^2 + 2bx + c = 0$ can be expressed as $\sqrt{c/a}\,\underline{/\pm\cos^{-1}(-b/\sqrt{ac}\,)}$.

6. (i) If $z = (2 - i)/(1 + i) - 2(3 + 4i)/(u + i)$ where u is a real number, find the values of u which make the complex number z lie on the line $x = y$ in the Argand diagram.

(ii) Find the complex number z such that $\arg z = \pi/4$ and $|z - 3 + 2i| = |z + 3i|$.

(iii) If x and y are real numbers and

$$\frac{2(x + iy)}{1 - i} + \frac{2(x - iy)}{i} = \frac{5(1 + i)}{2 - i}$$

find x and y.

7. Show that $z\bar{z}$ is real and equal to $|z|^2$.

8. Show that $\mathcal{R}(z) = \frac{1}{2}(z + \bar{z})$ and find an expression for $\mathcal{I}(z)$.

9. Show that $1/i = -i$, $1/(1 + i) = \frac{1}{2}(1 - i)$.

10. Evaluate $(3 + i)(3 - i)$, $(1 + 4i)(2 - 6i)$, $(5i + 6)/7i$, $(3 + 4i)/(4 + 3i)$

11. Find the modulus and argument of $-1 + \sqrt{3}\,i$, $3 + 4i$, $4 - 2i$; write each number in polar form.

12. Plot z, $\bar{z}$ on an Argand diagram and express $\arg \bar{z}$ in terms of $\arg z$.

13. Given that $|z - 3| = 4$ and $\arg z = \pi/4$, find z.

14. (i) Find the modulus and argument of each of the complex numbers which satisfy the equation $z^2 + 2z + 5 = 3i$.

(ii) What may one deduce if
(a) the quotient of two complex numbers is real,
(b) the product of two complex numbers is real and their difference is imaginary?

15. Given a complex number z, draw in an Argand diagram the positions of the points (i) iz,
(ii) $\bar{z}$
(iii) $1/z$
(iv) z^3.
Assume $|z| > 1$.

16. Prove the triangle inequality $|z_1 + z_2| \leqslant |z_1| + |z_2|$.

17. Evaluate $(2 + i)^3 - (2 - i)^3$, $(2 + i)^{-2} + (2 - i)^{-2}$.

18. Show that $\overline{(\bar{z})} = z$, $\overline{\left(\frac{1}{z}\right)} = 1/\bar{z}$, $(\bar{z})^n = \overline{(z^n)}$ by induction and the result $\overline{z_1 z_2} = \bar{z}_1\,\bar{z}_2$.

19. Show that $\arg(z_1/z_2) = \arg z_1 - \arg z_2$.

Regions and loci in the z-plane

Regions in the z-plane may be described in terms of complex numbers. For example, the equation $|z| = 2$ states that the length of the line joining the origin to z is always equal to 2. Hence the point represented by z must move on a circle, centre the origin, radius 2 [Figure 4.40(i)].

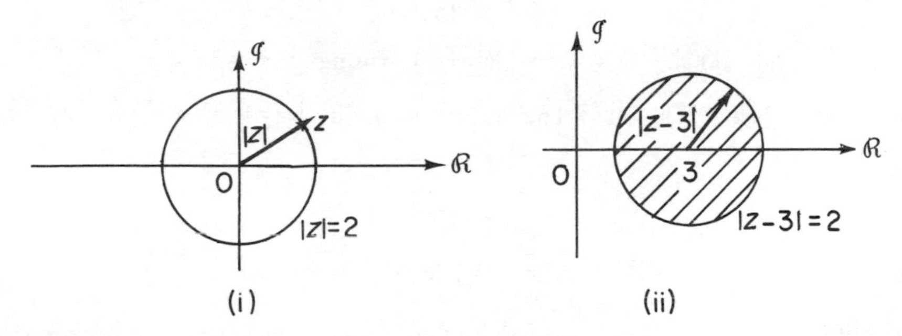

Figure 4.40

Similarly the equation $|z - 3| = 2$ can be thought of as $|z - (3 + \mathrm{i}0)| = 2$ and says that the distance between the point z and $(3, 0)$ is always equal to 2. Hence the locus of the point z is a circle, centre $(3, 0)$ and radius 2. The inequality $|z - 3| < 2$ is satisfied by all points inside this circle and so represents the shaded area in Figure 4.40(ii).

The inequality $\mathcal{R}(z) > 4$ represents the half-plane to the right of the line $x = 4$; the inequality $\mathcal{I}(z) \leqslant 3$ represents all points on and below the line $y = 3$. A straight line from the point z_0 at angle θ to the x-axis is described by $\arg(z - z_0) = \theta$ [Figure 4.41(i)]. To continue the line the other side of z_0 would require $\arg(z - z_0) = \theta + 180°$.

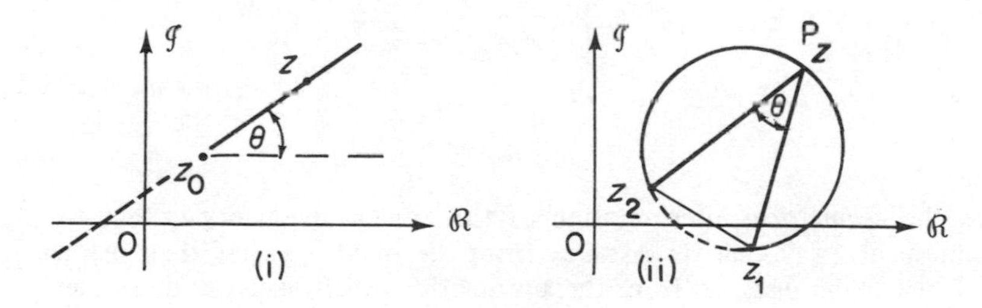

Figure 4.41

Consider the arc of a circle as shown in Figure 4.41(ii). The arc stands on the chord with end-points z_1 and z_2 and an angle θ is subtended at the circumference. If P, represented by z, is any point on the arc then we have $\arg[(z - z_1)/(z - z_2)] = \theta =$ constant, and this is the equation of the arc.

Note that the left-hand side is $\arg(z - z_1) - \arg(z - z_2)$ [Problem 19, page 178.] Try and interpret this from the diagram and convince yourself of the suitability of the polar form by writing the cartesian equivalent.

Finally, we see how to unravel an inequality by algebra and compare with a geometrical interpretation. Consider $|z - 1| \leqslant |z + 1|$. To see what domain in the complex plane this represents, we note that both sides of the inequality are positive and so we may square them to obtain

$$|z - 1|^2 \leqslant |z + 1|^2$$

i.e. $$(z - 1)\overline{(z - 1)} \leqslant (z + 1)\overline{(z + 1)} \quad (\text{using } |z|^2 = z\bar{z})$$

i.e. $$(z - 1)(\bar{z} - 1) \leqslant (z + 1)(\bar{z} + 1) \quad (\text{using } \overline{z_1 + z_2} = \bar{z}_1 + \bar{z}_2)$$

i.e. $$z\bar{z} - \bar{z} - z + 1 \leqslant z\bar{z} + \bar{z} + z + 1$$

i.e. $$-\bar{z} - z \leqslant \bar{z} + z$$

i.e. $$2(z + \bar{z}) \geqslant 0$$

therefore $$z + \bar{z} \geqslant 0$$

but $$z + \bar{z} = 2\mathcal{R}(z)$$

hence $$2\mathcal{R}(z) \geqslant 0$$

This represents all points on and to the right of the line $x = 0$ (see Figure 4.42).

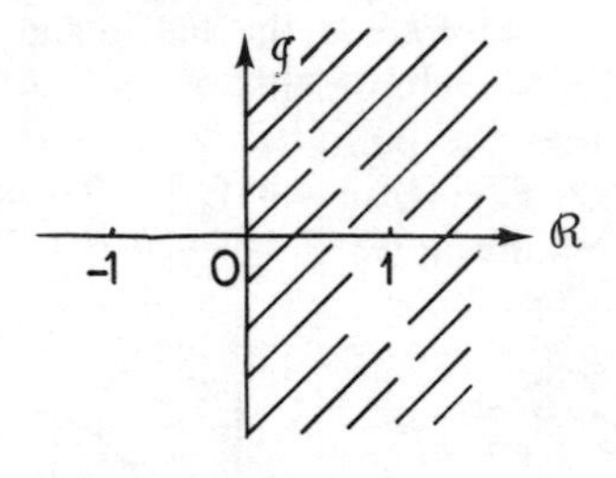

Figure 4.42

Now the *geometrical* interpretation of the original inequality is that z moves in such a way that its distance from the point 1 is less than or, at worst, equal to its distance from the point -1; which leads to the same region of the plane. Further $|z - 1| = |z + 1|$ gives the perpendicular bisector of the line segment joining $+1$ and -1 (that is, in the y-axis).

Note that we could have completed this problem by an alternative method. In this alternative method we write $z = x + iy$ and reduce the equation to x, y form. Thus

$$|z - 1| \leqslant |z + 1|$$

becomes

$$|x - 1 + iy| \leqslant |x + 1 + iy|$$

i.e.

$$\sqrt{(x - 1)^2 + y^2} \leqslant \sqrt{(x + 1)^2 + y^2}$$

i.e.

$$(x-1)^2+y^2 \leqslant (x+1)^2+y^2$$

i.e.

$$x^2-2x+1 \leqslant x^2+2x+1$$

i.e.

$$0 \leqslant 4x$$

Hence $x \geqslant 0$, as before .

This latter method is often the only one that can be used easily.

Multiplication and Division in Polar Form

We now turn to the multiplication and division of two complex numbers in polar form. First we observe that $z = (r, \theta) \equiv r \cos\theta + ir \sin\theta$.

Then if $z_1 = (r_1, \theta_1)$ and $z_2 = (r_2, \theta_2)$,

$$\begin{aligned} z_1 z_2 &= (r_1 \cos\theta_1 + ir_1 \sin\theta_1).(r_2 \cos\theta_2 + ir_2 \sin\theta_2) \\ &= r_1 r_2 \cos\theta_1 \cos\theta_2 + i^2 r_1 r_2 \sin\theta_1 \sin\theta_2 + \\ &\quad i\,[r_1 r_2 \cos\theta_1 \sin\theta_2 + r_1 r_2 \sin\theta_1 \cos\theta_2] \\ &= r_1 r_2 [\cos(\theta_1+\theta_2) + i \sin(\theta_1+\theta_2)] \end{aligned}$$

In other words, *to multiply two complex numbers, we* **multiply** *their* **moduli** *and* **add** *their* **arguments**. In shorthand $r_1 \underline{/\theta_1} \times r_2 \underline{/\theta_2}$ gives $r_1 r_2 \underline{/(\theta_1+\theta_2)}$.

Example

Consider $(1+i).[1/2+(\sqrt{3}/2)i] = 1/2 + (1/2)i + (\sqrt{3}/2)i + (\sqrt{3}/2)i^2 = (-\sqrt{3}+1)/2 + i[(1+\sqrt{3})/2]$.

We can do the same problem in polar form using the results just obtained; for $1+i = \sqrt{2}\,\underline{/45^\circ}$ and $1/2+(\sqrt{3}/2)i = 1\,\underline{/60^\circ}$.

Hence their product is $\sqrt{2}\,\underline{/45^\circ+60^\circ} = \sqrt{2}\,\underline{/105^\circ}$, which can be written

$$\sqrt{2}\cos 105^\circ + i\sqrt{2}\sin 105^\circ = -\sqrt{2}\cos 75^\circ + i\sqrt{2}\sin 75^\circ.$$

Convince yourself that this is the same as the result obtained when doing the problem in cartesian coordinates.

Division in polars follows a similar pattern:

$$\frac{z_1}{z_2} = \frac{r_1}{r_2}[\cos(\theta_1-\theta_2) + i\sin(\theta_1-\theta_2)] = \frac{r_1}{r_2}\underline{/(\theta_1-\theta_2)}$$

Hence in *division we* **divide** *the* **moduli**, *and* **subtract** *the* **arguments**. For example

$$\frac{1+i}{1/2+(\sqrt{3}/2)i} = \frac{\sqrt{2}\,\underline{/45^\circ}}{1\underline{/60^\circ}} = \frac{\sqrt{2}}{1}\underline{/(45^\circ-60^\circ)} = \sqrt{2}\,\underline{/(-15^\circ)}$$

These rules can obviously be combined, for example

$$\frac{(3\angle 60^\circ) \times (2\angle 40^\circ)}{4\angle 30^\circ} = \frac{3}{2}\angle 70^\circ$$

Two observations are worthy of attention:

1. $1/z_1 = \dfrac{1}{r_1}\angle -\theta_1, \quad z_1^2 = r_1^2\angle 2\theta_1, \quad z_1^3 = r_1^3\angle 3\theta_1$ etc.

2. It is generally true that multiplication and division are best carried out in polars.

Problems II

1. Simplify, giving answer in cartesian form:

   (i) $\dfrac{(\sqrt{2}\angle 3\pi/4)^2(2\angle -2\pi/3)^2}{(2\angle -\pi/6)}$ (ii) $\dfrac{(3\angle \pi/3)^3(2\angle -\pi/4)^5}{(4\angle \pi/2)}$

2. Simplify, giving answer in polar form:

   (i) $\dfrac{(3\angle 15^\circ)^2 \times (2\angle 40^\circ)^3}{6\angle 25^\circ \times 4\angle 35^\circ}$ (ii) $2\angle \pi/3 + 2\angle 5\pi/6$

   (iii) $\dfrac{(-1 + i)^2(-1 - i\sqrt{3})^2}{\sqrt{3} - i}$

3. (i) If $z_2 = \rho i z_1$ with ρ real and positive prove algebraically that $|z_1|^2 + |z_2|^2 = |z_1 - z_2|^2$ and interpret this result geometrically in an Argand diagram.

   (ii) If the point P represents the complex number z in an Argand diagram and $\arg[(z - 3)/(z - 1)] = \pm\pi/2$, show that P lies on the circle of unit radius with centre at the point (2, 0).

4. (i) The centre of an equilateral triangle is represented by $-1 + i$ and one vertex by $3 + 5i$. Find the complex numbers representing the other vertices.

   (ii) If the complex number z is such that the expression $(z - 1)/(z + i)$ is wholly imaginary, show that the locus of z in the Argand diagram is a circle with centre at the point $(1/2) - (1/2)i$ and radius $(1/2)\sqrt{2}$.

5. Plot on an Argand diagram $7 + 3i$, $6i$, $-3 - i$ and show that they are vertices of a square; find the fourth vertex.

6. Sketch the domains

   (i) $|z| \leqslant 4$ (ii) $|\arg z| < \pi/6$

   (iii) $(|z - 1|/|z - 2|) < 1$ (iv) $0 < \arg[(z - 2)/(z - 1)] < \pi/2$

7. Evaluate iz in polar form and show that the effect of multiplying a

complex number by i is to rotate the line joining it to the origin by 90° anticlockwise. Hence prove $i^2 = -1$.

8. If $z(\equiv x + iy)$ is a complex number and $w = (z - 2)/(z - i)$

   (a) show that when the point in the Argand diagram represented by w moves along the real axis, z traces a straight line through $(2, 0)$ and $(0, 1)$;

   (b) determine $|z - 1 - \frac{1}{2}i|$ when w lies on the imaginary axis. What is the locus of z as w moves along the imaginary axis?

9. Determine the region in the Argand plane defined by $|z - 1| + |z - i| < 4$.

10. Show that the equation of any circle in the complex plane may be written in the form $z\bar{z} + \alpha z + \bar{\alpha}\bar{z} + a = 0$ where a is a real constant and α is (in general) a complex constant. Express in this form the equation of the circle passing through the points $1 - i$, $2i$, $1 + i$ and find its radius and the position of its centre.

11. $(\cos 60^\circ + i \sin 60^\circ)$ is represented by a line segment OP. What number is represented by OP turned through 20° and halved in length?

12. Determine the locus $|z + 3i|^2 - |z - 3i|^2 = 12$.

Complex numbers as vectors

At this stage it will be worthwhile examining the consequences of the rule for addition and we represent this graphically in Figure 4.43.

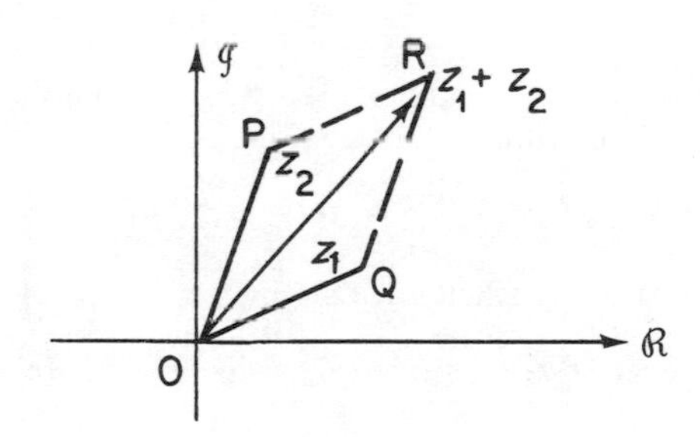

Figure 4.43

If OQ and OP represent the complex numbers z_1 and z_2 respectively, and OPRQ is a parallelogram, then it is not too hard to show by coordinate geometry that R has the coordinates of $(z_1 + z_2)$ and hence OR represents the sum $z_1 + z_2$.

We may regard a complex number as an ordered pair of real numbers (x, y) and define addition as $(x_1, y_1) + (x_2, y_2) = [(x_1 + x_2), (y_1 + y_2)]$.

Entities which obey this rule are called *two-dimensional vectors* and these are dealt with in the next section. It is important to bear in mind that complex numbers can be regarded as vectors for some purposes.

De Moivre's Theorem

An important consequence of the polar rule for multiplication is *de Moivre's Theorem* . This states that if p/q be a rational number, one value of

$$(\cos\theta + i\sin\theta)^{p/q} \quad \text{is} \quad (\cos\frac{p}{q}\theta + i\sin\frac{p}{q}\theta)$$

We shall prove this only for the special case p/q is a positive integer, n, using induction.

Clearly the result holds for $n = 1$.

Let us assume that $(\cos\theta + i\sin\theta)^m = \cos m\theta + i\sin m\theta$ for a positive integer m.

Multiplying both sides by $\cos\theta + i\sin\theta$ we get

$$\begin{aligned}(\cos\theta + i\sin\theta)^{m+1} &= (\cos m\theta + i\sin m\theta)(\cos\theta + i\sin\theta)\\ &= \cos m\theta\cos\theta + i\sin m\theta\cos\theta + i\cos m\theta\sin\theta\\ &\quad - \sin m\theta\sin\theta\\ &= \cos(m+1)\theta + i\sin(m+1)\theta\end{aligned}$$

Hence if the result is true for a positive integer m, it is also true for $m + 1$. But it is true for $m = 1$.

Hence, by induction, the result is true for all integers.

Example

We can use this result to evaluate $(1 - i)^8$. We know that $1 - i \equiv \sqrt{2}\underline{/-45^\circ}$. Therefore

$$(1 - i)^8 = (\sqrt{2})^8\underline{/-360^\circ} = 16\underline{/0^\circ} = 16$$

We note that $(\cos\theta + i\sin\theta)^{-r}$, can be expressed as

$$\cos(-r)\theta + i\sin(-r)\theta = \cos r\theta - i\sin r\theta$$

where r is any real rational number.
In particular,

$$(\cos\theta + i\sin\theta)^{-1} = \cos\theta - i\sin\theta$$

Hence, if $\cos\theta + i\sin\theta = z$, then $(\cos\theta + i\sin\theta)^{-1} = 1/z = \cos\theta - i\sin\theta$, and further, by addition, $\cos\theta = ½[z + (1/z)]$, and subtracting, $\sin\theta = (1/2i)[z - (1/z)]$.

Let us now investigate the effect of allowing n to increase without limit in the expression $[\cos(\theta/n) + i\sin(\theta/n)]^n$. By de Moivre's theorem we know that the expression takes the value $(\cos\theta + i\sin\theta)$ for any value of n and hence the limiting value must also be $(\cos\theta + i\sin\theta)$.

But we shall find later that $\cos\phi = 1 - (\phi^2/2!) + (\phi^4/4!) - \ldots\ldots$ and $\sin\phi = \phi \quad (\phi^3/3!) + (\phi^5/5!) - \ldots\ldots$. Hence

$$\cos(\theta/n) = 1 + 0(1/n)^2 \quad \text{and} \quad \sin(\theta/n) = \theta/n + 0(1/n^3)$$

where $0(1/n^2)$ means that the next term is of order $1/n^2$, similarly $0(1/n^3)$

Hence

$$\left(\cos\frac{\theta}{n} + i\sin\frac{\theta}{n}\right)^n = \left(1 + i\frac{\theta}{n} + 0\left(\frac{1}{n^2}\right)\right)^n$$

It may be shown that

$$\lim_{n\to\infty}\left(1 + i\frac{\theta}{n} + 0\left(\frac{1}{n^2}\right)\right)^n$$

$$= \lim_{n\to\infty}\left(1 + i\frac{\theta}{n}\right)^n = e^{i\theta}$$

where we have treated $i\theta$ in the same way as we did x in the definition of e^x. Hence

$$\cos\theta + i\sin\theta \equiv e^{i\theta} \tag{4.26}$$

which can be regarded as a definition of $e^{i\theta}$ as an alternative. The two expressions on either side of the identity sign behave identically for all values of θ.

Using $\cos\theta + i\sin\theta = e^{i\theta}$, de Moivre's theorem now becomes $(e^{i\theta})^{p/q} = e^{i(p\theta/q)}$.

Further, since $x + iy = r\cos\theta + ir\sin\theta = r(\cos\theta + i\sin\theta)$, we can write

$$z = re^{i\theta} \tag{4.27}$$

and this is one of the most useful forms of a complex number. It is called the **exponential form**. Also we have

$$e^{-i\theta} = \cos\theta - i\sin\theta \tag{4.28}$$

and so

$$\left.\begin{aligned} \cos\theta &= \tfrac{1}{2}(e^{i\theta} + e^{-i\theta}) \\ \sin\theta &= \tfrac{1}{2}i(e^{i\theta} - e^{-i\theta}) \end{aligned}\right\} \tag{4.29}$$

We can obtain expressions for $\cos n\theta$ and $\sin n\theta$ by using these results, e.g.

$$\begin{aligned} \cos 3\theta + i\sin 3\theta &= (\cos\theta + i\sin\theta)^3 \\ &= \cos^3\theta + 3\cos^2\theta\, i\sin\theta + 3\cos\theta\, i^2\sin^2\theta + i^3\sin^3\theta \\ &= \cos^3\theta - 3\cos\theta\sin^2\theta + i(3\cos^2\theta\sin\theta - \sin^3\theta) \end{aligned}$$

Hence picking out real and imaginary parts

$$\cos 3\theta = \cos^3\theta - 3\cos\theta\sin^2\theta = \cos^3\theta - 3\cos\theta(1-\cos^2\theta)$$
$$= 4\cos^3\theta - 3\cos\theta$$

and

$$\sin 3\theta = 3\cos^2\theta\sin\theta - \sin^3\theta = 3\sin\theta(1-\sin^2\theta) - \sin^3\theta$$
$$= 3\sin\theta - 4\sin^3\theta$$

Likewise, we can find expressions for $\cos^n\theta$, $\sin^n\theta$. For example,

$$\cos^4\theta = \frac{1}{16}(e^{i\theta} + e^{-i\theta})^4$$

$$= \frac{1}{16}(e^{4i\theta} + 4e^{3i\theta}e^{-i\theta} + 6e^{2i\theta}e^{-2i\theta} + 4e^{i\theta}e^{-3i\theta} + e^{-4i\theta})$$

$$= \frac{1}{16}(e^{4i\theta} + e^{-4i\theta}) + \frac{4}{16}(e^{2i\theta} + e^{-2i\theta}) + \frac{6}{16}$$

Now

$$\cos 4\theta + i\sin 4\theta = e^{4i\theta}$$

and

$$\cos 4\theta - i\sin 4\theta = e^{-4i\theta}$$

Hence

$$e^{4i\theta} + e^{-4i\theta} = 2\cos 4\theta$$

and so

$$\cos^4\theta = \frac{1}{16}.2\cos 4\theta + \frac{4}{16}.2\cos 2\theta + \frac{6}{16}$$

$$= \frac{1}{8}\cos 4\theta + \frac{1}{2}\cos 2\theta + \frac{3}{8}$$

Problems III

1. Use de Moivre's theorem to obtain $\cos 5\theta$, $\sin 5\theta$ as polynomials in $\cos\theta$, $\sin\theta$ respectively. Hence obtain surd expressions for $\cos 18°$ and $\sin 36°$.

2. Use the definitions of $\cos\theta$, $\sin\theta$ in terms of complex exponentials to show that
$$\cos^4\theta\sin^2\theta = \frac{1}{32}(2 + \cos 2\theta - 2\cos 4\theta - \cos 6\theta)$$

3. If $z = \cos\theta + i\sin\theta$, show that for positive integers n, $z^n + (1/z^n) = 2\cos n\theta$ and hence obtain the identity $4\cos^3\theta - 3\cos\theta = \cos 3\theta$.

4. (i) If a and b are real and $(a + ib)e^{i(5\pi/6)} + \sqrt{2}\,e^{-i(\pi/4)} = 2 - i$, find a, b.

   (ii) If $3c^{i\theta} - e^{-i\theta} = \sqrt{2} + ib$, where θ and b are real show that $\theta = 2k\pi \pm \pi/4$ (k an integer), $b = \pm 2\sqrt{2}$.

5. Solve the equation $\sin z = 2$.

6. Using de Moivre's theorem, or otherwise, show that

$$\frac{\sin 5\theta}{\sin \theta} = 16 \cos^4\theta - 12 \cos^2\theta + 1$$

For what values of θ is this result not true?

7. (i) Show, using de Moivre's theorem that

$$\cos 4\theta = \cos^4\theta - 6 \cos^2\theta \sin^2\theta + \sin^4\theta$$
$$\text{and } \sin 4\theta = 4 \sin\theta \cos\theta (\cos^2\theta - \sin^2\theta)$$

If, further, we define $T_n(x) = \cos n\theta$, where $x = \cos\theta$, obtain an expression for $T_4(x)$ in terms of x only, and derive the inverse relation $8x^4 = T_4(x) + 4T_2(x) + 3T_0(x)$.

(ii) Draw the curve represented by $|z| = a$ on an Argand diagram. If $Z = ze^{i\theta}$, what is the locus of the point Z as z moves on the above curve?
If also $W = z + (a^2/z)$, show that the point W will describe a straight line segment and indicate this on the Argand diagram.

8. $\cos\theta + i \sin\theta$ is sometimes written cis θ. Find $|\text{cis } 30°|$.

9. Using the exponential form of z, produce a suitable interpretation of $\log z$; what difficulties does your definition give?

10. Demonstrate that $e^{i\pi} = -1$. What is a suitable value for $e^{i\pi/2}$?

11. Show that $\sin(iz) = i \sinh z$, $\cos(iz) = \cosh z$. Find $\cosh(2 + i)$.

12. The current entering a telephone line is the real part of

$$\frac{\cos \omega t + i \sin \omega t}{\cosh(s + is)}$$

Express this in the form $A \sin(\omega t + \alpha)$.

13. By taking logarithms show that a suitable value for i^i is $e^{-\pi/2}$.

Roots of a complex number

Let us consider the problem of finding the cube roots of 1. Although we might expect three roots, we know there is only one real root. Let z be any root so that $z^3 = 1$. If z be (r, θ), then $z^3 = (r^3, 3\theta)$ and $1 \equiv (1, 0)$. Hence

$$r^3(\cos 3\theta + i \sin 3\theta) = 1(\cos 0° + i \sin 0°) \tag{4.30}$$

From this we see that $r^3 = 1$ and since r is real, $r = 1$.

Hence the root lies on the unit circle $|z| = 1$; $3\theta = 0$ and so the root z_1 is $(1, 0)$, which is the real root we knew already. But $\sin(360 + \phi) \equiv \sin\phi$ and $\cos(360 + \phi) \equiv \cos\phi$ and the number 1 may also be represented $(1, 360°)$; hence equation (4.30) may be replaced by

$$r^3(\cos 3\theta + i \sin 3\theta) = 1(\cos 360^\circ + i \sin 360^\circ)$$

which gives $r = 1$ and $\theta = 120^\circ$. We claim that a second root is

$$z_2 = \cos 120^\circ + i \sin 120^\circ = -\frac{1}{2} + \frac{\sqrt{3}}{2}i$$

You should check formally that $z_2^3 = 1$.

(4.30) may also be replaced by

$$r^3(\cos 3\theta + i \sin 3\theta) = 1(\cos 720^\circ + i \sin 720^\circ)$$

whence $r = 1$, $\theta = 240^\circ$ and $z_3 = -\frac{1}{2} - \frac{\sqrt{3}}{2}i$.

Let us plot these three roots on an Argand diagram (Figure 4.44).

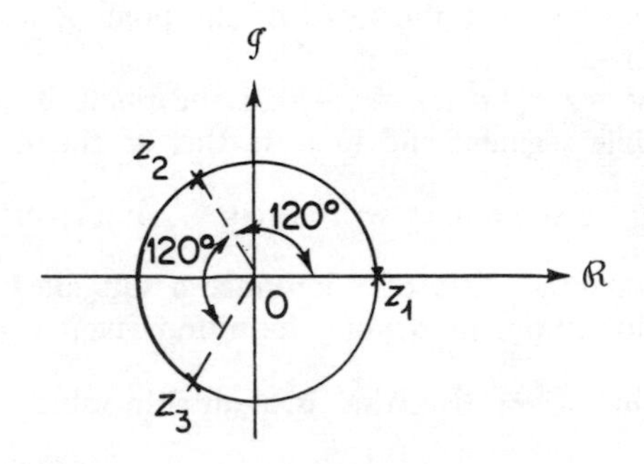

Figure 4.44

There are several features worthy of note.

1. The roots are equally spaced around the unit circle $|z| = 1$.
2. There are no more roots to be found; if we replaced (4.30) by $r^3(\cos 3\theta + i \sin 3\theta) = 1(\cos 1080^\circ + i \sin 1080^\circ)$ we would obtain $\theta = 360^\circ$ which coincides with z_1 and we would merely restart the cycle.
3. $\bar{z}_3 = z_2$ and $\bar{z}_2 = z_3$.
4. Since $z_3 = (1, 240^\circ)$ and $z_2 = (1, 120^\circ)$, we have $z_3 = z_2^2$.

We label the roots, 1, ω, ω^2 therefore, and the equation $z^3 - 1 = 0$ can be written as $(z - 1)(z - \omega)(z - \omega^2) = 0$. The left-hand side can be multiplied out to give

$$z^3 - z^2(1 + \omega + \omega^2) + z(\omega + \omega^2 + \omega^3) - \omega^3 = 0$$

Since $z_2 = \omega$ is a root of $z^3 = 1$, $\omega^3 = 1$ and as there is no z^2 term in the equation $z^3 - 1 = 0$, the sum of the roots $= 1 + \omega + \omega^2 = 0$. This also leads to a vanishing of the z term and the equation reduces to $z^3 = 1$.

It is useful to observe that we may cast the equation as

$$(z - 1)[z^2 - z(\omega + \omega^2) + \omega^3] = 0$$

i.e. $(z - 1)(z^2 + z + 1) = 0$ and we have the left-hand side as the product of real linear and real quadratic factors.

In fact there are two general rules which apply to polynomial equations

$$z^n + a_{n-1}z^{n-1} + \ldots\ldots + a_1 z + a_0 = 0$$

where a_r are *real* coefficients: any complex roots occur in *conjugate pairs* (z and $\bar{z}$) and the left-hand side can be decomposed into a product of real linear and real quadratic factors. You should show that the second result follows from the first.

Example

Let us find the fifth roots of $1 + i$. First, we write $1 + i$ as $\sqrt{2}[(1/\sqrt{2}) + i/\sqrt{2}]$ and hence in polar form as $\sqrt{2}\underline{/45^\circ}$.

De Moivre's theorem in its general form states that if $z = (r,\theta)$

$$z^{p/q} = r^{p/q}\left[\cos\frac{p(\theta + 2k\pi)}{q} + i\sin\frac{p(\theta + 2k\pi)}{q}\right]$$

where $k = 0, 1, 2, \ldots\ldots, (q - 1)$.

This may be remembered more easily by noting that all the roots are equally spaced around the circle of radius $r^{p/q}$. The radius of the circle we seek is $(\sqrt{2})^{1/5}$, i.e. $2^{1/10} \simeq 1.072$. One root, called the *principal root*, has argument $(45^\circ/5)$, i.e. 9°. Since the others are equally spaced, they must be $(360/5)^\circ$, i.e. 72° apart; i.e. they have arguments 81°, 153°, 225°, 297°; They are plotted on an Argand diagram, Figure 4.45.

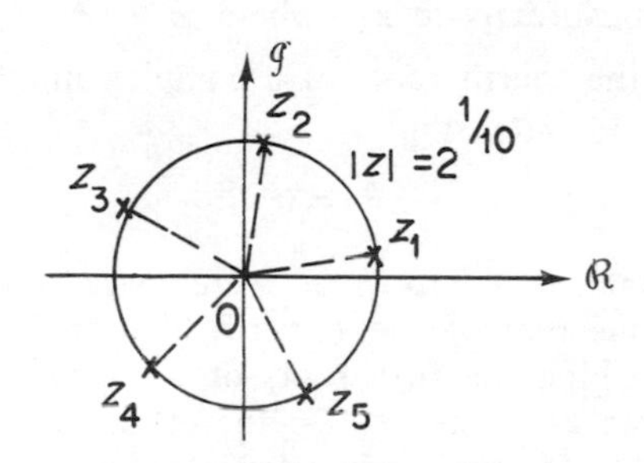

Figure 4.45

In this case no root is the conjugate of one of the others. Why not?

Summary of the method for finding the roots of a complex number

If we wish to find $\sqrt[n]{a + ib}$

(i) plot $a + ib$ on an Argand diagram
(ii) find $|a + ib| = r$ and $\arg(a + ib) = \theta$
(iii) find $\sqrt[n]{r}$ and draw a circle centre the origin of this radius.
(iv) the first root lies on this circle with argument θ/n
(v) find $360°/n$ and, starting at the first root, plot out the remaining roots on the circle by consistently increasing the argument by $360°/n$.

Problems IV

1. (i) Determine the cube roots of $\sqrt{3} - \mathrm{i}$, and express them in cartesian form.

   (ii) Determine the fourth roots of $z = \dfrac{(-2 + \mathrm{i})^3(-2 - 3\mathrm{i})^3}{(4 - 3\mathrm{i})^2}$ expressing them in the cartesian form.

2. (i) Find the complex number z which satisfies the equation

$$(1 + 2\mathrm{i})z + (3 + \mathrm{i})\bar{z} = \frac{5(1 - \mathrm{i})}{2 + \mathrm{i}}$$

   (ii) If $x = 1 + \mathrm{i}$ is a root of the quartic equation $x^4 + ax^3 + bx^2 + 8 = 0$, where a and b are real coefficients, find the other roots and determine the values of a and b.

3. Solve the equations (i) $z + z^5 = 0$
   (ii) $z^6 - z^3 - 12 = 0$

4. (i) Find the cube roots of $\sqrt{3} + \mathrm{i}$.
   (ii) Find the fourth roots of $3 + 4\mathrm{i}$.

5. (i) Show that the cube roots of a complex number may be written in the form z_1, ωz_1, $\omega^2 z_1$ (where $\omega = e^{2\pi \mathrm{i}/3}$).

   (ii) Show that the fourth roots of a complex number may be written as $\pm z_2$, $\pm \mathrm{i}z_2$.

6. Verify directly that $1 + \omega + \omega^2 = 0$

7. Using the fact that the nth roots of 1 are spaced equally round the unit circle with one root of $z = 1$, verify that the fourth roots of 1 are 1, i, -1, $-\mathrm{i}$. Find the fifth roots of 1. Label the first root anti-clockwise from $z = 1$ as γ. Do the other roots have a simple relationship to γ? Are there any relations akin to $1 + \omega + \omega^2 = 0$? Can you say anything in general about the nature of the nth roots of 1? Consider the cases n even and n odd.

Application – linear AC circuits

A periodic disturbance may be represented via complex variables. If a real quantity x varies harmonically with time as $x = a \cos(\omega t + \phi)$, a is called the *amplitude*, $\omega t + \phi$ the *phase*, $2\pi/\omega$ the *period* and $\omega/2\pi$ the *frequency* of the disturbance. We can **represent** the disturbance by $z = ae^{i(\omega t + \phi)} = a[\cos(\omega t + \phi) + i \sin(\omega t + \phi)]$; the *real* part of this complex quantity is the *disturbance* x. In applications to linear AC circuits, we shall need to attach a meaning to dz/dt. We can find dx/dt as $-\omega a \sin(\omega t + \phi)$; since we want this to be the real part of some complex expression we note that $i\omega ae^{i\omega t} = i\omega a(\cos \omega t + i \sin \omega t) = i\omega a \cos \omega t - \omega a \sin \omega t$ and hence dx/dt is the real part of $i\omega ae^{i(\omega t + \phi)}$.

If you have already done differentiation you will appreciate that this is consistent with saying that $dz/dt = i\omega ae^{i(\omega t + \phi)}$ which follows from the usual rules for differentiation.

Consider the circuit shown in Figure 4.46(i), where the applied e.m.f. $E = E_0 \cos \omega t$. When the circuit settles down to a steady oscillatory behaviour after the effective decay of any transient behaviour we can assume the alternating current to be of the same frequency as the applied e.m.f. but not necessarily in phase with it, i.e. $I = I_0 \cos(\omega t + \phi)$. It is customary in such problems to write the current representation as i and hence $\sqrt{-1}$ must be given the alternative symbol, j.

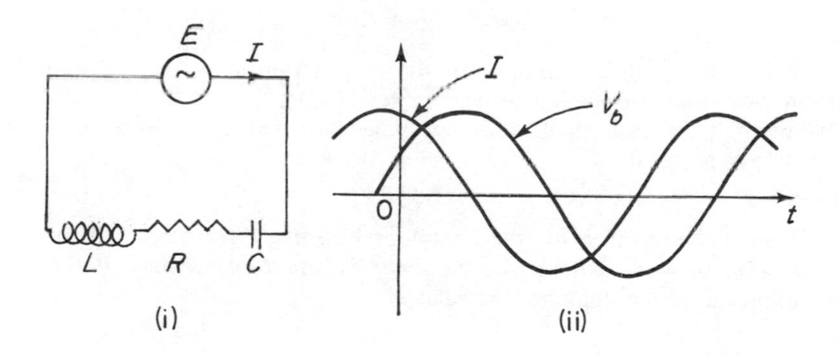

Figure 4.46

When a current I passes through an inductance coil of self-inductance L the back e.m.f. developed is $L(dI/dt) = V_b$.

If

$$I = I_0 \cos(\omega t + \phi),\ V_b = -\omega L I_0 \sin(\omega t + \phi)$$
$$= \omega L I_0 \cos(\omega t + \phi + \frac{\pi}{2})$$

In other words, the back e.m.f. is ahead of the current by 90°. See Figure 4.46(ii).

If we *represent* the current by $i = I_0\, e^{j(\omega t + \phi)}$ then the *representation* of the back e.m.f. is

$$v_b = \omega L I_0\, e^{j(\omega t + \phi + \pi/2)} = \omega L I_0\, e^{j(\omega t + \phi)} e^{j\pi/2}$$

$$= j\omega L i \qquad (\text{since } e^{j\pi/2} \equiv j)$$

Similarly, it can be shown that the *representations* v_R, v_C of the e.m.f.'s in the resistance and in the condenser are given by $v_R = Ri$ and $v_C = i/j\omega C = -ji/\omega C$. Now ωL and $1/\omega C$ are the *reactances* of the inductance and the capacitance; they play the same role as the resistance.

The *representation* of the total e.m.f. on the circuit,

$$e = i(j\omega L + R + \frac{-j}{\omega C}) \tag{4.31}$$

e/i is called the *complex impedance* of the circuit, often denoted by z.

Then $|z| = [R^2 + (\omega L - 1/\omega C)^2]^{1/2}$. If we let $\arg z = -\phi$ then

$$\tan \phi = (1/\omega C - \omega L)/R = (1 - \omega^2 LC)/R\omega C$$

Transforming back to real variables, and noting that $z = |z|\, e^{-j\phi}$

$$I = \mathcal{R}(i) = \mathcal{R}\{e/z\} = \mathcal{R}\{E_0\, e^{j\omega t} \times e^{j\phi} / [R^2 + (\omega L - \frac{1}{\omega C})^2]^{1/2}\}$$

$$= \frac{E_0 \cos(\omega t + \phi)}{[R^2 + (\omega L - 1/\omega C)^2]^{1/2}}$$

Note that equation (4.31) contains all the information about the solution and can yield this information after complex algebra.

We see that the concept of complex impedance is one which facilitates solution to a problem.

Problems V

1. Develop the solution for the circuit of Figure 4.47(i). You will find it easier to work in terms of the *complex admittance*, which is the reciprocal of the complex impedance.

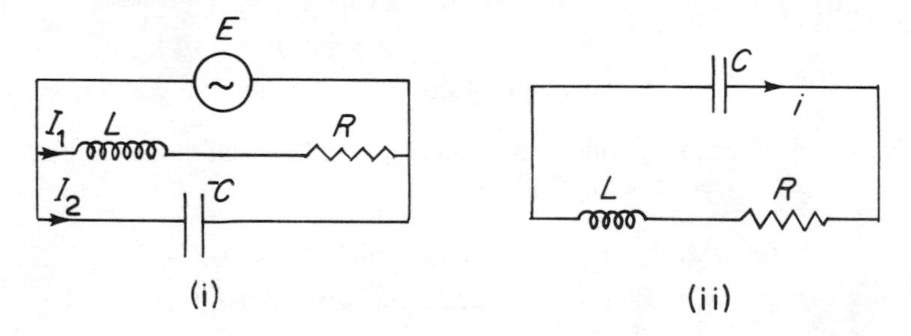

Figure 4.47

2. For a circuit shown in Figure 4.47(ii) with no e.m.f. generator, allow the current to be represented by $i = I_0 e^{j\omega t}$. Find an equation for ω by equating the total e.m.f. to zero. Write down the solution for i. The effect of the factor $e^{-Rt/2L}$ denotes attenuation; explain what happens if $R = 0$. Distinguish between the cases $4C - R^2 > 0, = 0, < 0$.

4.7 VECTOR ALGEBRA

In the section on complex numbers we introduced the idea of a complex number represented in the Argand diagram [Figure 4.48(i)] as being equivalent to an ordered pair of real numbers (x, y).

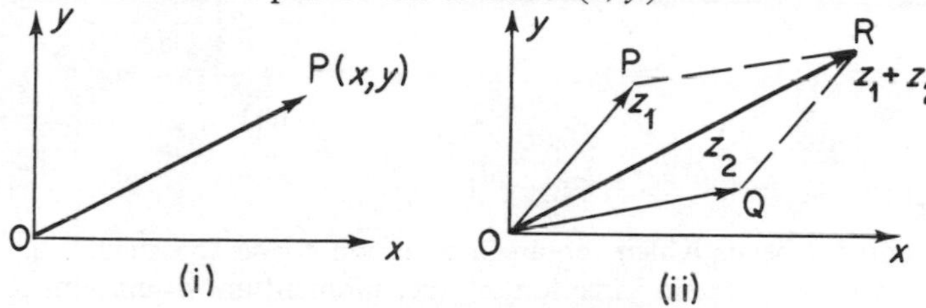

Figure 4.48

The addition of complex numbers was defined as the operation

$$(x_1, y_1) + (x_2, y_2) = [(x_1 + x_2), (y_1 + y_2)] \tag{4.32}$$

Entities which obey this rule are called *two-dimensional vectors* and the addition rule is equivalent to the *parallelogram rule* [Figure 4.48(ii)].

OP represents z_1, OQ represents z_2 and OR represents $z_1 + z_2$; x and y are called the **components** of the vector and you should notice that equation (4.32) says that we add the components in the x and y directions separately.

Now many quantities occur in science and engineering which obey the law of addition of the parallelogram rule. For example, you have probably done an experiment in physics similar to the following:

A weight W is suspended by two cords which pass over pulleys and carry weights W_1 and W_2. A third weight W is suspended from the cord as shown in Figure 4.49(i). It is found that the system takes up a position of equilibrium. The weights produce tensions in the strings and, at the point 0, these tensions must balance each other.

Experiment shows that if we complete the parallelogram in Figure 4.49(ii) (OP, OQ, OR being drawn so that they point along the strings and of lengths equivalent to W_1, W_2, W respectively) then the diagonal OR′ is *exactly* equal and opposite to OR, so balancing the downward force. Thus, the combined effect of W_1 and W_2 is represented by OR′ and addition of W_1 and W_2 is therefore defined in this way. The characteristic possessed by the quantities W_1, W_2 and W which makes addition different from normal algebra is that they have direction as well as magnitude and are therefore vectors. Thus we are led to define the addition of vectors in general in the same way. You will find that we do this on page 196.

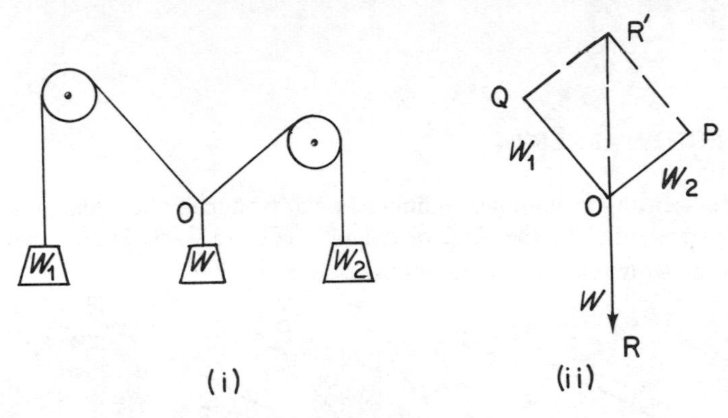

Figure 4.49

Many quantities occur which require a direction to be specified, e.g. velocity (a speed in a certain direction), force, momentum, displacement (a movement in a certain direction). Such quantities are called **vectors** as distinct from **scalar** quantities which require the specification of magnitude only.

It is clear that unless all the vectors we are considering lie in a plane, we shall not be able to represent them by complex numbers (or ordered pairs) since complex number representations are strictly two-dimensional in character.

We therefore have to extend our representations while retaining the idea that a vector is a directed length.

To distinguish between vector quantities and scalars we write vectors in bold-face type, such as **V**, **F** and **a**. Usually on paper or the blackboard we write $\underline{V}$, $\underline{F}$ and $\underline{a}$.

Now, just as we learn in algebra to add, subtract and multiply scalar quantities, we have to learn how to manipulate vector quantities and we shall go through the rules of addition, subtraction and multiplication to create an algebra of vectors.

Representation of a vector quantity

We can represent a vector quantity **F** graphically both in magnitude and direction by a line OP.

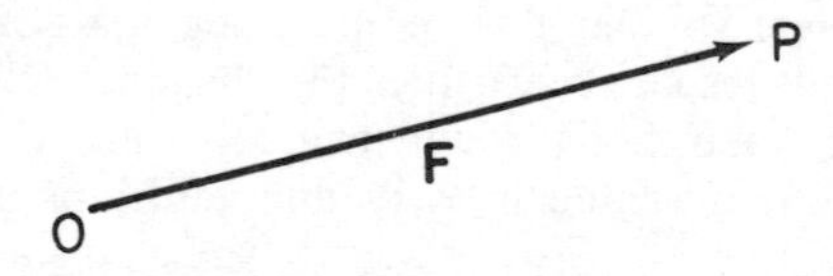

The magnitude of the vector is represented by the length of the line OP (using some convenient scale) and its direction in space is shown by an arrow on the line. Various notations are used in books when describing this vector. We could write for the vector

$$\mathbf{F} = \overline{OP} \quad \text{or} \quad \overrightarrow{OP}$$

and for its magnitude we would have

$$F = |\mathbf{F}| = |\overrightarrow{OP}| = OP$$

All these are possible notations that you will meet. Here we shall use

$$\mathbf{F} = \overrightarrow{OP} \quad \text{and} \quad F = |\mathbf{F}| = OP$$

It is obvious from our definition that if we multiply a vector by some scalar λ, then all we are doing is to multiply the length of the vector by λ: we are not changing the direction.

For example, if **F** is the directed length $\overrightarrow{OP}$

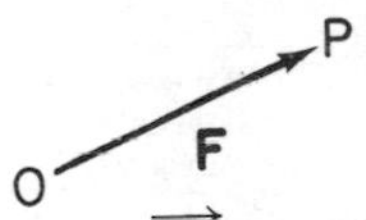

then 3**F** would be the vector $\overrightarrow{OR}$ as follows.

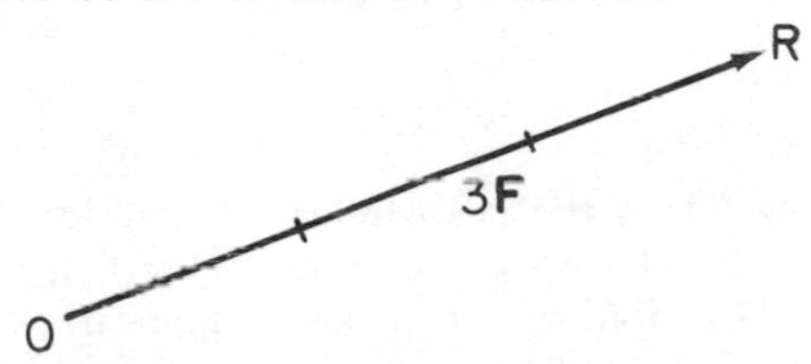

In particular if **f** is a vector of unit size (called a **unit vector**) in the direction of **F** then

$$\mathbf{F} = F\mathbf{f}$$

Equality of two vectors

If two vectors are parallel, pointing in the same direction, and of equal length, we say they are *equal*. Let $\mathbf{F} = \overrightarrow{OP} = \overrightarrow{QR}$.

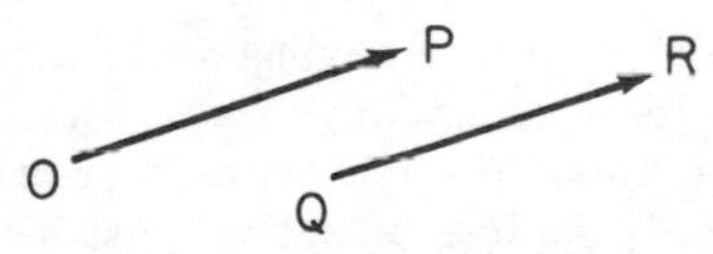

The points of application O and Q do not need to be the same point here. For example, it is common to say an aeroplane is flying due East with speed 250 m.p.h. without specifying its point in space. A vector where the point of application need not be specified is called a **free vector**.

If we have two vectors **F** and $\mathbf{F}_1$ which are parallel and of the same length but opposite in direction, we call the second vector $-\mathbf{F}$ since it is $(-1).\mathbf{F}$ and then $\mathbf{F}_1 = -\mathbf{F}$.

Addition and subtraction of vectors

To be consistent, we retain the parallelogram law of addition. Formally, if we want to find $\mathbf{E} + \mathbf{F}$ we draw $\mathbf{E}$ and $\mathbf{F}$ with the *same* point of application and complete the parallelogram. The diagonal of this parallelogram is then $\mathbf{E} + \mathbf{F}$ [Figure 4.50(i)]. $\mathbf{E} + \mathbf{F}$ is called the **resultant** of **E** and **F**.

It is obvious that $\mathbf{F} + \mathbf{E}$ must give the same resultant so that $\mathbf{E} + \mathbf{F} = \mathbf{F} + \mathbf{E}$. (Vector addition is therefore *commutative.*)

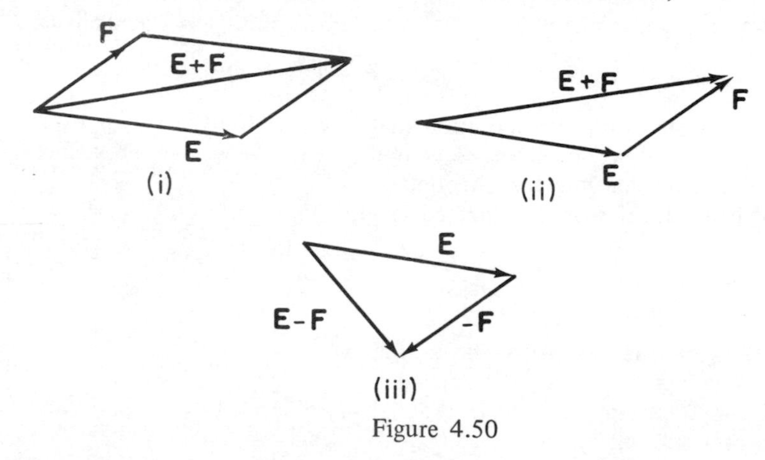

Figure 4.50

An alternative form of this is the **triangle law of addition** which is really half the above parallelogram. Here we draw the vector **E** first. At the end of **E** we place the vector **F**. Joining the point of application of **E** and the end of **F** gives the vector $\mathbf{E} + \mathbf{F}$ (directions of arrows are important). See Figure 4.50(ii).

Subtraction immediately follows for we can regard $\mathbf{E} - \mathbf{F}$ as meaning $\mathbf{E} + (-\mathbf{F})$ which gives Figure 4.50(iii). (Again note the directions of arrows.)

We can extend addition to more than two vectors and build up a polygon by placing the vectors successively end to end. For example, with three vectors, **A**, **B**, **C** (which need not necessarily be in one plane the sum $\mathbf{A} + \mathbf{B} + \mathbf{C}$ is represented by the closing side of the polygon shown in

---

Footnote A word of warning: when applying vector methods to physical problems it will be necessary at times to specify the line of action *and/or* the point of application. For example, if a force acts on a rigid body its line of action **must** be specified since a change in the line of action will alter the torque produced on the body. Similarly when dealing with a body which changes its shape when forces are applied we must specify the points of application of these forces. Vectors which have some *fixed* point of application are called **bound vectors.**

Figure 4.51(i).

The order of the addition is not important and we can easily see from Figures 4.51(ii), (iii) and (iv) that

$$(\mathbf{A} + \mathbf{B}) + \mathbf{C} = \mathbf{A} + (\mathbf{B} + \mathbf{C}) = (\mathbf{A} + \mathbf{C}) + \mathbf{B}$$

(combining the *associative* and *commutative* laws).

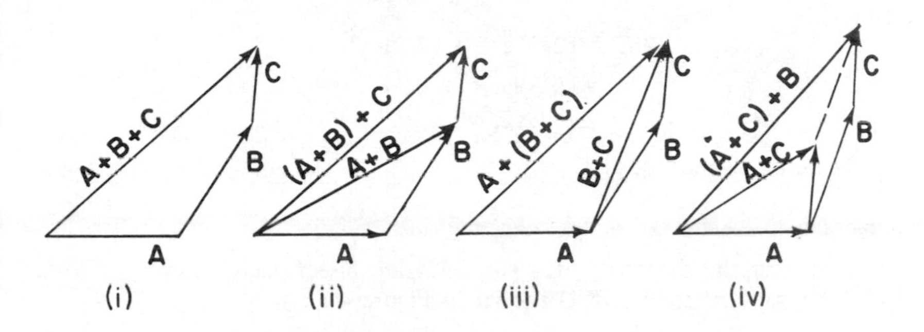

Figure 4.51

Similarly it is easy to show that $\lambda(\mathbf{A} + \mathbf{B}) = \lambda\mathbf{A} + \lambda\mathbf{B}$ and $(\lambda + \mu)\mathbf{A} = \lambda\mathbf{A} + \mu\mathbf{A}$ for scalars λ and μ.

Thus vector addition and subtraction obey all the laws of the normal algebra of real (or complex) numbers and we can treat linear vector equations in exactly the same way as in the algebra of real numbers. For example, if $\mathbf{A} + \mathbf{B} = \mathbf{C}$, then $\mathbf{A} = \mathbf{C} - \mathbf{B}$. We define further a **null vector** or **zero vector 0** as having no particular direction and zero length, so that we can then say in equations like the above – if $\mathbf{A} + \mathbf{B} = \mathbf{C}$, then $\mathbf{A} + \mathbf{B} - \mathbf{C} = \mathbf{0}$. Also $\mathbf{A} + \mathbf{0} = \mathbf{0} + \mathbf{A} = \mathbf{A}$.

Applications in plane geometry and mechanics

We now consider some applications in plane geometry and mechanics.

Example 1 (Refer to Figure 4.52)

Given DB = (2/5)AB and BE = (2/5)BC,
Prove DE || AC and DE = (2/5)AC.

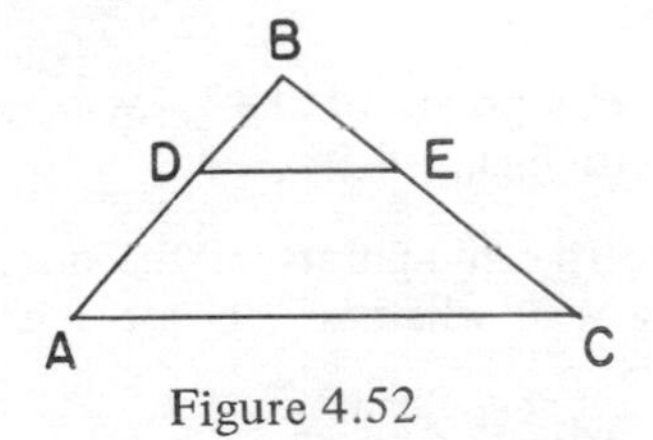

Figure 4.52

In vector language, $\overrightarrow{DB} = (2/5)\overrightarrow{AB}$ and $\overrightarrow{BE} = (2/5)\overrightarrow{BC}$ and we must prove that $\overrightarrow{DE} = (2/5)\overrightarrow{AC}$. (Since this means that DE = (2/5)AC and is also parallel to AC.)

Now $\overrightarrow{AB} + \overrightarrow{BC} = \overrightarrow{AC}$, $\overrightarrow{AD} = (3/5)\overrightarrow{AB}$, $\overrightarrow{EC} = (3/5)\overrightarrow{BC}$,
but

$$\overrightarrow{DB} + \overrightarrow{BE} = \overrightarrow{DE}$$

therefore

$$\begin{aligned}\overrightarrow{DE} &= (2/5)\overrightarrow{AB} + (2/5)\overrightarrow{BC}\\ &= (2/5)(\overrightarrow{AB} + \overrightarrow{BC})\\ &= (2/5)\overrightarrow{AC}\end{aligned}$$

Hence the result is proved.

Example 2

To prove that the diagonals of a parallelogram bisect each other. Consider the parallelogram ABCD shown in Figure 4.53.

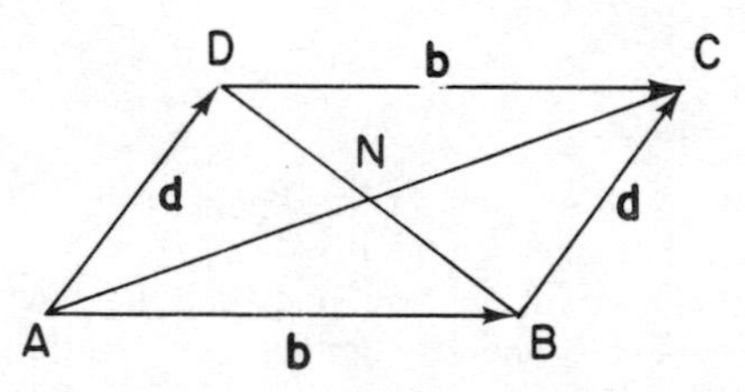

Figure 4.53

Let $\overrightarrow{AB} = \mathbf{b}$ and $\overrightarrow{AD} = \mathbf{d}$. We wish to show $\overrightarrow{BN} = \tfrac{1}{2}\overrightarrow{BD}$ and $\overrightarrow{AN} = \tfrac{1}{2}\overrightarrow{AC}$

Now

$$\overrightarrow{AC} = \mathbf{b} + \mathbf{d} \qquad \overrightarrow{BC} = \mathbf{d} \qquad \overrightarrow{DC} = \mathbf{b} \qquad \overrightarrow{BD} = \mathbf{d} - \mathbf{b}$$

But $\overrightarrow{BN} = \lambda\overrightarrow{BD}$ for some scalar λ; i.e. $\overrightarrow{BN} = \lambda(\mathbf{d} - \mathbf{b}) = \lambda\mathbf{d} - \lambda\mathbf{b}$
Also $\overrightarrow{AN} = \mu\overrightarrow{AC}$ for some scalar μ; i.e. $\overrightarrow{AN} = \mu(\mathbf{b} + \mathbf{d}) = \mu\mathbf{b} + \mu\mathbf{d}$
Also $\overrightarrow{AB} + \overrightarrow{BN} = \overrightarrow{AN}$, therefore $\mathbf{b} + \lambda\mathbf{d} - \lambda\mathbf{b} = \mu\mathbf{b} + \mu\mathbf{d}$. Comparing both sides, $\mu = \lambda$ and $1 - \lambda = \mu = \lambda$. Solving simultaneously we obtain $\lambda = \mu = \tfrac{1}{2}$. That is, the diagonals of a parallelogram bisect each other.

Example 3

To prove that the mid-points of the sides of any quadrilateral form a parallelogram. Refer to Figure 4.54.

Let E, F, G, H be the mid-points of the quadrilateral ABCD. Let $\overrightarrow{AB} = \mathbf{a}$, $\overrightarrow{BC} = \mathbf{b}$, $\overrightarrow{CD} = \mathbf{c}$, $\overrightarrow{DA} = \mathbf{d}$, where $\mathbf{a} + \mathbf{b} + \mathbf{c} + \mathbf{d} = \mathbf{0}$
Now

$$\overrightarrow{AE} = \tfrac{1}{2}\overrightarrow{AB} = \tfrac{1}{2}\mathbf{a};$$

therefore

$$\overrightarrow{EA} = -\tfrac{1}{2}\mathbf{a}$$

Also

$$\overrightarrow{HA} = \tfrac{1}{2}\mathbf{d}$$

therefore

$$\overrightarrow{AH} = -\tfrac{1}{2}\mathbf{d}$$

therefore

$$\overrightarrow{EH} = \overrightarrow{EA} + \overrightarrow{AH} = -\tfrac{1}{2}\mathbf{a} - \tfrac{1}{2}\mathbf{d}$$

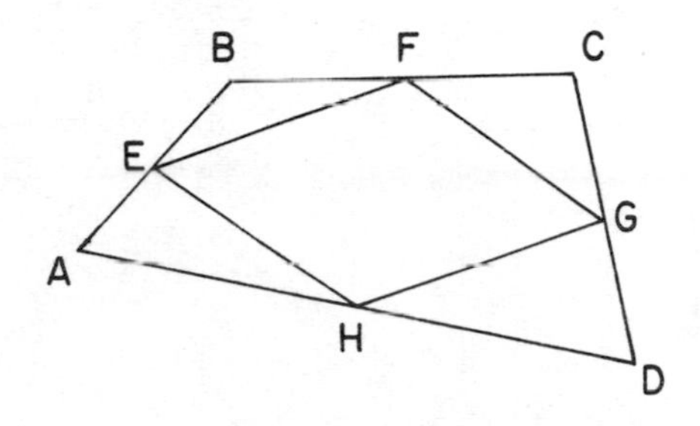

Figure 4.54

Further

$$\overrightarrow{FC} = \tfrac{1}{2}\mathbf{b} \quad \text{and} \quad \overrightarrow{CG} = \tfrac{1}{2}\mathbf{c}$$

therefore

$$\overrightarrow{FG} = \overrightarrow{FC} + \overrightarrow{CG} = \tfrac{1}{2}\mathbf{b} + \tfrac{1}{2}\mathbf{c}$$

But since

$$\mathbf{a} + \mathbf{b} + \mathbf{c} + \mathbf{d} = \mathbf{0}$$

$$\mathbf{b} + \mathbf{c} = \quad (\mathbf{a} + \mathbf{d})$$

Hence

$$\overrightarrow{FG} = \tfrac{1}{2}(\mathbf{b} + \mathbf{c}) = -\tfrac{1}{2}(\mathbf{a} + \mathbf{d}) = \overrightarrow{EH}$$

Since

$$\overrightarrow{FG} = \overrightarrow{EH}, \quad \text{FG} \parallel \text{EH and FG = EH.}$$

Similarly, $\overrightarrow{EF} = \overrightarrow{EB} + \overrightarrow{BF} = \tfrac{1}{2}\mathbf{a} + \tfrac{1}{2}\mathbf{b}$

$\overrightarrow{HG} = \overrightarrow{HD} + \overrightarrow{DG} = -\tfrac{1}{2}\mathbf{d} - \tfrac{1}{2}\mathbf{c} = \overrightarrow{EF}$

Hence HG = EF and HG $\parallel$ EF

The result is therefore proved.

Example 4

The equation of a straight line

(i) Suppose we are given a point on the line and the direction of the line [see Figure 4.55(i)].

Take an origin O and let A be the given point such that $\overrightarrow{OA} = \mathbf{a}$ and let the unit vector $\hat{\mathbf{u}}$ describe the direction of the line. We seek an equation which is satisfied by any point P on the line.
Now $\overrightarrow{AP} = t\hat{\mathbf{u}}$, t being a scalar which can vary, and $\overrightarrow{OP} = \overrightarrow{OA} + \overrightarrow{AP}$.
If $\overrightarrow{OP} = \mathbf{r}$, then

$$\mathbf{r} = \mathbf{a} + t\hat{\mathbf{u}}$$

and this is the desired equation. It gives for different values of t the value of $\mathbf{r}$ for all points on the line (e.g. $t = 0$ gives A).

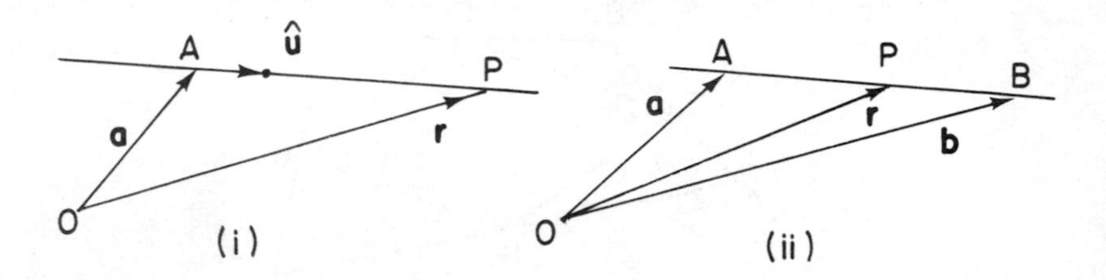

Figure 4.55

(ii) Suppose we are given two points on the line A and B . In Figure 4.55(ii) let $\overrightarrow{OA} = \mathbf{a}$ and $\overrightarrow{OB} = \mathbf{b}$, then $\overrightarrow{AB} = \mathbf{b} - \mathbf{a}$ and if P is any point on the line then $\overrightarrow{AP} = t(\mathbf{b} - \mathbf{a})$ for some value of t.
Now $\overrightarrow{OP} = \overrightarrow{OA} + \overrightarrow{AP}$, i.e. $\mathbf{r} = \mathbf{a} + t(\mathbf{b} - \mathbf{a})$, or

$$\mathbf{r} = (1 - t)\mathbf{a} + t\mathbf{b}$$

This is the required equation.

Note that $t = 0 \equiv$ A, $t = 1 \equiv$ B, $t < 0$ to left of A, $t > 0$ to right of B.

Example 5

Resultant of two forces.

Forces of magnitude 4 and 5 act along AB, AC respectively; find the resultant.

We refer to Figure 4.56.

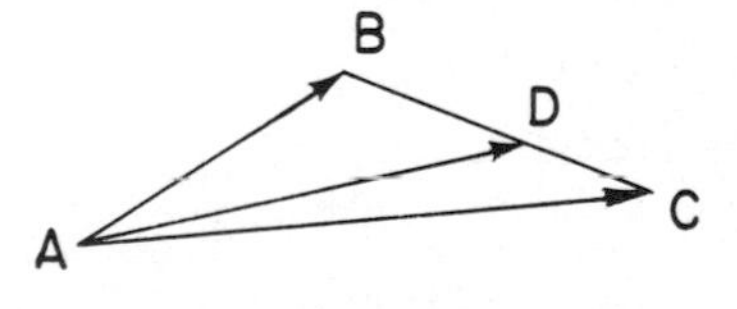

Figure 4.56

Let $\overrightarrow{AD}$ be the resultant direction.
Now $\qquad 4\overrightarrow{AB} = 4\overrightarrow{AD} + 4\overrightarrow{DB}$

and $\quad 5\overrightarrow{AC} = 5\overrightarrow{AD} + 5\overrightarrow{DC}$

therefore $\quad 4\overrightarrow{AB} + 5\overrightarrow{AC} = 9\overrightarrow{AD} + (4\overrightarrow{DB} + 5\overrightarrow{DC})$

Since the resultant acts along $\overrightarrow{AD}$, $4\overrightarrow{DB} + 5\overrightarrow{DC} = \mathbf{0}$. Therefore 4BD = 5DC, which locates D on the line BC.

The resultant force is of magnitude 9 acting along AD.

Example 6

Two forces act at the corner of a quadrilateral ABDC represented by $\overrightarrow{AB}$ and $\overrightarrow{AD}$; and two at C represented by $\overrightarrow{CB}$ and $\overrightarrow{CD}$. Show that their resultant is represented by $4\overrightarrow{PQ}$ where P and Q are the mid-points of AC and BD respectively.

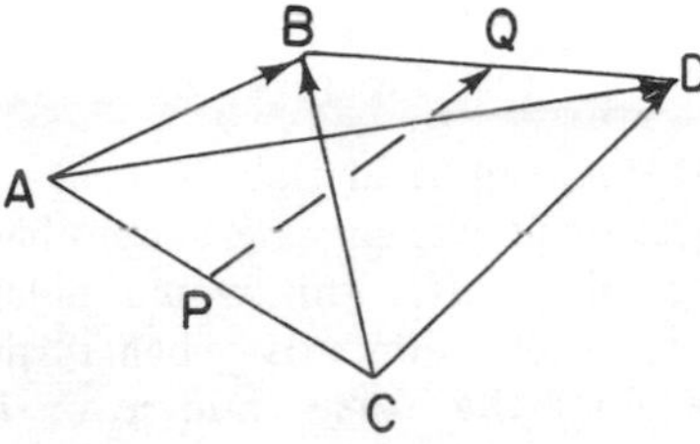

Figure 4.57

Referring to Figure 4.57 we see that the resultant equals

$$\begin{aligned}\overrightarrow{AB} + \overrightarrow{AD} + \overrightarrow{CB} + \overrightarrow{CD} &= (\overrightarrow{AP} + \overrightarrow{PQ} + \overrightarrow{QB}) + (\overrightarrow{AP} + \overrightarrow{PQ} + \overrightarrow{QD}) \\ &\quad + (\overrightarrow{CP} + \overrightarrow{PQ} + \overrightarrow{QB}) + (\overrightarrow{CP} + \overrightarrow{PQ} + \overrightarrow{QD}) \\ &= 2(\overrightarrow{AP} + \overrightarrow{CP}) + 2(\overrightarrow{QB} + \overrightarrow{QD}) + 4\overrightarrow{PQ}\end{aligned}$$

Now P is the mid-point of AC, therefore $\overrightarrow{AP} + \overrightarrow{CP} = 0$

Similarly, $\overrightarrow{QB} + \overrightarrow{QD} = 0$

Hence the result.

General components of a vector

From the definition of vector addition it is clear that any vector **F** can be written as the sum of other vectors, such as

$$\mathbf{F} = \mathbf{F}_1 + \mathbf{F}_2 + \ldots\ldots + \mathbf{F}_n$$

where the n vectors $\mathbf{F}_i$ and the vector $-\mathbf{F}$ when put head to tail form a closed polygon, as in Figure 4.58(i).

Notice this need not necessarily be a plane figure.

Further, $(n - 1)$ of the vectors $\mathbf{F}_i$ can be chosen arbitrarily but the nth vector must close the polygon.

$\mathbf{F}_1$, $\mathbf{F}_2$, ..., $\mathbf{F}_n$ are called **component vectors** of **F**. The most useful decomposition of a vector is into 3 components parallel to the 3 orthogonal axes of cartesian coordinates OX, OY and OZ, as in Figure 4.58(ii).

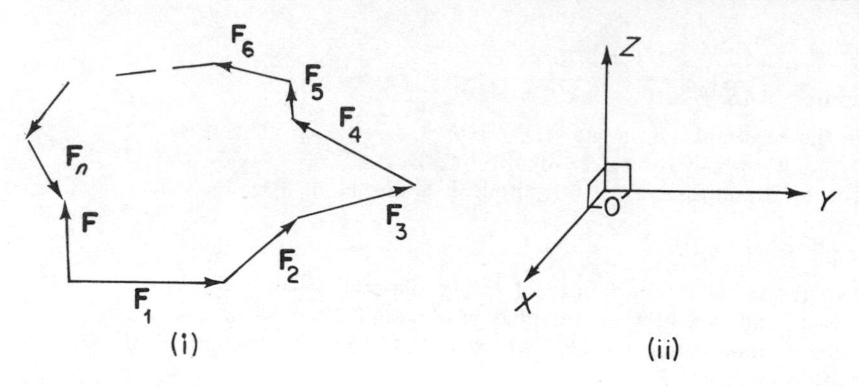

Figure 4.58

This figure shows the conventional right-handed system of coordinate axes – right-handed because if we imagine a right-handed screw placed at the origin pointing along the X-axis, this would advance along the X-axis when rotating OY towards OZ. Similarly when turning OZ to OX we would advance a screw along the Y-axis, and so on for the cyclic pattern

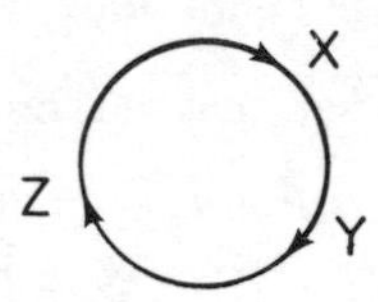

If now we take a vector **F** as $\overrightarrow{O'P}$ and draw lines through O′ parallel to the rectangular cartesian axes, we can complete the rectangular box having O′P as diagonal; this is shown in Figure 4.59.

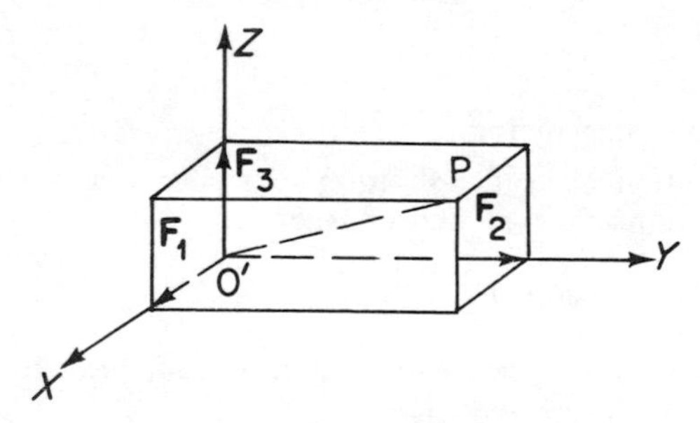

Figure 4.59

The intercepts on the lines O′X, O′Y and O′Z are $\mathbf{F}_1$, $\mathbf{F}_2$ and $\mathbf{F}_3$ respectively in the direction of the axes. Then

$$\mathbf{F} = \mathbf{F}_1 + \mathbf{F}_2 + \mathbf{F}_3$$

and $\mathbf{F}_1$, $\mathbf{F}_2$ and $\mathbf{F}_3$ are the **cartesian components** of the vector $\mathbf{F}$. Usually unit vectors in the directions of the X, Y, Z axes are denoted by **i, j, k** respectively.

Now $\mathbf{F}_1 = F_1\mathbf{i}$, $\mathbf{F}_2 = F_2\mathbf{j}$, $\mathbf{F}_3 = F_3\mathbf{k}$, so that

$$\mathbf{F} = F_1\mathbf{i} + F_2\mathbf{j} + F_3\mathbf{k}$$

We very often refer to F_1, F_2, F_3 as the first, second and third scalar components of the vector $\mathbf{F}$.

The *magnitude* of the vector $\mathbf{F}$ is easily shown by Pythagoras' Theorem to be $\sqrt{F_1^2 + F_2^2 + F_3^2}$. That is,

$$O'P = |\mathbf{F}| = F = \sqrt{F_1^2 + F_2^2 + F_3^2}$$

The vectors **i**, **j** and **k** are said to be **linearly independent** (this means that no one of the quantities **i**, **j**, **k** can be expressed as a *linear combination* of the other two). Thus, for addition and subtraction of vectors we have

$$\text{if} \quad \begin{cases} \mathbf{A} = A_1\mathbf{i} + A_2\mathbf{j} + A_3\mathbf{k} \\ \mathbf{B} = B_1\mathbf{i} + B_2\mathbf{j} + B_3\mathbf{k} \end{cases}$$

$$\text{then} \quad \mathbf{A} \pm \mathbf{B} = (A_1 \pm B_1)\mathbf{i} + (A_2 \pm B_2)\mathbf{j} + (A_3 \pm B_3)\mathbf{k} \qquad (4.33)$$

In other words, we may just add or subtract the respective components, which ties in with our ideas of two-dimensional vectors.

Example

If $\mathbf{A} = 3\mathbf{i} - 2\mathbf{j} + \mathbf{k}$, $\mathbf{B} = -\mathbf{i} + 2\mathbf{j} - \mathbf{k}$
then

$$\mathbf{A} + \mathbf{B} = 2\mathbf{i}$$

$$\mathbf{A} - \mathbf{B} = 4\mathbf{i} - 4\mathbf{j} + 2\mathbf{k} \qquad \mathbf{A} + 3\mathbf{B} = 4\mathbf{j} - 2\mathbf{k}$$

As an additional consequence, if two vectors are equal then the respective scalar components must be equal.

For example, if we have $\mathbf{A} = x\mathbf{i} + y\mathbf{j} + z\mathbf{k}$ and $\mathbf{A} = \mathbf{B} + \mathbf{C}$ where $\mathbf{B} = 2\mathbf{i} - \mathbf{j} + \mathbf{k}$ and $\mathbf{C} = 3\mathbf{i} + 5\mathbf{j} - \mathbf{k}$, then

$$x = 2 + 3 = 5, \quad y = -1 + 5 = 4, \quad z = 1 - 1 = 0$$

Position vector

A special case of the above component form is where we have a point P in space, with coordinates (x, y, z) referred to rectangular cartesian axes, as in Figure 4.60.

Then $\overrightarrow{OP}$ is the **position vector** of the point P, usually denoted by **r**.

The lengths along the coordinate axes are $OA = x$, $OB = y$, $OC = z$ so that $\mathbf{r} = x\mathbf{i} + y\mathbf{j} + z\mathbf{k}$.

The distance of this point from the origin is the length of OP and is

$$r = \sqrt{x^2 + y^2 + z^2}$$

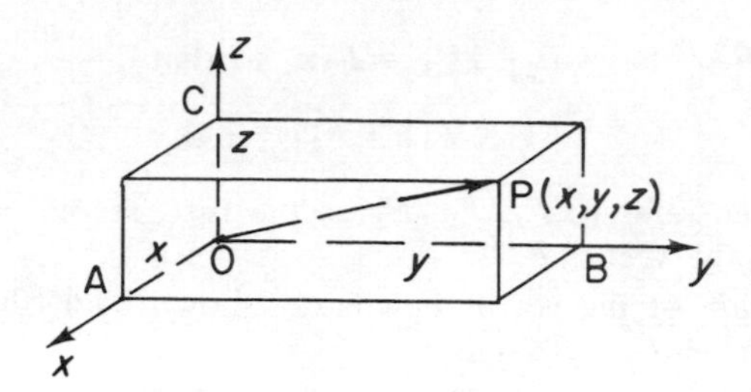

Figure 4.60

Direction cosines and direction ratios

Suppose a vector $\mathbf{F} = \overrightarrow{O'P}$, where $\mathbf{F} = F_1\mathbf{i} + F_2\mathbf{j} + F_3\mathbf{k}$.

The direction of O′P can be specified by the angles it makes with the O′X, O′Y, O′Z directions. These are usually designated by α, β, γ as shown.

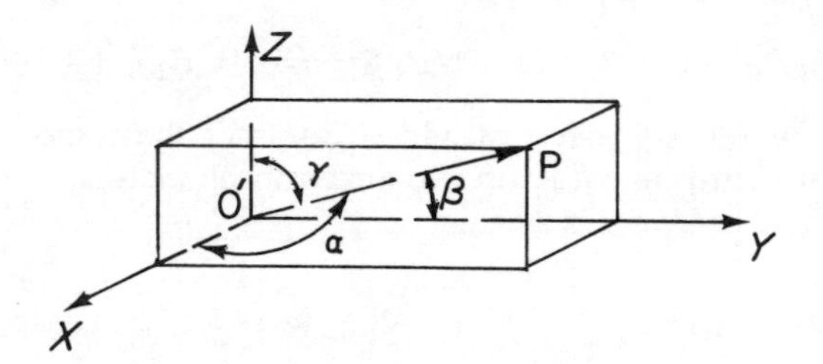

Figure 4.61

Now

$$\begin{aligned}\mathbf{F} &= F_1\mathbf{i} + F_2\mathbf{j} + F_3\mathbf{k} \\ &= \mathrm{O'P}\cos\alpha\mathbf{i} + \mathrm{O'P}\cos\beta\mathbf{j} + \mathrm{O'P}\cos\gamma\mathbf{k} \\ &= \mathrm{O'P}(\cos\alpha\mathbf{i} + \cos\beta\mathbf{j} + \cos\gamma\mathbf{k}) \qquad (4.34)\end{aligned}$$

The quantities $\cos\alpha$, $\cos\beta$, $\cos\gamma$ are called the **direction cosines** of the vector **F**.

We can see from equation (4.34) that, since O′P is the length of the vector **F**, $(\cos\alpha\mathbf{i} + \cos\beta\mathbf{j} + \cos\gamma\mathbf{k})$ must be a unit vector. That is, it has unit length so that

$$\cos^2\alpha + \cos^2\beta + \cos^2\gamma = 1$$

Notice that $\cos\alpha = F_1/F$, $\cos\beta = F_2/F$, $\cos\gamma = F_3/F$

Sometimes it is convenient to use quantities which are proportional to the direction cosines, but which do not have a sum of squares equal to 1. That is, we have three quantities $[l,m,n]$ which are such that

$$\frac{l}{\cos\alpha} = \frac{m}{\cos\beta} = \frac{n}{\cos\gamma}$$

They are called **direction ratios** and are convenient to use since they can be taken as integers. Indeed $[F_1, F_2, F_3]$ are direction ratios of O′P. We can easily convert them into direction cosines when needed because if we let

$$\frac{l}{\cos\alpha} = \frac{m}{\cos\beta} = \frac{n}{\cos\gamma} = \lambda$$

then

$$\cos\alpha = \frac{l}{\lambda}, \quad \cos\beta = \frac{m}{\lambda}, \quad \cos\gamma = \frac{n}{\lambda}$$

But

$$\cos^2\alpha + \cos^2\beta + \cos^2\gamma = 1$$

Hence

$$l^2 + m^2 + n^2 = \lambda^2$$

and the direction cosines are

$$\frac{l}{\sqrt{l^2 + m^2 + n^2}}, \quad \frac{m}{\sqrt{l^2 + m^2 + n^2}}, \quad \frac{n}{\sqrt{l^2 + m^2 + n^2}} \tag{4.35}$$

Example

Find the angles made with the coordinate directions by the line joining the points P(1,1,1) and Q(3,−2,4) in Figure 4.62

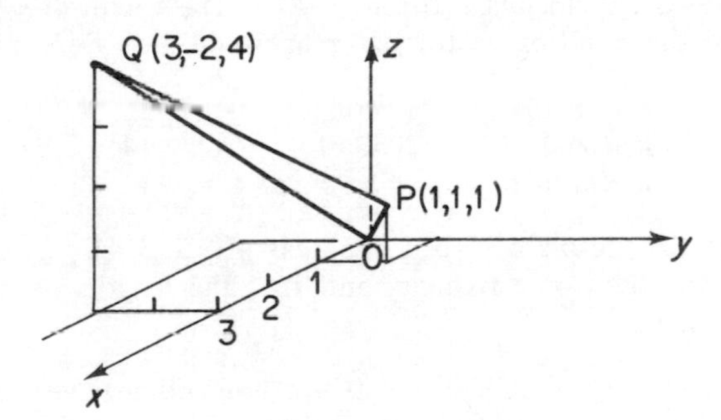

Figure 4.62

Now

$$\overrightarrow{OP} = \mathbf{i} + \mathbf{j} + \mathbf{k}$$
$$\overrightarrow{OQ} = 3\mathbf{i} - 2\mathbf{j} + 4\mathbf{k}$$
$$\overrightarrow{PQ} = \overrightarrow{OQ} - \overrightarrow{OP} = 2\mathbf{i} - 3\mathbf{j} + 3\mathbf{k}$$

therefore

$$PQ = \sqrt{2^2 + 3^2 + 3^2} = \sqrt{22}$$

hence

$$\overrightarrow{PQ} = PQ\left(\frac{2}{\sqrt{22}}\mathbf{i} - \frac{3}{\sqrt{22}}\mathbf{j} + \frac{3}{\sqrt{22}}\mathbf{k}\right)$$

The direction cosines are $\cos\alpha = \frac{2}{\sqrt{22}}$, $\cos\beta = -\frac{3}{\sqrt{22}}$, $\cos\gamma = \frac{3}{\sqrt{22}}$

Hence

$$\alpha = 64°17', \quad \beta = 129°46', \quad \gamma = 50°14'$$

Notice that we can say

$$\overrightarrow{PQ} = 2\mathbf{i} - 3\mathbf{j} + 3\mathbf{k}$$

Its direction ratios are [2, −3, 3].
Hence the direction cosines are

$$\left(\frac{2}{\sqrt{2^2 + 3^2 + 3^2}}, \frac{-3}{\sqrt{2^2 + 3^2 + 3^2}}, \frac{3}{\sqrt{2^2 + 3^2 + 3^2}}\right)$$

Problems I

1. If $\mathbf{r} = (2, -1, -1)$, $\mathbf{s} = (2, 1, 1)$, $\mathbf{t} = (-1, -1, 3)$
   (a) find the vector $\mathbf{u}$ such that $\mathbf{r} + 2\mathbf{s} + \mathbf{u} = (-1, -1, 3)$
   (b) find $|\mathbf{v}|$ if $2\mathbf{v} + 4\mathbf{r} + \mathbf{s} - 2\mathbf{t} = (1, 0, -3)$

2. Find a unit vector parallel to the resultant of the vectors $\mathbf{a} = (2, 4, -5)$ and $\mathbf{b} = (1, 2, 3)$.

3. The vector $(-2\mathbf{i} + 2\sqrt{3}\,\mathbf{j})$ in the (x, y) plane is multiplied by the scalar λ and rotated anticlockwise through 90°. The vector $(\mathbf{i} + \mathbf{j})$ is then added and the resulting vector has magnitude 4λ. Determine the value of λ.

4. If O is the origin and $\overrightarrow{OP} = (2, 3, -1)$, $\overrightarrow{OQ} = (4, -3, 2)$, find $\overrightarrow{PQ}$ and determine its magnitude.

5. Prove that the vectors $\mathbf{a} = (3, 1, -2)$, $\mathbf{b} = (-1, 3, 4)$, $\mathbf{c} = (4, -2, -6)$ can form the sides of a triangle, and find the lengths of the medians of this triangle.

6. Show geometrically that if $\mathbf{a}$ and $\mathbf{b}$ are non-collinear vectors in a certain plane, any other vector $\mathbf{c}$ in that plane can be represented as $(m\mathbf{a} + n\mathbf{b})$ for certain values of m and n.
   If $\mathbf{a} = (2\mathbf{i} - 5\mathbf{j})$, $\mathbf{b} = (-6\mathbf{i} + 4\mathbf{j})$ and $\mathbf{c} = (11\mathbf{i} - \mathbf{j})$, find m and n.

7. P, Q, R are the mid-points of the sides BC, CA, AB of the triangle ABC, and O is any other point. Show that

$$\overrightarrow{OP} + \overrightarrow{OQ} + \overrightarrow{OR} = \overrightarrow{OA} + \overrightarrow{OB} + \overrightarrow{OC}$$

Products of vectors – scalar product

Having defined vectors we should like to extend the algebra to include the products of vector quantities. Clearly, since vectors have both direction and magnitude, the mere multiplication of scalar magnitudes cannot be sufficient. It is not at all evident what form the product of vectors should take but if we look to physical problems we do find two particular kinds of multiplication occurring time and again. The first one occurs frequently in physics and gives a form AB cos θ for the combination of two vectors **A** and **B** where θ is the angle between them. For instance, the work done by a force **F** in moving a body through a displacement **d** is Fd cos θ and this is a scalar quantity: see Figure 4.63.

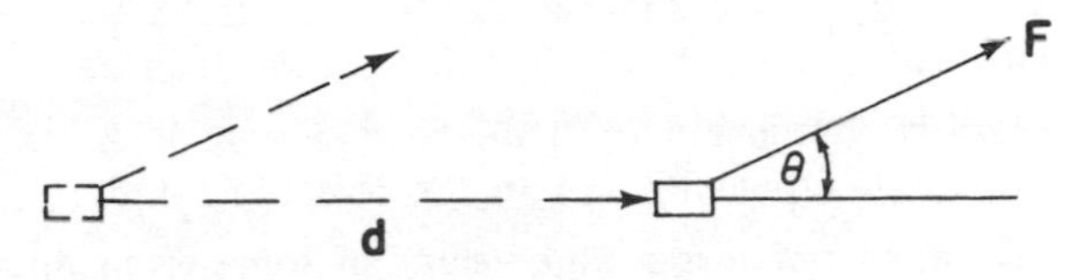

Figure 4.63

You have probably learnt that the work done is the displacement times the projection of the force along the direction of movement (the component of the force perpendicular to the direction of movement does no work).

Thus we are led to define the **scalar product** of two vectors in the following form.

Given two vectors **A** and **B**, θ being the angle between their directions $(0 \leqslant \theta \leqslant 180^\circ)$, then the *scalar product* of **A** and **B** is AB cos θ and is written **A . B** (read 'A dot B').

Now **A . B** = AB cos θ is a scalar quantity. Clearly **B . A** = BA cos θ = **A . B** and the product is **commutative**.

We have drawn the two vectors in Figure 4.64 as if they had a common point O. If they are in different planes then θ is the angle between two vectors parallel to **A** and **B** respectively drawn through any specified point.

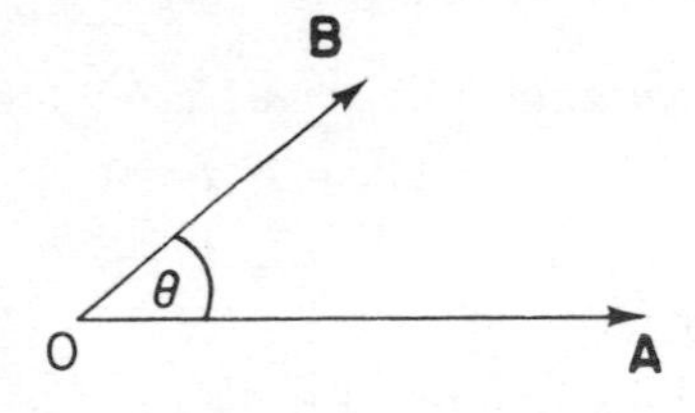

Figure 4.64

Properties of the scalar product

1. By definition, if the two vectors are *perpendicular* then $\theta = 90^\circ$ and $\cos 90^\circ = 0$ so that **A . B** = 0. This is the first major difference between vector and scalar algebra, for in scalars we have:

If $ab = 0$, then either $a = 0$ or $b = 0$.

However, we have for vectors:

If $\mathbf{A}.\mathbf{B} = 0$, then either $\mathbf{A} = 0$ or $\mathbf{B} = 0$ or $\mathbf{A}$ is perpendicular to $\mathbf{B}$. (This is consistent with the fact that work done by a force perpendicular to a displacement is zero.)

2. If **A** and **B** are parallel, then $\cos\theta = 1$ and $\mathbf{A}.\mathbf{B} = AB$. In particular if $\mathbf{B} = \mathbf{A}$, then

$$\mathbf{A}.\mathbf{A} = A^2 = (\text{magnitude of } \mathbf{A})^2$$

(Note: this may be used in reverse to find the magnitude of a vector.)

3. $\mathbf{A}.\mathbf{B} = A(B\cos\theta) = (A\cos\theta)B.$
From Figure 4.65(i) we can see that this product can be interpreted as

$A \times$ component of **B** in the direction of **A**, or
$B \times$ component of **A** in the direction of **B**.

Note, therefore, that if **n** is a unit vector in some given direction, then $\mathbf{A}.\mathbf{n} = A.1\cos\theta = A\cos\theta$ = component of **A** in the direction of **n**.

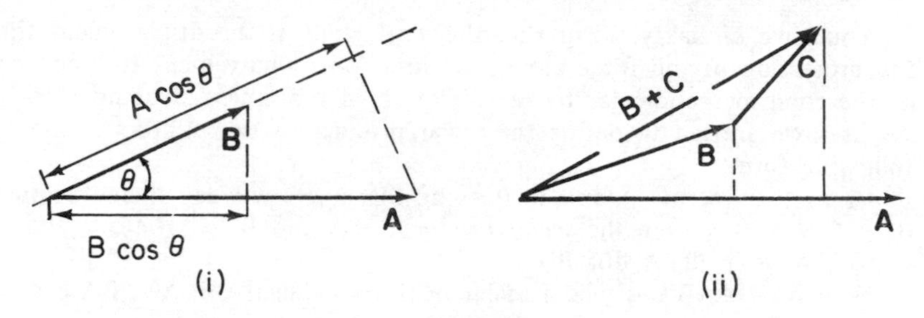

Figure 4.65

4. $\mathbf{A}.(\mathbf{B} + \mathbf{C}) = \mathbf{A}.\mathbf{B} + \mathbf{A}.\mathbf{C}$

Try to prove this by using Property 3 and considering Figure 4.65(ii).

5. Since **i**, **j**, **k** are *mutually orthogonal* vectors, then we have

$$\mathbf{i}.\mathbf{j} = \mathbf{j}.\mathbf{k} = \mathbf{k}.\mathbf{i} = 0$$

$$\mathbf{i}.\mathbf{i} = \mathbf{j}.\mathbf{j} = \mathbf{k}.\mathbf{k} = 1$$

Using these results we have

$$\mathbf{A}.\mathbf{B} = (A_1\mathbf{i} + A_2\mathbf{j} + A_3\mathbf{k}).(B_1\mathbf{i} + B_2\mathbf{j} + B_3\mathbf{k})$$

i.e. $$\mathbf{A}.\mathbf{B} = A_1B_1 + A_2B_2 + A_3B_3 \tag{4.36}$$

Notice that effectively only like components are multiplied. For example

$$(2\mathbf{i} - 3\mathbf{j} + \mathbf{k}).(4\mathbf{i} + 2\mathbf{j} - 2\mathbf{k}) = (2).(4) + (-3).(2) + (1).(-2) = 0$$

Neither of these vectors is zero. Hence they are perpendicular

Notice further that $\mathbf{A}.\mathbf{A} = A^2 = A_1^2 + A_2^2 + A_3^2$ which is consistent with our finding of the magnitude of **A**.

Notice that we can use the scalar product to find the angle between two vectors, for if we are given two vectors **A** and **B** then we can find $\mathbf{A}.\mathbf{B}$ But $\mathbf{A}.\mathbf{B} = |\mathbf{A}||\mathbf{B}| \cos\theta$ and since we can also find $|\mathbf{A}|$ and $|\mathbf{B}|$ then $\cos\theta$ is easily found giving θ the angle between the vectors.

Example

Find the angle between the vectors $\mathbf{A} = \mathbf{i} - \mathbf{j} - \mathbf{k}$ and $\mathbf{B} = 2\mathbf{i} + \mathbf{j} + 2\mathbf{k}$.
Now

$$\mathbf{A}.\mathbf{B} = 1.2 - 1.1 - 1.2 = -1$$
$$|\mathbf{A}| = \sqrt{1^2 + 1^2 + 1^2} = \sqrt{3}$$
$$|\mathbf{B}| = \sqrt{2^2 + 1^2 + 2^2} = 3$$

But $\mathbf{A}.\mathbf{B} = |\mathbf{A}||\mathbf{B}| \cos\theta$

Hence

$$\cos\theta = \frac{\mathbf{A}.\mathbf{B}}{|\mathbf{A}||\mathbf{B}|} = \frac{-1}{3\sqrt{3}}$$

Hence we can find θ as approximately $101°6'$.

Geometrical applications of scalar products

(i) *Pythagoras' Theorem*
From Figure 4.66(i) $\mathbf{a}.\mathbf{b} = 0$ and $\mathbf{a} + \mathbf{b} = \mathbf{c}$.
Now

$$\mathbf{c}.\mathbf{c} = \mathbf{c}.(\mathbf{a} + \mathbf{b}) = (\mathbf{a} + \mathbf{b}).(\mathbf{a} + \mathbf{b})$$
$$= \mathbf{a}.\mathbf{a} + \mathbf{b}.\mathbf{a} + \mathbf{a}.\mathbf{b} + \mathbf{b}.\mathbf{b}$$

therefore

$$c^2 = a^2 + b^2$$

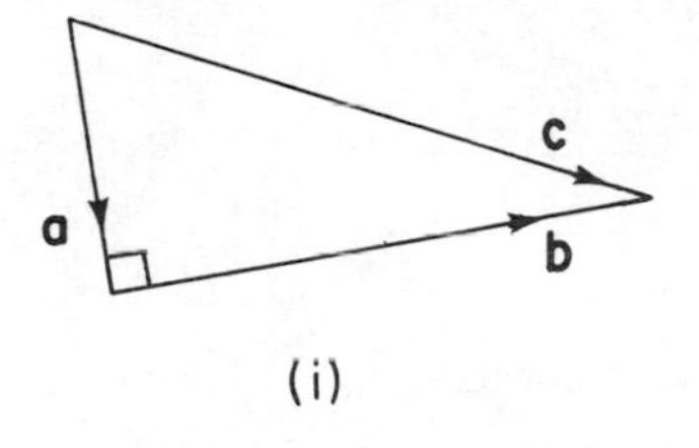

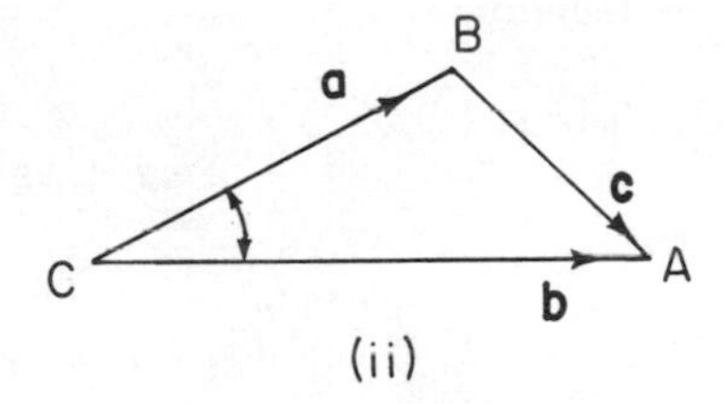

Figure 4.66

(ii) *Cosine Rule*
From Figure 4.66(ii) $\mathbf{b} = \mathbf{a} + \mathbf{c}$
therefore

$$\mathbf{c} = \mathbf{b} - \mathbf{a}$$

therefore

$$\mathbf{c}\,.\,\mathbf{c} = (\mathbf{b} - \mathbf{a})\,.\,(\mathbf{b} - \mathbf{a})$$

i.e.

$$c^2 = \mathbf{b}\,.\,\mathbf{b} - \mathbf{a}\,.\,\mathbf{b} - \mathbf{a}\,.\,\mathbf{b} + \mathbf{a}\,.\,\mathbf{a}$$
$$= b^2 + a^2 - 2ab \cos C$$

(iii) *The diagonals of a rhombus are perpendicular*
From Figure 4.67(i) $\mathbf{AC} = \mathbf{a} + \mathbf{b}$, $\mathbf{BD} = \mathbf{a} - \mathbf{b}$
therefore

$$\mathbf{AC}\,.\,\mathbf{BD} = (\mathbf{a} + \mathbf{b})\,.\,(\mathbf{a} - \mathbf{b}) = a^2 - b^2 = 0$$
since $a = b$ for a rhombus

Therefore

AC is perpendicular to BD.

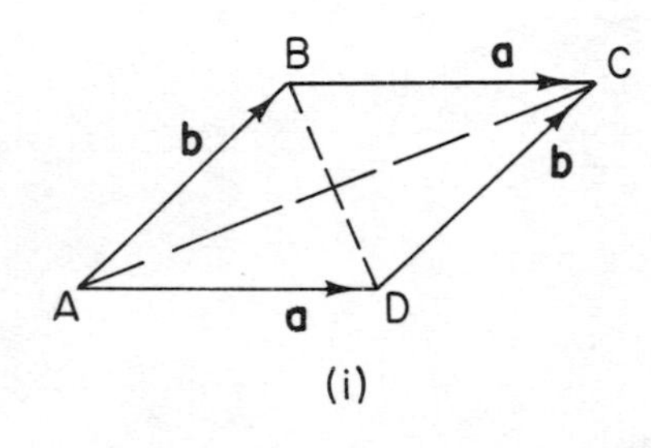

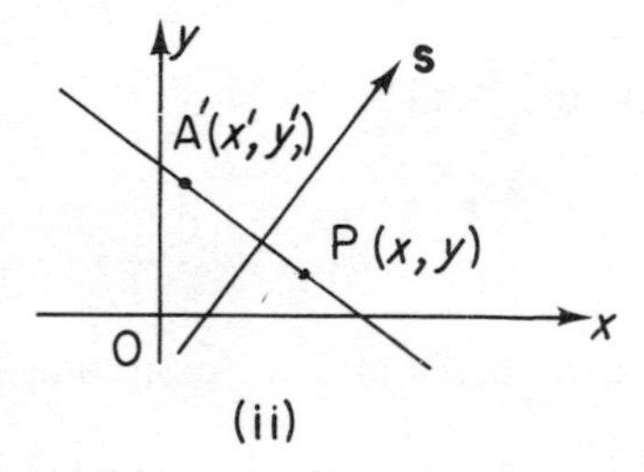

Figure 4.67

(iv) *Line perpendicular to a given line*
Let $\mathbf{s} = a\mathbf{i} + b\mathbf{j}$ be the given vector.
Let the required line pass through a point A′: Figure 4.67(ii).

$$\mathbf{A'P} = (x - x')\mathbf{i} + (y - y')\mathbf{j}$$

We require

$$\mathbf{s}\,.\,\mathbf{A'P} = 0$$

therefore

$$(x - x')\,a + (y - y')b = 0$$

hence

$$ax + by = ax' + by'$$

i.e.

$$ax + by = c$$

where $c = ax' + by' =$ constant

This is the equation of the line perpendicular to $\mathbf{s} = a\mathbf{i} + b\mathbf{j}$

(v) *Angle between two straight lines*
This angle is the same as the angle between the perpendiculars to the lines. If the lines are

$$x + 2y + 3 = 0$$
$$3x + 2y - 3 = 0$$

the perpendiculars are $\mathbf{P}_1 = \mathbf{i} + 2\mathbf{j}$; $\mathbf{P}_2 = 3\mathbf{i} + 2\mathbf{j}$ [using (iv)]
Then

$$\cos\theta = \frac{\mathbf{P}_1 \,.\, \mathbf{P}_2}{|\mathbf{P}_1|\,.\,|\mathbf{P}_2|} = \frac{7}{\sqrt{5}\,.\sqrt{13}}$$

(vi) *Perpendicular distance from a given point to a given line*
For example, we go from $\mathbf{A} \equiv (3, 1)$ to $3x + 2y - 6 = 0$ (see Figure 4.68).
The perpendicular to the line is $\hat{\mathbf{p}} = (3\mathbf{i} + 2\mathbf{j})/|3\mathbf{i} + 2\mathbf{j}|$
therefore

$$\hat{\mathbf{p}} = \frac{1}{\sqrt{13}}(3\mathbf{i} + 2\mathbf{j})$$

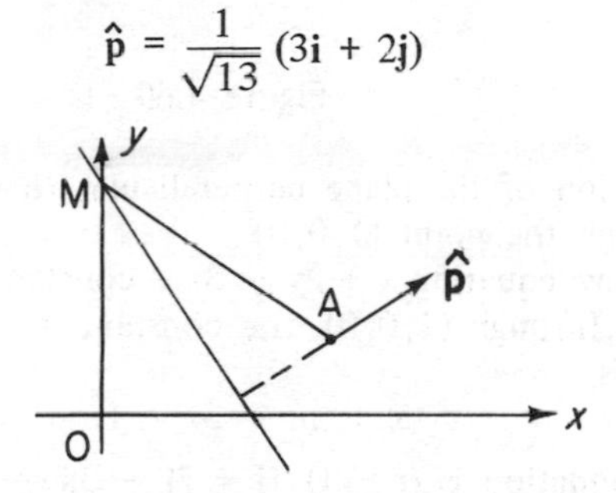

Figure 4.68

Take any point on the line, for example $M \equiv (0, 3)$, then

$$\overrightarrow{AM} = -3\mathbf{i} + 2\mathbf{j} \quad \text{and} \quad |\overrightarrow{AM}\,.\,\hat{\mathbf{p}}| = |-5(\sqrt{13}/13)| = 5/\sqrt{13}$$

is the required distance.

(vii) *Equation of a plane*
Suppose we are given a normal direction to the plane and a point B in the plane ($\overrightarrow{OB} = \mathbf{b}$) [see Figure 4.69(i)]. Let $\mathbf{a}$ be perpendicular to the plane and let P be *any* point in the plane ($\overrightarrow{OP} = \mathbf{r}$).
Now **BP** lies in the plane and is perpendicular to **a**.
Therefore

$$(\mathbf{r} - \mathbf{b})\,.\,\mathbf{a} = 0$$

If $\mathbf{r} = x\mathbf{i} + y\mathbf{j} + z\mathbf{k}$, $\mathbf{b} = b_1\mathbf{i} + b_2\mathbf{j} + b_3\mathbf{k}$ and $\mathbf{a} = a_1\mathbf{i} + a_2\mathbf{j} + a_3\mathbf{k}$, then we get

$$(x - b_1)a_1 + (y - b_2)a_2 + (z - b_3)a_3 = 0$$

i.e.

$$a_1x + a_2y + a_3z = a_1b_1 + a_2b_2 + a_3b_3 = \text{constant}$$

Notice that the normal direction (a_1, a_2, a_3) constitutes the coefficients of x, y and z in this equation.

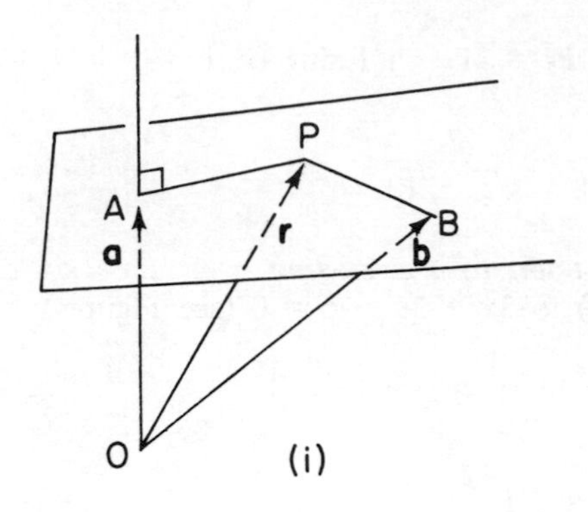

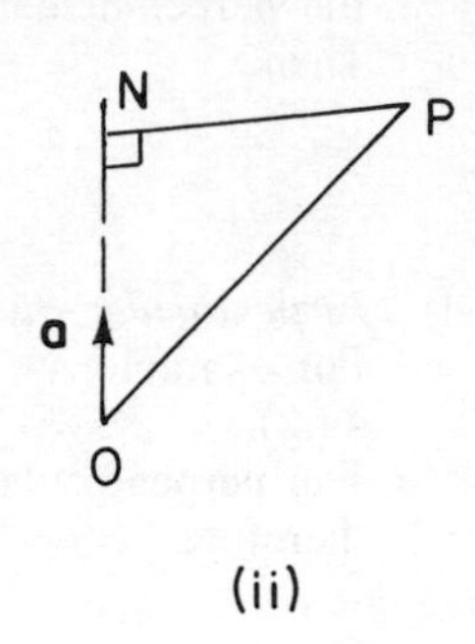

Figure 4.69

Example 1

Find the equation of the plane perpendicular to the direction $\mathbf{i} + 2\mathbf{j} - 3\mathbf{k}$ and passing through the point $(1, 0, 0)$.
The plane must have equation $x + 2y - 3z = \text{constant}$.

Since it passes through $(1, 0, 0)$, the constant $= 1 + 2.0 - 3.0 = 1$.
Hence the plane is

$$x + 2y - 3z = 1$$

Alternatively its equation is $(\mathbf{r} - \mathbf{i}).(\mathbf{i} + 2\mathbf{j} - 3\mathbf{k}) = 0$
i.e.

$$(x - 1) + (y - 0).2 + (z - 0)(-3) = 0$$

as before.

Example 2

Given the plane $2x + 3y + 2z = 4$, find a unit vector perpendicular to it. A perpendicular to the plane is $\mathbf{a} = 2\mathbf{i} + 3\mathbf{j} + 2\mathbf{k}$. Now $|\mathbf{a}| = \sqrt{17}$.
Hence the unit vector perpendicular to the plane is

$$\hat{\mathbf{a}} = \frac{2}{\sqrt{17}}\mathbf{i} + \frac{3}{\sqrt{17}}\mathbf{j} + \frac{2}{\sqrt{17}}\mathbf{k}$$

Let us find the perpendicular distance to the plane from O; refer to Figure 4.69(ii).
Take any point P in the plane – say $(2, 0, 0)$

Then $$\overrightarrow{OP} = 2\mathbf{i}$$

Perpendicular distance from O = ON = $\overrightarrow{OP}.\hat{\mathbf{a}} = 2\mathbf{i}.(2\mathbf{i} + 3\mathbf{j} + 2\mathbf{k})/\sqrt{17} = 4/\sqrt{17}$.

Problems II

1. Determine a unit vector perpendicular to the plane containing the vectors $2\mathbf{i} + \mathbf{j} - 3\mathbf{k}$, $\mathbf{i} - 2\mathbf{j} + \mathbf{k}$.

2. A force has components in the x and y directions respectively of 2 lb.wt. and 5 lb.wt. Its point of application moves from $(1, 5)$ to $(6, 8)$. Express the work done as a dot product and evaluate it, the unit of displacement being 1 ft.

3. Show that the vectors **a** = 3**i** − 2**j** + **k**, **b** = **i** − 3**j** + 5**k**, **c** = 2**i** + **j** − 4**k** form a right-angled triangle.

4. Find the work done in moving an object along the vector = (3, 2, 15) if the applied force is **F** = (2, −1, −1).

5. Show that the vector **a** = 5**i** − 20**j** + 35**k** is perpendicular to the plane containing the vectors **b** = **i** + 2**j** + **k** and **c** = 3**i** − **j** − **k**.

The vector product

We now look for a product of two vectors which results in a *vector* quantity.

We again look to physical situations to decide on the form that should be taken by this product. For instance, consider a force **F** acting at one end, Q, of a very light rod which is freely pivoted at its other end, P, as in Figure 4.70.

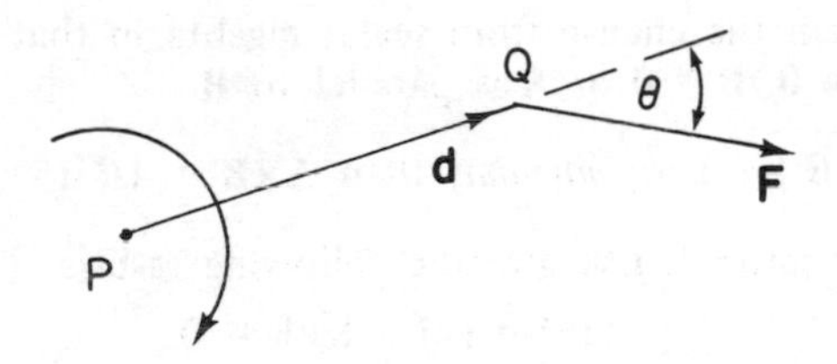

Figure 4.70

Let $\overrightarrow{PQ}$ = **d**

The moment of the force **F** about P will have magnitude $Fd \sin \theta$. The moment also has a direction since it is trying to turn the rod about an axis through P perpendicular to the plane of the rod and to the force **F** (the sense of rotation is shown about P by the curved arrow).

We therefore have the product of two vectors giving a vectorial result and are led to the following definition:

Given two vectors **A** and **B** with an included angle θ then the **vector product** of **A** and **B** is written **A**×**B** (sometimes **A**∧**B**) and is such that

(i) **A**×**B** is perpendicular to both **A** and **B**
(ii) the magnitude of **A**×**B** = $AB \sin \theta$
(iii) a *right-handed* rotation about the vector **A**×**B** through an angle θ would move **A** to the same direction as **B** (see Figure 4.71).

We can write **A**×**B** = $AB \sin \theta\, \hat{\mathbf{n}}$ where $\hat{\mathbf{n}}$ is a unit vector in the direction of **A**×**B**.

Properties of the vector product

1. By the definition, **B**×**A** = $BA \sin \theta\, (-\hat{\mathbf{n}})$ since the rotation is from **B**

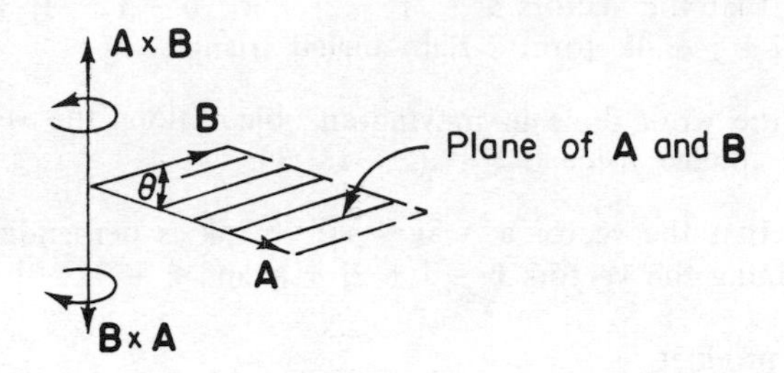

Figure 4.71

to **A** through an angle θ. Hence $\mathbf{B}\times\mathbf{A} = -\mathbf{A}\times\mathbf{B}$. Thus the vector product is not commutative and this vector product does not follow the usual rule of scalar algebra.

2. If **A** and **B** are *parallel*, then $\theta = 0$ and $\sin\theta = 0 \Rightarrow \mathbf{A}\times\mathbf{B} = 0$. Notice again the change from scalar algebra in that if $\mathbf{A}\times\mathbf{B} = 0$, then either $\mathbf{A} = \mathbf{0}$, $\mathbf{B} = \mathbf{0}$ or **A** is parallel to **B**.

3. If **A** and **B** are *perpendicular*, then $\mathbf{A}\times\mathbf{B} = AB\,\hat{\mathbf{n}}$

4. The unit vectors **i**, **j**, **k** give the following results

$$\mathbf{i}\times\mathbf{i} = \mathbf{j}\times\mathbf{j} = \mathbf{k}\times\mathbf{k} = \mathbf{0}$$

$$\mathbf{i}\times\mathbf{j} = \mathbf{k}, \quad \mathbf{j}\times\mathbf{k} = \mathbf{i}, \quad \mathbf{k}\times\mathbf{i} = \mathbf{j}$$

The last products, where the unit vectors are in cyclic order, are positive, but of course, if there is any change from cyclic order, a minus sign is introduced, as follows

$$\mathbf{i}\times\mathbf{k} = -\mathbf{j}, \quad \mathbf{k}\times\mathbf{j} = -\mathbf{i}, \quad \mathbf{j}\times\mathbf{i} = -\mathbf{k}$$

5. $\mathbf{A}\times(\mathbf{B}+\mathbf{C}) = \mathbf{A}\times\mathbf{B} + \mathbf{A}\times\mathbf{C}$
The proof of this law is quite difficult and we shall not give it here.

6. If we write **A** and **B** in component form, we get

$$\mathbf{A}\times\mathbf{B} = (A_1\mathbf{i} + A_2\mathbf{j} + A_3\mathbf{k})\times(B_1\mathbf{i} + B_2\mathbf{j} + B_3\mathbf{k})$$

Using the results of (4) and (5) we find

$$\begin{aligned}\mathbf{A}\times\mathbf{B} = {} & \cancel{A_1B_1\,\mathbf{i}\times\mathbf{i}} + A_1B_2\,\mathbf{i}\times\mathbf{j} + A_1B_3\,\mathbf{i}\times\mathbf{k} \\ & + A_2B_1\,\mathbf{j}\times\mathbf{i} + \cancel{A_2B_2\,\mathbf{j}\times\mathbf{j}} + A_2B_3\,\mathbf{j}\times\mathbf{k} \\ & + A_3B_1\,\mathbf{k}\times\mathbf{i} + A_3B_2\,\mathbf{k}\times\mathbf{j} + \cancel{A_3B_3\,\mathbf{k}\times\mathbf{k}}\end{aligned}$$

That is

$$\mathbf{A}\times\mathbf{B} = (A_2B_3 - A_3B_2)\mathbf{i} + (A_3B_1 - A_1B_3)\mathbf{j} + (A_1B_2 - A_2B_1)\mathbf{k}$$

This result can be written most conveniently in the form of a **determinant**. We shall not meet determinants until Chapter 5, but for completeness we quote the result in this form. It is

$$\mathbf{A}\times\mathbf{B} = \begin{vmatrix} \mathbf{i} & \mathbf{j} & \mathbf{k} \\ A_1 & A_2 & A_3 \\ B_1 & B_2 & B_3 \end{vmatrix}$$

$$= \mathbf{i}(A_2B_3 - A_3B_2) + \mathbf{j}(A_3B_1 - A_1B_3) + \mathbf{k}(A_1B_2 - A_2B_1) \quad (4.37)$$

7. The magnitude of $\mathbf{A}\times\mathbf{B}$ is $AB \sin\theta$. Looking at the plane of **A** and **B** and completing the parallelogram on **A** and **B** we see from Figure 4.72 that the area of this parallelogram is $AB \sin\theta$.

   Hence we have $|\mathbf{A}\times\mathbf{B}|$ = area of parallelogram with sides **A** and **B**

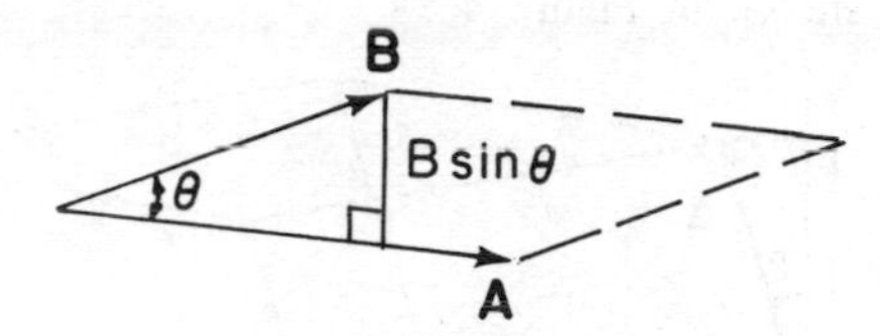

Figure 4.72

Problems III

1. Given $\mathbf{A} = 2\mathbf{i} - \mathbf{j} + \mathbf{k}$, $\mathbf{B} = 3\mathbf{i} + 5\mathbf{j} - \mathbf{k}$, find $\mathbf{A}\times\mathbf{B}$

2. Find the area of the triangle with vertices at the points $P(1,3,2)$, $Q(2,-1,1)$, $R(-1,2,3)$.

3. Show $(\mathbf{A} - \mathbf{B})\times(\mathbf{A} + \mathbf{B}) = 2(\mathbf{A}\times\mathbf{B})$

4. Prove the sine rule for a triangle.

5. If $\mathbf{a} = (12, 1, -13)$ and $\mathbf{c} = (8, -5, 7)$ find $\mathbf{b}$ so that $\mathbf{a}\times\mathbf{b} = \mathbf{c}$ and $\mathbf{a}.\mathbf{b} = 2$.

6. The magnetic induction **B** is defined by the *Lorentz force equation* $\mathbf{F} = q(\mathbf{V}\times\mathbf{B})$ where **F** is the force on a charge q moving with velocity **V**. In three experiments it was found that

   when $\mathbf{V} = \mathbf{i}$, $\mathbf{F}/q = -\mathbf{j} - \mathbf{k}$

   when $\mathbf{V} = \mathbf{j}$, $\mathbf{F}/q = \mathbf{i} - 2\mathbf{k}$

   when $\mathbf{V} = \mathbf{k}$, $\mathbf{F}/q = \mathbf{i} + 2\mathbf{j}$

   Using these results calculate **B**.

Products of three vectors

We shall not spend very much time on such products, but they do occur in physical situations often enough to deserve some attention. For example, they occur in advanced dynamics when considering the motion of a body.

The product of two vectors $\mathbf{B}$ and $\mathbf{C}$ can be either a scalar $\mathbf{B}.\mathbf{C}$ or a vector $\mathbf{B}\times\mathbf{C}$ and in each case can be multiplied by a third vector. Three types of product can occur.

(a) *Type* $(\mathbf{B}.\mathbf{C})\mathbf{A}$

Now $\mathbf{B}.\mathbf{C}$ is a scalar quantity and as such is simply a number ϕ, say. Then $(\mathbf{B}.\mathbf{C})\mathbf{A} = \phi\mathbf{A}$, which is just a multiple of $\mathbf{A}$.

(b) *Type* $\mathbf{A}.(\mathbf{B}\times\mathbf{C})$

Since $\mathbf{B}\times\mathbf{C}$ is a vector quantity, we can perform the scalar product of this with vector $\mathbf{A}$ to generate the product $\mathbf{A}.(\mathbf{B}\times\mathbf{C})$. The result is a scalar quantity and is called the **scalar triple product**.
Let us first of all construct the parallelepiped with three of its sides as $\mathbf{A}$, $\mathbf{B}$, $\mathbf{C}$ as shown in Figure 4.73.

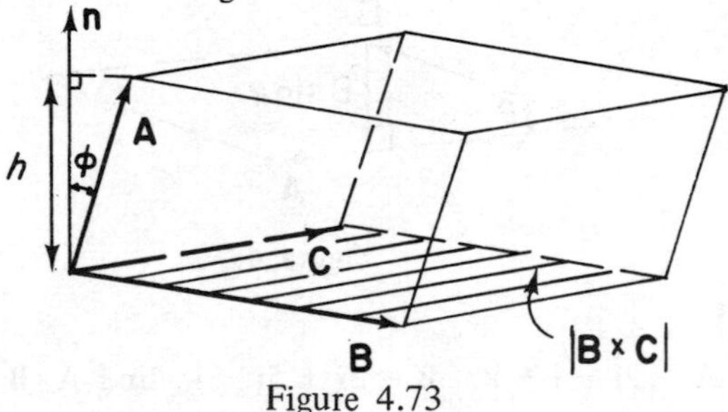

Figure 4.73

Now $\mathbf{B}\times\mathbf{C}$ is a vector in the direction of $\mathbf{n}$ with magnitude equal to the area of the shaded parallelogram.
Hence

$$\begin{aligned}\mathbf{A}.(\mathbf{B}\times\mathbf{C}) &= \mathbf{A}.\mathbf{n} \times \text{(area of parallelogram)}\\ &= A\cos\phi \times \text{(area of parallelogram)}\\ &= h \times \text{(area of parallelogram)}\end{aligned}$$

That is

$$\mathbf{A}.(\mathbf{B}\times\mathbf{C}) = \text{Volume of the parallelepiped}$$

Now in the above, the face constructed on $\mathbf{B}$ and $\mathbf{C}$ was taken as the base of the solid, but it is obvious that we could take any of the other faces as the base. It then follows that the expressions $\mathbf{C}.(\mathbf{A}\times\mathbf{B})$ and $\mathbf{B}.(\mathbf{C}\times\mathbf{A})$ also give the volume of the solid. (Notice that the cyclic order of $\mathbf{A}$, $\mathbf{B}$, $\mathbf{C}$ *must* be maintained to obtain the volume with positive sign.)
Hence

$$\mathbf{A}.(\mathbf{B}\times\mathbf{C}) = \mathbf{B}.(\mathbf{C}\times\mathbf{A}) = \mathbf{C}.(\mathbf{A}\times\mathbf{B})$$

If we interchange the last one of these expressions we get

$$\mathbf{A}.(\mathbf{B}\times\mathbf{C}) = (\mathbf{A}\times\mathbf{B}).\mathbf{C}$$

The order $\mathbf{A}$, $\mathbf{B}$, $\mathbf{C}$ is the same, but the . and $\times$ are interchanged.

It is the practice therefore in many books to write the scalar triple product as

$$[\mathbf{A}, \mathbf{B}, \mathbf{C}]$$

since it can be any of the expressions $\mathbf{A}.\mathbf{B}\times\mathbf{C}$, $\mathbf{A}\times\mathbf{B}.\mathbf{C}$, together with their cyclic permutations. Notice that $(\mathbf{A}.\mathbf{B})\times\mathbf{C}$ has no meaning (why is this so?) and so the position of the brackets is obvious.
In component form we would get the following expression

$$\mathbf{A}.(\mathbf{B}\times\mathbf{C}) = A_1(B_2C_3 - B_3C_2) + A_2(B_3C_1 - B_1C_3) + A_3(B_1C_2 - B_2C_1)$$

You should try to obtain this from direct expansion of the expression on the left. Again the most convenient form is as a *determinant*, which we quote here:

$$\begin{vmatrix} A_1 & A_2 & A_3 \\ B_1 & B_2 & B_3 \\ C_1 & C_2 & C_3 \end{vmatrix} \tag{4.38}$$

Important property of the scalar triple product
The triple scalar product will vanish if

(i) any of the vectors is zero
(ii) any two of the vectors are parallel (this makes the area of a face of the solid zero)
(iii) the three vectors are coplanar (the volume of the solid is zero).

It is noted that (ii) is really a special case of (iii).

Problems IV

1. $\mathbf{A} = \mathbf{i} + 2\mathbf{j} - \mathbf{k}$, $\mathbf{B} = 2\mathbf{i} + \mathbf{j} + 3\mathbf{k}$, $\mathbf{C} = \mathbf{i} - \mathbf{j} - 2\mathbf{k}$
   Find $\mathbf{A}.(\mathbf{B}\times\mathbf{C})$.

2. Show that $(\mathbf{A}\times\mathbf{B}).(\mathbf{C}\times\mathbf{D}) = (\mathbf{A}.\mathbf{C})(\mathbf{B}.\mathbf{D}) - (\mathbf{A}.\mathbf{D})(\mathbf{B}.\mathbf{C})$

3. Prove the vectors $\mathbf{A} = 2\mathbf{i} + \mathbf{j} + 2\mathbf{k}$, $\mathbf{B} = \mathbf{i} + 3\mathbf{j} - 7\mathbf{k}$, $\mathbf{C} = 3\mathbf{i} + 4\mathbf{j} - 5\mathbf{k}$ are coplanar and form a right-angled triangle.

4. $\mathbf{A} = \lambda\mathbf{B} + \mu\mathbf{C}$, show $\mathbf{A}.(\mathbf{B}\times\mathbf{C}) = 0$ and vice versa.

(c) *Type* $\mathbf{A}\times(\mathbf{B}\times\mathbf{C})$
Since $\mathbf{B}\times\mathbf{C}$ is a vector we can also multiply this vectorially by a third vector and obtain the expression

$$\mathbf{A}\times(\mathbf{B}\times\mathbf{C})$$

This is a vector and is called the **vector triple product**.
Looking at this expression geometrically in Figure 4.74 (i), we note that $\mathbf{B}\times\mathbf{C}$ is a vector in the direction perpendicular to $\mathbf{B}$ and $\mathbf{C}$.

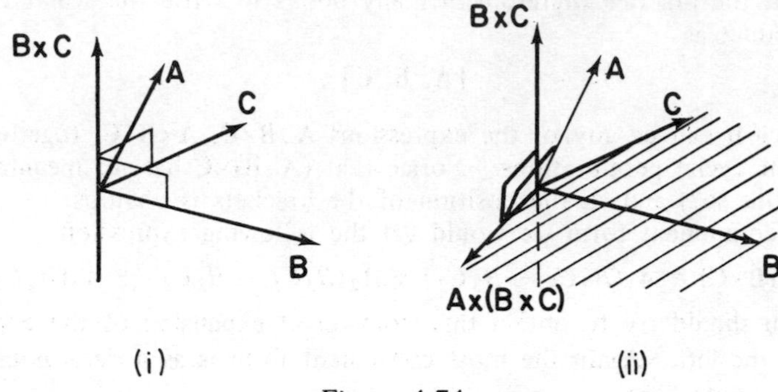

Figure 4.74

Now $\mathbf{A}\times(\mathbf{B}\times\mathbf{C})$ will be perpendicular to both $\mathbf{A}$ and $\mathbf{B}\times\mathbf{C}$. Since it is perpendicular to $\mathbf{B}\times\mathbf{C}$ it must lie in the plane of $\mathbf{B}$ and $\mathbf{C}$. [Figure 4.74 (ii)]. A vector in the plane of $\mathbf{B}$ and $\mathbf{C}$ must be of the form

$$\lambda\mathbf{B} + \mu\mathbf{C} \text{ for some } \lambda \text{ and } \mu$$

Hence $\mathbf{A}\times(\mathbf{B}\times\mathbf{C}) = \lambda\mathbf{B} + \mu\mathbf{C}$ where the λ and μ have to be determined. We can find λ and μ by several methods – see the problem that follows here for the method of expanding into cartesian components. More elegant methods do exist, but it is not felt necessary to pursue them here. What is important, is that you should know the result, for it has great uses in problems and in future derivations. The result is

$$\mathbf{A}\times(\mathbf{B}\times\mathbf{C}) = (\mathbf{A}.\mathbf{C})\mathbf{B} - (\mathbf{A}.\mathbf{B})\mathbf{C} \qquad (4.39)$$

Notice that the order and brackets are important here, for example,

$$\begin{aligned}(\mathbf{A}\times\mathbf{B})\times\mathbf{C} &= -\mathbf{C}\times(\mathbf{A}\times\mathbf{B})\\ &= -[(\mathbf{C}.\mathbf{B})\mathbf{A} - (\mathbf{C}.\mathbf{A})\mathbf{B}]\\ &= (\mathbf{A}.\mathbf{C})\mathbf{B} - (\mathbf{C}.\mathbf{B})\mathbf{A}\end{aligned}$$

an entirely different result from equation (4.39).

Problems V

1. If $\mathbf{A} = \mathbf{i} + \mathbf{j} + \mathbf{k}$, $\mathbf{B} = 2\mathbf{i} + \mathbf{j} - \mathbf{k}$, $\mathbf{C} = 3\mathbf{i} - 2\mathbf{j} + 2\mathbf{k}$, verify equation (4.39) by multiplying out left-hand side and the right-hand side and comparing results.

2. Choose axes so that $\mathbf{B} = B_1\mathbf{i}$, $\mathbf{C} = C_1\mathbf{i} + C_2\mathbf{j}$, $\mathbf{A} = A_1\mathbf{i} + A_2\mathbf{j} + A_3\mathbf{k}$. Expand $\mathbf{A}\times(\mathbf{B}\times\mathbf{C})$ and collect the terms, remembering that the result required is $\lambda\mathbf{B} + \mu\mathbf{C}$ or here

$\lambda B_1\mathbf{i} + \mu(C_1\mathbf{i} + C_2\mathbf{j})$. Hence show $\mathbf{A}\times(\mathbf{B}\times\mathbf{C}) = (\mathbf{A}.\mathbf{C})\mathbf{B} - (\mathbf{A}.\mathbf{B})\mathbf{C}$.

3. If $\mathbf{a} = (3,-1,2)$, $\mathbf{b} = (2,1,-1)$, $\mathbf{c} = (1,-2,2)$ find $(\mathbf{a}\times\mathbf{b})\times\mathbf{c}$ and $\mathbf{a}\times(\mathbf{b}\times\mathbf{c})$.

4. An electric charge q_1 moving with velocity $\mathbf{v}_1$ produces a magnetic induction $\mathbf{B}$ given by

$$\mathbf{B} = \frac{\mu}{4\pi} q_1 \frac{\mathbf{v}_1 \times \mathbf{r}}{r^2} \qquad \textit{(Biot-Savart Law)}$$

The magnetic force exerted on a second charge q_2, moving with velocity $\mathbf{v}_2$ is given by

$$\mathbf{F} = q_2 \mathbf{v}_2 \times \mathbf{B}$$

Show that

$$\mathbf{F} = \frac{\mu}{4\pi r^2} q_1 q_2 \; \mathbf{v}_2 \times (\mathbf{v}_1 \times \mathbf{r})$$

and find $\mathbf{F}$ if $\mathbf{v}_1 = (\mathbf{i} + \mathbf{j} - \mathbf{k})$, $\mathbf{v}_2 = -3\mathbf{i} + 2\mathbf{j}$, $\mathbf{r} = 2\mathbf{i} - \mathbf{j} + 3\mathbf{k}$, $q_1 = q_2 = 1$.

5. Show that $(\mathbf{A}\times\mathbf{B})\times(\mathbf{C}\times\mathbf{D}) = (\mathbf{A}.\mathbf{B}\times\mathbf{D})\mathbf{C} - (\mathbf{A}.\mathbf{B}\times\mathbf{C})\mathbf{D}$

SUMMARY

The common theme running throughout this chapter is that of an ordered pair of real numbers. We first present some results in coordinate geometry which you should retain. The representation of a point on a curve by cartesian, polar or parametric coordinates is an idea which it is essential to understand. You ought to be able to express in geometric terms an inequality given algebraically and vice-versa. The section on curve sketching is one which should not be neglected; it not only is a means of visually representing the main features of a function, but also it requires careful analytical thought.

Complex numbers are a useful device in some calculations and you should be conversant with the basic algebraic operations. It is helpful to regard a complex number as an ordered pair of real numbers, because this allows a geometrical interpretation. The ideas behind De Moivre's theorem should be understood as should the method of finding the roots of a complex number.

You must appreciate the difference between bound and free vectors. Since vector algebra is important as a background to understanding later work, you should learn how to perform it with ease.

Chapter Five

Solution of Equations

5.0 INTRODUCTION

In this chapter we study methods of solving simultaneous linear equations and single non-linear equations. We have put the solution of these two kinds of equation in the same chapter because the elegance of the methods of solution of linear equations contrasts sharply with the methods for non-linear equations and the relative simplicity of the former methods provides a clue as to why so many mathematical models are made linear (even at the expense of losing accuracy in their solution).

5.1 SIMULTANEOUS LINEAR EQUATIONS

Let us first of all consider the solution of two equations in two unknowns. These equations will take the form

$$\begin{cases} a_1x + b_1y = c_1 \\ a_2x + b_2y = c_2 \end{cases}$$

where x and y are unknown and the a's, b's and c's are known numbers.

We know from our earlier work that each of these equations will represent a straight line in the x-y plane. Thus the solution to two such simultaneous equations will be represented by the intersection of two straight lines. Three possible cases arise in theory, as shown in Figure 5.1.

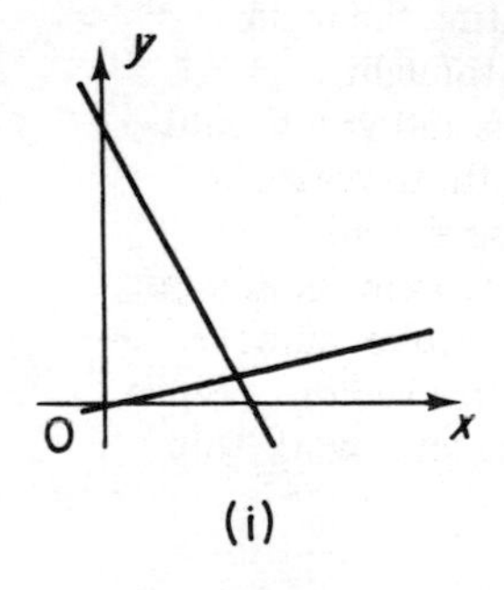

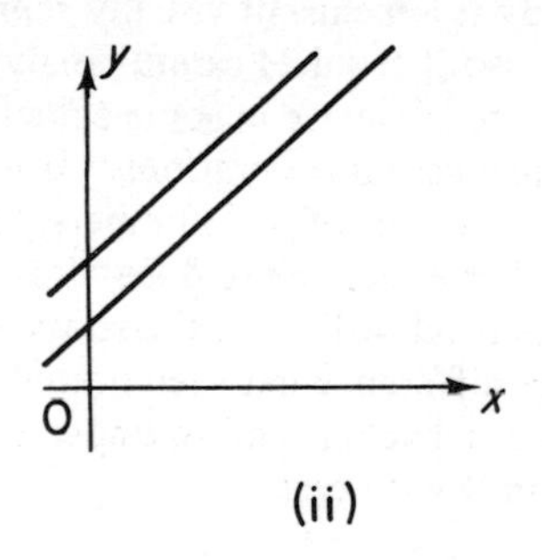

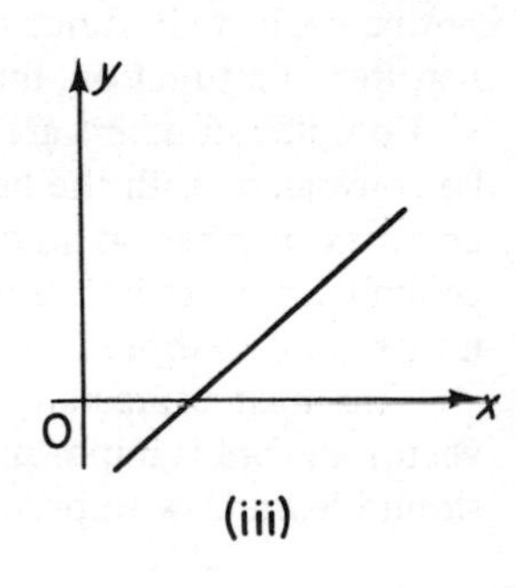

Figure 5.1

Case (i)
For example,

$$\begin{cases} 2x + y = 1 \\ x - 2y = 0.1 \end{cases}$$

Here the two lines intersect and there is a unique solution.

Case (ii)
For example,

$$\begin{cases} 2x - y = -1 \\ 2x - y = -2 \end{cases}$$

In this case the two lines are parallel and hence there is no solution.

Case (iii)
For example,

$$\begin{cases} x - y = 1 \\ 3x - 3y = 3 \end{cases}$$

Here the two lines coincide and there is, therefore, an infinite number of solutions.

From a practical point of view, there is a fourth possibility This occurs when the straight lines are almost parallel. Take, for example,

$$\begin{cases} x + 3y = 4 \\ 30x + 90.1y = 120.1 \end{cases}$$

These have a solution $x = 1$, $y = 1$; however, a change of +(1/9)% in the coefficient of y in the first equation yields a pair of equations which have no solution but, if this is accompanied by a change of +(1/12)% in the constant term, yields a pair of equations which are identical. These changes in the coefficients could easily be caused by inexact data. Alternatively, if the data is exact, the lack of precision in subsequent arithmetic could greatly affect the end result. We have said that a set of equations so finely balanced is **ill-conditioned** (or **unstable**) and needs special treatment.

Consider then the general equations

$$a_1x + b_1y = c_1 \tag{5.1}$$

$$a_2x + b_2y = c_2 \tag{5.2}$$

Multiplying (5.1) by b_2 and subtracting (5.2) multiplied by b_1 gives

$$(a_1b_2 - a_2b_1)x = c_1b_2 - c_2b_1$$

Similarly (5.2) $\times\, a_1$ $-$ (5.1) $\times\, a_2$ gives

$$(a_1b_2 - a_2b_1)y = c_2a_1 - c_1a_2$$

Hence if $a_1 b_2 \neq a_2 b_1$, we have a unique solution for x and y, namely

$$x = (c_1 b_2 - c_2 b_1)/(a_1 b_2 - a_2 b_1) \tag{5.3}$$

$$y = (c_2 a_1 - c_1 a_2)/(a_1 b_2 - a_2 b_1) \tag{5.4}$$

If $a_1 b_2 = a_2 b_1$, then, generally speaking, we have $a_1/b_1 = a_2/b_2$ which means the slopes of the straight lines are equal (and if this ratio $= c_1/c_2$, the lines are *coincident*). We say *generally* since there are special cases where $a_1 = b_1 = 0$ etc.

For shorthand we write

$$a_1 b_2 - a_2 b_1 \equiv \begin{vmatrix} a_1 & a_2 \\ b_1 & b_2 \end{vmatrix}$$

and this is an example of a **determinant** which we shall study in detail later.

Example

Solve

$$\begin{cases} 3x - y = 2 \\ x + y = 4 \end{cases}$$

Using (5.3) and (5.4) we find the solutions

$$x = \frac{2 + 4}{4} = 1.5$$

$$y = \frac{12 - 2}{4} = 2.5$$

Applying (5.3) and (5.4) you should now try to solve for x and y the examples given earlier. Notice how to spot that the solution has broken down.

We now turn our attention to three simultaneous linear equations in three unknowns and we shall develop methods of solution for these which can be extended to deal with n equations in n unknowns. Furthermore we should like these methods of solution to be suitable for computer treatment.

To allow generality we consider the equations in the form

$$\begin{cases} a_{11}x_1 + a_{12}x_2 + a_{13}x_3 = b_1 \\ a_{21}x_1 + a_{22}x_2 + a_{23}x_3 = b_2 \\ a_{31}x_1 + a_{32}x_2 + a_{33}x_3 = b_3 \end{cases}$$

where the a_{ij} (i, j can take values 1, 2 or 3) and the b_i ($i = 1$, 2 or 3) are constants.

It is noted that, just as with two equations in two unknowns, cases can arise where we get

(i) a unique solution
(ii) no solution, or
(iii) an infinite number of solutions.

We shall consider cases (ii) and (iii) later in more detail with the reasons for their occurrence, *but at the moment* let us assume that the equations have a *unique* answer and hence proceed on to develop for these a numerical method of solution.

5.2 DIRECT METHODS OF SOLUTION OF SIMULTANEOUS LINEAR EQUATIONS

Gauss' method of elimination

The basic method of Gauss reduces the equations to the triangular form

$$\begin{aligned} a_{11}x_1 + a_{12}x_2 + a_{13}x_3 &= b_1 \\ A_{22}x_2 + A_{23}x_3 &= B_2 \\ A_{33}x_3 &= B_3 \end{aligned}$$

(Note that the operations involved will change the coefficients in the second and third lines.)

The third equation then gives $x_3 = B_3/A_{33}$, assuming $A_{33} \neq 0$. Back substitution in the second equation now gives x_2, assuming $A_{22} \neq 0$. Finally, back substitution in the first equation gives x_1, assuming $a_{11} \neq 0$.

Let us work through an example.

Example 1

$$\begin{cases} 2x_1 - 4x_2 + x_3 = -9 & (5.5) \\ 4x_1 + 2x_2 - 3x_3 = 17 & (5.6) \\ x_1 - x_2 + 5x_3 = -11 & (5.7) \end{cases}$$

First of all we eliminate x_1 from equations (5.6) and (5.7) by

(i) subtracting $2 \times (5.5)$ from (5.6) and
(ii) subtracting $\frac{1}{2} \times (5.5)$ from (5.7).

This gives

$$2x_1 - 4x_2 + x_3 = -9 \quad (5.5)$$

$$10x_2 - 5x_3 = 35 \quad (5.8)$$

$$x_2 + (9/2)x_3 = -13/2 \quad (5.9)$$

For convenience divide equation (5.8) by the common factor 5 .

$$2x_1 - 4x_2 + x_3 = -9 \qquad (5.5)$$
$$2x_2 - x_3 = 7 \qquad (5.10)$$
$$x_2 + (9/2)x_3 = -13/2 \qquad (5.9)$$

We now eliminate x_2 from (5.9) by taking (5.9) − ½ × (5.10) giving

$$2x_1 - 4x_2 + x_3 = -9 \qquad (5.5)$$
$$2x_2 - x_3 = 7 \qquad (5.10)$$
$$5x_3 = -10 \qquad (5.11)$$

(5.11) $\Rightarrow x_3 = -2$

(5.10) $\Rightarrow x_2 = 7/2 + x_3/2 = 5/2$

(5.5) $\Rightarrow x_1 = -9/2 + 2x_2 - x_3/2 = -9/2 + 5 + 1 = 3/2$

You will notice that this method relies on the coefficients only and so we can isolate them for purposes of calculation. Also it would be worth having a check of some kind. One of the simplest is a row sum check. We sum all the constants in each equation and whatever operations are done on the equations should also check with the same operation on the row sums.

Let us do the above example again, isolating the coefficients and doing a row sum check. We would have

Row sum

$$\begin{bmatrix} 2 & -4 & 1 \\ 4 & 2 & -3 \\ 1 & -1 & 5 \end{bmatrix} \begin{bmatrix} -9 \\ 17 \\ -11 \end{bmatrix} \qquad \begin{bmatrix} -10 \\ 20 \\ -6 \end{bmatrix}$$

$$\begin{bmatrix} 2 & -4 & 1 \\ 0 & 10 & -5 \\ 0 & 1 & 9/2 \end{bmatrix} \begin{bmatrix} -9 \\ 35 \\ -13/2 \end{bmatrix} \qquad \begin{bmatrix} -10 \\ 40 \\ -1 \end{bmatrix} \begin{matrix} \\ \checkmark \;\; \text{Row 2} - (2 \times \text{Row 1}) \\ \checkmark \;\; \text{Row 3} - (\tfrac{1}{2} \times \text{Row 1}) \end{matrix}$$

$$\begin{bmatrix} 2 & -4 & 1 \\ 0 & 2 & -1 \\ 0 & 1 & 9/2 \end{bmatrix} \begin{bmatrix} -9 \\ 7 \\ -13/2 \end{bmatrix} \qquad \begin{bmatrix} -10 \\ 8 \\ -1 \end{bmatrix} \begin{matrix} \\ \checkmark \;\; \text{Row 2} \div 5 \\ \\ \end{matrix}$$

$$\begin{bmatrix} 2 & -4 & 1 \\ 0 & 2 & -1 \\ 0 & 0 & 5 \end{bmatrix} \begin{bmatrix} -9 \\ 7 \\ -10 \end{bmatrix} \qquad \begin{bmatrix} -10 \\ 8 \\ -5 \end{bmatrix} \begin{matrix} \\ \\ \checkmark \;\; \text{Row 3} - (\tfrac{1}{2} \times \text{Row 2}) \end{matrix}$$

The coefficient sums are added in each new row and checked against the new row sums produced by the operations.

An alternative check is to substitute the values obtained for x_1, x_2, x_3 into the second and third equations of the original set (why not the first?).

Hence, substitution in equation (5.6) gives

$$(4 \times 3/2) + (2 \times 5/2) - (3 \times -2) = 17 \quad \checkmark$$

and in equation (5.7) produces

$$(1 \times 3/2) - (1 \times 5/2) + (5 \times -2) = -11 \checkmark$$

Example 2

Let us work one more example using this method. Solve

$$\left\{\begin{aligned} 3x + 2y + z &= 4 \\ x - y + z &= 2 \\ -2x + 2z &= 5 \end{aligned}\right. \tag{5.12}$$

We have

$$\begin{bmatrix} 3 & 2 & 1 \\ 1 & -1 & 1 \\ -2 & 0 & 2 \end{bmatrix} \begin{bmatrix} 4 \\ 2 \\ 5 \end{bmatrix} \qquad \overset{\text{Row sum}}{\begin{bmatrix} 10 \\ 3 \\ 5 \end{bmatrix}}$$

$$\begin{bmatrix} 3 & 2 & 1 \\ 0 & -5/3 & 2/3 \\ 0 & 4/3 & 8/3 \end{bmatrix} \begin{bmatrix} 4 \\ 2/3 \\ 23/3 \end{bmatrix} \qquad \begin{bmatrix} 10 \\ -1/3 \\ 35/3 \end{bmatrix} \begin{matrix} \\ \checkmark \;\; \text{Row 2} - (1/3 \times \text{Row 1}) \\ \checkmark \;\; \text{Row 3} + (2/3 \times \text{Row 1}) \end{matrix}$$

$$\begin{bmatrix} 3 & 2 & 1 \\ 0 & -5/3 & 2/3 \\ 0 & 0 & 16/5 \end{bmatrix} \begin{bmatrix} 4 \\ 2/3 \\ 41/5 \end{bmatrix} \qquad \begin{bmatrix} 10 \\ -1/3 \\ 57/5 \end{bmatrix} \begin{matrix} \\ \\ \checkmark \;\; \text{Row 3} + (4/5 \times \text{Row 2}) \end{matrix}$$

$$z = \frac{41}{16}$$

$$y = -\frac{3}{5}\left(\frac{2}{3} - \frac{2}{3}\cdot\frac{41}{16}\right) = \frac{5}{8}$$

$$x = \frac{1}{3}\left(4 - \frac{41}{16} - 2\cdot\frac{5}{8}\right) = \frac{1}{16}$$

Check for yourself that these values satisfy the original equations (5.12).

Problems I

1. Solve the following simultaneous equations using Gauss' method working with exact arithmetic (fractions or whole numbers) and using a row sum check.

(a) $$\begin{cases} 5x - y + 2z - 3 = 0 \\ 2x + 4y + z - 8 = 0 \\ x + 3y - 3z - 2 = 0 \end{cases}$$

(b) $$\begin{cases} 25x_1 + 36x_2 + 36x_3 = 12 \\ 4x_1 + 81x_2 + 4x_3 = 27 \\ 9x_1 + 9x_2 + 64x_3 = 3 \end{cases}$$

(c) For the circuit shown in Figure 5.2, Kirchhoff's Laws produce the equations

$$\begin{aligned} i_1 + i_2 &= 3 \\ i_3 + i_5 &= 3 \\ i_1 - i_3 - i_4 &= 0 \\ i_2 + i_4 - i_5 &= 0 \\ 10i_1 + 10i_4 - 5i_2 &= 0 \\ 10i_4 + 5i_5 - 5i_3 &= 0 \end{aligned}$$

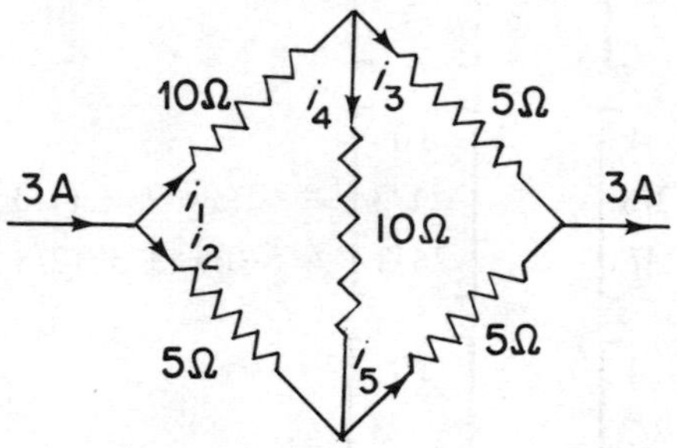

Figure 5.2

Show that the fourth equation can be removed without altering the problem and solve the resulting set of equations.

(d) *(Beware!)* $$\begin{cases} 20x - 12y - 14z = 20 \\ 8x + 4y - 6z = 16 \\ 3x - 4y - 2z = 2 \end{cases}$$

(e) $$\begin{cases} 3x - 4y + 7z = 16 \\ 2x + 3y + z = 11 \\ x - 5z = 14 \end{cases}$$

(f) $$\begin{cases} 2x + 3y + 4z = 3 \\ 6x - 3y + 8z = 4 \\ -2x + 6y - 12z = -2 \end{cases}$$

2. Solve by Gauss elimination method, where $i^2 = -1$,

(a) $(2 + i)x - iy = i - 2$
$3ix + (1 + i)y = -1$

(b) $(3 - 2i)x + (1 - 2i)y = 1 + 2i$
$(6 - 4i)x + (6 - 3i)y = -5 - 2i$

Round-off errors, Pivoting and Residuals

Up to now we have applied the Gauss elimination in situations where the original coefficients were whole numbers, and we have worked in fractions. This will not usually be the case for examples arising from physical situations and we may have to work with inexact arithmetic, rounding-off to the required number of decimal places. In such cases we try to arrange the rows so that the scale factors used in subtracting rows are less than 1 numerically. We can do this by **partially pivoting** – that is, at each stage we work with the largest coefficient of the variable to be eliminated (the largest coefficient is called the **pivot**).

Example 3

Consider the circuit in Figure 5.3 with resistances and batteries as shown; we choose the current directions arbitrarily at this stage.

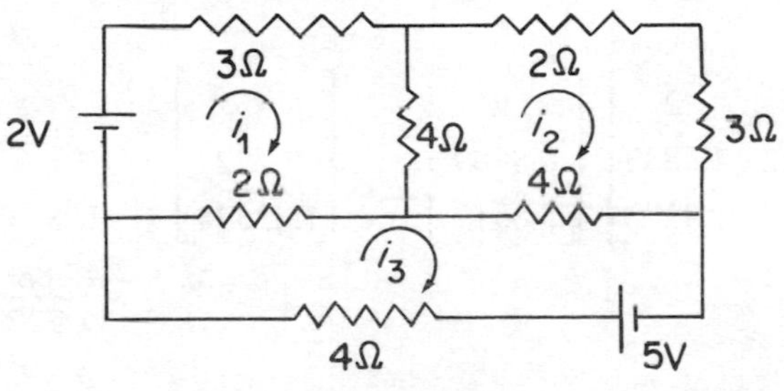

Figure 5.3

By applying Kirchhoff's Law, which states that the algebraic sum of the potential differences around any closed loop is zero, we find for the currents i_1, i_2, i_3

$$\begin{cases} 4i_3 + 2(i_3 - i_1) + 4(i_3 - i_2) - 5 = 0 \\ 3i_1 + 4(i_1 - i_2) + 2(i_1 - i_3) - 2 = 0 \\ 2i_2 + 3i_2 + 4(i_2 - i_3) + 4(i_2 - i_1) = 0 \end{cases}$$

or, collecting terms in i_1, i_2, i_3

$$\begin{cases} -2i_1 - 4i_2 + 10i_3 = 5 \\ 9i_1 - 4i_2 - 2i_3 = 2 \\ -4i_1 + 13i_2 - 4i_3 = 0 \end{cases} \tag{5.13}$$

The coefficient of i_1 with greatest magnitude is 9, so we rearrange the equations with this in the top left-hand position giving

$$\begin{bmatrix} 9 & -4 & -2 \\ -2 & -4 & 10 \\ -4 & 13 & -4 \end{bmatrix} \begin{bmatrix} 2 \\ 5 \\ 0 \end{bmatrix} \qquad \overset{\text{Row sum}}{\begin{bmatrix} 5 \\ 9 \\ 5 \end{bmatrix}}$$

Thus, working to four places of decimals

$$\begin{bmatrix} 9 & -4 & -2 \\ 0 & -4.8889 & 9.5556 \\ 0 & 11.2222 & -4.8889 \end{bmatrix} \begin{bmatrix} 2 \\ 5.4444 \\ 0.8889 \end{bmatrix} \quad \begin{bmatrix} 5 \\ 10.1111 \\ 7.2222 \end{bmatrix} \begin{matrix} \\ \checkmark \text{ Row 2} + (2/9 \times \text{Row 1}) \\ \checkmark \text{ Row 3} + (4/9 \times \text{Row 1}) \end{matrix}$$

The largest coefficient of i_2 in the second and third equations is 11.2222 so we rearrange the last two rows

$$\begin{bmatrix} 9 & -4 & -2 \\ 0 & 11.2222 & -4.8889 \\ 0 & -4.8889 & 9.5556 \end{bmatrix} \begin{bmatrix} 2 \\ 0.8889 \\ 5.4444 \end{bmatrix} \quad \begin{bmatrix} 5 \\ 7.2222 \\ 10.1111 \end{bmatrix}$$

giving

$$\begin{bmatrix} 9 & -4 & -2 \\ 0 & 11.2222 & -4.8889 \\ 0 & 0 & 7.4258 \end{bmatrix} \begin{bmatrix} 2 \\ 0.8889 \\ 5.8317 \end{bmatrix} \quad \begin{bmatrix} 5 \\ 7.2222 \\ 13.2574 \end{bmatrix} \begin{matrix} \\ \\ \checkmark \text{ Row 3} + \dfrac{4.8889}{11.2222} \times \text{Row 2} \end{matrix}$$

Thus

$$i_3 = 0.7853$$
$$i_2 = 0.4213$$
$$i_1 = 0.5840$$

Notice that the scale factors we used in this example were 2/9, 4/9 and 4.8889/11.2222, all of them less than 1 in magnitude, so limiting the spreading of round-off error as far as possible.

Our answers are likely to be more accurate than those obtained without pivoting.

Consider the set of equations rewritten as

$$\begin{cases} -2i_1 - 4i_2 + 10i_3 = 5 \\ -9i_1 + 4i_2 + 2i_3 = -2 \\ +4i_1 - 13i_2 + 4i_3 = 0 \end{cases}$$

The coefficient of i_1 which has greatest magnitude is -9, since the sign is irrelevant to the choice of pivot. Comparing with the example above, we see that in the elimination of i_1 from the other equations, the scale factors are $-2/9$ and $4/9$. The lesson to learn is that the sign of the coefficients is ignored when selecting the pivot.

Even so, we can still improve the accuracy of the answers as follows:

Suppose we have found values of i_1, i_2, i_3 which differ from the true values I_1, I_2, I_3 by i_1', i_2', i_3' so that

$$i_1 = 0.5840 = I_1 + i_1'$$
$$i_2 = 0.4213 = I_2 + i_2'$$
$$i_3 = 0.7853 = I_3 + i_3'$$

Now I_1, I_2, I_3 will satisfy the equations exactly, but the i_1, i_2, i_3 found will leave a small imbalance in each equation when substituted back into the equations in that the left-hand sides and right-hand sides will not be equal. These small imbalances are called **residuals**.

In the first equation, after rearrangement of equations (5.13), the residual r_1 is given by

$$r_1 = 9 \times 0.5840 - 4 \times 0.4213 - 2 \times 0.7853 - 2 = 0.0002$$

Likewise we find the residuals r_2 and r_3 as

$$r_2 = -0.0002 \quad \text{and} \quad r_3 = -0.0003$$

Now

$$\begin{aligned} 9i_1 - 4i_2 - 2i_3 &= 9(I_1 + i_1') - 4(I_2 + i_2') - 2(I_3 + i_3') \\ &= (9I_1 - 4I_2 - 2I_3) + 9i_1' - 4i_2' - 2i_3' \\ &= 2 + 9i_1' - 4i_2' - 2i_3' \end{aligned}$$

But the left-hand side $= 2 + r_1$

therefore

$$9i_1' - 4i_2' - 2i_3' = r_1$$

Similarly

$$-2i_1' - 4i_2' + 10i_3' = r_2$$
$$-4i_1' + 13i_2' - 4i_3' = r_3$$

We solve these equations for i_1', i_2', i_3'. Since the coefficients are the same as those in the original equations, the same sequence of operations will be followed and in effect we only have to perform these operations on the residuals.

This produces

$$i_1' = +0.00000 \qquad i_2' = -0.00003 \qquad i_3' = -0.00004$$

Hence more accurate answers are

$$i_1 = 0.58400$$
$$i_2 = 0.42133$$
$$i_3 = 0.78534 \quad \text{correct to 5 d.p.}$$

The process can be repeated as many times as is desired as long as the accuracy of the machine being used permits.

The Gaussian elimination method is suitable for use with more than 3 simultaneous equations and it should be clear how this is to be applied to n simultaneous equations in n unknowns. The method is also suitable for use on a digital computer and you should try drawing a flow chart for the procedure both with and without partial pivoting.

Problems II

1. Solve to 2 d.p. using Gaussian elimination with and without pivoting; work to 3 d.p.

(a) $$\begin{cases} -i_1 - 3i_2 + 7i_3 = 2 \\ 4i_1 - 2i_2 - i_3 = 10 \\ -2i_1 + 14i_2 - 3i_3 = 5 \end{cases}$$

(b) $$\begin{cases} 0.19x + 0.22y + 0.42z = 0.25 \\ 0.27x + 0.34y + 0.56z = 0.18 \\ 0.52x + 0.41y + 0.17z = 0.69 \end{cases}$$

(c) $$\begin{cases} 0.22x + 0.40y + 0.39z = 0.15 \\ 0.41x + 0.37y + 0.70z = 0.40 \\ 0.36x + 0.37y + 0.19z = 0.42 \end{cases}$$

(d) $$\begin{cases} 0.732x + 1.013y - 5.421z = 4.256 \\ 3.491x + 0.782y + 2.203z = 7.113 \\ 0.961x + 4.265y - 1.523z = 3.727 \end{cases}$$

2. Find the residuals in Problems 1(a) and 1(b). Using these, improve the accuracy of the solutions by one application of the process given in the above theory.

3. Solve, using Gaussian elimination, first using exact arithmetic and then working with 2 d.p.

$$\begin{cases} 5x_1 + 7x_2 + 6x_3 + 5x_4 = 23 \\ 7x_1 + 10x_2 + 8x_3 + 7x_4 = 32 \\ 6x_1 + 8x_2 + 10x_3 + 9x_4 = 33 \\ 5x_1 + 7x_2 + 9x_3 + 10x_4 = 31 \end{cases}$$

Gauss-Jordan Method

This is a slight variation of the Gauss elimination procedure which allows the values of x_1, x_2, ... to be read off directly.

The elimination procedure is extended to reduce the array of coefficients, for example, in a 3 X 3 system of equations, to the form

$$\begin{bmatrix} a_{11} & 0 & 0 \\ 0 & a'_{22} & 0 \\ 0 & 0 & a'_{33} \end{bmatrix}$$

Sometimes this is further extended to give

$$\begin{bmatrix} 1 & 0 & 0 \\ 0 & 1 & 0 \\ 0 & 0 & 1 \end{bmatrix}$$

Returning to

$$\begin{cases} 2x_1 - 4x_2 + x_3 = -9 & (5.5) \\ 4x_1 + 2x_2 - 3x_3 = 17 & (5.6) \\ x_1 - x_2 + 5x_3 = -11 & (5.7) \end{cases}$$

the procedure would be as before to write

$$\begin{bmatrix} 2 & -4 & 1 \\ 4 & 2 & -3 \\ 1 & -1 & 5 \end{bmatrix} \begin{bmatrix} -9 \\ 17 \\ -11 \end{bmatrix} \quad \overset{\text{Row sum}}{\begin{bmatrix} -10 \\ 20 \\ 6 \end{bmatrix}}$$

and then eliminating to obtain

$$\begin{bmatrix} 2 & -4 & 1 \\ 0 & 2 & -1 \\ 0 & 0 & 5 \end{bmatrix} \begin{bmatrix} -9 \\ 7 \\ -10 \end{bmatrix} \quad \begin{bmatrix} -10 \\ 8 \\ -5 \end{bmatrix}$$

Gauss-Jordan then extends this as follows:

$$\begin{bmatrix} 2 & -4 & 0 \\ 0 & 2 & 0 \\ 0 & 0 & 5 \end{bmatrix} \begin{bmatrix} -7 \\ 5 \\ -10 \end{bmatrix} \quad \begin{bmatrix} -9 \\ 7 \\ -5 \end{bmatrix} \begin{matrix} \checkmark \text{ Row 1} - (1/5 \times \text{Row 3}) \\ \checkmark \text{ Row 2} + (1/5 \times \text{Row 3}) \\ \end{matrix}$$

which gives

$$\begin{bmatrix} 2 & 0 & 0 \\ 0 & 2 & 0 \\ 0 & 0 & 5 \end{bmatrix} \begin{bmatrix} 3 \\ 5 \\ -10 \end{bmatrix} \quad \begin{bmatrix} 5 \\ 7 \\ -5 \end{bmatrix} \begin{matrix} \checkmark \text{ Row 1} + (2 \times \text{Row 2}) \\ \\ \end{matrix}$$

or

$$\begin{bmatrix} 1 & 0 & 0 \\ 0 & 1 & 0 \\ 0 & 0 & 1 \end{bmatrix} \begin{bmatrix} 3/2 \\ 5/2 \\ -2 \end{bmatrix} \quad \begin{bmatrix} 5/2 \\ 7/2 \\ -1 \end{bmatrix} \quad \begin{matrix} \text{Row } 1 \div 2 \\ \text{Row } 2 \div 2 \\ \text{Row } 3 \div 5 \end{matrix}$$

giving directly

$$x_1 = 3/2, \quad x_2 = 5/2, \quad x_3 = -2$$

Problems III

Solve the simultaneous equations given in Problems I and II of this section using Gauss-Jordan elimination.

Inconsistencies

There are *three* types of snags which can occur.

Case 1.

If we consider the solution of the equations

$$\begin{cases} 13x_1 + 4x_2 - x_3 = -14 \\ 4x_1 + 5x_2 - 3x_3 = 0 \\ 3x_1 + 2x_2 - x_3 = -2 \end{cases}$$

we get

$$\begin{bmatrix} 13 & 4 & -1 \\ 4 & 5 & -3 \\ 3 & 2 & -1 \end{bmatrix} \begin{bmatrix} -14 \\ 0 \\ -2 \end{bmatrix} \quad \overset{\text{Row sum}}{\begin{bmatrix} 2 \\ 6 \\ 2 \end{bmatrix}}$$

so that

$$\begin{bmatrix} 13 & 4 & -1 \\ 0 & \frac{49}{13} & -\frac{35}{13} \\ 0 & \frac{14}{13} & -\frac{10}{13} \end{bmatrix} \begin{bmatrix} -14 \\ \frac{56}{13} \\ \frac{16}{13} \end{bmatrix} \begin{bmatrix} 2 \\ \frac{70}{13} \\ \frac{20}{13} \end{bmatrix} \begin{matrix} \\ \checkmark \quad \text{Row } 2 - \left(\frac{4}{13} \times \text{Row } 1\right) \\ \checkmark \quad \text{Row } 3 - \left(\frac{3}{13} \times \text{Row } 1\right) \end{matrix}$$

and

$$\begin{bmatrix} 13 & 4 & -1 \\ 0 & \frac{49}{13} & -\frac{35}{13} \\ 0 & 0 & 0 \end{bmatrix} \begin{bmatrix} -14 \\ \frac{56}{13} \\ 0 \end{bmatrix} \begin{bmatrix} 2 \\ \frac{70}{13} \\ 0 \end{bmatrix} \begin{matrix} \\ \\ \checkmark \quad \text{Row } 3 - \left(\frac{14}{49} \times \text{Row } 2\right) \end{matrix}$$

The last equation has been eliminated completely and effectively we now have only 2 equations in 3 unknowns. We can, of course, assign an arbitrary value to $x_3 = \lambda$ (say) and we get

$$x_2 = \frac{8}{7} + \frac{5}{7}\lambda$$

$$x_1 = \frac{-10}{7} - \frac{\lambda}{7}$$

There is effectively an infinite number of solutions here.

Case 2

Consider now the solution of the equations

$$\begin{aligned} 3x_1 + 2x_2 + x_3 &= 4 \\ x_1 - x_2 + x_3 &= 2 \\ 9x_1 + x_2 + 5x_3 &= 10 \end{aligned}$$

We get

Row sum

$$\begin{bmatrix} 3 & 2 & 1 \\ 1 & -1 & 1 \\ 9 & 1 & 5 \end{bmatrix} \begin{bmatrix} 4 \\ 2 \\ 10 \end{bmatrix} \qquad \begin{bmatrix} 10 \\ 3 \\ 25 \end{bmatrix}$$

$$\begin{bmatrix} 3 & 2 & 1 \\ 0 & -\frac{5}{3} & \frac{2}{3} \\ 0 & -5 & 2 \end{bmatrix} \begin{bmatrix} 4 \\ \frac{2}{3} \\ -2 \end{bmatrix} \qquad \begin{bmatrix} 10 \\ -\frac{1}{3} \\ -5 \end{bmatrix} \begin{matrix} \\ \checkmark \text{ Row 2} - \left(\frac{1}{3} \times \text{Row 1}\right) \\ \checkmark \text{ Row 3} - (3 \times \text{Row 1}) \end{matrix}$$

$$\begin{bmatrix} 3 & 2 & 1 \\ 0 & -\frac{5}{3} & \frac{2}{3} \\ 0 & 0 & 0 \end{bmatrix} \begin{bmatrix} 4 \\ \frac{2}{3} \\ -4 \end{bmatrix} \qquad \begin{bmatrix} 10 \\ -\frac{1}{3} \\ -4 \end{bmatrix} \begin{matrix} \\ \\ \checkmark \text{ Row 3} - (3 \times \text{Row 2}) \end{matrix}$$

The last equation is **false** and thus our elimination procedure has led to a false result. Hence the equations have **no** solution.

The two examples above illustrate that sometimes we shall get no answer at all or perhaps an infinite number of solutions. Can we recognise when these situations will occur? In fact we can. You will remember that with 2 equations in two unknowns it was important that the *determinant* of the l.h.s. coefficients should not be zero. That is, for the equations

$$\begin{cases} a_1x + b_1y = c_1 \\ a_2x + b_2y = c_2 \end{cases}$$

to have a solution

$$\begin{vmatrix} a_1 & b_1 \\ a_2 & b_2 \end{vmatrix} \neq 0$$

For 3 equations in three unknowns the same criterion applies; that is, for the equations on page 222, the determinant

$$\begin{vmatrix} a_{11} & a_{12} & a_{13} \\ a_{21} & a_{22} & a_{23} \\ a_{31} & a_{32} & a_{33} \end{vmatrix} \neq 0$$

We do not yet know how to evaluate such determinants but we shall deal with this later in this chapter. However at the moment we state that

$$\begin{vmatrix} a_{11} & a_{12} & a_{13} \\ a_{21} & a_{22} & a_{23} \\ a_{31} & a_{32} & a_{33} \end{vmatrix} = \begin{array}{l} a_{11}a_{22}a_{33} + a_{12}a_{23}a_{31} + a_{13}a_{21}a_{32} \\ - a_{11}a_{23}a_{32} - a_{12}a_{21}a_{33} - a_{13}a_{31}a_{22} \end{array}$$

Looking at the two examples we have just done and considering the determinant of the left-hand side we find in the example of Case 1

$$\begin{vmatrix} 13 & 4 & -1 \\ 4 & 5 & -3 \\ 3 & 2 & -1 \end{vmatrix} = - 65 - 36 - 8 + 78 + 16 + 15 = 0$$

and for the example in Case 2

$$\begin{vmatrix} 3 & 2 & 1 \\ 1 & -1 & 1 \\ 9 & 1 & 5 \end{vmatrix} = - 15 + 18 + 1 - 3 - 10 + 9 = 0$$

It should be fairly obvious how to extend this check to 4×4, ..., $n \times n$ simultaneous linear equations.

We have then a straightforward check as to when snags discussed in Cases 1 and 2 will arise. *This is when the determinant of the coefficients of the equations is zero.*

Case 3

The third possibility is that of ill-conditioning which occurs when the determinant of coefficients is relatively small. You will remember the example of ill-conditioned equations in Section 5.1

$$\begin{cases} x + 3y = 4 \\ 30x + 90.1y = 120.1 \end{cases}$$

The determinant of coefficients is

$$\begin{vmatrix} 1 & 3 \\ 30 & 90.1 \end{vmatrix} = 0.1$$

which is relatively small compared with the exact answers of $x = 1, y = 1$.

Example

A well-known ill-conditioned set of 3 equations is

$$\begin{cases} \frac{1}{3}x_1 + \frac{1}{4}x_2 + \frac{1}{5}x_3 = 0 \\ \frac{1}{2}x_1 + \frac{1}{3}x_2 + \frac{1}{4}x_3 = 0 \\ x_1 + \frac{1}{2}x_2 + \frac{1}{3}x_3 = 1 \end{cases}$$

The determinant of the coefficients is

$$\begin{vmatrix} \frac{1}{3} & \frac{1}{4} & \frac{1}{5} \\ \frac{1}{2} & \frac{1}{3} & \frac{1}{4} \\ 1 & \frac{1}{2} & \frac{1}{3} \end{vmatrix} = -\frac{1}{27}$$

and the exact answers to this set of equations are

$$x_1 = 9$$
$$x_2 = -36$$
$$x_3 = 30$$

If we approximate 1/3 by 0.33 and solve the equations using Gauss elimination, we find

$$x_1 = 55.56$$
$$x_2 = -277.79$$
$$x_3 = 255.57$$

Thus for a small change in the coefficients, we produce a very large change in the solution. You will find more cases of ill-conditioning in the following problems.

Problems IV

1. Show, using Gaussian elimination, that each of the sets of equations

$$\begin{cases} 2x + 4y = 3 \\ 3x + 6y = 4.5 \end{cases} \qquad \begin{cases} 5x - 3y + 4z = 1 \\ 18x - 3y + 13z = 6 \\ 8x + 3y + 5z = 4 \end{cases}$$

$$\begin{cases} x + y + 2z = 3 \\ 4x - y - 2z = 4 \\ 2x - 3y - 6z = -2 \end{cases} \qquad \begin{cases} 2x - 3y - z = 2 \\ 8x - 23y - 7z = 3 \\ -x + 7y + 2z = 1.5 \end{cases}$$

has an infinite number of solutions.
Evaluate the determinant of the left-hand side coefficients in each case.

2. Show that each of the sets of equations

$$\begin{cases} 2x + 4y = 3 \\ 3x + 6y = 4.9 \end{cases} \qquad \begin{cases} 7x + y + 3z = 14 \\ 2x - 2y + 2z = 3 \\ 3x + 5y - z = 9 \end{cases}$$

$$\begin{cases} 2x - 3y - z = 2 \\ 8x - 23y - 7z = 4 \\ -2x + 14y + 4z = 6 \end{cases} \qquad \begin{cases} x + y + z = 1 \\ 2x - 2y + z = 4 \\ 3x - y + 2z = 6 \end{cases}$$

has no solution. Check that the determinant of l.h.s. coefficients is zero in each case.

3. Solve the equations

$$\begin{cases} x_1 - 2x_2 + x_3 = -2 \\ 3x_1 - 2x_2 + 4x_3 = 36 \\ 8x_1 - 20x_2 + 9x_3 = -38 \end{cases}$$

using exact arithmetic.
Solve the equations again when they are in the form

$$\begin{cases} x_1 - 2x_2 + x_3 = -2 \\ x_1 - 0.67x_2 + 1.33x_3 = 12 \\ x_1 - 2.50x_2 + 1.13x_3 = -4.75 \end{cases}$$

working to 2 d.p.
Give reasons for the inconsistencies.

4. For Problem 3 under the set of Problems II , try the effect of substituting for x_1, x_2, x_3 and x_4

(i) 14.6, −7.2, −2.5, 3.1
(ii) 2.36, 0.18, 0.65, 1.21.

Comment.

5.3 MATRICES

We saw in the solution of linear equations that the whole calculation could proceed using only arrays of coefficients. Such arrays are called **matrices** and are simply a convenient way of storing information. They have their own algebra for addition, subtraction and multiplication with which we have to become familiar.

It must be emphasised at this point that there is a fundamental difference between matrices and determinants. A matrix is an *array* of elements, each of which has its own distinct position in the array. A determinant is a *number* which is produced by combining these elements in a prescribed manner.

Looking back through our example on Gauss elimination (Section 5.2) you will see that we have met the matrices

$$\begin{bmatrix} 2 & -4 & 1 \\ 4 & 2 & -3 \\ 1 & -1 & 5 \end{bmatrix}$$

which is an example of a 3×3 matrix since it has 3 rows and 3 columns,

and

$$\begin{bmatrix} -9 \\ 17 \\ -11 \end{bmatrix}$$

which is an example of a 3×1 matrix as it has 3 rows and 1 column.

Generally, a matrix $\mathbf{A}$ is of order $m \times n$ if it has m rows and n columns and it would be written as

$$\mathbf{A} = \begin{bmatrix} a_{11} & a_{12} & a_{13} & \cdots & a_{1n} \\ a_{21} & a_{22} & a_{23} & \cdots & a_{2n} \\ \vdots & & & & \\ a_{m1} & a_{m2} & a_{m3} & \cdots & a_{mn} \end{bmatrix}$$

If $m = n$, the matrix is **square** .

If $m = 1$, the matrix is called a **row vector**, e.g. $(1, -1, 2, 4)$
If $n = 1$, the matrix is called a **column vector**, e.g. $\begin{bmatrix} 1 \\ 0 \\ -1 \end{bmatrix}$

If $m = n = 1$, the matrix is a single **number**, e.g. (3)
The $a_{11}, a_{12}, \ldots\ldots, a_{mn}$ are called **elements.**
A matrix having all its elements **zero** is called the **null** matrix. There are many such null matrices, e.g.

$$(0) \qquad \begin{bmatrix} 0 & 0 \\ 0 & 0 \end{bmatrix} \qquad (0 \quad 0 \quad 0 \quad 0)$$

A square matrix whose only non-zero elements are on the main diagonal (from the top left to the bottom right) is called a **diagonal** matrix. For example

$$\begin{bmatrix} -3 & 0 & 0 & 0 \\ 0 & 2 & 0 & 0 \\ 0 & 0 & 0 & 0 \\ 0 & 0 & 0 & -5 \end{bmatrix}$$

Transpose of a matrix A

This results if we interchange columns and rows of matrix **A** and is written $\mathbf{A}^T$. Examples:

$$\mathbf{A} = (1, 1, -1, 0) \qquad \mathbf{A}^T = \begin{bmatrix} 1 \\ 1 \\ -1 \\ 0 \end{bmatrix}$$

$$\mathbf{A} = \begin{bmatrix} 1 & 2 & 3 \\ 0 & -1 & 4 \end{bmatrix} \qquad \mathbf{A}^T = \begin{bmatrix} 1 & 0 \\ 2 & -1 \\ 3 & 4 \end{bmatrix}$$

Equality

Two matrices are **equal** if they are the same shape (i.e. have the same number of rows and columns) and if the corresponding elements are equal.

Addition

If matrices **A** and **B** are of the same shape, they can be added by adding corresponding elements and forming a matrix of the same shape as the original ones. For example,

$$\begin{bmatrix} 1 & 2 & 0 \\ 4 & -1 & 8 \end{bmatrix} + \begin{bmatrix} 0 & 1 & 4 \\ -1 & 1 & 4 \end{bmatrix} = \begin{bmatrix} 1 & 3 & 4 \\ 3 & 0 & 12 \end{bmatrix}$$

The resulting matrix is written $\mathbf{A} + \mathbf{B}$; observe that $\mathbf{A} + \mathbf{B} \equiv \mathbf{B} + \mathbf{A}$.

We cannot add $\begin{bmatrix} 1 & 2 \\ 3 & 4 \end{bmatrix}$ and $\begin{bmatrix} 1 & 1 & 1 \\ 1 & 1 & 1 \end{bmatrix}$; why?

Subtraction

Subtraction is obtained in the same way by subtraction of corresponding elements

e.g.

$$\begin{bmatrix} 1 & 2 & 0 \\ 4 & -1 & 8 \end{bmatrix} - \begin{bmatrix} 0 & 1 & 4 \\ -1 & 1 & 4 \end{bmatrix} = \begin{bmatrix} 1 & 1 & -4 \\ 5 & -2 & 4 \end{bmatrix}$$

Multiplication

(a) *Multiplication of a matrix by a scalar*

$$\text{If } \mathbf{A} \text{ is a matrix} = \begin{bmatrix} a_{11} & a_{12} & \cdots\cdots & a_{1n} \\ a_{21} & a_{22} & \cdots\cdots & a_{2n} \\ \vdots & & & \\ a_{m1} & a_{m2} & \cdots\cdots & a_{mn} \end{bmatrix}$$

$$\text{then } k\mathbf{A} = \begin{bmatrix} ka_{11} & ka_{12} & \cdots\cdots & ka_{1n} \\ ka_{21} & ka_{22} & \cdots\cdots & ka_{2n} \\ \vdots & & & \\ ka_{m1} & ka_{m2} & \cdots\cdots & ka_{mn} \end{bmatrix}$$

That is, every element is multiplied by the scalar k.

e.g.

$$3 \begin{bmatrix} 1 & 2 \\ 3 & 4 \end{bmatrix} = \begin{bmatrix} 3 & 6 \\ 9 & 12 \end{bmatrix}$$

(b) *Multiplication of two matrices*

Two matrices **A** and **B** can be multiplied in the order $\mathbf{A} \times \mathbf{B}$ if the number of columns in **A** = number of rows in **B**. If this condition does not hold, multiplication **cannot** be defined.

If **A** is an $m \times n$ and **B** is an $n \times p$ matrix, then the product **C** is an $m \times p$ matrix

$$\underset{m\times n}{\mathbf{A}} \times \underset{n\times p}{\mathbf{B}} = \underset{m\times p}{\mathbf{C}}$$

We say that **A** is post-multiplied by **B** or that **B** is premultiplied by **A**. The element of **C** in the kl position is obtained by multiplying and summing the corresponding elements of the kth *row* of **A** with the lth *column* of **B**. That is,

$$\text{if } \mathbf{A} = \begin{bmatrix} a_{11} & a_{12} & \cdots & a_{1n} \\ a_{21} & a_{22} & \cdots & a_{2n} \\ \vdots & & & \\ a_{k1} & a_{k2} & \cdots & a_{kn} \\ \vdots & & & \\ a_{m1} & a_{m2} & \cdots & a_{mn} \end{bmatrix} \text{ and } \mathbf{B} = \begin{bmatrix} b_{11} & b_{12} & \cdots & b_{1l} & \cdots & b_{1p} \\ b_{21} & b_{22} & \cdots & b_{2l} & \cdots & b_{2p} \\ \vdots & & & & & \\ b_{n1} & b_{n2} & \cdots & b_{nl} & \cdots & b_{np} \end{bmatrix}$$

then **C** =

l^{th} column ↓

k^{th} row ⟶ ○ ⟵ $a_{k1}b_{1l} + a_{k2}b_{2l} + \ldots\ldots a_{kn}b_{nl}$

The process will be clearer by studying two examples .

Example 1

$$\begin{bmatrix} 1 & -1 \\ 2 & 0 \end{bmatrix} \begin{bmatrix} 2 & -1 \\ 1 & 3 \end{bmatrix} = \begin{bmatrix} 1 \times 2 + (-1) \times 1 & 1 \times (-1) + (-1) \times 3 \\ 2 \times 2 + 0 \times 1 & 2 \times (-1) + 0 \times 3 \end{bmatrix}$$

$$= \begin{bmatrix} 1 & -4 \\ 4 & -2 \end{bmatrix}$$

Example 2

$$\underset{(3 \times 3)}{\begin{bmatrix} 1 & -1 & 2 \\ 0 & 3 & 4 \\ -2 & 5 & -1 \end{bmatrix}} \underset{(3 \times 2)}{\begin{bmatrix} 2 & 0 \\ -1 & 3 \\ 1 & 1 \end{bmatrix}} = \begin{bmatrix} 1\times2+(-1)\times(-1)+2\times1 & 1\times0+(-1)\times3+2\times1 \\ 0\times2+3\times(-1)+4\times1 & 0\times0+3\times3+4\times1 \\ -2\times2+5\times(-1)+(-1)\times1 & -2\times0+5\times3+(-1)\times1 \end{bmatrix}$$

$$= \underset{(3 \times 2)}{\begin{bmatrix} 5 & -1 \\ 1 & 13 \\ -10 & 14 \end{bmatrix}}$$

Notice that in the reverse order the product

$$\begin{bmatrix} 2 & 0 \\ -1 & 3 \\ 1 & 1 \end{bmatrix} \begin{bmatrix} 1 & -1 & 2 \\ 0 & 3 & 4 \\ -2 & 5 & -1 \end{bmatrix}$$

is not defined since it is a 3×2 times a 3×3 matrix. If the matrices are both square, then we can form $\mathbf{A} \times \mathbf{B}$ and $\mathbf{B} \times \mathbf{A}$ but the answers are not necessarily the same.

e.g.

$$\begin{bmatrix} 1 & 3 \\ 2 & 4 \end{bmatrix} \begin{bmatrix} 2 & 5 \\ 1 & 4 \end{bmatrix} = \begin{bmatrix} 5 & 17 \\ 8 & 26 \end{bmatrix}$$

and

$$\begin{bmatrix} 2 & 5 \\ 1 & 4 \end{bmatrix} \begin{bmatrix} 1 & 3 \\ 2 & 4 \end{bmatrix} = \begin{bmatrix} 12 & 26 \\ 9 & 19 \end{bmatrix}$$

That is, matrix multiplication is **not commutative.**
As in Example 2, note that

$$\begin{bmatrix} 1 & 2 \\ 2 & 2 \end{bmatrix} \begin{bmatrix} 1 & 1 & 1 \\ 1 & 1 & 1 \end{bmatrix} = \begin{bmatrix} 3 & 3 & 3 \\ 4 & 4 & 4 \end{bmatrix}$$

but

$$\begin{bmatrix} 1 & 1 & 1 \\ 1 & 1 & 1 \end{bmatrix} \begin{bmatrix} 1 & 2 \\ 2 & 2 \end{bmatrix} \text{ is } \textit{not} \text{ defined.}$$

A further difference from normal algebra is that we can have $\mathbf{A}.\mathbf{B} = \mathbf{0}$ = null matrix, without either $\mathbf{A} = \mathbf{0}$ or $\mathbf{B} = \mathbf{0}$; i.e. neither **A** nor **B** is a null matrix.

e.g.

$$\begin{bmatrix} 2 & 1 & 4 \\ 4 & 2 & 8 \end{bmatrix} \begin{bmatrix} 3 & 2 \\ 2 & 4 \\ -2 & -2 \end{bmatrix} = \begin{bmatrix} 0 & 0 \\ 0 & 0 \end{bmatrix}$$

However, with matrices of the correct shape for which multiplication can be defined,

$$\mathbf{A}.\mathbf{0} \equiv \mathbf{0}$$

Problems I

1. Find, where they exist, the matrices $[\mathbf{A} + \mathbf{B}]$, $[\mathbf{A} - \mathbf{B}]$, $[\mathbf{AB}]$, $[\mathbf{BA}]$, $[\mathbf{A} + \mathbf{B}][\mathbf{A} - \mathbf{B}]$ and $[\mathbf{A}^2 - \mathbf{B}^2]$, if:

   (i)
   $$\mathbf{A} = \begin{bmatrix} 4 & -5 \\ 2 & 0 \\ -6 & 3 \end{bmatrix} \qquad \mathbf{B} = \begin{bmatrix} -1 & 0 \\ 0 & 7 \\ 2 & -5 \end{bmatrix}$$

(ii)

$$\mathbf{A} = \begin{bmatrix} 2 & 3 & 4 \\ 1 & 5 & 6 \end{bmatrix} \qquad \mathbf{B} = \begin{bmatrix} 1 \\ 2 \\ 3 \end{bmatrix}$$

(iii)

$$\mathbf{A} = \begin{bmatrix} 3 & -4 \\ 1 & 5 \\ -2 & 2 \end{bmatrix} \qquad \mathbf{B} = \begin{bmatrix} 1 & 2 & 1 \\ 4 & 0 & 2 \end{bmatrix}$$

(iv)

$$\mathbf{A} = \begin{bmatrix} 1 & 2 & 3 \\ -4 & 0 & 6 \end{bmatrix} \qquad \mathbf{B} = \begin{bmatrix} 2 & 6 \\ 8 & 4 \\ -1 & 0 \\ 0 & 0 \end{bmatrix}$$

(v)

$$\mathbf{A} = \begin{bmatrix} 2 & -1 & 1 \\ 0 & 1 & 2 \\ 1 & 0 & 1 \end{bmatrix} \qquad \mathbf{B} = \begin{bmatrix} 0 & 3 & -5 \\ 2 & 0 & 6 \\ 1 & -1 & 4 \end{bmatrix}$$

2. If the dimensions of matrices **A**, **B** and **C** are such as to make both products possible, then $[\mathbf{AB}]\,\mathbf{C} = \mathbf{A}\,[\mathbf{BC}]$. Verify this for the case where

$$\mathbf{A} = \begin{bmatrix} 1 & 2 \\ -5 & 4 \\ 3 & -8 \end{bmatrix} \qquad \mathbf{B} = \begin{bmatrix} 1 & 0 & -1 & 4 \\ -5 & 6 & 7 & 0 \end{bmatrix} \qquad \mathbf{C} = \begin{bmatrix} 1 \\ 2 \\ 1 \\ 2 \end{bmatrix}$$

(This means that matrix multiplication is *associative*.)

3. For 2 X 2 and 3 X 3 matrices, show that $\mathbf{AA}^{\mathrm{T}}$ is a **symmetric** matrix – that is, a matrix of the general form

$$\begin{bmatrix} a_{11} & a_{12} & a_{13} & \cdot\ \cdot \\ a_{12} & a_{22} & a_{23} & \cdot\ \cdot \\ a_{13} & a_{23} & a_{33} & \cdot\ \cdot \\ \vdots & \vdots & \vdots & \ddots \end{bmatrix} \qquad (\text{or } a_{ij} = a_{ji})$$

Verify this for the matrix

$$\mathbf{A} = \begin{bmatrix} 1 & 0 & 7 \\ -3 & 5 & -4 \\ 0 & 2 & 1 \end{bmatrix}$$

Matrix notation for simultaneous equations

For simultaneous equations we now have a very simple way of writing down the system.

Take the first example we considered.

$$\begin{cases} 3x - y = 2 \\ x + y = 4 \end{cases}$$

We can write these equations as

$$\begin{bmatrix} 3 & -1 \\ 1 & 1 \end{bmatrix} \begin{bmatrix} x \\ y \end{bmatrix} = \begin{bmatrix} 2 \\ 4 \end{bmatrix}$$

Our next worked example was the solution of the equations

$$\begin{cases} 2x_1 - 4x_2 + x_3 = -9 \\ 4x_1 + 2x_2 - 3x_3 = 17 \\ x_1 - x_2 + 5x_3 = -11 \end{cases}$$

We could write this system of equations in the form

$$\begin{bmatrix} 2 & -4 & 1 \\ 4 & 2 & -3 \\ 1 & -1 & 5 \end{bmatrix} \begin{bmatrix} x_1 \\ x_2 \\ x_3 \end{bmatrix} = \begin{bmatrix} -9 \\ 17 \\ -11 \end{bmatrix}$$

Furthermore when using Gauss elimination we have been working with this matrix notation. We could write the next step in the process as

$$\begin{bmatrix} 2 & -4 & 1 \\ 0 & 10 & -5 \\ 0 & 1 & \frac{9}{2} \end{bmatrix} \begin{bmatrix} x_1 \\ x_2 \\ x_3 \end{bmatrix} = \begin{bmatrix} -9 \\ 35 \\ -\frac{13}{2} \end{bmatrix}$$

and the next as

$$\begin{bmatrix} 2 & -4 & 1 \\ 0 & 2 & -1 \\ 0 & 1 & \frac{9}{2} \end{bmatrix} \begin{bmatrix} x_1 \\ x_2 \\ x_3 \end{bmatrix} = \begin{bmatrix} -9 \\ 7 \\ -\frac{13}{2} \end{bmatrix}$$

with finally

$$\begin{bmatrix} 2 & -4 & 1 \\ 0 & 2 & -1 \\ 0 & 0 & 5 \end{bmatrix} \begin{bmatrix} x_1 \\ x_2 \\ x_3 \end{bmatrix} = \begin{bmatrix} -9 \\ 7 \\ -10 \end{bmatrix}$$

Unit Matrix

The special square matrices of the form

$$[1] \quad \begin{bmatrix} 1 & 0 \\ 0 & 1 \end{bmatrix} \begin{bmatrix} 1 & 0 & 0 \\ 0 & 1 & 0 \\ 0 & 0 & 1 \end{bmatrix} \begin{bmatrix} 1 & 0 & 0 & 0 \\ 0 & 1 & 0 & 0 \\ 0 & 0 & 1 & 0 \\ 0 & 0 & 0 & 1 \end{bmatrix}$$

and so on which have ones on the main diagonal and zeros everywhere else are called **unit matrices**. Any unit matrix is usually denoted by **I**. A unit matrix plays the same role in matrix algebra as "1" in ordinary arithmetic since, for appropriately shaped matrices

$$\mathbf{AI} = \mathbf{IA} = \mathbf{A} \qquad \mathbf{BI} = \mathbf{B}$$

e.g.

$$\begin{bmatrix} 1 & 1 & 1 \\ 2 & 0 & 4 \\ -1 & 4 & 7 \end{bmatrix} \begin{bmatrix} 1 & 0 & 0 \\ 0 & 1 & 0 \\ 0 & 0 & 1 \end{bmatrix} = \begin{bmatrix} 1 & 1 & 1 \\ 2 & 0 & 4 \\ -1 & 4 & 7 \end{bmatrix} = \begin{bmatrix} 1 & 0 & 0 \\ 0 & 1 & 0 \\ 0 & 0 & 1 \end{bmatrix} \begin{bmatrix} 1 & 1 & 1 \\ 2 & 0 & 4 \\ -1 & 4 & 7 \end{bmatrix}$$

$$\begin{bmatrix} 1 & 2 & 3 \\ -1 & 1 & 4 \end{bmatrix} \begin{bmatrix} 1 & 0 & 0 \\ 0 & 1 & 0 \\ 0 & 0 & 1 \end{bmatrix} = \begin{bmatrix} 1 & 2 & 3 \\ -1 & 1 & 4 \end{bmatrix}$$

IB is not defined in this case unless we choose **I** to be 2 × 2 when the result is

$$\begin{bmatrix} 1 & 0 \\ 0 & 1 \end{bmatrix} \begin{bmatrix} 1 & 2 & 3 \\ -1 & 1 & 4 \end{bmatrix} = \begin{bmatrix} 1 & 2 & 3 \\ -1 & 1 & 4 \end{bmatrix}$$

Also,

$$\begin{bmatrix} 1 & 2 \end{bmatrix} \begin{bmatrix} 1 & 0 \\ 0 & 1 \end{bmatrix} = \begin{bmatrix} 1 & 2 \end{bmatrix}$$

Notice that when we use Gauss-Jordan elimination the aim is to produce a unit matrix on the left hand side. Taking the example on page 231 further, we had reached the stage

$$\begin{bmatrix} 2 & -4 & 1 \\ 0 & 2 & -1 \\ 0 & 0 & 5 \end{bmatrix} \begin{bmatrix} x_1 \\ x_2 \\ x_3 \end{bmatrix} = \begin{bmatrix} -9 \\ 7 \\ -10 \end{bmatrix}$$

and you may remember that the Gauss-Jordan method then gave

$$\begin{bmatrix} 2 & -4 & 0 \\ 0 & 2 & 0 \\ 0 & 0 & 5 \end{bmatrix} \begin{bmatrix} x_1 \\ x_2 \\ x_3 \end{bmatrix} = \begin{bmatrix} -7 \\ 5 \\ -10 \end{bmatrix}$$

then

$$\begin{bmatrix} 2 & 0 & 0 \\ 0 & 2 & 0 \\ 0 & 0 & 5 \end{bmatrix} \begin{bmatrix} x_1 \\ x_2 \\ x_3 \end{bmatrix} = \begin{bmatrix} 3 \\ 5 \\ -10 \end{bmatrix}$$

and finally

$$\begin{bmatrix} 1 & 0 & 0 \\ 0 & 1 & 0 \\ 0 & 0 & 1 \end{bmatrix} \begin{bmatrix} x_1 \\ x_2 \\ x_3 \end{bmatrix} = \begin{bmatrix} \frac{3}{2} \\ \frac{5}{2} \\ -2 \end{bmatrix}$$

But the left-hand side is $\mathbf{I} \begin{bmatrix} x_1 \\ x_2 \\ x_3 \end{bmatrix}$ and this is just $\begin{bmatrix} x_1 \\ x_2 \\ x_3 \end{bmatrix}$

Therefore

$$\begin{bmatrix} x_1 \\ x_2 \\ x_3 \end{bmatrix} = \begin{bmatrix} \frac{3}{2} \\ \frac{5}{2} \\ 2 \end{bmatrix}$$

You can perhaps see now how useful are matrix notation and theory.

Problems II

1. Show that

$$\begin{bmatrix} 1 & 3 & 3 \\ 1 & 4 & 3 \\ 1 & 3 & 4 \end{bmatrix} \begin{bmatrix} 7 & -3 & -3 \\ -1 & 1 & 0 \\ -1 & 0 & 1 \end{bmatrix} = \mathbf{I}, \text{ the unit matrix.}$$

2. Show that

$$\text{if } \mathbf{A} = \begin{bmatrix} \frac{3}{13} & \frac{4}{13} & \frac{12}{13} \\ \frac{4}{5} & -\frac{3}{5} & 0 \\ \frac{36}{65} & \frac{48}{65} & -\frac{25}{65} \end{bmatrix} \text{ then } \mathbf{AA}^T = \mathbf{I}$$

Multiplication by Special Matrices

Special multiplications involving the columns of the unit matrix are most important. Consider for example

$$\begin{bmatrix} a_{11} & a_{12} & a_{13} \\ a_{21} & a_{22} & a_{23} \\ a_{31} & a_{32} & a_{33} \end{bmatrix} \begin{bmatrix} 1 \\ 0 \\ 0 \end{bmatrix}$$

Performing the multiplication we just get

$$\begin{bmatrix} a_{11} \\ a_{21} \\ a_{31} \end{bmatrix}$$

that is, the first column of the matrix **A**.

Similarly,

$$\begin{bmatrix} a_{11} & a_{12} & a_{13} \\ a_{21} & a_{22} & a_{23} \\ a_{31} & a_{32} & a_{33} \end{bmatrix} \begin{bmatrix} 0 \\ 1 \\ 0 \end{bmatrix} = \begin{bmatrix} a_{12} \\ a_{22} \\ a_{32} \end{bmatrix} = \text{second column of matrix } \mathbf{A}$$

and

$$\begin{bmatrix} a_{11} & a_{12} & a_{13} \\ a_{21} & a_{22} & a_{23} \\ a_{31} & a_{32} & a_{33} \end{bmatrix} \begin{bmatrix} 0 \\ 0 \\ 1 \end{bmatrix} = \begin{bmatrix} a_{13} \\ a_{23} \\ a_{33} \end{bmatrix} = \text{third column of matrix } \mathbf{A}$$

Although we have only illustrated work with columns of the 3×3 unit matrix it is obvious that it will be exactly the same for unit matrices of other sizes.

Problems III

1. Show that pre-multiplying the matrix

$$\mathbf{A} = \begin{bmatrix} 1 & 2 & 3 \\ 4 & 5 & 6 \\ 7 & 8 & 9 \end{bmatrix}$$

by the matrix

$$\begin{bmatrix} 1 & 0 & 0 \\ 0 & 0 & 1 \\ 0 & 1 & 0 \end{bmatrix}$$

interchanges rows two and three and that post-multiplying by

$$\begin{bmatrix} 0 & 0 & 1 \\ 0 & 1 & 0 \\ 1 & 0 & 0 \end{bmatrix}$$

interchanges rows one and three.

2. Show that the effect of post-multiplying a matrix with three columns by the matrix

$$\begin{bmatrix} 1 & 0 & 0 \\ 0 & 1 & 0 \\ 2 & 0 & 1 \end{bmatrix}$$

is to replace the first column by (the first column + twice the third column).

Generally speaking, the rule is: do to the unit matrix that which you want to do to the given matrix, post-multiplying if the effect is to be on columns, pre-multiplying if the effect is to be on rows.

3. Find matrices which will perform each of the following operations on a 3×3 matrix and verify your choice by application to matrix **A** of Problem 1:
   (a) add three times row 3 to row 1;
   (b) halve column 2;
   (c) subtract row 3 from row 2.

Inverse of a Matrix

Let the matrix **A** be square, then if we can find a matrix **B** which is such that **AB** = **I** = **BA**, then **B** is said to be the **inverse** of **A** and it is written $\mathbf{A}^{-1}$. (Notice **A** must be square for us to be able to form both **AB** and **BA**.)
e.g.

$$\mathbf{A} = \begin{bmatrix} 2 & 3 & -1 \\ 0 & 1 & 1 \\ -1 & 2 & 1 \end{bmatrix} \qquad \mathbf{B} = \frac{1}{6}\begin{bmatrix} 1 & 5 & -4 \\ 1 & -1 & 2 \\ -1 & 7 & -2 \end{bmatrix}$$

$$\mathbf{AB} = \frac{1}{6}\begin{bmatrix} 6 & 0 & 0 \\ 0 & 6 & 0 \\ 0 & 0 & 6 \end{bmatrix} = \begin{bmatrix} 1 & 0 & 0 \\ 0 & 1 & 0 \\ 0 & 0 & 1 \end{bmatrix} = \mathbf{I}$$

and

$$\mathbf{BA} = \frac{1}{6}\begin{bmatrix} 6 & 0 & 0 \\ 0 & 6 & 0 \\ 0 & 0 & 6 \end{bmatrix} = \mathbf{I}$$

Thus $\mathbf{B} = \mathbf{A}^{-1}$

If the inverse exists, then it is unique.

Suppose there are two matrices **B** and **C** which are such that

$$\mathbf{AB} = \mathbf{I} = \mathbf{BA}$$

and

$$\mathbf{AC} = \mathbf{I} = \mathbf{CA}$$

Then

$$\mathbf{AB} - \mathbf{AC} = \mathbf{0}$$

Premultiplication by **B** gives $\mathbf{BA}(\mathbf{B} - \mathbf{C}) = \mathbf{B0} = \mathbf{0}$
That is,

$$\mathbf{I}(\mathbf{B} - \mathbf{C}) = \mathbf{0}$$

or

$$\mathbf{B} - \mathbf{C} = \mathbf{0}$$

(since multiplying by **I** leaves the matrix unchanged)

Hence

$$\mathbf{B} = \mathbf{C}$$

To find the inverse of a matrix

Consider the set of simultaneous equations

$$\begin{bmatrix} a_{11} & a_{12} & a_{13} \\ a_{21} & a_{22} & a_{23} \\ a_{31} & a_{32} & a_{33} \end{bmatrix} \begin{bmatrix} x_1 \\ x_2 \\ x_3 \end{bmatrix} = \begin{bmatrix} b_1 \\ b_2 \\ b_3 \end{bmatrix}$$

This can be represented in matrix notation as

$$\mathbf{AX} = \mathbf{B}$$

First of all, if we use Gauss-Jordan elimination we transform this to

$$\mathbf{IX} = \mathbf{B}'$$

Looking at the equation $\mathbf{AX} = \mathbf{B}$ from a different viewpoint, if we pre-multiply both sides of the equation by $\mathbf{A}^{-1}$ we get

$$\mathbf{A}^{-1}\mathbf{AX} = \mathbf{A}^{-1}\mathbf{B}$$

or

$$\mathbf{IX} = \mathbf{A}^{-1}\mathbf{B}$$

Thus

$$\mathbf{B}' = \mathbf{A}^{-1}\mathbf{B}$$

By picking **B** as the special column

$$\begin{bmatrix} 1 \\ 0 \\ 0 \end{bmatrix}$$

we know that $\mathbf{A}^{-1}\mathbf{B}$ will be the first column of $\mathbf{A}^{-1}$. Similarly by choosing **B** as

$$\begin{bmatrix} 0 \\ 1 \\ 0 \end{bmatrix}$$

we shall obtain the second column of $\mathbf{A}^{-1}$ and finally if **B** is chosen as

$$\begin{bmatrix} 0 \\ 0 \\ 1 \end{bmatrix}$$

we get the third column of $\mathbf{A}^{-1}$.

We can thus find all the columns of $\mathbf{A}^{-1}$.

Example 3

Let us take an example: find the inverse of

$$\mathbf{A} = \begin{bmatrix} 2 & 3 & -1 \\ 0 & 1 & 1 \\ -1 & 2 & 1 \end{bmatrix}$$

Consider the equation

$$\begin{bmatrix} 2 & 3 & -1 \\ 0 & 1 & 1 \\ -1 & 2 & 1 \end{bmatrix} \begin{bmatrix} x_1 \\ x_2 \\ x_3 \end{bmatrix} = \begin{bmatrix} 1 \\ 0 \\ 0 \end{bmatrix}$$

We get

$$\begin{bmatrix} 2 & 3 & -1 \\ 0 & 1 & 1 \\ 0 & \frac{7}{2} & \frac{1}{2} \end{bmatrix} \begin{bmatrix} x_1 \\ x_2 \\ x_3 \end{bmatrix} = \begin{bmatrix} 1 \\ 0 \\ \frac{1}{2} \end{bmatrix} \qquad \text{Row 3} + (\tfrac{1}{2} \times \text{Row 1})$$

$$\begin{bmatrix} 2 & 3 & -1 \\ 0 & 1 & 1 \\ 0 & 0 & -3 \end{bmatrix} \begin{bmatrix} x_1 \\ x_2 \\ x_3 \end{bmatrix} = \begin{bmatrix} 1 \\ 0 \\ \frac{1}{2} \end{bmatrix} \qquad \text{Row 3} - (7/2 \times \text{Row 2})$$

$$\begin{bmatrix} 2 & 3 & 0 \\ 0 & 1 & 0 \\ 0 & 0 & -3 \end{bmatrix}\begin{bmatrix} x_1 \\ x_2 \\ x_3 \end{bmatrix} = \begin{bmatrix} \frac{5}{6} \\ \frac{1}{6} \\ \frac{1}{2} \end{bmatrix} \quad \begin{matrix} \text{Row 1} - (1/3 \times \text{Row 3}) \\ \text{Row 2} + (1/3 \times \text{Row 3}) \\ \end{matrix}$$

$$\begin{bmatrix} 2 & 0 & 0 \\ 0 & 1 & 0 \\ 0 & 0 & -3 \end{bmatrix}\begin{bmatrix} x_1 \\ x_2 \\ x_3 \end{bmatrix} = \begin{bmatrix} \frac{2}{6} \\ \frac{1}{6} \\ \frac{1}{2} \end{bmatrix} \quad \text{Row 1} - (3 \times \text{Row 2})$$

and finally

$$\begin{bmatrix} 1 & 0 & 0 \\ 0 & 1 & 0 \\ 0 & 0 & 1 \end{bmatrix}\begin{bmatrix} x_1 \\ x_2 \\ x_3 \end{bmatrix} = \begin{bmatrix} \frac{1}{6} \\ \frac{1}{6} \\ -\frac{1}{6} \end{bmatrix} \quad \begin{matrix} \text{Row 1} \div 2 \\ \\ \text{Row 3} \div (-3) \end{matrix}$$

Thus the first column of $\mathbf{A}^{-1}$ is

$$\begin{bmatrix} \frac{1}{6} \\ \frac{1}{6} \\ -\frac{1}{6} \end{bmatrix}$$

We now consider

$$\begin{bmatrix} 2 & 3 & -1 \\ 0 & 1 & 1 \\ -1 & 2 & 1 \end{bmatrix}\begin{bmatrix} x_1 \\ x_2 \\ x_3 \end{bmatrix} = \begin{bmatrix} 0 \\ 1 \\ 0 \end{bmatrix}$$

We shall perform exactly the same operations to reduce the l.h.s. to **IX**. Hence we get for r.h.s.

$$\begin{bmatrix} 0 \\ 1 \\ 0 \end{bmatrix} \qquad \begin{bmatrix} 0 \\ 1 \\ 0 \end{bmatrix} \qquad \begin{bmatrix} 0 \\ 1 \\ -\frac{7}{2} \end{bmatrix} \qquad \begin{bmatrix} \frac{7}{6} \\ -\frac{1}{6} \\ -\frac{7}{2} \end{bmatrix}$$

Row 3 + (½ Row 1) Row 3 − (7/2 × Row 2) Row 1 − (1/3 × Row 3)
Row 2 + (1/3 × Row 3)

$$\begin{bmatrix} \frac{10}{6} \\ -\frac{1}{6} \\ -\frac{7}{2} \end{bmatrix} \qquad \begin{bmatrix} \frac{5}{6} \\ -\frac{1}{6} \\ \frac{7}{6} \end{bmatrix}$$

Row 1 − (3 × Row 2)   Row 1 ÷ 2, Row 3 ÷ (−3)

This is the 2nd column of $\mathbf{A}^{-1}$

Similarly for the third column, giving

$$\begin{bmatrix} -\frac{4}{6} \\ \frac{2}{6} \\ -\frac{2}{6} \end{bmatrix}$$

Hence

$$\mathbf{A}^{-1} = \frac{1}{6}\begin{bmatrix} 1 & 5 & -4 \\ 1 & -1 & 2 \\ -1 & 7 & -2 \end{bmatrix}$$

Now since we are performing the same operations each time with different right-hand sides we could set up the system in tabloid form dealing with all of the right-hand sides simultaneously as follows.

$$\begin{bmatrix} 2 & 3 & -1 \\ 0 & 1 & 1 \\ -1 & 2 & 1 \end{bmatrix}\begin{bmatrix} 1 & 0 & 0 \\ 0 & 1 & 0 \\ 0 & 0 & 1 \end{bmatrix}$$

$$\begin{bmatrix} 2 & 3 & -1 \\ 0 & 1 & 1 \\ 0 & \frac{7}{2} & \frac{1}{2} \end{bmatrix}\begin{bmatrix} 1 & 0 & 0 \\ 0 & 1 & 0 \\ \frac{1}{2} & 0 & 1 \end{bmatrix} \quad \text{Row 3 + (½ × Row 1)}$$

$$\begin{bmatrix} 2 & 3 & -1 \\ 0 & 1 & 1 \\ 0 & 0 & -3 \end{bmatrix}\begin{bmatrix} 1 & 0 & 0 \\ 0 & 1 & 0 \\ \frac{1}{2} & -\frac{7}{2} & 1 \end{bmatrix} \quad \text{Row 3 − (7/2 × Row 2)}$$

$$\begin{bmatrix} 2 & 3 & 0 \\ 0 & 1 & 0 \\ 0 & 0 & -3 \end{bmatrix}\begin{bmatrix} \frac{5}{6} & \frac{7}{6} & -\frac{1}{3} \\ \frac{1}{6} & -\frac{1}{6} & \frac{1}{3} \\ \frac{1}{2} & -\frac{7}{2} & 1 \end{bmatrix} \quad \begin{matrix} \text{Row 1 − (1/3 × Row 3)} \\ \text{Row 2 + (1/3 × Row 3)} \\ \end{matrix}$$

$$\begin{bmatrix} 2 & 0 & 0 \\ 0 & 1 & 0 \\ 0 & 0 & -3 \end{bmatrix} \begin{bmatrix} \frac{2}{6} & \frac{10}{6} & -\frac{8}{6} \\ \frac{1}{6} & -\frac{1}{6} & \frac{1}{3} \\ \frac{1}{2} & -\frac{7}{2} & 1 \end{bmatrix} \quad \text{Row 1} - (3 \times \text{Row 2})$$

$$\begin{bmatrix} 1 & 0 & 0 \\ 0 & 1 & 0 \\ 0 & 0 & 1 \end{bmatrix} \begin{bmatrix} \frac{1}{6} & \frac{5}{6} & -\frac{4}{6} \\ \frac{1}{6} & -\frac{1}{6} & \frac{2}{6} \\ -\frac{1}{6} & \frac{7}{6} & -\frac{2}{6} \end{bmatrix} \quad \begin{matrix} \text{Row 1} \div 2 \\ \\ \text{Row 3} \div (-3) \end{matrix}$$

$$\uparrow$$

$$A^{-1}$$

Notice that **A** has become **I** and **I** has become $\mathbf{A}^{-1}$. We have assumed that $\mathbf{A}^{-1}$ does in fact exist – there are cases when we cannot find $\mathbf{A}^{-1}$ and we shall be able to recognise these cases after the next section. Also, there is a direct way of finding $\mathbf{A}^{-1}$ if it exists which we can show you after the next section which is on determinants.

The computer program for Gauss-Jordan elimination can be modified to invert a matrix. There is no column of constants to be catered for, but there are two arrays **A** and **B**. The necessary operations on **A** are repeated on **B**.

Problems IV

1. Find the inverses where possible of the following matrices

(i) $\begin{bmatrix} -2 & 3 \\ 1 & 4 \end{bmatrix}$ (ii) $\begin{bmatrix} 2 & 3 \\ 1.8 & 2.7 \end{bmatrix}$

(iii) $\begin{bmatrix} 1 & 2 & 3 \\ 2 & -2 & -4 \\ 3 & -4 & -3 \end{bmatrix}$ (iv) $\begin{bmatrix} 1 & 3 & 3 \\ 1 & 4 & 3 \\ 1 & 3 & 4 \end{bmatrix}$

(v) $\begin{bmatrix} 1 & 3 & 3 \\ 1 & 4 & 3 \\ 1 & 5 & 3 \end{bmatrix}$ (vi) $\begin{bmatrix} 1 & -1 & 2 & 1 \\ -1 & 0 & 3 & 2 \\ 2 & 1 & 0 & -1 \\ 2 & -2 & 1 & 3 \end{bmatrix}$

5.4 DETERMINANTS

We have already met determinants of order two of the form

$$\begin{vmatrix} a_{11} & a_{12} \\ a_{21} & a_{22} \end{vmatrix}$$

The order of a determinant is the number of rows (or columns) it has. The value of this determinant is $a_{11}a_{22} - a_{12}a_{21}$. Here again we emphasise this is simply a *number* in contrast to the corresponding *matrix* which is an array.

We mentioned briefly in the solution of linear equations the third order determinant

$$D = \begin{vmatrix} a_{11} & a_{12} & a_{13} \\ a_{21} & a_{22} & a_{23} \\ a_{31} & a_{32} & a_{33} \end{vmatrix}$$

If the corresponding matrix

$$\begin{bmatrix} a_{11} & a_{12} & a_{13} \\ a_{21} & a_{22} & a_{23} \\ a_{31} & a_{32} & a_{33} \end{bmatrix}$$

is written **A**, this determinant is written $|A|$, pronounced 'det A'.

We wish to study this and higher order determinants a little more fully.

Formally, the value of $|A|$ is given by

$$|A| = a_{11}\begin{vmatrix} a_{22} & a_{23} \\ a_{32} & a_{33} \end{vmatrix} - a_{12}\begin{vmatrix} a_{21} & a_{23} \\ a_{31} & a_{33} \end{vmatrix} + a_{13}\begin{vmatrix} a_{21} & a_{22} \\ a_{31} & a_{32} \end{vmatrix} \tag{5.14}$$

which is called *expanding along the top row*. Notice that each element in the top row is multiplied by the determinant of the elements left when the column and row through that element are struck out. For example, looking at the element a_{12}

$$\begin{vmatrix} a_{11} - - & a_{12} & - - a_{13} \\ a_{21} & a_{22} & a_{23} \\ a_{31} & a_{32} & a_{33} \end{vmatrix}$$

the determinant of elements left over is

$$\begin{vmatrix} a_{21} & a_{23} \\ a_{31} & a_{33} \end{vmatrix}$$

These determinants left are called **minors**.

So

$$\begin{vmatrix} a_{21} & a_{23} \\ a_{31} & a_{33} \end{vmatrix}$$

is the **minor** of a_{12}.

You will notice in relation to (5.14) that a minus sign is allocated to a_{12}; this arises purely from position according to the pattern of signs

$$\begin{matrix} + & - & + \\ - & + & - \\ + & - & + \end{matrix}$$

For any size determinant we always start with a + in the top left position and from this position work both horizontally and vertically with alternating signs

$$\begin{matrix} + & \rightarrow - & \rightarrow + \\ \downarrow & & \\ - & & \\ \downarrow & & \\ + & & \end{matrix}$$

If we write down a minor and multiply this by its sign from the position pattern we then call it a **cofactor.** So

$$-\begin{vmatrix} a_{21} & a_{23} \\ a_{31} & a_{33} \end{vmatrix}$$

is the **cofactor** of a_{12}, and is written A_{12}. In general, the cofactor of a_{ij} is written A_{ij}.

The rules for a larger order determinant are exactly the same.

e.g.

$$\begin{vmatrix} a_{11} & a_{12} & a_{13} & a_{14} \\ a_{21} & a_{22} & a_{23} & a_{24} \\ a_{31} & a_{32} & a_{33} & a_{34} \\ a_{41} & a_{42} & a_{43} & a_{44} \end{vmatrix} = a_{11}\begin{vmatrix} a_{22} & a_{23} & a_{24} \\ a_{32} & a_{33} & a_{34} \\ a_{42} & a_{43} & a_{44} \end{vmatrix} - a_{12}\begin{vmatrix} a_{21} & a_{23} & a_{24} \\ a_{31} & a_{33} & a_{34} \\ a_{41} & a_{43} & a_{44} \end{vmatrix}$$

$$+ a_{13}\begin{vmatrix} a_{21} & a_{22} & a_{24} \\ a_{31} & a_{32} & a_{34} \\ a_{41} & a_{42} & a_{44} \end{vmatrix} - a_{14}\begin{vmatrix} a_{21} & a_{22} & a_{23} \\ a_{31} & a_{32} & a_{33} \\ a_{41} & a_{42} & a_{43} \end{vmatrix}$$

Notice the sign pattern is

$$\begin{matrix} + & - & + & - \\ - & + & - & + \\ + & - & + & - \\ - & + & - & + \end{matrix}$$

We then expand the 3rd order determinants as before.

In the two following examples, we shall evaluate two determinants.

Example 1

$$\begin{vmatrix} 1 & -1 & 3 \\ 0 & 2 & 5 \\ -2 & 1 & 6 \end{vmatrix} = 1\begin{vmatrix} 2 & 5 \\ 1 & 6 \end{vmatrix} - (-1)\begin{vmatrix} 0 & 5 \\ -2 & 6 \end{vmatrix} + 3\begin{vmatrix} 0 & 2 \\ -2 & 1 \end{vmatrix}$$

$$= 7 + 10 + 12 \quad = 29$$

Example 2

$$\begin{vmatrix} 1 & 2 & 1 & 2 \\ 2 & 1 & 2 & 1 \\ 0 & 1 & 1 & 0 \\ 0 & 4 & -1 & 3 \end{vmatrix} = 1\begin{vmatrix} 1 & 2 & 1 \\ 1 & 1 & 0 \\ 4 & -1 & 3 \end{vmatrix} - 2\begin{vmatrix} 2 & 2 & 1 \\ 0 & 1 & 0 \\ 0 & -1 & 3 \end{vmatrix} + 1\begin{vmatrix} 2 & 1 & 1 \\ 0 & 1 & 0 \\ 0 & 4 & 3 \end{vmatrix} - 2\begin{vmatrix} 2 & 1 & 2 \\ 0 & 1 & 1 \\ 0 & 4 & -1 \end{vmatrix}$$

$$= 1\left\{1\begin{vmatrix} 1 & 0 \\ -1 & 3 \end{vmatrix} - 2\begin{vmatrix} 1 & 0 \\ 4 & 3 \end{vmatrix} + 1\begin{vmatrix} 1 & 1 \\ 4 & -1 \end{vmatrix}\right\}$$

$$- 2\left\{2\begin{vmatrix} 1 & 0 \\ -1 & 3 \end{vmatrix} - 2\begin{vmatrix} 0 & 0 \\ 0 & 3 \end{vmatrix} + 1\begin{vmatrix} 0 & 1 \\ 0 & -1 \end{vmatrix}\right\}$$

$$+ 1\left\{2\begin{vmatrix} 1 & 0 \\ 4 & 3 \end{vmatrix} - 1\begin{vmatrix} 0 & 0 \\ 0 & 3 \end{vmatrix} + 1\begin{vmatrix} 0 & 1 \\ 0 & 4 \end{vmatrix}\right\}$$

$$- 2\left\{2\begin{vmatrix} 1 & 1 \\ 4 & -1 \end{vmatrix} - 1\begin{vmatrix} 0 & 1 \\ 0 & -1 \end{vmatrix} + 2\begin{vmatrix} 0 & 1 \\ 0 & 4 \end{vmatrix}\right\}$$

$$= 3 - 6 - 5 - 12 + 6 + 20$$

$$= 6$$

Properties of Determinants (no proofs given)

(i) If two rows (or two columns) are interchanged, $|A| \to -|A|$, i.e. the value of the determinant is multiplied by -1.

(ii) If the rows and columns are interchanged (transposed) the value is not changed; $|A^T| = |A|$.

(iii) If two rows or columns are identical, $|A| = 0$.

(iv) If a row or column has a common factor we proceed thus:

$$\begin{vmatrix} a_{11} & a_{12} & a_{13} \\ ka_{21} & ka_{22} & ka_{23} \\ a_{31} & a_{32} & a_{33} \end{vmatrix} = k \begin{vmatrix} a_{11} & a_{12} & a_{13} \\ a_{21} & a_{22} & a_{23} \\ a_{31} & a_{32} & a_{33} \end{vmatrix} = k|\mathrm{A}|$$

(v) $$\begin{vmatrix} a_{11} + ka_{12} + la_{13} & a_{12} + na_{13} & a_{13} \\ a_{21} + ka_{22} + la_{23} & a_{22} + na_{23} & a_{23} \\ a_{31} + ka_{32} + la_{33} & a_{32} + na_{33} & a_{33} \end{vmatrix} = |\mathrm{A}|$$

This means that we can add multiples of corresponding elements of one column to another. (Similarly for rows.)

(vi) $$\begin{vmatrix} a_{11} + \alpha & a_{12} & a_{13} \\ a_{21} + \beta & a_{22} & a_{23} \\ a_{31} + \gamma & a_{32} & a_{33} \end{vmatrix} = \begin{vmatrix} a_{11} & a_{12} & a_{13} \\ a_{21} & a_{22} & a_{23} \\ a_{31} & a_{32} & a_{33} \end{vmatrix} + \begin{vmatrix} \alpha & a_{12} & a_{13} \\ \beta & a_{22} & a_{23} \\ \gamma & a_{32} & a_{33} \end{vmatrix}$$

These properties hold for determinants of any order. Usually, what is true for columns is true for rows and vice versa. Property (i) implies that we can expand a determinant along any row and together with property (ii) it implies that we can expand along any row or column as long as we take account of the sign positions.

We have for instance

$$\begin{vmatrix} a_{11} & a_{12} & a_{13} \\ a_{21} & a_{22} & a_{23} \\ a_{31} & a_{32} & a_{33} \end{vmatrix} = -a_{21} \begin{vmatrix} a_{12} & a_{13} \\ a_{32} & a_{33} \end{vmatrix} + a_{22} \begin{vmatrix} a_{11} & a_{13} \\ a_{31} & a_{33} \end{vmatrix}$$

$$-a_{23} \begin{vmatrix} a_{11} & a_{12} \\ a_{31} & a_{32} \end{vmatrix}$$ (Expand along second row)

$$= a_{13} \begin{vmatrix} a_{21} & a_{22} \\ a_{31} & a_{32} \end{vmatrix} - a_{23} \begin{vmatrix} a_{11} & a_{12} \\ a_{31} & a_{32} \end{vmatrix} + a_{33} \begin{vmatrix} a_{11} & a_{12} \\ a_{21} & a_{22} \end{vmatrix}$$

(Expand along 3rd column).

Property (v) is probably the most useful since it allows us to simplify a determinant by introducing zeros. We shall take one of our previous cxamplcs.

Example 3

$$\begin{vmatrix} 1 & -1 & 3 \\ 0 & 2 & 5 \\ -2 & 1 & 6 \end{vmatrix}$$

Property (v) says that we can add any multiple of one row (or column) to another and not change the value of the determinant. We use this as follows

$$D = \begin{vmatrix} 1 & -1 & 3 \\ 0 & 2 & 5 \\ -2 & 1 & 6 \end{vmatrix} = \begin{vmatrix} 1 & 0 & 3 \\ 0 & 2 & 5 \\ -2 & -1 & 6 \end{vmatrix} \quad \text{add column 1 to column 2}$$

$$= \begin{vmatrix} 1 & 0 & 0 \\ 0 & 2 & 5 \\ -2 & -1 & 12 \end{vmatrix} \quad \text{add column 1} \times (-3) \text{ to column 3}$$

Notice that we could have saved writing by doing both these operations together.

Now if we expand along the top row, we get

$$D = 1\begin{vmatrix} 2 & 5 \\ -1 & 12 \end{vmatrix} - 0\begin{vmatrix} 0 & 5 \\ -2 & 12 \end{vmatrix} + 0\begin{vmatrix} 0 & 2 \\ -2 & -1 \end{vmatrix}$$

$$= \begin{vmatrix} 2 & 5 \\ -1 & 12 \end{vmatrix}$$

So we have effectively reduced the 3rd order determinant to a second order determinant giving

$$D = 12 \times 2 - 5 \times (-1) = 29$$

We could, of course, have obtained zeros in any other row or column. It would have been easier to work on the first column since there is already one zero there as follows.

$$D = \begin{vmatrix} 1 & -1 & 3 \\ 0 & 2 & 5 \\ -2 & 1 & 6 \end{vmatrix} = \begin{vmatrix} 1 & -1 & 3 \\ 0 & 2 & 5 \\ 0 & -1 & 12 \end{vmatrix} \quad \text{Add } (2 \times \text{Row 1}) \text{ to Row 3}$$

If we now expand down the first column we get

$$D = 1\begin{vmatrix} 2 & 5 \\ -1 & 12 \end{vmatrix} - 0\begin{vmatrix} \cdot & \cdot \\ \cdot & \cdot \end{vmatrix} + 0\begin{vmatrix} \cdot & \cdot \\ \cdot & \cdot \end{vmatrix}$$

giving the same result as before.

Another useful device is to reduce the determinant to one with zeros everywhere below the main diagonal **(upper triangular)** or to one with zeros everywhere above the main diagonal **(lower triangular)**.

$$\begin{vmatrix} 1 & -1 & 3 \\ 0 & 2 & 5 \\ -2 & 1 & 6 \end{vmatrix} = \begin{vmatrix} 1 & -1 & 3 \\ 0 & 2 & 5 \\ 0 & -1 & 12 \end{vmatrix} \quad \text{Add } (2 \times \text{Row } 1) \text{ to Row } 3$$

$$= \begin{vmatrix} 1 & -1 & 3 \\ 0 & 2 & 5 \\ 0 & 0 & \frac{29}{2} \end{vmatrix} \quad \text{Add } (\tfrac{1}{2} \times \text{Row } 2) \text{ to Row } 3$$

The value of such a determinant is simply the product of its diagonal elements, i.e. $1 \times 2 \times 29/2 = 29$, as before. Try reducing the original determinant to a lower triangular form and check the value thus obtained.

Let us now take an example of a fourth order determinant:

Example 4

$$D = \begin{vmatrix} -3 & 2 & 2 & 5 \\ 3 & -1 & 1 & 4 \\ 6 & -2 & 3 & 7 \\ 8 & -3 & -4 & 2 \end{vmatrix}$$

It will be easiest to avoid fractions by fixing our attention on the "1" present in row 2, column 3, and producing zeros in the row or column in which it lies. We get

$$D = \begin{vmatrix} -3 & 2 & 2 & 5 \\ 3 & -1 & \textcircled{1} & 4 \\ 6 & -2 & 3 & 7 \\ 8 & -3 & -4 & 2 \end{vmatrix}$$

$$= \begin{vmatrix} -9 & 4 & 0 & -3 \\ 3 & -1 & \textcircled{1} & 4 \\ -3 & 1 & 0 & -5 \\ 20 & -7 & 0 & 18 \end{vmatrix} \quad \begin{array}{l} \text{Add } (-2 \times \text{Row } 2) \text{ to Row } 1 \\ \\ \text{Add } (-3 \times \text{Row } 2) \text{ to Row } 3 \\ \text{Add } (4 \times \text{Row } 2) \text{ to Row } 4 \end{array}$$

Now if we expand down the 3rd column we effectively reduce the determinant to one of 3rd order. Remembering the position signs we get

$$D = -1 \begin{vmatrix} -9 & 4 & -3 \\ -3 & 1 & -5 \\ 20 & -7 & 18 \end{vmatrix}$$

There is a "1" here again so we work to manufacture zeros in the row or column containing this. We proceed:

$$D = -1 \begin{vmatrix} 3 & 4 & 17 \\ 0 & 1 & 0 \\ -1 & -7 & -17 \end{vmatrix} \quad \begin{array}{l} \text{Add } (3 \times \text{Column 2}) \text{ to Column 1} \\ \\ \text{Add } (5 \times \text{Column 2}) \text{ to Column 3} \end{array}$$

$$= -1 \begin{vmatrix} 3 & 17 \\ -1 & -17 \end{vmatrix} \quad (\text{Expand along } 2^{\text{nd}} \text{ Row})$$

$$= 34$$

If there is not a "1" present, then we can manufacture one by dividing a whole row or column by one of its elements, taking this outside as a factor. Thus

$$\begin{vmatrix} 3 & 6 & -5 & 4 \\ 5 & 5 & 3 & 5 \\ 9 & -7 & 8 & 4 \\ 8 & -11 & 14 & 19 \end{vmatrix} = 3 \begin{vmatrix} 1 & 6 & -5 & 4 \\ \frac{5}{3} & 5 & 3 & 5 \\ 3 & -7 & 8 & 4 \\ \frac{8}{3} & -11 & 14 & 19 \end{vmatrix}$$

and so on.

Problems I

1. Check properties (i), (ii), (iii) of determinants (see page 255) for a general 3×3 determinant.

2. Evaluate the following determinants:

(i) $\begin{vmatrix} 3 & 1 & -2 \\ 8 & -5 & 7 \\ 4 & 0 & 1 \end{vmatrix}$ (ii) $\begin{vmatrix} 2{\cdot}7 & 3{\cdot}6 & 4{\cdot}1 \\ 2{\cdot}5 & 3{\cdot}9 & 4{\cdot}5 \\ 2{\cdot}9 & 3{\cdot}3 & 4{\cdot}2 \end{vmatrix}$

(iii) $\begin{vmatrix} 2 & 1+i & 4 \\ 1-i & 3 & 2-i \\ 4 & 2+i & 1 \end{vmatrix}$ $(i^2 = -1)$ (iv) $\begin{vmatrix} 2 & 4 & 5 & 7 \\ 3 & 5 & 9 & 2 \\ 4 & 1 & 7 & 3 \\ 2 & 11 & 7 & 13 \end{vmatrix}$

3. Show that

$$\begin{vmatrix} 1 & a & b \\ a & 1 & b \\ a & b & 1 \end{vmatrix} = (a-1)(b-1)(a+b+1)$$

4. Show that

$$\begin{vmatrix} 1 & a & bc \\ 1 & b & ca \\ 1 & c & ab \end{vmatrix} = (a-b)(b-c)(c-a)$$

5. For third order determinants *only* we can use a special rule called **Sarrus Rule** which says: to evaluate

$$\Delta = \begin{vmatrix} a_1 & a_2 & a_3 \\ b_1 & b_2 & b_3 \\ c_1 & c_2 & c_3 \end{vmatrix}$$

repeat the first 2 columns as follows and multiply the diagonals with 3 elements. The arrows pointing down give positive multiplications $a_1b_2c_3$, $a_2b_3c_1$ and $a_3b_1c_2$. Those with arrows pointing up give negative multiplications $-c_1b_2a_3$, $-c_2b_3a_1$, $-c_3b_1a_2$.

$$\begin{vmatrix} a_1 & a_2 & a_3 & a_1 & a_2 \\ b_1 & b_2 & b_3 & b_1 & b_2 \\ c_1 & c_2 & c_3 & c_1 & c_2 \end{vmatrix}$$

Then $\Delta = (a_1b_2c_3 + a_2b_3c_1 + a_3b_1c_2) - (c_1b_2a_3 + c_2b_3a_1 + c_3b_1a_2)$
Using this rule repeat Problems 2(i), 2(ii) and 2(iii).

6. For the determinant

$$D = \begin{vmatrix} 4 & 1 & 1 \\ 1 & 2 & 3 \\ 3 & 1 & 2 \end{vmatrix}$$

find the cofactor of each element and show that

$$D = a_{11}A_{11} + a_{12}A_{12} + a_{13}A_{13}$$
$$D = a_{11}A_{11} + a_{21}A_{21} + a_{31}A_{31}$$
$$D = a_{31}A_{31} + a_{32}A_{32} + a_{33}A_{33}$$

What other expressions can you find which also equal D?

Use of Determinants to solve Simultaneous Equations

We shall not prove the following results for, although the proofs are straightforward, they are tedious.

We have for a set of 2 equations

$$\begin{cases} a_{11}x_1 + a_{12}x_2 = b_1 \\ a_{21}x_1 + a_{22}x_2 = b_2 \end{cases}$$

the result
$$\frac{x_1}{\begin{vmatrix} b_1 & a_{12} \\ b_2 & a_{22} \end{vmatrix}} = \frac{x_2}{\begin{vmatrix} a_{11} & b_1 \\ a_{21} & b_2 \end{vmatrix}} = \frac{1}{\begin{vmatrix} a_{11} & a_{12} \\ a_{21} & a_{22} \end{vmatrix}}$$

This is known as **Cramer's rule** for two simultaneous equations.

This is easily extended to apply to the set of 3 equations

$$\begin{cases} a_{11}x_1 + a_{12}x_2 + a_{13}x_3 = b_1 \\ a_{21}x_1 + a_{22}x_2 + a_{23}x_3 = b_2 \\ a_{31}x_1 + a_{32}x_2 + a_{33}x_3 = b_3 \end{cases}$$

The solution is

$$\frac{x_1}{\begin{vmatrix} b_1 & a_{12} & a_{13} \\ b_2 & a_{22} & a_{23} \\ b_3 & a_{32} & a_{33} \end{vmatrix}} = \frac{x_2}{\begin{vmatrix} a_{11} & b_1 & a_{13} \\ a_{21} & b_2 & a_{23} \\ a_{31} & b_3 & a_{33} \end{vmatrix}} = \frac{x_3}{\begin{vmatrix} a_{11} & a_{12} & b_1 \\ a_{21} & a_{22} & b_2 \\ a_{31} & a_{32} & b_3 \end{vmatrix}} = \frac{1}{\begin{vmatrix} a_{11} & a_{12} & a_{13} \\ a_{21} & a_{22} & a_{23} \\ a_{31} & a_{32} & a_{33} \end{vmatrix}}$$

You can now see why we get inconsistencies if the determinant of coefficients is zero, because we have

$$\frac{x_1}{\begin{vmatrix} \cdot & \cdot \\ \cdot & \cdot \end{vmatrix}} = \frac{x_2}{\begin{vmatrix} \cdot & \cdot \\ \cdot & \cdot \end{vmatrix}} = \frac{x_3}{\begin{vmatrix} \cdot & \cdot \\ \cdot & \cdot \end{vmatrix}} = \frac{1}{0}$$

and division by zero is not defined.

Moreover if this determinant is not zero but small compared with the other determinants involved we are likely to get problems of ill-conditioning.

Problems II

Solve the sets of equations (using determinants).

1. (i) $\begin{cases} 2x - y = 3 \\ x + 5y = 2 \end{cases}$ (ii) $\begin{cases} x - 3y = -2 \\ x + y = 4 \end{cases}$

2. (i) $\begin{cases} 2x_1 - 4x_2 + x_3 = -9 \\ 4x_1 + 2x_2 - 3x_3 = 17 \\ x_1 - x_2 + 5x_3 = -11 \end{cases}$ (ii) $\begin{cases} 25x_1 + 24x_2 + 36x_3 = 12 \\ 4x_1 + 54x_2 + 4x_3 = 27 \\ 9x_1 + 6x_2 + 64x_3 = 3 \end{cases}$

3. (i) $\begin{cases} 5x - y + 2z - 3 = 0 \\ 2x + 4y + z - 8 = 0 \\ x + 3y - 3z - 2 = 0 \end{cases}$ (ii) $\begin{cases} 3x - 4y + 7z = 16 \\ 2x + 3y + z = 11 \\ x - 5z = 14 \end{cases}$

(iii) $$\begin{cases} 4x + 2y + 3z = 3 \\ 8x + 6y - 3z = 4 \\ -12x - 2y + 6z = -2 \end{cases}$$

5.5 FORMAL EVALUATION OF THE INVERSE OF A MATRIX

We have seen how to find the inverse of a matrix by the use of elementary operations. As mentioned then, there is an *alternative method of finding the inverse.*

This procedure is as follows:

(i) Take the matrix **A** and replace each element by its cofactor.
(ii) Transpose the result to form the **adjoint** matrix **adj A**.
(iii) Find the determinant of $A = |A|$.
(iv) Then the inverse matrix $A^{-1} = \mathbf{adj\ A}/|A|$, provided $|A| \neq 0$.

Example 1

$$\mathbf{A} = \begin{bmatrix} 3 & -1 \\ 2 & -2 \end{bmatrix}$$

Replacing each element by its cofactor, we get (Rule of signs $\begin{smallmatrix} + & - \\ - & + \end{smallmatrix}$)

$$\mathbf{adj\ A} = \begin{bmatrix} -2 & -2 \\ 1 & 3 \end{bmatrix}^{T} = \begin{bmatrix} -2 & 1 \\ -2 & 3 \end{bmatrix}$$

$$|A| = -4$$

therefore

$$A^{-1} = -\frac{1}{4}\begin{bmatrix} -2 & 1 \\ -2 & 3 \end{bmatrix}$$

Check

$$\mathbf{AA^{-1}} = -\frac{1}{4}\begin{bmatrix} 3 & -1 \\ 2 & -2 \end{bmatrix}\begin{bmatrix} -2 & 1 \\ -2 & 3 \end{bmatrix} = -\frac{1}{4}\begin{bmatrix} -4 & 0 \\ 0 & -4 \end{bmatrix} = \mathbf{I}$$

and

$$\mathbf{A^{-1}A} = -\frac{1}{4}\begin{bmatrix} -2 & 1 \\ -2 & 3 \end{bmatrix}\begin{bmatrix} 3 & -1 \\ 2 & -2 \end{bmatrix} = -\frac{1}{4}\begin{bmatrix} -4 & 0 \\ 0 & -4 \end{bmatrix} = \mathbf{I}$$

Example 2

$$\mathbf{A} = \begin{bmatrix} 3 & 1 & 2 \\ 5 & -4 & 1 \\ -1 & 2 & 1 \end{bmatrix}$$

Replacing each element by its cofactor and, remembering the rule of signs

$$\begin{matrix} + & - & + \\ - & + & - \\ + & - & + \end{matrix}$$

we have

$$\begin{bmatrix} \begin{vmatrix} -4 & 1 \\ 2 & 1 \end{vmatrix} & -\begin{vmatrix} 5 & 1 \\ -1 & 1 \end{vmatrix} & \begin{vmatrix} 5 & -4 \\ -1 & 2 \end{vmatrix} \\ -\begin{vmatrix} 1 & 2 \\ 2 & 1 \end{vmatrix} & \begin{vmatrix} 3 & 2 \\ -1 & 1 \end{vmatrix} & -\begin{vmatrix} 3 & 1 \\ -1 & 2 \end{vmatrix} \\ \begin{vmatrix} 1 & 2 \\ -4 & 1 \end{vmatrix} & -\begin{vmatrix} 3 & 2 \\ 5 & 1 \end{vmatrix} & \begin{vmatrix} 3 & 1 \\ 5 & 4 \end{vmatrix} \end{bmatrix}$$

$$= \begin{bmatrix} -6 & -6 & 6 \\ 3 & 5 & -7 \\ 9 & 7 & -17 \end{bmatrix}$$

Thus

$$\text{adj } \mathbf{A} = \begin{bmatrix} -6 & -6 & 6 \\ 3 & 5 & -7 \\ 9 & 7 & -17 \end{bmatrix}^{T} = \begin{bmatrix} -6 & 3 & 9 \\ -6 & 5 & 7 \\ 6 & -7 & -17 \end{bmatrix}$$

$$|\mathbf{A}| = \begin{vmatrix} 3 & 1 & 2 \\ 5 & -4 & 1 \\ -1 & 2 & 1 \end{vmatrix} = \begin{vmatrix} 5 & -3 & 0 \\ 6 & -6 & 0 \\ -1 & 2 & 1 \end{vmatrix} \quad \begin{matrix} \text{Row 1} - (2 \times \text{Row 3}) \\ \text{Row 2} - \text{Row 3} \\ \end{matrix}$$

$$= \begin{vmatrix} 5 & -3 \\ 6 & -6 \end{vmatrix}$$

$$= -12$$

A quicker method of finding $|\mathbf{A}|$ in these problems is to remember that $|\mathbf{A}| = a_{11}A_{11} + a_{12}A_{12} + a_{13}A_{13}$ and notice that A_{11}, A_{12}, A_{13} have already been evaluated. Therefore,

$$\begin{aligned} |\mathbf{A}| &= 3 \times (-6) + 1 \times (-6) + 2 \times 6 \\ &= -12, \text{ as before.} \end{aligned}$$

Thus,

$$\mathbf{A}^{-1} = \frac{-1}{12}\begin{bmatrix} -6 & 3 & 9 \\ -6 & 5 & 7 \\ 6 & -7 & -17 \end{bmatrix}$$

Check

$$\begin{aligned} \mathbf{AA}^{-1} &= \frac{-1}{12}\begin{bmatrix} 3 & 1 & 2 \\ 5 & -4 & 1 \\ -1 & 2 & 1 \end{bmatrix}\begin{bmatrix} -6 & 3 & 9 \\ -6 & 5 & 7 \\ 6 & -7 & -17 \end{bmatrix} \\ &= \frac{-1}{12}\begin{bmatrix} -12 & 0 & 0 \\ 0 & -12 & 0 \\ 0 & 0 & -12 \end{bmatrix} \\ &= \mathbf{I} \end{aligned}$$

Similarly $\mathbf{A}^{-1}\mathbf{A} = \mathbf{I}$. Remember that *both* $\mathbf{AA}^{-1}$ *and* $\mathbf{A}^{-1}\mathbf{A}$ must equal **I**.

Singular Matrix

Since, in finding $\mathbf{A}^{-1}$ we divide **adj A** by $|\mathbf{A}|$, we shall not be able to find $\mathbf{A}^{-1}$ if $|\mathbf{A}| = 0$. Such a matrix is called a **singular matrix**. Hence a *non-singular* matrix *has* an inverse, a *singular* matrix does *not.*

Solution of simultaneous equations using the inverse matrix

We have considered the general set of equations

$$\begin{cases} a_{11}x_1 + a_{12}x_2 + a_{13}x_3 + \ldots\ldots + a_{1n}x_n = b_1 \\ a_{21}x_1 + a_{22}x_2 + \ldots\ldots\ldots\ldots\ldots\ldots + a_{2n}x_n = b_2 \\ \vdots \qquad\qquad\qquad\qquad\qquad\qquad \vdots \qquad\quad \vdots \\ a_{n1}x_1 + a_{n2}x_2 + \ldots\ldots\ldots\ldots\ldots\ldots + a_{nn}x_n = b_n \end{cases}$$

We know that we can write this system of equations in matrix form as

$$\mathbf{AX} = \mathbf{B}$$

where

$$\mathbf{A} = \begin{bmatrix} a_{11} & \cdots & a_{1n} \\ \vdots & & \\ a_{n1} & \cdots & a_{nn} \end{bmatrix}, \quad \mathbf{X} = \begin{bmatrix} x_1 \\ \vdots \\ x_n \end{bmatrix}, \quad \mathbf{B} = \begin{bmatrix} b_1 \\ b_2 \\ \vdots \\ b_n \end{bmatrix}$$

If we now pre-multiply both sides of this matrix equation by $\mathbf{A}^{-1}$, then

$$\mathbf{A}^{-1}\mathbf{A}\mathbf{X} = \mathbf{A}^{-1}\mathbf{B}$$

That is,

$$\mathbf{I}\mathbf{X} = \mathbf{A}^{-1}\mathbf{B}$$

or

$$\mathbf{X} = \mathbf{A}^{-1}\mathbf{B}$$

Thus we can solve the equations by multiplying the vector of the right-hand sides by $\mathbf{A}^{-1}$.

Problems

1. Find the inverse of the matrix

$$\begin{bmatrix} 5 & -1 & 2 \\ 2 & 4 & 1 \\ 1 & 3 & -3 \end{bmatrix}$$

and hence solve the simultaneous equations

$$\begin{cases} 5x - y + 2z = 3 \\ 2x + 4y + z = 8 \\ x + 3y - 3z = 2 \end{cases}$$

2. Find the inverse of

$$\mathbf{A} = \begin{bmatrix} 1 & 3 & 3 \\ 1 & 4 & 3 \\ 1 & 3 & 4 \end{bmatrix}$$

and check that $\mathbf{A}\mathbf{A}^{-1} = \mathbf{A}^{-1}\mathbf{A} = \mathbf{I}$.

3. Use the formal method of finding an inverse matrix to check your answers obtained to Problems (i) and (iii) on page 252 and use these to solve the equations

$$\begin{cases} -2x_1 + 3x_2 = 1 \\ x_1 + 4x_2 = 3 \end{cases}$$

$$\begin{cases} x_1 + 2x_2 + 3x_3 = 12 \\ 2x_1 - 2x_2 - 4x_3 = -14 \\ 3x_1 - 4x_2 - 3x_3 = -4 \end{cases}$$

5.6 EFFICIENCY OF METHODS OF SOLUTION OF SIMULTANEOUS LINEAR EQUATIONS

It is of interest to compare the methods outlined above for solving simultaneous linear equations.

For n simultaneous equations, using the direct method of solution involving determinants requires $\sim (1.72 \times n \times n!)$ multiplications and/or divisions (additions being regarded as taking negligible time compared with multiplication and division). For two equations the number is exactly 8, and for 100 equations it is $\sim 1.6 \times 10^{160}$, which would take a fast digital computer $\sim 10^{147}$ years!

The Gauss elimination method requires $[(n^3/3) + n^2 - (n/3)]$ multiplications and/or divisions; for two equations this becomes 6 operations and for 100 equations 343,333 operations – a considerable saving, relative to the determinant method.

Finding the inverse matrix directly requires $n^3 + n^2$ operations. Thus for large sets of simultaneous equations (> 3) it is best to use Gaussian elimination.

5.7 ITERATIVE SOLUTION OF SIMULTANEOUS LINEAR EQUATIONS

Consider the equations

$$5x + 3y = 6 \qquad (5.15)$$

$$4x + 7y = 8 \qquad (5.16)$$

We may arrange them as follows

$$x = 6/5 - (3/5)y$$

$$y = 8/7 - (4/7)x$$

and produce the iterative scheme

$$x_{n+1} = 6/5 - (3/5)y_n \qquad (5.17)$$

$$y_{n+1} = 8/7 - (4/7)x_n \qquad (5.18)$$

Now we guess as the solution $x_0 = 0, y_0 = 0$ which we substitute into (5.17) and (5.18).

These equations then yield $x_1 = 6/5 = 1.20$, and $y_1 = 8/7 = 1.14$ (2 d.p.). Substituting these into (5.17) and (5.18), $x_2 = 6/5 - 3/5(1.14) = 0.52$, and $y_2 = 8/7 - 4/7(1.2) = 0.46$ (2 d.p.). In Table 5.1 we show the successive approximations (working to 2 d.p.)

In fact, the solution (to 2 d.p.) is $x = 0.78$ and $y = 0.70$

The oscillation is due to rounding off to 2 d.p. and had we worked to 3 d.p. we should have obtained Table 5.2. This emphasises the need to work to one more decimal place than the accuracy required in the answer. This iterative method is known as **Jacobi's** method.

Table 5.1

n	x_n	y_n
0	0	0
1	1.2	1.14
2	0.52	0.46
3	0.92	0.85
4	0.69	0.62
5	0.83	0.75
6	0.75	0.67
7	0.79	0.71
8	0.77	0.69
9	0.79	0.70
10	0.78	0.69

Table 5.2

n	x_n	y_n
0	0	0
1	1.2	1.143
2	0.514	0.457
3	0.926	0.856
4	0.686	0.614
5	0.832	0.751
6	0.749	0.667
7	0.800	0.715
8	0.771	0.688
9	0.787	0.702
10	0.778	0.693
11	0.784	0.698
12	0.781	0.695
13	0.783	0.697

Suppose this time we obtain x from (5.16) and y from (5.15), i.e.

$$x = 8/4 - (7/4)y$$
$$y = 6/3 - (5/3)x$$

to produce the iterative scheme

$$x_{n+1} = 8/4 - (7/4)y_n \qquad (5.19)$$
$$y_{n+1} = 6/3 - (5/3)x_n \qquad (5.20)$$

Repeating the Jacobi procedure gives Table 5.3

Table 5.3

n	x_n	y_n
0	0	0
1	2.00	2.00
2	−1.50	1.33
3	4.33	4.50
4	−5.88	−5.23
5	11.14	11.80
⋮		

This time we see that the solutions do **not** converge. In general, if we have the two equations

$$a_1x + b_1y + c_1 = 0$$
$$a_2x + b_2y + c_2 = 0$$

then the condition for Jacobi's method to converge is that

$$|a_1b_2| > |a_2b_1|$$

This means geometrically that the initial substitution is made into the equation with the lowest absolute value of slope.

In the first arrangement $|a_1b_2| = 35$, $|a_2b_1| = 12$. In the second rearrangement we did not get convergence; there, the condition did not hold†.

The above method is also known as the method of *simultaneous* displacements. An alternative is the **Gauss-Seidel** method which displaces values *successively*. That is, it uses the latest guess for x to find y and then that new value for y to find a new value for x, and so on. Equations (5.17) and (5.18) become

$$x_{n+1} = 6/5 - (3/5)y_n \tag{5.17a}$$

and

$$y_{n+1} = 8/7 - (4/7)x_{n+1} \tag{5.18a}$$

Applying this method to equations (5.17) and (5.18), $x_0 = 0$, $y_0 = (8 - 4 \times 0)/7 = 1.14$, $x_1 = (6 - 3 \times 1.14)/5 = 0.52$, $y_1 = (8 - 4 \times 0.52)/7 = 0.85$ and so on. It can be seen that we are cutting down on the number of steps at each stage compared with Jacobi's method.

Table 5.4 shows the approximations

Table 5.4

n	x_n	y_n	x_n	y_n
0	0.00	1.14	0	1.143
1	0.52	0.85	0.514	0.856
2	0.69	0.75	0.686	0.751
3	0.75	0.71	0.749	0.715
4	0.77	0.70	0.771	0.702
5	0.78	0.70	0.778	0.698
6	0.78	0.70	0.781	0.697
	(working to 2 d.p.)		(working to 3 d.p.)	

† All we have said is that there will be convergence if the condition holds. If it does not, we shall be unable to conclude anything.

The graphical approach is shown in Figure 5.4

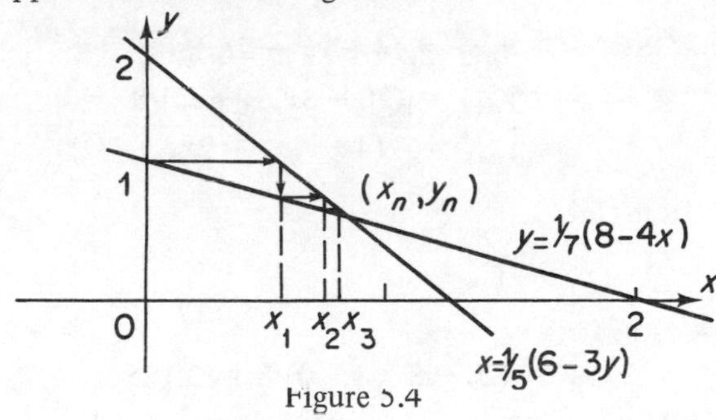

Figure 5.4

Trying rearrangements (5.19) and (5.20) we obtain Table 5.5

Table 5.5

n	x_n	y_n
0	0.00	2.00
1	−1.50	4.50
2	−5.88	11.80
3	−18.65	33.08
⋮		

Figure 5.5 shows the geometrical equivalent.

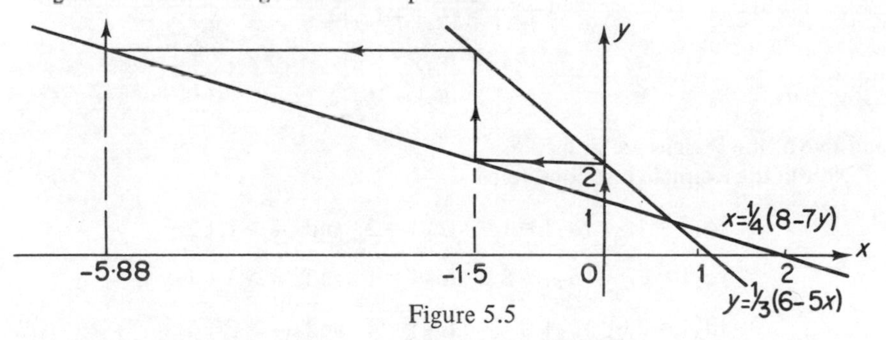

Figure 5.5

The Gauss-Seidel method can be extended to more than two equations.

Example

Consider
$$\begin{cases} 4x + y + 2z = 4 \\ 3x + 8y - z = 20 \\ 2x - y - 4z = 4 \end{cases}$$

We rearrange them and form the iterative scheme

$$x_{n+1} = (4 - y_n - 2z_n)/4 \tag{5.21}$$

$$y_{n+1} = (20 - 3x_{n+1} + z_n)/8 \tag{5.22}$$

$$z_{n+1} = -(4 + y_{n+1} - 2x_{n+1})/4 \tag{5.23}$$

If $y_0 = z_0 = 0$, from (5.21)

$$x_1 = 1$$

then from (5.22)

$$y_1 = (20 - 3 \times 1 + 0)/8 = 2.125$$

finally, from (5.23)

$$z_1 = -(4 + 2.125 - 2)/4 = -1.03$$

Now we use y_1, z_1 in (5.21) to calculate x_2
x_2, z_1 in (5.22) to calculate y_2
x_2, y_2 in (5.23) to calculate z_2 and so on.

Try to continue the iterations to $x = 1, y = 2, z = -1$, which is the true solution.

In general, for the equations

$$\begin{cases} a_1x + b_1y + c_1z + d_1 = 0 \\ a_2x + b_2y + c_2z + d_2 = 0 \\ a_3x + b_3y + c_3z + d_3 = 0 \end{cases}$$

the iterative method will converge if

$$|a_1| \geqslant |b_1| + |c_1|$$
$$|b_2| \geqslant |a_2| + |c_2|$$
$$|c_3| \geqslant |a_3| + |b_3|$$

and two of the $\geqslant$ signs are in fact $>$.

Now in the example just considered

$$|a_1| = 4, \quad |b_1| = 1, \quad |c_1| = 2 \quad \text{and} \quad 4 > 1 + 2$$

$$|a_2| = 3, \quad |b_2| = 8, \quad |c_2| = 1 \quad \text{and} \quad 8 > 3 + 1$$

$$|a_3| = 2, \quad |b_3| = 1, \quad |c_3| = 4 \quad \text{and} \quad 4 > 2 + 1$$

Hence convergence is assured.

Comparison of Iteration and Direct Methods

Gauss Elimination works for *any* non-singular set of n equations in n unknowns.

The Jacobi and Gauss-Seidel methods work only for *special* systems where convergence is assured. However, when iteration works it should be used because it can be shown that the number of multiplications for Gauss-Seidel iteration is proportional to n^2 per iteration; for Gauss elimination this number is proportional to n^3; also the round-off error is generally smaller for Gauss-Seidel.

Further, if the matrix is **sparse** (i.e. it contains several zero elements) elimination tends to destroy this sparseness.

Computer Program for Jacobi or Gauss-Seidel methods

The flow chart (Figure 5.6) is suitable for *either* the Jacobi *or* the Gauss-Seidel method; the choice depends on the statements used in the box

Compute new X(I)

To check whether convergence has occurred, we calculate the difference between new and old values of each X(I) and if **all** these differences are less than the specified tolerance TOL we say that convergence has taken place.

A is the coefficient matrix, **B** the vector of constants; ITCOUN counts the number of iterations performed.

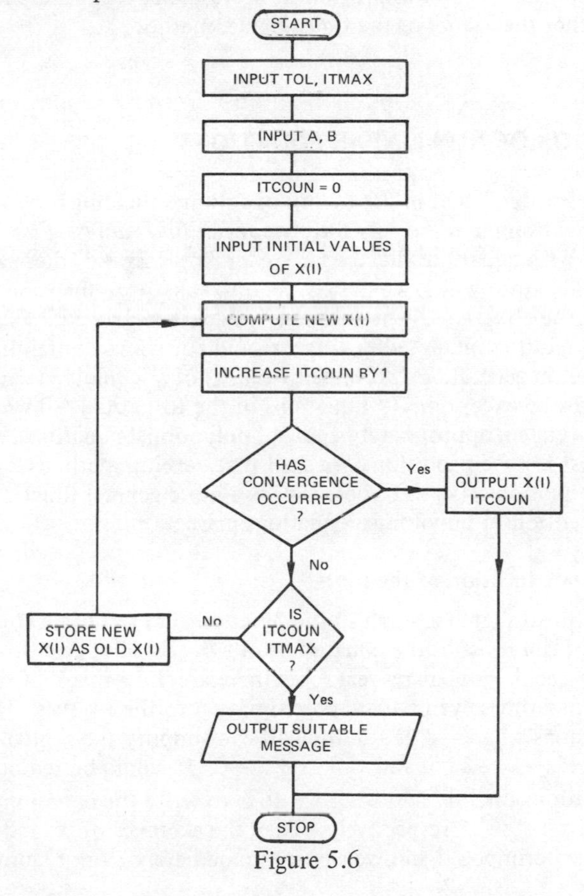

Figure 5.6

Problems

1. Consider each of the following systems of three simultaneous equations

(i) $\begin{cases} 3x_1 - 3x_2 + 7x_3 = 18 \\ x_1 + 6x_2 - x_3 = 10 \\ 10x_1 - 2x_2 + 7x_3 = 27 \end{cases}$ (ii) $\begin{cases} 4x_1 + x_2 + 2x_3 = 16 \\ x_1 + 3x_2 + x_3 = 10 \\ x_1 + 2x_2 + 5x_3 = 12 \end{cases}$

(a) Without rearranging the equations, try to find the solutions iteratively using both the Jacobi and the Gauss-Seidel methods starting with values (0, 0, 0), (1, 1, 1) and (1.01, 2.01, 3.01) for (x_1, x_2, x_3).

(b) Rearrange the equations if necessary to satisfy the convergence criteria on page 270 and repeat (a).

(c) Check your solutions in the original equations.

2. Write and run a computer program to solve the equations in Problem 1 using either the Jacobi or the Gauss-Seidel method.

5.8 SOLUTION OF NON-LINEAR EQUATIONS

We saw in Chapter 3 that in the theory of column buckling it was necessary to solve the equation $\tan u = u$. By *solve* we mean *find values of u which satisfy the equation.* With equations like $3x + 2 = 6$ or $3x^2 + 2x = 6$, there are general formulae available into which we merely need to substitute the values of the coefficients in the equations. The solution of equations like $3x^3 + 2x^2 = 6$ or $3x^4 + 2x^3 = 6$ requires much tedious algebra and equations containing higher powers of x are, in general, *not* solvable by means of a formula. (Can you see why there might be exceptions?) Equations of the form $p(x) = 0$ where $p(x)$ is a polynomial are called, appropriately enough, **polynomial equations**, and special techniques exist for their solution. We shall first develop methods which are applicable to equations $f(x) = 0$ where $f(x)$ is a more general function of x and pay special attention to polynomial equations in Section 5.9.

First approximate location of the roots

Often it helps to sketch a graph of the function $f(x)$ to obtain some idea as to the *nature* of the roots of the equation $f(x) = 0$. Usually, even from a rough sketch, we can decide how many real roots there are, how many of these are positive and, sometimes even a first approximation to their values. However, there are occasions when a little forethought can simplify the sketching. Consider two examples $x^3 - x - 1 = 0$ and $x^3 - x + 1 = 0$. It would be tedious to sketch these two functions directly and what we do is to write the equations as $x^3 = x + 1$ and $x^3 = x - 1$ respectively. Then the sketches of x^3 and $x + 1$, of x^3 and $x - 1$ are superimposed and we can see immediately from Figure 5.7 that the first equation has one real positive root and the second has one real negative root.

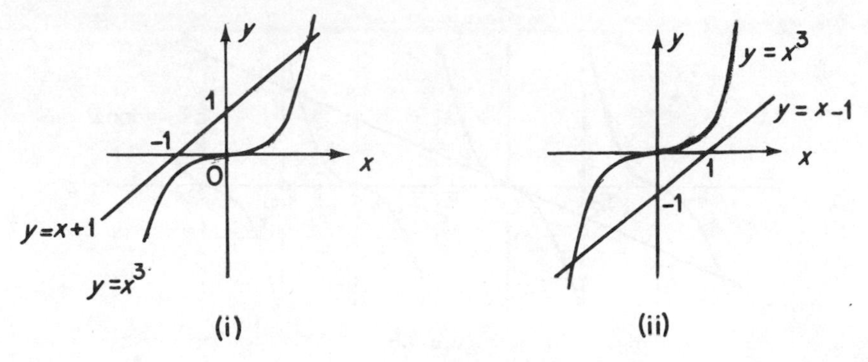

Figure 5.7

A pair of reasonably accurate sketches would give us first approximations for these roots of 1¼ and −1¼ respectively. Alternatively, we may tabulate each function near the likely value of its zero [a **zero** of $f(x)$ is another name for a root of $f(x) = 0$].

In Table 5.6 we consider the equation $x^3 - x - 1 = 0$.

Table 5.6

x	0	1	2	1½
x^3	0	1	8	3⅜
$x + 1$	1	2	3	2½
$x^3 - x - 1$	−1	−1	5	1⅞

We can see that the root lies between 1 and 1½. You should check the second equation in a similar fashion.

Given merely a table of values for $f(x)$ we must be careful that we do not jump to conclusions. We might say that if $f(a) < 0$ and $f(b) > 0$ where $a < b$, then there is at least one value of x in the interval (a, b) for which $f(x) = 0$. This is true only if $f(x)$ is a continuous function in the interval (a, b). It is clear that $x^3 - x - 1$ is continuous for all x; and this allows us to use the above method. However, for example, $\tan u - u$ is positive for $u = 1$, and negative for $u = 2$, yet there is no root of $\tan u - u = 0$ in the interval $(1, 2)$ since the function $(\tan u - u)$ is discontinuous at $\pi/2$. Care must therefore be exercised in such cases.

Example

The equation $\tan u = u$ is already in a form which suggests the superimposition of two graphs (we shall cheat a little and assume that we know the slope of $\tan u$ is 1 at $u = 0$); refer to Figure 5.8.

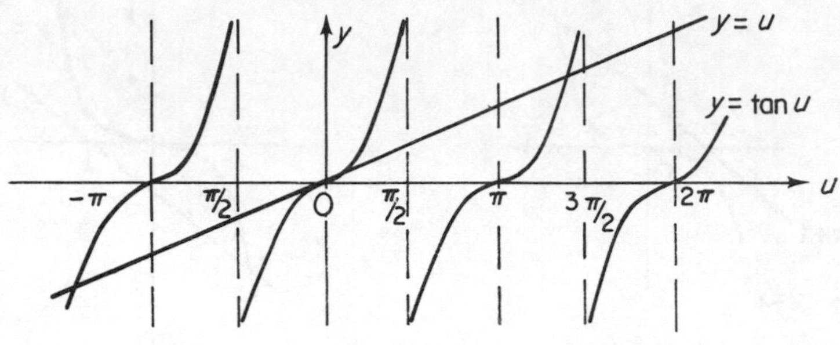

Figure 5.8

Several features are worthy of note:

(i) there is an *infinite number* of roots;

(ii) putting $v = -u$ gives $\tan(-v) = -v$, i.e. $-\tan v = -v$, and this means that for each *positive* root there is a corresponding *negative* root of equal magnitude;

(iii) there is a root at $u = 0$;

(iv) apart from that root, reasonable approximations to the others are $(2n + 1)(\pi/2)$ for $n = 1, 2, 3, \ldots\ldots$ and the larger the value of u, the better the approximation.

Problems I

1. By sketching suitable functions, obtain information about the number, the nature and, where possible, the location of the roots of the following equations.

(i) $x^3 + 2x^2 + 1 = 0$ (ii) $x^3 - 2x^2 - 1 = 0$

(iii) $100 - x - (2/x) = 0$ (iv) $7 - x - (9/x) = 0$

(v) $\sin x - e^{-2x} = 0$ (vi) $x - 1 - \sin 2x = 0$

Improving the approximation

How can we obtain a better estimate for a root? One way is to extend the idea behind Table 5.6. Realising that the root lay between $x = 1$ and $x = 2$, we chose $x = 1\frac{1}{2}$ as our next approximation simply because it was mid-way between those two values. Worthy of note is the fact that the original interval of uncertainty (as to where the root might be) was of length 1 and after the evaluation of $f(1\frac{1}{2})$ this interval was reduced to $\frac{1}{2}$. Since we know that the root lies between $x = 1$ and $x = 1\frac{1}{2}$ we may evaluate $f(1\frac{1}{4}) = -19/64 = -0.297$ (3 d.p.) and conclude that the root lies in the interval (1.25, 1.5). Of course, all we need to know about $f(1\frac{1}{4})$ is that is is negative. Our interval of uncertainty is now of length $\frac{1}{4}$ and this gives rise to the naming of the technique as the **method of successive bisection**.

You should study the flow chart of this process, in Figure 5.9, which produces the root correct to 4 d.p.

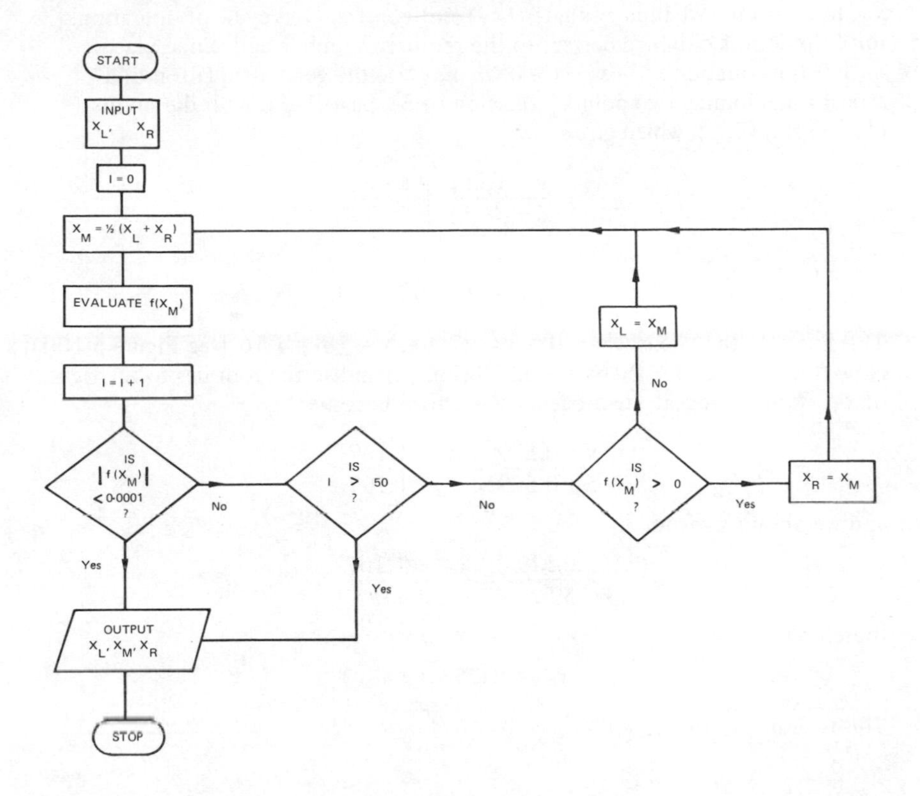

Figure 5.9

Comment on this flow chart; it is assumed that $f(X_L) < 0$ and $f(X_R) > 0$: how would you modify the flow chart if this were not necessarily the case?

If possible, write and run the program. But certainly carry out a few more steps by hand to get a feel for the method. The method is suitable for a digital computer but it may not be the most efficient technique and it can certainly be tedious for hand calculation. After one step the interval of uncertainty was ½, after two steps it was $1/2^2$ and after n steps it will be $1/2^n$. Bearing in mind that $2^{10} = 1024$ we see that it will take ten steps to reduce the interval to less than 0.001 and hence achieve 3 d.p. accuracy. We look for faster methods and in the rest of this section we study three of the best-known.

(i) Rule of False Position

Referring back to Table 5.6 on page 273 we notice that the value of $f(1)$ is smaller in magnitude than $f(2)$ and this would suggest that it is more sensible to look for the root nearer to $x = 1$ than $x = 2$. The rule of **false position** *(regula falsi)* makes use of this; in effect, we join the points $[1, f(1)]$ and $[2, f(2)]$ by a chord and take the intersection with the x-axis as the next approximation, x_3, to the root. We then evaluate $f(x_3)$ and continue the cycle of operations until the root is obtained correct to the required number of decimal places.

For the equation $x^3 - x - 1 = 0$ we may use the general equation of a straight line joining two points [equation (4.3), page 143], with the points $(1, -1)$ and $(2, 5)$, which gives

$$\frac{y-(-1)}{5-(-1)} = \frac{x-1}{2-1}$$

that is,

$$y = 6x - 7$$

This line cuts the x-axis where $0 = 6x - 7$, i.e. $x = 7/6 = 1.1\dot{6}$ [see Figure 5.10(i)]. Now $f(7/6) \triangleq 1.585 - 2.164 = -0.579$ (3 d.p.) and so the root lies to the right of x_3. We now repeat the process. The chord becomes

$$\frac{y + 0.579}{5 + 0.579} = \frac{x - 1.167}{2 - 1.167}$$

and we obtain x_4 from

$$\frac{0.579}{5.579} = \frac{x_4 - 1.167}{0.833}$$

therefore

$$x_4 = 1.253 \text{ (3 d.p.)}$$

This is shown in Figure 5.10(ii).

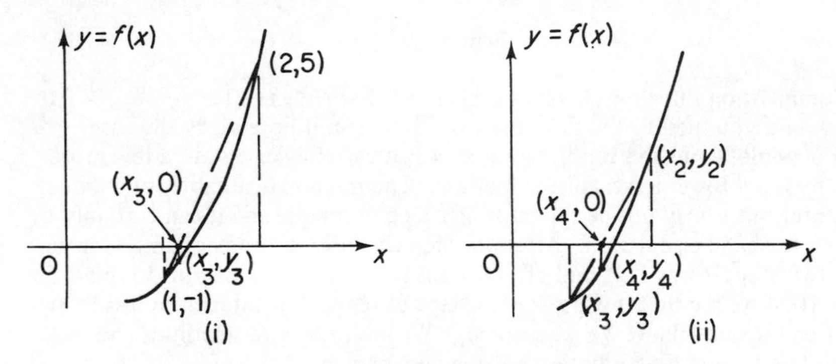

Figure 5.10

Applying the method repeatedly we obtain the results shown in Table 5.7.

We can see that we have obtained the root correct to 3 d.p. after nine applications compared to the ten needed for successive bisection.

Table 5.7

n	x_n (3 d.p.)	$f(x_n)$ (3 d.p.)
1	1	−1
2	2	5
3	1.167	−0.579
4	1.253	−0.285
5	1.293	−0.129
6	1.311	−0.057
7	1.311	−0.024
8	1.322	−0.010
9	1.324	−0.002
10	1.325	−0.000
11	1.325	−0.000

A variation on this method is to keep the point (x_1, y_1) and successively replace (x_2, y_2) by (x_3, y_3), (x_4, y_4) etc. This works if the curve is concave (or convex) between the initial points and can be defined by

$$x_{n+1} = x_n - y_n \frac{(x_n - x_1)}{(y_n - y_1)}, \quad n \geqslant 2 \tag{5.24}$$

Problem

Try writing a flow chart for the above variation and running a computer program on the example we have been using.

(ii) **Basic Iteration**

So far, the two methods we have examined have shown no signs of failure and there is always a danger that methods which sometimes work faster than these may not *always* work. In examining other methods we must be on the look-out for ways in which they may not reach the desired result.

The method of **Basic Iteration** relies on writing the equation to be solved in the form $x = F(x)$ and then using the companion iterative formula $x_{n+1} = F(x_n)$; in other words, the latest approximation to the root under consideration is substituted into the right-hand side of the formula and the result interpreted as the next approximation. We have no reason to suppose that this Basic Iteration will always work and so we shall test its behaviour on simple equations whose roots we know by other means.

We start with the equation $x^2 - 5x + 4 = 0$ (which has roots of 1 and 4) and we note that we can rearrange it to $x = (1/5)(x^2 + 4)$, to $x = (5x - 4)^{1/2}$, to $x = x^2 - 4x + 4$ [i.e. $x = (x - 2)^2$] to $x = x^{1/2} + 2$ and to $x = 2 - x^{1/2}$. Some of these rearrangements are obvious, others more subtle; does it matter which we choose?

In Tables 5.8, 5.9, 5.10 and 5.11 we see the effect of different rearrangements and different initial guesses. The results were obtained with a desk calculator and are rounded off to 3 d.p. in the tables.

Table 5.8

$$x_{n+1} = 0.2(x_n^2 + 4)$$

n	x_n	x_n	x_n
0	5	3	0
1	5.8	2.6	0.8
2	7.531	2.152	0.928
3	12.142	1.726	0.972
4		1.396	0.989
5		1.190	0.996
6		1.083	0.998
7		1.035	0.999
8		1.014	1.000
9		1.006	
10		1.002	
11		1.001	
12		1.000	
13		1.000	

Table 5.9

$$x_{n+1} = (5x_n - 4)^{1/2}$$

n	x_n	x_n	x_n	x_n
0	5	3	0	2
1	4.583	3.317	No real value	2.449
2	4.349	3.547		2.872
3	4.212	3.706		3.218
4	4.131	3.812		3.477
5	4.081	3.881		3.659
6	4.050	3.925		3.781
7	4.031	3.953		3.860
8	4.019	3.970		3.912
9	4.012	3.981		3.944
10	4.007	3.988		3.965
11	4.005	3.993		3.978

Table 5.10

$x_{n+1} = x_n^{1/2} + 2$

n	x_n	x_n
0	9	0
1	5	2
2	4.236	3.414
3	4.058	3.850
4	4.014	3.962
5	4.003	3.991
6	4.001	3.997
7	4.000	3.999
8	4.000	4.000

Table 5.11

$x_{n+1} = 2 - x_n^{1/2}$

n	x_n	x_n	x_n
0	0	9	3
1	2	−1	0.268
2	0.59		1.423
3	1.23		0.817
4	0.891		1.096
5	1.056		
6	0.997		
7	1.002		
8	0.999		
9	1.000		

What conclusions can we draw from these tables? It is clear that not all these earrangements provide corresponding convergent iterative formulae.

From Table 5.8 we see that, in an attempt to locate the root 4, if we make an nitial guess too high the successive approximations get rapidly worse; on the other hand, if the initial guess is too low the iterations converge, but to the root 1. However, attempting to locate the root 1 directly is successful with the initial guess either too low or too high. Clearly, a different rearrangement will be required to locate the root 4. In Table 5.9 we see a similar state of affairs but with convergence to the root $x = 4$; notice that $x_0 = 0$ gives trouble.

Tables 5.10 and 5.11 show that the slightly simpler formulae converge to the roots more quickly, yet again each seems to converge to only one root.

A little investigation shows that $x = 4$ satisfies the rearrangement behind the formula of Table 5.10, but $x = 1$ does not. The opposite is true for the rearrangement leading to Table 5.11. Yet both these values of x satisfy the rearrangements associated with Tables 5.8 and 5.9, so there must be a deeper reason for the temperamental behaviour of these formulae.

Before we study these causes further consider $x^2 - x - 2 = 0$ which has roots 2 and −1, and $x^2 + x - 2 = 0$ with roots −2 and 1. Attempts to locate these roots are shown in Tables 5.12 and 5.13.

Notice that, although both roots satisfy the respective rearrangements, convergence to these roots does not automatically follow.

In order to study the underlying cause of convergence or non-convergence we examine Tables 5.10 to 5.13 in more detail. First we try and obtain a pictorial representation of the behaviour of the successive approximations in each case.

Table 5.12

$x_{n+1} = x_n^2 - 2$

n	x_n	x_n
0	1	2.1
1	−1	2.41
2	−1	3.81

Table 5.13

$x_{n+1} = 2 - x_n^2$

n	x_n
0	1.2
1	0.56
2	1.69
3	−0.86
4	1.26
5	−0.41
6	1.83
7	−1.35
8	0.18
9	1.74
10	−1.03
11	0.94
12	1.12
13	0.75
14	1.44

In Figure 5.11(i) we plot schematically the effect of applying the formula $x_{n+1} = F(x_n) = x_n^{1/2} + 2$ three times. Since $x_1 = F(x_0)$ and $x_2 = F(x_1)$, we may symbolise x_2 as $(F_oF)(x_0)$. We see clearly the *homing-in* on $x = 4$.

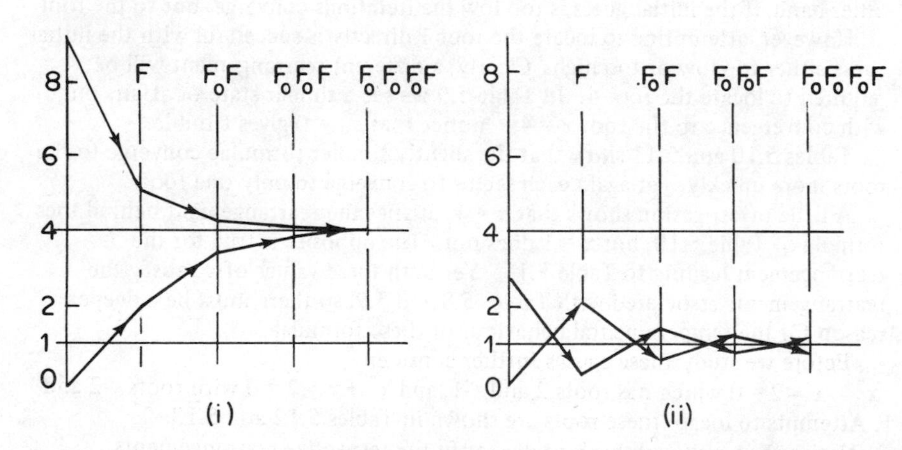

Figure 5.11

Let us analyse this situation algebraically. Suppose the true root is x and let x_n be our n^{th} approximation to it so that the error is ϵ_n, that is, $x_n = x + \epsilon_n$. Now

$$
\begin{aligned}
F(x_n) &= x_n^{1/2} + 2 \\
&= (x + \epsilon_n)^{1/2} + 2 \\
&= x^{1/2}[1 + (\epsilon_n/x)]^{1/2} + 2 \\
&\simeq x^{1/2}[1 + \tfrac{1}{2}(\epsilon_n/x)] + 2 \qquad \text{(Binomial Expansion)} \\
&= x^{1/2} + 2 + (\epsilon_n/2x^{1/2}) \qquad \text{(This is } \textit{exactly} \text{ equal to the expression on the previous line.)}
\end{aligned}
$$

Therefore

$$x_{n+1} \simeq x^{1/2} + 2 + (\epsilon_n/2x^{1/2})$$

But $x = x^{1/2} + 2$ exactly, therefore

$$x_{n+1} \simeq x + (\epsilon_n/2x^{1/2})$$

that is,

$$\epsilon_{n+1} \simeq \epsilon_n/2x^{1/2}$$

The approximate effect of applying the formula once is to multiply the error in the approximation by a **scale factor (S.F.)** of $1/2x^{1/2}$.

This approximate effect is near enough to the truth if ϵ_n/x is small. Now at $x = 4$, the scale factor is $1/4$, and at $x = 1$ the S.F. is $1/2$. In both cases, if we are near either root the error should continually decrease. (The fact that the root $x = 1$ does not satisfy the rearrangement leading to this iterative formula means that the error, ϵ_n, must be taken to be from $x = 4$ and the value of the scale factor at $x = 1$ is not relevant here.)

For $x_{n+1} = 2 - x_n^{1/2}$, Figure 5.11(ii) shows the homing-in on $x = 1$ and the oscillatory nature of the approximations about this value is reflected in the scale factor of $-1/2x^{1/2}$ which you should try to obtain.

The rearrangement of Table 5.12 has a scale factor of $2x$ and hence is larger than 1 for both roots; the divergent nature of the approximations for $x_0 = 2.1$ is seen in Figure 5.12(i) but there *is* convergence at $x = -1$. It would seem that the basic iteration *may* work even if the scale factor is larger than 1. Finally, the rearrangement of Table 5.13 has a S.F. of $-2x$ and the haphazard behaviour is shown in Figure 5.12(ii).

Figure 5.13 shows pictorially how convergence or divergence is obtained in these four examples. In each case graphs of $y = x$ and $y = F(x)$ are superimposed. The ordinate through x_n intersects the curve $y = F(x)$ to give $F(x_n)$ and the abscissa through $F(x_n)$ intersects $y = x$ to give x_{n+1}. It would seem that where the slope of $y = F(x)$ near the root $x = a$ is small, the method converges. We shall see in Chapter 6 that the scale factor is, in fact, $F'(a)$ and we may state the rule:

A **sufficient** *condition for the iterative formula $x_{n+1} = F(x_n)$ to converge to a real root $x = a$ of $x = F(x)$ is that $|F'(x)| < 1$* **near the root.**

Since, in general, we do not know a, we cannot evaluate $F'(a)$, but if $F(x)$

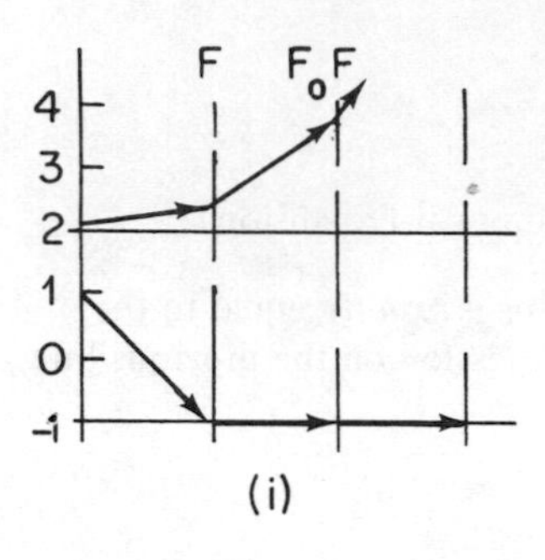

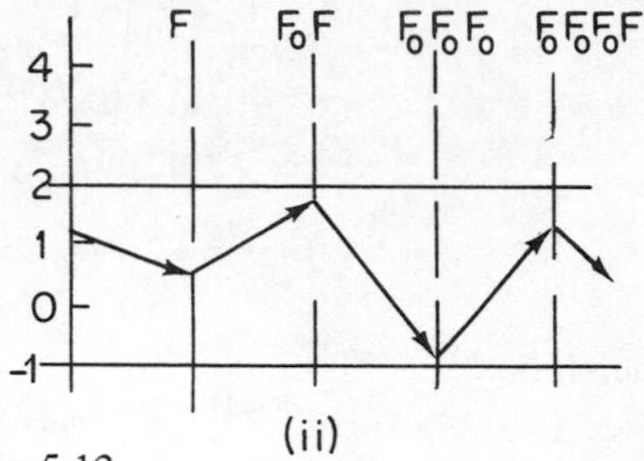

Figure 5.12

is a reasonably-behaved function, $F'(x)$ will be close enough to $F'(a)$, provided x is reasonably close to a.

Note: If $|F'(x)|$ is less than 1 near the root in question, we are **guaranteed** convergence to that root; on the other hand, it is **not certain** (but *still possible*) that convergence will occur if $|F'(x)| \geqslant 1$ near the root.

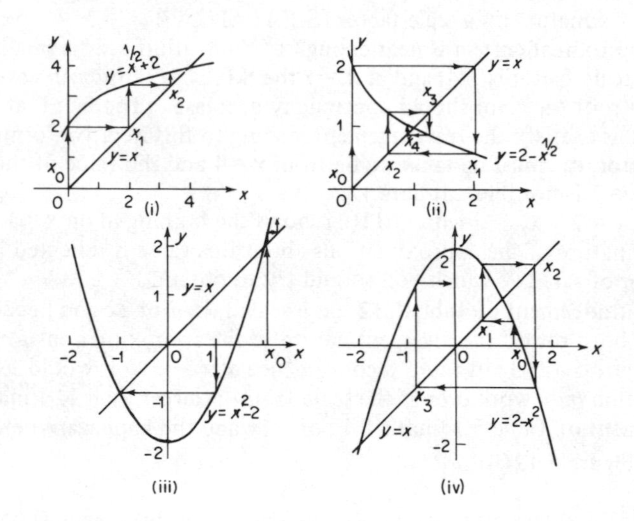

Figure 5.13

Problems II

1. Use the method of successive bisection and the rule of false position on the equations behind Tables 5.8 and 5.9 and compare the relative speeds of convergence.

2. Carry out an analysis of the situations featured in Tables 5.8 and 5.9. Are your predictions borne out in practice?

3. Show that $x = (x^2 + 2)/(2x - 1)$ is a rearrangement of the equation $x^2 - x - 2 = 0$. Use the companion iterative formula with $x_0 = -2, 0, 1, 3$ and comment on the results.

4. Find algebraically the roots of $x^2 - 3x + 2 = 0$.
Find four rearrangements of the equation and, where possible, the scale factor associated with each. Hence predict the behaviour of the companion iterative sequences. Construct a flow chart for the basic iteration method in general and write a computer program for this particular problem. Run the program to check your predictions.

5. Solve, using basic iteration, the equation $xe^x = 4$ (the derived function of e^x is e^x and that of $\log_e x$ is $1/x$). Then solve $xe^x = 2$, $xe^x = 10$.

6. The formula $x_{n+1} = 2x_n - Ax_n^2$ is a candidate for finding the inverse of A, $1/A$. Show that if the formula converges, then it does converge to $1/A$ and determine the limits on the initial guess, x_0, in order that it should converge. Test your conclusions on the cases $A = 9$ with $x_0 = 0.2, 1.0$ respectively.

7. Find the positive root of the equation $\sinh x = 5x$ by iteration and the other roots by inspection.

8. Consider the equation $x^4 + x^2 = 20$. Three rearrangements are
(i) $x = \sqrt[4]{20 - x^2}$; $F'(x) = -x/[2(20 - x^2)^{3/4}]$
(ii) $x = \sqrt{20 - x^4}$; $F'(x) = -2x^3/\sqrt{20 - x^4}$
(iii) $x = \sqrt{20/(1 + x^2)}$; $F'(x) = -\sqrt{20}x/[(1 + x^2)^{3/2}]$
Predict the behaviour of the accompanying iterative formulae and check by carrying out a few iterations with each. Then solve the equation algebraically.

(iii) Newton-Raphson method

We have said that the smaller $|F'(a)|$ the better and we might ask if it is possible to find a rearrangement for which $|F'(a)|$ is zero. Such a method, starting from a different approach, is the one here described. This time the equation to be solved is cast in the form $f(x) = 0$. Consider Figure 5.14(i).

Starting with an approximation x_0, the tangent to the curve, $y = f(x)$, at the point $[x_0, f(x_0)]$ is drawn and meets the x-axis at x_1, the next approximation. The tangent to the curve at $[x_1, f(x_1)]$ is drawn and this meets the x-axis at a third approximation, x_2; the cycle is repeated until $|x_{n+1} - x_n|$ is less than a specified amount. To produce a formula for the method, examine Figure 5.14(ii); we assume x_n has been used to produce x_{n+1}. The slope of CB is $f'(x_n)$ and since the length of AB is $f(x_n)$ we see that the length CA is

$$\frac{f(x_n)}{f'(x_n)}$$

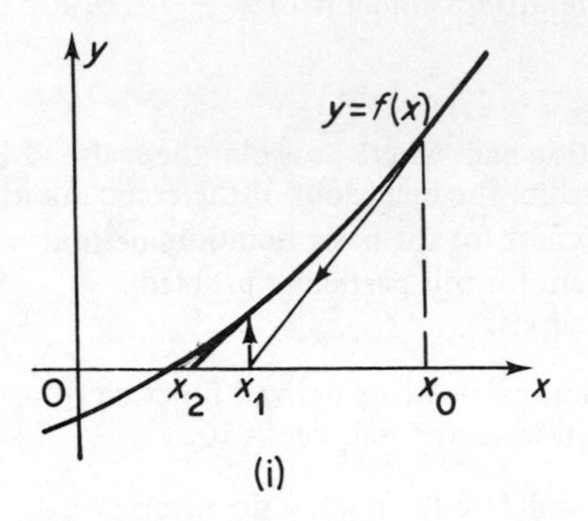

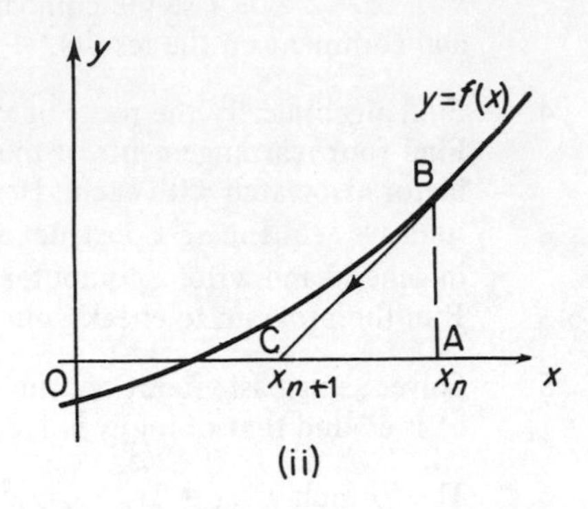

Figure 5.14

Hence OC = OA − CA, i.e.

$$x_{n+1} = x_n - \frac{f(x_n)}{f'(x_n)} \tag{5.25}$$

This is the **Newton-Raphson formula.** [Note that the same result holds if $f'(x_n) < 0$.]

Let us apply the method to the equation $x^2 - 5x + 4 = 0$. Since

$$f(x) = x^2 - 5x + 4, \quad f'(x) = 2x - 5$$

and (5.25) becomes

$$x_{n+1} = x_n - \frac{(x_n^2 - 5x_n + 4)}{2x_n - 5}$$

that is,

$$x_{n+1} = \frac{x_n^2 - 4}{2x_n - 5}$$

Notice that $x = (x^2 - 4)/(2x - 5)$ is another rearrangement of the equation whose roots we seek.

If we start with $x_0 = 5$,

$$x_1 = \frac{25 - 4}{10 - 5} = 4.2$$

and so

$$x_2 = \frac{(4.2)^2 - 4}{8.4 - 5} = 4.012 \text{ (3 d.p.)}$$

and

$$x_3 = \frac{(4.012)^2 - 4}{8.024 - 5} = 4.000 \ \text{(3 d.p.)}$$

Comparing these results with those of Table 5.9 we see how much more rapidly we have obtained the root to 3 d.p. (It is worth restating that the inaccuracy in x_2 does not matter in the calculation of x_3 since iterative methods are usually *self-correcting.*) To be complete we must apply the formula once more to obtain

$$x_4 = \frac{(4.000)^2 - 4}{8.000 - 5} = 4.000 \ \text{(3 d.p.)}$$

Table 5.14 shows the effect of different initial approximations on the Newton-Raphson method.

Table 5.14

n	x_n	x_n	x_n	x_n
0	5	3	0	2
1	4.2	5	0.8	0.8
2	4.012	4.2	0.988	etc.
3	4.000	4.012	0.999	
4	4.000	4.000	1.000	
5		4.000	1.000	

It is clear that *when* the Newton-Raphson method works it works very well; however, there *are* cases of failure and we illustrate two of these below. Unfortunately there is no simple criterion for convergence as there was for Basic Iteration.

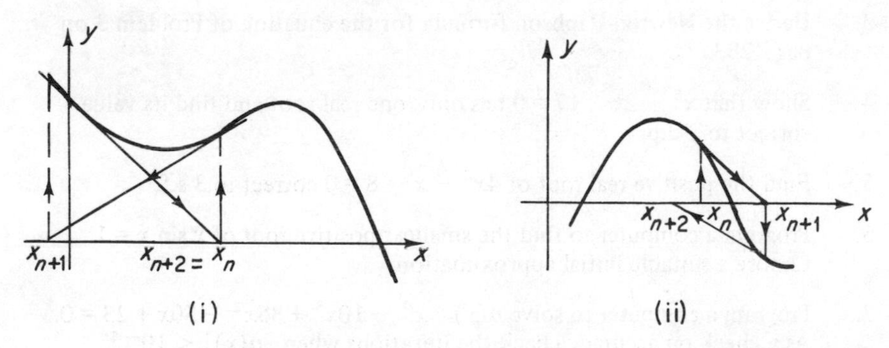

Figure 5.15

In Figure 5.15(i) there is a local minimum near the root and we get into an infinite looping situation. In Figure 5.15(ii) there are two roots close together and this causes a close initial approximation to take us further away from either root.

When the technique does work it is said to have *second-order convergence* as opposed to the *first-order convergence* of the Basic Iteration method. This means $\epsilon_{n+1} \propto \epsilon_n^2$ as opposed to $\epsilon_{n+1} \propto \epsilon_n$. In Chapter 6 we shall prove this result and it is worth now showing that the Newton-Raphson method is a special case of Basic Iteration. The formula (5.25) is of the form

$$x_{n+1} = F(x_n)$$

if we identify $F(x)$ as

$$x - \frac{f(x)}{f'(x)}$$

In Chapter 6 we see that $F'(x)$ in this case can be written

$$\frac{f(x)f''(x)}{[f'(x)]^2}$$

and at the root a, $f(a) = 0$, hence $|F'(a)| = 0$.

Problems III

(See page 301 for a table of derivatives.)

1. Use the Newton-Raphson method on the equation $x^2 - A = 0$ in an attempt to find $\sqrt{A}$. What do you notice about the iterative formula? Repeat for $x^3 - A = 0$, $x^r - A = 0$ (r any positive integer).

2. The equation $\tan u = u$ was discussed earlier. Find the first non-zero root to 3 d.p. (The derived function of $\tan u$ is $\sec^2 u$.)

3. Derive the Newton-Raphson formula for the equation of Problem 3 on page 283.

4. Show that $x^3 - 2x - 17 = 0$ has only one real root and find its value correct to 2 d.p.

5. Find the positive real root of $4x^4 - x - 8 = 0$ correct to 3 s.f.

6. Program a computer to find the smallest positive root of $x \sin x = 1$. Choose a suitable initial approximation.

7. Program a computer to solve $p(x) = x^4 - 10x^3 + 35x^2 - 50x + 23 = 0$. As a check on accuracy check the iterations when $|p(x)| < 10^{-4}$.

8 Which would you prefer for computing by hand the real root of $x^3 - 972 = 0$

(a) successive bisection
(b) Basic Iteration
(c) Newton-Raphson?

9. Kepler's equation, used for computing orbits of satellites, is

$$M = x - E \sin x$$

Given $E = 0.2$, solve this when

(i) $M = 0.5$
(ii) $M = 0.8$

correct to 4 decimal places.

10. Solve approximately the equation $\sin \omega t - e^{-at} = 0$ (arising from the motion of a planetary gear system used in automatic transmission). In particular, determine the smallest root correct to 4 d.p., taking $\omega = 0.573$ and $a = 0.01$.

11. Freudenstein's equation relating the output crank angle ϕ to the input crank angle θ of a 4-bar mechanism is

$$\frac{D}{C} \cos \theta - \frac{D}{A} \cos \phi + \frac{(D^2 + A^2 - B^2 + C^2)}{2AC} - \cos (\theta - \phi) = 0$$

where A, B, C and D are the lengths of the input crank, coupling link output crank and fixed link respectively.

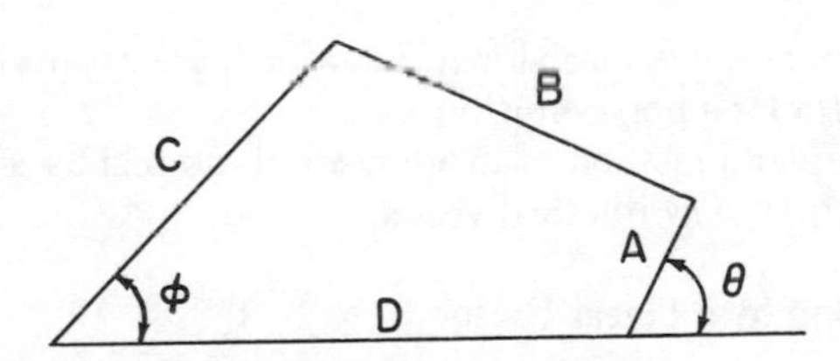

Figure 5.16

If $A = 5$ cm, $B = 10$ cm, $C = 10$ cm, $D = 10$ cm, program a computer to produce a table of values of ϕ against θ. [The derived function with respect to ϕ of $\cos (\theta - \phi)$ is $+ \sin (\theta - \phi)$].

12. A sphere of density ρ_1 and radius a floats on a fluid of density ρ_0. If it is submerged to a depth h ($< a$) it can be shown that the volume submerged is $(\pi/3)(3ah^2 - h^3)$.

If $\rho_0 = 1$ and $\rho_1 = 0.4$ find a suitable approximation to the ratio $\alpha = h/a$ and use the Newton-Raphson method to find α to 2 d.p.

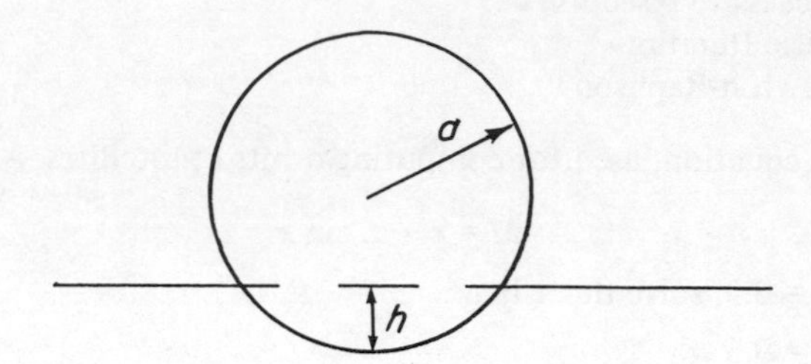

Figure 5.17

13. Multiple roots are a hazard; so are two roots of $f(x) = 0$ which are close together. A technique for resolving two close roots is as follows: having located their position approximately, solve $x = x + f'(x) - 1$ by Newton-Raphson for x^*; then evaluate

$$d = \sqrt{\frac{2[x^* - f(x^*)]}{f''(x^*)}}$$

and take $x^* - d$ and $x^* + d$ as starting approximations for the two original roots. For the equation $(x - 2)(x - 1.001)(x - 0.999) = 0$ try the usual Newton-Raphson method and then use the above technique.

5.9 POLYNOMIAL EQUATIONS

We now examine some special methods which are applicable to equations $f(x) = 0$ where $f(x)$ is a polynomial in x.

First we consider a method of dividing a polynomial by a linear or a quadratic factor: the method of **synthetic division.**

Synthetic Division by a Linear Factor

A brief return to the *nesting* of polynomials is required. The polynomial $p(x) = a_0x^3 + a_1x^2 + a_2x + a_3$ can be nested as

$$[(a_0x + a_1)x + a_2]x + a_3 \tag{5.26}$$

and this latter form requires only 3 multiplications and 3 additions, as opposed to the $3 + 2 + 1 = 6$ multiplications and 3 additions of the first form. This is clearly more efficient and less round-off error is likely.

If we consider the division of the polynomial by a factor $(x - x_1)$ in its original form we would have

$$p(x) = (x - x_1)(b_0x^2 + b_1x + b_2) + b_3$$

Multiplying out the R H.S. and comparing coefficients of like powers of x we have

$$
\begin{aligned}
x^3:\ & a_0 = b_0 && \Rightarrow b_0 = a_0 \\
x^2:\ & a_1 = b_1 - b_0 x_1 && \Rightarrow b_1 = a_1 + b_0 x_1 \\
x:\ & a_2 = b_2 - b_1 x_1 && \Rightarrow b_2 = a_2 + b_1 x_1 \\
& a_3 = b_3 - b_2 x_1 && \Rightarrow b_3 = a_3 + b_2 x_1
\end{aligned}
$$

Hence, apart from b_0, all the coefficients may be generated from the *recurrence formula*

$$b_r = a_r + b_{r-1} x_1$$

Compare the formulae with the result of putting $x = x_1$ in (5.26) giving

$$[(a_0 x_1 + a_1)x_1 + a_2]x_1 + a_3$$

or

$$[(b_0 x_1 + a_1)x_1 + a_2]x_1 + a_3$$

or

$$[(b_1 x_1 + a_2)x_1] + a_3$$

or

$$b_2 x_1 + a_3$$

You should see that the processes are the same.

Example 1

In order to make use of this synthetic division we first carry out, in traditional fashion, the division of $2x^3 + 3x^2 - 10x + 6$ by $x - 2$

$$
\begin{array}{r@{\,}l}
 & 2x^2 + 7x + 4 \\
x - 2 \,\big) & 2x^3 + 3x^2 - 10x + 6 \\
 & \underline{2x^3 - 4x^2} \\
 & \quad 7x^2 - 10x \\
 & \quad \underline{7x^2 - 14x} \\
 & \qquad 4x + 6 \\
 & \qquad \underline{4x - 8} \\
 & \qquad\quad 14
\end{array}
$$

We see that the *quotient* is $2x^2 + 7x + 4$ and the *remainder* is 14. In general terms

$$p(x) = (x - 2)(2x^2 + 7x + 4) + 14$$

therefore

$$p(2) = 14$$

and so if

$$p(x) = (x - a)q(x) + r$$

then

$$p(a) = r$$

You should notice that the only essential quantities in the process are the coefficients in bold-face type. We now lay out these numbers so that their roles are apparent.

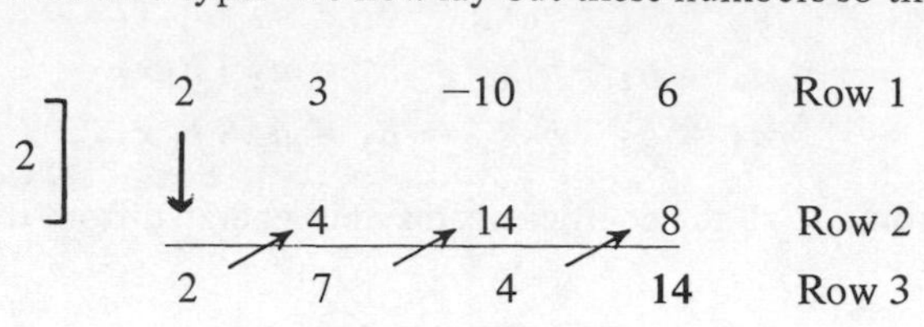

We proceed as follows. First the coefficients of the polynomial are written in Row 1, the number **−2** in the divisor is reversed in sign and placed to the left of the symbol] in Row 2 and the coefficient of x^3 is repeated in Row 3. Then this number is multiplied by the *reversed sign number* **2** from Row 2 to give the **4** in Row 2. This is added to the number above to give the entry **7** in Row 3 and the steps repeated until the *remainder* **14** is reached in Row 3. The first three numbers in that row are the coefficients in the *quotient*.

Example 2

We work through the example $(3x^3 + 2x - 6) \div (x + 2)$

	3	0	2	−6
−2]		−6	12	−28
	3	−6	14	−34

This gives a quotient of $3x^2 - 6x + 14$ and a remainder of −34.

It may be shown that, having divided $p(x)$ by $(x - a)$ to obtain a quotient $q(x)$ and a remainder $r = p(a)$ we may divide $q(x)$ by $(x - a)$ to obtain $p'(a)$. (See Chapter 6.)

This may be applied with the Newton-Raphson method as in the following example.

Example 3

To solve $p(x) = x^3 + 9x^2 + 23x + 14 = 0$, given that one root is near $x_0 = -0.7$. (We divide by $x + 0.7$.)

	1	9	23	14	
−0.7]		−0.7	−5.81	−12.033	
	1	8.3	17.19	[1.967	$= p(x_0)$]
−0.7]		−0.7	−5.32		
	1	7.6	[11.87	$= p'(x_0)$]	

$$x_1 = x_0 - \frac{p(x_0)}{p'(x_0)} = -0.7 - \frac{1.967}{11.87} = -0.866 \quad (3 \text{ d.p.})$$

Repetition of this process yields

$$x_2 = -0.88485 \quad (5 \text{ d.p.}) \qquad \text{and } x_3 = -0.88509 \quad (5 \text{ d.p.})$$

Bearing in mind the small correction we would suspect this last approximation to be a very good one, a verification coming from the evaluation of $p(x_3) = 0.000001$ (6 d.p.).

We may then divide $p(x)$ by the factor $(x + 0.88509)$ to give $(x^2 + 8.11491x + 15.81756)$ and this can be solved by the usual formula to yield the other roots as -4.86081 and -3.25410 (5 d.p.); their accuracy may be checked likewise, using nested multiplication (of course) and the values of $p(x)$ are -0.000009 and 0.000011 respectively, both to 6 d.p.

The above scheme is sometimes referred to as the **Birge-Vieta** method.

Finally in Figure 5.18 we flow chart the process to divide a polynomial by $(x + a)$.

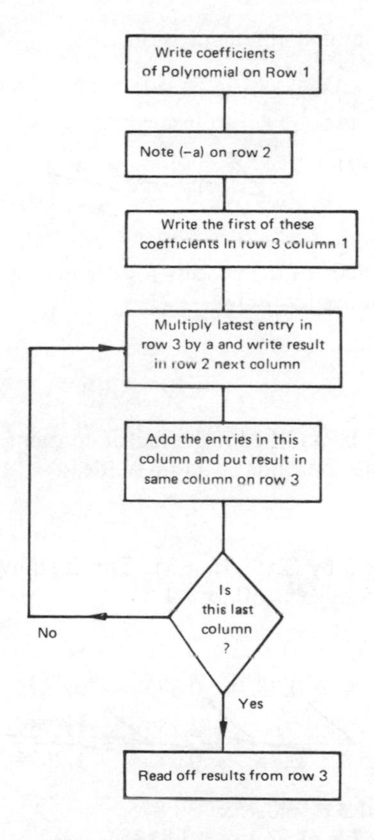

Figure 5.18

It is worth noting that if we wanted to divide, say, $3x^3 + x - 6$ by $(2x - 3)$ we we should divide $1.5x^3 + 0.5x - 3$ by $(x - 1.5)$ and double the resulting remainder (Why?).

Division by a Quadratic Factor

In much the same way as for a linear factor we may divide a polynomial by a quadratic factor. We divide $x^4 + 3x^3 - x^2 + 2x + 6$ by $(x^2 + x + 2)$ first by the traditional method.

$$\begin{array}{r|l} & x^2 + 2x - 5 \\ \hline x^2 + x + 2 & x^4 + 3x^3 - x^2 + 2x + 6 \\ & x^4 + x^3 + 2x^2 \\ \hline & \quad 2x^3 - 3x^2 + 2x \\ & \quad 2x^3 + 2x^2 + 4x \\ \hline & \qquad -5x^2 - 2x + 6 \\ & \qquad -5x^2 - 5x - 10 \\ \hline & \qquad\qquad 3x + 16 \end{array}$$

Now we apply the method of synthetic division

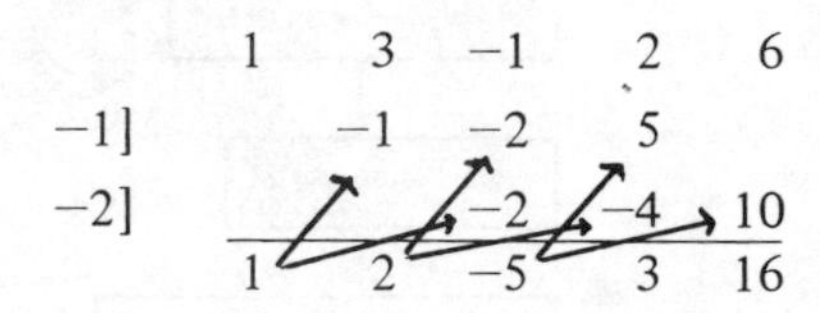

Figure 5.19 is a flow chart of the process for division by $x^2 + cx + d$ and you should follow the steps above by using it.

As before, the process is modified if the factor contains a coefficient of x^2 other than 1. An example consolidates all these ideas.

Example

Divide $3x^4 + 2x^2 - 4x + 5$ by $(2x^2 - x + 3)$. This is equivalent to the division of $1.5x^4 + x^2 - 2x + 2.5$ by $(x^2 - 0.5x + 1.5)$.

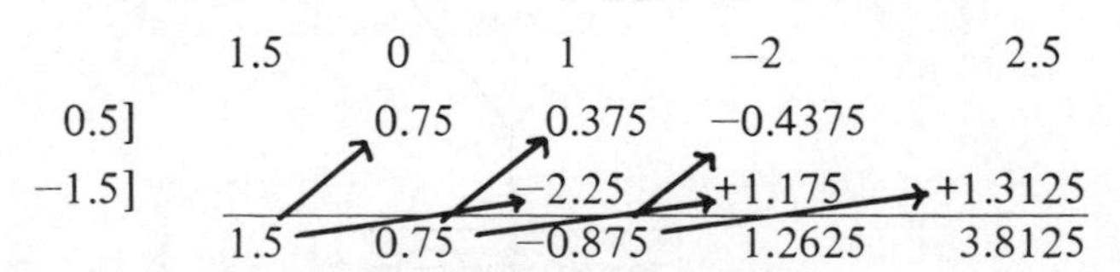

Hence we have quotient $1.5x^2 + 0.75x - 0.875$
and remainder $2 \times (1.2625x + 3.8125)$

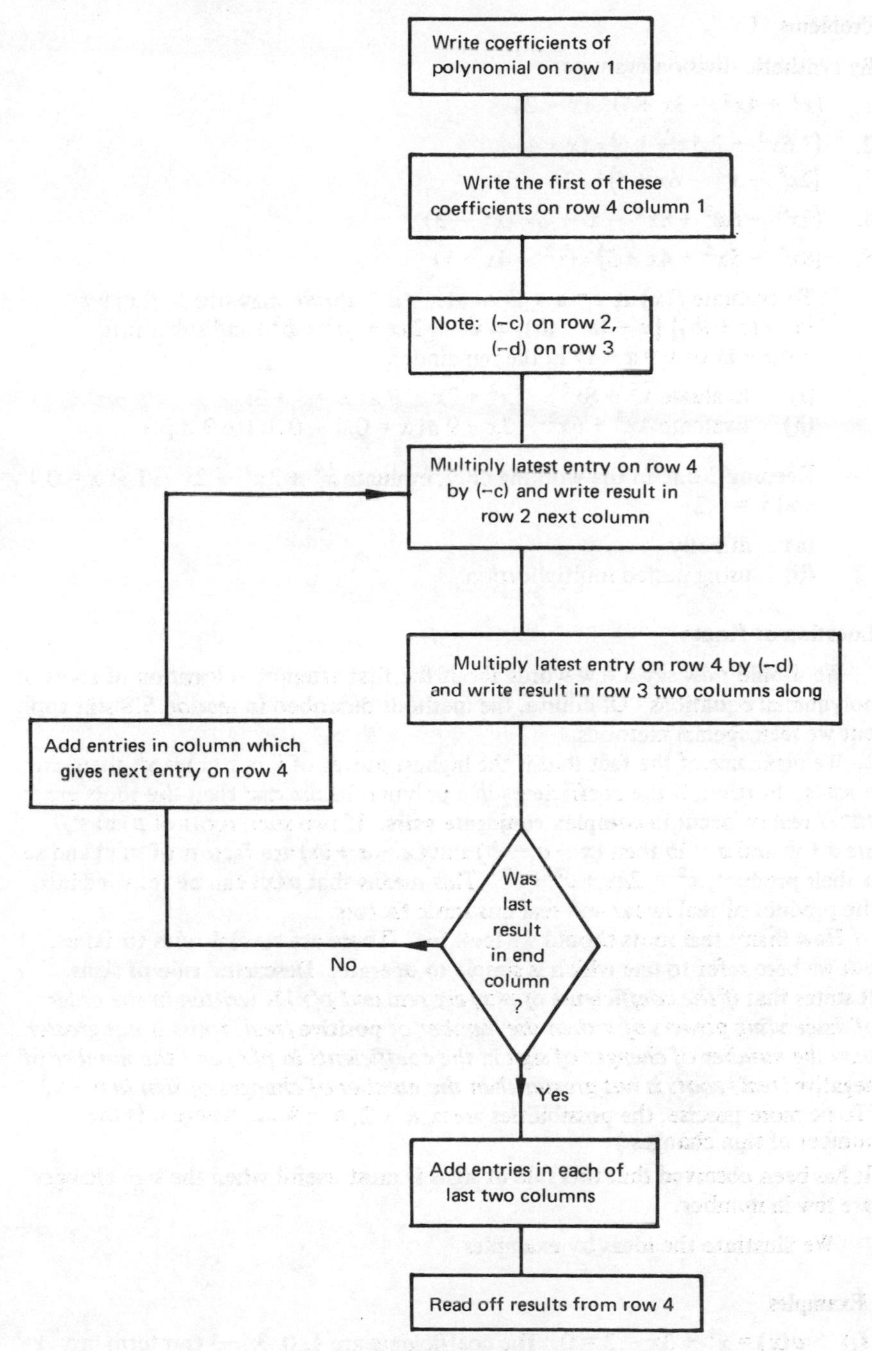
Write coefficients of polynomial on row 1
Write the first of these coefficients on row 4 column 1
Note: (−c) on row 2, (−d) on row 3
Multiply latest entry on row 4 by (−c) and write result in row 2 next column
Multiply latest entry on row 4 by (−d) and write result in row 3 two columns along
Add entries in column which gives next entry on row 4
Was last result in end column ?
No
Yes
Add entries in each of last two columns
Read off results from row 4

Figure 5.19

Problems I

By synthetic division evaluate:

1. $(x^3 + 4x^2 - 3x + 7) \div (x - 2)$
2. $(3.6x^2 + 2.5x + 1.6) \div (x + 4)$
3. $(2x^3 - x^2 - 6x + 5) \div (2x - 5)$
4. $(2x^4 - 6x^3 + 8x^2 - x - 7) \div (x^2 - 3)$
5. $(8x^3 - 5x^2 + 4x + 5) \div (x^2 - 4x + 5)$
6. To evaluate $f(x)$ at $x = a + ib$ or at $x = a - ib$, we may divide $f(x)$ by $[x - (a + ib)]\,[x - (a - ib)] \equiv x^2 - 2ax + (a^2 + b^2)$ and substitute $x = a + ib$ or $x = a - ib$ in the remainder.
   (i) Evaluate $x^4 + 8x^3 - 5x^2 + 7x - 7$ at $x = 3 + 2\text{i}$
   (ii) Evaluate $4x^3 + 6x^2 - 3x + 9$ at $x = 0.2 - 0.3\text{i}$ (to 3 d.p.)
7. Keeping 2 d.p. in the working only, evaluate $x^4 + 3x^2 - 2x - 1$ at $x = 0.1$ and $x = 1.2$
   (a) directly
   (b) using nested multiplication.

Location of Roots

We should now say a few words about the first attempt at location of roots of polynomial equations. Of course, the methods described in section 5.8 still apply but we seek special methods.

We make use of the fact that if the highest power of x in $p(x)$ is x^n there are n roots. Further, if the coefficients in a polynomial are *real* then the roots are *either* **real** *or* occur in **complex conjugate pairs**. If two such roots of $p(x) = 0$ are $a + ib$ and $a - ib$ then $(x - a - ib)$ and $(x - a + ib)$ are factors of $p(x)$ and so is their product, $x^2 - 2ax + a^2 + b^2$. This means that $p(x)$ can be resolved into the product of real linear and real quadratic factors.

How many real roots should we look for? There are several rules to aid us, but we here refer to one which is simple to operate: **Descartes' rule of signs.** It states that *if the coefficients of p(x) are real and p(x) is written in the order of descending powers of x then the number of* **positive** *(real) roots is not greater than the number of changes of sign in the coefficients in p(x) and the number of* **negative** *(real) roots is not greater than the number of changes of sign in p(−x).* (To be more precise, the possibilities are n, $n - 2$, $n - 4$ ..... where n is the number of sign changes.)

It has been observed that this rule of signs is most useful when the sign changes are few in number.

We illustrate the ideas by examples.

Examples

(i) $p(x) = x^3 + 3x - 3 = 0$. The coefficients are 1, 0, 3, −3 (no term in x^2) and the one change of sign (between 3 and −3) implies at most one positive root and since $f(0) < 0 < f(1)$ the root does exist. Now, if we replace x by $-x$,

$$p(-x) = -x^3 - 3x - 3$$

and there are no sign changes in the coefficients $-1, 0, -3, -3$ and hence no negative roots.

(ii) $x^3 - 2x^2 - 2x + 4 = 0$. $p(x)$ has 2 sign changes and $p(-x) \equiv -x^3 - 2x^2 + 2x + 4$ has one sign change. The possibilities are 1 negative root and either 2 positive roots or no positive roots. (If there are no positive roots then there will be two complex conjugate roots.)

In the following try to check on the number of roots by applying the rule of signs.

(iii) $x^3 - 3x^2 - 3 = 0$ has one positive root.

(iv) $x^3 - 1 = 0$ has one real root.

(v) $x^4 - 1 = 0$ has at most 1 positive root and 1 negative root.

(vi) $x^4 + 2x^2 + 4 = 0$ has no real roots.

(vii) $x^4 - 3x^2 + 6x + 2 = 0$. There are at most two positive roots and at most two negative roots. A rough tabulation is called for.

x	-3	-2	-1	0	1	2
$p(x)$	38	-6	-6	2	6	18

It is clear that there is one negative root between -3 and -2 (nearer the latter) and one between -1 and 0 (nearer the latter). The problem of the positive roots is not yet resolved and we decide to examine the slope function, $p'(x) = 4x^3 - 6x + 6$.

You should convince yourself that Figure 5.20(i) represents the behaviour of $p'(x)$.

This figure tells us that the slope of $p(x)$ is negative until about -1.6, zero there and for ever afterwards positive. Since $p(0) = 2$ we conclude that there are no positive roots [see Figure 5.20(ii)].

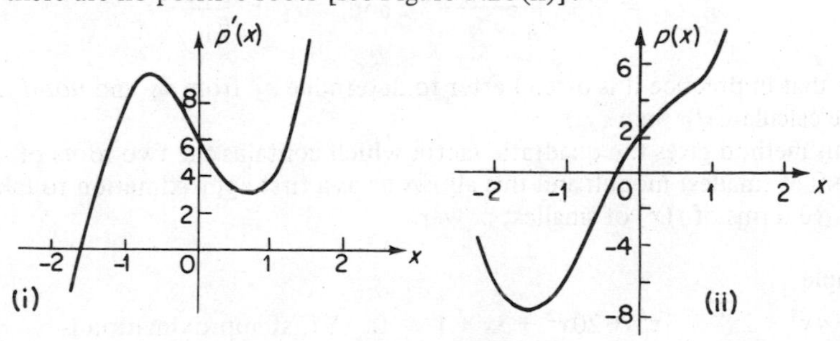

Figure 5.20

Evaluation of Quadratic Factors

Suppose we divide $p(x)$ by $x^2 + cx + d$ where $p(x) = a_0x^n + a_1x^{n-1} + \ldots + a_n$ giving a remainder $R(x)$, i.e.

$$p(x) = (x^2 + cx + d)q(x) + R(x) \tag{5.27}$$

where

$$q(x) = b_0x^{n-2} + b_1x^{n-3} + \ldots + b_{n-2}$$

and

$$R(x) = b_{n-1}(x + d) + b_n$$

In general, $R(x)$ will not vanish unless $q(x)$ is an *exact* factor of $p(x)$. Comparing coefficients in (5.27) we have

$$\begin{aligned}
b_0 &= a_0 \\
b_1 &= a_1 - cb_0 \\
b_2 &= a_2 - cb_1 - db_0 \\
&\vdots \\
b_r &= a_r - cb_{r-1} - db_{r-2} \\
b_{n-1} &= a_{n-1} - cb_{n-2} - db_{n-3} &\quad (5.28) \\
b_n &= a_n - cb_{n-1} - db_{n-2} &\quad (5.29)
\end{aligned}$$

We require that b_{n-1} and b_n should vanish *simultaneously*, and we consider two methods of achieving this. Unfortunately we have not the expertise to be able to derive the second method but we shall develop these techniques in Chapter 7. For the sake of completeness, the method is included at this stage.

Lin's Method

Let $q(x) = x^2 + c_0x + d_0$ be a first approximation to the required quadratic factor. A second approximation can be obtained by ignoring b_{n-1} and b_n in the relationships (5.28) and (5.29); then we have

$$c_1 = \frac{a_{n-1} - d_0b_{n-3}}{b_{n-2}} \quad \text{and} \quad d_1 = \frac{a_n}{b_{n-2}}$$

(Note that in practice it is often better to determine c_1 from d_1 and *not* d_0, i.e. we calculate d_1 then c_1.)

This method gives the quadratic factor which contains the two roots of $f(x) = 0$ of smallest moduli and this allows us as a first approximation to take the three terms of $f(x)$ of smallest power.

Example

$5x^6 + 4x^5 + 2x^4 + 3x^3 + 20x^2 + 3x + 1 = 0$. A first approximation is $20x^2 + 3x + 1 = 0$, i.e. $x^2 + (3/20)x + (1/20)$ giving $c_0 = 0.15$, $d_0 = 0.05$.

You should check by synthetic division that $b_3 = 2.6481$, $b_4 = 19.5397$ (4 d.p.). Hence

$$d_1 = \frac{1}{19.5397} = 0.0512 \quad \text{and} \quad c_1 = \frac{3 - 0.0512 \times 2.0481}{19.5397} = 0.1466$$

Using these values we repeat the cycle of operations to obtain

$$d_2 = 0.05120 \quad \text{and} \quad c_2 = 0.1465$$

giving the factor

$$19.5471x^2 + 2.8655x + 1$$

Bairstow's Method

We consider this method via an example.

$$x^4 + 5x^3 + 12x^2 + 14x + 8 = 0$$

First, we carry out a synthetic division using the first approximation $x^2 + 1.970x + 1.965$ to obtain a quotient of $q(x) = x^2 + 3.03x + 4.0659$ and a remainder of $0.0362x + 0.0105$ (i.e. $rx + s$).

We then take $xq(x)$ and $q(x)$ and synthetically divide these by $x^2 + 1.970x + 1.965$ as follows:

	$x\,q(x)$				$q(x)$		
	1	3.03	4.066	0	1	3.03	4.066
−1.970]		−1.970	−2.088			−1.970	
−1.965]			−1.965	−2.083			−1.965
	1	1.060	0.013	−2.083	1	1.060	2.101
			‖	‖		‖	‖
			α	β		γ	δ

We then solve the equations

$$+r = +\alpha\Delta c + \gamma\Delta d$$
$$s = \beta\Delta c + \delta\Delta d$$

for the corrections Δc, Δd to our estimates c_0, d_0.

In our example we solve

$$0.0362 = 0.013\Delta c + 1.06\Delta d$$
$$0.0105 = -2.083\Delta c + 2.101\Delta d$$

giving $\Delta c = 0.0291$, $\Delta d = 0.0337$, therefore

$$c_1 = 1.970 + 0.0291 = 1.9991$$
$$d_1 = 1.965 + 0.337 = 1.9987$$

and a better approximation of $x^2 + 1.9991x + 1.9987$.

Compare this method on the example that was used to illustrate Lin's method.

Note that $q(x)$ is obtained ready for further quadratic factors to be found.

The problem often arises as to how to find a suitable first approximation.

Problems II

1. Find the number and nature of the roots of

   (a) $3x^3 - 2x^2 + 4x - 1 = 0$

   (b) $3x^3 + 6x^2 - 2x - 4 = 0$

   (c) $3x^3 - 8x^2 - 7x - 1 = 0$

   (d) $x^4 - 8x^3 - 8x^2 + 7x + 6 = 0$

2. Using (a) Lin's method, (b) Bairstow's method, find the roots of

   (i) $x^4 - 2x^2 - x + 3 = 0$

   (ii) $x^4 + x^2 - x + 1 = 0$

3. Using Lin's method twice find an approximate quadratic factor of $x^4 - 8x^3 + 27x^2 - 38x + 26 = 0$, starting with $x^2 - 1.5x + 1$ as a first approximation.

4. Given that $x^2 - 3.9x + 4.8$ is an approximate factor of $x^4 - 4x^3 + 4x^2 + 4x - 5 = 0$, use Bairstow's method to improve the approximation.

5. In pipe-flow problems one frequently has to solve the equation $c_5 D^5 + c_1 D + c_0 = 0$. If $c_5 = 1000$, $c_1 = -3$, $c_0 = 9.04$, find a first root using the Newton-Raphson method and then apply either the Lin or the Bairstow method to find one quadratic factor.

6. Given a polynomial equation

$$a_n x^n + a_{n-1} x^{n-1} + \ldots + a_1 x + a_0 = 0$$

show that the root of maximum modulus, $x^*_{\max}$, satisfies the inequality

$$|x^*_{\max}| \leqslant \sqrt{\left[\frac{a_{n-1}}{a_n}\right]^2 - 2\left[\frac{a_{n-2}}{a_n}\right]}$$

SUMMARY

You should now be in a position to solve simultaneous linear equations by the Gauss Elimination method and to understand the ideas of pivoting and residuals. You should recognise the relative merits of direct methods and iterative methods of solution.

The concept of a matrix is very important; you should be able to perform matrix algebra without difficulty, including methods of finding the inverse of a given matrix. The application of matrix methods to the solution of simultaneous equations should be appreciated. You should be able to test whether a set of simultaneous equations has a unique solution and realise the dangers of ill-conditioning.

Those iterative methods for finding the roots of a non-linear equation which are suitable for a digital computer are worth adding to your stock of techniques, but you must understand their fallibility. The *special* methods for polynomial equations should be noted, for you must recognise that such special methods (though by their very nature, of limited applicability) are often more efficient in their particular sphere.

Chapter Six

Differentiation

6.1 A SIMPLE PROBLEM

Consider the mechanism shown in Figure 6.1 (i). It is a piston connected to a crank OQ by a connecting rod PQ; when the crank rotates about O with constant angular velocity ω the piston moves in a straight line. We wish to discover how the piston moves. To do this we must first find a relationship between the linear displacement of the piston and the angle turned through by the crank. We use the schematic diagram of Figure 6.1 (ii).

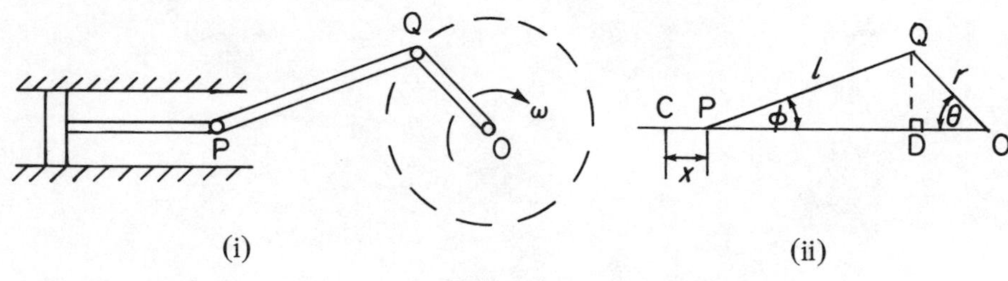

Figure 6.1

When PQO is a straight line ($\theta = \phi = 0$) then the piston is as far to the left as it will go and P is in the position C. This is called the *dead centre* position. The length PQO is then $(l + r)$ and, at that stage, $x = 0$; hence $CO = l + r$. As θ increases, so the displacement x increases until θ reaches π when x has a maximum of $2r$; thereafter x decreases to zero when $\theta = 2\pi$, whence the motion repeats every 2π.

In Figure 6.1 (ii), the distance PO is $l \cos \phi + r \cos \theta$ and so $x = l + r - (l \cos \phi + r \cos \theta)$. However, from triangles PQD and OQD we can see that $l \sin \phi = r \sin \theta$ therefore

$$\cos \phi = \sqrt{1 - \frac{r^2}{l^2} \sin^2 \theta}$$

It is convenient at this point to let the ratio l/r be denoted by m; usually, in practice, $l \gg r$. Then we have the equation

$$x = r(1 - \cos \theta + m - \sqrt{m^2 - \sin^2 \theta}) \tag{6.1}$$

If we wish to determine the velocity dx/dt and acceleration d^2x/dt^2 of the piston for the purpose of estimating the force on it we must clearly have the expertise to differentiate the right-hand side of equation (6.1). For this purpose, we need some general techniques of differentiation.

In Chapter 3 we discussed the concept of differentiation but we only learned how to find the derivatives of simple expressions. Now we shall study further techniques of differentiation, returning to the piston problem on pages 303 and 313.

6.2 TECHNIQUES OF DIFFERENTIATION

First, we provide a table of standard derivatives, Table 6.1; you should familiarise yourselves with the results therein. They can all be derived from first principles, but in the problems at the end of this section some other ways of obtaining them are indicated.

Table 6.1

Table of standard derivatives

$f(x)$	$\frac{df(x)}{dx} \equiv f'(x)$	$f(x)$	$\frac{df(x)}{dx} \equiv f'(x)$
x^n	nx^{n-1}	$\sinh x$	$\cosh x$
$\sin x$	$\cos x$	$\cosh x$	$\sinh x$
$\cos x$	$-\sin x$	$\tanh x$	$\text{sech}^2 x$
$\tan x$	$\sec^2 x$	$\log x$	$\frac{1}{x}$
$\text{cosec}\, x$	$-\text{cosec}\, x \cot x$		
$\sec x$	$\sec x \tan x$	$\log [f(x)]$	$f'(x)/f(x)$
$\cot x$	$-\text{cosec}^2 x$	e^x	e^x
$\sin^{-1} x$	$\frac{1}{\sqrt{1-x^2}}$	$\sinh^{-1} x$	$\frac{1}{\sqrt{x^2+1}}$
$\cos^{-1} x$	$\frac{-1}{\sqrt{1-x^2}}$	$\cosh^{-1} x$	$\frac{1}{\sqrt{x^2-1}}$
$\tan^{-1} x$	$\frac{1}{1+x^2}$		

In the case of inverse functions the range is restricted - see section 2.8.

Rules for combinations of functions

These rules come directly from the corresponding rules for limits of sequences; they should really be applied at a point x_0, but for practical purposes we need to assume that the functions involved are *differentiable in some interval of* x. Also we should confine the term *derivative* to a point, e.g. $f'(x_0)$ and speak of the *derived function* $f'(x)$ but we shall sometimes, for brevity, call $f'(x)$ the *derivative of* $f(x)$.

For these rules we assume that $f(x)$ and $g(x)$ are differentiable functions in some interval I and that α is some number. Then

(i)
$$\frac{\mathrm{d}}{\mathrm{d}x}[f(x) \pm g(x)] = f'(x) \pm g'(x) \tag{6.2}$$

e.g.

$$y = x^2 + \sin x; \quad \frac{\mathrm{d}y}{\mathrm{d}x} = \frac{\mathrm{d}}{\mathrm{d}x}(x^2) + \frac{\mathrm{d}}{\mathrm{d}x}(\sin x) = 2x + \cos x$$

(ii)
$$\frac{\mathrm{d}}{\mathrm{d}x}[\alpha f(x)] = \alpha f'(x) \tag{6.3}$$

e.g.

$$y = \sqrt{\pi}\cos x; \quad \frac{\mathrm{d}y}{\mathrm{d}x} = \sqrt{\pi}\,\frac{\mathrm{d}}{\mathrm{d}x}(\cos x) = -\sqrt{\pi}\sin x$$

(iii)
$$\frac{\mathrm{d}}{\mathrm{d}x}[f(x).g(x)] = f'(x).g(x) + f(x).g'(x) \tag{6.4}$$

e.g.

$$y = \mathrm{e}^x \tan x; \quad \frac{\mathrm{d}y}{\mathrm{d}x} = \frac{\mathrm{d}}{\mathrm{d}x}(\mathrm{e}^x)\tan x + \mathrm{e}^x\,\frac{\mathrm{d}}{\mathrm{d}x}(\tan x) = \mathrm{e}^x(\tan x + \sec^2 x)$$

(iv)
$$\frac{\mathrm{d}}{\mathrm{d}x}\left(\frac{f(x)}{g(x)}\right) = \frac{f'(x)g(x) - f(x)g'(x)}{[g(x)]^2} \tag{6.5}$$

provided $g(x) \neq 0$ in I,

e.g.

$$\frac{\mathrm{d}}{\mathrm{d}x}[(\log x)/x] = [(1/x).x - \log x.1]/x^2 = (1 - \log x)/x^2$$

provided I does not include $x = 0$.

Try to obtain the special case of this **Quotient Rule** when $f(x) \equiv 1$.

Example

Let us demonstrate these rules for the function $h(x) = \dfrac{(x+2)(x^2 - 2x + 1)}{(4x - 4)}$

By rules (6.2) and (6.3) the derivative of $(x + 2)$ is 1, of $(x^2 - 2x + 1)$ is $(2x - 2)$ and of $(4x - 4)$ is 4.

Then, by rule (6.4) the derivative of the numerator is

$$1.(x^2 - 2x + 1) + (x + 2)(2x - 2)$$

that is

$$(3x^2 - 3)$$

Finally, using rule (6.5) we obtain

$$h'(x) = \frac{(3x^2 - 3)(4x - 4) - (x + 2)(x^2 - 2x + 1).4}{(4x - 4)^2}$$

$$= \frac{12(x^2 - 1)(x - 1) - 4(x + 2)(x - 1)^2}{16(x - 1)^2}$$

$$= \frac{3(x + 1) - (x + 2)}{4}$$

$$= \frac{2x + 1}{4}$$

Of course, this result could have been, and should have been, obtained more easily had we noticed that $h(x) = \frac{1}{4}(x + 2)(x - 1) = \frac{1}{4}(x^2 + x - 2)$. (For what value of x is our result invalid?)

Chain Rule

When we are dealing with the composition of two functions the chain rule is applied. This rule is as follows: *if y is a differentiable function of u and u is a differentiable function of x, then y is a differentiable function of x and*

$$\frac{dy}{dx} = \frac{dy}{du} \cdot \frac{du}{dx} \tag{6.6}$$

Example 1

Consider $y = (2x^2 + 3)^3 + 2$

If we put $u = 2x^2 + 3$ then $y = u^3 + 2$ so that $du/dx = 4x$ and $dy/du = 3u^2$. The rule tells us that

$$\frac{dy}{dx} = 3u^2 . 4x = 3(2x^2 + 3)^2 \times 4x = 12x(2x^2 + 3)^2$$

You should first see how this result could have been obtained in one step and then check the answer by expanding the original expression for y and differentiating.

Example 2

We now consider $h(x) = \sin^2(3x^3 + 1)$. We write $u = 3x^3 + 1$ and note that $\sin^2 u \equiv (\sin u)^2$. Then $dh/du = 2 \sin u . \cos u$ by one application of the rule and by another $dh/dx = 2 \sin u \cos u . 9x^2 = (\sin 2u) . 9x^2 = [\sin 2(3x^3 + 1)] . 9x^2$

Example 3

Finally, we return to the motion of the piston discussed at the beginning of this chapter. We have equation (6.1), i.e. $x = r(1 - \cos\theta + m - \sqrt{m^2 - \sin^2\theta})$ so that x is a function of θ where θ itself is a function of time, t.

Then by the chain rule $dx/dt = dx/d\theta \,.\, d\theta/dt$, where $d\theta/dt$ is the angular velocity which has constant magnitude, ω. The tricky part in finding $dx/d\theta$ is to differentiate $\sqrt{m^2 - \sin^2\theta} \equiv (m^2 - \sin^2\theta)^{1/2}$; applying the chain rule, the derivative of this is

$$\frac{1}{2} \cdot \frac{-2 \sin\theta \,.\, \cos\theta}{(m^2 - \sin^2\theta)^{1/2}} \equiv \frac{-\sin 2\theta}{2\sqrt{m^2 - \sin^2\theta}}$$

Hence the velocity of the piston

$$v \equiv \frac{dx}{dt} = r\left(\sin\theta + \frac{\frac{1}{2}\sin 2\theta}{\sqrt{m^2 - \sin^2\theta}}\right) . \,\omega \qquad (6.7)$$

Try to sketch a graph of v against θ; when $\theta = \pi/2$, $v = \omega r$ and you could show that this is the maximum value of v. (When is v a minimum?) It should be observed that ω is measured in radians/sec; a dimensional check on equation (6.7) will verify this. We consider this problem further on page 313 .

Note that the chain rule can be extended to more than one stage: see Problem 9.

Problems I

1. Using the standard table of derivatives, Table 6.1, differentiate the following functions with respect to x (use the product, quotient and chain rules as appropriate)

   (i) $x^3 \sin x$ (ii) $e^x \cos x$

   (iii) $x^4 \log x$ (iv) $x^2/(1 + x)$

   (v) $(\sec x)/x$ (vi) $(3 - 2x)^5$

   (vii) $\sqrt{1 - x^2}$ (viii) $\sqrt{1 - x^2}\,\sin^{-1} x$

   (ix) $\sin 3x$ (x) $\sin^2 4x$

   (xi) $\cos(\sin 3x)$ (xii) e^{3x+2}

   (xiii) $x^2 \tan^2 2x$ (xiv) $\log(1 - x + x^2)$

   (xv) $\log[(1 + x)/(1 - x)]$ (xvi) $\log(\cot x)$

   (xvii) $e^{2x} \log(1 + x^2)$

   (xviii) $(\sin x)/(\cos x)$ (compare with the result in the table for $\tan x$)

   (xix) $\operatorname{cosec} x = (\sin x)^{-1}$ (compare with Table 6.1)

   (xx) $\cosh 2x$ (check from first principles using $\cosh 2x = [e^{2x} + e^{-2x}]/2$)

2. The distance of a moving particle from its starting point is $s = 3 + 8t - 7t^2$ (cm) where t is the time in seconds. Find the speed and acceleration of the particle after 3 seconds.

3. The velocity of a body moving through a resisting medium is $20(1 - e^{-0.02t})$; find the acceleration when $t = 5$.

4. A spring moves according to $x = e^{-0.2t}(0.6 \cos 4t - 0.4 \sin 4t)$ (Damped Motion). Sketch $x(t)$ and find the velocity at any time.

5. Show that $y = (\operatorname{cosec} x)/[\sinh(1/x)]$ satisfies the relation

$$\frac{dy}{dx} = y\left[-\cot x + \frac{1}{x^2}\coth\left(\frac{1}{x}\right)\right] \qquad \text{(C.E.I.)}$$

6. A spherical bubble expands. The rate at which its radius is increasing is 3 cm/sec. Find the rate of increase of its volume when the radius is 25 cm.

7. The ends of a rod PQ of length a are constrained to move as shown in Figure 6.2. Q is made to move such that $x = 4 \sin 3t$.

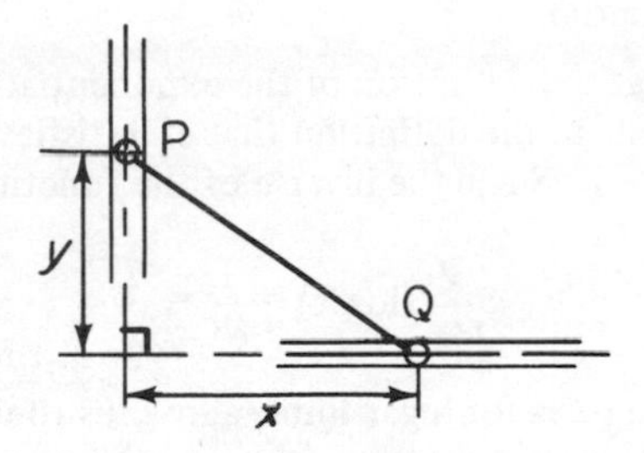

Figure 6.2

Find the speed of the end P at any instant.

8. Soil falls into a conical heap such that $h = (2/3)r$. Given that the soil falls on to the heap at the rate of 10 ml/sec, find the rate at which h increases.

9. The chain rule can be extended to the result

$$\frac{dy}{dx} = \frac{dy}{du} \cdot \frac{du}{dv} \cdot \frac{dv}{dx}$$

Check this for $y = \sqrt{u}$, $u = v(1 + v)$, $v = x^3$, by finding dy/dx by two methods.

Inverse Functions

In the special case of the chain rule where y is a function of u and u is a function of y we have $1 = (dy/du).(du/dy)$ and so

$$\frac{du}{dy} = 1\bigg/\frac{dy}{du} \qquad (6.8)$$

This is **not** a case of treating dy/du as a fraction; nor was this so for the general chain rule.

Examples

(i) If $y = u^3$, $u = y^{1/3}$ and u is a function of y so that

$$\frac{du}{dy} = \frac{1}{3u^2} = \frac{1}{3y^{2/3}}$$

as we might expect.

(ii) If $y = \sin 3u$, $u = (1/3)\sin^{-1} y$ and if we restrict the function y to the domain $-\pi/6 \leqslant u \leqslant \pi/6$, u is a function of y with $du/dy = 1/(3\cos 3u) = 1/3\sqrt{1-y^2}$.

(iii) If $y = u^2$, $u = y^{1/2}$ and $1/(dy/du) = 1/2u$.

Logarithmic Differentiation

It can be shown that the definition of the exponential function e^x given in Section 2.8 is equivalent to the definition that e^x satisfies the identity $f'(x) = f(x)$ with $f(0) = 1$. Since the inverse of the function $x = e^u$ is the function $u = \log x$ it follows that

$$\frac{d}{dx}(\log x) = \frac{1}{e^u} = \frac{1}{x}$$

This, together with the rules for logarithms, allows us to differentiate products and quotients of functions more easily. We need the two additional results

$$\frac{d}{dx}[\log f(x)] = \frac{f'(x)}{f(x)} \quad \text{and} \quad \frac{d}{dx}\log y = \left(\frac{d}{dy}\log y\right)\frac{dy}{dx} = \frac{1}{y}\frac{dy}{dx}$$

both of which follow from the chain rule, as you can verify.

Example

$$y = \frac{(2x+3)^2(x-4)^3}{(7x-2)^2(x^2+2x+2)}$$

Now,

$$\begin{aligned}\log y &= \log(2x+3)^2 + \log(x-4)^3 - \log(7x-2)^2 - \log(x^2+2x+2)\\ &= 2\log(2x+3) + 3\log(x-4) - 2\log(7x-2) - \log(x^2+2x+2)\end{aligned}$$

Differentiating both sides of this equation with respect to x we obtain

$$\frac{1}{y}\frac{dy}{dx} = 2\cdot\frac{2}{2x+3} + \frac{3}{x-4} - 2\cdot\frac{7}{7x-2} - \frac{2x+2}{x^2+2x+2}$$

Hence we may obtain dy/dx. Check the example on page 302 by this method.

Implicit Differentiation

Often we have an equation of the form $f(x, y) = 0$ where it would be difficult or impossible to write it as $y = g(x)$ or where perhaps so doing would lead to a complicated expression. In such cases, it is better to differentiate the equation as it stands.

Example 1

$$y^2 + 2ay + xy + \sin y = 4$$

Differentiating with respect to x,

$$2y\frac{dy}{dx} + 2a\frac{dy}{dx} + 1 . y + x\frac{dy}{dx} + \cos y\frac{dy}{dx} = 0$$

therefore

$$\frac{dy}{dx} = \frac{-y}{2y + 2a + x + \cos y}$$

Example 2

The circle $x^2 + y^2 = a^2$ gives on differentiating w.r.t. x

$$2x + 2y\frac{dy}{dx} = 0$$

Therefore

$$\frac{dy}{dx} = -\frac{x}{y}$$

This result gives the gradient of the tangent to the circle at any point. Take a point on the circle from each of the four quadrants and check that the slope has the right sign; qualitatively check the slope as x moves from 0 to a in the first quadrant.

The slope at the point (x_1, y_1) is $(-x_1/y_1)$ and the equation of the tangent to the circle at that point is $y - y_1 = (-x_1/y_1)(x - x_1)$, that is,

$$yy_1 - y_1^2 + x_1x - x_1^2 = 0$$

or

$$yy_1 + xx_1 = x_1^2 + y_1^2 = a^2$$

which is the standard equation.

Parametric Differentiation

When a curve is given in parametric form, e.g. $x = x(t)$, $y = y(t)$, it is often not wise to eliminate the parameter t (even if this is possible) before differentiating to find dy/dx. We consider first the parabola $y^2 = 4ax$, which in parametric form is $x = at^2$, $y = 2at$. Now it is straightforward to find $2y(dy/dx) = 4a$ and hence $dy/dx = 2a/y$.

But $dx/dt = 2at$ and $dy/dt = 2a$, therefore

$$\frac{dy}{dx} = \frac{dy}{dt} \times \frac{dt}{dx} = \frac{dy}{dt} \bigg/ \frac{dx}{dt} = \frac{2a}{2at} = \frac{1}{t}$$

we see that we get the same result as before; however, t is **not** a function of x and we need care in interpreting this result. In this case there should be little difficulty.

Example

The equations for a cycloid are $x = a(\theta - \sin\theta)$, $y = a(1 - \cos\theta)$. Here we have

$$\frac{dx}{d\theta} = a(1 - \cos\theta), \quad \frac{dy}{d\theta} = +a\sin\theta \quad \text{and} \quad \frac{dy}{dx} = \frac{\sin\theta}{1 - \cos\theta} = \cot\frac{\theta}{2}$$

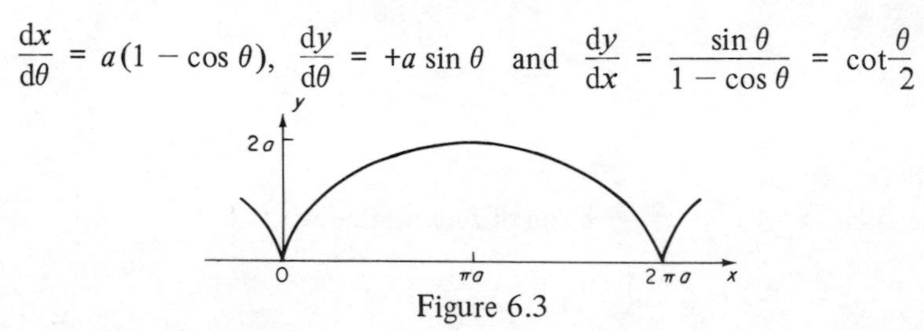

Figure 6.3

It would not be easy to express y as a function of x and less easy to differentiate the resulting expression. Is our result for dy/dx subject to misinterpretation? Compare the slope from Figure 6.3 with the formula for dy/dx. In this example the parameter θ has a physical significance. A **cycloid** is the path traced out by a point on the circumference of a wheel of radius a if the point is initially in contact with the ground; θ is the angle turned through by the wheel.

Problems II

1. Find dy/dx for each of the following functions
   (i) $y = \sin^{-1}(\cos x)$, $0 < x < \pi$
   (ii) $y = \tan^{-1}(\log x)$ (iii) $y = \sin[(x+y)^2]$
   (iv) $y = (1+x)(25 - x^2)^{1/2}$ (v) $y = [x + (1/x)]^x$
   (vi) $x = a\sin t - b\sin(at/b)$, $y = a\cos t - b\cos(at/b)$
   (vii) $x = \cos 2\theta$, $y = 2\theta + \sin 2\theta$
   (viii) $x = 2y - \tan^{-1}y$ (ix) $x^2 + y^2 + 3x + 2y - 5 = 0$
   (x) $y = (2 - x)\sqrt{\dfrac{3-x}{1+x}}$ (xi) $y = \sin^{-1}\left(\dfrac{1 + 2\sin x}{2 + \sin x}\right)$
   (xii) $y = \sin^{-1}(2x\sqrt{1 - x^2})$

2. By considering $\sin y = x$ and using implicit differentiation, derive the standard result

$$\frac{d}{dx}(\sin^{-1}x) = \frac{1}{\sqrt{1 - x^2}}$$

3. Show that
$$\frac{d}{dx}\left(\frac{1}{a}\tan^{-1}\frac{x}{a}\right) = \frac{1}{x^2 + a^2}$$

4. Find
(i) $\dfrac{d}{dx}\left(\sqrt{\dfrac{2x+3}{2x-3}}\right)$ (ii) $\dfrac{d}{dx}[e^x(1+x)^2/\sqrt{(1-x)^3}]$

5. Find dy/dx if
(i) $x^2 + 3xy + y^2 = 2$ (ii) $x^3/y^2 = 5$
(iii) $\sin(x+y) = 3xy$ (iv) $\log y = \sin x + \cos y$
(v) $e^{x-y} - 2x = 0$

6. Find dy/dx if
(i) $x = \sin t,\ y = \sin \lambda t$ (ii) $x = a\cos\theta,\ y = b\sin\theta$
(iii) $x = t^3,\ y = 2t^2 - 1$ (iv) $x = a\cos 2\theta,\ y = \cos\theta + 1$

7. Using logarithmic differentiation, show that
(i) $\dfrac{d}{dx}(10^x) = 10^x.\log_e 10$ (ii) $\dfrac{d}{dx}(x^x) = (1 + \log_e x)x^x$

General Problems

1. Differentiate with respect to x
(i) $\tan^{-1}\dfrac{4\sin x}{3 + 5\cos x}$ (ii) $\log\left(\dfrac{x^2-1}{x^2+1}\right)$
(iii) $e^{-2x}(2x^2 + 2x - 1)$ (iv) $\tan^{-1}\left(\dfrac{2x}{1-x^2}\right)$
expressing each answer in its simplest form. (L.U.)

2. Find the values when $x = 1$ of the differential coefficients with respect to x of the following functions
(i) $x^2 e^x \sin(\pi x)$ (ii) $\tan^{-1}(2x)$
(iii) $\ln\left(\dfrac{3x^2-2}{2x^2+3}\right)$ (L.U.)

3. Find the value, when $x = 1$, of the derivative with respect to x of each of the following expressions
(i) $e^{-2x}\cos\pi x$ (ii) $\ln\left(\dfrac{x^2}{1+x^3}\right)$
(iii) $\cot^{-1}x + \cot^{-1}(1/x)$ (iv) x^x (L.U.)

4. (i) If

$$y = \left(\frac{1+x}{1-x}\right)^k$$

prove that

$$(1-x^2)\frac{dy}{dx} = 2ky$$

(ii) If

$$y = \sqrt{ax - x^2} - a\tan^{-1}\sqrt{\frac{a-x}{x}}$$

prove that

$$x\left(\frac{dy}{dx}\right)^2 = a - x$$

(iii) If $y = \log(\sec\theta + \tan\theta) - (2 - \sec\theta)\tan\theta$, where $0 < \theta < \frac{1}{2}\pi$, show that $dy/d\theta$ is positive and deduce that y is always positive in this range. (L.U.)

5. (i) If

$$y = \frac{\sin^{-1}x}{\sqrt{(1-x^2)}}$$

prove that

$$(1-x^2)\frac{dy}{dx} = xy + 1$$

(ii) Find dy/dx in its simplest form if

$$y = \frac{(2x-1)(3-x)^3}{(2-x)^2}$$

(L.U.)

6. If

$$y = \tan^{-1}\left(\frac{ae\sin x}{b + a\cos x}\right)$$

where $a^2e^2 = a^2 - b^2$, show that

$$\frac{dy}{dx} = \frac{ae}{a + b\cos x}$$

(L.U.)

7. (i) Find the value when $x = \frac{1}{2}$ of the differential coefficient with respect to x of

(a) $x\sin^{-1}x + \sqrt{(1-x^2)}$ (b) $\cosh^{-1}(1/x)$ (c) x^{2x}

(ii) If $f(x) = x - \log(1+x)$, find $f'(x)$ and hence show that $f(x) > 0$ for $x > 0$. (L.U.)

8. Given that the implicit parametric equations of a particular curve representing a particular streamline in a two-dimensional flow pattern are

$$\cos ax + \sin b\theta = x\theta, \quad \sin ay + \cos b\theta = y\theta,$$

find dy/dx in terms of x, y and θ. (C.E.I.)

6.3 MAXIMUM AND MINIMUM VALUES OF A FUNCTION

We gave the definitions of *increasing* and *decreasing* regions of a function and also of a *local minimum* in Section 4.3.. For completeness we state the following rules:

(i) If $f'(x)$ is zero at a point x_0, then x_0 is said to be a **turning point** of $f(x)$.

(iia) If $f'(x_0-) < 0$ and $f'(x_0+) > 0$ the turning point is a **local minimum** [e.g. $f(x) = x^2$ at $x_0 = 0$].
[Remember that $f'(x_0-)$ means the derivative of $f(x)$ at the point $(x_0 - \epsilon)$ where ϵ is a small positive number.]

(iib) If $f'(x_0-) > 0$ and $f'(x_0+) < 0$ the turning point is a **local maximum** [e.g. $f(x) = 1 - x^2$ at $x_0 = 0$].

(iic) If $f'(x_0-)$ and $f'(x_0+)$ have the same sign then the turning point is a **Point of Inflection** [e.g. $f(x) = x^3$ at $x_0 = 0$].

We shall return to the case of points of inflection in detail in the next section.

Example

Consider $f(x) = x^3 - 12x + 4$. We can easily find $f'(x) = 3x^2 - 12$. It should be obvious that $f'(x) = 0$ when $x = \pm 2$, so these are the turning points. Since $f(x)$ is *continuous* we can take x_0- and x_0+ at convenient points between these two values of x without having to worry about being too close to ± 2 (we shall not lose local effects here). Now $f'(-3) > 0$ and $f'(-1) < 0$ hence $x = -2$ is a *local maximum*; likewise $f'(1) < 0$ and $f'(3) > 0$ so that $x = 2$ is a *local minimum*.

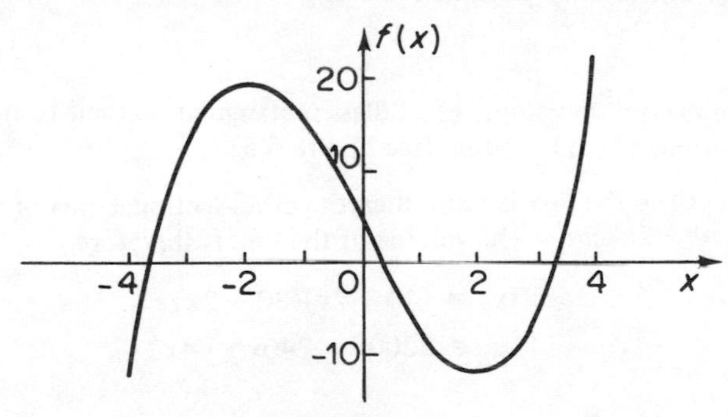

Figure 6.4

The values of $f(x)$ at the turning points are called **stationary values** and in this example they are $f(-2) = 20$ and $f(2) = -12$. A sketch of $f(x)$ is shown in Figure 6.4. What other information was needed for the sketch?

It should be borne in mind that we have so far mentioned *local* maxima and *local* minima; these are classed as *local* **extrema** and may not coincide with the *overall* extreme values of the function. In the example above, if we seek the greatest and least values of $f(x)$ in the domain $[-3, 4]$ we find these occur at $x = 4$ and $x = -2$ respectively; in this case, the local minimum is also the overall minimum in the interval $[-3, 4]$ but the overall maximum is achieved at an end-point of that interval. It is a general result that if an overall maximum (or minimum) does not coincide with its local counterparts then it will occur at an end-point of the interval in question, if this is finite. (Why should this be so? What happens if the interval is not finite? Try to find some relevant examples.)

In general, to find overall extreme values, it is advisable to evaluate $f(x)$ at points where $f'(x) = 0$ and at the end-points of the relevant interval and also at points where $f'(x)$ does not exist (remember $f(x) = |x|$?); then we sift through all these values of $f(x)$ to find the greatest and the least ones. (We deal with an example of this in Section 6.4.)

There are two situations where we can cut down on the calculus required. The first is where algebraic considerations help us to locate an overall extreme.

Example 1

Find the greatest value of $\dfrac{1}{13 - 6x + x^2}$

We can rearrange the expression to

$$\frac{1}{4 + (x-3)^2}$$

and it will take its greatest value when the denominator is least; this being the sum of two positive quantities is least when $(x - 3)^2 = 0$, i.e. when $x = 3$. The value of the original expression is then $1/4$.

A second situation is where the determination of the nature of any turning point can be achieved by physical reasoning.

Example 2

Find the maximum volume of a lidless rectangular box cut from a sheet of metal measuring 80 cm by 40 cm (see Figure 6.5).

If the depth of the box is x cm, then the cross-sectional area of its base is $(40 - 2x)(80 - 2x)\text{cm}^2$. The volume of the box is therefore

$$\begin{aligned} V(x) &= (40 - 2x)(80 - 2x)x \\ &= 3200x - 240x^2 + 4x^3 \end{aligned}$$

Hence

$$V'(x) = 3200 - 480x + 12x^2$$

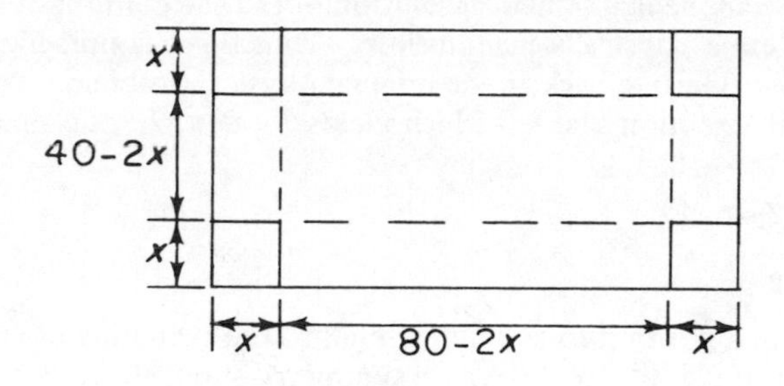

Figure 6.5

Now $V'(x) = 0$ when $4(3x^2 - 120x + 800) = 0$, i.e. when $x = 20[1 \pm (1/\sqrt{3})]$. On physical grounds we reject the + sign since this would mean $40 - 2x < 0$ and we note that this is impossible by reference to Figure 6.4. It is easy to see that x is confined to the range $0 \leqslant x \leqslant 20$ and, since $x = 0$ or $x = 20$ give zero volume, it follows that $x = 20[1 - (1/\sqrt{3})]$ will provide the maximum volume.

Example 3

Let us return to equation (6.7): $v = \omega r(\sin\theta + \frac{1}{2}\sin 2\theta/\sqrt{m^2 - \sin^2\theta})$; we see that when $\theta = 0$ or π or 2π, $v = 0$ and hence x has local extreme values; check that these correspond to extreme positions of the piston. In fact, $v = 0$ when

$$\sin\theta + \frac{\frac{1}{2}\sin 2\theta}{\sqrt{m^2 - \sin^2\theta}} = 0, \text{ i.e. } \sqrt{m^2 - \sin^2\theta} \,.\, \sin\theta + \sin\theta\cos\theta = 0$$

which reduces to

$$\sin\theta\,[\sqrt{m^2 - \sin^2\theta} + \cos\theta] = 0$$

Thus *either*

$$\sin\theta = 0$$

or

$$\sqrt{m^2 - \sin^2\theta} = -\cos\theta$$

i.e.

$$m^2 - \sin^2\theta = \cos^2\theta$$

i.e.

$$m^2 = \cos^2\theta + \sin^2\theta$$

i.e.

$$m^2 = 1$$

If $m = -1$ then, remembering that $m = l/r$, we find that $l = -r$, which is nonsense since l and r are lengths, and must be positive. If $m = 1$, then $l = r$. *However*, it can be reasoned on physical grounds that m must be at least 2. The case $m = -1$ has arisen because at one stage we squared an equation in order to obtain a solution for m. This is a spurious root arising from algebraic manipulation. On the other hand, the case $m = 1$ is a genuine mathematical solution but in the context of this particular problem is physically inadmissible. Thus this example highlights the danger of not referring back to the original physical problem. We are left with the only solution $\sin\theta = 0$ which yields $\theta = 0, \pi, 2\pi$ as claimed earlier.

Curve Sketching

We are now in a position to add more detail to the sketching of curves. Consider, for example, $y^2 = x^5$. We have symmetry about the x-axis; x cannot be negative; the only intersection with either axis is at the origin; for large values of x, y is larger than x. Now we find that

$$\frac{dy}{dx} = \pm\frac{5}{2}x^{3/2}$$

which means that as x increases, the slope increases in magnitude from its value of zero at the origin. The feature there is called a **cusp**, as shown in Figure 6.6.

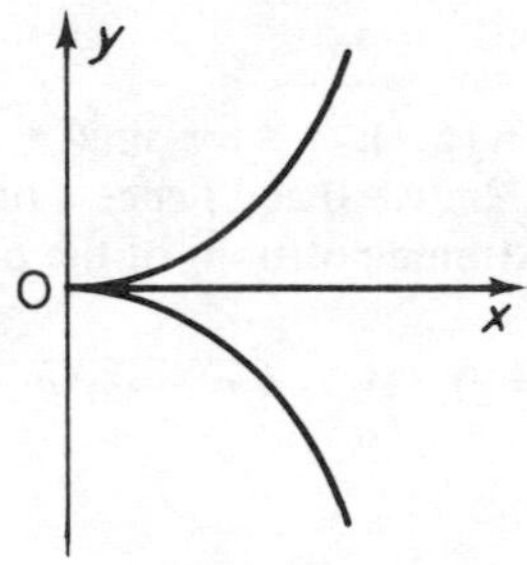

Figure 6.6

Problems

1. (i) Show that the greatest and least values of $f(x) = x^2 - 3x + 2$ in the interval $[0, 4]$ are 6 and $-\frac{1}{4}$ respectively. Repeat for the intervals $[-3, 3]$, $[-3, 0]$ and $[1, 2]$.

   (ii) Repeat the above for $f(x) = 2 + x - x^2$.

2. Examine the following for maximum and minimum values:

   (i) $(x - 4)^2(x + 3)^3$ (ii) $x^4 - 8x^3 + 18x^2 - 14$

   (iii) $x^2 + (54/x)$ (iv) $x^{2/3}(6x - x^2 - 8)^{1/3}$

   (v) $\theta \sin\theta + (1 + h)\cos\theta$ (consider the cases $h < 1$ and $h \geqslant 1$ separately)

3. The velocity of a signal at a distance r along a cable is $v = kr \log 1/r$ where $0 < r < 1$; at what distance is the velocity greatest? Sketch the graph of $v(r)$.

4. The velocity of a wave in a deep channel is

$$v = c\sqrt{\frac{\lambda}{b} + \frac{b}{\lambda}}$$

where λ is the wavelength, b and c constants. What is the least velocity?

5. The transmitted power of a belt can be represented as $H = Cv(T_0 - wv^2/g)$ where T_0 is the initial tension in the belt, w its weight/unit length, v its running speed and C a constant. Find the speed at which maximum power is transmitted.

6. In an LCR circuit, the impedance is given by

$$Z = \sqrt{R^2 + [\omega L - (1/\omega C)]^2}$$

At what frequency ω is the impedance least?

7. The torque exerted by an induction motor is given by $T = k^2 R/(R^2 + k^2)$ where R is the rotor resistance, and k is a constant. Find the maximum torque.

8. A high-frequency coaxial feeder has internal radius r_1 and external radius r_2. To determine its optimum dimensions we must find the value of r_1, given r_2, for which attenuation is least; this requires the least value of

$$\left(\frac{1}{r_1} + \frac{1}{r_2}\right) \Big/ \log \frac{r_2}{r_1}$$

Write $x = r_2/r_1$ and find an equation for x. Show that $3 < x < 4$ and use the Newton-Raphson method to obtain x to 3 d.p.

9. Sketch the curves

(i) $y^2(2 + x) = x^2(2 - x)$ (ii) $x^3 + y^3 - 9x^2 = 0$

10. Show that the equation $x^4 - x^3 - 6x^2 + 4x + 8 = 0$ has two coincident roots. Hence find the four roots of the equation.

11. Sketch and discuss the curve $y^4 = x^2 y - 4$, showing clearly the salient features, including any asymptotes and extrema. (C.E.I.)

12. Sketch and discuss the curve $y^2(x + 1) = x^2(1 - x)$.
Include in your discussion the maximum and minimum points and the asymptotes. (C.E.I.)

13. (i) If $f(x) = \ln(1 - x) + x/(1 - x)$, find $f'(x)$ and deduce that $f(x) > 0$ for $0 < x < 1$.

(ii) Find the maximum and minimum values of the function $g(x) = (x^2 + 2x - 1)e^{-2x}$.
Show that the curve $y = g(x)$ has points of inflection and sketch the curve. (L.U.)

14. Find the stationary values of

$$y = x + 5 + \frac{3}{x} - \frac{1}{x^2}$$

and determine its finite maximum and minimum values, if any. Show that $\mathrm{d}y/\mathrm{d}x$ is not negative if x is positive or less than -2.
Prove that

$$\lim_{x \to \infty} \frac{\mathrm{d}y}{\mathrm{d}x} = 1$$

and give a careful sketch of the curve represented by the given equation. (L.U.)

6.4 HIGHER DERIVATIVES

We have seen that a tabulated function has second, third and higher differences, so a function given by the rule $f(x)$ has higher derivatives $f''(x)$, $f'''(x)$, etc. *if they exist.* The nth order derivative is denoted $f^{(n)}(x)$ or $\mathrm{d}^n y/\mathrm{d}x^n$.

Some functions are *infinitely differentiable,* i.e. they possess derivatives of all orders. Examples are

(i) finite polynomials which at some stage have a zero derivative and further differentiation yields zero on each subsequent occasion;

(ii) $\sin x$ for which the derivatives have a cyclic pattern: $\cos x, -\sin x, -\cos x, \sin x, \ldots$

There are examples where each time we differentiate a function it becomes more badly behaved and at some stage we may obtain a non-differentiable function: for example, $f(x) = |x|$ gives $f'(x)$ which is not defined at $x = 0$.

Many of the techniques we have developed so far carry through for higher derivatives.

Examples

(i) We found on differentiating implicitly the equation $x^2 + y^2 = a^2$ that $2x + 2y(\mathrm{d}y/\mathrm{d}x) = 0$. If we repeat the process we obtain

$$2 + 2\left(\frac{\mathrm{d}y}{\mathrm{d}x}\right)\frac{\mathrm{d}y}{\mathrm{d}x} + 2y\frac{\mathrm{d}^2y}{\mathrm{d}x^2} = 0$$

On substituting for $\mathrm{d}y/\mathrm{d}x$,

$$2 + 2\left(\frac{-x}{y}\right)^2 + 2y\frac{d^2y}{dx^2} = 0$$

whence

$$\frac{d^2y}{dx^2} = -\frac{1 + (x^2/y^2)}{y} = -\frac{a^2}{y^3}$$

It would have been more tedious to have found

$$\frac{dy}{dx} = \frac{-x}{\sqrt{a^2 - x^2}}$$

and differentiated this equation to find d^2y/dx^2.

(ii) For the parabola $y^2 = 4ax$, in parametric form we had $dy/dx = 1/t$. In general, if we seek d^2y/dx^2 we note that this is $(d/dx)(dy/dx)$ but since dy/dx is a function of t, we apply the chain rule so that

$$\frac{d^2y}{dx^2} = \frac{d}{dt}\left(\frac{dy}{dx}\right)\cdot\frac{dt}{dx} \tag{6.9}$$

Hence

$$\frac{d^2y}{dx^2} = \frac{d}{dt}\left(\frac{1}{t}\right)\cdot\frac{dt}{dx} = \left(\frac{-1}{t^2}\right)\cdot\frac{1}{2at} = -\frac{1}{2at^3}$$

[A common error is to say that

$$\frac{d^2y}{dx^2} = \frac{d^2y}{dt^2}\bigg/\frac{d^2x}{dt^2}$$

which is **absolutely wrong** - we emphasise that d^2y/dx^2 means *differentiate y w.r.t. x twice.*]

Again we must take care in interpreting the result.

(iii) What interpretation can we put on the second derivative? A physical example is where x represents displacement and t time; then d^2x/dt^2 is the rate of change of dx/dt with time, i.e. it is an acceleration. With the piston example we may differentiate equation (6.7) to obtain

$$\frac{d^2x}{dt^2} = \omega^2 r\left[\cos\theta + \frac{\sin^4\theta + m^2\cos 2\theta}{(m^2 - \sin^2\theta)^{3/2}}\right] \tag{6.9a}$$

Check this for yourself. Note that the knowledge of this acceleration would allow us to evaluate the forces involved by using Newton's Laws.

Geometrical Interpretation of the Second Derivative

Geometrically d^2y/dx^2 represents the rate of change of slope of $y = f(x)$ with x [since $d^2y/dx^2 = d/dx(dy/dx)$].

To consider the applications of this idea see Figure 6.7.

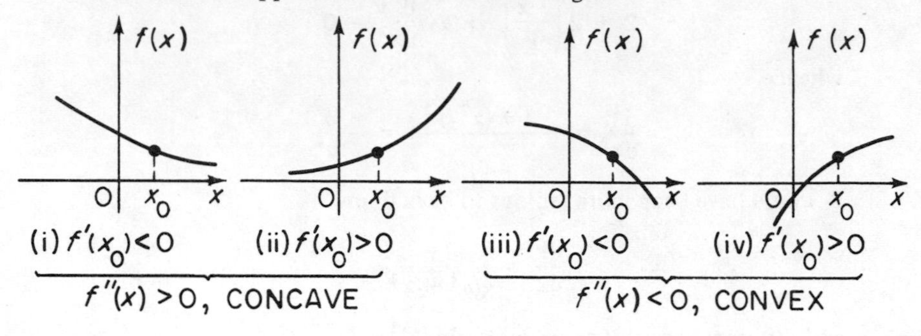

Figure 6.7

We see that $f''(x)$ gives the *local* sense of *concavity* of a function. By matching (i) and (ii) we see that at a local minimum, $f''(x_0) \geqslant 0$ and by matching (iii) and (iv) at a local maximum, $f''(x_0) \leqslant 0$.

These provide alternative rules for determination of a turning point.

But in the case (iic) on page 311 we had a point where $f'(x)$ kept the same sign as it passed through the turning point. To see what implication this has on $f''(x_0)$ consider the following examples.

Example 1

Consider $f(x) = x^3$. Now $f'(x) = 3x^2$, $f''(x) = 6x$. At $x = 0$, $f'(x) = 0$; hence we have a turning point. If we now look at the expression for $f''(x)$ we find this is also zero at $x = 0$. The graph of $f(x) = x^3$, Figure 6.8 (i), shows that $f'(x) = 0$ at the origin. Comparing with the diagrams of Figure 6.7 we see that the graph is made up from the two parts shown in Figure 6.8 (ii). As we pass through the origin, the sense of concavity changes and $f''(x)$ changes from negative to positive. Thus, at the origin we must have $f''(x) = 0$ as we have found algebraically.

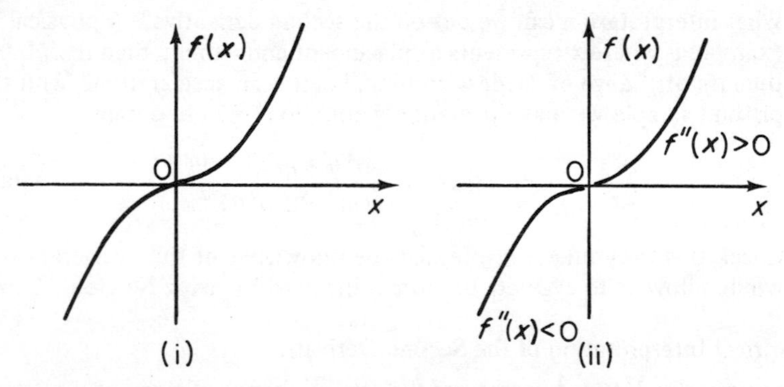

Figure 6.8

We define a *point of inflection* as being a point *where the concavity changes* and at such a point $f''(x) = 0$. It is not essential for $f'(x)$ to be zero at a point of inflection – see Figure 6.9.

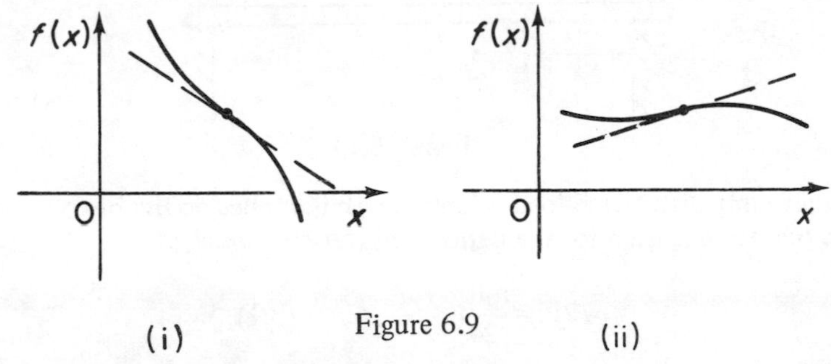

Figure 6.9

It is to be noted that where $f'(x) = 0$ at the same point, trouble can occur in identifying the nature of the turning point.

Example 2

Consider $f(x) = x^4$; then $f'(x) = 4x^3$ and $f''(x) = 12x^2$. Now both $f'(x)$ and $f''(x)$ vanish at $x = 0$, yet from Figure 6.10 we can see that, in fact, there is a local minimum there.

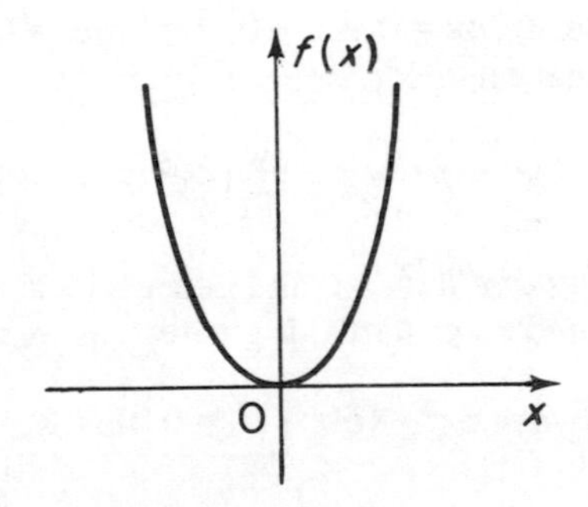

Figure 6.10

For the time being, we advise that where both first and second derivatives vanish at a value of x it is best to examine $f'(x)$ on either side of this value; we return with a better explanation of the situation in section 6.12.

Example 3

Figure 6.11 shows an encastre beam which is fixed into a supporting structure so that the slope at each end is zero. We denote distance from the left-hand wall by x and the downwards deflection of that point from the level by y.

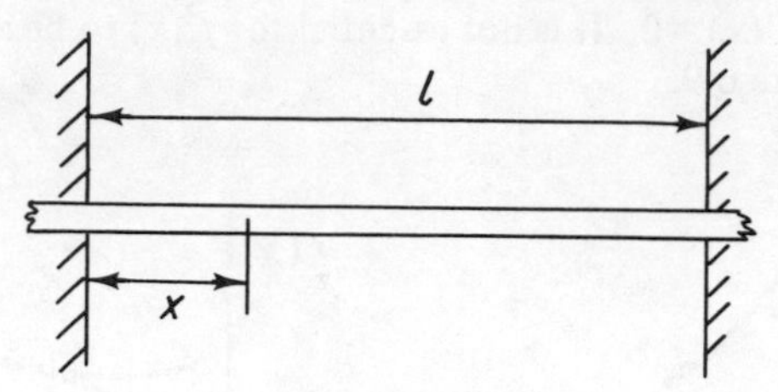

Figure 6.11

A uniformly distributed load w/unit length is applied to the beam. It can be shown that the equation for the deflection, *assumed small,* is

$$EIy = \frac{w}{24}(x^4 - 2lx^3 + l^2x^2)$$

where EI is the (constant) flexural rigidity of the beam.

It is easy to check that this formula gives $y(0) = y(l) = 0$.

On differentiating, we find that

$$EI\frac{dy}{dx} = \frac{w}{24}(4x^3 - 6lx^2 + 2l^2x) = \frac{2w}{24}x(2x^2 - 3lx + l^2)$$

Hence $dy/dx = 0$ at $x = 0$ or where $2x^2 - 3lx + l^2 = 0$, i.e. $x = 0$ or $(x - l)(2x - l) = 0$, i.e. $dy/dx = 0$ at $x = 0, x = l$ or $x = l/2$.

A second differentiation yields

$$-M = EI\frac{d^2y}{dx^2} = \frac{w}{12}(6x^2 - 6lx + l^2)$$

which at $x = l/2$ implies $d^2y/dx^2 > 0$ and hence a local minimum; this is where the deflection (measured downwards) is greatest, as we might have expected from symmetry.

Now $d^2y/dx^2 = 0$ when $6x^2 - 6lx + l^2 = 0$, that is,

$$x = \frac{l}{6}(3 \pm \sqrt{3})$$

$$\simeq 0.79l \quad \text{and} \quad 0.21l$$

These values give the points of inflection and are points where, *under the assumption of small deflection,* the bending moment, M, is zero. (Note they are symmetrically placed about $x = l/2$.) A further differentiation will give

$$-F = EI\frac{d^3y}{dx^3} = \frac{w}{2}(2x - l)$$

where F is the shear force. Finally,

$$EI\frac{d^4y}{dx^4} = w$$

which is a statement about the load at any point.

In practice, we would usually start from the bending moment equation and undo the differentiation to obtain the deflection.

Example 4

Find the greatest and least values of $f(x) = e^{-x}\sin x$ in $[0, \infty)$. We find $f'(x) = e^{-x}(-\sin x + \cos x)$ and this vanishes where $\tan x = +1$, that is, when

$$x = (4n + 1)\frac{\pi}{4}; \quad n = 0, 1, 2, \ldots$$

Further,

$$f''(x) = e^{-x}(\sin x - 2\cos x - \sin x) = -2\cos x\, e^{-x};$$

at $x = \pi/4, 9\pi/4, 17\pi/4\ldots$ $f''(x) < 0$, indicating a local maximum, whereas at $x = 5\pi/4, 13\pi/4, 21\pi/4, \ldots$ $f''(x) > 0$, indicating a local minimum.

However, the absolute values of these stationary points are decaying with increasing x. The first local maximum of $f(x)$, viz. $(1/\sqrt{2})e^{-\pi/4}$, is the greatest value we seek and the first local minimum, $(-1/\sqrt{2})e^{-5\pi/4}$ is the least value.

Problems

1. (i) If $y = A\log_e x + B$, prove that $x\dfrac{d^2y}{dx^2} + \dfrac{dy}{dx} = 0$.

   (ii) If $pV^\gamma = c$, show $V^2\dfrac{d^2p}{dV^2} = \gamma(\gamma + 1)p$ and $p^2\dfrac{d^2V}{dp^2} = \dfrac{\gamma + 1}{\gamma^2}V$

2. Find d^2y/dx^2 for each of the parts of Questions 5 and 6 of Problems II on page 309.

3. The profit P in a certain manufacturing process is given by

$$P = px - [F + V(x/k)^{3/2}]$$

   where x is the number produced, p is the sales price and V, F and k are constants of cost. Find the maximum profit.

4. The potential energy of one atom due to another is given by

$$V = \frac{A}{r^{12}} - \frac{B}{r^6} \quad (A, B > 0)$$

   where r is the separation of the atoms.
   Find where $dV/dr = 0$ (the equilibrium separation) and show that $d^2V/dr^2 > 0$ at this value of r.

5. If $y = \sin(2\sin^{-1}x)$, show that

$$(1-x^2)\frac{d^2y}{dx^2} - x\frac{dy}{dx} + 4y = 0$$

6. Prove that

$$\frac{d^2x}{dy^2} + \left(\frac{dx}{dy}\right)^3 \frac{d^2y}{dx^2} = 0 \qquad \text{(C.E.I.)}$$

7. (i) If

$$x = a\left(t + \frac{1}{t}\right) \quad \text{and} \quad y = a\left(t - \frac{1}{t}\right)$$

prove that

$$y^3\frac{d^2y}{dx^2} + 4a^2 = 0$$

(ii) When

$$y = \frac{x}{x + \sqrt{(1+x^2)}}$$

show that

$$\sqrt{1+x^2}\,\frac{dy}{dx} = \frac{y^2}{x^2} \qquad \text{(L.U.)}$$

8. (i) Given that $x = 1/(1+t^2)$, $y = t^3/(1+t^2)$, obtain expressions for dy/dx and d^2y/dx^2 in terms of t.

(ii) By considering the stationary value of $f(x) = e^{ax} - x$, where a is real and positive, prove that the equation $f(x) = 0$ has two and only two real roots when $a < e^{-1}$. Show that, when a is small, the smaller of these roots is

$$1 + a + \frac{3}{2}a^2 + O(a^3) \qquad \text{(L.U.)}$$

9. If $y = (1-x^2)^{1/2}\sin^{-1}x$, prove that

$$(1-x^2)\frac{d^2y}{dx^2} - x\frac{dy}{dx} + 2x + y = 0 \qquad \text{(L.U.)}$$

10. *Leibnitz' Theorem*

Let u and v be functions of x which possess derivatives of any order. Starting from

$$\frac{d}{dx}(uv) = u\frac{dv}{dx} + \frac{du}{dx}v$$

show that

$$\frac{d^2}{dx^2}(uv) = u\frac{d^2v}{dx^2} + 2\frac{du}{dx}\cdot\frac{dv}{dx} + \frac{d^2u}{dx^2}v$$

and deduce that

$$\frac{d^3}{dx^3}(uv) = u\frac{d^3 v}{dx^3} + 3\frac{du}{dx}\cdot\frac{d^2 v}{dx^2} + 3\frac{d^2 u}{dx^2}\frac{dv}{dx} + \frac{d^3 u}{dx^3}v$$

By induction show that

$$\frac{d^n}{dx^n}(uv) = \sum_{r=0}^{n} {}^nC_r \frac{d^r u}{dx^r}\frac{d^{n-r}v}{dx^{n-r}}$$

where $\dfrac{d^0 u}{dx^0} \equiv u, \quad \dfrac{d^0 v}{dx^0} \equiv v$

11. Using Leibnitz' Theorem for the nth derivative of the product of two functions, show that

$$\frac{d^n}{dx^n}(x \log x) = (-1)^n \frac{(n-2)!}{x^{n-1}} \qquad \text{(C.E.I.)}$$

12. In Mechanics, a point of equilibrium is said to be *stable* if the potential energy V is a minimum at that point and *unstable* if V is a maximum there; otherwise it is in *neutral equilibrium.* Investigate the stability of the points of equilibrium of the motion described by

$$V(x) = \begin{cases} c & \text{for } |x| > a \\ c - \lambda \exp\{-x^2/(a^2 - x^2)\} & \text{for } |x| \leqslant a \end{cases} \qquad \text{(C.E.I.)}$$

13. (i) Find the sixth derivative with respect to x of $x^2 \log x$.

(ii) If $y = \exp\{x + \sqrt{(1 + x^2)}\}$, find d^2y/dx^2 when $x = 0$. (L.U.)

14 State Leibnitz' Theorem on the nth differential coefficient of a product.

If $y = \sin x/(1 - x^2)$, show that

(i) $(1 - x^2)\dfrac{d^2 y}{dx^2} - 4x\dfrac{dy}{dx} - (1 + x^2)y = 0$

(ii) $y_{n+2} - (n^2 + 3n + 1)y_n - n(n-1)y_{n-2} = 0$

where y_n is the value of $d^n y/dx^n$ when $x = 0$. (L.U.)

15. Show that $(x^2 - 1)^n$ satisfies the equation $(x^2 - 1)\dfrac{dy}{dx} - 2nxy = 0$ and hence deduce that

$$(x^2 - 1)\frac{d^{n+2}}{dx^{n+2}}(x^2 - 1)^n + 2x\frac{d^{n+1}}{dx^{n+1}}(x^2 - 1)^n - n(n+1)\frac{d^n}{dx^n}(x^2 - 1)^n = 0$$

(L.U.)

16. If $f(x) = \sin(k \sin^{-1} x)$ show that

$$(1 - x^2)f'' - xf' + k^2 f = 0$$

and

$$(1 - x^2)f^{(n+2)} - (2n+1)xf^{(n+1)} + (k^2 - n^2)f^{(n)} = 0 \qquad \text{(L.U.)}$$

6.5 CURVATURE

The curvature of an arc AB of the curve $y = f(x)$ is the angle through which the tangent moves as its point of contact traverses the arc from A to B. The angle is shown as $\delta\psi$ and if the length of the arc AB is δs (both assumed small) we have an average rate of curvature of $\delta\psi/\delta s$. (See Figure 6.12)

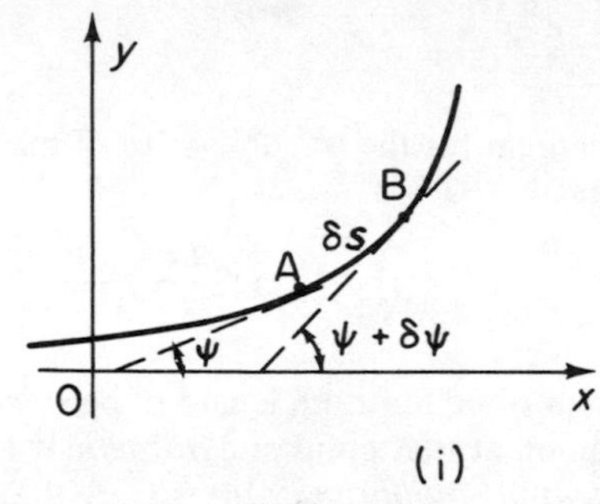

Figure 6.12

We define **curvature at a point** as

$$\kappa = \frac{d\psi}{ds} \tag{6.10}$$

Clearly the larger κ, the more rapid the *bending* of the curve.

Related to this concept is that of **radius of curvature.** The radius of curvature at a point is given by

$$\rho = \frac{1}{\kappa} = \frac{ds}{d\psi} \tag{6.11}$$

The basic idea stems from that of a circle whose radius of curvature is constant (and equal to its radius). Figure 6.13 shows us that $\delta s \triangleq r\delta\psi$ (exact equality occurs for a circle) and hence $r \triangleq \delta s/\delta\psi$, which gives us ρ.

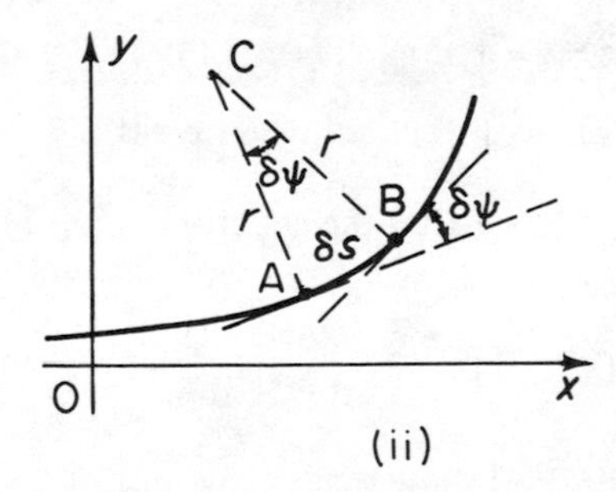

Figure 6.13

It follows that the *larger* ρ is the *flatter* the curve and the *smaller* ρ is the *sharper* the bending.

Locally, the arc AB can be taken to be the arc of a circle of radius r.

C is called the local **centre of curvature.**

To derive the formula for ρ in Cartesian coordinates we refer to Figure 6.14. If the arc AB is *small*, it can be approximated by a straight line and we have, in the limit, the relationships

$$\frac{dy}{dx} = \tan\psi \quad \text{and} \quad \left(\frac{ds}{dx}\right)^2 = 1 + \left(\frac{dy}{dx}\right)^2 \tag{6.12}$$

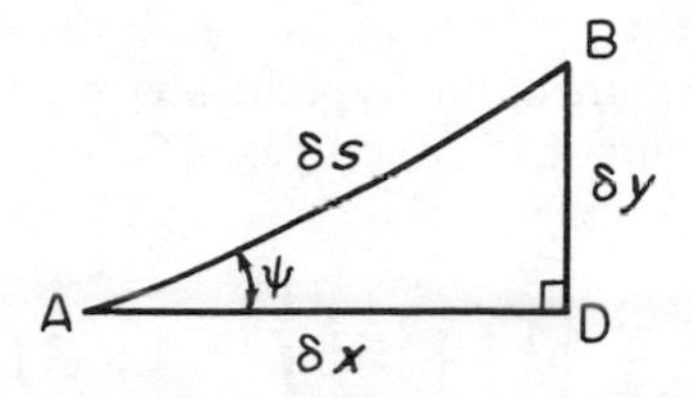

Figure 6.14

We attempt to find $d\psi/ds$ and to do this we begin with

$$\frac{d}{ds}\left(\frac{dy}{dx}\right) = \frac{d}{ds}(\tan\psi) = \frac{d}{d\psi}(\tan\psi)\,.\,\frac{d\psi}{ds} = \sec^2\psi\frac{d\psi}{ds}$$

$$= (1+\tan^2\psi)\frac{d\psi}{ds} = \left\{1 + \left(\frac{dy}{dx}\right)^2\right\}\frac{d\psi}{ds}$$

But

$$\frac{d}{ds}\left(\frac{dy}{dx}\right) = \frac{d}{dx}\left(\frac{dy}{dx}\right)\frac{dx}{ds} = \frac{d^2y}{dx^2}\,.\cos\psi$$

$$= \frac{d^2y}{dx^2}\,\frac{1}{\sqrt{1+\tan^2\psi}}$$

Hence, comparing,

$$\rho = \frac{\left\{1 + \left(\dfrac{dy}{dx}\right)^2\right\}^{3/2}}{\dfrac{d^2y}{dx^2}} \tag{6.13}$$

[Notice that if dy/dx is small then ρ is approximately $1/(d^2y/dx^2)$, hence the approximation $M = -EI(d^2y/dx^2)$ for small deflections of beams.]

Example 1

Find the radius of curvature of the curve $y = \log x$.

We have

$$\frac{dy}{dx} = \frac{1}{x} \quad \text{and} \quad \frac{d^2y}{dx^2} = -\frac{1}{x^2}$$

Hence

$$\rho = \frac{\left\{1+\frac{1}{x^2}\right\}^{3/2}}{-\frac{1}{x^2}} = \frac{-(x^2+1)^{3/2}}{x}$$

Check this result against the graph of $\log x$ on page 84.

Example 2

Find the radius of curvature of the hyperbola $xy = 1$ at the point (1, 1). (It will help to refer to Figure 4.15 (ii) on page 156.)

Noting that

$$y = \frac{1}{x}, \quad \rho = \frac{\left\{1+\left(\frac{-1}{x^2}\right)^2\right\}^{3/2}}{2/x^3} = \frac{x^3}{2}\left\{1+\frac{1}{x^4}\right\}^{3/2}$$

At (1, 1) $\rho = \tfrac{1}{2}\{2\}^{3/2} = \sqrt{2}$

Observe that at the point (−1, −1) $\rho = -\sqrt{2}$ and the negative sign merely indicates the sense of bending. For large x, ρ is approximately $\tfrac{1}{2}(x^3)$ and the curve flattens out as $x \to \infty$; for small x, ρ is approximately $x^3/2x^6 = 1/2x^3$ and as $x \to 0$ the curve again flattens out.

If the centre of curvature C has coordinates (x_c, y_c) then it can be shown (see Problem 1 following) that

$$x_c = x - \rho \sin \psi$$
$$y_c = y + \rho \cos \psi$$

or

$$\left.\begin{aligned} x_c &= x - \frac{\left\{1+\left(\frac{dy}{dx}\right)^2\right\}\frac{dy}{dx}}{\frac{d^2y}{dx^2}} \\ y_c &= y + \frac{\left\{1+\left(\frac{dy}{dx}\right)^2\right\}}{\frac{d^2y}{dx^2}} \end{aligned}\right\} \qquad (6.14)$$

Hence for the hyperbola $xy = 1$

$$x_c = x + \frac{x}{2}\left(1+\frac{1}{x^4}\right); \quad y_c = \frac{1}{x} + \frac{x^3}{2}\left(1+\frac{1}{x^4}\right)$$

i.e.

$$x_c = \frac{1}{2}\left(3x+\frac{1}{x^3}\right); \quad y_c = \frac{1}{2}\left(x^3+\frac{3}{x}\right)$$

As we move along a curve the centre of curvature will trace out a locus called the **evolute** of the original curve. The evolute of the hyperbola $xy = 1$ is shown in Figure 6.15. For the circle $x^2 + y^2 = a^2$ you should be able to show that $(x_c, y_c) \equiv (0, 0)$ and therefore the evolute here reduces to a *single point.*

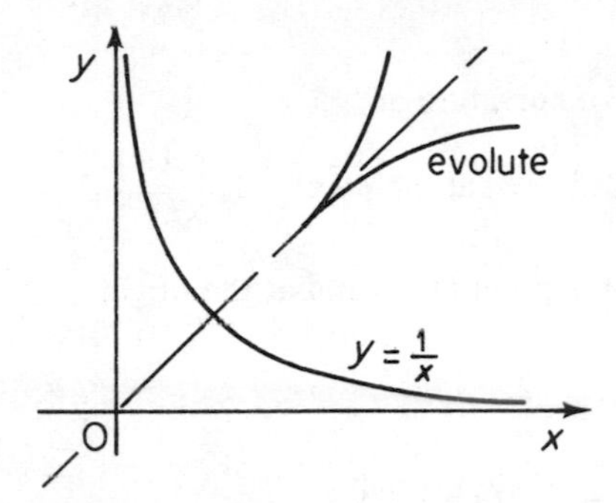

Figure 6.15

Problems

1. By considering Figure 6.16, show that for the centre of curvature C

$$x_c = x - \rho \sin \psi$$
$$y_c = y + \rho \cos \psi$$

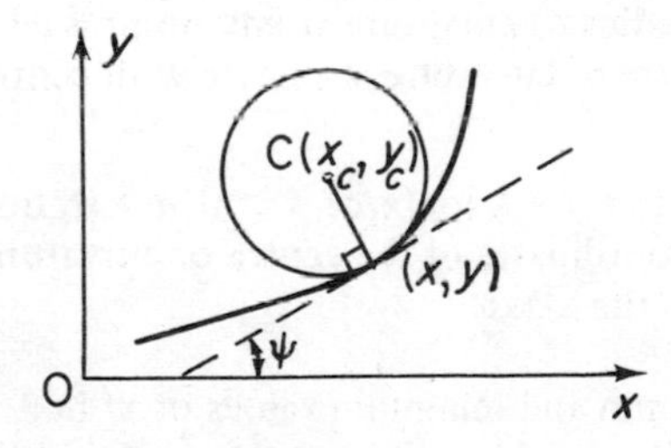

Figure 6.16

Given that $\tan \psi = \mathrm{d}y/\mathrm{d}x$ and

$$\rho = \frac{\left\{1 + \left(\frac{\mathrm{d}y}{\mathrm{d}x}\right)^2\right\}^{3/2}}{\frac{\mathrm{d}^2 y}{\mathrm{d}x^2}}$$

show that

$$x_c = x - \left\{1 + \left(\frac{\mathrm{d}y}{\mathrm{d}x}\right)^2\right\} \frac{\mathrm{d}y}{\mathrm{d}x} \Big/ \frac{\mathrm{d}^2 y}{\mathrm{d}x^2}$$

$$y_c = y + \left\{1 + \left(\frac{dy}{dx}\right)^2\right\} \Big/ \frac{d^2y}{dx^2}$$

2. Find the radius of curvature ρ of the catenary $y = c \cosh(x/c)$ and show that $\rho = y^2/c$.

3. Find the radius of curvature of

   (a) $y = \dfrac{1}{\sqrt{3}}(2x^2 - x)$ at the point $\left(1, \dfrac{1}{\sqrt{3}}\right)$

   (b) $y = x^3$ at the point (1, 1) and at the origin.

   (c) $y = \dfrac{1}{1-x}$ at (2, −1).

   (d) $\dfrac{x^2}{25} + \dfrac{y^2}{12} = 1$ at (5/2, 3)

4. Obtain the radius of curvature at the point (x_1, y_1) on the curve $y = 1 + 2x - 4x^2$. Sketch the curve. Show that the circle of curvature at (1/4, 5/4) is given by the equation $16(x^2 + y^2) - 8x - 36y + 21 = 0$. (L.U.)

5. A curve is given in terms of the parameter θ by the equations

$$x = a(\cos\theta + \theta\sin\theta), \quad y = a(\sin\theta - \theta\cos\theta)$$

   Show that the radius of curvature at any point is $a\theta$. Show also that the locus of the centre of curvature is a circle with centre at the origin and radius a. (L.U.)

6. Given the function $y = k\log(x/a)$, $k > 0$, $a > 0$, find the curvature of its graph and the coordinates of the centre of curvature, for the point where the graph meets the x-axis. (L.U.)

7. Find the maximum and minimum values of $x^2(x-2)^3$. Find also the radius of curvature of the curve $y = x^2(x-2)^3$ at the point (1, −1). (L.U.)

6.6 FINITE DIFFERENCES AND INTERPOLATION

In Section 2.9 we introduced the idea of differences of a tabulated function, the notation for forward differences:

$$\Delta f(x_0) = f(x_0 + h) - f(x_0)$$

and the first steps in interpolation. We now re-examine these differences and bear in mind their obvious similarity to differentiation. Since engineers often have to estimate values of a quantity for which we have only a certain number of values tabulated, it is important to obtain estimates which are as accurate as the data permits. Table 6.2 is a typical difference table.

Table 6.2

x	$f(x)$	Δ	Δ^2	Δ^3
x_0	$f(x_0)$			
		$\Delta f(x_0)$		
$x_0 + h$	$f(x_0 + h)$		$\Delta^2 f(x_0)$	
		$\Delta f(x_0 + h)$		$\Delta^3 f(x_0)$
$x_0 + 2h$	$f(x_0 + 2h)$		$\Delta^2 f(x_0 + h)$	
		$\Delta f(x_0 + 2h)$		$\Delta^3 f(x_0 + h)$
$x_0 + 3h$	$f(x_0 + 3h)$		$\Delta^2 f(x_0 + 2h)$	
		$\Delta f(x_0 + 3h)$		
$x_0 + 4h$	$f(x_0 + 4h)$			

Some books write $\Delta_h f(x_0)$ to emphasise the step size, h.

An alternative notation is shown in Table 6.3.

Table 6.3

x	$f(x)$	Δ	Δ^2	Δ^3
x_0	f_0			
		Δf_0		
x_1	f_1		$\Delta^2 f_0$	
		Δf_1		$\Delta^3 f_0$
x_2	f_2		$\Delta^2 f_1$	
		Δf_2		$\Delta^3 f_1$
x_3	f_3		$\Delta^2 f_2$	
		Δf_3		
x_4	f_4			

Since $\Delta f(x_0) = f(x_0 + h) - f(x_0)$ we have

$$f(x_0 + h) = f(x_0) + \Delta f(x_0) \tag{6.15a}$$

and

$$\begin{aligned} f(x_0 + 2h) &= f(x_0 + h) + \Delta f(x_0 + h) \\ &= f(x_0) + \Delta f(x_0) + \Delta f(x_0 + h) \end{aligned} \tag{6.15b}$$

[We have $f(x) = x^2$ as a *function* taking values of x and producing values x^2; differentiation is an **operation** taking a function and producing a new function and $\mathrm{d}/\mathrm{d}x$ is called an **operator.**]

Now we treat Δ as an *operator* on $f(x)$; this means it can be manipulated by the usual rules of algebra, provided it precedes the entity on which it operates. Therefore we can operate on the equation (6.15a) to obtain

$$\Delta f(x_0 + h) = \Delta f(x_0) + \Delta[\Delta f(x_0)] = \Delta f(x_0) + \Delta^2 f(x_0)$$

and so

$$f(x_0 + 2h) = f(x_0) + 2\Delta f(x_0) + \Delta^2 f(x_0)$$

In a similar way we can show that

$$f(x_0 + 3h) = f(x_0) + 3\Delta f(x_0) + 3\Delta^2 f(x_0) + \Delta^3 f(x_0)$$

and by induction that

$$f(x_0 + nh) = f(x_0) + n\Delta f(x_0) + \frac{n(n-1)}{2!}\Delta^2 f(x_0) + \ldots + \Delta^n f(x_0) \quad (6.16)$$

where n is a positive integer. Note the strong similarity between (6.16) and the binomial theorem. Can we extend this result for non-integer values of n to give us a possible formula for interpolation?

Interpolation using forward differences

First we approach the problem of interpolation by a different route. We know that the linear interpolation formula $f(x_0 + ph) = f(x_0) + p\Delta f(x_0)$ is *exact* for a *linear* function $f(x) = bx + c$. If we wish to interpolate a quadratic function $f(x) = ax^2 + bx + c$ using this formula we would expect an error since we cannot approximate accurately a quadratic function by a straight line. To find this error we consider simply $f(x) = ax^2$ (why?) and find that

$$\Delta f(x_0) = a(x_0 + h)^2 - ax_0^2 \equiv 2ahx_0 + ah^2$$

and that

$$\Delta^2 f(x_0) = 2ah^2$$

Linear interpolation gives

$$f(x_0 + ph) = ax_0^2 + p(2ahx_0 + ah^2)$$

However, the true value is

$$a(x_0 + ph)^2 \equiv ax_0^2 + 2aphx_0 + ap^2h^2$$

and so the error is

$$ah^2(p - p^2)$$

(You should be able to show that the maximum error occurs when $p = \frac{1}{2}$ and is then $\frac{1}{4}ah^2$.) If we add on this correction term which can be written $\frac{1}{2}p(p-1)\Delta^2 f(x_0)$, we have a formula exact for a quadratic polynomial, viz.

$$f(x_0 + ph) = f(x_0) + p\Delta f(x_0) + \tfrac{1}{2}p(p-1)\Delta^2 f(x_0)$$

A similar exercise that you could try is to establish the correction term to make the formula correct for a cubic polynomial: the correction is

$$\frac{1}{6}p(p-1)(p-2)\Delta^3 f(x_0)$$

Therefore, for a cubic polynomial

$$f(x_0 + ph) = f(x_0) + p\Delta f(x_0) + \frac{1}{2!}p(p-1)\Delta^2 f(x_0) + \frac{1}{3!}p(p-1)(p-2)\Delta^3 f(x_0)$$

Notice that the formula (6.16) seems to be true, at least as far as the term in $\Delta^3 f(x_0)$, if we substitute p instead of n and we therefore believe that (6.16) is also true for non-integral values of n.

A proof of the formula can be accomplished using the idea of operators. We introduce the operator E where $\mathrm{E}f(x_0) = f(x_0 + h)$ with the obvious extension

$$\mathrm{E}^2 f(x_0) = \mathrm{E}[\mathrm{E}f(x_0)] = \mathrm{E}[f(x_0 + h)] = f(x_0 + 2h)$$

and so on.

Now

$$(1 + \Delta)f(x_0) = f(x_0) + \Delta f(x_0) = f(x_0) + f(x_0 + h) - f(x_0) = f(x_0 + h)$$

and we see that the operator $1 + \Delta$ has the same effect as the operator E. This allows us to identify them by an operator equation: $1 + \Delta = \mathrm{E}$. It can be shown directly (try it) that

$$(1 + \Delta)^2 = 1 + 2\Delta + \Delta^2 = \mathrm{E}^2$$

and so on.

It is fairly clear what interpretation to put on $\mathrm{E}^n f(x_0)$ if n is a positive integer but what about E^{-1} or $\mathrm{E}^{1/2}$? In fact we *assume* that

$$\mathrm{E}^{1/2}[\mathrm{E}^{1/2} f(x_0)] \equiv \mathrm{E}f(x_0) \quad \text{and} \quad \mathrm{E}^{-1}[\mathrm{E}f(x_0)] = \mathrm{E}[\mathrm{E}^{-1}f(x_0)] = f(x_0)$$

Interpretations for other values of n are similar.

Then for any *rational* value of p

$$f(x_0 + ph) = \mathrm{E}^p f(x_0) = (1 + \Delta)^p f(x_0)$$

$$= \left\{ 1 + p\Delta + \frac{p(p-1)}{1.2}\Delta^2 + \ldots + \frac{p(p-1)\ldots(p-r+1)}{1.2 \;\ldots\ldots\ldots\; r}\Delta^r + \ldots\ldots\ldots \right\} f(x_0) \quad \text{(by the binomial expansion)}$$

that is,

$$f(x_0 + ph) = f(x_0) + p\Delta f(x_0) + \frac{p(p-1)}{2!}\Delta^2 f(x_0) + \frac{p(p-1)(p-2)}{3!}\Delta^3 f(x_0) + \ldots \tag{6.17}$$

Equation (6.17) is the **Newton-Gregory forward difference interpolation formula** which can be used in interpolating a function at values between tabulated points. Let us apply the formula to a specific example.

Example

The angle θ, turned through by a shaft, was measured and the results shown in Table 6.4.

Table 6.4

t (secs)	0	0.2	0.4	0.6	0.8	1.0
θ (rads)	−0.002	0.058	0.149	0.283	0.470	0.717

We first form a difference table, Table 6.5.

Table 6.5

t	θ	Δ	Δ^2	Δ^3	Δ^4
0	−0.002				
		0.060			
0.2	0.058		0.031		
		0.091		0.012	
0.4	0.149		0.043		−0.002
		0.134		0.010	
0.6	0.283		0.053		−0.003
		0.187		0.007	
0.8	0.470		0.060		
		0.247			
1.0	0.717				

It would seem unwise to carry the tabulation further.

We estimate θ when

(i) $t = 0.3$, (ii) $t = 0.15$, (iii) $t = 0.7$, (iv) $t = 0.9$

(i) 0.3 lies half way between 0.2 and 0.4, and using $x_0 = 0.2$, $h = 0.2$ then $0.3 = x_0 + ph$ gives $p = \frac{1}{2}$. Using the Newton-Gregory forward difference formula we obtain

$$\theta(0.2 + \tfrac{1}{2} \times 0.2) \simeq 0.058 + \tfrac{1}{2} \times 0.091 + \frac{\frac{1}{2}(-\frac{1}{2})}{2} \times 0.043 + \frac{\frac{1}{2}(-\frac{1}{2})(-\frac{3}{2})}{6} \times (0.010)$$

$$= 0.058 + 0.0455 - 0.005375 + 0.000625$$

i.e. $\theta(0.3) = 0.099$ (3 d.p.)

We quote the result to 3 d.p. since we cannot hope to achieve an accuracy greater than the tabulated values.

(ii) Here we take $x_0 = 0$ and since $h = 0.2$, then $p = \frac{3}{4}$. A similar calculation gives

$$\theta(0 + \tfrac{3}{4} . 0.2) \simeq -0.002 + 0.0435 - 0.00290625 + 0.0004687 5$$

i.e. $\theta(0.15) = 0.039$ (3 d.p.)

(iii) This time take $x_0 = 0.6$ with $p = \frac{1}{2}$. Here, we can only pick up the first and second differences and

$\theta(0.6 + \frac{1}{2} \, . \, 0.2) = 0.283 + 0.0935 - 0.0075 = 0.369$ (3 d.p.)

i.e. $\theta(0.7) = 0.369$ (3 d.p.)

(iv) $h = 0.2$, $x_0 = 0.8$, $p = \frac{1}{2}$. This time we are restricted to linear interpolation:

$\theta(0.9) \simeq 0.470 + \frac{1}{2}(0.247) = 0.5593(5)$ (3 d.p.)

This clearly demonstrates that the formula ceases to be of use at the end of a table and we wonder if a similar formula exists which allows us to go back into a table to pick up values.

Backward Differences

If we refer back to Table 6.2 we can label the entries in terms of so-called **backward differences.** The backward difference $f(x_0) - f(x_0 - h)$ can be written as

$$\nabla f(x_0) \quad \text{or} \quad \nabla f_0 \tag{6.18}$$

It is to be emphasised that the entry values are unchanged; it is the labels which differ. Table 6.6 shows the changes.

Table 6.6

x	$f(x)$	∇	∇^2	∇^3
x_0	f_0			
		∇f_1		
$x_0 + h$	f_1		$\nabla^2 f_2$	
		∇f_2		$\nabla^3 f_3$
$x_0 + 2h$	f_2		$\nabla^2 f_3$	
		∇f_3		$\nabla^3 f_4$
$x_0 + 3h$	f_3		$\nabla^2 f_4$	
		∇f_4		
$x_0 + 4h$	f_4			

Now $E^{-1} f(x_0) = f(x_0 - h)$ and therefore

$$(1 - E^{-1}) f(x_0) = f(x_0) - f(x_0 - h) \equiv \nabla f(x_0)$$

Hence

$$\nabla = 1 - E^{-1}$$

and so

$$E^{-1} = 1 - \nabla, \quad E = (1 - \nabla)^{-1}$$

and

$$f(x_0 + ph) = (1 - \nabla)^{-p} f(x_0)$$

This gives the **Newton-Gregory backward difference interpolation formula**

$$f(x_0 + ph) = f(x_0) + p\nabla f(x_0) + \frac{p(p+1)}{2!} \nabla^2 f(x_0) + \frac{p(p+1)(p+2)}{3!} \nabla^3 f(x_0) + \ldots \tag{6.19}$$

(The differences used if we choose $x_0 + 3h$ or $x_0 + 4h$ as the base are shown in Table 6.6 by dashed lines.)

Example

We can therefore apply formula (6.19) to case (iv) of the previous example where the forward interpolation formula reduced to linear interpolation; we have $h = 0.2$, $x_0 = 0.8$, $p = \frac{1}{2}$.

Then

$$\begin{aligned}\theta(0.9) &= 0.470 + \tfrac{1}{2} \times 0.187 + \frac{\frac{1}{2}\cdot\frac{3}{2}}{2} \times 0.053 + \frac{\frac{1}{2}\cdot\frac{3}{2}\cdot\frac{5}{2}}{6} \times 0.010 \\ &= 0.470 + 0.0935 + 0.019875 + 0.003125 \\ &= 0.586(5) \quad \text{(3 d.p.)}\end{aligned}$$

Alternatively, we may take $x_0 = 1.0$ whence $p = -\frac{1}{2}$, to obtain

$$\begin{aligned}\theta(0.9) &= 0.717 + (-\tfrac{1}{2}) \times 0.247 + \frac{(-\frac{1}{2})(\frac{1}{2})}{2} \times 0.060 + \frac{(-\frac{1}{2})(\frac{1}{2})(\frac{3}{2})}{6} \times 0.007 \\ &= 0.586(5) \quad \text{(3 d.p.) which is the same result as before.}\end{aligned}$$

Central Differences

Earlier we have seen that a central average gave more accurate results for numerical differentiation and we might expect a similar result of affairs for interpolation.

We define the **central difference**

$$\delta f(x_0 + \tfrac{1}{2}h) \equiv f(x_0 + h) - f(x_0)$$

and sometimes write it as $\delta f_{\frac{1}{2}}$. Then if we shift the centre of attention from x_0 to $x_r = x_0 + rh$,

$$\delta f[x_0 + (r + \tfrac{1}{2})h] \equiv f[x_0 + (r+1)h] - f(x_0 + rh)$$

which is written $\delta f_{r+\frac{1}{2}}$. It follows that $\delta f_{r+\frac{1}{2}} \equiv \Delta f_r - \nabla f_{r+1}$. Table 6.7 shows the same entries as Tables 6.2 and 6.6 with their new labels.

Table 6.7

x	$f(x)$	δ	δ^2	δ^3
x_0	f_0			
		$\delta f_{\frac{1}{2}}$		
$x_0 + h$	f_1		$\delta^2 f_1$	
		$\delta f_{\frac{3}{2}}$		$\delta^3 f_{\frac{3}{2}}$
$x_0 + 2h$	f_2		$\delta^2 f_2$	
		$\delta f_{\frac{5}{2}}$		$\delta^3 f_{\frac{5}{2}}$
$x_0 + 3h$	f_3		$\delta^2 f_3$	
		$\delta f_{\frac{7}{2}}$		
$x_0 + 4h$	f_4			

Since $\delta^2 f_r \equiv \delta(\delta f_r)$, it follows that

$$\delta^2 f_1 \equiv \delta(f_{r+½} - f_{r-½}) \equiv (f_{r+1} - f_r) - (f_r - f_{r-1}) \equiv f_{r+1} - 2f_r + f_{r-1}$$

Check this result against the values of Table 6.5. Observe how the suffices change as we move from column to column in Table 6.7.

We now state without proof two interpolation formulae which use central differences.

Bessel's Formula

$$f(x_0 + ph) = f_0 + p\delta f_{1_2} + \frac{1}{2.2!}p(p-1)(\delta^2 f_0 + \delta^2 f_1) + \frac{1}{3!}p(p-1)(p-½)\delta^3 f_{½}$$
$$+ \frac{1}{2.4!}(p+1)p(p-1)(p-2)(\delta^4 f_0 + \delta^4 f_1) + \ldots\ldots \qquad (6.20)$$

Everett's Formula

$$f(x_0 + ph) = q\,[f_0 + \frac{1}{3!}(q^2-1)\delta^2 f_0 + \frac{1}{5!}(q^2-1)(q^2-4)\delta^4 f_0 + \ldots]$$
$$+ p\,[f_1 + \frac{1}{3!}(p^2-1)\delta^2 f_1 + \frac{1}{5!}(p^2-1)(p^2-4)\delta^4 f_1 + \ldots] \qquad (6.21)$$

where $q = 1 - p$.

Example

If we return to Table 6.5 we shall apply both formulae to the evaluation of $\theta(0.5)$. You should compare the results with estimates from the Newton-Gregory formulae.

(i) *Bessel's Formula*

Here $x_0 = 0.4$ and $p = ½$. Then

$$\theta(0.5) = 0.149 + \frac{1}{2} \times 0.134 + \frac{1}{4}\cdot\frac{1}{2}\left(-\frac{1}{2}\right)(0.043 + 0.053)$$
$$+ \frac{1}{6}\cdot\frac{1}{2}\cdot\left(-\frac{1}{2}\right)(0).(0.010) + \frac{1}{48}\cdot\left(\frac{3}{2}\right)\cdot\left(\frac{1}{2}\right)\cdot\left(-\frac{1}{2}\right)\cdot\left(-\frac{3}{2}\right)(-0.002 - 0.003)$$
$$= 0.149 + 0.067 - 0.006 + 0 - 0.0000585$$
$$= 0.210 \quad \text{(3 d.p.)}$$

The way differences are picked up for use follows the pattern below. We have also included this pattern on Table 6.5.

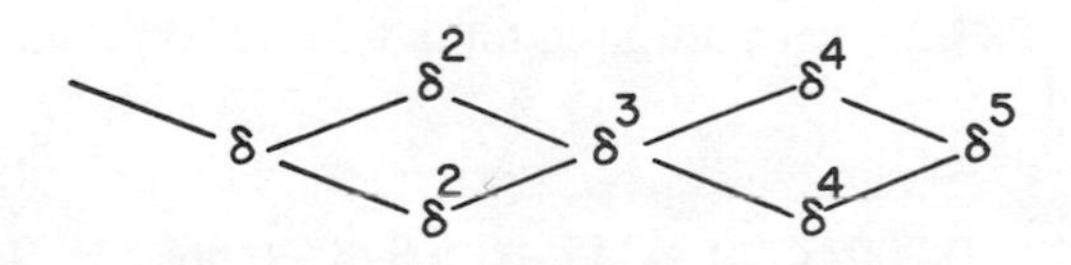

(ii) *Everett's Formula*

Here $x_0 = 0.4$, $p = ½$, $q = ½$. Then

$$\theta(0.5) = \frac{1}{2}\left[0.149 + \frac{1}{6}\left(-\frac{3}{4}\right)(0.043) + \frac{1}{120}\left(-\frac{3}{4}\right)\cdot\left(-\frac{15}{4}\right)(-0.002)\right]$$

$$+\frac{1}{2}\left[0.283 + \frac{1}{6}\left(-\frac{3}{4}\right)(0.053) + \frac{1}{120}\left(-\frac{3}{4}\right)\cdot\left(-\frac{15}{4}\right)(-0.003)\right]$$

$$= \frac{1}{2}\left[0.149 - 0.005375 - 0.000046875\right] + \frac{1}{2}\left[0.283 - 0.006625 - 0.0000703125\right]$$

$$= 0.0717891 + 0.1381523 \quad (3 \text{ d.p.})$$

$$= 0.210 \quad (3 \text{ d.p.})$$

Choice of Formulae

It would seem advisable to give some guidance as to which formula to use. It is clear that the central difference based formulae produce as accurate an estimation as the Newton-Gregory formulae using fewer terms. In general, the central difference formulae should be used where possible because of the rapid decay of the successive terms; however, at the beginning of a table, the Newton-Gregory forward difference formula is used and its backward difference counterpart is used at the end of a table. To emphasise this, if we wanted to find $f(0.1)$, $f(0.3)$, $f(0.55)$, $f(0.93)$ from the table

x	0.0	0.2	0.4	0.6	0.8	1.0
$f(x)$	0.0	0.008	0.064	0.216	0.516	1.000

we would use Newton-Gregory forward difference, central difference, central difference and Newton-Gregory backward difference formulae respectively.

Problems I

1. Given the following table for a function $f(x)$ find $f(3.15)$.

x	3.0	3.2	3.4	3.6	3.8	4.0
$f(x)$	3.0103	3.4242	3.8021	4.1497	4.4716	4.7712

Estimate the accuracy of your answer. (C.E.I.)

2. Values of $f(x) = \tan x$ are given for $x = 35°(2°)45°$. Find tan 36° and tan 43°12′.

$x°$	35	37	39	41	43	45
tan x	0.70021	0.75355	0.80978	0.86929	0.93252	1.00000

3. Given the following table find sin 37°30′ using both Bessel and Everett interpolation formulae.

x	0	10	20	30	40	50	60	70
sin x	0.00000	0.17365	0.34202	0.50000	0.64279	0.76604	0.86603	0.93969

4. Use the following table of $f(x) = \tan x$ to interpolate tan 85°30′ and tan 87°30′

$x°$	85	86	87	88	89
$f(x)$	11.430	14.301	19.081	28.636	57.290

Note how poorly the values compare with tabular values 12.706 and 22.904. In cases like this where $f(x)$ changes very rapidly, the reciprocal function $g(x) = 1/[f(x)]$ will change more slowly. Form a difference table for $g(x)$, obtain $g(85°30')$ and $g(87°30')$, and hence obtain more accurate values of tan 85°30′ and tan 87°30′.

5. The following table gives the values of a function $f(x)$ for values of x between 1.0 and 1.5.

x	$f(x)$	x	$f(x)$
1.0	1.54308	1.3	1.97091
1.1	1.66852	1.4	2.15090
1.2	1.81066	1.5	2.35241

Calculate the values of $f(1.05)$ and $f(1.47)$.
(Any interpolation formula may be used, but it should be clearly stated and all symbols defined.) (L.U.)

6. A fourth degree polynomial is tabulated as follows

x	0	0.1	0.2	0.3	0.4
y	1.0000	0.9208	0.6928	0.3448	−0.0752

x	0.5	0.6	0.7	0.8	0.9
y	−0.5000	−0.8452	−0.9992	−0.8432	−0.2312

Show from a difference table that there is an error and use the corrected table with the Stirling interpolation formula

$$f_p = f_0 + \tfrac{1}{2}p(\delta f_{1/2} + \delta f_{-1/2}) + \tfrac{1}{2}p^2\delta^2 f_0 + \frac{p(p^2-1)}{2.3!}(\delta^3 f_{1/2} + \delta^3 f_{-1/2}) + \frac{p^2(p^2-1)}{4!}\delta^4 f_0 + \ldots$$

to find the value of y when $x = 0.45$.

7. The values of $f(x)$, a low degree polynomial are given in the table

x	2	3	4	5	6	7	8	9	10
$f(x)$	15	40	85	165	259	400	585	820	1111

It is suspected that there is a transposition error in one of the values of $f(x)$.
By differencing, locate and correct this error.
Hence, by using an appropriate interpolation formula, evaluate $f(x)$ when $x = 2.5$. (L.U.)

8. Define the shift operator E and the forward difference operator Δ when applied to the function $f(x) = f(x_0 + ph)$. Hence obtain Newton's forward interpolation formula

$$f_p = f_0 + p\Delta f_0 + \frac{p(p-1)}{2}\Delta^2 f_0 + \ldots + \binom{p}{r}\Delta^r f_0 + \ldots$$

A polynomial function is given by the following table.

x	0	1	2	3	4	5	6
f	0	3	14	39	84	155	258

Make a difference table and explain how the correctness of the arithmetic may be checked.

Use this table to find f when $x = 1.5$ and when $x = 7$. (L.U.)

Lagrange Interpolation

Two questions should spring to mind: if in a formula we neglect differences higher than, say, the fourth, is this equivalent to assuming we can approximate the curve by a fourth order polynomial? What happens if the tabulated values are not equally spaced? The answer to the first question is *Yes* and this leads us to develop a different method of interpolation to overcome the difficulty raised by the second question. We make use of the result that there is a unique polynomial of Nth degree passing through $(N + 1)$ specified (different) points. We quote without proof the **Lagrange polynomial** through the N points

$$(x_1, f_1), (x_2, f_2), \ldots, (x_N, f_N)$$

This polynomial is

$$\sum_{r=1}^{N} f_r \prod_{\substack{k=1\\k\neq r}}^{N} \frac{(x - x_k)}{(x_r - x_k)} \tag{6.22}$$

where $\prod_{\substack{k=1\\k\neq r}}^{N}$ means the product of terms like the one quoted where k takes values $1, 2, \ldots, N$ but excluding $k = r$. For example, the *three-point formula* is

$$\frac{(x - x_2)(x - x_3)}{(x_1 - x_2)(x_1 - x_3)}f_1 + \frac{(x - x_1)(x - x_3)}{(x_2 - x_1)(x_2 - x_3)}f_2 + \frac{(x - x_1)(x - x_2)}{(x_3 - x_1)(x_3 - x_2)}f_3$$

Observe that this uses only function values and does not require a table of differences (which would require that these function values be equally spaced).

Example

Suppose we return to Table 6.5 and use the values $\theta(0.2)$, $\theta(0.4)$, $\theta(0.6)$ to estimate $\theta(0.5)$, then $x_1 = 0.2$, $x_2 = 0.4$, $x_3 = 0.6$, $f_1 = 0.058$, $f_2 = 0.149$, $f_3 = 0.283$ and $x = 0.5$. Then we have the approximation

$$\theta(0.5) \simeq \frac{(0.5-0.4)(0.5-0.6)}{(-0.2)(-0.4)}(0.058) + \frac{(0.5-0.2)(0.5-0.6)}{(0.2)(-0.2)}(0.149) + $$

$$+ \frac{(0.5-0.2)(0.5-0.4)}{(0.4)(0.2)}(0.283)$$

$$= -0.00725 + 0.11175 + 0.106125$$

$$= 0.211 \quad \text{(3 d.p.)}$$

Bearing in mind the fact that the third differences are not quite constant, it might have been better to take the four-point formula: you should try to derive it in general and apply it to this example.

Disadvantages

The disadvantages of Lagrange interpolation are

(i) that there is no easy check on its accuracy as can be the case when a table of differences is formed;

(ii) that we have to decide in advance what degree of polynomial to fit;

(iii) that to go to a higher degree polynomial we have to start from scratch whereas in the case of a difference-based formula we simply have to add on extra terms.

Before leaving this section we note the hazards involved in extrapolation.

Example

The difference table (Table 6.8) is given

Table 6.8

x	$f(x)$	∇	∇^2	∇^3
0.0	0.0			
		0.4		
0.25	0.4		−0.8	
		−0.4		1.0
0.50	0.0		+0.2	
		−0.2		
0.75	−0.2			

We wish to estimate $f(0.9)$ and $f(1.0)$ using (i) all points and cubic interpolation, and (ii) the last three points and quadratic interpolation.

(i) For the cubic interpolation, in order to obtain $f(0.9)$ we use the Newton-Gregory backward formula with $x_0 = 0.75$, $p = 0.6$. Then

$$f(0.9) \simeq -0.2 + (0.6)(-0.2) + \frac{(0.6)(1.6)}{2}(0.2) + \frac{(0.6)(1.6)(2.6)}{6}(1.0)$$

$$\simeq -0.2 - 0.12 + 0.096 + 0.416 = 0.19 \quad \text{(2 d.p.)}$$

Similarly $\quad f(1.0) = -0.2 - 0.2 + 0.2 + 1.0 = 0.80 \quad$ (2 d.p.)

[We could have found $f(1.0)$ by continuing the difference table.]

(ii) For quadratic interpolation, we merely omit the last term in each case to obtain $f(0.9) = -0.22$ (2 d.p.) and $f(1.0) = -0.20$ (2 d.p.)

Now suppose we try Lagrange interpolation on (0.5, 0.0), (0.75, −0.2) and an additional point (1.25, 0.4). You should be able to show that revised estimates are $f(0.9) = -0.16$ (2 d.p.) and $f(1.0) = -0.067$ (3 d.p.).

Clearly something is amiss. In fact, the tabular values came from a sine-type function and the attempts at extrapolation can be seen to fail by superimposing their predictions on the curve as shown in Figure 6.17.

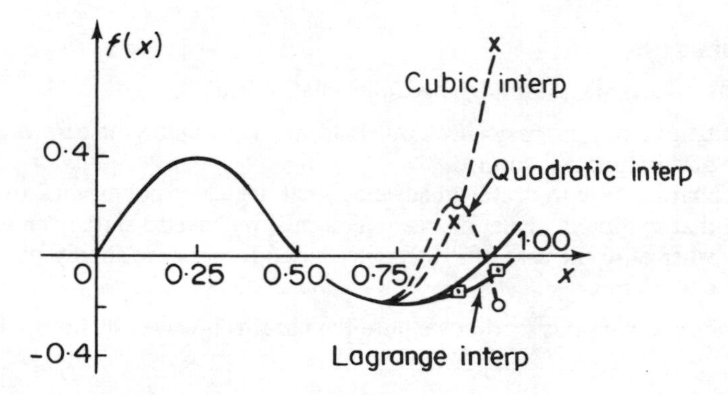

Figure 6.17

This example emphasises two points:

(i) extrapolation is always risky

(ii) if the data is badly approximated by low degree polynomials then a higher-order extrapolation formula is not necessarily any better.

Problems II

1. Given the following data

x	2	3	4	5	6
$f(x)$	9.00	4.00	2.25	1.44	1.00

evaluate $f(4.5)$ using

(a) 3-point formula with $x_1 = 3,\ x_2 = 4,\ x_3 = 5$

(b) 3-point formula with $x_1 = 4,\ x_2 = 5,\ x_3 = 6$

(c) 4-point formula with $x_1 = 3,\ x_2 = 4,\ x_3 = 5,\ x_4 = 6$

2. The values of a function $f(x)$ are given for $x = x_i$ $(i = 1, 2, ..., n)$. Assuming Lagrange's interpolation polynomial of degree $(n - 1)$ is of the form

$$\sum_{i=1}^{n} l_i(x) f(x_i)$$

prove that

$$\sum_{i=1}^{n} l_i(x) \equiv 1$$

A function $f(x)$ is given by the following table of values

x	-2	0	3	4
$f(x)$	-14	8	-4	4

Determine $f(1)$ and $f(2)$. (L.U.)

6.7 NUMERICAL DIFFERENTIATION

In Section 3.5 we developed ad hoc some formulae for differentiation from a table of values. We can, in fact, derive such formulae and more accurate ones by differentiating suitable interpolation formulae.

For example, if we differentiate successive terms in (6.17) with respect to p, we obtain the equation

$$hf'(x_0 + ph) = 0 + \Delta f(x_0) + \frac{2p - 1}{2!} \Delta^2 f(x_0) + \frac{3p^2 - 6p + 2}{3!} \Delta^3 f(x_0) + ...$$

You should ask whether such differentiation is acceptable since the terms on the right-hand side are theoretically infinite in number; our only reply at present is that provided the terms still decay reasonably rapidly we shall be satisfied with the results. You might even ask whether we can even contemplate differentiating with respect to p; but, once the table of values is given, the differences are automatically determined and all that can cause $f(x_0 + ph)$ to vary is a change in p.

In the special case $p = 0$ we obtain

$$f'(x_0) \simeq \frac{1}{h}\left\{\Delta f(x_0) - \frac{1}{2} \Delta^2 f(x_0) + \frac{1}{3} \Delta^3 f(x_0) - \frac{1}{4} \Delta^4 f(x_0) \; ...\right\} \quad (6.23)$$

which gives the derivative *at a tabulated point* x_0.

If we put $p = ½$ we produce

$$f'(x_0 + ½h) \simeq \frac{1}{h}\left\{\Delta f(x_0) - \frac{1}{24} \Delta^3 f(x_0) + \frac{1}{24} \Delta^4 f(x_0) \; ...\right\} \quad (6.24)$$

which gives the derivative *mid-way between tabulated points*. You should notice that the term in $\Delta^2 f(x_0)$ has vanished.

As might be expected, these formulae are best suited at the beginning of a table. At the end of a table we use the corresponding backward differences formula, viz.

$$f'(x_0) \triangleq \frac{1}{h}\left\{\nabla f_0 + \frac{1}{2}\nabla^2 f_0 + \frac{1}{3}\nabla^3 f_0\right\} \tag{6.25}$$

or

$$f'(x_0 - \tfrac{1}{2}h) \triangleq \frac{1}{h}\left\{\nabla f_0 - \frac{1}{24}\nabla^3 f_0 - \frac{1}{24}\nabla^4 f_0\right\} \tag{6.26}$$

Similarly, by differentiating central difference formulae we find that

$$f'(x_0) \triangleq \frac{1}{h}\left\{\mu\delta f_0 - \frac{1}{6}\mu\delta^3 f_0 + \frac{1}{30}\mu\delta^5 f_0\right\} \tag{6.27}$$

$$f'(x_0 + \tfrac{1}{2}h) \triangleq \frac{1}{h}\left\{\delta f_{1/2} - \frac{1}{24}\delta^3 f_{1/2} + \frac{3}{640}\delta^5 f_{1/2}\right\} \tag{6.28}$$

[Here μ is the averaging operator defined by $\mu f_p = \frac{1}{2}(f_{p+1/2} + f_{p-1/2})$]

If we rewrite the first two terms on the right-hand side of (6.27) we obtain

$$f'(x_0) \triangleq \frac{1}{2h}\left\{f(x_0 + h) - f(x_0 - h)\right\} - \frac{1}{6h}\mu\delta^3 f_0$$

and we see that we have a correction term to our central formula of Section 3.5.

You should check that (6.23) and (6.25) produce similar results.

We saw in Problem 11 on page 141 that repeated application of numerical differentiation yielded successively less accurate values. Hence care must be exercised in the use of numerical differentiation.

We quote some approximate formulae for higher derivatives:

$$f''(x_0) \triangleq \frac{1}{h^2}\left\{\delta^2 f_0 - \frac{1}{12}\delta^4 f_0\right\} = \frac{f(x_0 + h) - 2f(x_0) + f(x_0 - h)}{h^2} - \frac{1}{12h^2}\delta^4 f_0$$

$$f'''(x_0) \triangleq \frac{1}{h^3}\mu\delta^3 f_0 \tag{6.29}$$

$$f^{(iv)}(x_0) \triangleq \frac{1}{h^4}\delta^4 f_0$$

Example 1

We apply the above results to Table 6.9, which is part of a table of reciprocals so that we can check our results.

We endeavour to find the first four derivatives of $f(x)$ at $x = 1.0$; we expect values of $-1, 2, -6, 24$.

Using (6.23) we have, ignoring the terms in δ^4 (why?),

$$f'(1.0) \triangleq \frac{1}{0.1}\left\{-0.0909 - \frac{1}{2}(+0.0151) + \frac{1}{3}(-0.0034)\right\} = -0.986 \text{ (3 d.p.)}$$

Table 6.9

x	$f(x)$	δ	δ^2	δ^3	δ^4
0.7	1.4290				
		−0.1790			
0.8	1.2500		+0.0401		
		−0.1389		−0.0123	
0.9	1.1111		+0.0278		+0.0047
		−0.1111		−0.0076	
1.0	1.0000		+0.0202		+0.0025
		−0.0909		−0.0051	
1.1	0.9091		+0.0151		+0.0017
		−0.0758		−0.0034	
1.2	0.8333		+0.0117		
		−0.0641			
1.3	0.7692				

From (6.25)

$$f'(1.0) \simeq \frac{1}{0.1}\left\{-0.1111 + \frac{1}{2}(+0.0278) + \frac{1}{3}(-0.0123)\right\} = -1.013 \text{ (3 d.p.)}$$

From (6.27)

$$f'(1.0) \simeq \frac{1}{0.1}\left\{\frac{1}{2}\left[-0.0909 - 0.1111\right] - \frac{1}{6}\cdot\frac{1}{2}\left[-0.0051 - 0.0076\right]\right\} = -0.999 \text{ (3 d.p.)}$$

To find the higher derivatives, we use (6.29)

$$f''(1.0) \simeq \frac{1}{0.01}\left[+0.0202 - \frac{1}{12}(+0.0025)\right] = 2.00 \quad \text{(2 d.p.)}$$

$$f'''(1.0) \simeq \frac{1}{0.001}\left\{\frac{1}{2}(-0.0076 - 0.0051)\right\} = -6.35 \text{ (2 d.p.)}$$

$$f^{(iv)}(1.0) \simeq \frac{1}{0.0001}(+0.0025) = 25.0$$

It is evident that the central difference formulae are the most accurate and that the higher the derivative the fewer decimal places we can achieve and the less the accuracy.

We need some way of estimating the error in these interpolation and differentiation formulae and to do this we would have to resort to calculus as is so often the case.

Problems

1. The following is a table of $f(x) = \sin x$ (x in radians).

x	0.7	0.8	0.9	1.0	1.1	1.2	1.3
$\sin x$	.644218	.717356	.783327	.841471	.891207	.932039	.963558

Using formulae (6.23), (6.25) and (6.27), show that at $x = 1.0$, $f'(x) = 0.54052,\ 0.54011,\ 0.54030$ respectively. Look up cos (1.0) and compare with the above.

Use (6.29) to show at $x = 1.0$, $f'' = -0.8415$
$f''' = -0.538$
$f^{(iv)} = 0.85$

What answers should we get? Note how accuracy is lost in numerical differentiation.

Find $(d/dx)(\sin x)$ at $x = 1.05$ and compare with $\cos(1.05) = 0.497571$.

2. Given the following table for $\cos x$

x	0.499	0.500	0.501
$\cos x$	0.878062	0.877563	0.877103

determine the value of $d/dx(\cos x)$ at $x = 0.5$. Estimate the possible error in your answer. (C.E.I.)

3. For the data of Problem 1 on page 336 find d^2f/dx^2 at $x = 3.6$. Estimate the accuracy of your answer. (C.E.I.)

6.8 THE MEAN VALUE THEOREM

This theorem has many applications as it stands and is also a basis for other theorems. Before we state it we need a basic result used in its proof.

Rolle's Theorem

If a function $g(x)$ is continuous in the closed interval $[a,b]$ and differentiable in the open interval (a,b) and if, further, $g(a) = 0 = g(b)$ then there is at least one point ξ in the interval $a < \xi < b$ such that $g'(\xi) = 0$.

Geometrically this means that if a continuous curve whose formula has a continuous first derivative intersects the x-axis at two points, it has a horizontal tangent at some intermediate point (see Figure 6.18).

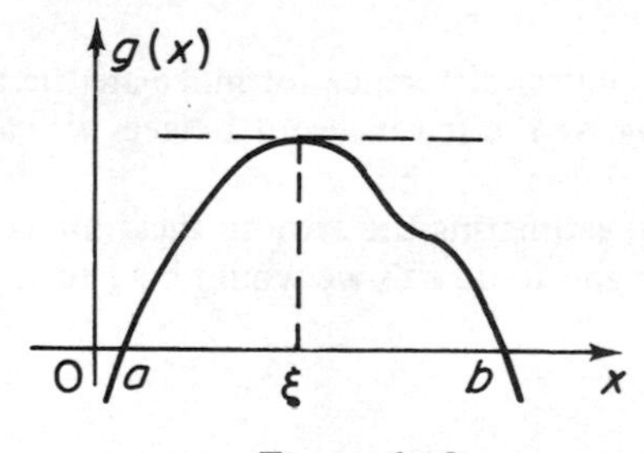

Figure 6.18

(This can be compared with the **Intermediate Value Theorem,** touched on in section 3.4; *if $f(x)$ be continuous in $[a,b]$ then for any value c between $f(a)$ and $f(b)$ there is at least one point value ξ in the interval $a \leqslant \xi \leqslant b$ such that $f(\xi) = c$.*)

What Rolle's Theorem does not tell us is where the point ξ is precisely. In individual examples, we may be able to establish this directly, by differentiation.

Proof of Rolle's Theorem

We assume the result that if M and m are the least upper bound and greatest lower bound respectively of a continuous function in some closed interval then the function achieves those values somewhere in the interval. Clearly if $m = M$ then the function $g(x)$ takes a constant value in the interval $[a, b]$ and $g'(x) \equiv 0$ there. Otherwise, we have at least one of m and M non-zero. We prove the result for the case $M \neq 0$; let $M > 0$ then there is a point ξ in (a,b) at which $g(\xi) = M$ (we have excluded $\xi = a$ and $\xi = b$; (why?). Now $g(x) \leqslant g(\xi)$ for all $x \in [a,b]$ and so for any small enough number $h > 0$, $g(\xi + h) - g(\xi) \leqslant 0$ and $g(\xi - h) - g(\xi) \leqslant 0$. Hence $[g(\xi + h) - g(\xi)]/h \leqslant 0$ and $[g(\xi) - g(\xi - h)]/h \geqslant 0$; if we take the limiting cases as $h \to 0$ then the first quotient stays negative or zero and the second stays positive or zero. But, in the limit both approach $g'(\xi)$ and so we conclude that $g'(\xi) = 0$.

Mean Value Theorem (MVT)

If $f(x)$ is a function continuous in $[a,b]$ and differentiable in (a,b) then there is at least one value ξ where $a < \xi < b$ for which $f'(\xi) = [f(b) - f(a)]/(b - a)$.

Geometrically, this means that there is a point ξ at which the tangent to the curve of $f(x)$ is parallel to the chord joining the end-points as shown in Figure 6.19. Again we are told nothing as to the precise location of the point ξ.

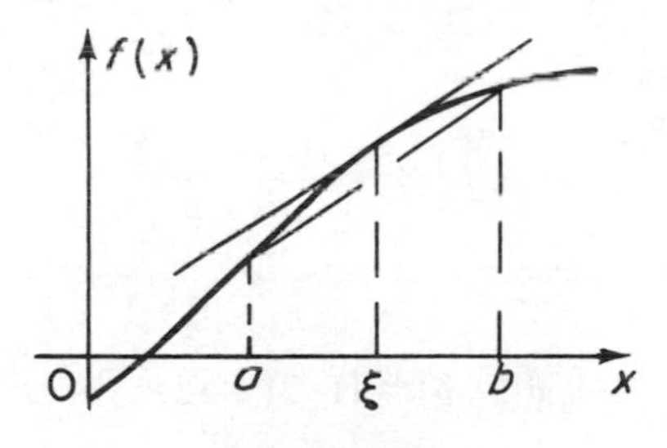

Figure 6.19

As an analogy imagine two cars setting off from the same place at the same time to arrive at a second place at the same later time. One car travels at a constant speed, the other is free to go at what speed it likes, stopping if it so desires. Then the MVT says that at some time the speed of the second car is equal to that of the first.

Proof of MVT

The proof of the theorem comes by applying Rolle's theorem to the function

$$g(x) = f(x) - \left\{ f(a) + \frac{f(b) - f(a)}{b - a}(x - a) \right\}$$

The function $g(x)$ represents the vertical distance PQ shown in Figure 6.20.

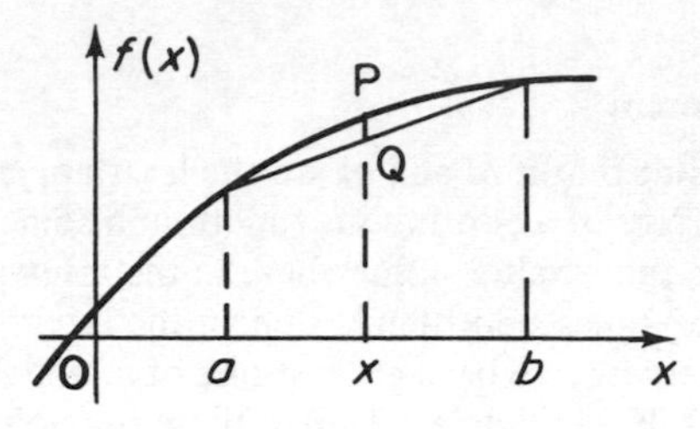

Figure 6.20

Now this function satisfies the first two conditions of Rolle's theorem because we are given that $f(x)$ does and so also does the other function in the braces { }. Further,

$$g(a) = f(a) - \{f(a) + 0\} = 0$$

and

$$g(b) = f(b) - \{f(a) + f(b) - f(a)\} = 0$$

Then, by Rolle's theorem, there is a value ξ where $a < \xi < b$ with $g'(\xi) = 0$. But

$$g'(x) = f'(x) - \left\{\frac{f(b) - f(a)}{b - a}\right\}$$

hence

$$0 = g'(\xi) = f'(\xi) - \left\{\frac{f(b) - f(a)}{b - a}\right\}$$

Thus

$$f'(\xi) = \frac{f(b) - f(a)}{b - a}$$

and the existence of ξ is established.

Example 1

(i) Let $f(x) = x^3 + 3$ and $[a, b] = [1, 2]$. Then

$$\frac{f(b) - f(a)}{b - a} = \frac{11 - 4}{2 - 1} = 7$$

But $f'(x) = 3x^2$ and so ξ is such that $3\xi^2 = 7$, which gives $\xi = \pm 1.5275$ (4 d.p.) and only the value $+1.5275$ lies inside the interval [sketch $f(x)$ and check].

(ii) Let $f(x) = |x|$ and $[a, b] = [-1, 2]$. Then

$$\frac{f(b) - f(a)}{b - a} = \frac{2 - 1}{2 - (-1)} = \frac{1}{3}$$

yet nowhere in $[-1, 2]$ does $f'(x) = 1/3$. Why should this be? Check whether $f(x)$ obeys the conditions of MVT.

Example 2

One application of MVT is to estimate the maximum truncation error in linear interpolation. If we replace $f(x)$ in $[a, b]$ by

$$\phi(x) = f(a) + \left\{\frac{f(b) - f(a)}{b - a}\right\}(x - a)$$

then $\phi(a) = f(a)$, and $\phi(b) = f(b)$ and ϕ is of the form $\alpha + \beta x$.
The error at any point in $[a, b]$ is $f(x) - \phi(x)$

$$= (x - a)\left[\frac{f(x) - f(a)}{x - a} - \left\{\frac{f(b) - f(a)}{b - a}\right\}\right]$$

Now, by MVT the right hand side is equal to

$$(x - a)[f'(\xi_1) - f'(\xi_2)]$$

where both ξ_1 and $\xi_2 \in (a, b)$.

Applying MVT to the expression

$$\frac{[f'(\xi_1) - f'(\xi_2)]}{\xi_1 - \xi_2}$$

we find it to be equal to $f''(\psi)$ where ψ lies between ξ_1 and ξ_2,[and therefore in (a, b)].

Then $f(x) - \phi(x) = (x - a)f''(\psi)(\xi_1 - \xi_2)$. Now, the worst that $|\xi_1 - \xi_2|$ can be is $(b - a)$ as can $|x - a|$; therefore $|f(x) - \phi(x)| \leqslant |f''(\psi)|(b - a)^2$. If we can find a number M such that $|f''(x)| \leqslant M$ for all $x \in (a, b)$ then

$$|f(x) - \phi(x)| \leqslant M(b - a)^2 \tag{6.30}$$

Since (6.30) is a general result it is quite likely that the actual error is much less than $M(b - a)^2$ but at least we have a starting point. Unless we have a means of determining $f''(x)$ this estimation is of little use, anyway.

Example 3

We have seen that $(\sin\theta)/\theta \to 1$ as $\theta \to 0$, and so we may approximate $\sin\theta$ by θ. Suppose we regard this approximation as satisfactory if it is at most 0.01 in error. Let $a = 0$ and note that, since $f''(x) = -\sin x$, $M = 1$; then the maximum error in the approximation is b^2 and if we require this to be at most 0.01 we must restrict b to be $\leqslant 0.1$. By symmetry we conclude that the acceptable interval is $[-0.1, 0.1]$. In fact, $\sin 0.38^c = 0.3709$ (4 d.p.) and so the approximation is still acceptable here.

Problems

1. For small values of x we can approximate $e^x \simeq 1 + x$. What will be the maximum error in this truncation over the interval $0 < x < 0.1$?

2. Use the MVT to show that for $a > b \geqslant 0$

$$\frac{a - b}{1 + b} > \log\left(\frac{1 + a}{1 + b}\right) > \frac{a - b}{1 + a}$$

and hence find upper and lower bounds for the value of log(1.01).

(C.E.I.)

For interpolation given only tabulated values, errors may be estimated from tables designed for the purpose. See Interpolation and Allied Tables (HMSO).

6.9 APPROXIMATION OF FUNCTIONS

We turn our attention to the approximation of a function $f(x)$ by a straight line so that we may predict values of $f(x)$ without too much calculation. Certainly, from the point of view of using a computer the calculation of polynomial values is straightforward and therefore polynomials are the most commonly used approximating functions (a straight line is, of course, just a first-order polynomial).

The Tangent Approximation

The equation of the chord to the function $f(x)$ shown in Figure 6.21 (i) is

$$y = f(x_0) + \left\{\frac{f(x_0 + h) - f(x_0)}{h}\right\}(x - x_0)$$

and we see that provided h is small and the slope $f'(x)$ does not vary too much in the interval $[x_0, x_0 + h]$, then the chord is a reasonable approximation to the curve. Note that to draw the chord we need to know two values of $f(x)$

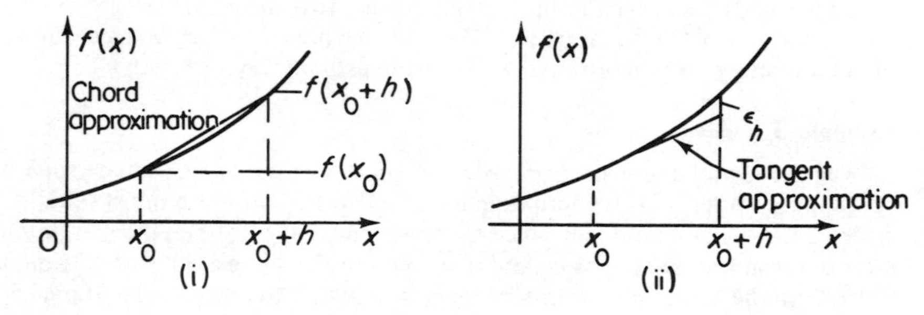

Figure 6.21

If we let $h \to 0$ then the equation of the straight line becomes

$$y = f(x_0) + f'(x_0)(x - x_0) \tag{6.31}$$

which is called the **Tangent Approximation** to $f(x)$ at x_0 [see Figure 6.21 (ii)] and requires knowledge of the function at the point x_0 only. Note the special case where $x_0 = 0$, viz.

$$y = f(0) + xf'(0)$$

We can use the MVT to put an upper bound on the error involved in this approximation. The discrepancy in the values at $(x_0 + h)$, [see Figure 6.21, (ii)] is $\epsilon_h \equiv f(x_0 + h) - [f(x_0) + f'(x_0)h]$.

Now

$$\frac{\epsilon_h}{h} = \frac{f(x_0 + h) - f(x_0)}{h} - f'(x_0)$$

$$= f'(\xi) - f'(x_0)$$

[by MVT where ξ is in the interval $(x_0, x_0 + h)$].

A second application of MVT yields

$$\frac{\epsilon_h}{h(\xi - x_0)} = \frac{f'(\xi) - f'(x_0)}{\xi - x_0} = f''(\psi)$$

where ψ lies between x_0 and ξ and hence $\psi \in (x_0, x_0 + h)$.

If we replace $f''(\psi)$ by M, the greatest value of $|f''(x)|$ in $[x_0, x_0 + h]$, then $|\epsilon_h| \leqslant Mh|\xi - x_0| < Mh^2$.

We shall now apply the tangent approximation to a specific function and see how realistic is this upper bound on $|\epsilon_h|$.

Example

We wish to approximate $y = \log x$ with $x_0 = 2$ as base; we quote the results in Table 6.10 to 3 d.p. The tangent approximation (6.31) becomes

$$y = \log 2 + \frac{1}{2}(x - 2)$$

Table 6.10

x	$\log x$ from tables	$x-2$	Tangent Approxn.	Error	Predicted Maximum Error	M
1.7	0.531	−0.3	0.543	0.012	0.031	0.346
1.8	0.588	−0.2	0.593	0.005	0.012	0.309
1.9	0.642	−0.1	0.643	0.001	0.003	0.277
2.0	0.693	0	0.693	0	0	Not relevant
2.1	0.742	0.1	0.743	0.001	0.003	0.25
2.2	0.789	0.2	0.793	0.004	0.010	0.25
2.3	0.833	0.3	0.843	0.010	0.023	0.25

In this example $f''(x) = -(1/x^2)$ and the maximum value of $|f''(x)|$ occurs at the left-hand end of each interval. A schematic view of the approximation is shown in Figure 6.22.

We see that the error is in all cases $\leqslant \frac{1}{2}Mh^2$ rather than Mh^2 which was the value quoted. It is clear, as expected, that the larger h, the larger the error. The

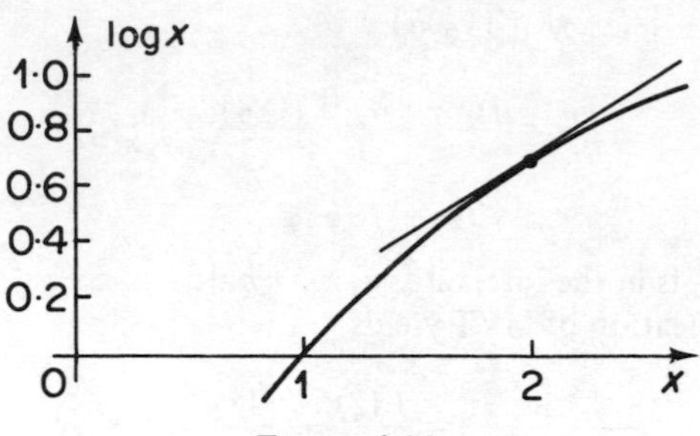

Figure 6.22

error seems to be roughly proportional to h^2; this means that the error becomes rapidly worse as we increase h. We have, in effect, matched the value $f(x_0)$ and slope $f'(x_0)$; we would expect to obtain a better approximation by making $f''(x_0)$ also match up.

Applications of the Tangent Approximation

(i) *Basic Iteration*

We may apply the Tangent Approximation to verify the convergence criterion for Basic Iteration stated in Section 5.8. Suppose we have the equation $x = F(x)$, with root $x = a$ which we attempt to solve by the iterative formula $x_{n+1} = F(x_n)$. Using the tangent approximation we obtain $F(x_n) = F(a) + F'(a)(x_n - a)$; therefore, since $a = F(a)$, $\epsilon_{n+1} = x_{n+1} - a = F'(a)(x_n - a)$ and so

$$|\epsilon_{n+1}| = |F'(a)|.|x_n - a| = |F'(a)|.|\epsilon_n|$$

If the iterative method is to converge, $|\epsilon_{n+1}|$ must be less than $|\epsilon_n|$, and this requires $|F'(a)| < 1$ or, if the function is well-behaved, $|F'(x)| < 1$ near $x = a$.

(ii) *Newton-Raphson iterative method*

Let $f(a + h) = 0$ and let h be small, then

$$0 = f(a + h) \simeq f(a) + hf'(a)$$

therefore, $h \simeq - f(a)/f'(a)$ and hence $a - f(a)/f'(a)$ is, in general, a better approximation to $a + h$ than was a.

The Quadratic Approximation

Let $q(x) \equiv \alpha + \beta(x - x_0) + \gamma(x - x_0)^2$ where $\alpha, \beta, \gamma, x_0$ are constants; we choose this form for $q(x)$ by analogy with the tangent approximation: it is still a perfectly general quadratic. Given the function $f(x)$ we require that if $q(x)$ is to approximate $f(x)$ near $x = x_0$ then $q(x_0) = f(x_0)$, $q'(x_0) = f'(x_0)$ and $q''(x_0) = f''(x_0)$.

Applying these constraints we find that

$$\alpha = f(x_0) \qquad \beta = f'(x_0) \qquad 2\gamma = f''(x_0)$$

Hence

$$f(x) \simeq f(x_0) + f'(x_0)(x - x_0) + \tfrac{1}{2} f''(x_0)(x - x_0)^2 \qquad (6.32)$$

is the required quadratic approximation to $f(x)$ at $x = x_0$.

Note the special case when $x_0 = 0$:

$$f(x) \simeq f(0) + xf'(0) + \frac{x^2}{2} f''(0) \qquad (6.32a)$$

Example

For $f(x) = \log x$ with $x_0 = 2$ as base we have

$$f(x) \simeq \log 2 + \frac{1}{2}(x - 2) + \frac{1}{2} \cdot (-\frac{1}{4})(x - 2)^2$$

In effect we have added on a correction term to the tangent approximation. Results are quoted in Table 6.11 to 4 d.p. Refer to Figure 6.23.

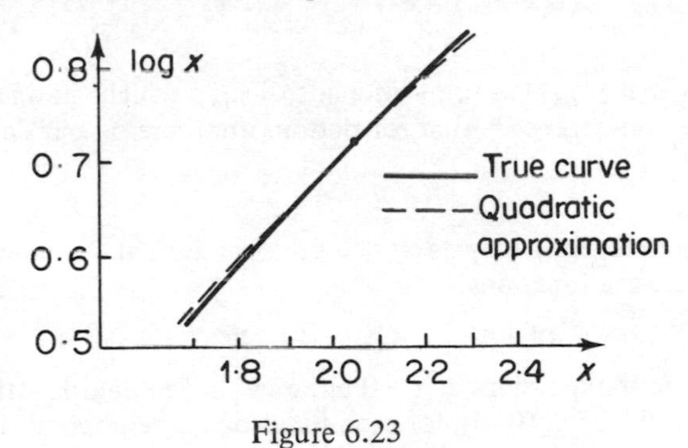

Figure 6.23

Table 6.11

x	$\log x$	$x-2$	$(x-2)^2$	Quadratic Approxn	Error
1.7	0.5306	−0.3	0.09	0.5320	0.0014
1.8	0.5878	−0.2	0.04	0.5881	0.0003
1.9	0.6419	−0.1	0.01	0.6420	−0.0001
2.0	0.6931	0	0	0.6931	0
2.1	0.7419	0.1	0.01	0.7418	−0.0001
2.2	0.7885	0.2	0.04	0.7881	−0.0004
2.3	0.8329	0.3	0.09	0.8320	−0.0009

We see clearly the improved accuracy as compared with Table 6.10, but it is not too easy to estimate the error in general; in fact, it turns out to be proportional to h^3.

Application

We can use the quadratic approximation to show the speed of convergence of the Newton-Raphson method.

Let the root of $f(x) = 0$ that we seek be $x = a$, i.e. $f(a) = 0$. Then the error at the $(n+1)^{\text{th}}$ step is given by

$$\epsilon_{n+1} = a - x_{n+1} = a - \left\{ x_n - \frac{f(x_n)}{f'(x_n)} \right\} = a - x_n - \left\{ \frac{f(a) - f(x_n)}{f'(x_n)} \right\}$$

From the quadratic approximation, we have

$$f(a) - f(x_n) \simeq f'(x_n)(a - x_n) + \frac{1}{2} f''(x_n)(a - x_n)^2$$

and hence

$$\epsilon_{n+1} \simeq -\frac{1}{2}\frac{f''(x_n)}{f'(x_n)}(a - x_n)^2 = -\frac{1}{2}\frac{f''(x_n)}{f'(x_n)}\epsilon_n^2$$

It follows that $|\epsilon_{n+1}|$ is proportional to $|\epsilon_n|^2$ which shows the quadratic nature of the convergence. What restrictions are there on our demonstration?

Problems

1. Find the tangent and quadratic approximations at the points $x = 0$ and $x = x_0$ to the functions

   (a) e^x (b) $\log(1 + x)$ (c) $\cos x$ (d) $x^3 - 2x^2 + 7x - 5$

2. Using your expansions at $x = 0$ obtained in Problem 1, estimate $f(0.1)$, $f(0.25)$, $f(0.5)$, for each function and compare with the estimates obtained from your approximations based at the points $x = 0.2$, $x = 0.3$ and $x = 0.6$ respectively. Comment on your results.

We might suppose that by matching higher derivatives of $f(x)$ at x_0, we could obtain a succession of better approximations. This is so for a suitable function $f(x)$, and in such a case we would extend the argument to an **infinite series**

$$a_0 + a_1 x + a_2 x^2 + a_3 x^3 + \ldots$$

where the coefficients a_i have to be determined in each case. We must now turn our attention to such series to see what conditions are imposed on $f(x)$ if it is to be approximated by such a power series.

6.10 INFINITE SERIES

The entity $1 + 2 + 3 + 4 + 5 + \ldots + 101$ is called a **finite series** and the expression $1 + 2x + 13x^2 + \ldots + 0.76x^{12}$ is called a **finite power series**: in essence, we are adding a collection of terms in a definite order. Such series are easy to handle and to evaluate. But what happens if there is an infinite number of terms to be added together?

Consider the expression

$$1 + \frac{1}{2} + \frac{1}{4} + \frac{1}{8} + \frac{1}{16} + \ldots$$

Ignoring for the moment the first term, we can see that the remaining terms can be geometrically interpreted as the area formed by taking half a square of unit area (Figure 6.24) then adding half of the remainder, then adding half of what is left, and so on. No matter how many times we perform these operations there will always be a small area left to divide into two parts. In this way we can see that

$$\frac{1}{2} + \frac{1}{4} + \frac{1}{8} + \frac{1}{16} + \ldots$$

can be interpreted as 1, in the sense that if we add enough of the terms together the result is as close to 1 as desired. We can construct a sequence of **partial sums** (i.e. the sum of the first term, of the first two terms, of the first three terms, etc) and study their behaviour. For the series

$$\frac{1}{2} + \frac{1}{4} + \frac{1}{8} + \frac{1}{16} + \ldots$$

the partial sums are

$$S_1 = \frac{1}{2}, \quad S_2 = \frac{3}{4}, \quad S_3 = \frac{7}{8}, \quad S_4 = \frac{15}{16} \text{ etc}$$

and we can see that $S_n = (2^n - 1)/2^n$. Now, since $\{S_n\} \equiv \{(2^n - 1)/2^n\} \to 1$ as $n \to \infty$ we say that the **sum** of the series

$$\frac{1}{2} + \frac{1}{4} + \frac{1}{8} + \frac{1}{16} + \ldots$$

is 1.

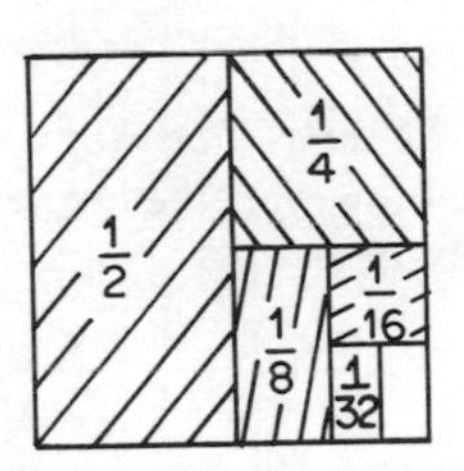

Figure 6.24

In general, *if the sequence* $\{S_n\}$ *of partial sums of an infinite series converges to a value, S, the series is said to have a sum S or to converge to the sum S.* It remains to establish suitable tests on the terms of a series in order to determine whether it converges.

For the moment, we hope that you are familiar with arithmetic and geometric progressions. An application of a geometric progression occurs in an electrical context.

Example (see Figure 6.25)

A feedback amplifier operates by taking a fraction β of its total amplified output and feeding it back. Let the amplifying factor be A then the input voltage v is amplified to Av; of this βAv is fed back into the amplifier and hence $Av - \beta Av = (1-\beta)Av$ is allowed to be output. The feedback βAv is amplified to $\beta A^2 v$ of which $\beta^2 A^2 v$ is fed back and $(1-\beta)\beta A^2 v$ is allowed to be output. Continuing in this way, the total useful output voltage V is seen to be $(1-\beta)Av + (1-\beta)\beta A^2 v + (1-\beta)\beta^2 A^3 v + \ldots$, that is,

$$V = (1-\beta)Av(1 + \beta A + \beta^2 A^2 + \ldots)$$

$$= \frac{(1-\beta)Av}{(1-\beta A)} \quad \text{since } 0 < \beta < 1$$

The effective amplification is $(1-\beta)A/(1-\beta A)$ which is called the *gain* of the amplifier. What happens if $\beta = 0$ or if $\beta = 1$?

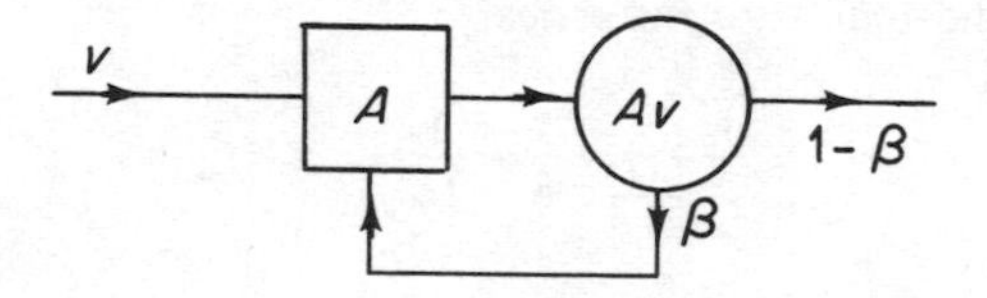

Figure 6.25

Convergence Tests

If we consider $1 + 2 + 3 + 4 + \ldots$ it is clear that the partial sums increase without limit and the infinite series is said to **diverge.** The series $1 - 1 + 1 - 1 + 1 \ldots$ is said to **oscillate finitely** (its partial sums are alternatively 1 and 0); the series $1 - 2 + 3 - 4 + 5 \ldots$ **oscillates infinitely.** We need some tests to determine into which of these categories to place a particular series.

(i) *Divergence criterion*

If $u_n \not\to 0$ as $n \to \infty$ the series does **not** *converge.* For example, the series $1 + 2 + 3 + 4 + 5 + \ldots$. However, the converse does **not** follow : if we consider the **harmonic series**

$$1 + \frac{1}{2} + \frac{1}{3} + \frac{1}{4} + \frac{1}{5} + \ldots$$

the individual terms are decreasing but the series is known to diverge. It is clearly not enough that the terms get smaller: they must do so sufficiently rapidly.

(ii) *Comparison Tests* (for series whose terms are non-negative)

(a) *If the series under test is such that each term is less than the corresponding term in a series which is known to converge, then the test series converges.* (The sum of the test series is unknown.)

Example

The series

$$1 + \frac{1}{2^2} + \frac{1}{3^3} + \frac{1}{4^4} + \ldots$$

may be compared with the series

$$1 + \frac{1}{2^2} + \frac{1}{2^3} + \frac{1}{2^4} + \ldots$$

which is known to converge; in fact, the sum of the second series is

$$\frac{1}{2} + \frac{1}{2^1} + \frac{1}{2^2} + \frac{1}{2^3} + \frac{1}{2^4} + \ldots$$

which, by the sum of a G.P., gives

$$\frac{1}{2} + \frac{\frac{1}{2}}{1 - \frac{1}{2}} = 1\frac{1}{2}$$

Term for term, the test series is smaller than the second series and hence it converges (to what, we cannot say at the moment).

(b) *If the series under test is such that each term is larger than the corresponding term in a series which is known to be divergent, then the test series diverges.*

Example

The series

$$1.01 + \frac{1.01}{2} + \frac{1.01}{3} + \frac{1.01}{4} + \ldots$$

when compared with the divergent series

$$1 + \frac{1}{2} + \frac{1}{3} + \frac{1}{4} + \ldots$$

shows divergence.

The tests are shown schematically in Figure 6.26; the test series is shown by crosses, and the known series by circles.

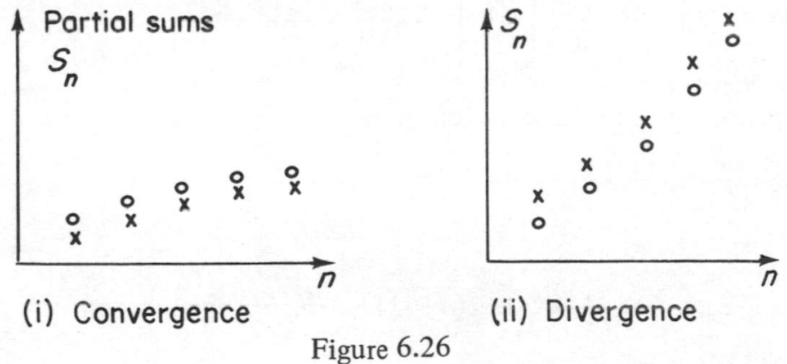

Figure 6.26

The drawback to these tests is in choosing which series to use as a comparison; it is of no use, for example, knowing that our test series is term for term less than a known divergent series. There is a 'grey' area in which these tests are not sensitive enough to pick up the convergence or otherwise of a series. However a useful class of series for comparison purposes is

$$\sum_{r=1}^{\infty} \frac{1}{r^p} = 1 + \frac{1}{2^p} + \frac{1}{3^p} + \ldots$$

where there is convergence for $p > 1$ and divergence for $p \leqslant 1$.

It is not always a straightforward matter to effect a comparison and it is preferable often to compare successive terms of a series with themselves.

(iii) *D'Alembert's Ratio Test*

Let the series be written $u_1 + u_2 + u_3 + \ldots + u_n + u_{n+1} + \ldots$. If we can find an expression for u_n (and hence u_{n+1}) we form the ratio

$$\left|\frac{u_n}{u_{n+1}}\right|$$

and look at its long-term behaviour (as with sequences, it is only the *tail* which matters).

If

$$\left|\frac{u_n}{u_{n+1}}\right| \to l$$

as $n \to \infty$ (where l is a number) then $l > 1 \Rightarrow$ convergence of the series and $l < 1 \Rightarrow$ divergence. If $l = 1$, the test is not sufficiently sensitive to help.

Example 1

(i) $1 + \frac{3}{2} + \frac{5}{2^2} + \frac{7}{2^3} + \frac{9}{2^4} + \ldots$

Here,

$$u_n = \frac{2n-1}{2^{n-1}} \text{ and } u_{n+1} = \frac{2n+1}{2^n}$$

therefore

$$\left|\frac{u_n}{u_{n+1}}\right| = \frac{2(2n-1)}{2n+1} \to 2$$

as $n \to \infty$.

Hence $l = 2$ and the series converges.

(ii) $1 + \frac{1}{2.1} + \frac{1}{2.2} + \frac{1}{2.3} + \ldots$

Now,

$$u_n = \frac{1}{2(n-1)}, \quad u_{n+1} = \frac{1}{2n}$$

and so

$$\left|\frac{u_n}{u_{n+1}}\right| = \frac{2n}{2n-1} \to 1$$

as $n \to \infty$.

We can conclude nothing about the series.

Other more sensitive (and more complicated) tests exist but we shall not develop them here.

If each term u_n in a series is replaced by $|u_n|$ and the new series converges, the first series is called **absolutely convergent**. An absolutely convergent series is also convergent, but the converse is not always true, and when it is not the series is called **conditionally convergent.**

Example 2

(i) The series

$$1 - \frac{1}{2} + \frac{1}{3} - \frac{1}{4} + \frac{1}{5} - \dots$$

can be shown to be convergent, but the series

$$1 + \frac{1}{2} + \frac{1}{3} + \frac{1}{4} + \frac{1}{5} \dots$$

is divergent.

(ii) The series

$$1 + \frac{1}{2^2} + \frac{1}{3^2} + \frac{1}{4^2} + \dots$$

is convergent and therefore the series

$$1 - \frac{1}{2^2} + \frac{1}{3^2} - \frac{1}{4^2} + \dots.$$

is absolutely convergent.

Problems I

1. Use the ratio test to show the convergence or divergence of the following series

(i) $\frac{1}{1!} + \frac{1}{2!} + \dots + \frac{1}{n!} + \dots$

(ii) $\frac{3}{1} + \frac{3^2}{2} + \dots + \frac{3^n}{n} + \dots$

(iii) $\frac{1}{\sqrt{3}} + \frac{3}{3} + \frac{5}{(\sqrt{3})^3} + \dots + \frac{(2n-1)}{(\sqrt{3})^n} + \dots$

(iv) $1 + \frac{1.2}{1.3} + \frac{1.2.3}{1.3.5} + \dots + \frac{1.2.3.\dots\dots\dots\dots n}{1.3.5\dots(2n-1)} + \dots$

2. Show that the ratio test is inconclusive for the following series and use the comparison test to show convergence or divergence.

(i) $1+\frac{3}{2.4}+\frac{7}{4.9}+\ldots+\frac{2^r-1}{2^{r-1}.r^2}+\ldots$ (compare with $\sum\frac{2}{r^2}$)

(ii) $\frac{1}{3^2}+\frac{1}{5^2}+\frac{1}{7^2}+\ldots+\frac{1}{(2n+1)^2}+\ldots$ (compare with $\sum\frac{1}{r^2}$)

(iii) $1+\frac{2^2+1}{2^3+1}+\frac{3^2+1}{3^3+1}+\frac{4^2+1}{4^3+1}+\ldots+\frac{n^2+1}{n^3+1}+\ldots$

(iv) $\frac{\log_e 2}{2}+\frac{\log_e 3}{3}+\ldots+\frac{\log_e n}{n}+\ldots$

3. Show that the series

$$\sum_{n=1}^{\infty}\frac{(-1)^{n+1}}{2n-1}$$

converges. How many terms of the series are needed in order to obtain in the sum an error which does not exceed 0.002 in magnitude? (C.E.I.)

Power Series

Consider the three power series

(i) $1+x+2x^2+3x^3+4x^4+\ldots$

(ii) $1+x+\frac{1}{2}x^2+\frac{1}{3}x^3+\frac{1}{4}x^4+\ldots$

(iii) $1+x+\frac{1}{2}x^2+\frac{1}{2^2}x^3+\frac{1}{2^3}x^4+\ldots$

Two factors affect the way the terms grow or decrease: the coefficients and the increasing powers of x. Different values of x affect whether a series converges or not. If $|x|>1$, the values of the powers of x grow in magnitude and if $|x|<1$ they diminish. We have already seen that if the terms of a series do not decrease then the series must diverge and hence for a power series with such terms as coefficients, $|x|$ must be <1 if the power series is to stand a chance of converging. Suppose in the above three series that we put $x = 1$, then we know that, of the resulting series the first two diverge, and it is easy to show that the third converges. If $x = 1.5$, we cannot be sure that the third series still converges, but it is certain that the first two do not. If $x = ½$ the third series will converge and we need to check the first two again. It should be clear that the bigger $|x|$ the less likely the series is to converge and in fact we define a **radius of convergence** for a power series $a_0+a_1x+a_2x^2+a_3x^3+\ldots$ as that *largest* value ρ such that, for all x which satisfy $|x|<\rho$ the series converges and if $|x|>\rho$ the series diverges; notice that this range of x is symmetrical about $x = 0$. If a series converged for $-1<x<2$ and diverged elsewhere, the radius of convergence would be taken as 1. Refer to Figure 6.27.

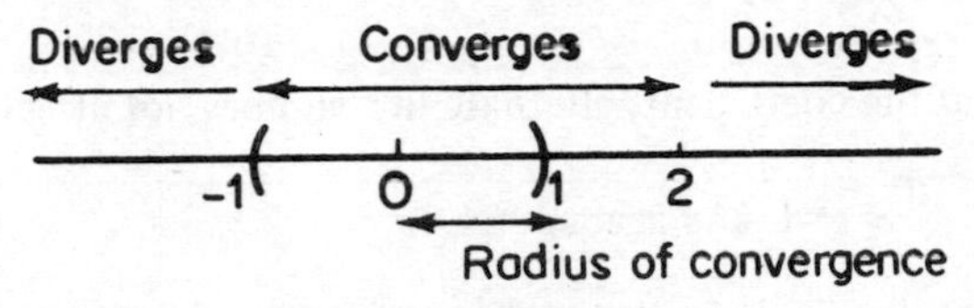

Figure 6.27

In effect, the coefficients determine how far we can allow x to stray from $x = 0$ before the terms fail to decay rapidly enough for convergence.

We test for convergence of a power series by the ratio test.

Consider the series

$$a_0 + a_1 x + a_2 x^2 + a_3 x^3 + \dots$$

Now

$$u_n = a_{n-1} x^{n-1}, \quad u_{n+1} = a_n x^n$$

and therefore

$$\left|\frac{u_n}{u_{n+1}}\right| = \left|\frac{a_{n-1} x^{n-1}}{a_n x^n}\right| = \left|\frac{a_{n-1}}{a_n}\right| \cdot \frac{1}{|x|}$$

If $\left|\dfrac{a_{n-1}}{a_n}\right| \to l$ as $n \to \infty$ then in order for $\left|\dfrac{u_n}{u_{n+1}}\right|$ to approach a limit $l > 1$ we require $|x| < l$. In this case therefore we have convergence for $|x| < l$.

If $|x| > l$ we have divergence, but if $|x| = l$ we must make a special check.

Examples

(i) $1 + x + 2x^2 + 3x^3 + 4x^4 + \dots$

$$\left|\frac{a_{n-1}}{a_n}\right| = \frac{n-1}{n} \to 1$$ and hence all we require is $|x| < 1$, a somewhat surprising result, since $x = 1$ clearly gives a divergent series.

(ii) $1 + x + \frac{1}{2}x^2 + \frac{1}{3}x^3 + \frac{1}{4}x^4 + \dots$

$$\left|\frac{a_{n-1}}{a_n}\right| = \frac{n}{n-1} \to 1$$ and again $\rho = 1$.

(iii) $1 + x + \frac{1}{2}x^2 + \frac{1}{2^2}x^3 + \frac{1}{2^3}x^4 + \dots$

$$\left|\frac{a_{n-1}}{a_n}\right| = \frac{2^{n-1}}{2^{n-2}} = 2 \to 2$$ and so $|x| < 2 \Rightarrow$ convergence.

When $x = 2$, the series is $1 + 2 + 2 + 2 + 2 + \dots$ which in fact diverges.

Two further examples will help consolidate ideas.

(iv) $1 - x + \frac{1}{2}x^2 - \frac{1}{3}x^3 + ...$

(The fact that the coefficients alternate in sign does not affect the test.)

$$\left|\frac{a_{n-1}}{a_n}\right| = \frac{n}{n-1} \to 1 \text{ and again } \rho = 1.$$

We know that if $x = -1$ we have a divergent series, but if $x = 1$ it can be shown that the resulting series converges.

(v) $1 + x + \frac{x^2}{2!} + \frac{x^3}{3!} + ...$

$$\left|\frac{a_{n-1}}{a_n}\right| = \frac{n!}{(n-1)!} = n \text{ and so } \left|\frac{u_n}{u_{n+1}}\right| = \frac{n}{|x|}$$

and as $n \to \infty$, this ratio tends to a value > 1 no matter what the (fixed) value of x.

Hence this series converges for all x and has an *infinite* radius of convergence.

Practical Convergence of Series

It is all very well to say that a power series will converge for a particular value of x, but if we want to compute the sum of that series to within 1% of its true value we may need to calculate so many terms as to render the operation tedious to say the least. (There are sometimes algebraic methods of speeding up the convergence.)

In Figure 6.28 we flow chart a general method of calculating the sum of a series correct to within a specified amount, EPS. If this has not been achieved after NMAX terms have been added together, a suitable message is output. For a particular series it is simetimes possible to calculate each term from its predecessor by a general formula.

Problems II

1. Find the interval of convergence of

(i) $\frac{x-2}{1} + \frac{(x-2)^2}{2} + \frac{(x-2)^3}{3} + ... + \frac{(x-2)^n}{n} + ...$

(ii) $\frac{x+1}{\sqrt{1}} + \frac{(x+1)^2}{\sqrt{2}} + \frac{(x+1)^3}{\sqrt{3}} + ... + \frac{(x+1)^n}{\sqrt{n}} + ...$

(iii) $x - \frac{x^3}{3} + \frac{x^5}{5} - \frac{x^7}{7} + ...$

(iv) $x - \frac{x^2}{4} + \frac{x^3}{9} - \frac{x^4}{16} + ...$

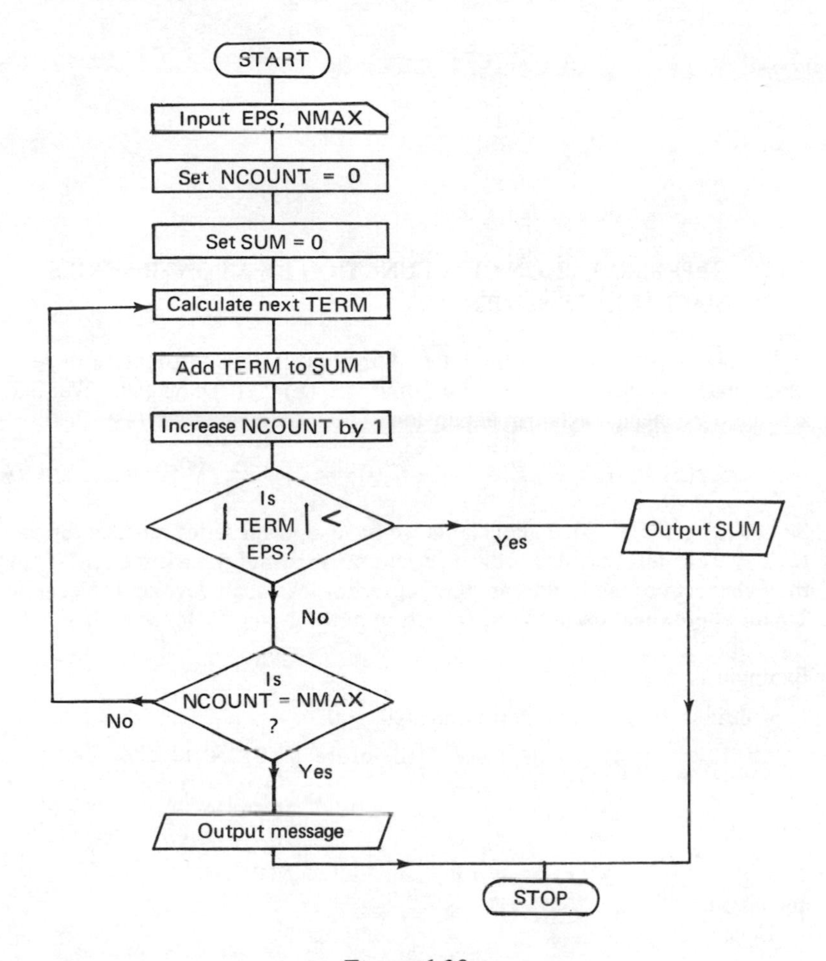

Figure 6.28

2. Find the sum of any one of the above series for a selection of values of x using a computer program.

3. An infinite series $S(x)$ is defined by

$$S(x) = 1 + \frac{x^2}{2} + \frac{x^4}{2.4} + \frac{x^6}{2.4.6} + \ldots$$

For what values of x is the series convergent? (C.E.I.)

4. Compare the two series below as methods of calculating π. Write a computer program to help in your investigations.

(i) $\frac{\pi}{4} = 1 - \frac{1}{3} + \frac{1}{5} - \frac{1}{7} + \ldots$

(ii) $\frac{\pi^2}{6} = 1 + \frac{1}{2^2} + \frac{1}{3^2} + \frac{1}{4^2} + \ldots$

6.11 REPRESENTATION OF A FUNCTION BY A POWER SERIES: MACLAURIN'S SERIES

We saw in the section on the tangent and quadratic approximations to a function the special case where the function $f(x) \simeq f(0) + xf'(0)$. We shall assume the general **Maclaurin expansion**

$$f(x) = f(0) + xf'(0) + \frac{x^2}{2!}f''(0) + \ldots + \frac{x^n}{n!}f^{(n)}(0) + \ldots. \qquad (6.33)$$

Since $f(0), f'(0), f''(0)$ are constants we have a power series representation of $f(x)$. We can test that such power series converge and we **assume** that if they do, they converge to the functions they represent. We shall develop the general **Taylor approximation** in the next section and this will tie loose ends together.

Example 1: Sine Series

Consider $f(x) = \sin x$. Note that $f(0) = 0$.

$$\begin{aligned} f'(x) &= \cos x & \text{therefore } f'(0) &= +1 \\ f''(x) &= -\sin x & f''(0) &= 0 \\ f'''(x) &= -\cos x & f'''(0) &= -1 \\ f^{\text{iv}}(x) &= \sin x & f^{\text{iv}}(0) &= 0 \\ f^{\text{v}}(x) &= \cos x & f^{\text{v}}(0) &= 1 \end{aligned}$$

and so on.

Hence

$$\sin x = x - \frac{x^3}{3!} + \frac{x^5}{5!} - \ldots + (-1)^{n-1}\frac{x^{2n-1}}{(2n-1)!} \ldots. \qquad (6.34)$$

and thus $\sin x \simeq x$ for small values of x (alternatively, $(\sin x)/x \to 1$ as $x \to 0$). (Note the effect of replacing x by $-x$ on both sides.)

We now ask for what values of x the series converges.

The ratio

$$\left|\frac{u_n}{u_{n+1}}\right| = \frac{(2n+1)!}{(2n-1)!}\cdot\left|\frac{-x^{2n-1}}{x^{2n-1}}\right| = \frac{(2n+1)(2n)}{|x|^2} \to \infty \quad \text{as } n \to \infty$$

therefore the series converges for all values of x.

Example 2: Binomial Expansion

Here,

$$f(x) = (1 + x)^s \qquad ; \qquad f(0) = 1$$

Now

$$f'(x) = s(1+x)^{s-1} \quad ; \quad f'(0) = s$$
$$f''(x) = s(s-1)(1+x)^{s-2}; \quad f''(0) = s(s-1)$$

Similarly, $f'''(0) = s(s-1)(s-2)$; $f^{iv}(0) = s(s-1)(s-2)(s-3)$ etc.
Then

$$(1+x)^s = 1 + sx + \frac{s(s-1)}{1.2}x^2 + \frac{s(s-1)(s-2)}{1.2.3}x^3 + \frac{s(s-1)(s-2)(s-3)}{1.2.3.4}x^4 + \ldots \qquad (6.35)$$

This is the so-called **binomial expansion.**

The coefficient of x^r is

$$\frac{s(s-1)(s-2)\ldots(s-r+1)}{1.2.3\ldots r} \quad \text{or} \quad \frac{s!}{r!(s-r)!}$$

which is sometimes written

$$^sC_r \quad \text{or} \quad \binom{s}{r}.$$

To test for convergence we form

$$\left|\frac{u_n}{u_{n+1}}\right| = \left|\frac{s!}{n!(s-n)!} \quad \frac{(n+1)!(s-n-1)!}{s!}\right| \frac{1}{|x|}$$

$$= \left|\frac{n+1}{s-n}\right| \cdot \frac{1}{|x|} \to \frac{1}{|x|} \quad \text{as } n \to \infty$$

The expansion is valid for $|x| < 1$.

If $x = -1$, $(1+x)^s = 0$, but the right-hand side of (6.35) is

$$1 - s + \frac{s(s-1)}{1.2} - \frac{s(s-1)(s-2)}{1.2.3} + \ldots$$

Now if $s = -2$, the series becomes

$$1 + 2 + 3 + 4 + \ldots$$

which clearly diverges, showing no representation for $x = -1$.

What do you think happens if $x = +1$?

Series representations for other functions can be derived directly or by some algebraic devices. Other series may be formed by simple manipulations.

Example 3: Cosine Series

The series for $\cos x$ can be obtained by differentiating (6.34) term by term, therefore

$$\cos x = 1 - \frac{x^2}{2!} + \frac{x^4}{4!} - \ldots \qquad (6.36)$$

Differentiating tends to destroy good behaviour and it is necessary to check directly that this series converges; it can be shown by the ratio test that for all values of x it does converge.

Example 4: Exponential Series

$$e^{ix} = \cos x + i \sin x$$

$$= (1 - \frac{x^2}{2!} + \frac{x^4}{4!} - ...) + i(x - \frac{x^3}{3!} + \frac{x^5}{5!} - ...)$$

(assumes addition valid)

$$= (1 + ix - \frac{x^2}{2!} - \frac{ix^3}{3!} + \frac{x^4}{4!} + \frac{ix^5}{5!} - ...)$$

$$= 1 + ix + \frac{(ix)^2}{2!} + \frac{(ix)^3}{3!} + \frac{(ix)^4}{4!} + \frac{(ix)^5}{5!} + ...$$

If we replace ix by x we obtain the exponential series :

$$e^x = 1 + x + \frac{x^2}{2!} + \frac{x^3}{3!} + \frac{x^4}{4!} + \frac{x^5}{5!} + ... \qquad (6.37a)$$

Similarly,

$$e^{-x} = 1 - x + \frac{x^2}{2!} - \frac{x^3}{3!} + \frac{x^4}{4!} - \frac{x^5}{5!} + ... \qquad (6.37b)$$

Both series converge for all values of x.

Example 5: Hyperbolic Series

$$\cosh x = ½(e^x + e^{-x})$$

therefore

$$\cosh x = 1 + \frac{x^2}{2!} + \frac{x^4}{4!} ... \qquad (6.38)$$

Similarly, or by differentiation

$$\sinh x = x + \frac{x^3}{3!} + \frac{x^5}{5!} + ... \qquad (6.39)$$

Example 6: Logarithmic Series

Since log(0) is not defined we cannot expand $\log x$ by a Maclaurin's series. Instead, we expand $\log(1 + x)$. You should show that

$$\log(1 + x) = x - \frac{x^2}{2} + \frac{x^3}{3} - \frac{x^4}{4} ... \qquad (6.40)$$

and that the series converges for $|x| < 1$. We know it cannot converge for $x = -1$ but it does in fact converge for $x = 1$.

In each of these examples you should try to write down the nth term of the series.

Example 7

As well as providing an approximate formula for a function which we shall use in applications a little later, a Maclaurin's series is useful for calculating approximate values of that function for small values of x (provided that these values are within the range of convergence). The following three calculations illustrate the ideas.

(i) $\sin 0.1^c = 0.1 - \dfrac{(0.1)^3}{6} + \dfrac{(0.1)^5}{120} + \ldots$

$= 0.1 - 0.000167 + 0.0000001 - \ldots$

$= 0.0998$ (4 d.p.) (From tables, this is correct to 4 d.p.)

(ii) $\sin 0.5^c = 0.5 - \dfrac{(0.5)^3}{6} + \dfrac{(0.5)^5}{120} - \ldots$

$= 0.5 - 0.02083 + 0.00006 - \ldots$

$= 0.4798$ (4 d.p.) (Tables give 0.4794 to 4 d.p.)

(iii) $\sin 1^c = 1 - \dfrac{1}{6} + \dfrac{1}{120} - \ldots$

$= 1 - 0.1667 + 0.0083 - \ldots$

$= 0.8416$ (4 d.p.) (0.8415 by tables)

We see that $\sin 0.5^c$ takes more terms than $\sin 0.1^c$ to achieve comparable accuracy yet $\sin 1^c$ seems to give a better result with only three terms. However, the first term neglected is $-(1/7!) = -0.0002$ and including this will not improve matters. What is happening is that this last series is converging more slowly than the others and it will take many more terms than the others to achieve 5 d.p. accuracy.

In Figure 6.29 we see the way in which successive Maclaurin approximations give a greater range of agreement.

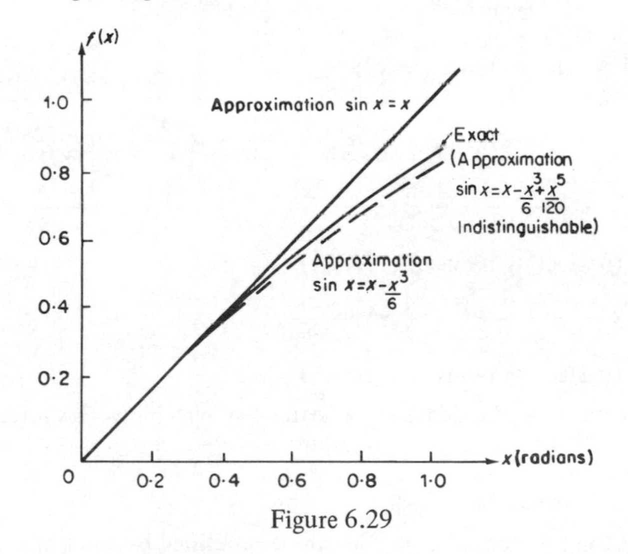

Figure 6.29

We conclude with an example to show the manipulations often necessary to obtain a Maclaurin approximation.

Example 8

Find the Maclaurin expansion for $f(x) = e^{\tan^{-1}x}$.

It would be unwise to attempt to differentiate the function as it stands. We first tackle therefore the expansion of $u = \tan^{-1}x$; again we would be rash to differentiate u many times. Instead note that since $\mathrm{d}u/\mathrm{d}x = 1/(1+x^2)$ we have $(1+x^2)(\mathrm{d}u/\mathrm{d}x) = 1$. Differentiating this equation we obtain

$$2x\frac{\mathrm{d}u}{\mathrm{d}x} + (1+x^2)\frac{\mathrm{d}^2u}{\mathrm{d}x^2} = 0$$

and, repeating the operation,

$$2\frac{\mathrm{d}u}{\mathrm{d}x} + 4x\frac{\mathrm{d}^2u}{\mathrm{d}x^2} + (1+x^2)\frac{\mathrm{d}^3u}{\mathrm{d}x^3} = 0$$

$$6\frac{\mathrm{d}^2u}{\mathrm{d}x^2} + 6x\frac{\mathrm{d}^3u}{\mathrm{d}x^3} + (1+x^2)\frac{\mathrm{d}^4u}{\mathrm{d}x^4} = 0 \quad \text{etc.}$$

Now, $u(0) = 0$, $u'(0) = 1$ and hence $2.0.u'(0) + 1.u''(0) = 0$ whence $u''(0) = 0$; also $2.1 + 0 + 1.u'''(0) = 0$, therefore $u'''(0) = -2$. Similarly $u^{(\text{iv})}(0) = 0$.

Proceeding in this way we obtain

$$\tan^{-1}x = x - \frac{x^3}{3} + \frac{x^5}{5} \ldots \tag{6.41}$$

Suppose we decide in advance that we want to keep the terms of the expansion of $f(x)$ as far as x^4 only.

We have

$$f(x) = \mathrm{e}^u \simeq 1 + u + \frac{u^2}{2!} + \frac{u^3}{3!} + \frac{u^4}{4!}$$

and we need no further terms (why?).

Then

$$\mathrm{e}^{\tan^{-1}x} \simeq 1 + \left(x - \frac{x^3}{3}\right) + \frac{1}{2}\left(x - \frac{x^3}{3}\right)^2 + \frac{1}{6}\left(x - \frac{x^3}{3}\right)^3 + \frac{1}{24}\left(x - \frac{x^3}{3}\right)^4 + \ldots$$

$$\simeq 1 + x - \frac{x^3}{3} + \frac{1}{2}x^2\left(1 - \frac{2x^2}{3} + \frac{x^4}{9}\right) + \frac{1}{6}x^3(1) + \frac{1}{24}x^4(1)$$

(neglecting terms of order x^5)

$$\simeq 1 + x + \frac{1}{2}x^2 - \frac{1}{6}x^3 - \frac{7}{24}x^4$$

(neglecting terms of order x^5)

This expansion may be derived in a neater way which involves integration.

Applications of Maclaurin's Series

(i) The piston problem of page 300 can be simplified by using approximations to equations (6.1), (6.7) and (6.9a) if m is large. The expression for

$$\cos\phi = \left(1 - \frac{\sin^2\theta}{m^2}\right)^{1/2} \simeq 1 - \frac{\sin^2\theta}{2m^2} - \frac{\sin^4\theta}{8m^4}$$

Now when $m = 4$ the last term is $(-\sin^4\theta)/2048$ which has magnitude no greater than 1/2048 and can thus be neglected. Hence, we have a simplified form for x using (6.1) and you should show that

$$x = r\left[(1 - \cos\theta) + \frac{1}{2m} - \frac{\cos^2\theta}{2m}\right]$$

On differentiating we have

the velocity $\quad v \simeq \omega r\left(\sin\theta + \dfrac{\sin 2\theta}{2m}\right)$, and

the acceleration $\quad f \simeq \omega^2 r\left(\cos\theta + \dfrac{\cos 2\theta}{m}\right)$

This last form can be got by neglecting $\sin^4\theta$ and $\sin^2\theta$ in (6.9a).

(ii) Webb's approximation for $\sec\theta$ which arises in the theory of buckling of struts (where the equation $\theta = \sec\theta$ has to be solved) follows from the Maclaurin approximation.

$$\sec\theta = \frac{1}{\cos\theta} \simeq \frac{1}{1 - \frac{1}{2}\theta^2} \simeq 1 + \tfrac{1}{2}\theta^2$$

Webb's approximation is

$$\sec\theta \simeq \frac{1 + \dfrac{4\theta^2}{\pi^2} \times 0.26}{1 - \dfrac{4\theta^2}{\pi^2}}$$

You should check its accuracy for $\theta = 0(15^\circ)90^\circ$ by writing an appropriate computer program.

Problems

1. Obtain the following Maclaurin expansions:

   (i) $\cos^2 x = 1 - x^2 + \dfrac{1}{3}x^4 - \dfrac{2}{45}x^6 + \ldots$

   (ii) $\sin^2 x = x^2 - \dfrac{1}{3}x^4 + \dfrac{2}{45}x^6 - \dfrac{1}{315}x^8 + \ldots$

   (iii) $2^x = 1 + x\log 2 + \dfrac{1}{2}(x\log 2)^2 + \dfrac{1}{6}(x\log 2)^3 + \ldots$

   (iv) $a^{mx} = 1 + mx\log a + (mx\log a)^2/2! + (mx\log a)^3/3! + \ldots$
   (a and m constants)

   (v) $\log(1 + \cos x) = \log 2 - \dfrac{1}{4}x^2 - \dfrac{1}{96}x^4 - \ldots$

2. Show that if c is large compared with x, $s = c\sinh(x/c)$ may be replaced approximately by $s = x + (x^3/6c^2)$, and $y = c\cosh(x/c)$ becomes approximately the equation of a parabola.

3. State the expansions, in ascending powers of x, of $\log(1+x)$, $\sin x$ and e^x. Deduce the following results:

(i) $\log(1-x^3) = -x^3 - \dfrac{x^6}{2} - \dfrac{x^9}{3} - \ldots$

(ii) $\log(1+x+x^2) = x + \dfrac{x^2}{2} - \dfrac{2}{3}x^3 + \dfrac{1}{4}x^4 + \ldots$

(iii) $\log(1+\sin x) = x - \dfrac{x^2}{2} + \dfrac{x^3}{6} - \ldots$

(iv) $e^x \log(1+x) = x + \dfrac{x^2}{2!} + \dfrac{2x^3}{3!} + \dfrac{9x^5}{5!} + \ldots$

4. If $y = \log \cos x$ prove that

$$\frac{d^3y}{dx^3} + 2\frac{d^2y}{dx^2}\frac{dy}{dx} = 0$$

Hence, or otherwise, obtain the Maclaurin expansion of y as far as the term in x^4, and, by the substitution $x = \pi/4$, deduce the approximate relation

$$\log 2 \simeq \frac{\pi^2}{16}(1 + \frac{\pi^2}{96})$$

Show that

$$\lim_{x\to 0} \frac{\log \cos x}{x^2} = -\frac{1}{2}$$

5. Use Maclaurin's Theorem to expand $\log(1+x)$, $\sin x$ and $\cos x$ in ascending powers of x. Hence find the series for $\log[(1/x)\sin x]$ and $\log \cos x$ as far as the terms in x^4, and show that if x is small, $\tan x \simeq x\, e^{x^2/3}$

6. (i) Show that

$$1 - x^2 + x^4 + \ldots + (-1)^{n-1}x^{2n-2} + \frac{(-1)^n x^{2n}}{1+x^2} = \frac{1}{1+x^2}$$

(ii) Hence expand $\tan^{-1} y$ as an infinite series in powers of y. For what range of values of y is the series convergent? (C.E.I.)

Hint: $\dfrac{d}{du}\tan^{-1}u = \dfrac{1}{1+u^2}$

7. Expand

$$\sqrt{(1+ax+bx^2+cx^3)} - \sqrt{(1+ax+bx^2)}$$

in ascending powers of x as far as the term in x^5. (L.U.)

8. State the Maclaurin expansion for a function $f(x)$ and use it to obtain the series for $\cos x$ in powers of x. Hence or otherwise, show that for sufficiently

small values of x

$$\sec x = 1 + \frac{x^2}{2!} + \frac{5x^4}{4!} + \ldots$$

Use this expansion to obtain the value of sec 2° correct to five decimal places. (L.U.)

9. Use Maclaurin's theorem to obtain the series expansion for $\log(1+x)$ in terms of x and state the range of values of x for which this expansion is valid. Deduce that

$$\log\left(1 + \frac{1}{n}\right) = 2\left(\frac{1}{2n+1} + \frac{1}{3(2n+1)^3} + \frac{1}{5(2n+1)^5} + \ldots\right)$$

and hence calculate log 10 correct to five significant figures given that $\log 3 = 1.09860$. (L.U.)

10. If $y = \sin^{-1}x$ show that $(1 - x^2)y_{n+2} - (2n+1)xy_{n+1} - n^2 y_n = 0$ where y_n denotes $d^n y/dx^n$, and obtain the Maclaurin series

$$y = x + \frac{1^2 x^3}{3!} + \frac{1^2.3^2.x^5}{5!} + \ldots$$ (C.E.I.)

11. Show that $x = \log 2 + a/8$ is an approximate solution of

$$2e^x \sinh x = 3 + a$$

where a is small. Find a better approximation by including terms in a^2. (C.E.I.)

12. (i) Expand $\sinh^{-1}x$ as a power series as far as the term in x^3.

(ii) A quantity $\alpha_1 = \sqrt{RG}$ is estimated as α_2 by a mathematical model via the equation

$$\sinh^2 \frac{\alpha_2}{2n} = \frac{RG}{4n^2}$$

R, G and n being constants. Show that, if $\sqrt{RG}/2n$ is small the difference between α_1 and α_2 is approximately $\alpha_1^3/24n^2$.

6.12 THE TAYLOR SERIES

When using a Maclaurin series which is based on $x = 0$, if we estimate the function at values of x far away from $x = 0$, we usually have to take more terms to achieve a particular accuracy than were we to make the estimation at a value of x nearer $x = 0$. Indeed, we cannot take values of x outside the range of convergence. It would seem reasonable to attempt a shift of base from $x = 0$ to another value x_0. This could suggest a generalisation of the tangent and quadratic approximations met earlier.

We can turn the tangent approximation into an exact equality thus:

$$f(x_0 + h) = f(x_0) + hf'(x_0) + \frac{h^2}{2!}f''(\xi)$$

where $x_0 < \xi < x_0 + h$.

The quadratic approximation can be converted to

$$f(x_0 + h) = f(x_0) + hf'(x_0) + \frac{h^2}{2!}f''(x_0) + \frac{h^3}{3!}f'''(\xi)$$

where $x_0 < \xi < x_0 + h$.

Then by analogy (the rigorous proof involves a generalisation of the MVT) we have the result that if $f(x)$ is continuous in $[x_0, x_0 + h]$ and possesses derivatives up to the $(n + 1)^{\text{th}}$ order in $(x_0, x_0 + h)$ then

$$f(x_0 + h) = f(x_0) + hf'(x_0) + \frac{h^2}{2!}f''(x_0) + \ldots + \frac{h^n}{n!}f^{(n)}(x_0) + \frac{h^{n+1}}{(n + 1)!}f^{(n+1)}(\xi)$$

with $x_0 < \xi < x_0 + h$ (6.42)

This is known as **Taylor's Theorem**. The corresponding Maclaurin's Theorem is deduced by putting $x_0 = 0$. We can estimate the maximum error in truncating a Taylor series when estimating a particular value as in the following.

Example

Consider $f(x) = 1/(1 - x) = 1 + x + x^2 + x^3 + \ldots$ when expanded about $x_0 = 0$.

If we truncate the series at the term shown we commit an error of magnitude $(x^4/4!)f^{(iv)}(\xi)$. Now $f^{(iv)}(x) = 4!/(1 - x)^5$ and if we are interested in estimating $f(0.3)$ then $0 < \xi < 0.3$. The largest value of $f^{(iv)}(x)$ occurs at $x = 0.3$ when it has the value $4!/(0.7)^5$; then the magnitude of the error committed in estimating $f(0.3)$ as $1 + 0.3 + 0.09 + 0.027 = 1.417$ is at most

$$\frac{(0.3)^4}{4!} \cdot \frac{4!}{(0.7)^5} = 0.0481$$

In fact, $f(0.3) = 1/0.7 = 1.429$ (3 d.p.) and the error committed is actually 0.012 which is well within the maximum quoted. Were we unable to do the above analysis we should have had to quote our estimate as 1.417 ± 0.012.

Let us repeat the calculations where the expansion is based on $x = 0.1$.

The expansion is $f(x) = 1.1111 + 1.235(x - 0.1) + 1.372(x - 0.1)^2 + 1.524(x - 0.1)^3$ where coefficients are kept to 3 d.p. This estimates $f(0.3)$ as 1.425 (3 d.p.) with a smaller error of 0.004. The maximum error from the formula is

$$\frac{(0.2)^4}{4!} \frac{4!}{(0.7)^5} = 0.0095$$

It should scarcely be necessary to remark that the error

$$R_{n+1} = \frac{h^{n+1}}{(n + 1)!}f^{(n+1)}(\xi)$$

must decrease in magnitude as $n \to \infty$ if the series representation is to be exact. There are cases where R_{n+1} increases indefinitely as $n \to \infty$ and then the series is of little use. This really brings us back to the idea of radius of convergence except that we must remember that the general Taylor expansion (6.42) can be written

$$f(x) = f(x_0) + (x - x_0)f'(x_0) + \frac{(x - x_0)^2}{2!} f''(x_0) + \ldots$$

$$+ \frac{(x - x_0)^{n+1}}{(n + 1)!} f^{(n+1)}(\xi) \qquad (6.43a)$$

We can also write the general Taylor expansion as an infinite series

$$f(x) = f(x_0) + (x - x_0)f'(x_0) + \frac{(x - x_0)^2}{2!} f''(x_0) + \ldots$$

$$+ \frac{(x - x_0)^n}{n!} f^{(n)}(x_0) + \ldots \qquad (6.43b)$$

Problems I

1. Expand each of the functions below about the point given
   (a) e^x about $x = 1$ (b) $2/(1 + x)$ about $x = 2$
   (c) $\sin x$ about $x = \pi/6$
   State in each case for what range of x the expansion is permissible.

2. (i) Prove that

   $$\sin x = \sin a + (x - a)\cos a - \frac{(x - a)^2 \sin a}{2!} - \frac{(x - a)^3 \cos \xi}{3!}$$

   where $a \leqslant \xi \leqslant x$.

   (ii) Use the relationship in (i) to evaluate $\sin 51^\circ$ given that $\sin 45^\circ = 1/\sqrt{2}$. To how many decimal places is the answer accurate?

   (C.E.I.)

3. For small values of x the approximations $e^x \simeq 1 + x$, $\sin x \simeq x$ are sometimes employed. Use the error term from Taylor's expansion to estimate how large a value of x can be used such that the error in the approximation is less than 0.01.

4. The hydraulic radius χ used in open channel flow is defined by χ = cross-sectional area of water flowing/wetted perimeter. In a rectangular channel the depth d is small compared with the width w. Show that $\chi \simeq d - 2d^2/w$ and estimate the error involved in using this formula when $d = 0.1$ m and $w = 5$ m.

5. Given that $\pi/90 = 0.03491$ use four terms of Taylor's series to estimate $\tan 43^\circ$ correct to four decimal places.

6. Show that

$$\sqrt{a+h} = \sqrt{a} + \frac{1}{2}\frac{h}{\sqrt{a}} - \frac{1}{8}\frac{h^2}{a\sqrt{a}} + \frac{1}{16}\frac{h^3}{a^2\sqrt{a}} - \frac{5}{128}\frac{h^4}{a^3\sqrt{a}}$$

and hence, with $a = 9$, $h = 1$, show that $\sqrt{10} = 3.16228$ correct to five decimal places.

7. Use Taylor's expansion to show that :

(i) $y = \dfrac{\log(2-x)}{x^3 - 3x + 2}$ behaves like $-\dfrac{1}{3}\dfrac{1}{(x-1)}$ for x near 1

(ii) $y = \dfrac{(1+\cos x)^2}{\sin x}$ behaves like $-\dfrac{1}{4}(x-\pi)^3$ for x near π

Hence sketch a graph of y against x near the relevant value in each case.

Applications of Taylor's Series

(i) *Numerical Differentiation*

Suppose we wish to estimate $f'(a)$ and $f''(a)$. By Taylor's series we have

$$f(a+h) = f(a) + hf'(a) + \frac{h^2}{2!}f''(a) + \frac{h^3}{3!}f'''(a) + \ldots \quad (6.44a)$$

$$f(a-h) = f(a) - hf'(a) + \frac{h^2}{2!}f''(a) - \frac{h^3}{3!}f'''(a) + \ldots \quad (6.44b)$$

Subtracting (6.44b) from (6.44a), (is this justified?)

$$f(a+h) - f(a-h) = 2hf'(a) + \frac{2h^3}{3!}f'''(a) + \ldots$$

$$= 2hf'(a) + 0(h^3)$$

therefore

$$\frac{f(a+h) - f(a-h)}{2h} = f'(a) + 0(h^2)$$

Similarly, adding (6.44a) and (6.44b) gives

$$\frac{f(a+h) - 2f(a) + f(a-h)}{h^2} = f''(a) + 0(h^2)$$

Higher accuracy formulae can be found by expanding $f(a+2h)$, $f(a-2h)$ etc. and taking suitable combinations of these expansions.

(ii) *Newton's formula for curvature at the origin*

Suppose the curve of $y = f(x)$ touches the x-axis at the origin so that $f(0) = 0 = f'(0)$. Then the radius of curvature of the curve at the origin is $\{1 + [f'(0)]^2\}^{3/2}/f''(0) = 1/f''(0)$. If the curve is given in parametric form, the following manipulations help.

$$y = f(0) + xf'(0) + \tfrac{1}{2}x^2 f''(0) + 0(x^3) = \tfrac{1}{2}x^2 f''(0) + 0(x^3)$$

[since $f(0)$ and $f'(0)$ are both zero]

Hence

$$\frac{2y}{x^2} = f''(0) + 0(x)$$

and

$$\lim_{x \to 0} \frac{2y}{x^2} = f''(0)$$

and so the radius of curvature at the origin is $\lim_{x \to 0} \frac{x^2}{2y}$, which is **Newton's formula for curvature at the origin.** (6.45)

Example

The cycloid $x = 2(\theta + \sin\theta)$, $y = 2(1 - \cos\theta)$ touches the x-axis at the origin since, when $\theta = 0$, $x = 0$, $y = 0$, and

$$\frac{dy}{dx} = \left(\frac{dy}{d\theta}\right) \bigg/ \left(\frac{dx}{d\theta}\right) = 0$$

Then the curvature at the origin

$$= \lim_{x \to 0} \frac{4(\theta + \sin\theta)^2}{4(1 - \cos\theta)}$$

$$= \lim_{x \to 0} \frac{\left(1 + \frac{\sin\theta}{\theta}\right)^2 . \theta^2}{2\sin^2\frac{\theta}{2}}$$

$$= \lim_{x \to 0} \left[2(1 + \frac{\sin\theta}{\theta})^2 \Big/ (\frac{\sin \frac{1}{2}\theta}{\frac{1}{2}\theta})^2\right]$$

$$= 2.2^2/1$$

$$= 8$$

(iii) *Indeterminate forms: L'Hôpital's Rule*

In the above example we had a potential limiting form 0/0; equally awkward is the form ∞/∞. Although there are various methods of dealing with such forms we shall here develop a simple rule.

We suppose that for two differentiable functions, $f(x)$ and $g(x)$, $f(0) = g(0) = 0$. Then

$$\frac{f(x) - f(0)}{x} \cdot \frac{x}{g(x) - g(0)} = \frac{f(x)}{g(x)}$$

If we proceed to the limit as $x \to 0$ we obtain

$$\lim_{x \to 0} \frac{f(x)}{g(x)} = \frac{f'(0)}{g'(0)}$$

Which is **L'Hôpital's Rule.** (What case do we have to beware?)

Examples

(i) $\lim_{x \to 0} \frac{1 - e^x}{x}$

Here $1 - e^0 = 0$ and so the conditions of the rule are obeyed. $f'(x) = -e^x$ and $g'(x) = 1$ so that the limit is $-e^0/1 = -1$. We could have deduced the same result by expanding the numerator as a Maclaurin's series.

(ii) $\lim_{x \to 0} \dfrac{x - \sin x}{x^2}$

Here $f(0) = 0 = g(0)$, but $f'(x) = 1 - \cos x$, and so $f'(0) = g'(0) = 0$ and an application of L'Hôpital's rule gives 0/0. The basic principle is, *when in doubt apply the rule again* (**provided** the differentiations can be done). Hence $f''(x) = \sin x$, $g''(x) = 2$ and so the limit is

$$\frac{f''(0)}{g''(0)} = 0$$

The generalisation of L'Hôpital's rule follows from an extension of MVT.

(iv) *Small Errors*

Suppose $\delta f \equiv f(a + \delta x) - f(a)$, then by Taylor's expansion, if h is small,

$$\delta f \simeq \delta x f'(a) \tag{6.46}$$

If any small error, δx, has been made in the measurement of x then (6.46) gives the approximate error in $f(x)$. We call $\delta f/f(a)$ the **relative error**; it is approximately

$$\delta x \cdot \frac{f'(a)}{f(a)}$$

the **percentage error** is

$$\left\{\frac{\delta f}{f(a)} \times 100\right\}\% \simeq \left\{\delta x \frac{f'(a)}{f(a)} \times 100\right\}\%$$

Example 1

A tower 50 metres high stands on one bank of a river, the elevation of the top from a point on the opposite bank being 30° (see Figure 6.30). Find the percentage error in the width of the river as measured via the angle which is subject to an error of 1′.

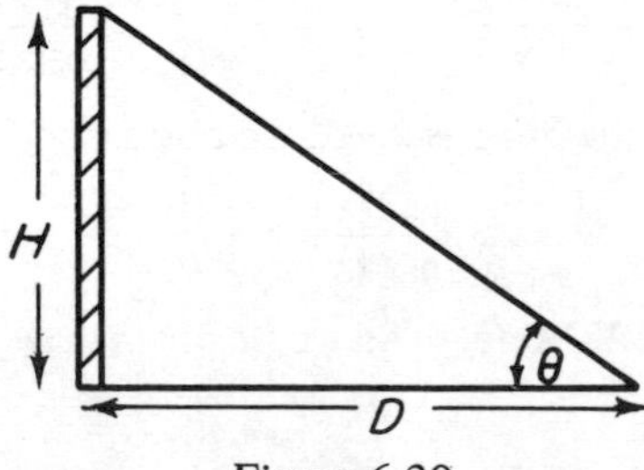

Figure 6.30

If D is the width of the river, H the height of the tower and θ the angle of elevation, then

$$D(\theta) = H \cot \theta$$

Since

$$D'(\theta) = -H \operatorname{cosec}^2 \theta$$

the percentage error is

$$\frac{-H \operatorname{cosec}^2 \theta}{H \cot \theta} . \delta\theta \times 100$$

$$-\frac{100}{\cos \theta \sin \theta} \delta\theta$$

and since $\theta = 30°$, this error

$$= \frac{-100}{\frac{\sqrt{3}}{2} . \frac{1}{2}} . \frac{1}{60} . \frac{\pi}{180} = \frac{-400\pi}{10800\sqrt{3}}$$

$$= -0.067\% \text{ (3 d.p.)}$$

This result shows the scaling of the error by the function $D(\theta)$.

Example 2

The period of a simple pendulum is $T = 2\pi\sqrt{l/g}$. We use this to estimate g; suppose a small percentage error of +0.2% is made in measuring the length and −0.1% in measuring T and we seek the effect on g.

Now $g = (4\pi^2/T^2)l$ and it helps to take logarithms to obtain $\log g = \log 4\pi^2 + \log l - 2 \log T$. Differentiating both sides w.r.t. g

$$\frac{1}{g} = 0 + \frac{1}{l}\frac{dl}{dg} - \frac{2}{T}\frac{dT}{dg}$$

and replacing derivatives by their approximations

$$\frac{\delta g}{g} \simeq \frac{\delta l}{l} - \frac{2\delta T}{T}$$

and hence

$$\frac{dg}{g} \times 100 \simeq 0.2 - 2(-0.1) = 0.4\%$$

This suggests that T must be measured twice as accurately as l to have as little effect on the value of g.

Optimal Values: A revisit

We saw in section 6.4 the difficulty encountered with x^3 and x^4 in the determination of their turning points. Remember that the definition of a turning point of $f(x)$ is that $f'(x) = 0$. Then, if h is small, Taylor's expansion (6.42) becomes

$$\delta f = f(x_0 + h) - f(x_0) = \cancel{hf'(x_0)} + \frac{h^2}{2!} f''(x_0) + ... \simeq \frac{h^2}{2!} f''(x_0)$$

Now, if $f''(x_0) > 0$, $\delta f > 0$ irrespective of whether h is positive or negative, and so, *locally*, $f(x_0)$ is the smallest value; conversely, if $f''(x_0) < 0$, $f(x_0)$ is a local maximum. If $f''(x_0) = 0$ then the approximation is too coarse to determine the details of the behaviour at x_0. We have to take the next term of the Taylor series to get

$$\delta f \simeq \frac{h^3}{3!} f'''(x_0)$$

Here a different state of affairs obtains: if $f'''(x_0) > 0$ then δf takes opposite signs on either side of x_0 [likewise if $f'''(x_0) < 0$] which implies a point of inflection. If $f'''(x_0) = 0$ we have to proceed further along the Taylor series to obtain $\delta f \simeq (h^4/4!) f^{(iv)}(x_0)$. It is clear that behaviour will continue to alternate. We have the rule:

if the first non-vanishing derivative of $f(x)$ at $x = x_0$ is of even order then there is a local maximum or minimum at x_0 (local minimum if the value of that derivative is positive, local maximum if it is negative); if the first non-vanishing derivative is of odd order there is a point of inflection at x_0.

Note the flatter behaviour near $x = 0$ of x^4 than x^2 as shown in Figure 6.31.

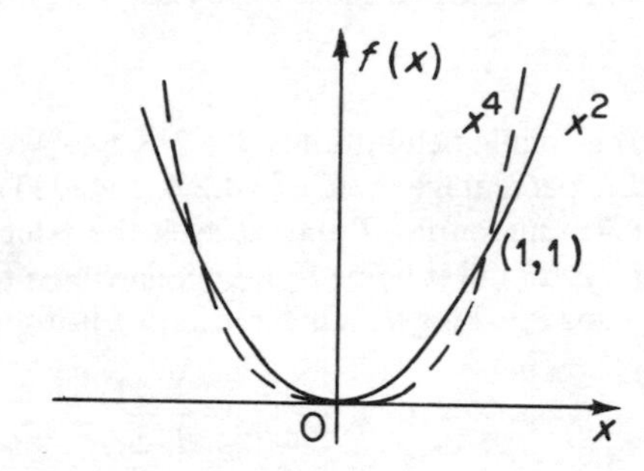

Figure 6.31

Problems II

1. Use L'Hôpital's rule or other methods to establish:

(i) $\lim_{x\to 0} \frac{\sin^2 x}{x}$ (ii) $\lim_{x\to 0} \frac{\log(1+x)}{x}$

(iii) $\lim_{x\to 1/2} \frac{\log 2x}{2x-1}$ (iv) $\lim_{x\to 0} \frac{\sec x - 1}{x \sin x}$

(v) $\lim_{\lambda\to -1} \frac{a^{\lambda+1} - 1}{\lambda + 1} = \log a \quad (a > 0)$

(vi) $\lim_{x\to 1} \frac{\log \cos 2\pi x}{1 + \cos \pi x} = -4$ (vii) $\lim_{\theta\to 0} \frac{\tan 2\theta - 2\tan\theta}{\sin 2\theta - 2\sin\theta} = -2$

2. The measurement of the side of a cube is in error by -1%. What is the error in the estimated volume of the cube?

3. The density of a spherical ball-bearing is estimated by measuring the radius as 3.00 ± 0.02 cm. and weighing the sphere as 30 ± 0.01 gm. What is a reasonable estimate of the density?

4. The stress on a shaft of diameter d under a torque T is given by $S = 16T/\pi d^3$. Find the approximate % error in the calculated stress if the measured value of d is liable to an error of 2%.

5. Find the curvature at the origin of the curves

(i) $y = \dfrac{x^2}{1+x^2}$ (ii) $y^2 = 4ax$

(iii) $\begin{cases} x = \cos\theta \\ y = 2(\sin\theta - 1) \end{cases}$

6. Obtain the expansion of $\tan\theta$ in powers of θ as far as the term in θ^5. A column of length l has a vertical load P and a horizontal load F at the top. The transverse deflection is

$$\delta = \frac{Fl}{P}\left[\frac{\tan ml}{ml} - 1\right]$$

where $m^2 = P/EI$. Show that as $P \to 0$, $\delta \to Fl^3/3EI$ and that a small value of P leads to an increase of this by about $(40Pl^2/EI)\%$ (L.U.)

7. The current i in an electrical circuit at time t is given by

$$i = \frac{E}{R}(1 - e^{-Rt/L})$$

Use L'Hôpital's rule to derive a suitable formula for i when R is negligibly small. (C.E.I.)

8. Show that, as far as the term in x^5

$$\log\left[\frac{1+\tan(\frac{1}{2}x)}{1-\tan(\frac{1}{2}x)}\right] = x + \frac{x^3}{6} + \frac{x^5}{24} + \ldots$$

Hence, or otherwise, prove that

$$\lim_{x\to 0}\left\{\frac{\log\dfrac{1+\tan(\frac{1}{2}x)}{1-\tan(\frac{1}{2}x)} - \sin x}{x^3}\right\} = \frac{1}{3}$$ (L.U.)

6.13 SIMPLE SEARCH PROCEDURES FOR OPTIMAL VALUES

In certain problems we may be faced with the possibility of being able to evaluate $f(x)$ only at a few points of our choice and basing a decision as to the location of an optimum value of the function solely on the evidence of those values. (This occurs particularly with functions of several variables where the only reasonable method is a search procedure with the help of a computer.) In this section we illustrate some simple methods for finding a local minimum of a function of one variable. Local maxima can be found by examining $[-f(x)]$ for local minima and overall optima by also checking values at the end-points of relevant intervals.

Exhaustive search

The simplest technique is **exhaustive search** where $f(x)$ is evaluated at specified (usually equally spaced) points, and the least of these values taken. The danger is that the spacing may be too large. See Figure 6.32. If the values of x are selected by some random number generator the technique is called **random search**.

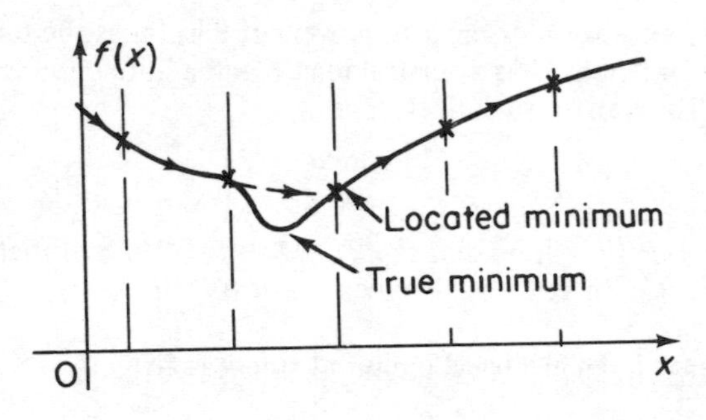

Figure 6.32

Example

$f(x) = (x - 20)^4$ (which we know has a minimum at $x = 20$ of 0).

$$
\begin{aligned}
f(0) &= 160000 \\
f(5) &= 50625 \\
f(10) &= 10000 \\
f(15) &= 625 \\
f(20) &= 0 \quad \leftarrow \quad \text{minimum} \\
f(25) &= 625 \\
f(30) &= 10000
\end{aligned}
$$

Grid search

Here we assume that the domain of $f(x)$ is limited to the interval $[a, b]$; this may be because of prior knowledge or, sometimes, because of physical considerations. We start by choosing a suitable grid size h so that $a = x_0$, $b = x_n$, where $n = (b - a)/h$. We then evaluate $f(x)$ at $x_0, x_1, x_2, \ldots$ as long as $f(x_r) > f(x_{r+1})$. At some stage, it might be that $f(x_s) < f(x_{s+1})$. Then we know that the local minimum lies in the interval $[x_{s-1}, x_{s+1}]$ and we can recommence at x_{s-1} with a smaller grid size, say $h/10$. We can repeat this process until the range of uncertainty is within acceptable limits. A flow chart for this is given in Figure 6.33. See whether you can modify it to include some sophistications.

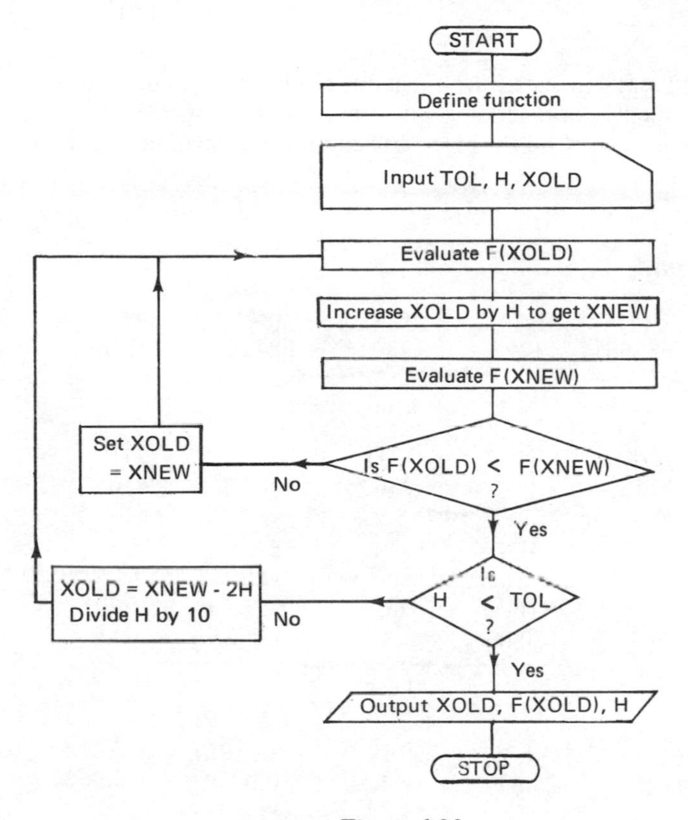

Figure 6.33

Equal Interval search

Suppose the function has been already studied so that its minimum is known to lie in the range (0, 1). Then, the function is evaluated at 1/4, 1/2, 3/4 and one of the three situations shown in Figure 6.34 will obtain.

In each case we can, by comparing these three function values, decide which interval to keep and notice that this cuts the *interval of uncertainty* by half. We are then ready to repeat the process but the benefit which makes the division of the initial interval into four equal parts is that we only need two further function evaluations, for example, in Figure 6.34 (i) we need evaluate only at

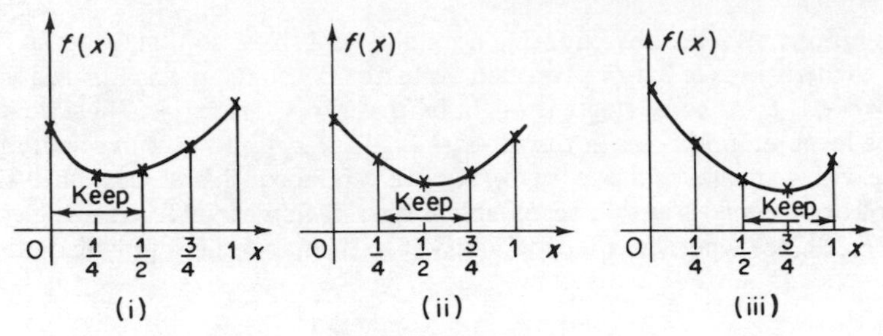

Figure 6.34

1/8 and 3/8. By M stages after the initial one the interval of uncertainty is $(\frac{1}{2})^M$ and the number of function evaluations has been $(2M + 3)$.

Draw up a flow chart for the procedure and run a program to check the following results.

Example

$f(x) = (x - 20)^4$

$f(4)$	$f(14)$	$f(24)$	$f(34)$	$f(44)$
65536	1296	256	38416	331776

keep (from $f(14)$ to $f(34)$)

$f(14)$	$f(19)$	$f(24)$	$f(29)$	$f(34)$
1296	1	256	6561	38416

keep (from $f(14)$ to $f(24)$)

$f(14)$	$f(16.5)$	$f(19)$	$f(21.5)$	$f(24)$
1296	150.06	1	5.0625	256

keep (from $f(16.5)$ to $f(21.5)$)

$f(16.5)$	$f(17.75)$	$f(19)$	$f(20.25)$	$f(21.5)$
150.06	25.63	1	0.0039	5.0625

keep (from $f(19)$ to $f(21.5)$)

$f(19)$	$f(19.625)$	$f(20.25)$	$f(20.875)$	$f(21.5)$
1	0.0198	0.0039	0.5862	5.0625

keep (from $f(19.625)$ to $f(20.875)$)

$f(19.625)$	$f(19.9375)$	$f(20.25)$	$f(20.5625)$	$f(20.875)$
0.0198	0.000015	0.0039	0.0986	0.5862

keep (from $f(19.625)$ to $f(20.25)$)

Problems

1. Use the method of Equal Interval search on $x^4 + 3x^2 - 5$; check the result by algebra.

2. Given that the minimum of $f(x)$ is in $[x_{s-1}, x_{s+1}]$, with a best estimate of x_s, fit a quadratic curve to these three points and take the minimum of this quadratic as the minimum of $f(x)$. Apply the method to the examples in the text.

3. **Dichotomous search** evaluates $f(x)$ at a small distance ϵ on either side of the mid-point of the interval of uncertainty and rejects just less than half the interval. Sketch the possibilities to see what would be rejected. Try out the method on $x^2 - 6x + 2$ given $0 \leqslant x \leqslant 10$ and check by calculus. Run the method on a digital computer. (Don't forget to write a flow chart first.)

4. Use equal interval search on the text examples, with a division of the intervals into three parts.

5. If the minimum is some way from the starting point, many function evaluations may be needed to get a first location. A routine that is useful is to double the step length h every time a success is encountered, i.e. $f(x_r + h) < f(x_r)$. If $f(x_r + h) > f(x_r)$, then the step size is halved and we start again from x_r until another failure occurs; the cycle being repeated until the required accuracy is reached. Draw a flow chart to cover this technique and cater for the case $f(x_r + h) = f(x_r)$.

SUMMARY

This chapter contains much important material with which it is necessary to become conversant. Having understood the idea of differentiation from Chapter 3, you are now expected to be expert in the techniques of differentiation and their application to problems of locating optimum values of a function and to problems of curvature. Of course, if the function is specified by a table of values, you will have to use numerical methods for finding approximate derivatives and for locating maximum and minimum values. The approximation of a tabulated function by a finite difference formula (e.g. Newton-Gregory) is important and you should appreciate the strengths and weaknesses of the various formulae. The approximation of a function by a power series is of prime importance and must be understood. You should become familiar with the ways of deriving Maclaurin's and Taylor's series together with their application and with the notion of convergence of an infinite series. We do not expect you to become an expert at juggling with infinite series, but you ought to be able to cope with simple cases.

Chapter Seven

Partial Differentiation

7.0 INTRODUCTION

So far we have dealt primarily with the behaviour of functions of one variable; in this chapter we start with a practical example where the need to determine the nature of such a functional relationship from experimental results leads to the study of functions of more than one variable. Most of the results we derive for functions of two variables can be easily extended to more than two variables but we concentrate on cases which involve two variables because (a) we have the possibility of geometrical interpretation of results, and (b) the significant step in our thinking is from one variable to two variables, the steps from two to three or more being relatively simple.

7.1 FITTING RELATIONSHIPS TO EXPERIMENTAL DATA

Figure 7.1 shows a set of plotted results from an experiment. We wish to fit a particular relationship to the data; for example, a straight line $y = a_0 + a_1x$ or a parabola $y = a_0 + a_1x + a_2x^2$.

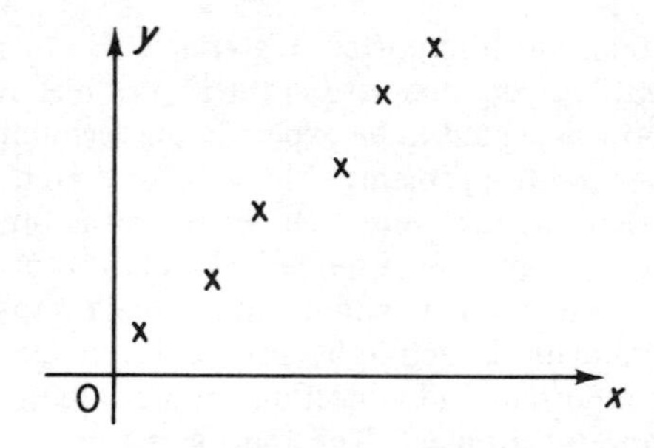

Figure 7.1

Once we have decided on the *kind* of relationship we wish to fit, then we have to decide on the coefficients that determine the particular curve of its kind which fits the data best. But what is *best* in this context? If we were to fit a straight line to the data of Figure 7.1, we could use a transparent straight-edge and move it on the paper until it seemed to split the data points roughly equally above and below the line. However, our idea of best fit and someone else's might not coincide and if the line of best fit were to be used to make predictions about y for non-observed values of x, the results could be significantly different.

To overcome this subjective element, we need an algebraically defined criterion and we embark on this using Figure 7.2. We generalise the situation to a relationship $y(x)$ being fitted to n data points and we examine the behaviour of $y(x)$ near two of them. We assume that the experiment is one in which we can regard the measurement of x values to be *exact* (or nearly so) and the measurement of y values to be *subject to error* (e.g. x might be time, and y voltage). Then for each observed x_i, $i = 1, 2, ..., n$, there is an observed y_i and, if we assume a particular relationship $y(x)$, a calculated value y_i^c where $y_i^c = y(x_i)$. We measure the discrepancy, e_i, between the observed and calculated quantities in each case; $e_i = y_i - y_i^c$. In Figure 7.2 $e_1 > 0$, $e_2 < 0$. In an ideal case, where all the data points lie *exactly* on the curve of a known relationship $y(x)$, $e_i = 0$ for all points. However, in general we must expect a discrepancy at each point and we seek that curve of a chosen kind for which the *total* discrepancy (i.e. the *sums* of discrepancies at each data point) is least.

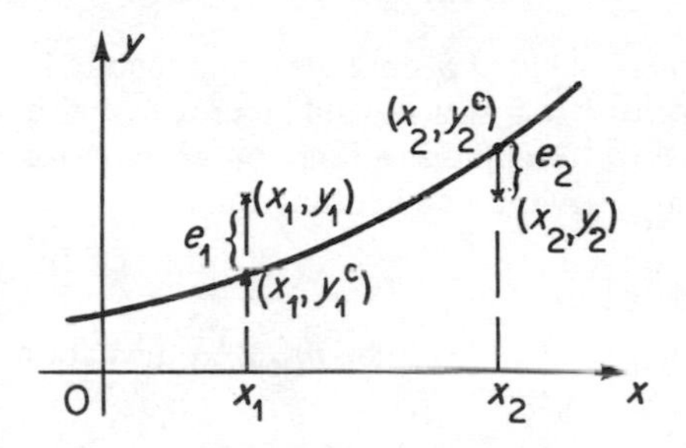

Figure 7.2

The algebraic sum of discrepancies is a poor choice since it permits partial cancellation between positive and negative quantities and makes the situation seem better than it is. It would be wrong to say that the data of Figure 7.3 fitted the straight line well yet the sum of the discrepancies is nearly zero.

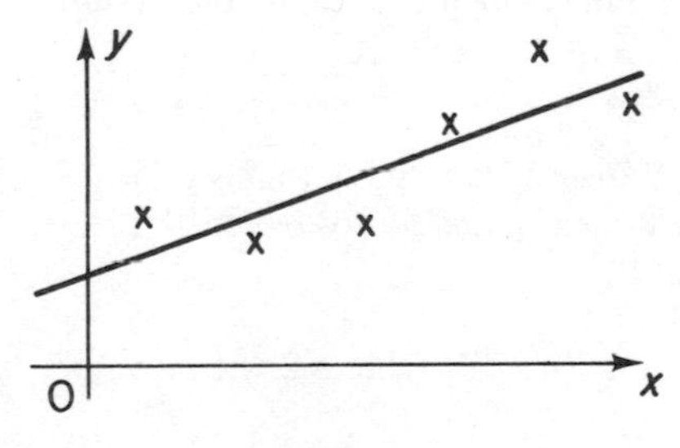

Figure 7.3

For simplicity with later calculations the **least squares criterion** is used: this calculates

$$S = \sum_{i=1}^{n} e_i^2 \tag{7.1}$$

for the n data points and seeks that curve of its class which *minimises* S. We note that this gives undue emphasis to points which are well separated from the majority but this is a risk which has to be weighed against the objectiveness of the criterion and its easy adaptation to programming for a computer (Problem 1 at the end of this section). Problem 2 shows that the attempt to fit a straight line $y = a_0 + a_1 x$ to given data points provides a value for S which depends on a_0 and a_1; we merely note in passing that under most circumstances we shall always be able to find a_0 and a_1 no matter how badly the data is fitted by a straight line and promise to return to this matter in section 7.5. For the moment, we state that, in general, S will be a *function* of a_0 and a_1, viz. $S = S(a_0, a_1)$ and we seek to find those values of a_0 and a_1 which minimise S. Problem 4 at the end of this section shows that for the straight line fit

$$S = \sum_{i=1}^{n} (y_i - a_0 - a_1 x_i)^2$$

[There is no maximum value of S since we could merely keep the slope a_1 fixed and increase (or decrease) a_0, which would increase S without limit.] We need therefore to know how to minimise a function of two variables but we first need to study the general behaviour of such functions.

Problems

1. A straight line $y = 3x - 2$ is to be fitted to the data points

x	0	1	2	3
y	-1.9	0.9	4.4	6.9

Plot the points and draw the straight line and notice the *outlier* – the point much further from the line than the others. Tabulate for each point e_i, $|e_i|$ and e_i^2 and find the sums of each quantity. Then remove the outlier from the calculations and repeat. Comment on your results.

2. Fit the straight line $y = a_0 + a_1 x$ to the data points

x	-1	0	1
y	2	3	4

by calculating S from (7.1). Show that $S = 2(a_1 - 1)^2 + 3(a_0 - 3)^2$ and hence find the values a_0 and a_1 for which S is least. Verify your conclusions graphically.

3. Repeat the calculations of Problem 2 with y values 1.9, 3.1, 4.1.

4. For a straight line $y = a_0 + a_1 x$, write down an expression for e_i and deduce a form of S when the data points are $(x_1, y_1), ..., (x_n, y_n)$.

7.2 FUNCTIONS OF TWO VARIABLES

We now turn our attention to the study of functions of two variables to see what useful results we can obtain. Just as a function of one variable can be represented by a line on a piece of paper, so a function of two variables $f(x, y)$ can be represented by a surface in three dimensions. We often write $z = f(x, y)$ and we see from Figure 7.4 how the three-dimensional representation of a point $z_0 = f(x_0, y_0)$ ties up with the idea of the point (x_0, y_0, z_0) on the surface of $f(x, y)$.

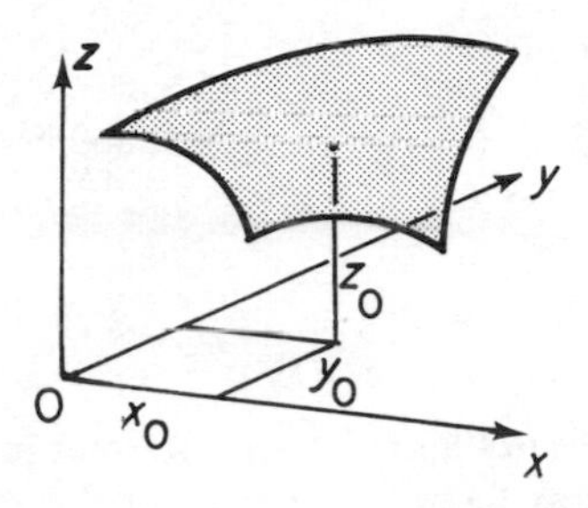

Figure 7.4

Formally we say that *if to each ordered pair of values of two independent variables x and y, each defined over some range, there corresponds* **one and only one value** *of a third (dependent) variable z, then z is a* **function** *of x and y in that range.*

For example (i) $z = \sqrt{x^2 + y^2}$ defined for all values of x and y

(ii) $z = \sqrt{x^2 + y^2 - 1}$ defined for all values of x and y satisfying $x^2 + y^2 \geqslant 1$.

Great care would be needed in interpreting the function $z = \tan^{-1}(y/x)$.

You can perhaps visualise the surface representation more clearly if you imagine that at each point (x, y) of the x-y plane, within the ranges defined, a thin matchstick is placed upright with the height of the matchstick equal to the value of the function at that point. The tops of the matchsticks form the surface representing $f(x, y)$.

Evaluation of the function at a point is straightforward.
For example, if

$$f(x, y) = x^2 - 2y^2 + 3x + 1$$

then

$$f(1, 2) = 1 - 8 + 3 + 1 = -3$$

and

$$f(2, -1) = 4 - 2 + 6 + 1 = 9$$

The above are all examples of an **explicit** definition of the function. An example of an **implicit** definition is $x^2 + xz + y^2 + \sin z = 0$. (Compare page 6)

Surface representation of f(x, y)

It is not always easy to draw the surface representing $f(x, y)$: one class of surfaces which is simple to sketch is the class of planes [see Figures 7.5 (i) and (ii)].

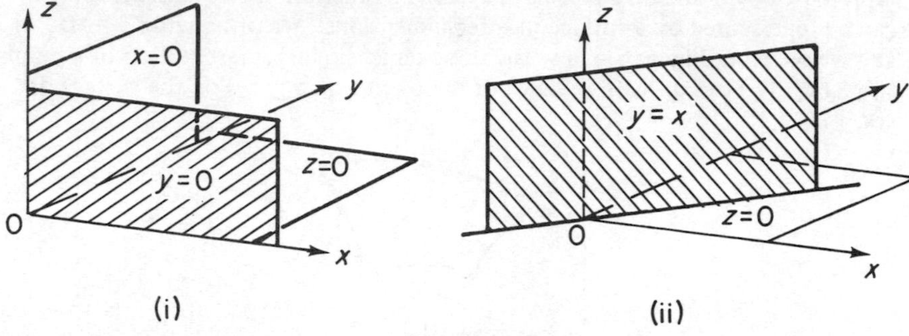

Figure 7.5

Notice that the two planes in Figure 7.5 (ii) intersect in a straight line (this is generally so – see Problems 4 and 5 at the end of this section). The specification of the plane $y = x$ must not be confused with the intersection line $y = x$ which, in three dimensions, should strictly be $y = x, z = 0$.

A point free to roam in three dimensions has 3 **degrees of freedom** (d.o.f.). The one equation confining it to a surface/plane means it is reduced to 2 degrees of freedom and to restrict it to a curve/line with 1 d.o.f., needs a *second* equation.

You have doubtless seen contour lines on maps which are two-dimensional representations of a three-dimensional piece of country-side. For a function of two variables the contour lines are curves joining points of equal height above sea-level and are formed by horizontal planes intersecting the surface, representing the function; see Figures 7.6(i) and (ii). The equation of such a horizontal plane is $z = c$, where c is a constant.

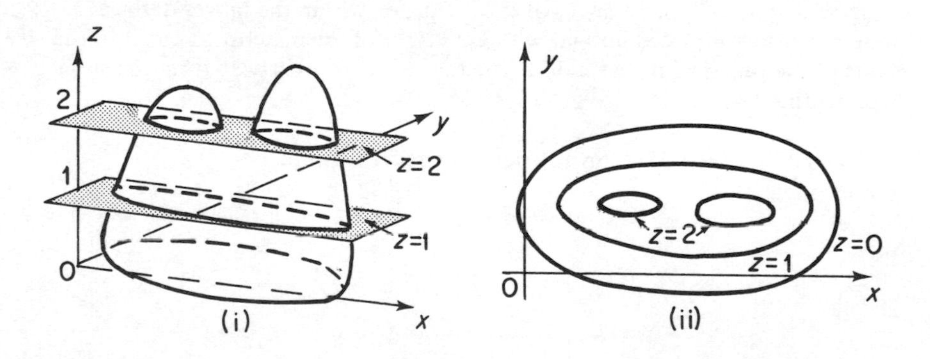

Figure 7.6

In the example of the cone $z = \sqrt{x^2 + y^2}$ the contour lines are concentric circles, $x^2 + y^2 = c^2$; see Figures 7.7 (i) and (ii).

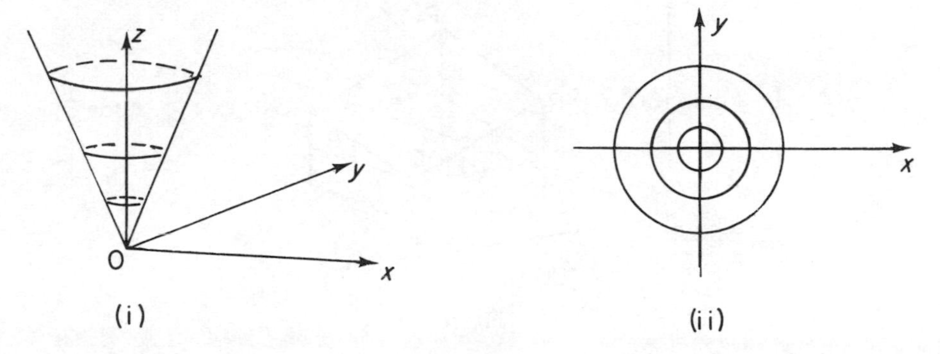

Figure 7.7

Note that if we intersect a surface with a vertical plane then the intersection is a curve representing a function of one variable: see Figure 7.8.

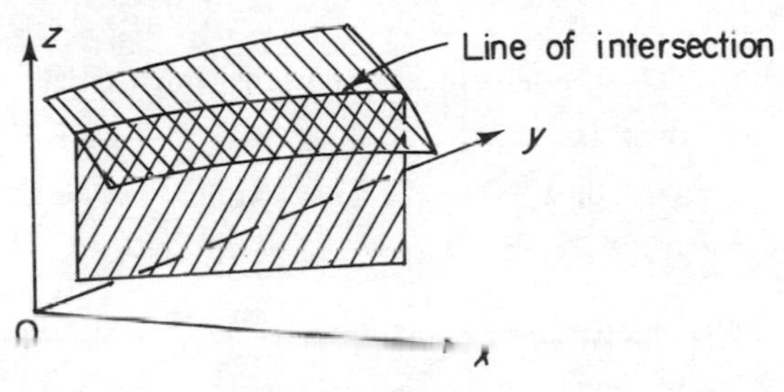

Figure 7.8

Equation of a plane

We produce the equation of a plane inclined at α to the x-axis and β to the y-axis: see Figure 7.9. From this equation you should be able to see the special cases $\alpha \to \pi/2$, $\beta \to \pi/2$.

The point Q is (a, b, c) and we seek a relationship connecting the coordinates (x, y, z) of P. To get to P we can first rise to R by an amount $(a - x)\tan(180^\circ - \alpha)$ or first to S by an amount $(y - b)\tan \beta$. Similar triangles show that the total rise to P, $(z - c)$, is the sum of these and, in general, if $\alpha \neq \pi/2 \neq \beta$, we obtain the equation

$$\left.\begin{array}{c} z = c + (x - a)\tan \alpha + (y - b)\tan \beta \\ \text{or for suitable coefficients } A, B, C, D \\ Ax + By + Cz = D \end{array}\right\} \quad (7.2)$$

This is the *general equation of a plane.* Note that a direct comparison between

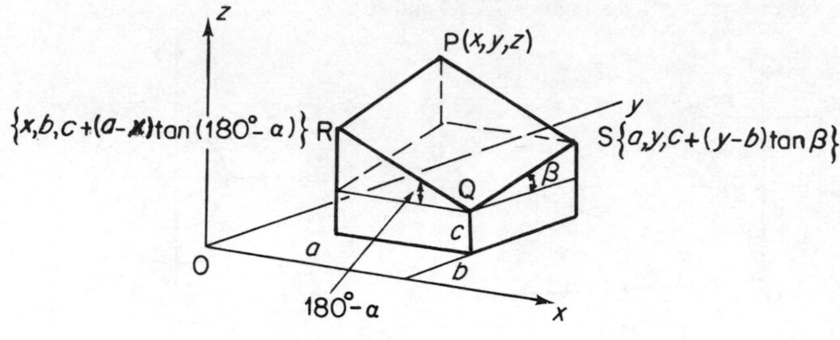

Figure 7.9

these last two equations would give $A = \tan \alpha$, $B = \tan \beta$, $C = -1$, $D = -c + a \tan \alpha + b \tan \beta$. However, the equation of a plane is often given in the latter form and in any particular problem it may not be necessary to determine α and β.

Problems

1. Sketch the sets of points in 3 dimensions which satisfy

   (i) $3x - y = 0$ (ii) $2x + y = 0$

   (iii) $2x + y = 0,\ z = 0$ (iv) $x^2 + y^2 = 36$

   (v) $x^2 + y^2 = 36,\ z = 2$

2. Describe geometrically the functions

   (i) $z = x^2 - y^2$ (ii) $z = \sqrt{x^2 + y^2 - 2x + 1}$

   (iii) $z = \sqrt{x^2 + y^2 - 2x + 2}$ (iv) $z = 1 + \sqrt{4x - x^2 - y^2}$

   On what range of values is each of these functions defined?

3. Evaluate $f(1, 2)$, $f(-2, -1)$, $f(0, 0)$ for

   (i) $f(x, y) = x^2 + 3y^2 - x + 1$ (ii) $f(x, y) = xy/(x^2 + y^2)$

   (iii) $f(x, y) = \log(1 + x^2 + 2y)$ (iv) $f(x, y) = 3x\,e^y + y^2 e^{2x}$

4. Find the points which lie on the intersection of

   (i) $x = 2,\ y = 3$ (ii) $x = 0,\ z = 0$

   (iii) $x + y = 2,\ x + z = 1$

5. Let two planes have equations

$$ax + by + cz = d$$
$$a'x + b'y + c'z = d'$$

   Under what conditions do they intersect? Interpret geometrically.

6. Sketch the contours $z = 0(1)4$ of the surface $z = x^2 + 2y^2 - 3x + 1$ and hence describe the function.

7. Consider the intersection of $z = x^2 + y^2 + 2y + 1$ with the planes $x = 1$, $x = 2, x = a$; $y = 1, y = 2, y = b$; $x + y + z = 1$.

8. Where does the plane $ax + by + cz = d$ intersect the three axes?

9. The density of a gas at the point (x, y, z) at time t is given by $\rho(x, y, z, t) = (2x + 3y - z)\sin t$. Evaluate $\rho(1, -2, 1, 4)$.

7.3 PARTIAL DIFFERENTIATION

Recalling the problem of curve fitting by least squares, we seek the minimum of $S(a_0, a_1)$ and this would indicate some kind of differentiation, which in its turn would suggest the need for limit of a function. This calls for caution since we can approach the point (x_0, y_0, z_0) from a point (x, y, z) along the surface of $f(x, y)$ by a *whole host* of routes, not just by two as for a function of one variable. (See Problems 1 and 2 at the end of this section.)

We can say that the condition for continuity of a function at (a, b) is

$$\lim_{(x, y)\to(a, b)} f(x, y) = f(a, b)$$

and we take the symbol '$\to$' to imply *by any route whatsoever*; [we assume that $f(a, b)$ exists and is finite]. Geometrically, for a function which is continuous over some range of values of x and y, the surface which represents it has no holes in that range nor any vertical cliff faces.

Regarding differentiation as being bound up with some slope, we ask: what slope do we look for at a point on the surface? If we take a short needle and place its centre on the point in question then it can indicate an infinite number of slopes depending on its direction relative to the x and y axes. Since the function is defined in terms of x and y we take two principal slopes at a point: the first obtained by keeping x fixed and varying y, and the second by keeping y fixed and varying x.

In Figure 7.10(i), by proceeding along the surface keeping our value of x fixed we have effectively a function of y only and we can define the slope at (x_0, y_0) in the obvious way as follows: *assuming that the limit exists, the* **partial derivative** *at (x_0, y_0) of $f(x, y)$ with respect to y* [written $f_y(x_0, y_0)$ or

$$\left.\frac{\partial f}{\partial y}\right|_{(x_0, y_0)}$$

is equal to

$$\lim_{k\to 0} \frac{f(x_0, y_0 + k) - f(x_0, y_0)}{k} \tag{7.3a}$$

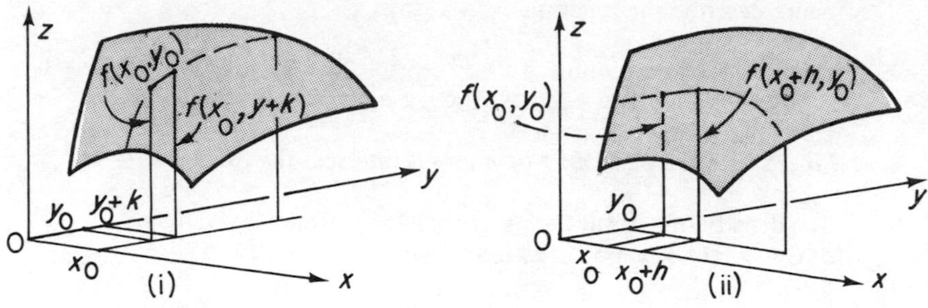

Figure 7.10

In a similar way, from Figure 7.10(ii) we may define $f_x(x_0, y_0)$ as

$$\lim_{h \to 0} \frac{f(x_0 + h, y_0) - f(x_0, y_0)}{h} \tag{7.3b}$$

You should check out this definition on Figure 7.10(ii).

In practice, to differentiate partially with respect to x, say, we treat y, wherever it occurs, as a constant. Conversely, to differentiate partially with respect to y, we treat x everywhere as a constant.

Examples

1. $z = ax^2 + 2hxy + by^2 + 2gx + 2fy + c$

$$f_x \equiv z_x \equiv \frac{\partial z}{\partial x} = 2ax + 2hy + 2g$$

$$f_y \equiv z_y \equiv \frac{\partial z}{\partial y} = 2hx + 2by + 2f$$

2. $z = x^2\sqrt{x^2 + y^2}$

$$\frac{\partial z}{\partial x} = 2x\sqrt{x^2 + y^2} + x^2 \cdot \frac{1}{2} \frac{2x}{\sqrt{x^2 + y^2}}$$

$$z_y \equiv \frac{\partial z}{\partial y} = x^2 \cdot \frac{1}{2} \frac{2y}{\sqrt{x^2 + y^2}} \equiv f_y(x, y)$$

$$\text{e.g.} \quad f_y(1, 2) = 1 \cdot \frac{1}{2} \cdot \frac{2 \cdot 2}{\sqrt{5}} = \frac{2}{\sqrt{5}}$$

Continuity Theorem

An application of the previous result is the **continuity theorem** of fluid flow. On the macro-scale the theorem says that in some region of the field of flow, the amount of fluid entering the volume of the region in a given time must balance that amount which flows out of that volume in the same time (unless build-up of fluid occurs). On the small scale, it gives a necessary relationship between components of velocity. Let the flow be two-dimensional, i.e. no flow in the

z-direction and let the x and y components of velocity be $u(x, y)$ and $v(x, y)$ respectively. We take an imaginary cuboidal region (as shown in Figure 7.11) of the space in which the fluid is flowing and consider the flow through its faces.

We know there is no flow through the top and bottom faces and we first concentrate on the flow in the x-direction; refer to Figure 7.11 (i).

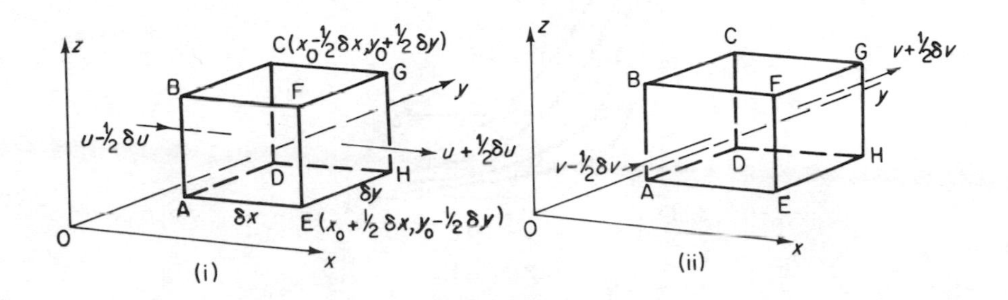

Figure 7.11

The centre of the cuboid is at (x_0, y_0) and it is of height 1. We assume that the dimensions δx, δy are small. Then the flow into the region through the face ABCD (of area $\delta y\,.\,1$) depends on the x-component of velocity at that face; the x-coordinate of points on ABCD is $(x_0 - \frac{1}{2}\delta x)$ but the y-coordinates vary slightly. We assume that the average velocity in the x-direction is $u(x_0 - \frac{1}{2}\delta x, y_0)$, i.e. that at the centre of the face. Then the flow rate through ABCD is $u(x_0 - \frac{1}{2}\delta x, y_0).\delta y.1$; likewise the flow rate out through EFGH is $u(x_0 + \frac{1}{2}\delta x, y_0)\delta y.1$, which gives a net flow out of the cuboid in the x-direction of $[u(x_0 + \frac{1}{2}\delta x, y_0) - u(x_0 - \frac{1}{2}\delta x, y_0)]\,\delta y.1$. A similar argument applies to the flow rate in the y-direction [Figure 7.11(ii)] and you should see that we can obtain the net flow rate out of the cuboid (which is, of course, zero) as

$$[u(x_0 + \tfrac{1}{2}\delta x, y_0) - u(x_0 - \tfrac{1}{2}\delta x, y_0)]\,.\delta y.1\ +$$
$$[v(x_0, y_0 + \tfrac{1}{2}\delta y) - v(x_0, y_0 - \tfrac{1}{2}\delta y)]\,.\delta x.1 = 0$$

Dividing by $\delta x \delta y\,.1$ (the volume of the cuboid) we obtain the equation

$$\frac{u(x_0 + \frac{1}{2}\delta x, y_0) - u(x_0 - \frac{1}{2}\delta x, y_0)}{\delta x} + \frac{v(x_0, y_0 + \frac{1}{2}\delta y) - v(x_0, y_0 - \frac{1}{2}\delta y)}{\delta y} = 0$$

If we now take the limiting case when δx and $\delta y \to 0$ simultaneously (what happens to the cuboid?) we obtain $u_x(x_0, y_0) + v_y(x_0, y_0) = 0$ or, more generally,

$$\frac{\partial u}{\partial x} + \frac{\partial v}{\partial y} = 0 \qquad (7.4)$$

This equation means that the velocity components u and v are not independent and knowledge of one of them helps to determine the other.

Example†

For a particular flow of a frictionless fluid past a corner (see Figure 7.12) it is known that the velocity in the x-direction is $u = 3x$.

What is the value of the total speed of the fluid at any point?

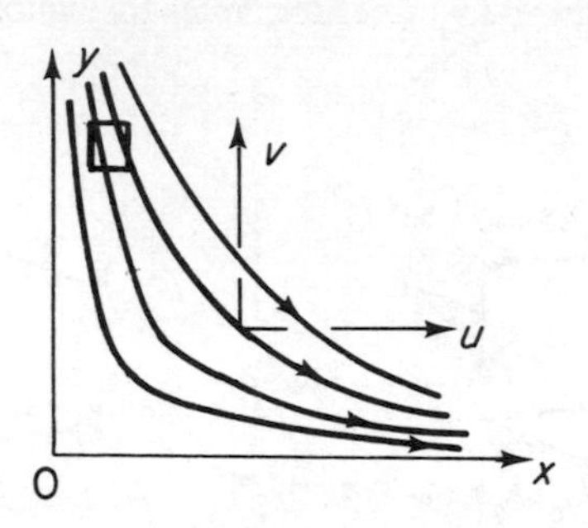

Figure 7.12

We have $\partial u/\partial x = 3$ and therefore, by (7.4), $\partial v/\partial y = -3$.

Now, integrating this last equation it is not enough for us to say that $v = -3y$ as $v = -3y + 2$ or $v = -3y + x + 2$ or, indeed, $v = -3y + f(x)$, where $f(x)$ is any (differentiable) function of x, will all give $\partial v/\partial y = -3$. However, for the flow specified, it can be shown that $v = -3y$ and the *total speed* is the magnitude of the velocity vector $\mathbf{w} \equiv (u, v) = (3x, -3y)$, that is, $\sqrt{9x^2 + 9y^2} = 3(x^2 + y^2)^{1/2}$.

Deduce some properties of the flow from the information to hand by considering a rectangular element of fluid approaching the corner from the position shown in Figure 7.12.

Higher Derivatives

By differentiating z_x partially with respect to x, we obtain

$$z_{xx} \text{ or } \frac{\partial^2 z}{\partial x^2} \equiv \frac{\partial}{\partial x}\left(\frac{\partial z}{\partial x}\right)$$

(provided the necessary limits exist). In a similar fashion we can obtain the other second derivatives z_{xy}, z_{yx}, z_{yy} where the first suffix refers to the first differentiation; the alternative notations are

$$\frac{\partial^2 f}{\partial y \partial x}, \frac{\partial^2 f}{\partial x \partial y}, \frac{\partial^2 f}{\partial y^2}$$

In **most** cases $z_{xy} \equiv z_{yx}$. Likewise, in most cases, there are *four* third partial derivatives,

$$z_{xxx}, z_{yxx}, z_{xyy}, z_{yyy} \quad \text{or} \quad \frac{\partial^3 z}{\partial x^3}, \frac{\partial^3 z}{\partial x^2 \partial y}, \frac{\partial^2 z}{\partial y^2 \partial x}\ \frac{\partial^3 z}{\partial y^3}$$

---

† If you have not yet studied integration, postpone this example until after reading Chapter 8.

Notation

In thermodynamics, partial differentiation plays an important role. The **entropy** S of a gas is given by

$$S = C_v \log p + C_p \log v + A \tag{7.5a}$$

where C_p, C_v and A are constants. We can substitute for v from the gas law $pv = RT$ to obtain

$$S = (C_v - C_p)\log p + C_p \log T + B \tag{7.5b}$$

where B is a constant.

Suppose we calculate $\partial S/\partial p$ from equation (7.5a), then v is held constant and we obtain

$$\frac{\partial S}{\partial p} = \frac{C_v}{p}$$

To find $\partial S/\partial p$ from equation (7.5b) means that T is held constant and we obtain

$$\frac{\partial S}{\partial p} = \frac{C_v - C_p}{p}$$

The difficulty here is that the notation $\partial S/\partial p$ is not sufficiently precise. For equation (7.5a) we write

$$\left(\frac{\partial S}{\partial p}\right)_v$$

to indicate that v is held constant and for equation (7.5b) we write

$$\left(\frac{\partial S}{\partial p}\right)_T$$

Examples

1. $z = ax^2 + 2hxy + by^2 + 2gx + 2fy + c$

   From page 390 we find $z_{xx} = 2a,\ z_{xy} = 2h = z_{yx},\ z_{yy} = 2b$

2. $z = x^2\sqrt{x^2 + y^2}$

   From page 390

$$z_x = (x^2 + y^2)^{-1/2}(3x^3 + 2y^2x)$$
$$z_y = x^2y(x^2 + y^2)^{-1/2}$$

Now

$$\begin{aligned} z_{xx} &= (x^2 + y^2)^{-1/2}(9x^2 + 2y^2) - \tfrac{1}{2}(x^2 + y^2)^{-3/2}.2x(3x^3 + 2y^2x) \\ &= (x^2 + y^2)^{-3/2}[(9x^2 + 2y^2)(x^2 + y^2) - 3x^4 - 2x^2y^2] \\ &= (x^2 + y^2)^{-3/2}(6x^4 + 9x^2y^2 + 2y^4) \end{aligned}$$

Likewise, as you can verify,

$$z_{xy} = xy(x^2 + 2y^2)(x^2 + y^2)^{-3/2} = z_{yx}$$
$$z_{yy} = x^4(x^2 + y^2)^{-3/2}$$

Problems

1. (i) Find $\displaystyle\lim_{(x,y)\to(1,2)} \frac{x}{x^2+y^2+2}$

   (ii) Find $\displaystyle\lim_{(x,y)\to(1,2)} \frac{(x-1)^3}{(x-1)^2+(y-2)}$

   (a) by first letting $x \to 1$ then letting $y \to 2$
   (b) by reversing the order of taking limits.

2. Find $\displaystyle\lim_{(x,y)\to(0,0)} \frac{x^2-y^2}{x^2+y^2}$ by
   (a) $x \to 0$ then $y \to 0$
   (b) $y \to 0$ then $x \to 0$
   (c) along line $y = x$
   (d) along $y = \frac{1}{2}x$
   (e) along $y = x^2$
   (f) put $x = r\cos\theta$, $y = r\sin\theta$ and let $r \to 0$.

3. By converting to polar coordinates find the limits as $(x, y) \to (0, 0)$ of

   (i) $\dfrac{x^2y^2}{x^2+y^2}$ (ii) $\dfrac{x+2y}{xy}$

   (iii) $\dfrac{1}{x^2+y^2}$ (iv) $(x^2+y^2)/xy$

4. Investigate the continuity of

   (i) $\dfrac{x}{x^2+y^2+1}$ (ii) $\dfrac{x^2y^2}{x^2+y^2}$

   (iii) $\dfrac{x-y}{x+y}$ (iv) $\dfrac{x+y}{x-y}$

   (v) $f(x, y) = \begin{cases} \dfrac{x^3y^3}{x^2+y^2} & (x, y) \neq (0, 0) \\ 10 & (x, y) = (0, 0) \end{cases}$

   What value could be given to $f(0, 0)$ in part (ii) to make $f(x, y)$ continuous?

5. Find the first partial derivatives of

   (i) $x^2+y^3+2z^2$ (ii) x^2y^3/z
   (iii) $(1+x^2y)e^{3z}$ (iv) $\sin(xz+y)$
   (v) $e^{(xy+2y^2)}$ (vi) $\frac{1}{2}xy\sin z$
   (vii) $\tan^{-1}(x/y)$ (viii) $(x/y^2)-(y/x^2)$

6. Find the first and second partial derivatives of the following functions and evaluate them at $(0, 0)$, $(1, 2)$, $(-1, 0)$.

(i) $x^3 y^4$ (ii) x^2/y^3

(iii) $2x \cos y$ (iv) $\sin(x + y)$

(v) $2y$ (vi) $e^x + \cos y$

(vii) $(1 + x^2 y)e^{3y}$ (viii) $\tan^{-1}(y/3x)$

(ix) $\sqrt{x^2 - y^2}$

7. (i) Show that $\phi = Ae^{-\frac{1}{2}kt} \sin pt \cos qx$ satisfies the equation

$$\frac{\partial^2 \phi}{\partial x^2} = \frac{1}{c^2}\left(\frac{\partial^2 \phi}{\partial t^2} + k\frac{\partial \phi}{\partial t}\right)$$

provided that $p^2 = c^2 q^2 - \frac{1}{4}k^2$

(ii) Show that $V = (Ar^n + Br^{-n})\cos(n\theta - \alpha)$, where A, B, n, α are constants, satisfies the equation

$$\frac{\partial^2 V}{\partial r^2} + \frac{1}{r}\frac{\partial V}{\partial r} + \frac{1}{r^2}\frac{\partial^2 V}{\partial \theta^2} = 0$$

(iii) Find values of the parameter n so that

$$V = r^n(3\cos^2\theta - 1)$$

satisfies

$$\frac{\partial}{\partial r}\left(r^2 \frac{\partial V}{\partial r}\right) + \frac{1}{\sin\theta}\frac{\partial}{\partial \theta}\left(\sin\theta \frac{\partial V}{\partial \theta}\right) = 0$$

8. (i) If $y = \phi(x - ct) + \psi(x + ct)$ where ϕ, ψ are arbitrary functions, show that

$$\frac{\partial^2 y}{\partial x^2} = \frac{1}{c^2}\frac{\partial^2 y}{\partial t^2}$$

(ii) If $z = xF(y/x) + f(y/x)$, F and f being arbitrary functions, show that

$$x\frac{\partial z}{\partial x} + y\frac{\partial z}{\partial y} = z - f$$

and deduce that

$$x^2 z_{xx} + 2xyz_{xy} + y^2 z_{yy} = 0$$

9. (i) If $V = \dfrac{x^3 y^3}{x^3 + y^3}$, show that

(a) $x\dfrac{\partial V}{\partial x} + y\dfrac{\partial V}{\partial y} = 3V$

(b) $x^2\dfrac{\partial^2 V}{\partial x^2} + 2xy\dfrac{\partial^2 V}{\partial x \partial y} + y^2\dfrac{\partial^2 V}{\partial y^2} = 6V$

(ii) If $z = xf(y/x) + g(x/y)$, show that $x\dfrac{\partial z}{\partial x} + y\dfrac{\partial z}{\partial y} = xf(y/x)$. (L.U.)

10. Find $T\left(\dfrac{\partial P}{\partial T}\right)_V - P$ for

(a) the ideal gas law $PV = RT$; R constant

(b) van der Waal's law $[P + (a/V^2)]\,[V - b] = RT$; a, b and R constants.

11. At time t, the displacement y of a point at distance x from one end of a vibrating string is given by $y_{tt} = c^2 y_{xx}$ where c is a constant. Show that this equation is satisfied by

$$y = B \sin px \,.\, \sin(cpt + a)$$

where B, p and a are constants.

7.4 STATIONARY POINTS ON THE SURFACE REPRESENTING f(x, y)

Suppose, in theory, we take a small abstract coin (small in relation to surface changes) and move it on the surface representing $f(x, y)$. At the tops of rounded hills (*local maxima*) and the bottoms of rounded valleys (*local minima*), it will be horizontal; it will be above the surface at a local maximum and below it at a local minimum. We need to find a necessary condition for the coin to be horizontal at a point (x_0, y_0); this is that the slopes at (x_0, y_0), $\partial f/\partial x$ and $\partial f/\partial y$, are **both** zero. If we are concerned with small movements on the surface away from (x_0, y_0) we can approximate the function by a plane which is tangential to the surface at (x_0, y_0). In Figure 7.9 we may take Q to be (x_0, y_0, z_0) and recast equation (7.2) to give

$$f(x_0+h, y_0+k) \simeq f(x_0, y_0) + h\left.\frac{\partial f}{\partial x}\right|_{(x_0,\, y_0)} + k\left.\frac{\partial f}{\partial y}\right|_{(x_0,\, y_0)} \tag{7.6}$$

This is the **tangent plane approximation** to $f(x, y)$ at (x_0, y_0) and has obvious similarity to the one-variable case.

We shall be interested in the change in $f(x, y)$ as we move the small distance from (x_0, y_0) to (x_0+h, y_0+k). This change is

$$\delta f \equiv f(x_0+h, y_0+k) - f(x_0, y_0)$$

and we therefore have

$$\delta f \simeq hf_x(x_0, y_0) + kf_y(x_0, y_0) \tag{7.7}$$

We return to this idea in section 7.6.

For the moment, note that if

$$\left.\frac{\partial f}{\partial x}\right|_{(x_0,\,y_0)} = 0 = \left.\frac{\partial f}{\partial y}\right|_{(x_0,\,y_0)}$$

then the tangent plane is horizontal at (x_0, y_0). (Try for yourself by experiment.) A necessary condition for (x_0, y_0) to be a stationary point of $f(x, y)$ is that

$$f_x(x_0, y_0) = 0 = f_y(x_0, y_0) \qquad (7.8)$$

Notice the connection between different notations.

A *local maximum* occurs when, holding x_0 fixed, the intersection of the plane $x = x_0$ with the surface shows a local maximum for the resulting function of y, **and** the intersection of the plane $y = y_0$ shows a local maximum for the resulting function of x. A *local minimum* is defined similarly.

There is a third case where we obtain a local maximum in one direction and a local minimum in the other; such a point is known as a **saddle point**, a name which you will see is clearly justified if you examine a saddle or, failing that, a camel's back. (There are other possible unusual features which satisfy $z_x = z_y = 0$ but we shall not deal with them.)

Example 1

Show that the function $f(x, y) = 2x^2 + 2xy - y^3$ has stationary points at $(0, 0)$ and $(\frac{1}{6}, -\frac{1}{3})$ and determine their nature.

Now $\qquad f_x = 4x + 2y; \quad f_y = 2x - 3y^2$

Stationary points occur where $4x + 2y = 0$ and $2x - 3y^2 = 0$; then $y = -2x$, and, substituting in the second relationship we obtain $2x - 12x^2 = 0$, i.e. $2x(1 - 6x) = 0$ from which the stationary points follow (always be careful not to miss any).

Now keep

$$x = \frac{1}{6}$$

so that

$$f\left(\frac{1}{6}, y\right) = \frac{1}{18} + \frac{1}{3}y - y^3 = g(y), \text{ say}$$

Then $g'(y) = (1/3) - 3y^2$ and this is zero when $y = -1/3$; further, $g''(y) = -6y$ and hence $g''(-\frac{1}{3}) > 0$ indicating a local minimum. Now put $y = -1/3$ so that

$$f\left(x, -\frac{1}{3}\right) = 2x^2 - \frac{2}{3}x + \frac{1}{27} = h(x), \text{ say.}$$

Then $h'(x) = 4x - (2/3)$, $h''(x) = 4 > 0$ indicating a local minimum. Hence $(\frac{1}{6}, -\frac{1}{3})$ is a local minimum; but *you* see what troubles arise with $(0, 0)$.

A criterion which involves second derivatives is stated without proof in Problem 3; the criterion follows from the Taylor's series for $f(x, y)$.

Example 2

$f(x, y) = 3(x^2 + y^2)^{1/2}$

Now $\sqrt{x^2 + y^2}$ represents the distance of the point (x, y) on the x-y plane from

the origin. This is clearly least when $x = y = 0$ but has **no** maximum value.
For reference

$$f_x = \frac{3x}{(x^2 + y^2)^{1/2}}, \quad f_y = \frac{3y}{(x^2 + y^2)^{1/2}}$$

What implications does this have in respect of the flow in a corner problem on page 392.

Problems

1. Find the equation of the tangent plane to the surface $x^2yz + 3y^2 = 2xz^2 - 8z$ at the point $(1, 2, -1)$. Show that the equation of the normal to the surface at the same point can be expressed parametrically in the form $x = 1 - 6t$, $y = 2 + 11t$, $z = 14t - 1$. (C.E.I.)

2. Find the location and nature of the stationary points of
   (i) $z = 34x^2 - 24xy + 41y^2$ (ii) $z = x^2 + y^2 + 6x - 4y + 25$
   (iii) $z = x^3 + 4x^2 + 3y^2 + 5x - 6y$ (iv) $z = x^4 + y^4 + 4xy$
   (v) $z = \log(x^2 + y^2) - x - 2y$ (vi) $z = x + xy^2 - y - x^2y$
   (vii) $z = 1/(x^2 + y^2 + 4)$ (viii) $z = xy(4x + 2y + 1)$
   (ix) $z = x^4 + y^4 + 2x^2y^2$

3. The criterion for determining the nature of stationary points at (a, b) may be stated in terms of

$$D \equiv f_{xx}(a, b)\, f_{yy}(a, b) - [f_{xy}(a, b)]^2$$

   $D > 0 \Rightarrow$ local maximum or local minimum
   $f_{xx}(a, b) > 0 \Rightarrow$ local minimum, $f_{xx}(a, b) < 0 \Rightarrow$ local maximum
   $D < 0 \Rightarrow$ saddle point
   $D = 0$ needs further investigation.

   Carry out these tests on (i), (ii), (iii), (iv), (vi), (viii), (ix) of Problem 2.

4. (i) Find the stationary values of $x^3 + ay^2 - 6axy$ where $a > 0$ and determine whether they are maxima, minima or saddle points.
   (ii) Prove that the function $xy(3 - x - y)$ has a maximum when $x = y = 1$.
   (iii) Find the greatest value of the function $(x^2 - y^2)\exp(-x^2 - 2y^2)$. (L.U.)

5. Derive the expression

$$\frac{|ax_1 + by_1 + cz_1 + d|}{\sqrt{(a^2 + b^2 + c^2)}}$$

   for the shortest distance, D, of the point (x_1, y_1, z_1) from the plane

$ax + by + cz + d = 0$.

[Hint: Consider $D^2 = (x - x_1)^2 + (y - y_1)^2 + (z - z_1)^2$; use the equation of the plane to substitute for z and obtain an expression $g(x, y)$. Find the least value of $g(x, y)$.]

Find the point nearest to the origin which lies on the plane $x - 2y + 2z = 16$.

6. (i) The trough of uniform cross-section shown in Figure 7.13 (i) is to have cross-sectional perimeter l. Find the dimensions for maximum cross-sectional area.

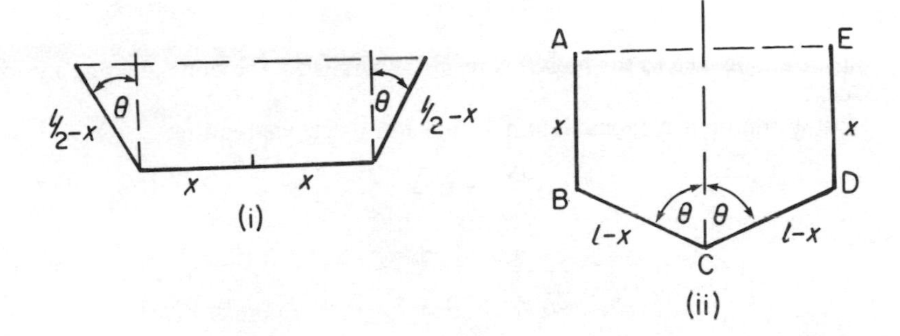

Figure 7.13

(ii) A rectangular sheet of metal of width $2l$ is bent to form a trough without ends. The cross-section is a polygon ABCDE as shown in Figure 7.13 (ii). Prove that as x and θ vary, the maximum value of the cross-sectional area is $l^2/\sqrt{3}$.

7.5 RETURN TO LEAST SQUARES CURVE FITTING

Referring back to page 384, to fit a straight line $y = a_0 + a_1 x$ to n pairs of points (x_i, y_i) we had an overall squared error

$$S = \sum_1^n (y_i - a_0 - a_1 x_i)^2; \quad S = S(a_0, a_1)$$

We use the result that the derivative of a sum is the sum of the separate derivatives to find the values of a_0 and a_1 which minimise S.

Now

$$\frac{\partial S}{\partial a_0} = \sum_1^n 2(y_i - a_0 - a_1 x_i).(-1).$$

and

$$\frac{\partial S}{\partial a_1} = \sum_1^n 2(y_i - a_0 - a_1 x_i).(-x_i)$$

Consider the case $n = 2$, then

$$S = (y_1 - a_0 - a_1 x_1)^2 + (y_2 - a_0 - a_1 x_2)^2$$

and

$$\frac{\partial S}{\partial a_1} = 2(y_1 - a_0 - a_1 x_1)(-x_1) + 2(y_2 - a_0 - a_1 x_2)(-x_2)$$

You should be able to see how this result generalises; the same ideas apply for $\partial S/\partial a_0$.

For a minimum S (a maximum S does not exist) we require

$$\frac{\partial S}{\partial a_0} = 0 = \frac{\partial S}{\partial a_1}$$

i.e.

$$\sum_1^n (y_i - a_0 - a_1 x_i) = 0 = \sum_1^n x_i(y_i - a_0 - a_1 x_i)$$

Hence

$$\left.\begin{aligned} \sum y_i &= \sum a_0 + \sum a_1 x_i \\ \sum x_i y_i &= \sum x_i a_0 + \sum a_1 x_i^2 \end{aligned}\right\}$$

or

$$\left.\begin{aligned} \sum y_i &= n a_0 + a_1 \sum x_i \\ \sum x_i y_i &= a_0 \sum x_i + a_1 \sum x_i^2 \end{aligned}\right\} \qquad (7.9)$$

These are called the **normal equations.**

All we need to do is to evaluate each of the sums to produce numbers and then solve these two equations for a_0 and a_1.

Similar analysis shows that the normal equations for a quadratic fit are

$$\left.\begin{aligned} \sum y_i &= n a_0 + a_1 \sum x_i + a_2 \sum x_i^2 \\ \sum x_i y_i &= a_0 \sum x_i + a_1 \sum x_i^2 + a_2 \sum x_i^3 \\ \sum x_i^2 y_i &= a_0 \sum x_i^2 + a_1 \sum x_i^3 + a_2 \sum x_i^4 \end{aligned}\right\} \qquad (7.10)$$

Example 1

The readings in Table 7.1 were obtained for the specific heat of ethyl alcohol

Table 7.1

x temperature °C	0	10	20	30	40	50
y specific heat	0.50680	0.54544	0.56617	0.58743	0.62984	0.66330

Now

$$\sum y_i = 0.50680 + \ldots + 0.66330 = 3.49898$$

$$\sum x_i = 150$$

$$\sum x_i y_i = 92.7593$$

$$\sum x_i^2 = 5500$$

Hence equations (7.9) give $\begin{cases} 3.49898 = 6a_0 + 150a_1 \\ 92.7593 = 150a_0 + 5500a_1 \end{cases}$

and solving we find

$$a_0 = 0.50766; \quad a_1 = 0.00311$$

Thus the straight line fit is

$$y = 0.50766 + 0.00311x$$

If we now find a quadratic fit $y = a_0 + a_1 x + a_2 x^2$ then

$$\sum x_i^3 = 225\,000$$

$$\sum x_i^2 y_i = 3474.693$$

$$\sum x_i^4 = 979 \times 10^4$$

which gives from equations (7.10)

$$\begin{cases} 3.48898 = 6a_0 + 150a_1 + 5500a_2 \\ 92.7593 = 150a_0 + 5500a_1 + 225\,000a_2 \\ 3474.693 = 5500a_0 + 225\,000a_1 + 979 \times 10^4 a_2 \end{cases}$$

Solving,

$$\begin{cases} a_0 = 0.50807 \\ a_1 = 0.00247 \\ a_2 = 0.00001 \end{cases}$$

Hence

$$y = 0.50807 + 0.00247x + 0.00001x^2$$

We can similarly fit a cubic to obtain

$$y = 0.50705 + 0.00294x - 0.00001x^2 + 0.00003x^3$$

Remember that we shall *always* be able to find a curve of best fit (no matter how poor a fit it is) for each class of curve.

Example 2

Let us investigate other troubles that may arise.

Suppose we take the relationship $y = 1 + 2x + x^2$ and generate the data points of Table 7.2.

Table 7.2

	x	y	x^2	xy	x^2y	x^3	x^4
	1	4	1	4	4	1	1
	2	9	4	18	36	8	16
	3	16	9	48	144	27	81
	4	25	16	100	400	64	256
	5	36	25	180	900	125	625
$\sum$	15	90	55	350	1484	225	979

Fitting a straight line $y = a_0 + a_1 x$ we obtain

$$\begin{cases} 90 = 5a_0 + 15a_1 \\ 350 = 15a_0 + 55a_1 \end{cases}$$

and hence

$$y = -6 + 8x$$

Fitting a quadratic $y = a_0 + a_1 x + a_2 x^2$ yields

$$\begin{cases} 90 = 5a_0 + 15a_1 + 55a_2 \\ 350 = 15a_0 + 55a_1 + 225a_2 \\ 1484 = 55a_0 + 225a_1 + 979a_2 \end{cases}$$

and hence

$$y = 1.00134 + 1.99145x + 1.01105x^2$$

[compare the true polynomial $y = 1 + 2x + x^2$]

Fitting a cubic provides

$$y = 1.00085 + 2.00125x + 1.00134x^2 + 0.00003x^3$$

and fitting a quartic

$$y = 1.00102 + 1.99913x + 0.99935x^2 + 0.00013x^3 + 0.00007x^4$$

Were we to try to fit a quintic we should expect that since we have only *five data points* to find *six coefficients* we would get infinitely many answers, but from the process we would, in fact, get *one* answer.

The trouble here is **round-off error.** If we had worked with *exact* arithmetic we should have got *exactly* $y = 1 + 2x + x^2$.

A flowchart is given in Figure 7.14 for the fitting of a linear relationship $y = a + bx$ to n pairs of observed x and y values.

We read the data into two one-dimensional arrays, calculate the values of a and b and write them out. Arrays are used in case the x and y values might be needed for calculations (these are omitted in this discussion).

The sum Σx will be represented by SX, Σy by SY, Σx^2 by SXX, Σxy by SXY.

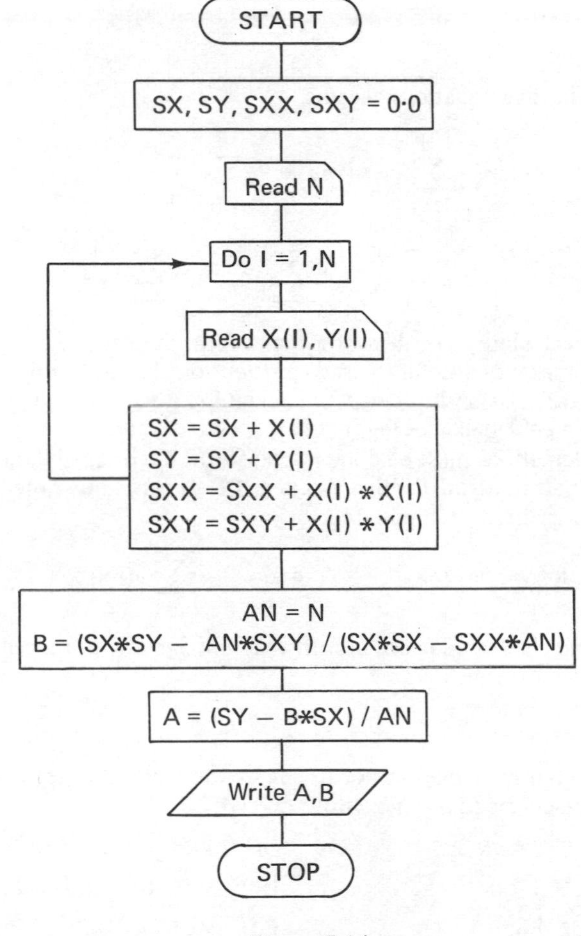

Figure 7.14

Now write the Fortran program from this flow chart.

Often we need to fit curves other than polynomials. The normal equations are then more difficult to calculate. Suppose, for example, we wish to fit the curve $y = a + (b/x)$ to n data points (x_i, y_i). The quantity to be minimised is

$$S = \sum_{i=1}^{n} (y_i - a - b/x_i)^2$$

Now

$$\frac{\partial S}{\partial a} = -\sum_{i=1}^{n} 2(y_i - a - b/x_i)$$

and

$$\frac{\partial S}{\partial b} = -\sum_{i=1}^{n} \frac{2}{x_i}(y_i - a - b/x_i)$$

and hence the normal equations become

$$\sum_{i=1}^{n} y_i = na + b\sum_{i=1}^{n} \frac{1}{x_i}$$

$$\sum_{i=1}^{n} \frac{y_i}{x_i} = a\sum_{i=1}^{n} \frac{1}{x_i} + b\sum_{i=1}^{n} \frac{1}{x_i^2}$$

We must then calculate y_i/x_i, $1/x_i$, $1/x_i^2$ and form their sums.

It is sometimes more useful to use *logarithms* on the relationship to be fitted. For example, the relationship $y = ax^n$ becomes $\log y = \log a + n \log x$ and the relationship $y = ae^{bx}$ becomes $\log_e y = \log_e a + bx$.

Clearly, calculations must be done to transform the original data into a suitable form before fitting the line Y = A + BX. It should be noted that in these cases the quantity being minimised is

$$\sum_{i=1}^{n} (\log y_i - \log a - n \log x_i)^2 \qquad \text{and} \qquad \sum_{i=1}^{n} (\log y_i - \log a - bx_i)^2$$

respectively and the least squares best fit may not agree with that obtained from fitting the curve directly.

Problems

1. Find the normal equations for fitting by the method of least squares the following curves to n observations (x_i, y_i):

   (i) $y = a + bx^3$ (ii) $y = a + bx^2 + cx^3$

   (iii) $y = 1 + ax$ (iv) $y = a\mathrm{e}^{bx}$, a fixed

   (v) $y = \log(a + bx)$ (vi) $y = a/x$

2. When unloaded reinforced concrete columns dry, compressive stresses occur due to shrinkage. The following table shows drying shrinkage in 10^{-6}cm/cm, y, measured against % reinforcement x.

x	0	2	4.5	8.5
y	475	380	190	75

Fit a least squares straight line and a quadratic.

3. The Highway Code gives the values of the distance D travelled by a car before coming to rest from a speed V

V (miles/hr)	20	30	40	50	60
D (ft)	45	75	120	175	240

Find a relationship of the form $D = a + bV + cV^2$ which represents the given data. Comment upon the validity of the expression for speeds below 20 miles/hr. Suggest a simple relation between D and V which could be applied for such speeds. (C.E.I.)

4. Fit by least squares a straight line to the data

x	0.5	0.6	0.7	0.8
y	3.8	2.1	−0.3	−1.7

Reject the point furthest from this line (called the *regression* line of y on x) and determine a new line to fit the remaining data.

5. Write down the appropriate normal equations and hence fit by least squares the following:

(i) $y = ax + \dfrac{b}{x}$ to the data

x	2	3	4	5
y	5.3	7.2	9.0	10.6

(ii) $y = ax^2 + b\sqrt{x}$ to the data

x	1	2	3	4
y	−5.2	−2.0	6.0	17.5

(iii) $y = a/x^2$ to

x	1	2	3	4	5
y	0.0224	0.0055	0.0030	0.0015	0.0010

(a) by direct application
(b) by transforming the equation into a linear relationship.

6. Linearise the relationship $y = a\mathrm{e}^{b/x}$ and fit this to the data

x	1	2	3	4	5
y	30.3	7.41	4.12	3.18	2.73

7. Use finite differences to find the degree of the polynomial which would best fit the data given. Fit a polynomial of this degree by the method of least squares; find and estimate the closeness of fit.

x	1	2	3	4	5
y	2.00	1.18	1.20	2.05	3.75

8. The following pairs of values of S and T were obtained experimentally.

T	−2	−1	0	1	2
S	0	2	4	5	7

Assuming the values of S only to be in error, use the method of least squares to fit to the data

(i) $S = a + bT + cT^2$ (ii) $S = a + bT$

(iii) $S = a + bT + cT^2 + dT^3$

To determine which of these curves fits the data best, evaluate for each curve

$$\Omega = \frac{S_{\min}}{n - m}$$

where $S_{\min}$ is the minimised sum of squares, n is the number of observations and m is the number of coefficients in the relationship governing the curve. The least value of Ω implies the best fit.

9. It is believed that the variables s and t, whose corresponding values are given by the table below, are connected by a relation of the form $s = \lambda t^k$, where λ and k are constants. Use the method of least squares to determine the best values of λ and k, with as much accuracy as the data justifies.

t	1.0	1.2	1.4	1.6	1.8
s	84.45	68.64	55.76	45.33	36.80

(L.U.)

10. Compare least squares fitting of a straight line with linear interpolation.

7.6 SMALL INCREMENTS AND DIFFERENTIALS

If we rewrite relationship (7.7), page 396 in a different notation we have the approximation

$$\delta f \simeq \frac{\partial f}{\partial x}\,\delta x + \frac{\partial f}{\partial y}\,\delta y \tag{7.11}$$

We may imagine that the change δx causes a change in $f(x, y)$, the rate of response being $\partial f/\partial x$; likewise the change δy gives rise to a second change in $f(x, y)$ with response rate $\partial f/\partial y$, and the total change δf is the sum of these two.

We can derive this result in a somewhat different way. Let $u = f(x, y)$ and let

x, y be changed by small amounts, δx, δy; we seek the resulting (small change) in u, i.e. δu.

$$\begin{aligned} \delta u &= f(x+\delta x,\ y+\delta y) - f(x,y) \\ &= f(x+\delta x,\ y+\delta y) - f(x,\ y+\delta y) + f(x,\ y+\delta y) - f(x,y) \\ &= f_x(x,\ y+\delta y).\delta x + f_y(x,y).\delta y + 2^{\text{nd}} \text{ order terms in } \delta x, \delta y \\ &= [f_x(x,y) + 0(\delta y)]\,\delta x + f_y(x,y)\delta y + 2^{\text{nd}} \text{ order terms} \end{aligned}$$

therefore

$$\delta u \simeq f_x \delta x + f_y \delta y$$

We call the quantity

$$\mathrm{d}u = \frac{\partial f}{\partial x}\delta x + \frac{\partial f}{\partial y}\delta y \tag{7.12}$$

the **differential** for the increments δx, δy (sometimes called the **total differential**). We shall see its uses a little later on. For the moment we use the approximate formula (7.12) to estimate errors. Notice the difference between δu and $\mathrm{d}u$.

Example

The area S of triangle ABC is calculated from the formula $S = \frac{1}{2}bc \sin A$; the length b, c and the angle A are measured as 3.00 cm, 4.00 cm and 30° respectively. Find the error and percentage error in S as calculated from the measurements, given that b and c each has a maximum error of ± 0.05 cm and A one of $\pm 0.1°$.

First, we have the result

$$S = \tfrac{1}{2}\,.\,3\,.\,4\,.\,\tfrac{1}{2} = 3 \text{ sq.cm}$$

Now,

$$\delta S = \frac{\partial S}{\partial b}\delta b + \frac{\partial S}{\partial c}\delta c + \frac{\partial S}{\partial A}\delta A \tag{7.13}$$

and

$$\frac{\partial S}{\partial b} = \tfrac{1}{2}c \sin A, \qquad \frac{\partial S}{\partial c} = \tfrac{1}{2}b \sin A, \qquad \frac{\partial S}{\partial A} = \tfrac{1}{2}bc \cos A$$

We compute these rates at the measured values of b, c and A, therefore

$$\frac{\partial S}{\partial b} = \tfrac{1}{2}\,.\,4\,.\,\tfrac{1}{2} = 1, \qquad \frac{\partial S}{\partial c} = \frac{3}{4}, \qquad \frac{\partial S}{\partial A} = 3\sqrt{3}$$

Hence

$$|\delta S| = 1 \times 0.05 + \frac{3}{4}\,.\,0.05 + 3\sqrt{3}\ 0.1(\pi/180)$$

$$= 0.05 \times \frac{7}{4} + 3\sqrt{3} \times \frac{\pi}{1800} \simeq 0.097$$

and this is sufficient precision for our purposes.

Then $S = (3.000 \pm 0.097)$ sq.cm.

To find the percentage error, $\delta S/S \times 100$ we may manipulate (7.13) or take logarithms.

Let

$$u = \log S = \log \tfrac{1}{2} + \log b + \log c + \log \sin A$$

then

$$\delta u \left\{ \simeq \frac{du}{dS}\delta S = \frac{1}{S}\delta S \right\} = 0 + \frac{1}{b}\delta b + \frac{1}{c}\delta c + \frac{\cos A}{\sin A}\delta A$$

Hence

$$\frac{\delta S}{S} \times 100 \simeq \frac{\delta b}{b} \times 100 + \frac{\delta c}{c} \times 100 + \cot A\, \delta A \times 100$$

$$\begin{aligned} \frac{\delta S}{S} \times 100 &\simeq \frac{0.05}{3.00} \times 100 + \frac{0.05}{4.00} \times 100 + \sqrt{3} \times [0.1 \,.\, (\pi/180)] \times 100 \\ &\simeq 1.667 + 1.250 + 0.302 \\ &\simeq \underline{3.219\%} \end{aligned}$$

Differentials and their applications

Given increments δx, δy the differential given by (7.12) is the change in u predicted by the tangent plane approximation. Before proceeding further, we note that $\partial x/\partial x = 1$, $\partial x/\partial y = 0$ if x, y are independent and hence, regarding x as a function of x and y, (equation 7.12) gives

$$dx = \frac{\partial x}{\partial x}\delta x + \frac{\partial x}{\partial y}\delta y = \delta x$$

Similarly we may show $dy = \delta y$, and we may rewrite (7.12) as

$$du = \frac{\partial u}{\partial x}dx + \frac{\partial u}{\partial y}dy \tag{7.14}$$

Example 1

The ideal gas law in thermodynamics is $pV = RT$ where R is constant. Hence $p = RT/V = p(T, V)$. This gives

$$dp = \frac{\partial p}{\partial T}dT + \frac{\partial p}{\partial V}dV = \frac{R}{V}dT - \frac{RT}{V^2}dV$$

Along an isobar, p is constant and hence $dp = 0$ irrespective of changes in V, T and so

$$0 = \frac{R}{V}dT - \frac{RT}{V^2}dV$$

It follows that $dV/dT = V/T$ and this is the differential equation for the family of isobars. (*Please beware:* we have **not** treated dV/dT as a *fraction:* the ideas go much deeper.)

Example 2

A **streamline** is an imaginary line in a fluid across which no flow takes place, i.e. the velocity vector at a point is tangential to the streamline passing through that point [Figure 7.15(i)].

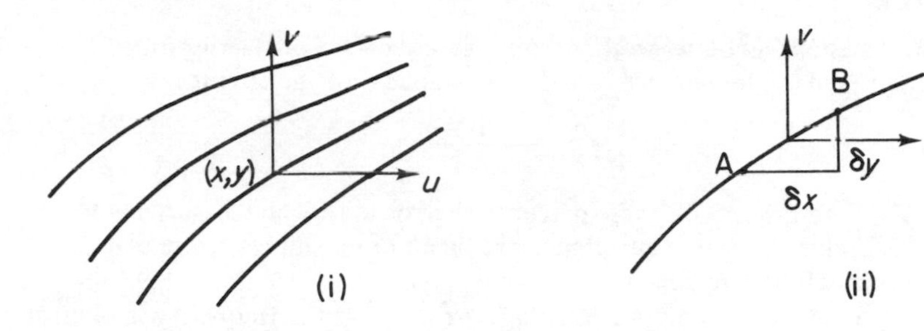

Figure 7.15

In Figure 7.15 (ii) we consider a small portion AB of the streamline, small enough to assume it to be a straight line. The average velocity in the x-direction across AB can be taken as $u(x, y)$, by application of the *mean-value theorem,* and that in the y-direction as $v(x, y)$. In order that the *total* flow rate across AB should be zero, we require $u(x, y)\delta y - v(x, y)\delta x = 0$.

Now suppose the streamline to be a contour line specified by some equation $\psi(x, y) = C$, a constant; then $\psi_B - \psi_A = 0$ and $d\psi = (\partial\psi/\partial x)dx + (\partial\psi/\partial y)dy = 0$. By ψ_B we mean the value of $\psi(x, y)$ evaluated at B.

Now if we compare this with the relationship $u dy - v dx = 0$ it is possible to choose ψ so that $\partial\psi/\partial x = -v$, $\partial\psi/\partial y = u$. (An alternative convention is $\partial\psi/\partial x = v$, $\partial\psi/\partial y = -u$.)

Then ψ is called a *stream function* for the flow.

For the problem of flow in a corner we may take $\psi(x, y) = 3xy$ (show this from the given forms for u and v) and hence the fluid will flow along curves $3xy$ = constant; these are rectangular hyperbolae. It is found that this is in reasonable agreement with experiment, provided we do not go close to the origin.

In general it is then possible under certain restrictions to find an equation satisfied by ψ. We have

$$\frac{\partial u}{\partial x} = \frac{\partial^2 \psi}{\partial x \partial y} \quad \text{and} \quad \frac{\partial v}{\partial y} = -\frac{\partial^2 \psi}{\partial y \partial x}$$

which from the equation of continuity (7.4) ties in with the equality of the mixed second derivatives.

Further,

$$\frac{\partial u}{\partial y} = \frac{\partial^2 \psi}{\partial y^2} \quad \text{and} \quad \frac{\partial v}{\partial x} = -\frac{\partial^2 \psi}{\partial x^2}$$

For flows which are *irrotational*, that is, particles of fluid do not spin about their axes, $\partial u/\partial y = \partial v/\partial x$ (consult a textbook on Fluid Dynamics to see why) and this leads to **Laplace's equation**

$$\frac{\partial^2 \psi}{\partial x^2} + \frac{\partial^2 \psi}{\partial y^2} = 0 \tag{7.15}$$

Problems I

1. The relationship between the resonant frequency f of a series-timed circuit and the inductance L and capacitance C of the circuit is given by

$$f = \frac{1}{2\pi\sqrt{LC}}$$

   (a) Find the approximate percentage error in f if the measurements of inductance and capacitance are liable to maximum errors of 1.25% and 0.75% respectively.
   (b) Find the maximum percentage error in L if the frequency is in error by ±0.5% and the capacitance is liable to an error of ±1.5%.

2. A cylindrical hole of diameter 6 in and height 4 in is to be cut in a block of metal by a process in which the maximum error in diameter is 0.003 in and in height is 0.002 in. What is the largest possible error in the volume of the cavity? (C.E.I.)

3. The breaking weight W of a cantilever beam is given by the formula $Wl = kbd^2$, where b is the breadth, l the length, d the depth and k a constant depending on the material of the beam. If the length is increased by 1% and the breadth by 5%, how much should the depth be altered to keep the breaking weight unchanged?

4. Find the total differential du given that $u = f(x, y)$ where $f(x, y)$ is

   (i) $\frac{1}{x^2 y} + xy$ (ii) $e^{-(x+y)} \sin(x + y)$
   (iii) $\sqrt{x^2 + y^2} \tan^{-1}(y/x)$ (iv) $x^3 y + 3xy^4$
   (v) $e^{(x^2 + y^2)}$ (vi) $\cos(x + 2y)$

5. (i) If $u = \partial\psi/\partial y$ and $v = -\partial\psi/\partial x$ where $u = x + y$ and $v = x - y$, write down the differential $d\psi$. Show that if $d\psi = 0$, then a possible expression for ψ is $y^2/2 + xy - x^2/2$. Sketch the streamlines for the flow represented by the velocities u and v.

   (ii) For the stream function in part (i) show that

$$\frac{\partial^2 \psi}{\partial x^2} + \frac{\partial^2 \psi}{\partial y^2} = 0$$

6. The power consumed in an electrical resistor is given by $P = E^2/R$ watts. If $E = 100$ volts and $R = 10$ ohms, by approximately how much does P change if E is decreased by 2 volts and R is increased by 0.5 ohms? Compare your result with the exact answer.

7. The rate of flow of gas in a pipe is given by $v = Cd^{1/2}\, T^{-5/6}$, where C is a constant, d is the pipe diameter and T is the aboslute gas temperature.

The measurement of d is subject to a maximum error of ±1.6% and that of T to one of ±1.2%. Find approximately the maximum percentage error in the value of v.

Chain Rule

If x and y are both functions of the single variable t then, from basic calculus, $\mathrm{d}x = (\mathrm{d}x/\mathrm{d}t)\mathrm{d}t$ and $\mathrm{d}y = (\mathrm{d}y/\mathrm{d}t)\mathrm{d}t$ and $z(x, y)$ will also be a function of t, therefore $\mathrm{d}z = (\mathrm{d}z/\mathrm{d}t)\mathrm{d}t$.

Then we have the **chain rule**

$$\frac{\mathrm{d}z}{\mathrm{d}t} = \frac{\partial z}{\partial x}\frac{\mathrm{d}x}{\mathrm{d}t} + \frac{\partial z}{\partial y}\frac{\mathrm{d}y}{\mathrm{d}t} \tag{7.16}$$

This yields

$$\mathrm{d}z = \frac{\partial z}{\partial x}\frac{\mathrm{d}x}{\mathrm{d}t}\mathrm{d}t + \frac{\partial z}{\partial y}\frac{\mathrm{d}y}{\mathrm{d}t}\mathrm{d}t = \frac{\partial z}{\partial x}\mathrm{d}x + \frac{\partial z}{\partial y}\mathrm{d}y$$

showing that a similar relationship to (7.12) holds in this case.

We may regard (7.16) in the following light. A change in t causes a change in x with a response rate $\mathrm{d}x/\mathrm{d}t$; this change in x causes a change in z with a response rate $\partial z/\partial x$. Likewise the change in t causes a change in y with a response rate $\mathrm{d}y/\mathrm{d}t$ and this causes a separate change in z with response rate $\partial z/\partial y$. The total response rate of z to the change in t is given by (7.16). We call $\mathrm{d}z/\mathrm{d}t$ the **total derivative** of z with respect to t.

The chain rule may be extended in cases where x and y are both functions of two further variables s and t. For example, suppose $z = x^2 + 2y^2$ where $x = 2s + 3t$, $y = s - 4t^3$. Then

$$\begin{aligned}\frac{\partial z}{\partial s} &= \frac{\partial z}{\partial x}\cdot\frac{\partial x}{\partial s} + \frac{\partial z}{\partial y}\cdot\frac{\partial y}{\partial s}\\ &= 2x.2 + 4y.1\\ &= 12s + 12t - 16t^3\end{aligned}$$

Notice that since $z = z(x, y)$, $x = x(s, t)$ and $y = y(s, t)$, all the derivatives are partial derivatives.

Similarly,

$$\begin{aligned}\frac{\partial z}{\partial t} = \frac{\partial z}{\partial x}\cdot\frac{\partial x}{\partial t} + \frac{\partial z}{\partial y}\cdot\frac{\partial y}{\partial t} &= 2x.3 + 4y(-12t^2)\\ &= 12s + 18t - 48st^2 + 192t^5\end{aligned}$$

Example

The radius, r, of a cylinder decreases at the rate of 0.3 cm/sec and the height increases at the rate of 0.4 cm/sec. Find the rate of change of the volume of the cylinder *at the instant* when $r = 6$ cm, $h = 10$ cm.

Now $V = \pi r^2 h$ and so

$$\frac{\mathrm{d}V}{\mathrm{d}t} = 2\pi rh\frac{\mathrm{d}r}{\mathrm{d}t} + \pi r^2\frac{\mathrm{d}h}{\mathrm{d}t}$$

$$= 2\pi . 6 . 10(-0.3) + \pi . 36 [0.4]$$
$$= -36\pi + 14.4\pi$$
$$= -21.6\pi$$

Hence the instantaneous *decrease* of volume is 21.6π c.c/sec.

Problems II

1. Show that the total surface area of a cone of base radius r and height h is given by
$$S = \pi r^2 + \pi r\sqrt{r^2 + h^2}$$
Find the rate at which the surface area is increasing when $h = 5$ cm and $r = 4$ cm if h and r are both increasing at a rate of 0.5 cm/sec.

2. If x increases at a rate of 2 cm/sec at the instant when $x = 2$ cm and $y = 1$ cm, find the rate at which y must be changing in order that $u = (x^2 + y^2)/(x + y)$ shall be neither increasing nor decreasing when x and y have these values.

3. The equation $e_N = 2\sqrt{kTBR}$ arises in the study of thermal noise. If T, B and R are all functions of time t, find the formula for de_N/dt.

4. In a triode valve the anode current i is given by $i = C(V_a + \mu V_g)^{3/2}$ where V_g is the grid voltage and V_a the anode voltage. If V_g increases at the rate of 0.1 volt/s and V_a at 0.3 volt/s, find the rate of change of i when $V_g = 8$ volts and $V_a = 240$ volts.

5. A gas obeys the law $PV = RT$ where P is its pressure in N/m^2, V its volume in m^3 and T its temperature in °K. If the volume decreases by $0.4\ m^3/s$ and the temperature increases by 4°K/s, find the rate of increase of the pressure; take $R = 8$ Nm/°K.

Implicit Differentiation

Suppose x and y are not independent, but connected by some equation $f(x, y) = 0$. If $\partial f/\partial x$ and $\partial f/\partial y$ exist, then since $df = 0$, it can be shown that

$$\frac{dy}{dx} = -\left(\frac{\partial f}{\partial x}\right) \Big/ \left(\frac{\partial f}{\partial y}\right) \tag{7.17}$$

The proof follows from the approximation

$$df \simeq \frac{\partial f}{\partial x}dx + \frac{\partial f}{\partial y}dy$$

Example 1

Consider $f(x, y) = x^3 + y^3 - 2xy = 0$

We use the chain rule in the form

$$\frac{df}{dx} = \frac{\partial f}{\partial x} + \frac{\partial f}{\partial y} \cdot \frac{dy}{dx}$$

to obtain

$$\frac{df}{dx} = 3x^2 - 2y + (3y^2 - 2x)\frac{dy}{dx}$$

But, since $f(x, y) = 0$, $\frac{df}{dx} = 0$ and therefore $\frac{dy}{dx} = -\frac{3x^2 - 2y}{3y^2 - 2x}$

Example 2

Consider the ellipse $x^2 + 3y^2 = 4$

Let

$$f(x, y) = x^2 + 3y^2 - 4 = 0$$

then

$$\frac{\partial f}{\partial x} = 2x \quad \text{and} \quad \frac{\partial f}{\partial y} = 6y$$

Hence

$$\frac{dy}{dx} = -\frac{2x}{6y}$$

Compare this with the (so far unjustified) technique of implicit differentiation quoted in section 6.2, page 306 .

An application of this idea is the **envelope** to a family of curves.

Any family of curves (e.g. streamlines) will have an arbitrary constant, C, in their defining equation, $f(x, y, C) = 0$. The envelope is found by eliminating C from that equation and the equation

$$\frac{\partial}{\partial C} f(x, y, C) = 0$$

Example 3

A family of straight lines is given by

$$y = Cx - C^2 \tag{7.18}$$

Then

$$f(x, y, C) = Cx - C^2 - y \quad \text{and} \quad \frac{\partial f}{\partial C} = x - 2C$$

We eliminate C between (7.18) and the equation $x - 2C = 0$, to obtain

$$y = \frac{x^2}{2} - \frac{x^2}{4} = \frac{x^2}{4}$$

The situation is depicted in Figure 7.16.

Each line of the family is a tangent to the parabola at some point and at every point of the parabola one of the family is a tangent there.

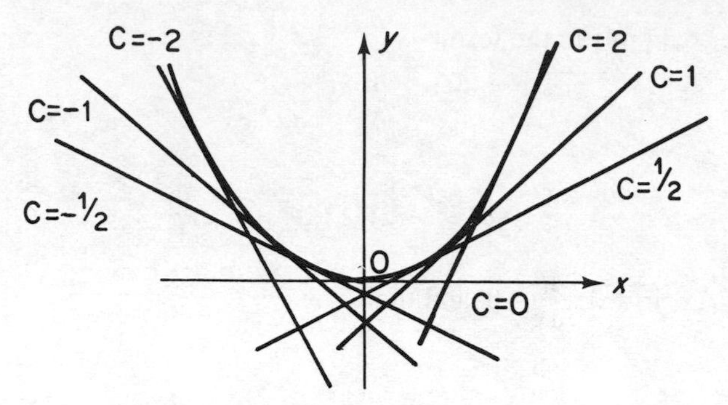

Figure 7.16

A second application is in **orthogonal trajectories** which arise in connection with stream functions. Let the streamlines for flow in a vortex be

$$x^2 + y^2 = a^2 \tag{7.19}$$

Now $f(x, y, a) = x^2 + y^2 - a^2$ and

$$\frac{\mathrm{d}f}{\mathrm{d}x} = \frac{\partial f}{\partial x} + \frac{\partial f}{\partial y} \cdot \frac{\mathrm{d}y}{\mathrm{d}x}$$

$$= 2x + 2y\frac{\mathrm{d}y}{\mathrm{d}x} = 0 \tag{7.20}$$

This gives a slope of

$$\frac{\mathrm{d}y}{\mathrm{d}x} = -\frac{x}{y}$$

Curves which intersect the family (7.19) everywhere at right angles have slopes $+y/x$ and thus have equation

$$x\frac{\mathrm{d}y}{\mathrm{d}x} - y = 0$$

It is seen that the family of straight lines $y = Cx$ satisfies this last equation. These *orthogonal curves* represent the velocity potential for the flow (Figure 7.17). Note: had (7.20) contained the parameter a then it would have been necessary to eliminate a between (7.19) and (7.20) before determining $\mathrm{d}y/\mathrm{d}x$.

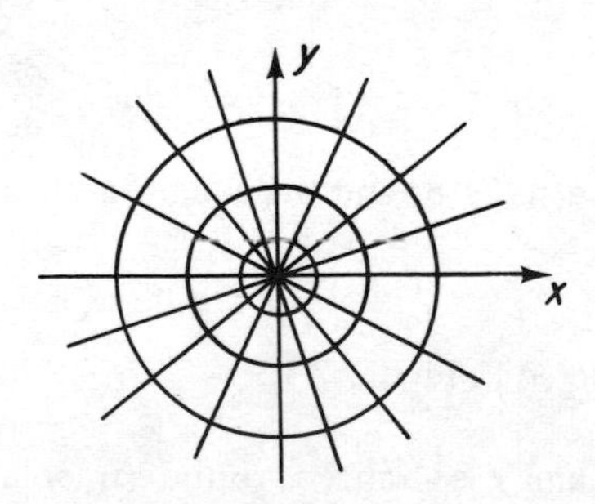

Figure 7.17

Problems III

1. Find dy/dx from the following implicit relationships

   (i) $x^3 + y^3 = 8$ (ii) $(x + y)\sin xy = 1$

   (iii) $x \cos y + y \cos x = 2$ (iv) $e^{y/x} \log(x + y) = 6$

2. Find the envelope of

   (i) the straight line $y = cx + c^3$

   (ii) the curve $\dfrac{x^2}{\alpha} + \dfrac{y^2}{1 - \alpha} = 1$

3. Show that the envelope of the family of curves with parameter α defined by $x \sin \alpha + y \cos \alpha = 4$ is a circle.

4. The path of a projectile is given by

$$y = x \tan \theta - \frac{gx^2}{2V^2} \sec^2 \theta$$

where θ is the angle of projection and V is the velocity of the projectile (see Figure 7.18).

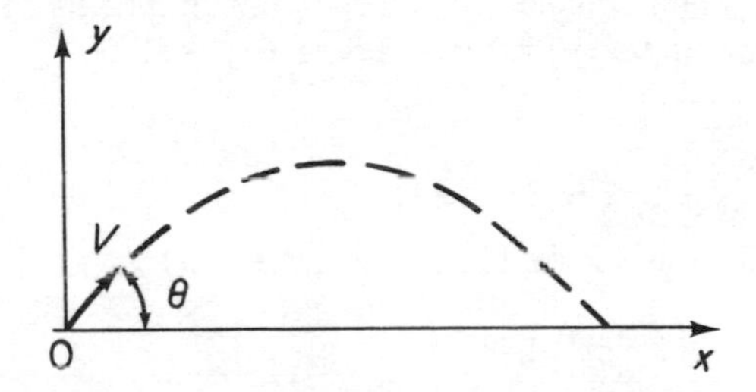

Figure 7.18

Regarding θ as a parameter, show that the enveloping parabola of projectiles projected from the origin with constant velocity V is $g^2x^2 = V^4 - 2gV^2y$. This is sometimes called the **parabola of safety** since no projectile can penetrate beyond it, no matter what value of θ is taken.

5. A long concrete wedge whose cross-section is a circular sector of angle α and radius R has one of its plane faces kept at temperature θ_1 and the other at θ_2 while the curved surface is insulated. See Figure 7.19. Except near the ends, it can be shown that the steady-state temperature is given by

$$\theta = \theta_1 + \frac{\theta_2 - \theta_1}{\alpha} \tan^{-1}(y/x)$$

Find the equation of the isotherms (constant temperature-lines) and show that these are the orthogonal trajectories of the lines of flow $x^2 + y^2 = k^2$.

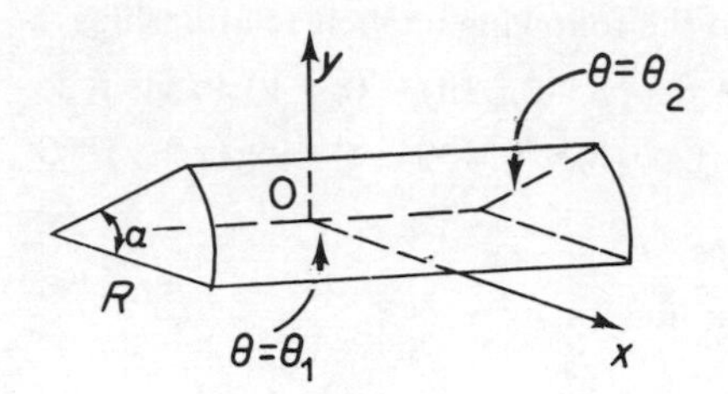

Figure 7.19

Draw both families of curves.

6. Show that in the following fluid flow problems the streamlines $\psi(x, y) = c$ and the equipotentials $\phi(x, y) = k$ form orthogonal trajectories.

   (i) $\phi = x^2 - y^2, \quad \psi = 2xy$

   (ii) ϕ and ψ are given by $\phi + i\psi = -Uz$ where $z = x + iy$

   (iii) ϕ and ψ are given by $\phi + i\psi = U(z + 4/z)$.

7. Find $\partial z/\partial x$ and $\partial z/\partial y$ given that z is defined implicitly in terms of x and y by the equation $f(x, y) = 2x^2 - 3xy^2z + z^3 = C$. (C.E.I.)

8. Find dv/dt when

   (i) $v = xy + 2, \ x = t^2, \ y = 1/t$

   (ii) $v = x^2 + y^2, \ x = t/(1 + t), \ y = t^2/(1 + t)$.

9. Find $\partial z/\partial x$ and $\partial z/\partial y$ when

   (i) $x^2 + 2y^2 - z^2 = 4$ (ii) $x + y + z = \log z$

   (iii) $z = x^2 + 8xy + 2$ (iv) $x^2 + 2yz + 2zx = 2$

   Hint: write each equation in the form $F(x, y, z) = 0$ and note that $dF/dx = 0 = dF/dy$.

10. A beam of light, parallel to the x-axis and coming from $x = -\infty$, hits the circular mirror $r = a, -\pi/3 \leqslant \theta \leqslant \pi/3$ and is reflected. Show that the envelope of the reflected ray (the **caustic curve**) is given by the equation

$$x = \tfrac{1}{2}a \cos\theta.(3 - 2\cos^2\theta), \quad y = a\sin^3\theta$$

Sketch the caustic curve.

SUMMARY

You have now been exposed to the additional problems involved with functions of more than one independent variable. We emphasise that the most difficulty occurs in extending results for one variable to cases of two variables; the further extension to three or more variables is relatively simple.

On the one hand, we develop methods of handling functions of two variables, particularly the techniques of partial differentiation, and then study the application to small increments and differentials; you should aim to be conversant with these.

The method of least squares curve fitting is the basis of some statistical methods of linear model-building. Consequently you should understand the underlying ideas and become familiar with the main techniques.

Chapter Eight

Integration

8.0 INTRODUCTION

In this chapter we consider two classes of problem:

(i) that of estimating areas, volumes and other geometrically-based quantities;

(ii) that of *undoing* differentiation. For example, given an equation for the velocity of the piston in Section 6.2, determining its displacement at any time, or given the bending moment equation for a loaded beam (Section 6.4), finding its deflected profile.

We shall show in Section 8.3 that the underlying process involved in classes (i) and (ii) is fundamentally the same.

8.1 ESTIMATION OF AREAS

It is desired to level a piece of uneven ground by taking earth from the higher portions and placing it in the lower regions: the method of *cut and fill*. This clearly requires the estimation of the volume of soil in an irregularly-shaped piece of land. As a simpler problem let us consider a cross-section through such a piece of land and try to estimate the cross-sectional area. We superimpose suitable axes on the profile, taking the lowest point as being at zero height. The typical measurements which might be taken (each subject to error ± 0.005 m) are tabulated in Table 8.1 and graphed in Figure 8.1.

Table 8.1

x (metres)	y (metres)
0	0
2	1.62
4	5.24
6	9.31
8	10.29
10	10.83

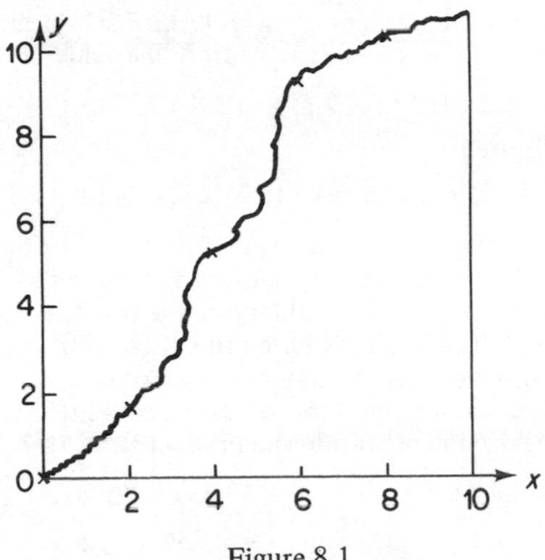

Figure 8.1

It would be of little use to say that the area of the cross-section is between 0 and $108.3\,\mathrm{m}^2$. A sensible thing to do would be to use the measurements and divide the area into strips as shown in Figure 8.2. We assume that the function representing the profile is monotonically increasing. How to modify the argument in other cases is dealt with later.

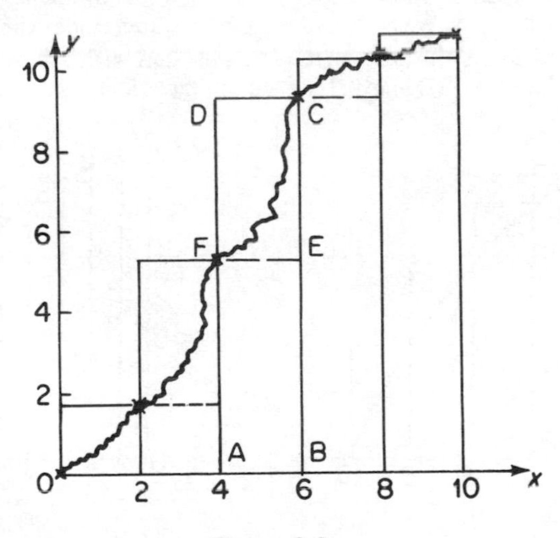

Figure 8.2

Since our basic idea of area is in terms of squares or rectangles we can estimate the area of each strip by the rectangles shown. If we choose the larger rectangles (ABCD is a typical one), their sum is an **overestimate** for the required cross-section area and the smaller rectangles (ABEF is typical) provide an **underestimate.** The value of the overestimate is, from the values in Table 8.1,

$$(2 \times 1.62) + (2 \times 5.24) + (2 \times 9.31) + (2 \times 10.29) + (2 \times 10.83) = 74.58\,\text{m}^2$$

and the underestimate is

$$(2 \times 0) + (2 \times 1.62) + (2 \times 5.24) + (2 \times 9.31) + (2 \times 10.29) = 52.92\,\text{m}^2$$

Notice that the difference is $(2 \times 10.83) - (2 \times 0) = 21.66\,\text{m}^2$, which is the sum of the differences of the rectangles for each strip (and in this case is the area of the largest rectangle: why?). We could quote the results as $\frac{1}{2}(74.58 + 52.92) \pm \frac{1}{2}21.66 = (63.75 \pm 10.83)\text{m}^2$. We have not yet taken into account the inexact measurements which will cause errors.

The worst that could happen is that all errors could reinforce; we know from Chapter 1 that the maximum error in the product is

$$(0.005)(1.62 + 5.24 + 9.31 + 10.29 + 10.83) + 0.005(2 + 2 + 2 + 2 + 2)$$
$$= 0.236 \quad (3 \text{ d.p.})$$

Therefore we must modify our estimate to $(63.75 \pm 11.07)\text{m}^2$.

From now on, we ignore the aspect of inaccurate measurements and assume that the strips we take are narrow enough to ensure that the function is monotonic in each strip. If the function is not monotonic overall, then the error in the result will be much less than the estimate. (We can always split the area into monotonic regions.) We hope that we can develop a method which will ideally reduce the error in our estimate and we choose a simple function on which to develop our method; in addition, we keep exact arithmetic as far as possible. Consider $y = 100 - x^2$. We seek to estimate the area under the curve from $x = 0$ to $x = 10$, first with 5 strips, then with 10, and then with 20, 40, 100, and finally 1000. [See Figures 8.3(i) and 8.3(ii), which have 5 and 10 strips respectively.]

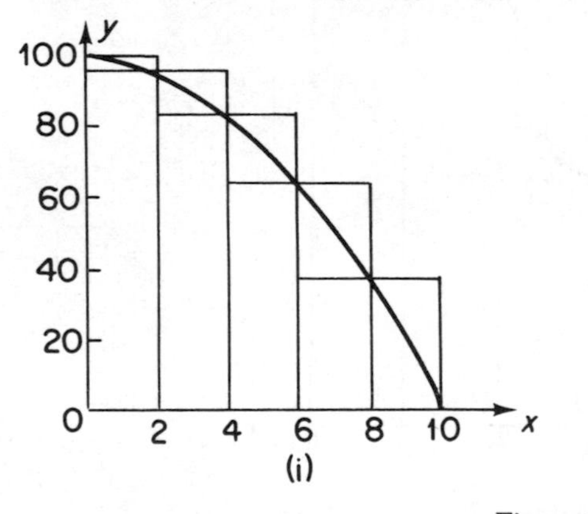

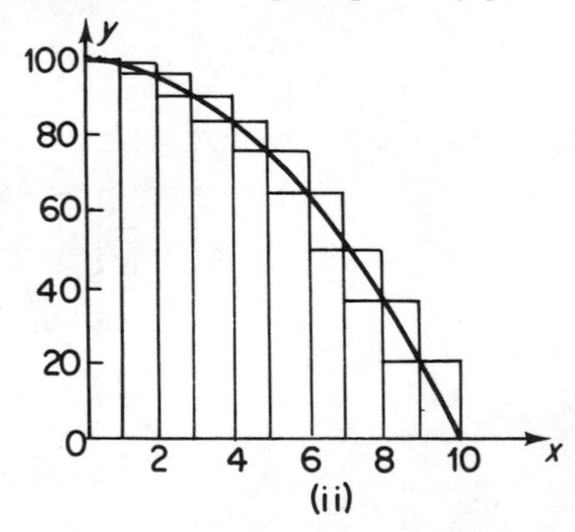

Figure 8.3

We first construct a table of values (see Table 8.2)

Table 8.2

x	0	1	2	3	4	5	6	7	8	9	10
y	100	99	96	91	84	75	64	51	36	19	0

For five strips the overestimate of the area is

$$(2 \times 100) + (2 \times 96) + (2 \times 84) + (2 \times 64) + (2 \times 36) = 760$$

and the underestimate is

$$(2 \times 96) + (2 \times 84) + (2 \times 64) + (2 \times 36) + (2 \times 0) = 560$$

(Note that the difference is again the area of the largest rectangle: why?) We therefore quote the result as

$$660 \pm 100$$

For ten strips, you can check that the result could be quoted as 665 ± 50.

Notice that halving the strip width halves the error bound. The results for other strip widths programmed on a computer are shown in tabular form in Table 8.3; try to write such a program yourself. It seems that the error bound is directly proportional to the strip width h (round-off error spoils things for a large number of strips, n).

Table 8.3

n	20	40	100	1000	10000
h	0.5	0.25	0.1	0.01	0.001
Overestimate	691.25	679.0623	671.6500	667.1665	666.7167
Underestimate	641.2500	654.0625	666.6500	666.1665	666.6167
Estimated value	666.2500	666.5625	666.6500	666.6665	666.6667
Error bound	25.0000	12.5000	5.0000	0.5000	0.0500

Results are quoted to 4 d.p.

We shall see in Section 8.3 that the exact area is $666\frac{2}{3}$ and so our quoted estimates are much nearer the truth than the size of the error bounds would have us believe. Clearly we shall need a better method of approximation and we examine a few of these in Section 8.3.

For the moment, we continue with this method and note that for each value of n the difference between the over- and underestimates (twice the error bound) is the area of the largest rectangle $(100 \times h) = 1000/n$. Hence if we want an estimate that we can be absolutely sure is correct to within $\pm\,\epsilon$ we simply choose n such that $1000/n < \epsilon$ and hope that n is not so large that round-off error spoils things. For example, to get an estimate correct to 1 d.p., i.e. $\pm\,0.05$, we need $1000/n < 1/20$, i.e. $n > 20\,000$.

Problems

1. Develop over- and underestimates for the area under the curve for $y = 80 - x^3$ in the range $x = 0$ to $x = 4$ with 4 strips, and 8 strips. If possible, write and run a computer program to cope with 16, 32 and 64 strips.
   What conclusions can you draw?

2. Repeat for $y = 400 - x^4$ over the range $x = 0$ to $x = 4$. How many strips would you need for an accuracy of 1 d.p., 2 d.p.?
   Comment.

8.2 THE DEFINITE INTEGRAL

Round-off apart, it would seem that if n is increased indefinitely then the estimate of the area becomes progressively better. It is tempting to think that in the limit as $n \to \infty$ the error will tend to zero and that the estimate will become exact. The dangers inherent in such an idea are pointed out in Problem 1 at the end of this section. Meanwhile we **assume** that the above result is true for **continuous** functions.

We generalise matters to consider any continuous function, $f(x)$, in some interval $[a, b]$. We seek the shaded area shown in Figure 8.4.

We divide the interval $[a, b]$ into n equal† strips of width $h = (b - a)/n$.

We denote by $\overline{S}_n$ the total area of the larger rectangles and $\underline{S}_n$ the total area of the smaller rectangles. Formally,

$$\overline{S}_n = \sum_{i=1}^{n} \overline{f}(x_i)\delta x$$

where

$$\overline{f}(x_i) \text{ is } \max_{x \in [x_{i-1}, x_i]} f(x)$$

and

$$\delta x = x_i - x_{i-1}$$

$\underline{S}_n$ is defined similarly.

Then the true area A is sandwiched between the two values: $\underline{S}_n < A < \overline{S}_n$. $\underline{S}_n$ is called the lower sum and $\overline{S}_n$ the upper sum. By what we have seen earlier, if we increase the number of strips the estimates improve (see Figure 8.5), i.e.

$$\underline{S}_n < \underline{S}_{n+1} < \ldots\ldots\ldots < A < \ldots\ldots\ldots < \overline{S}_{n+1} < \overline{S}_n$$

The true area A is sandwiched between two sequences $\{\underline{S}_n\}$ and $\{\overline{S}_n\}$.

---

† equality is not essential, it merely simplifies the subsequent algebra.

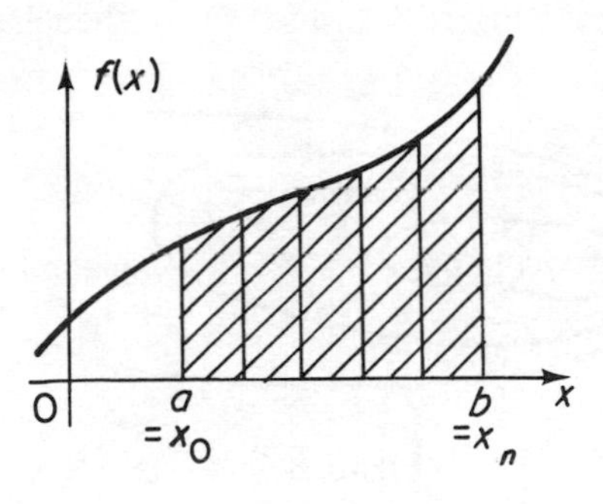

Figure 8.4

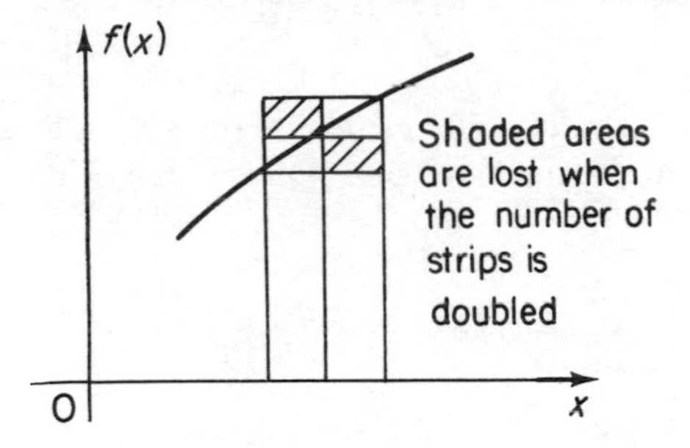

Figure 8.5

If each of these sequences tends to a limiting value and the value of the limit is the same in each case, the function is said to be **integrable** *over* the interval $[a, b]$ and the common limit is the **definite integral** of $f(x)$ over $[a, b]$ written

$$\int_a^b f(x)\mathrm{d}x \ \dagger \tag{8.1}$$

$f(x)$ is called the **integrand.**

We shall assume the result that all continuous functions are integrable over the the region of continuity. In the same way that differentiability is a stronger condition than continuity, this latter is stronger than integrability.

....

† the $\int$ is a corruption of S for **sum** and indicates the limit of $\underline{S}_n$ and $\overline{S}_n$. Some times it is written as

$$\int_a^b x \mapsto f, \text{ e.g. } \int_0^1 x \mapsto x^2 \qquad \text{for the function } x^2$$

We represent this idea in Figure 8.6 which shows the relationship between the three properties.

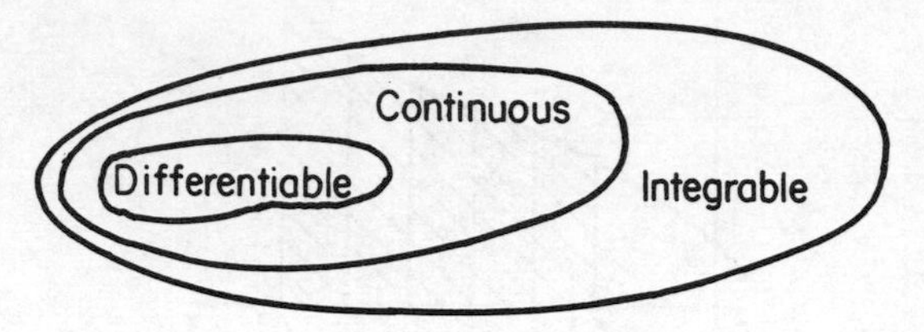

Figure 8.6

We consider the limiting process on an example,

$$\int_a^b x \; dx$$

The function x is a monotonically increasing function and in each of the n strips of width h, the maximum value of $f(x)$ is at the right-hand end, i.e. $a + rh$, giving an upper sum of

$$\sum_{r=1}^{n} h(a + rh) = nah + h^2 . \frac{n(n+1)}{2}$$

Likewise, the lower sum is

$$nah + h^2 . \frac{(n-1)n}{2}$$

Since $nh = (b - a)$ we may write

$$\overline{S_n} = a(b-a) + \frac{n(n+1)(b-a)^2}{2n^2} = a(b-a) + \frac{(b-a)^2}{2}\left(1 + \frac{1}{n}\right)$$

and

$$\underline{S_n} = a(b-a) + \frac{(b-a)^2}{2}\left(1 - \frac{1}{n}\right)$$

By the rules for limits, as $n \to \infty$,

$$\overline{S_n} \to a(b-a) + \frac{(b-a)^2}{2} = \frac{b^2}{2} - \frac{a^2}{2}$$

Similarly,

$$\underline{S_n} \to \frac{b^2}{2} - \frac{a^2}{2}$$

Hence

$$\int_a^b x \, dx = \frac{b^2}{2} - \frac{a^2}{2}$$

(We could have derived this by geometry. Try this yourself.)

In a similar way we can show

$$\int_a^b 1\,\mathrm{d}x \equiv \int_a^b x^0\,\mathrm{d}x = b - a$$

and

$$\int_a^b x^2\,\mathrm{d}x = \frac{b^3}{3} - \frac{a^3}{3}$$

It would seem reasonable to guess that

$$\int_a^b x^n\,\mathrm{d}x = \frac{b^{n+1}}{n+1} - \frac{a^{n+1}}{n+1} \tag{8.2}$$

(unless, of course, $n = -1$).

We could prove the results by an extension of the method for $f(x) = x$, but it would be cumbersome and we clearly need a more practicable technique.

Properties of the definite integral

Consequent upon the rules for limits in Section 3.1 we have the following results:

(i) $$\int_a^b \lambda f(x)\mathrm{d}x = \lambda \int_a^b f(x)\mathrm{d}x \tag{8.3}$$

(ii) $$\int_a^b \{f(x) + g(x)\}\mathrm{d}x = \int_a^b f(x)\mathrm{d}x + \int_a^b g(x)\mathrm{d}x \tag{8.4}$$

(iii) $$\int_a^c f(x)\mathrm{d}x + \int_c^b f(x)\mathrm{d}x = \int_a^b f(x)\mathrm{d}x \tag{8.5}$$

It is assumed that $f(x)$ and $g(x)$ are continuous in the interval $[a, b]$, that λ is any constant (note the case $\lambda = -1$) and that $a < c < b$. If at any part of $[a, b]$ $f(x) < 0$ then the integral of that region is taken as negative. The unfortunate consequences of this are shown in Example 1 given on page 428.

We have assumed that $a < b$ so far but in the course of manipulating integrals it may happen that $a > b$; we then use the result

(iv) $$\int_b^a f(x)\mathrm{d}x = -\int_a^b f(x)\mathrm{d}x \tag{8.6}$$

If for all x in $[a, b]$, $g(x) \leqslant f(x) \leqslant h(x)$ where the three functions are integrable over $[a, b]$ then

$$\int_a^b g(x)\mathrm{d}x \leqslant \int_a^b f(x)\mathrm{d}x \leqslant \int_a^b h(x)\mathrm{d}x \tag{8.7}$$

In particular, if the maximum and minimum values of $f(x)$ on $[a, b]$ are M and m respectively then

$$m(b-a) \leqslant \int_a^b f(x)\mathrm{d}x \leqslant M(b-a) \tag{8.8}$$

Now

$$m \leqslant \frac{1}{b-a}\int_a^b f(x)\mathrm{d}x \leqslant M$$

But if $f(x)$ is continuous on $[a, b]$ then, by the intermediate value theorem, there is a point ξ where $a \leqslant \xi \leqslant b$ such that

$$\frac{1}{b-a}\int_a^b f(x)\mathrm{d}x = f(\xi) \tag{8.9}$$

This is the **First Mean Value Theorem** for integrals and it implies that there is a point of average height (in the sense indicated) in $[a, b]$.

Problems

1. Assume that

$$\int_a^b \frac{1}{(x-2)^2}\,\mathrm{d}x = \frac{-1}{b-2} - \frac{-1}{a-2}$$

If $a = 1, b = 3$, use the method of Section 8.1 with 3, 5, 7, 9 strips and compare results. What is your conclusion?

2. From properties given in (8.3) to (8.6), prove

(i) $\displaystyle\int_a^b f(x)\mathrm{d}x = \int_a^b f(a+b-x)\mathrm{d}x$ (Hint: put $a+b-x=t$)

(ii) $\displaystyle\int_{-a}^a f(x)\mathrm{d}x = \begin{cases} 0 \text{ if } f \text{ is odd} \\ 2\int_0^a f(x)\mathrm{d}x \quad \text{if } f \text{ is even} \end{cases}$

(iii) $\displaystyle\int_0^{2a} f(x)\mathrm{d}x = \int_0^a [f(x)+f(2a-x)]\,\mathrm{d}x$

3. Find bounds on the following:

$$\int_0^{\pi/2} \cos x \, dx, \qquad \int_0^{1/2} (1 - x^2) dx, \qquad \int_0^3 \frac{1}{1+x^2} \, dx$$

4. Find the point ξ in property (8.9) for

$$\int_0^1 x \, dx, \qquad \int_0^1 x^2 \, dx, \qquad \int_0^4 x^3 \, dx$$

8.3 PRIMITIVE FUNCTIONS AND THE FUNDAMENTAL THEOREM OF THE CALCULUS

Was it coincidence that the results in Section 8.2 could all be written in the form $F(b) - F(a)$? We shall show in this section that the answer is 'no'.

First, a definition: $F(x)$ is called **an indefinite integral** or a **primitive function** of $f(x)$ if

$$\int_a^b f(x) dx = F(b) - F(a) \tag{8.10}$$

We say *an* indefinite integral rather than *the* indefinite integral since there is more than one such integral for a given function $f(x)$.

It would be fairly obvious that an indefinite integral of x^2 is $x^3/3$, but so is $(x^3/3) + 6$, $(x^3/3) - 8.2$ or $(x^3/3) + C$, where C is any constant, since the constant cancels out on subtraction of $F(a)$ from $F(b)$. We have then a whole family of indefinite integrals differing from each other only by a constant, i.e. the graph of one can be shifted into that of another by moving it parallel to the y-axis (Figure 8.7).

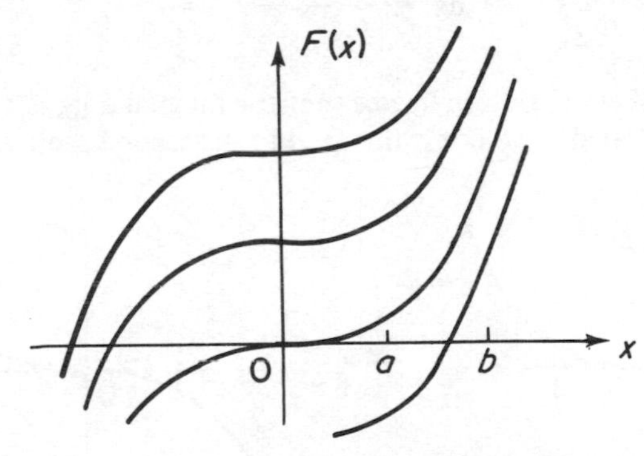

Figure 8.7

Every point whose x-coordinate is between a and b lies on one (and only one) of these curves. It should be fairly clear that there is certainly one curve which crosses the x-axis at $x = a$.

We denote the primitive function which takes a zero value at $x = a$ by $G(x)$; then

$$\int_a^b f(x)\mathrm{d}x = G(b) - G(a) = G(b)$$

If we hold the lower limit fixed and vary b, then

$$\int_a^x f(x)\mathrm{d}x = G(x)$$

We can therefore represent the indefinite integral as a function.

There is a danger in applying result (8.2) if the function $f(x)$ takes negative values on the range of integration.

Example 1

Suppose we want

$$\int_{-1}^1 x^3\,\mathrm{d}x$$

From equation (8.2) this would be

$$\frac{(1)^4}{4} - \frac{(-1)^4}{4} = 0$$

Again,

$$\int_{-1}^2 x^3\,\mathrm{d}x = \frac{2^4}{4} - \frac{(-1)^4}{4} = \frac{15}{4}$$

A look at the first sketch below indicates that the integral does not give the true value of the total shaded area; nor is this true in the second sketch, though this is not so obvious.

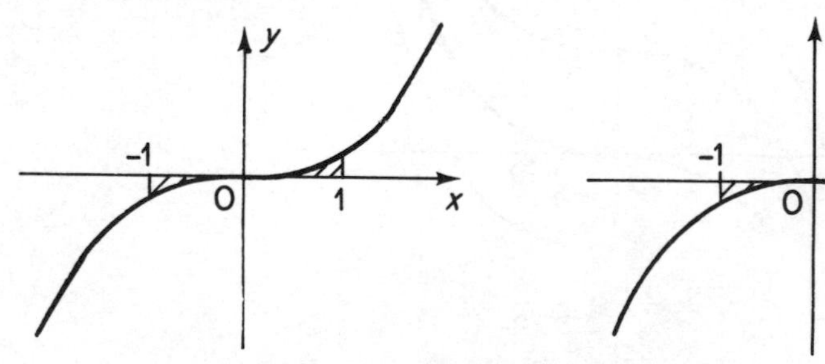

We now show two important results.

I. If $F(x)$ is a primitive function of $f(x)$, then $F'(x) = f(x)$ (8.11a)

II. Any two primitives of $f(x)$ differ only by a constant (8.11b)

The proof of the first is as follows:

Now

$$F(a+h) - F(a) = \int_a^{a+h} f(x)\,dx$$

and so

$$\frac{F(a+h) - F(a)}{h} = \frac{1}{h}\int_a^{a+h} f(x)\,dx$$

$$= f(\xi) \text{ where } a \leqslant \xi \leqslant a+h \quad \text{(by 8.9)}$$

If we take the limit of both sides as $h \to 0$ we have

$$F'(a) = f(a) \quad \text{since } \xi \text{ is squeezed down to } a.$$

Since 'a' is a general point, the result that $F'(x) = f(x)$ follows if we assume x to be any point in the interval over which $f(x)$ is integrable.

The second result is necessary since all we have shown is that if $F(x)$ is a primitive, its derivative is $f(x)$; it does **not** follow that if $F'(x) = f(x)$ then $F(x)$ is a primitive of $f(x)$. (The pair of statements 'All men are fools', 'X is a fool' does not imply X is a man since some women could be fools also.)

A sketch of the proof of the second result now follows.

Let $F(x)$, $H(x)$ both have derived functions $f(x)$. Then define $Q(x) \equiv F(x) - H(x)$. We have $Q'(x) \equiv F'(x) - H'(x) \equiv f(x) - f(x) = 0$. Since $Q'(x) = 0$ it follows that $Q(x)$ is a constant function. [We are in danger of going round in circles: how do we know there is not a non-constant function whose derivative is identically zero? You must take this result on trust since we are not proving the result (8.11b) rigorously, merely giving a plausible argument. Rigorous proofs can be found in the books listed in the bibliography.]

To sum up, we know that a primitive function exists and any function which differs from it by a constant is also a primitive; hence any function whose derivative is $f(x)$ is also a primitive.

Theorems (8.11a) and (8.11b) constitute the **Fundamental Theorem of the Calculus**, the link between differentiation and integration. We see that all we need for integration is to find a primitive of $f(x)$ to *undo* the differentiation that led to $f(x)$. Since $x^3/3$ differentiates to x^2, then $x^3/3$ is a primitive of x^2 sometimes called **antiderivative** or more loosely an **indefinite integral**. The process of finding a primitive is called **(indefinite) integration.** We usually write this simply as

$$\int f(x)\,dx.$$

To revert to the definite integral, we notice that

$$\int_a^b f(x)\mathrm{d}x \equiv \int_a^b f(u)\mathrm{d}u$$

This follows since, if $f(x) = x^2$, $f(u) = u^2$ and

$$\int_a^b f(x)\mathrm{d}x = F(b) - F(a)$$

Hence

$$F(x) = \frac{x^3}{3}$$

and therefore

$$F(u) = \frac{u^3}{3}$$

Alternatively,

$$\int_a^b f(u)\mathrm{d}u = \left[\frac{u^3}{3}\right]_a^b = \frac{b^3}{3} - \frac{a^3}{3}$$

We call u a **dummy variable**; it does not really matter what letter is used in the integrand, but

$\int_a^x f(u)\mathrm{d}u$ is neater than $\int_a^x f(x)\mathrm{d}x$ and avoids confusion.

Example 2

The parabolic integral

$$\int_0^{10} (100 - x^2)\mathrm{d}x = \int_0^{10} 100\mathrm{d}x - \int_0^{10} x^2\mathrm{d}x$$

$$= 100\int_0^{10} 1\,\mathrm{d}x - \int_0^{10} x^2\mathrm{d}x$$

$$= 100\,.\,[x]_0^{10} - \left[\frac{x^3}{3}\right]_0^{10}$$

$$= 100\,.\,10 - 1000/3$$

$$= 666\tfrac{2}{3}$$

This was the result promised on page 421.

Problems

1. Verify the following indefinite integrals by means of differentiation, C being any constant.

(i) $$\int \frac{dx}{a^2 - x^2} = \frac{1}{2a}\log\left(\frac{a+x}{a-x}\right) + C$$

(ii) $$\int \frac{dx}{\sqrt{a^2 - x^2}} = \sin^{-1}\frac{x}{a} + C$$

(iii) $$\int \sin x \cos x \, dx = -\frac{1}{4}\cos 2x + C$$

(iv) $$\int \frac{3x dx}{x^2 + 2} = \frac{3}{2}\log(x^2 + 2) + C$$

(v) $$\int (e^{-3x} + e^{2x}) dx = -\frac{1}{3}e^{-3x} + \frac{1}{2}e^{2x} + C$$

(vi) $$\int \cos x \sqrt{\sin x} \, dx = \frac{2}{3}(\sin x)^{3/2} + C$$

(vii) $$\int \frac{dx}{x\sqrt{x^2 - 4}} = \frac{1}{2}\cos^{-1}\frac{2}{x} + C$$

(viii) $$\int x^2(x^3 + 1)^3 dx = \frac{1}{12}(x^3 + 1)^4 + C$$

(ix) $$\int \frac{x^2 dx}{\sqrt{1 - x^2}} = -\frac{x}{2}\sqrt{1 - x^2} + \frac{1}{2}\sin^{-1}x + C$$

(x) $$\int x \sin x \, dx = \sin x - x \cos x + C$$

(xi) $$\int x e^{3x} dx = \frac{e^{3x}}{3}\left(x - \frac{1}{3}\right) + C$$

(xii) $$\int x^2 \sinh x \, dx = (x^2 + 2)\cosh x - 2x \sinh x + C$$

8.4 EXAMPLES OF INDEFINITE INTEGRATION

Here we have $dy/dx = f(x)$. This is an example of an ordinary differential equation (discussed in the next chapter), and our task is to find y as a function of x; e.g. $dy/dx = 4x^2$. We might be tempted to write $y = (4/3)x^3$ which does indeed satisfy the given equations, but so also does $y = (4/3)x^3 + 1$; as does $y = (4/3)x^3 + 476.2$ and, in general, $y = (4/3)x^3 + C$, where C is any constant, often called an **arbitrary constant.**

To decide on the particular value of C we need a **boundary condition** such as $y = 3$ when $x = 2$; then

$$3 = \frac{4}{3}(2)^3 + C$$

therefore

$$C = 3 - \frac{32}{3} = \frac{-23}{3}$$

and the particular solution is

$$y = \frac{4}{3}x^3 - \frac{23}{3}$$

[Note that if y is a function of time, $y = y(t)$, and the condition is given for $t = 0$, it is known as an **initial condition.**]

In fact, all ordinary differential equations require the use of integration for their solution. We can make a start at these solutions by writing down a list of standard results from differentiation and working back to form a list of standard integrals.

In the following C is the arbitrary constant.

$$\frac{d}{dx}(x^n) = nx^{n-1} \qquad \int x^n dx = \frac{x^{n+1}}{n+1} + C \quad (n \neq -1)$$

$$\frac{d}{dx}(\log_e x) = \frac{1}{x} \qquad \int \frac{1}{x} dx = \log_e x + C$$

$$\frac{d}{dx}[\log_e f(x)] = \frac{f'(x)}{f(x)} \qquad \int \frac{f'(x)dx}{f(x)} = \log_e f(x) + C$$

$$\frac{d}{dx}(e^{kx}) = ke^{kx} \quad (k \text{ constant}) \qquad \int e^{kx} dx = \frac{e^{kx}}{k} + C$$

$$\frac{d}{dx}(\sin x) = \cos x \qquad \int \cos x \, dx = \sin x + C$$

$$\frac{d}{dx}(\cos x) = -\sin x \qquad \int \sin x \, dx = -\cos x + C$$

$$\frac{d}{dx}(\tan x) = \sec^2 x \qquad \int \sec^2 x \, dx = \tan x + C$$

$$\frac{d}{dx}(\cosh x) = \sinh x \qquad \int \sinh x \, dx = \cosh x + C$$

$$\frac{d}{dx}(\sinh x) = \cosh x \qquad \int \cosh x \, dx = \sinh x + C$$

$$\frac{d}{dx}(\sin^{-1} x) = \frac{1}{\sqrt{1-x^2}} \qquad \int \frac{1}{\sqrt{1-x^2}} \, dx = \sin^{-1} x + C$$

We use some of these entries with the rules of Section 8.2.

Examples

1. $$\int 2 \cos x \, dx = 2 \int \cos x \, dx \qquad \text{by (8.3)}$$

$$= 2 \sin x + 2C$$

But if C is any constant, $2C$ can be rewritten as C without loss of generality.

2. $$\int (x^4 + e^{2x}) dx = \int x^4 dx + \int e^{2x} dx \qquad \text{by (8.4)}$$

$$= \frac{x^5}{5} + \frac{e^{2x}}{2} + C$$

Here we have *absorbed* the two arbitrary constants into one.

3. $$\int -e^{4x} dx = -\int e^{4x} dx \qquad \text{by (8.3)}$$

$$= -\frac{e^{4x}}{4} + C$$

The need to develop sophisticated techniques for integration is clear if we look again at the crank problem of Sections 6.1 and 6.2. If we were given equation (6.7) for the velocity of the piston and had, from that, to deduce its displacement (6.1) by integration, we should be at a loss at the moment. We could try for a *formula approach* or a *numerical technique.*

8.5 NUMERICAL TECHNIQUES

It would be tedious if we always had to resort to guesswork using tables of derivatives and the general rules (8.3) to (8.6). As we shall see in Section 8.6 there are methods – *tricks* – which save us some of the tedium. However, perhaps we may not be able to spot which trick to employ and perhaps there is no simple form for the indefinite integral expressible in terms of elementary functions. Hence we might resort to the expansion of the **integrand** $f(x)$ as a *Taylor series* and integrate the value we want; in general, since integration smooths out irregularities in a function we expect that convergence will take place. But it may happen that the Taylor's series does not converge for all values of x in the range of integration; even if it does it may converge too slowly to be of practical use.

In such cases, and, of course, in the case when the function $f(x)$ is given merely as a set of tabulated values then we must resort to a *numerical approximation* technique. We consider basically *two* methods and indicate other possibilities.

Method 1 – Trapezoidal Rule

The method of Section 8.1 used the over- and underestimates in each strip to find an area. Figure 8.8 indicates that we can obtain an improved estimate by taking the area of the trapezium whose *top* is the chord joining the end values of $f(x)$.

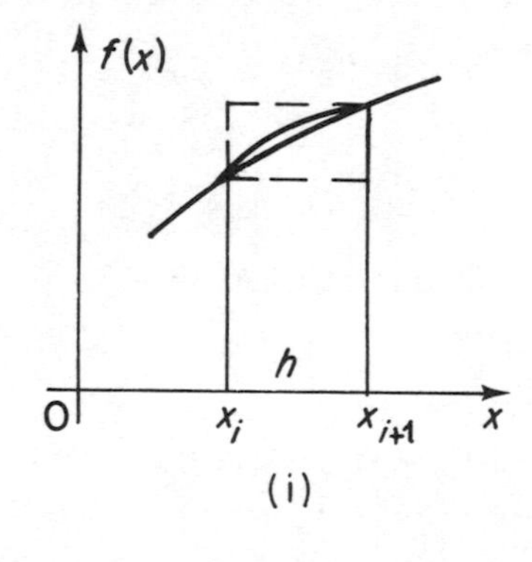

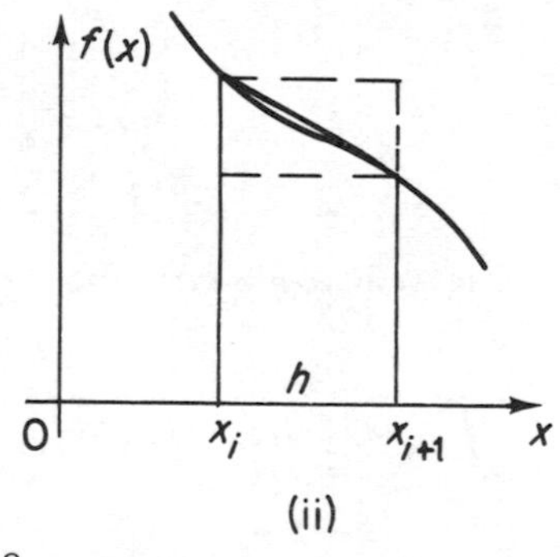

Figure 8.8

The area of the trapezium is $(x_{i+1} - x_i) \cdot \frac{1}{2}[f(x_i) + f(x_{i+1})]$, and if we denote the strip width by h and $f(x_i)$ by f_i etc. the area can be written $\frac{1}{2}h(f_i + f_{i+1})$.

It should be easy to see that if the interval $[a, b]$ is divided into n equal strips of width h then

$$\int_a^b f(x)\mathrm{d}x$$

may be approximated as the trapezoidal approximation to the integral I_T

$$I_T = \frac{h}{2}\{f_0 + 2f_1 + 2f_2 + ... + 2f_{n-1} + f_n\} \tag{8.12}$$

What we have done is to approximate the function in each strip by a straight line. Let us first see what sort of approximation we get and then try to estimate the error of the approximation.

Example 1

$$\int_0^{10} (100 - x^2)\mathrm{d}x$$

We know that the true result is 666⅔ and Table 8.3 (page 421) gives the estimates by the approximating rectangles technique. Let us first take *five* strips between $x = 0$ and $x = 10$; then $h = 2$ and from the values in Table 8.2, we may deduce

$$I_T = \frac{2}{2}\{100 + 2 \times 96 + 2 \times 84 + 2 \times 64 + 2 \times 36 + 0\} = 660$$

With *ten* strips we obtain

$$I_T = \frac{1}{2}\{100 + 2 \times 99 + 2 \times 96 + 2 \times 91 + 2 \times 84 + 2 \times 75 +$$

$$2 \times 64 + 2 \times 51 + 2 \times 36 + 2 \times 19 + 0\} = 665$$

We can see that these estimates are indeed better than the corresponding results by the rectangle method and, as before, the error decreases with decreasing strip width. This time the error ϵ_T has been reduced from 20/3 to 5/3, i.e. it is quartered by halving the strip width; it seems that the error $\epsilon_T \propto h^2$.

Table 8.4, obtained with the aid of a digital computer seems to confirm this view.

Table 8.4

n	20	40	100	1000	10000
h	0.5	0.25	0.1	0.01	0.001
I_T	666.2500	666.5625	666.6500	666.6665	666.6667
$\lvert\epsilon_T\rvert$	0.4167	0.1042	0.0167	0.0002	0.0000

Figure 8.9 shows a flow chart for the *Trapezoidal method* or *rule*. N is the number of strips; A and B the limits of integration, H the strip width. I is a dummy counter.

Modify it, as you think fit, and write the appropriate program.

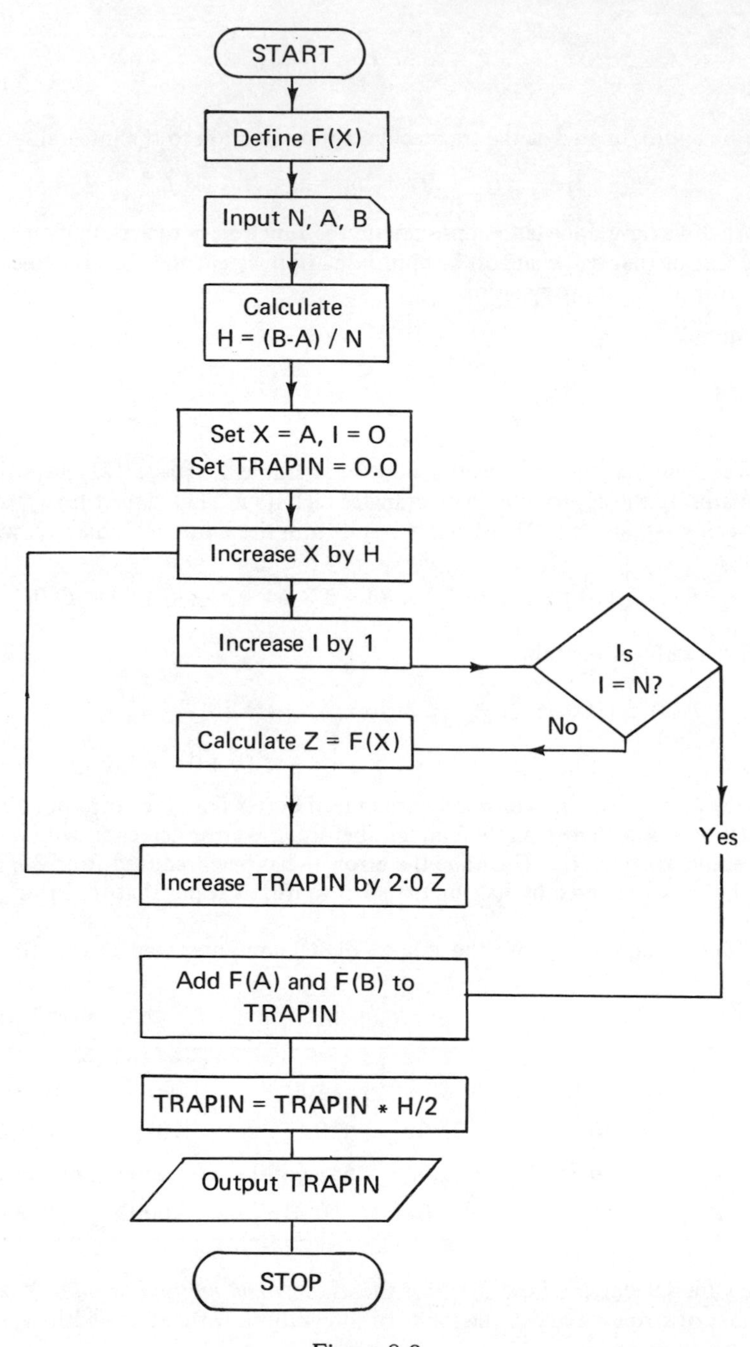
START
Define F(X)
Input N, A, B
Calculate
H = (B-A) / N
Set X = A, I = O
Set TRAPIN = O.O
Increase X by H
Increase I by 1
Is
I = N?
No
Yes
Calculate Z = F(X)
Increase TRAPIN by 2·0 Z
Add F(A) and F(B) to
TRAPIN
TRAPIN = TRAPIN * H/2
Output TRAPIN
STOP

Figure 8.9

If we seek the definite integral to a particular accuracy, we must find an expression for the error which will allow us to calculate the number of strips needed to achieve that accuracy, hoping that round-off error will not spoil things for us.

First, we estimate the error over one strip: $[x_i, x_{i+1}]$. The Trapezoidal rule gives

$$\frac{h}{2}\{f_i + f_{i+1}\}$$

if we expand $f(x)$ as a Taylor's series about x_i we obtain

$$f(x) = f_i + (x - x_i)f_i' + \frac{1}{2}(x - x_i)^2 f_i'' + \frac{1}{3!}(x - x_i)^3 f_i''' + \ldots$$

and so

$$\int_{x_i}^{x_{i+1}} f(x)\mathrm{d}x = \left[xf_i + \frac{1}{2}(x - x_i)^2 f_i' + \frac{1}{2.3}(x - x_i)^3 f_i'' + \frac{1}{4!}(x - x_i)^4 f_i''' + ..\right]_{x_i}^{x_{i+1}}$$

$$= (x_{i+1} - x_i)f_i + \frac{1}{2!}(x_{i+1} - x_i)^2 f_i' + \frac{1}{3!}(x_{i+1} - x_i)^3 f_i'' +$$

$$\frac{1}{4!}(x_{i+1} - x_i)^4 f_i''' + \ldots\ldots$$

$$= hf_i + \frac{1}{2!}h^2 f_i' + \frac{h^3}{3!}f_i'' + \frac{h^4}{4!}f_i''' + \ldots\ldots \tag{8.13a}$$

Now, we had from Section 6.12 that

$$\frac{f_{i+1} - f_i}{h} = f_i' + \frac{h}{2}f_i'' + 0(h^2)$$

and so

$$\frac{h^2}{2!}f_i' = \frac{h(f_{i+1} - f_i)}{2} - \frac{h^3}{4}f_i'' + 0(h^4)$$

Inserting this in equation (8.13a) we obtain

$$\int_{x_i}^{x_{i+1}} f(x)\mathrm{d}x = \frac{h}{2}(f_{i+1} + f_i) - \frac{h^3}{4}f_i'' + \frac{h^3}{6}f_i'' + 0(h^4)$$

$$= \frac{h}{2}(f_{i+1} + f_i) - \frac{h^3}{12}f_i'' + 0(h^4)$$

Hence

$$\epsilon_T = \frac{h^3}{12}f_i'' + 0(h^4) \tag{8.13b}$$

By application of the mean-value theorem we can state

$$\epsilon_T = \frac{h^3}{12}f''(\xi), \quad x_i \leqslant \xi \leqslant x_{i+1} \tag{8.13c}$$

But, if $|f''(x)| \leqslant M_i$ for $x_i \leqslant x \leqslant x_{i+1}$ we can estimate

$$|\epsilon_T| \leqslant \frac{h^3}{12} M_i \tag{8.13d}$$

Over the n strips from x_0 to x_n we *could* estimate the error in each strip, but at the risk of loss of accuracy we weaken our estimate for the sake of ease in computation so that if M is the largest absolute value of $f''(x)$ over the interval $x_0 \leqslant x \leqslant x_n$, i.e.

$$M = \max_{x \in [x_0, x_n]} |f''(x)|$$

then

$$|\epsilon_T| \leqslant \frac{nh^3}{12} M = (x_n - x_0)\frac{h^2}{12} M \tag{8.13e}$$

For example,

$$\int_0^{10} (100 - x^2)\,dx \quad \text{with 5 strips gives [since } f''(x) = -2]$$

$$|\epsilon_T| \leqslant 10 \cdot \frac{2^2}{12} \cdot 2 = 6\frac{2}{3}$$

The actual error was $6\frac{2}{3}$. With 10 strips we estimate $|\epsilon_T| \leqslant 10 \cdot (1^2/12) \cdot 2 = 5/3$, which was the actual error. This is the best (or worst) to be hoped for, since we have a constant second derivative.

Example 2

$$\int_0^{\pi/2} \cos x \,dx = \Big[\sin x\Big]_0^{\pi/2} = 1$$

If we apply the Trapezoidal rule with three strips ($h = \pi/6$) we obtain

$$I_T = \frac{\pi}{12}\left(\cos 0 + 2\cos\frac{\pi}{6} + 2\cos\frac{2\pi}{6} + \cos\frac{\pi}{2}\right)$$

$$\simeq \frac{\pi}{12}\left(1 + 1.732 + 1 + 0\right) = \frac{3.732\pi}{12} = 0.977 \quad (3\text{ d.p.})$$

The estimate of the error is given by

$$|\epsilon_T| \leqslant \frac{\pi}{2} \cdot \frac{\pi^2}{36} \cdot \frac{1}{12} = 0.036 \quad (3\text{ d.p.})$$

since $f''(x) = -\cos x$ and $|f''(x)| \leqslant 1$

It would seem that we could only quote the result as 1.00 ± 0.04, which is true, but the error estimate is far too generous when compared with the actual error of 0.023 (3 d.p.). Table 8.5 shows the estimates, actual errors and estimated errors with differing strip widths. We want an accuracy of 0.01 in the

Trapezoidal estimate; we know that for n strips, $h = \pi/2n$ and

$$|\epsilon_T| \leqslant \frac{\pi}{2} \cdot \frac{\pi^2}{4n^2} \cdot \frac{1}{12}$$

We therefore require

$$\frac{\pi^3}{12 \,.\, 8n^2} < \frac{1}{100} \quad \text{i.e.} \quad n^2 > \frac{100\pi^3}{96} \simeq 33$$

Hence we could take $n = 6$ and this is a convenient number with which to compute. Table 8.5 shows that $n = 6$ is, in fact, more than enough.

The one danger is that if the error estimate gives a large number of strips then round-off error may come in to the picture when in practice a smaller number of strips would suffice with negligible round-off.

We have not yet considered what would happen to this estimation if the function to be integrated was given only by a table of values. We postpone this until later and meanwhile examine a more accurate method.

Table 8.5

Number of strips	Estimate	Error	Estimated Error
1	.7854	.2146	.3230
2	.9481	.0519	.0807
3	.9770	.0230	.0359
4	.9871	.0129	.0202
5	.9918	.0082	.0129
6	.9943	.0057	.0090
7	.9958	.0042	.0066
8	.9968	.0032	.0050
9	.9975	.0025	.0040
10	.9979	.0021	.0032
11	.9983	.0017	.0027
12	.9986	.0014	.0022
13	.9988	.0012	.0019
14	.9990	.0010	.0016
15	.9991	.0009	.0014
20	.9995	.0005	.0008

Method 2 – Simpson's Rule

If we interpolate a parabola (see Figure 8.10) through three points (x_i, f_i), (x_{i+1}, f_{i+1}) and (x_{i+2}, f_{i+2}), we would expect to get a more accurate representation of

$$\int_{x_i}^{x_{i+2}} f(x)\,\mathrm{d}x$$

from the area under the parabola than by the application of the Trapezoidal rule to the two strips $[x_i, x_{i+1}]$ and $[x_{i+1}, x_{i+2}]$.

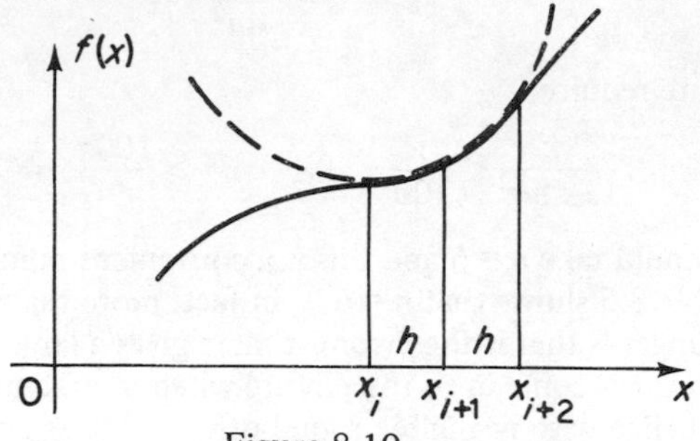

Figure 8.10

Let the quadratic approximation to $f(x)$ be $y = ax^2 + bx + c$; it clearly will not affect the value of the area concerned if we shift the origin to x_{i+1} and relabel the points as $(-h, f_{-1})$, $(0, f_0)$ and (h, f_1). Then, since the values of y (from the quadratic) and $f(x)$ coincide at these points, we have

$$f_{-1} = ah^2 - bh + c \qquad f_0 = c \qquad f_1 = ah^2 + bh + c$$

which give $c = f_0$, $b = (f_1 - f_{-1})/2h$, and $a = (f_1 - 2f_0 + f_{-1})/2h^2$

Now

$$\int_{-h}^{h} y \, dx = \left[a\frac{x^3}{3} + b\frac{x^2}{2} + cx \right]_{-h}^{h} = 2a\frac{h^3}{3} + 2ch$$

$$= \frac{h}{3}(f_1 - 2f_0 + f_{-1}) + 2hf_0$$

$$= \frac{h}{3}(f_1 + 4f_0 + f_{-1})$$

Making the obvious extension to the estimation of

$$\int_{x_0}^{x_{2n}} f(x) dx$$

by $2n$ strips we have the Simpson approximation, I_S, to the integral

$$I_S = \frac{h}{3}\{f_0 + 4f_1 + 2f_2 + 4f_3 + 2f_4 + \ldots\ldots + 4f_{2n-1} + f_{2n}\} \tag{8.14}$$

Example 3

We now estimate

$$\int_0^{\pi/2} \cos x \, dx \quad \text{with two strips } (h = \pi/4)$$

$$
\begin{aligned}
I_S &= \frac{\pi}{12}\{\cos 0 + 4\cos\frac{\pi}{4} + \cos\frac{\pi}{2}\} \\
&\triangleq \frac{\pi}{12}\{1 + 4 \times 0.707 + 0\} \\
&= \frac{3.828\pi}{12} \\
&= 1.002 \quad \text{(3 d.p.)}
\end{aligned}
$$

Thus the error is .002 and we see that the Simpson formula with 3 function evaluations is more accurate than the Trapezoidal rule with four evaluations.

Similarly, using 4 strips we obtain an estimate

$$
\begin{aligned}
I_S &= \frac{\pi}{24}\{\cos 0 + 4\cos\frac{\pi}{8} + 2\cos\frac{\pi}{4} + 4\cos\frac{3\pi}{8} + \cos\frac{\pi}{2}\} \\
&\triangleq \frac{\pi}{24}\{1 + 4 \times 0.9239 + 2 \times 0.7071 + 4 \times 0.3827 + 0\} \\
&= 1.0001(5)
\end{aligned}
$$

This is in error by 0.0001(5) which is considerably less than the corresponding error obtained when using the Trapezoidal rule.

Simpson's rule is, of course, exact for a quadratic function, but Problem 5 of this Section shows that it is *exact for a cubic*; this means we have hit on a very accurate formula for the simplicity it possesses. By methods similar to those employed for the Trapezoidal rule, we can show that the truncation error ϵ_S is such that

$$
|\epsilon_S| \leqslant \frac{(x_{2n} - x_0)h^4}{180} M \tag{8.15}
$$

where

$$
M = \max_{x \in [x_0, x_{2n}]} |f^{\text{iv}}(x)|
$$

i.e. the maximum absolute value that $f^{\text{iv}}(x)$ achieves in the interval $x_0 \leqslant x \leqslant x_{2n}$.

For example, with

$$
\int_0^{\pi/2} \cos x \, dx, \quad f^{\text{iv}}(x) = \cos x
$$

which has maximum absolute value $M = 1$, with 2 strips

$$
|\epsilon_S| \leqslant \frac{\pi}{2} \cdot \frac{\pi^4}{256} \cdot \frac{1}{180} \cdot 1 = 0.003 \quad \text{(3 d.p.)}
$$

and this is over-generous.

The accuracy achieved and the maximum error estimated for differing strip widths are shown in Table 8.6.

Table 8.6

Number of strips	Estimate	Error	Estimated error
2	1.0023	0.2E–02	0.3E–02
4	1.0001	0.1E–03	0.2E–03
6	1.0000	0.3E–04	0.4E–04
8	1.0000	0.8E–05	0.1E–04
10	1.0000	0.3E–05	0.5E–05
12	1.0000	0.2E–05	0.3E–05
14	1.0000	0.9E–06	0.1E–05
16	1.0000	0.5E–06	0.8E–06
18	1.0000	0.3E–06	0.5E–06
20	1.0000	0.2E–06	0.3E–06

(Remember 0.3E–02 means 0.3×10^{-2})

Example 4

$$I = \int_{-2}^{2} e^{-x^2/2}\,dx \quad \text{[The value obtained by other methods = 2.3925 (4 d.p.)]}$$

(a) *Trapezoidal rule, 4 strips,* $h = 1.0$

$$I_T = \frac{1.0}{2}\left(e^{-(-2)^2/2} + 2e^{-(-1)^2/2} + 2e^{-(0)^2/2} + 2e^{-(1)^2/2} + e^{-(2)^2/2}\right)$$

$= 2.3484$ (4 d.p.) (Error = 0.0441)

Truncation error $\sim [nh^3/12] f''(\xi)$

$f'(x) = -xe^{-x^2/2}$

$f''(x) = e^{-x^2/2}(x^2 - 1)$

$f'''(x) = e^{-x^2/2}(-x^3 + x + 2x) = 0$ when $x = 0$ or $x^2 - 3 = 0$

Now $f''(0) = -1$, $f''(\sqrt{3}) = 2e^{-3/2}$, $f''(-\sqrt{3}) = 2e^{-3/2}$

therefore

$$\text{Max}\, f''(x) = 2e^{-3/2} = 0.4462$$

but

$$\text{Max}\, |f''(x)| = 1$$

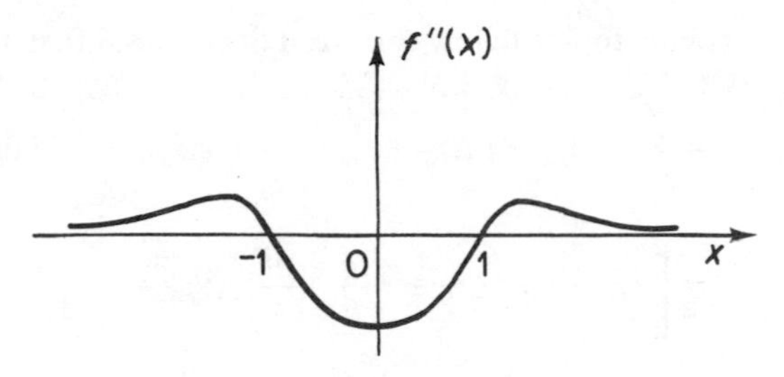

therefore

$$\text{Max error} = 4 \cdot \frac{1^3}{12} = 0.3333 \quad (4 \text{ d.p.})$$

(b) *Simpson's rule, with 2 double strips*

$$I_S = \frac{1.0}{3}\left(e^{-(-2)^2/2} + 4e^{-(-1)^2/2} + 2e^{-(0)^2/2} + 4e^{-(1)^2/2} + e^{-(2)^2/2}\right)$$

$$= 2.3743$$

Error = 0.0182.
Here the truncation error would be too tedious to estimate.

A way of ensuring that the approximation obtained is reasonable in cases where the evaluation of M is itself a complicated and even intractable problem, without resort to numerical approximations, is to halve the step size and repeat until two successive estimates agree to the accuracy required. Even so, this approach has its dangers.

If we wish to develop other formulae we need not resort to these ad hoc approaches; instead, there is a more general line of attack.

Integration of finite difference interpolation formulae

Since both the Simpson and Trapezoidal methods have, in effect, interpolated the function either by parabola or by straight line, it would seem reasonable to try to produce other formulae by integrating interpolation formulae; hopefully, we shall have an estimate of the error when using such a formula.

We illustrate the idea by one example and quote other results. The Newton-Gregory forward difference interpolation formula was given as

$$f(x_0 + ph) = f(x_0) + p\Delta f(x_0) + \tfrac{1}{2}p(p-1)\Delta^2 f(x_0) + \ldots\ldots \qquad (6.17)$$

Given the tabulated values (and hence the spacing, h) we may regard $f(x_0 + ph)$ as a function of p and we rewrite (6.17) as

$$f_p = f_0 + p\Delta f_0 + \tfrac{1}{2}p(p-1)\Delta^2 f_0 + \ldots\ldots \qquad (8.16)$$

Now

$$\int_{x_0}^{x_0+nh} f(x)\mathrm{d}x \equiv \int_0^n f_p \,.\, h\mathrm{d}p \qquad (8.17)$$

since $x = x_0 + ph$. (If you do not follow this, read Section 8.6 first.)

$$\int_0^n f_p \,\mathrm{d}p = \int_0^n [f_0 + p\Delta f_0 + \tfrac{1}{2}p(p-1)\Delta^2 f_0 + \ldots\ldots]\,\mathrm{d}p$$

$$= \left[pf_0 + \tfrac{1}{2}p^2\Delta f_0 + \left(\frac{p^3}{6} - \frac{p^2}{4}\right)\Delta^2 f_0 + \ldots\ldots\right]_0^n$$

$$= nf_0 + \tfrac{1}{2}n^2\Delta f_0 + \frac{n^2}{12}(2n-3)\Delta^2 f_0 + \ldots\ldots \qquad (8.18)$$

If we take $n = 1$ we have

$$\int_{x_0}^{x_0+h} f(x)\,\mathrm{d}x = h\{f_0 + \tfrac{1}{2}\Delta f_0 - \frac{1}{12}\Delta^2 f_0 + \ldots\ldots\} \qquad (8.19)$$

and if we write Δf_0 as $(f_1 - f_0)$ we obtain

$$\int_{x_0}^{x_0+h} f(x)\,\mathrm{d}x = h\left\{\frac{f_0 + f_1}{2} - \frac{1}{12}\Delta^2 f_0 + \ldots\ldots\right\} \qquad (8.20)$$

Hence if we can ignore second-order differences we have the Trapezoidal rule; if we can only ignore third-order and higher differences, we have an estimate of the correction to the Trapezoidal rule as

$$-\frac{h}{12}\Delta^2 f_0$$

(In fact we can show that the third order term is

$$+\frac{1}{24}\Delta^3 f_0\,)$$

To estimate the total error using the Trapezoidal rule, we simply add up the contributions from each strip. Of course, we should be in some difficulty near the end of a table : see the problems at the end of this section. This allows us to quote the error in estimating (8.17) by the Trapezoidal rule as

$$\epsilon_T = \frac{1}{12}(\Delta f_0 - \nabla f_n) - \frac{1}{24}(\Delta^2 f_0 + \nabla^2 f_n) \qquad (8.21)$$

Example 5

Evaluate $\int_0^1 x^2\,\mathrm{d}x$ using the Trapezoidal rule with 4 strips. (We know the true value to be 1/3.)

Table 8.7, given below, is the difference table.

Table 8.7

x	$f(x)$	Δ	Δ^2	Δ^3
0.00	0.0000			
		0.0625		
0.25	0.0625		0.1250	
		0.1875		0
0.50	0.2500		0.1250	
		0.3125		0
0.75	0.5625		0.1250	
		0.4375		
1.00	1.0000			

Here the Trapezoidal rule gives

$$\frac{1}{8}\{0 + 2 \times 0.0625 + 2 \times 0.2500 + 2 \times 0.5625 + 1.000\}$$

$$= \frac{1}{8} \times 2.75$$

$$= 0.344 \quad \text{(3 d.p.)}$$

The estimate of the error is

$$\frac{1}{12}(0.0625 - 0.4375) - \frac{1}{24}(0.1250 + 0.1250)$$

$$= \frac{1}{12}(-0.375) - \frac{1}{24}(0.25)$$

$$= -0.031 - 0.010$$

$$= -0.041 \quad \text{(3 d.p.)}$$

In fact, we need not have used the second differences.

Of course, in practice we would apply (8.19) to each strip as far as the last significant difference.

Reverting to (8.18) we can put $n = 2$ to obtain eventually

$$\int_{x_0}^{x_0+2h} f(x)\mathrm{d}x = \frac{h}{3}\{f_0 + 4f_1 + f_2\} - \frac{h}{90}\Delta^4 f_0 + \ldots\ldots \tag{8.22}$$

(try to derive this yourself) and we see that we have recovered Simpson's rule, and we can quote the approximate error as

$$\frac{h}{90}\Delta^4 f_0$$

In the centre of a table we have the errors in

$$\int_{x_r}^{x_r+h} f(x)\mathrm{d}x$$

by the Trapezoidal rule as

$$\frac{h}{24}(\delta^2 f_r + \delta^2 f_{r+1}) \tag{8.23}$$

and to

$$\int_{x_r}^{x_r+2h} f(x)\mathrm{d}x$$

by Simpson's rule as

$$\frac{h}{90}\delta^4 f_r \tag{8.24}$$

In Problems 10 and 11 we show other integration formulae; they are all in the form

$$h\sum_{r=0}^{n} w_r f_r$$

where w_r are weights such that

$$\sum_{r=0}^{n} w_r = n$$

In the Trapezoidal rule the weights are 1/2, 1/2 with $n = 1$ and in the Simpson rule the weights are 1/3, 4/3, 1/3 with $n = 2$.

Problems

1. Given that the mean value of $f(x)$ over (a, b) is

$$\frac{1}{b-a}\int_a^b f(x)\mathrm{d}x$$

calculate the mean of $6x - x^2$ over (0, 6) by

(i) the Trapezoidal rule
(ii) Simpson's rule } with 6 intervals

Why is (i) in error and (ii) exact?

2. The following table gives values of the Fresnel integral which occurs in optics and is defined by

$$S(x) = \int_0^x \sin\left(\frac{\pi}{2}t^2\right)\mathrm{d}t$$

x	0.60	0.64	0.68	0.72	0.76	0.80	0.84	0.88	0.92	0.96
$S(x)$	0.111	0.133	0.158	0.176	0.216	0.249	0.284	0.321	0.359	0.398

It is known that all the values for $S(x)$ in the table are correct apart from one isolated error. Form a difference table and locate this error. Correct the error and check your result by performing a numerical integration over one step, noting that the Trapezoidal rule may be used for three figure accuracy.

3. Draw a flow chart for Simpson's rule and write the corresponding computer program. Where appropriate,use this program to evaluate some of the integrals occurring in the following problems.

4. Evaluate

$$\int_1^3 \frac{dt}{t}$$

using Simpson's rule with 5 ordinates.
Estimate the error and compare this with the actual error by evaluating the integral analytically. How many ordinates would be needed to be sure of making the error less than 0.001?
Evaluate the integral by Simpson's rule to three decimal places using this number of ordinates.

5. Evaluate

$$\int_0^4 x^3\,dx$$

using Simpson's rule and 4 strips. Compare with the analytical result.
Now evaluate

$$\int_a^b x^3\,dx$$

by Simpson's rule and compare with the analytical result (use 4 strips).

6. The function $J(t)$ is defined by

$$J(t) = \frac{2}{\pi}\int_0^{\pi/2} \cos(t \sin x)dx$$

(i) Using Simpson's rule with 4 intervals, evaluate $J(0.5)$.

(ii) Write a computer program, giving any data which may be required, to evaluate $J(t)$ for $t = 1, 2, 3, 4, 5$, using Simpson's rule with 32 intervals.

7. What error is involved when

$$\int_0^1 \frac{dx}{1 + x^2}$$

is evaluated to 4 d.p. by Simpson's rule dividing the interval into four equal parts? (C.E.I.)

8. Determine numerically

$$\int_0^{\pi/4} \sin x \, dx$$

by using Simpson's rule with a step length so chosen that the truncation error is less than 0.001. Find a bound for the round-off error in your answer. (C.E.I.)

9. Evaluate

$$\int_0^1 \sqrt{x} \, dx$$

by using Simpson's rule with 4 strips.
Put $x = \sin^2\theta$ and evaluate the new integral, again with 4 strips. Why might you expect the second result to be better than the first?

10. (i) Expand $\sqrt{1 - e^2 \sin^2\theta}$ in powers of $\sin\theta$. Hence determine the Elliptic Integral

$$\int_0^{\pi/2} \sqrt{1 - \tfrac{1}{2}\sin^2\theta} \, d\theta$$

(ii) Compare the above with the result of integration over 6 strips using **Weddle's rule**

$$\int_{x_0}^{x_6} f(x)dx \simeq \frac{3h}{10}(f_0 + 5f_1 + f_2 + 6f_3 + f_4 + 5f_5 + f_6)$$

11. Given the following table

x	−0.2	−0.1	0	0.1	0.2	0.3	0.4	0.5	0.6	0.7
e^{-x^2}	0.9608	0.9900	1.0000	0.9900	0.9608	0.9139	0.8521	0.7788	0.6977	0.6126

evaluate to 4 d.p.

$$\int_{-0.2}^{0.7} e^{-x^2} \, dx$$

(i) by use of the Weddle rule [see Problem 10(ii)] and **Three-eighths rule**:

$$\int_{x_0}^{x_3} f(x)\mathrm{d}x = \frac{3h}{8}(f_0 + 3f_1 + 3f_2 + f_3)$$

(ii) by expanding e^{-x^2} and integrating term by term.

12. A particle moves along a straight line so that at time t its distance s from a fixed point of the line is given by

$$\frac{\mathrm{d}s}{\mathrm{d}t} = t\sqrt{8 - t^3}$$

Use Simpson's rule with 8 strips to calculate the approximate distance travelled by the particle from $t = 0$ to $t = 2$. (L.U.)

13. Apply Simpson's rule with 8 strips to evaluate the integral

$$\int_0^{0.4} \sqrt{1 + x^3}\,\mathrm{d}x$$

using four figure tables in your calculations.
Obtain an approximate value of this integral by using the binomial expansion of $(1 + x^3)^{1/2}$ as far as the term in x^6 and integrating this expansion. (L.U.)

14. Form the difference table of the function given in Problem 11. Of the difference formulae (i) – (iv) given in this problem, which is the most suitable in the evaluation of

$$\int_0^{0.1} e^{-x^2}\,\mathrm{d}x$$

(i) using the existing data
(ii) if the table was available for $x \geqslant 0$ only
(iii) if the table was available for $x \leqslant 0.1$ only?

By successively applying formula (iv), generate a table of values of the Normal Probability Integral

$$\frac{1}{\sqrt{2\pi}} \int_{-\infty}^{x} e^{-\frac{1}{2}t^2}\,\mathrm{d}t$$

to 3 d.p. for $x = 0.1, 0.2, 0.3$, assuming

$$\frac{1}{\sqrt{2\pi}} \int_{-\infty}^{0} e^{-\frac{1}{2}t^2}\,\mathrm{d}t = 0.5$$

(i) $$\int_{x_0}^{x_1} f(x)\mathrm{d}x = h\,\{f_0 + \frac{1}{2}\Delta_0 - \frac{1}{12}\Delta_0{}^2 + \frac{1}{124}\Delta_0{}^3 - \frac{19}{720}\Delta_0{}^4\}$$

(ii) $$= h\{f_1 - \frac{1}{2}\nabla_1 - \frac{1}{12}\nabla_1{}^2 - \frac{1}{24}\nabla_1{}^3 - \frac{19}{720}\nabla_1{}^4\}$$

(iii) $$= h\{f_0 + \frac{1}{2}\nabla_0 + \frac{5}{12}\nabla_0{}^2 + \frac{3}{8}\nabla_0{}^3 + \frac{251}{720}\nabla_0{}^4\}$$

(iv) $$= h\{\mu f_{1/2} - \frac{1}{12}\mu\delta_{1/2}{}^2 + \frac{11}{720}\mu\delta_{1/2}{}^4 - \frac{191}{60480}\mu\delta_{1/2}{}^6\}$$

15. (i) Find

$$\int_0^{1/2} \log_e(1+x)\,dx$$

by expanding $\log_e(1+x)$ as a Taylor's series.

(ii) Repeat (i) with $\int_0^1 \log_e(1+x)\,dx$ and $\int_0^4 \log_e(1+x)\,dx$

(iii) Use series expansions to evaluate $\int_0^4 \frac{e^x - e^{-x}}{2x}\,dx$

16. The quarter-perimeter of the ellipse $\frac{x^2}{a^2} + \frac{y^2}{b^2} = 1$ is

$$\int_0^a \sqrt{\frac{a^2 - m^2x^2}{a^2 - x^2}}\,dx$$

where $m^2 = 1 - b^2/a^2$.
Use infinite series to evaluate the integral in the case when $b = 0.8a$.

17. Evaluate approximately

$$\int_0^{1/2} e^{-x}\,dx$$

(i) by expanding the integrand in powers of x
(ii) using Simpson's rule with eleven ordinates.

In each case give your answer correct to three places of decimals. (L.U.)

18. If $f(x) = Ax^2 + Bx + C$, where A, B and C are constants, prove that

$$\int_a^b f(x)\,dx = \frac{1}{6}(b-a)\left\{f(a) + 4f\left(\frac{a+b}{2}\right) + f(b)\right\}$$

Deduce Simpson's rule for the approximate evaluation of a definite integral.

Apply the rule to obtain an approximate value of

$$\int_1^2 (x-1)^2 \log_{10} x \, dx$$

by dividing up the range of integration into 10 equal parts. (L.U.)

8.6 ANALYTICAL METHODS OF INTEGRATION

In this section we develop some of the tricks of the trade of calculus-based methods of solution. It must be emphasised that the only way to become proficient in this kind of approach is to obtain practice on many examples; the more familiar one becomes with the various methods and lines of attack the more readily one can solve further problems.

We have already given a table of some standard integrals on page 432 and we now extend these and introduce some techniques which are analogous to those used in differentiation.

(i) **Substitution or Change of Variable**

In differentiation we have the *chain rule*

$$\frac{dy}{dx} = \frac{dy}{du} \cdot \frac{du}{dx}$$

Suppose we require $F(x) = \int f(x)\,dx$

so that $F'(x) = f(x)$

Now if x is a continuous function of u, say $x = \phi(u)$, then $F\{\phi(u)\}$ will be a continuous function of u (for an appropriate interval) and

$$\frac{dF}{du} = \frac{dF}{dx} \cdot \frac{dx}{du} = f(x) \,.\, \phi'(u) = f\{\phi(u)\}\phi'(u)$$

Hence $F(x) = F\{\phi(u)\} = \int f\{\phi(u)\}\phi'(u)\,du$

Let us take examples to see how this is used.

Example 1

$$F(x) = \int \frac{dx}{\sqrt{a^2 - x^2}}$$

Let $x = \phi(u) = a \sin u$.

Then

$$\frac{dx}{du} = \phi'(u) = a \cos u$$

and

$$f\{\phi(u)\} = \frac{1}{\sqrt{a^2 - a^2 \sin^2 u}} = \frac{1}{a\sqrt{1 - \sin^2 u}}$$

$$= \frac{1}{a \cos u}$$

$$\text{Hence } F(x) = \int \frac{1}{a \cos u} \cdot a \cos u \, du$$

$$= \int du$$

$$= u + C$$

$$= \sin^{-1} \frac{x}{a} + C$$

which is one of our standard forms.

Example 2

$$F(x) = \int \frac{dx}{(1 + x^2)^{3/2}}$$

Let $x = \phi(u) = \tan u$. Then

$$\frac{dx}{du} = \phi'(u) = \sec^2 u$$

and

$$f\{\phi(u)\} = \frac{1}{(1 + \tan^2 u)^{3/2}}$$

$$= \frac{1}{\sec^3 u}$$

$$\text{Hence } F(x) = \int \frac{1}{\sec^3 u} \cdot \sec^2 u \, du = \int \cos u \, du = \sin u + C$$

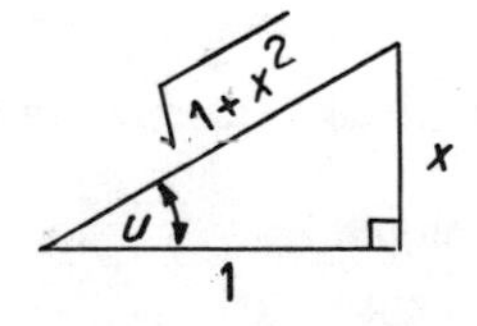

Now $x = \tan u$, hence $\sin u = \dfrac{x}{\sqrt{1 + x^2}}$

therefore

$$\int \frac{dx}{(1 + x^2)^{3/2}} = \frac{x}{\sqrt{1 + x^2}} + C$$

We usually shorten the writing in the following way

Example 3

$$I = \int \frac{dx}{\sqrt{2 + 3x}}$$

Put $u = 2 + 3x$ (or $x = \frac{u - 2}{3}$)

therefore

$$\frac{du}{dx} = 3 \quad \text{or} \quad \frac{dx}{du} = \frac{1}{3}$$

therefore

$$I = \int \frac{dx}{\sqrt{2 + 3x}} = \int \frac{1}{\sqrt{u}} \frac{dx}{du} du = \int \frac{1}{\sqrt{u}} \cdot \frac{1}{3} du$$

$$= \frac{1}{3} \int u^{-1/2} du$$

$$= \frac{1}{3} \frac{u^{1/2}}{(1/2)} + C$$

$$= \frac{2}{3} \sqrt{2 + 3x} + C$$

Example 4

$$I = \int \frac{dx}{x + 1 - \sqrt{2x + 1}}$$

Put $u^2 = 2x + 1$ (or $x = \frac{u^2 - 1}{2}$)

therefore

$$2u = 2 \frac{dx}{du}$$

Hence

$$I = \int \frac{1}{1/2(u^2 - 1) + 1 - u} u du$$

$$= 2 \int \frac{u du}{(u - 1)^2}$$

We have obtained an integral which is not yet a standard form. However, we can use a further substitution.

Let us put $v = (u - 1)$ or $u = v + 1$

therefore

$$1 = \frac{du}{dv}$$

Hence

$$I = 2\int \frac{(v+1)}{v^2} \cdot 1 \cdot dv$$

$$= 2\int \left(\frac{1}{v} + \frac{1}{v^2}\right) dv$$

$$= 2\left[\log_e v - \frac{1}{v}\right] + C$$

$$= 2\left[\log_e (u-1) - \frac{1}{(u-1)}\right] + C$$

$$= 2\left[\log_e (\sqrt{2x+1} - 1) - \frac{1}{\sqrt{2x+1} - 1}\right] + C$$

If we are given a definite integral we do not have to revert to x in the integration – we change to corresponding limits of integration as shown in the next example.

Example 5

$$I = \int_{\pi/4}^{\pi/2} \frac{\cos\theta}{\sin^3\theta}\, d\theta$$

Put $\sin\theta = u$

therefore

$$\cos\theta \frac{d\theta}{du} = 1$$

therefore

$$\frac{d\theta}{du} = \frac{1}{\cos\theta}$$

Hence

$$I = \int_{\theta=\pi/4}^{\theta=\pi/2} \frac{\cos\theta}{u^3} \cdot \frac{1}{\cos\theta} \cdot du = \int_{\theta=\pi/4}^{\theta=\pi/2} \frac{1}{u^3}\, du$$

Now when θ goes over the interval $[\pi/4, \pi/2]$, u will take corresponding values in the interval $[\sin \pi/4, \sin \pi/2]$, that is, $[1/\sqrt{2}, 1]$.
Hence

$$I = \int_{1/\sqrt{2}}^{1} \frac{1}{u^3}\, du = \left[-\frac{1}{2u^2}\right]_{1/\sqrt{2}}^{1} = \left(-\frac{1}{2}\right) - \left(\frac{-1}{2\,.\,½}\right)$$

$$= 1/2$$

Problems I

1. Using appropriate substitutions obtain the standard forms

$$\int \frac{1}{\sqrt{x^2 - a^2}}\,dx = \cosh^{-1}\frac{x}{a} + C$$

$$\int \frac{1}{\sqrt{x^2 + a^2}}\,dx = \sinh^{-1}\frac{x}{a} + C$$

$$\int \frac{1}{x^2 + a^2}\,dx = \frac{1}{a}\tan^{-1}\frac{x}{a} + C$$

$$\int \frac{1}{a^2 - x^2}\,dx = \frac{1}{a}\tanh^{-1}\frac{x}{a} + C$$

$$\int \frac{1}{x^2 - a^2}\,dx = -\frac{1}{a}\coth^{-1}\frac{x}{a} + C$$

2. Evaluate

(a) $\displaystyle\int 5x\sqrt{1 + 2x^2}\,dx$ (b) $\displaystyle\int_0^1 \frac{4x\,dx}{(2 - x^2)^{3/2}}$

(c) $\displaystyle\int_0^{\pi/3} \frac{\sin x\,dx}{(1 + \cos x)^2}$ (d) $\displaystyle\int_0^{\pi/2} \frac{\cos x\,dx}{(1 + \sin x)}$

(e) $\displaystyle\int_1^2 \frac{dx}{x^2\sqrt{x^2 + 1}}$ (put $x = 1/u$)

(f) $\displaystyle\int \sqrt{\frac{x}{1 - x}}\,dx$ (put $x = \sin^2 u$)

(ii) **Integration by Parts**

This stems from the product rule for differentiation. If u and v are functions of x, then we know that

$$\frac{d}{dx}(u.v) = u\frac{dv}{dx} + v\frac{du}{dx}$$

Integrate with respect to x to obtain

$$\int \frac{d}{dx}(uv)\,.\,dx = \int u\frac{dv}{dx}\,.\,dx + \int v\frac{du}{dx}\,.\,dx$$

therefore

$$\left[uv\right] = \int u\frac{dv}{dx}\,.\,dx + \int v\frac{du}{dx}\,dx$$

Hence

$$\int u\frac{dv}{dx}\,dx = \left[uv\right] - \int \frac{du}{dx}\,.\,v\,.\,dx$$

or, *in words,* the integral of a product comprises two terms: the first × (integral of the second) minus the integral of [derivative of the first × (integral of the second)].

Example 1

$$I = \int x\sin x\,dx = x(-\cos x) - \int(-\cos x)(1)dx$$

$$= -x\cos x + \int \cos x\,dx$$

$$= -x\cos x + \sin x + C$$

Notice that the power of x has been reduced in the operation and the integral has become *unmixed.*

Example 2

$$I = \int x^2 e^x dx = x^2 e^x - \int 2x e^x dx$$

($x^2 e^x$: integral of second; $2x e^x$: derivative of the first)

$$= x^2 e^x - \left[2x e^x - \int 2e^x dx\right] \quad \text{applying the rule once more}$$

$$= x^2 e^x - 2x e^x + 2e^x + C$$

Note that

$$\int e^x x^2 dx = e^x\,.\,\frac{x^3}{3} - \int e^x\,.\,\frac{x^3}{3}\,dx \quad !!$$

so it is necessary to choose carefully the order of writing the terms in the integrand to lead to an integral which we can evaluate.

Example 3

$$I = \int 3\log_e x\,dx = \int \log_e x\,.\,3dx = \log_e x\,.\,3x - \int \frac{1}{x}3x\,dx$$

$$= 3x\log_e x - 3x + C$$

In these examples, we suppress the arbitrary constant until the last integral is evaluated.

Problems II

Evaluate

(i) $\int x \sin x \, dx$ (ii) $\int_0^1 x \, e^{-x} dx$

(iii) $\int \sqrt{x} \log x \, dx$

(iv) $\int \log x \, dx$ (write as $\int 1 . \log x \, dx$)

(v) $\int e^x \sin 2x \, dx$ (integrate twice by parts and form an equation for the integral)

(vi) $\int_0^1 \sin^{-1} x \, dx$

(vii) $\int \tan^{-1} x \, dx$

(iii) **Integration of rational functions**

A *rational function* is of the form $P(x)/Q(x)$ where $P(x)$ and $Q(x)$ are polynomials in x.

(a) $P(x) = 1$

Example 1

$$\int \frac{dx}{(x+3)^2} = -\frac{1}{(x+3)} + C \quad (\text{put } x + 3 = u)$$

Example 2

$$\int \frac{dx}{4x - x^2} = \frac{1}{4} \int \frac{dx}{x} + \frac{1}{4} \int \frac{dx}{4 - x} \quad (\text{by partial fractions})$$

$$= \frac{1}{4} \log x - \frac{1}{4} \log (4 - x) + C$$

$$= \frac{1}{4} \log \left(\frac{x}{4-x}\right) + C$$

$$= \log A \left(\frac{x}{4-x}\right)^{1/4}, \quad \text{where } \log A = C$$

Example 3

$$\int \frac{dx}{x^2 - 6x + 12} = \int \frac{dx}{(x-3)^2 + 3} \quad (\text{put } u = x - 3)$$

$$= \int \frac{du}{u^2 + 3} = \frac{1}{\sqrt{3}} \tan^{-1} \frac{u}{\sqrt{3}} + C$$

$$= \frac{1}{\sqrt{3}} \tan^{-1} [(x-3)/\sqrt{3}] + C$$

(b) *$Q(x)$ factorises*, $P(x) \neq 1$

Here we use partial fractions.

Example

$$I = \int \frac{x+1}{x^2 - 3x + 2} dx$$

Now if $\dfrac{x+1}{x^2 - 3x + 2} \equiv \dfrac{A}{x-2} + \dfrac{B}{x-1}$

$$x + 1 = A(x-1) + B(x-2)$$

therefore comparing coefficients,

$$A + B = 1, \; -A - 2B = 1$$

and hence

$$A = 3, \quad B = -2$$

Therefore,

$$I = \int \frac{3}{x-2} dx - \int \frac{2dx}{x-1}$$

$$= 3 \log(x-2) - 2 \log(x-1) + C$$

$$= \log(x-2)^3 - \log(x-1)^2 + C$$

$$= \log\left\{\frac{(x-2)^3}{(x-1)^2}\right\} + C$$

(c) *Q(x) does* **not** *factorise*

Example 1

$$I = \int \frac{x^3 + 2}{x^2 + 1}\,dx = \int \frac{x(x^2 + 1) - x + 2}{x^2 + 1}\,dx$$

(divide out to leave remainder where the numerator is of lower degree than that of the denominator)

$$= \int x\,dx - \int \frac{x - 2}{x^2 + 1}\,dx$$

$$= \frac{x^2}{2} - \frac{1}{2}\int \frac{2x - 4}{x^2 + 1}\,dx \quad \text{(try to write numerator as multiple of the derivative of } x^2 + 1\text{)}$$

$$= \frac{1}{2}x^2 - \frac{1}{2}\int \frac{2x}{x^2 + 1}\,dx + 2\int \frac{dx}{x^2 + 1} \quad \text{(separate out terms)}$$

$$= \frac{1}{2}x^2 - \frac{1}{2}\log(x^2 + 1) + 2\tan^{-1}x + C$$

We have divided out as far as we can and then attempted to write the numerator of the remainder as a multiple of the derivative of the denominator, leaving a standard form.

Let us take one more example.

Example 2

$$I = \int \frac{x^2 - 8x + 20}{x^2 + 8x + 20}\,dx$$

$$= \int \frac{(x^2 + 8x + 20) - 16x}{x^2 + 8x + 20}\,dx \quad \text{(divide out)}$$

$$= \int \left(1 - \frac{16x}{x^2 + 8x + 20}\right)dx$$

$$= x - 8\int \frac{2x + 8 - 8}{x^2 + 8x + 20}\,dx \quad \left[\text{force } \frac{d}{dx}(x^2 + 8x + 20) \text{ on to numerator}\right]$$

$$= x - 8\log(x^2 + 8x + 20) + 64\int \frac{dx}{(x + 4)^2 + 4} \quad \text{(divide out and complete the square)}$$

$$= x - 8\log(x^2 + 8x + 20) + 64\,.\frac{1}{2}\tan^{-1}\left(\frac{x+4}{2}\right) + C$$

Problems III

Evaluate

(i) $\displaystyle\int_2^3 \frac{dx}{x^2 - 1}$ (ii) $\displaystyle\int \frac{x\,dx}{x^2 - 3x + 2}$

(iii) $\displaystyle\int \frac{16x + 1}{8x^2 - 14x + 3}\,dx$ (iv) $\displaystyle\int \frac{x + 5}{(x^3 - 3x^2 + 4x - 12)}\,dx$

(v) $\displaystyle\int \frac{(8 - x)\,dx}{(x - 2)^2(x + 1)}$ (vi) $\displaystyle\int \frac{dx}{x^4 - 1}$

(vii) $\displaystyle\int \frac{dx}{(x^2 + 8x - 25)}$ (viii) $\displaystyle\int \frac{dx}{(11 - 4x - 2x^2)}$

(ix) $\displaystyle\int_0^1 \left(\frac{x^2 + 4x + 6}{x^2 + 2x + 2}\right)dx$ (x) $\displaystyle\int \frac{(4x + 2)\,dx}{x(x^2 - 4x + 13)}$

(iv) **Integration of irrational functions**

(a) *$Q(x)$ irrational, $P(x) = 1$,* $I = \displaystyle\int \frac{dx}{Q(x)}$

Example 1

$$I = \int \frac{dx}{\sqrt{2x^2 - 7x + 6}} = \frac{1}{\sqrt{2}}\int \frac{dx}{\sqrt{x^2 - (7/2)x + 3}}$$

$$= \frac{1}{\sqrt{2}}\int \frac{dx}{\sqrt{(x - 7/4)^2 - (1/4)^2}}$$

(Put $x - \dfrac{7}{4} = \dfrac{1}{4}\cosh\theta$, $\dfrac{dx}{d\theta} = \dfrac{1}{4}\sinh\theta$)

therefore

$$I = \frac{1}{\sqrt{2}}\int \frac{\frac{1}{4}\sinh\theta\,d\theta}{\frac{1}{4}\sinh\theta}$$

$$= \frac{1}{\sqrt{2}}\cosh^{-1} 4(x - \tfrac{7}{4}) + C$$

Example 2

$$I = \int \frac{\mathrm{d}x}{\sqrt{x^2 + a^2}}$$

By putting $x = a \sinh u$ we have $I = \sinh^{-1} \dfrac{x}{a}$

But we may use low cunning, as follows,

$$I = \int \frac{1 + \dfrac{x}{\sqrt{x^2 + a^2}}}{x + \sqrt{x^2 + a^2}}\, \mathrm{d}x$$

$$= \log(x + \sqrt{x^2 + a^2}) + C$$

But $(\sqrt{x^2 + a^2} - x)(\sqrt{x^2 + a^2} + x) = a^2$, therefore

$$I = \log[a^2/(\sqrt{x^2 + a^2} - x)] + C$$

(To reconcile the two results, read again pages 89 and 90 and modify the argument, replacing x by x/a). How do you explain the disparity of $\log a^2$?

(b) $P(x) = px + q$, $Q(x)$ *irrational*

$$I = \int \frac{px + q}{\sqrt{ax^2 + bx + c}}\, \mathrm{d}x$$

Example

$$I = \int \frac{3x + 2}{\sqrt{3x^2 + 5x + 3}}\, \mathrm{d}x$$

Note that $\dfrac{\mathrm{d}}{\mathrm{d}x}(3x^2 + 5x + 3) = 6x + 5$

$$\text{Now } I = \int \frac{3x + \frac{5}{2}}{\sqrt{3x^2 + 5x + 3}}\, \mathrm{d}x - \int \frac{\frac{1}{2}\, \mathrm{d}x}{\sqrt{3x^2 + 5x + 3}}$$

Put $u = 3x^2 + 5x + 3$ in first integral

$$I = \int \frac{\frac{1}{2}\mathrm{d}u}{\sqrt{u}} - \frac{1}{\sqrt{3}} \int \frac{\frac{1}{2}\mathrm{d}x}{\sqrt{(x + 5/6)^2 + 11/36}}$$

$$= \sqrt{u} - \frac{1}{2\sqrt{3}} \sinh^{-1}\left(\frac{x + 5/6}{\sqrt{11/36}}\right) + C$$

$$= (3x^2 + 5x + 3)^{\frac{1}{2}} - \frac{1}{2\sqrt{3}} \sinh^{-1}\left(\frac{6x + 5}{\sqrt{11}}\right) + C$$

(c) *P(x) irrational,* $Q(x) = 1$

$$I = \int \sqrt{ax^2 + bx + c}\,\mathrm{d}x$$

Example

$$I = \int \sqrt{3x^2 + 12x + 39}\,\mathrm{d}x$$

$$= \sqrt{3}\int \sqrt{x^2 + 4x + 13}\,\mathrm{d}x$$

$$= \sqrt{3}\int \sqrt{(x+2)^2 + 3^2}\,\mathrm{d}x$$

(Put $x + 2 = 3\sinh\theta,\quad \dfrac{\mathrm{d}x}{\mathrm{d}\theta} = 3\cosh\theta$)

$$= \sqrt{3}\int (\sqrt{9\sinh^2\theta + 9})\,3\cosh\theta\,\mathrm{d}\theta$$

$$= 9\sqrt{3}\int \cosh^2\theta\,\mathrm{d}\theta \quad = \frac{9\sqrt{3}}{2}\int (1 + \cosh 2\theta)\,\mathrm{d}\theta$$

$$= \frac{9\sqrt{3}}{2}\left[\theta + \frac{1}{2}\sinh 2\theta\right] + C \quad \text{(Begin to } \textit{undo} \text{ the substitution)}$$

$$= \frac{9\sqrt{3}}{2}\left[\sinh^{-1}\left(\frac{x+2}{3}\right) + \sinh\theta\cosh\theta\right] + C$$

$$= \frac{9\sqrt{3}}{2}\left[\sinh^{-1}\left(\frac{x+2}{3}\right) + \frac{x+2}{3}\sqrt{1 + \left(\frac{x+2}{3}\right)^2}\right] + C$$

$$= \frac{9\sqrt{3}}{2}\left[\sinh^{-1}\left(\frac{x+2}{3}\right) + \frac{x+2}{9}\sqrt{x^2 + 4x + 13}\right] + C$$

Problems IV

1. Evaluate:

(i) $\displaystyle\int \sqrt{x^2 - 2}\,\mathrm{d}x$ (ii) $\displaystyle\int \frac{\mathrm{d}x}{\sqrt{x^2 - 3x + 1}}$

(iii) $\displaystyle\int \frac{(2x+5)}{\sqrt{x^2+5x-6}}\,dx$ (iv) $\displaystyle\int \sqrt{1+4x+3x^2}\,dx$

(v) $\displaystyle\int \frac{x^2+x+1}{\sqrt{x^2-x-1}}\,dx$

(v) **Trigonometric Integrals**

Example 1

$$\int \sin^4 x \cos^3 x \, dx = \int \sin^4 x(1-\sin^2 x)\,d(\sin x)^{\dagger}$$

[Put $s = \sin x$]

$$\equiv \int (s^4 - s^6)\,ds = \frac{\sin^5 x}{5} - \frac{\sin^7 x}{7} + C$$

Example 2

$$I = \int \sin 4x \,.\, \cos 2x \, dx = \frac{1}{2}\int \left[\sin 6x + \sin 2x\right] dx$$

[We have used the trigonometric identity: $2 \sin A \cos B \equiv \sin(A+B) + \sin(A-B)$]

therefore

$$I = \frac{1}{2}\left[-\frac{\cos 6x}{6} - \frac{\cos 2x}{2}\right] + C$$

Use of $t = \tan\theta$ *and* $t = \tan\theta/2$

(a) For a function of $\cos^2\theta$ and/or $\sin^2\theta$ we use the substitution $t = \tan\theta$

Example 1

$$I = \int \frac{1}{a + b\cos^2\theta + c\sin^2\theta}\,d\theta$$

Put $t = \tan\theta$, therefore

---

† Some authors would object to this notation $d(\sin x)$ but it suffices as long as it is clear that we are effectively substituting $u = \sin x$

$$\sin\theta = \frac{t}{\sqrt{1+t^2}}, \quad \cos\theta = \frac{1}{\sqrt{1+t^2}}$$

$$\frac{dt}{d\theta} = \sec^2\theta = 1 + t^2$$

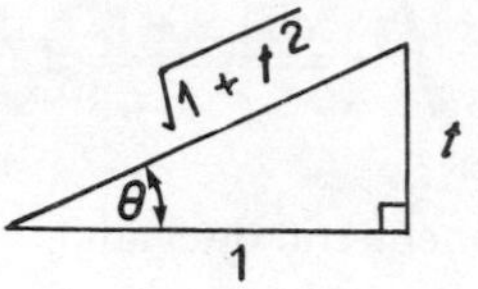

therefore

$$\frac{d\theta}{dt} = \frac{1}{1+t^2}$$

therefore

$$I = \int \frac{dt}{a(1+t^2) + b + ct^2}$$

which can be evaluated by our previous techniques.

(b) For a function of $\sin\theta$ and $\cos\theta$ we use the substitution $t = \tan\theta/2$

Example 2

$$I = \int \frac{1}{a + b\cos\theta + c\sin\theta}\, d\theta$$

Put $t = \tan \tfrac{1}{2}\theta$, therefore

$$\sin\frac{\theta}{2} = \frac{t}{\sqrt{1+t^2}}, \quad \cos\frac{\theta}{2} = \frac{1}{\sqrt{1+t^2}}$$

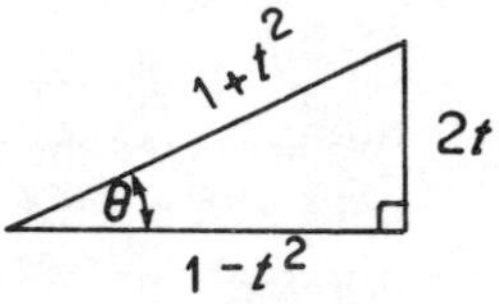

therefore

$$\sin\theta = \frac{2t}{1+t^2}, \quad \cos\theta = \frac{1-t^2}{1+t^2}, \quad \frac{d\theta}{dt} = \frac{2}{1+t^2}$$

therefore

$$I = \int \frac{2dt}{(1+t^2)\left(a + b\dfrac{(1-t^2)}{(1+t^2)} + c\,.\,\dfrac{2t}{1+t^2}\right)}$$

$$= \int \frac{2dt}{a(1+t^2) + b(1-t^2) + 2ct}$$

$$= \int \frac{2dt}{t^2(a-b) + 2ct + (a+b)}$$

which can be evaluated by techniques already considered.

Example 3

Assume the earth is a sphere of radius R, producing a force on a particle outside it such that the acceleration of the particle varies inversely as the square of its distance from the earth's centre. The value of this acceleration at the earth's surface is denoted by g. The particle is released from rest at a height R above the earth: when it is at height x above the earth its speed is $\sqrt{[gR(R-x)/(R+x)]}$. Integrate by using $x = R\cos\theta$ to find the time taken to reach the earth.

Now $\dfrac{dx}{dt} = -\left(\dfrac{gR(R-x)}{x+R}\right)^{1/2}$. Let T be time to reach earth.

Therefore

$$T = -\int_R^0 \left(\frac{R+x}{gR(R-x)}\right)^{1/2} dx$$

Put $x = R\cos\theta$; $\quad \dfrac{dx}{d\theta} = -R\sin\theta$

therefore $R + x = R(1+\cos\theta) = R.2\cos^2\dfrac{\theta}{2}$

and

$$R - x = R(1-\cos\theta) = R.2\sin^2\frac{\theta}{2}$$

When $x = 0$, $\cos\theta = 0 \Rightarrow \theta = \dfrac{\pi}{2}$

$x = R$, $\cos\theta = 1 \Rightarrow \theta = 0$

(continuity of substitution)

therefore

$$T = -\int_0^{\pi/2} \left(\frac{R.2\cos^2\theta/2}{gR.R.2\sin^2\theta/2}\right)^{1/2}(-R\sin\theta)d\theta$$

$$= -\sqrt{\frac{R}{g}}\int_0^{\pi/2} -\frac{\cos\frac{\theta}{2}.\sin\theta}{\sin\frac{\theta}{2}}d\theta$$

$$= \sqrt{\frac{R}{g}}\int_0^{\pi/2} \frac{\cos\frac{\theta}{2}.2\sin\frac{\theta}{2}.\cos\frac{\theta}{2}}{\sin\frac{\theta}{2}}d\theta$$

$$= \sqrt{\frac{R}{g}}\int_0^{\pi/2} 2\cos^2\frac{\theta}{2}\,d\theta$$

$$= \sqrt{\frac{R}{g}} \int_0^{\pi/2} (1 + \cos\theta)\mathrm{d}\theta$$

$$= \sqrt{\frac{R}{g}} \left[\theta + \sin\theta\right]_0^{\pi/2}$$

$$= \sqrt{\frac{R}{g}} \left(\frac{\pi}{2} + 1\right)$$

If you noticed, the integrand became infinite at $x = R$. We shall return to this problem later.

Problems V

Evaluate

(i) $\displaystyle\int \frac{\mathrm{d}x}{3 + 5\cos x}$ (ii) $\displaystyle\int \frac{\mathrm{d}x}{1 - \cos x + \sin x}$

(iii) $\displaystyle\int \frac{\mathrm{d}t}{1 + \sin^2 t}$ (iv) $\displaystyle\int \frac{\mathrm{d}t}{25 - 24\sin^2 t}$

(v) $\displaystyle\int \frac{2 + \sin x}{5\sin x - 4}\,\mathrm{d}x$

(vi) Infinite Integrals

We now consider cases where there is a *singularity* in the range of integration or where the range of integration is *infinite.*

(a) *Infinite Limits*

$$\int_a^{\infty} f(x)\mathrm{d}x = \lim_{x\to\infty} \int_a^{x} f(x)\mathrm{d}x$$

Example 1

$$\int_1^{\infty} \frac{1}{x^3}\,\mathrm{d}x = \lim_{x\to\infty} \int_1^{x} \frac{1}{x^3}\mathrm{d}x = \lim_{x\to\infty} \left[-\frac{1}{2x^2}\right]_1^x$$

$$= \lim_{x\to\infty} \left[\frac{1}{2} - \frac{1}{2x^2}\right]$$

$$= \frac{1}{2}$$

Example 2

$$\int_0^\infty x^{1/2}\, dx = \lim_{x\to\infty} \int_0^x x^{1/2}\, dx = \lim_{x\to\infty} \left[\frac{2}{3} x^{3/2}\right]_0^x$$

$$= \lim_{x\to\infty} \left[\frac{2}{3} x^{3/2}\right] \qquad \to \text{diverges}$$

Likewise

$$\int_{-\infty}^{\infty} = \lim_{x\to\infty} \int_a^x + \lim_{x\to\infty} \int_{-x}^a \qquad \textit{if both exist}$$

(b) *Suppose $f(x)$ goes infinite in the range of integration*

Example 1

$$\int_0^2 \frac{dx}{\sqrt{2-x}} = \lim_{\epsilon\to 0} \int_0^{2-\epsilon} \frac{dx}{\sqrt{2-x}} = \lim_{\epsilon\to 0} \left[-2(2-x)^{1/2}\right]_0^{2-\epsilon}$$

(since when $x = 2$ the integrand is infinite)

$$= 2 \lim_{\epsilon\to 0} \left[2^{1/2} - \epsilon^{1/2}\right] = 2 . \sqrt{2}$$

Example 2

$$\int_{-1}^1 x^{-1/3}\, dx = \lim_{\epsilon\to 0} \int_{-1}^{-\epsilon} x^{-1/3}\, dx + \lim_{\epsilon\to 0} \int_{\epsilon}^1 x^{-1/3}\, dx$$

(since when $x = 0$ the integrand is infinite)

$$= \lim_{\epsilon\to 0} \left[\frac{3}{2} x^{2/3}\right]_{-1}^{-\epsilon} + \lim_{\epsilon\to 0} \left[\frac{3}{2} x^{2/3}\right]_{\epsilon}^1$$

$$= \lim_{\epsilon\to 0} \left[\frac{3}{2} \epsilon^{2/3}\right] - \frac{3}{2} + \lim_{\epsilon\to 0} \left[\frac{3}{2} - \frac{3}{2} \epsilon^{2/3}\right]$$

$$= -\frac{3}{2} + \frac{3}{2}$$

$$= 0$$

Problems VI

1. Consider $\int_{\epsilon}^1 x \log x\, dx$ and hence evaluate $\int_0^1 x \log x\, dx$

2. Evaluate $\int_0^\infty x\,e^{-ax}\,dx \quad (a > 0)$

3. Find (i) $\int_0^1 \frac{dx}{\sqrt{1-x^2}}$ (ii) $\int_0^1 \frac{dx}{\sqrt{x}}$

(iii) $\int_0^\infty e^{-2\theta} \sin\theta \, d\theta$

4. Given that $\int_0^\infty e^{-x^2}\,dx = \frac{1}{2}\sqrt{\pi}$,

show that $\int_0^\infty x^{-1/2} e^{-x}\,dx = \sqrt{\pi}$

(C.E.I.)

Problems VII (General Problems)

1. Use suitable trigonometric substitutions to evaluate:

(i) $\int_0^1 \frac{dx}{(x^2+1)^{3/2}}$ (ii) $\int_0^1 \sqrt{4-x^2}\,dx$

2. Find $\int x e^{(1+i)x}\,dx$ and hence obtain

$\int x e^x \cos x\,dx$ and $\int x e^x \sin x\,dx$

3. Find $\int (x^2-4)^{1/2}\,dx$ and $\int x^3(x^2-4)^{1/2}\,dx$

4. Evaluate (i) $\int x \tan^{-1} x\,dx$ (ii) $\int_0^{\pi/2} \frac{dx}{3\cos x + 4\sin x + 5}$

5. (i) Evaluate the integrals

(a) $\int_0^1 \frac{(x+1)\,dx}{x^2-2x+4}$ (b) $\int_0^\pi e^x \sin(x/2)\,dx$

(c) $\displaystyle\int_0^1 \frac{e^x\,dx}{2+e^x}$

(ii) In determining the fugacity of a gas which obeys Van der Waal's equation we must find

$$\int\left[\frac{2a}{RTV^2} - \frac{V}{(V-b)^2}\right]dV$$

where R, a, b and T are constants. Evaluate this, using the substitution $z = (V - b)$ in the second part of the integrand.

6. A spherical surface of radius a metres, with centre O, carries an electric charge of surface density σ coulombs/square metre, symmetrically distributed with respect to an axis OX. The potential at a point P on this axis, such that OP $= x$ metres, is given by

$$\int_0^\pi \frac{a^2\,\sigma \sin\theta}{2\epsilon_0(a^2 + x^2 - 2ax\cos\theta)^{1/2}}\,d\theta \text{ volts}$$

Evaluate this integral, with ϵ_0 and σ constant, in the case when $x > a$.

7. Evaluate $\displaystyle\int \sqrt{x}\,\log_e x\,dx$

8. (i) Evaluate (a) $\displaystyle\int_1^2 \frac{dx}{\sqrt{1+2x-x^2}}$ (b) $\displaystyle\int_0^{\pi/2} \cos^2 x \sin^5 x\,dx$

(ii) Find $C = \displaystyle\int xe^{-x}\cos x\,dx$ and $S = \displaystyle\int xe^{-x}\sin x\,dx$

(Hint: Consider $C + iS$)

9. Find the indefinite integrals of

$$\frac{x}{(x^2+x+1)^{1/2}} \quad\text{and}\quad \frac{x}{(x^2+x+1)}$$

10. Evaluate (i) $\displaystyle\int_2^3 \frac{x}{(x-1)(x+2)}\,dx$ (ii) $\displaystyle\int_{-2}^0 \frac{1}{\sqrt{x^2+4x+5}}\,dx$

11. Integrate with respect to x:

(i) $x^2 e^{-x}$ (ii) $\dfrac{\sqrt{1-x^2}}{1+x}$

12. Evaluate (i) $\displaystyle\int_0^1 \frac{x+2}{x^2-2x+4}\,dx$ (ii) $\displaystyle\int \sec^3 x\,dx$

13. Write down the expressions for $\cos\theta$ and $\sin\theta$ in terms of $e^{i\theta}$ and $e^{-i\theta}$ and use these forms to expand $\sin^4 x$ in cosines of multiples of x.

Hence evaluate $\displaystyle\int_0^{\pi/4} \sin^4 x\,dx$

14. The force on a body distant x from a fixed point is $k(x^2+a^2)^{-3/2}$. Find the work done in moving the body from $x = a$ to $x = 2a$.

15. Evaluate the integrals

(i) $\displaystyle\int_0^{\pi} e^{-x}\cos 3x\,dx$ (ii) $\displaystyle\int \frac{1}{\sqrt{2x^2+3x-1}}\,dx$

(iii) $\displaystyle\int \sinh^3 x\,dx$

16. Show that

(i) $\displaystyle\int_0^a f(x)\,dx = \int_0^a f(a-x)\,dx$

(ii) $\displaystyle\int_0^{\pi/2} \log\sin x\,dx = \int_0^{\pi/2} \log\cos x\,dx$

$$= \frac{1}{2}\int_0^{\pi/2} \log\sin 2x\,dx - \frac{\pi}{4}\log 2$$

Deduce that $\displaystyle\int_0^{\pi/2} \log\sin x\,dx = -\frac{\pi}{2}\log 2$

(L.U.)

8.7 APPLICATIONS OF DEFINITE INTEGRATION

(i) Area under a curve

We saw in the early parts of this chapter that the value of

$$\int_a^b f(x)\,dx$$

is the area under the curve $y = f(x)$ between the ordinates $x = a$ and $x = b$ (Figure 8.11).

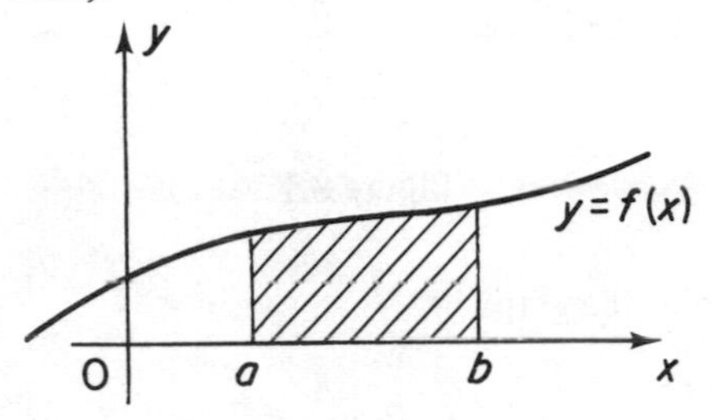

Figure 8.11

The integral is really the limit of the sum of the areas of the strips of width $\Delta_k x$ which can be drawn under the curve - a typical element of area being shown in Figure 8.12(i).

$$\int_a^b f(x)\,dx = \lim_{n\to\infty} \sum_{k=1}^{n} f(x_k).\Delta_k x$$

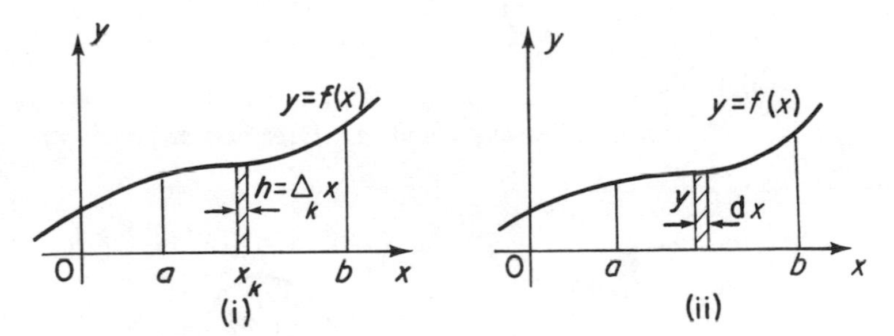

Figure 8.12

We can therefore say that

$$\int_a^b y\,dx$$

is the area under the curve $y = f(x)$ between $x = a$ and $x = b$ and can think of it as being made up of the limit of the sum of the elements of area y dx between a and b as shown in Figure 8.12(ii).

Example 1

A propeller blade has profile $y = \sin 2x - \sqrt{3}\,\sin x$ between $x = 0$ and $x = \pi/6$. Find the area of cross-section.

The graph of this function is sketched in Figure 8.13.

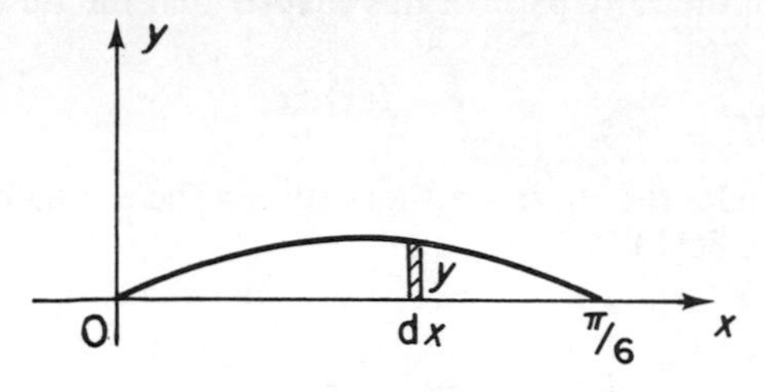

Figure 8.13

$$\text{The area under the graph} = \int_0^{\pi/6} y\,dx = \int_0^{\pi/6} (\sin 2x - \sqrt{3}\sin x)dx$$

$$= \left[-\frac{1}{2}\cos 2x + \sqrt{3}\cos x\right]_0^{\pi/6}$$

$$= \frac{7}{4} - \sqrt{3}$$

Note: As pointed out earlier – we have to be careful that the area lies entirely above the x-axis in using the integral to evaluate the area. For a function having a graph as in Figure 8.14 then

$$\int_a^b f(x)\,dx$$

would give

(the area between c and b) – (area between a and c)

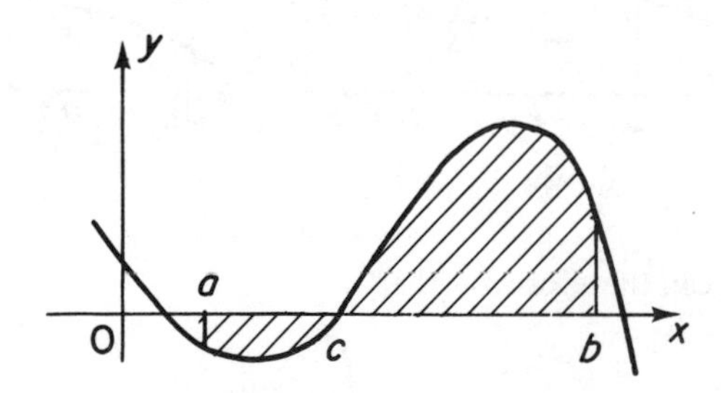

Figure 8.14

(ii) Volume of revolution

The idea of a definite integral being the limit of a sum of elements is used here. As shown in Figure 8.15 (i) we think of the area under the curve $y = f(x)$ as being revolved through 360° about the x-axis and the resulting volume is made up of elemental discs of volume $(\pi y^2)\,dx$.

$$\text{Hence the volume} = \int_a^b \pi y^2\,dx$$

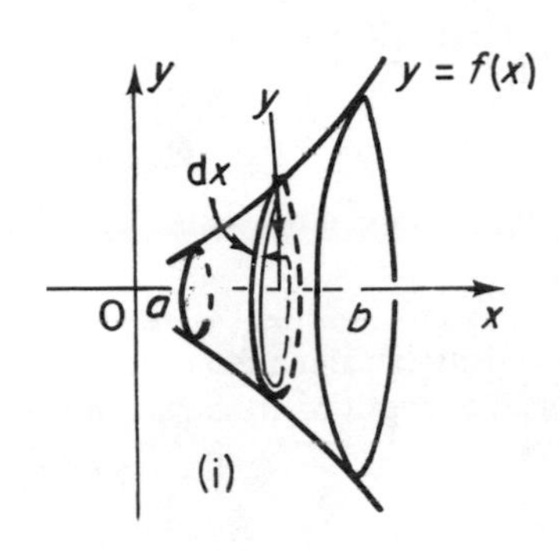

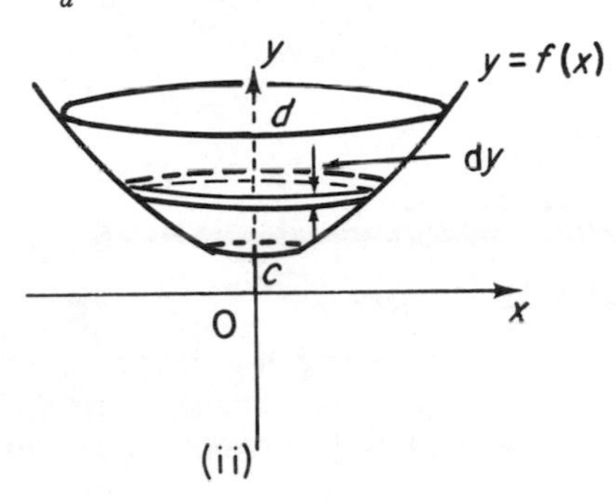

Figure 8.15

Similarly we have for revolution about the y-axis [Figure 8.15 (ii)]

$$\text{Volume} = \int_c^d \pi x^2\,dy$$

Example 2

Find the volume of the cap of a sphere as depicted in Figure 8.16.

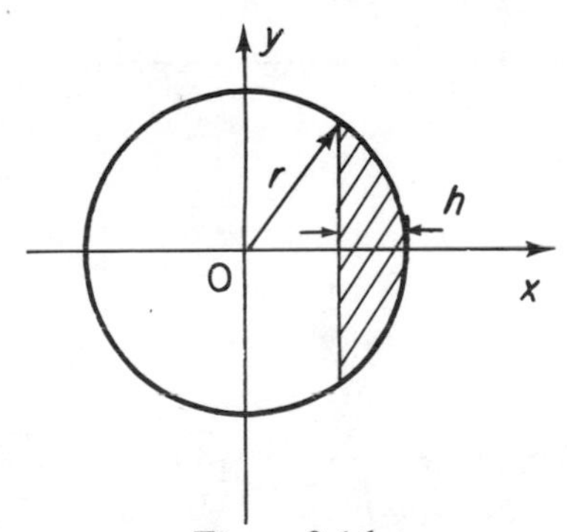

Figure 8.16

Now the circle is $x^2 + y^2 = r^2$, that is, $y^2 = r^2 - x^2$ is the equation of the circumference.

$$\text{Hence the volume} = \pi \int_{r-h}^{r} y^2 \, \mathrm{d}x$$

$$= \pi \int_{r-h}^{r} (r^2 - x^2)\, \mathrm{d}x$$

$$= \pi \left[r^2 x - \frac{x^3}{3} \right]_{r-h}^{r}$$

$$= \pi \left[r^3 - \frac{r^3}{3} - r^2 (r-h) + \frac{(r-h)^3}{3} \right]$$

$$= \frac{\pi h^2}{3} (3r - h)$$

(iii) Perimeter

The perimeter or length of arc of a curve is easily obtained from first principles. Looking at Figure 8.17 we see that the length of arc between A and B is the limit of the sum of small arcs δs.

That is

$$\text{Perimeter} = \int_{A}^{B} \mathrm{d}s$$

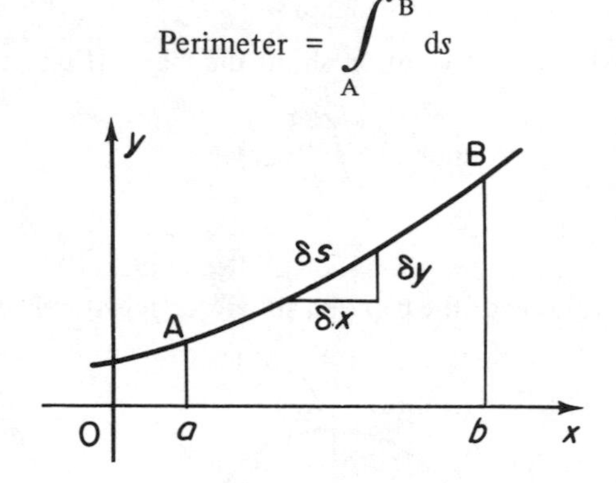

Figure 8.17

But

$$(\delta s)^2 = (\delta x)^2 + (\delta y)^2 \quad \text{(for small enough arcs)}$$

or

$$\left(\frac{\delta s}{\delta x}\right)^2 = 1 + \left(\frac{\delta y}{\delta x}\right)^2$$

Hence

$$\frac{\delta s}{\delta x} = \sqrt{1 + \left(\frac{\delta y}{\delta x}\right)^2}$$

Similarly

$$\frac{\delta s}{\delta y} = \sqrt{1 + \left(\frac{\delta x}{\delta y}\right)^2}$$

Hence

$$\int_A^B ds = \int_a^b \frac{ds}{dx}\,dx = \int_a^b \sqrt{1 + \left(\frac{dy}{dx}\right)^2}\,dx$$

(We can obtain a similar expression for the perimeter as an integral over y.)

Example 3

Find the length of the curve $y = \log_e x$ from $x = 1$ to $x = 2\sqrt{2}$

$$\text{The length, } l = \int_{x=1}^{x=2\sqrt{2}} \sqrt{1 + \left(\frac{1}{x}\right)^2}\,dx$$

$$= \int_1^{2\sqrt{2}} \frac{\sqrt{x^2+1}}{x}\,dx$$

We must now use the techniques we have learned to evaluate this integral.

Put

$$u^2 = x^2 + 1, \quad \frac{du}{dx} = \frac{x}{u}$$

Hence

$$l = \int_{\sqrt{2}}^{3} \frac{u}{x} \cdot \frac{u}{x}\,du = \int_{\sqrt{2}}^{3} \frac{u^2}{u^2 - 1}\,du$$

$$= \int_{\sqrt{2}}^{3} \left(1 + \frac{1}{u^2 - 1}\right) du$$

$$= \frac{1}{2} \int_{\sqrt{2}}^{3} \left(2 + \frac{1}{u-1} - \frac{1}{u+1}\right) du$$

$$= \frac{1}{2}\left[2u + \log \frac{u-1}{u+1}\right]_{\sqrt{2}}^{3}$$

$$= 3 - \sqrt{2} - \frac{1}{2}\log 2 - \frac{1}{2}\log\left(\frac{\sqrt{2}-1}{\sqrt{2}+1}\right)$$

(iv) Area of surface of revolution (Figure 8.18)

Area of curved surface of elemental disc $= (2\pi y)\delta s$

$$\text{Surface area} = \int_{x=a}^{b} 2\pi y \, \mathrm{d}s$$

Figure 8.18

Example 4

The arc of the catenary $y = c \cosh (x/c)$ from $x = 0$ to $x = c$ rotates about the axis of x. Find the area of surface formed.

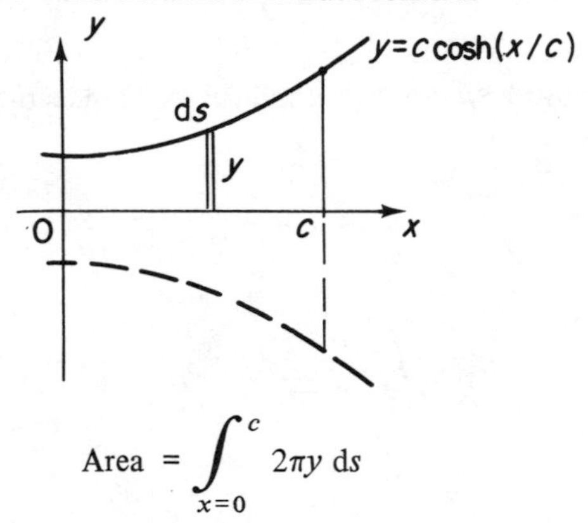

$$\text{Area} = \int_{x=0}^{c} 2\pi y \, \mathrm{d}s$$

Now

$$\frac{\mathrm{d}s}{\mathrm{d}x} = \sqrt{1 + \left(\frac{\mathrm{d}y}{\mathrm{d}x}\right)^2} = \sqrt{1 + \sinh^2 \frac{x}{c}} = \cosh \frac{x}{c}$$

therefore

$$\text{Area} = \int_0^c 2\pi y \frac{\mathrm{d}s}{\mathrm{d}x} \mathrm{d}x = 2\pi c \int_0^c \cosh^2 \frac{x}{c} \mathrm{d}x$$

$$= \pi c \int_0^c \left(1 + \cosh \frac{2x}{c}\right) \mathrm{d}x$$

$$= \pi c\left[x + \frac{c}{2}\sinh\frac{2x}{c}\right]_0^c$$

$$= \pi c(c + \frac{c}{2}\sinh 2) - 0$$

$$= \pi c^2 [1 + \frac{1}{4}(e^2 - \frac{1}{e^2})]$$

(v) **Centroid**

We shall leave you to show from first principles the following results:

(1) Centroid of a perimeter

$$\bar{x} = \left[\int_A^B x\,ds\right] \Big/ S_{AB}, \qquad \bar{y} = \left[\int_A^B y\,ds\right] \Big/ S_{AB}$$

where S_{AB} is the perimeter length.

(2) Centroid of area under a curve

$$\bar{x} = \left[\int_a^b yx\,dx\right] \Big/ \text{Area}, \qquad \bar{y} = \left[\int_a^b \frac{y^2}{2}\,dx\right] \Big/ \text{Area}$$

(3) Centroid of volume of revolution

$$\bar{x} = \pi\left[\int_a^b y^2 x\,dx\right] \Big/ \text{Volume}, \qquad \bar{y} = 0$$

(4) Centroid of surface of revolution

$$\bar{x} = 2\pi\left[\int_A^B yx\,ds\right] \Big/ \text{Surface Area}, \qquad \bar{y} = 0$$

The essential technique used in the application of integration is the building up of the formula from an elemental part. We do not usually try to remember such formulae.

(Some of the following problems require reference to the Appendix on page 728).

Problems

1. Find the surface area of the paraboloid of revolution formed by rotating about the y-axis that portion of the parabola $y = x^2/a$ between $x = 0$ and $x = a$.

2. Show by integration that the volume of a hemisphere of radius a is $(2/3)\pi a^3$ and determine the position of its centroid.

3. A segment of height h is taken from a sphere of radius r as shown.

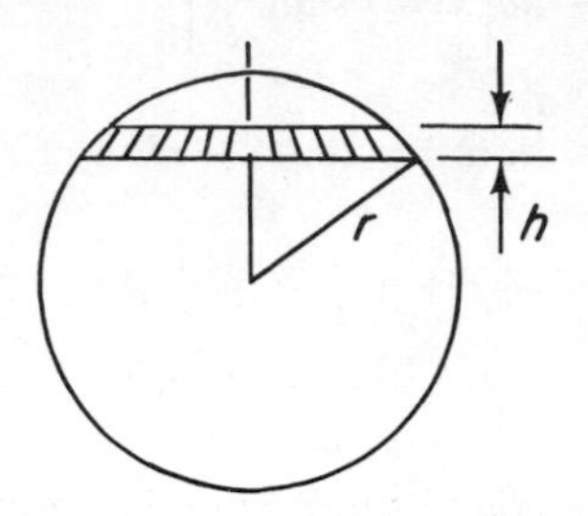

Find the curved surface area of this segment.

4. (i) Determine the area enclosed between the curves $y^2 = 4x$, $y^2 = 4(8 - x)$ and find the x-coordinate of the centroid of this same area.

   (ii) Sketch the cycloid $x = a(\theta - \sin\theta)$, $y = a(1 - \cos\theta)$. Prove that one arch of the cycloid is of length $8a$.

5. (i) Sketch the curve $y^2 = x^2(4 - x)$ and find the finite area which is completely enclosed by part of this curve.

   (ii) The portion of the curve $y^2 = 9/(2x + 5)$ between $x = 1$ and $x = 2$ is rotated about the x-axis to sweep out a solid of revolution. Find the coordinates of the centre of gravity of this solid.

6. (i) Find the coordinates of the centroid of the area enclosed by the curve $y = 8/(x^2 + 4)$, the x-axis and the ordinates $x = -2\sqrt{3}$ and $x = +2\sqrt{3}$.

   (ii) Sketch the curve $ay^2 = x^2(a - x)$ $(a > 0)$ and find the area of its loop.

7. A parabolic mirror is formed by revolving the part of the parabola $y^2 = 4x$ from $x = 0$ to $x = 2$ about Ox. Find the surface area of the mirror so formed.

8. Show that for the catenary $y = c\cosh(x/c)$, the area under the curve between any two ordinates is c times the arc length of the curve between the corresponding points.

9. Sketch the curve $3ay^2 = x^2(a - x)$, a being a positive constant, and find the area and length of its loop.

10. The profile of an impeller blade is bounded by the lines $x = 0.1$, $y = 2x$, $y = e^{-x}$, $x = 1$ and the x-axis. The blade thickness t varies linearly with x, thus $t = (1.1 - x)\tau$, where τ is a constant. Find to 2 d.p. from tables,

or otherwise, the point given by the equation $2x = e^{-x}$.
Hence determine the volume of the blade, showing the result to be approximately $\pi/4$. (C.E.I.)

11. (i) Find the area of the loop of the curve whose equation is $ay^2 = (x - a)(x - 5a)^2$.

    (ii) Determine the distance of the mass centre of the above loop from the y-axis.

12. Find the area bounded by the curves $y^2 = 4x$ and $x^2 = 4y$.

13. Find the area under the curve given parametrically by $x = a(\theta - \sin\theta)$, $y = a(1 - \cos\theta)$ between the points $\theta = 0$ and $\theta = \pi$.

14. Find the length of the curve $x = \log\sec y$ from $y = 0$ to $y = \pi/3$.

15. Find the volume obtained by revolving the following area about the axes Ox, Oy:

    The area under $y = \sin x$ from $x = 0$ to $x = \pi$

16. The area bounded by $y^2x = 4a^2(2a - x)$ and the ordinates $x = a$, $x = 2a$ revolves around Ox. Show that the volume generated is $4\pi a^3(2\log 2 - 1)$.

17. If $e = 24\cos pt + 4\cos 3pt$ and $i = 4\sin pt + \sin 3pt$, prove that $ei = 52\sin 2pt + 20\sin 4pt + 2\sin 6pt$. Hence show that the mean value of ei over the range $t = 0$ to $t = \pi/2p$ is 33.5.

18. Prove that the volume of a segment of a sphere of height h and base radius c is $\pi h(3c^2 + h^2)/6$.

19. Sketch the curve $3y^2 = x^2(x + 1)$. Find the length and area of its loop.

20. (i) Find the length of the curve $y = \log\cos x$ from $x = 0$ to $x = \pi/3$.

    (ii) The area enclosed by the parabola $y^2 = 6ax + 16a^2$ and the ordinates $x = 0$ and $x = 4a$ is rotated about the x-axis to form a solid of revolution. Show that the area of the curved surface of the solid is $436\pi a^2/9$.

21. Find, by integration, the form factor (i.e. R.M.S. Value/Mean Value) of the wave $e = E_1\sin\omega t + E_3\sin 3\omega t$ over the range $t = 0$ to $t = \pi/\omega$.

22. Find the second moment (a) about the x-axis, (b) about the y-axis of the area enclosed between the curves $y = x^{1/2}$ and $y = x$. Find also the volume of the solid of revolution obtained by rotating this area through 2π radians about the x-axis. (L.U.)

23. Find the first two points (starting at $x = 0$) at which the curve $y = e^x \sin x$ crosses the axis of x. Find the height above the axis of x, of the centroid of the area bounded by the curve and the axis of x between these points.

24. The density at any point P of a cone of height h is proportional to the square of the perpendicular distance of that point from the base of the cone. Show that the centre of gravity of the cone is at a distance $\frac{1}{2}h$ from the vertex.

25. Find the coordinates of the centroid of the area between the parabola $y = x^2 - 7x + 12$ and the axes of x and y.

26. The area bounded by the curve $y^2 = 4(2 - x)$ and the y-axis is rotated through two right angles about the x-axis. Show that the volume of the solid of revolution is 8π and find the radius of gyration of this solid about the y-axis.

27. A rope is suspended between two vertical poles of equal height and takes the shape of the curve $y = 5 \cosh x/5$, where the origin O is mid-way between the feet of the poles and y is the height above the ground. If the rope is 450 metres long show that the distance between poles is 45 metres to the nearest metre. Find the height of the centroid of the rope above the ground to the nearest metre. (L.U.)

28. Sketch the graph of $r = a(3 - 2 \cos \theta)$ where r, θ are polar coordinates and a is constant.
Find the area enclosed by this curve and the position of the centroid of the area. (L.U.)

29. A uniform lamina of mass M is in the shape of the area bounded by the curve $y = c \cosh (x/c)$, the coordinate axes and the line $x = c$. Find the distance of the centre of gravity of the lamina from the y-axis, and show that the moment of inertia of the lamina about the y-axis is $Mc^2(3 - 2 \coth 1)$. (L.U.)

30. Sketch the curve whose equation is $y^2 = x^2(1 - x^2)$ and find the area enclosed by one loop of the curve.
If this loop is rotated about the x-axis to generate a surface of revolution, show that this surface encloses a volume $2\pi/15$. (L.U.)

31. Sketch the curve $ay^2 = x^2(a - x)$, where $a > 0$, and show that the tangents at the origin to this curve are perpendicular. Calculate the area A enclosed by the x-axis and that part of the curve which lies in the first quadrant. Calculate also the volume swept out when the area A is rotated through 2π about (i) the x-axis, (ii) the y-axis.
Deduce the coordinates of the centroid of the area A. (L.U.)

8.8 REDUCTION FORMULAE

We give now a few examples of where it is possible to obtain a formula which can be repeatedly used to evaluate an integral.

Example 1

$$I_n = \int_0^{\pi/2} \sin^n\theta \, d\theta$$

$$= \int_0^{\pi/2} \sin^{n-1}\theta \sin\theta \, d\theta$$

Integrating this by parts,

$$I_n = \Big[-\sin^{n-1}\theta\cos\theta\Big]_0^{\pi/2} - \int_0^{\pi/2} (n-1)\sin^{n-2}\theta\cos\theta(-\cos\theta)\,d\theta$$

$$= 0 - 0 + (n-1)\int_0^{\pi/2} \sin^{n-2}\theta(1-\sin^2\theta)\,d\theta$$

$$= (n-1)\int_0^{\pi/2} \sin^{n-2}\theta\,d\theta - (n-1)\int_0^{\pi/2} \sin^n\theta\,d\theta$$

that is,

$$I_n = (n-1)I_{n-2} - (n-1)I_n$$

Therefore

$$I_n = \frac{(n-1)}{n}I_{n-2}$$

Notice that the power of n in the integral has been reduced to $n-2$. Thus repeated use of this **Reduction formula** will reduce the power of n by 2 each time.

Replacing n by $(n-2)$ gives

$$I_{n-2} = \left(\frac{n-3}{n-2}\right)I_{n-4} \quad \text{etc.}$$

Note: $I_1 = \int_0^{\pi/2} \sin\theta\,d\theta = \Big[-\cos\theta\Big]_0^{\pi/2} = -(-1) = 1$

$$I_0 = \int_0^{\pi/2} d\theta = \frac{\pi}{2}$$

Hence, by repeated use of the formula,

if n is **even**, $$I_n = \frac{(n-1)}{n} \cdot \frac{(n-3)}{(n-2)} \cdots\cdots \frac{3}{4} \cdot \frac{1}{2} I_0$$

that is,

$$I_n = \frac{(n-1)}{n} \cdot \frac{(n-3)}{(n-2)} \cdots\cdots \frac{3}{4} \cdot \frac{1}{2} \cdot \frac{\pi}{2}$$

if n is **odd**, $$I_n = \frac{(n-1)}{n} \cdot \frac{(n-3)}{(n-2)} \cdots\cdots \frac{4}{5} \cdot \frac{2}{3} \cdot 1$$

Thus, for example,

$$\int_0^{\pi/2} \sin^7\theta \, d\theta = \frac{6}{7} \cdot \frac{4}{5} \cdot \frac{2}{3} \cdot 1 = \frac{16}{35}$$

and

$$\int_0^{\pi/2} \sin^8\theta \, d\theta = \frac{7}{8} \cdot \frac{5}{6} \cdot \frac{3}{4} \cdot \frac{1}{2} \cdot \frac{\pi}{2} = \frac{35}{256}\pi$$

Note: $$I_n = \int_0^{\pi/2} \sin^n\theta \, d\theta = \int_0^{\pi/2} \cos^n\theta \, d\theta$$

[Put $\theta = \frac{\pi}{2} - t$ then $\sin\theta = \cos t$ and $d\theta = -dt$, therefore

$$I_n = \int_{\pi/2}^{0} -dt \, . \cos^n t = \int_0^{\pi/2} \cos^n t \, dt = \int_0^{\pi/2} \cos^n\theta \, d\theta$$]

Example 2

$$I_{m,n} = \int_0^{\pi/2} \sin^m\theta \cos^n\theta \, d\theta$$

$$= -\int_0^{\pi/2} \sin^{m-1}\theta \cos^n\theta \, d(\cos\theta) \quad \text{(Now use integration by parts)}$$

$$= -\left[\sin^{m-1}\theta \, . \frac{\cos^{n+1}\theta}{n+1}\right]_0^{\pi/2} + \int_0^{\pi/2} \left[(m-1)\sin^{m-2}\theta \times \cos\theta \frac{\cos^{n+1}\theta}{n+1}\right] d\theta$$

$$= 0 - 0 + \frac{(m-1)}{(n+1)} \int_0^{\pi/2} \sin^{m-2}\theta \cos^{n+2}\theta \, d\theta$$

$$= \left(\frac{m-1}{n+1}\right) \int_0^{\pi/2} \sin^{m-2}\theta(1 - \sin^2\theta).\cos^n\theta \, d\theta$$

therefore

$$I_{m,n} = \left(\frac{m-1}{n+1}\right) I_{m-2,n} - \left(\frac{m-1}{n+1}\right) I_{m,n}$$

therefore

$$I_{m,n}\left(\frac{m-1}{n+1} + \frac{n+1}{n+1}\right) = \left(\frac{m-1}{n+1}\right) I_{m-2,n}$$

therefore

$$I_{m,n} = \left(\frac{m-1}{m+n}\right) I_{m-2,n} = \left(\frac{n-1}{m+n}\right) I_{m,n-2}$$

Note:

$$I_{1,0} = \int_0^{\pi/2} \sin\theta \, d\theta = \left[-\cos\theta\right]_0^{\pi/2} = 1$$

$$I_{0,1} = \int_0^{\pi/2} \cos\theta \, d\theta = \left[\sin\theta\right]_0^{\pi/2} = 1$$

$$I_{0,0} = \int_0^{\pi/2} d\theta = \frac{\pi}{2}$$

Thus, if m and n are both even,

$$I_{m,n} = \left\{\frac{(m-1)(m-3)(m-5) \,\ldots\ldots\, (n-1)(n-3)(n-5) \,\ldots\ldots}{(m+n)(m+n-2) \,\ldots\ldots}\right\}\frac{\pi}{2}$$

In all other cases,

$$I_{m,n} = \left\{\frac{(m-1)(m-3)(m-5) \,\ldots\ldots\, (n-1)(n-3)(n-5) \,\ldots\ldots}{(m+n)(m+n-2) \,\ldots\ldots}\right\}.1$$

For example

$$\int_0^{\pi/2} \sin^6\theta \cos^4\theta \, d\theta = \frac{5.3.1.3.1}{10.8.6.4.2}\frac{\pi}{2}$$

$$= \frac{3}{512}\pi$$

and

$$\int_0^{\pi/2} \sin^7\theta \cos^5\theta \, d\theta = \frac{6.4.2.4.2}{12.10.8.6.4.2}.1 = \frac{1}{120}$$

Example 3

$$I_n = \int x^n e^{ax}\,dx$$

therefore

$$I_n = x^n \frac{e^{ax}}{a} - \frac{n}{a}\int x^{n-1} e^{ax}\,dx$$

therefore

$$I_n = \frac{x^n e^{ax}}{a} - \frac{n}{a} I_{n-1}$$

Hence

$$I_5 = \int x^5 e^{ax}\,dx = \frac{x^5 e^{ax}}{a} - \frac{5}{a}[I_4]$$

$$= \frac{x^5 e^{ax}}{a} - \frac{5}{a}\left[\frac{x^4 e^{ax}}{4} - \frac{4}{a}\left\{\frac{x^3 e^{ax}}{a} - \frac{3}{a}\left(\frac{x^2 e^{ax}}{a} - \frac{2}{a}\left[\frac{xe^{ax}}{a} - \frac{1}{a^2} e^{ax}\right]\right)\right\}\right]$$

Problems

1. Evaluate $\displaystyle\int_0^{\pi/2} \cos^2 x \sin^5 x\,dx$

2. If $\displaystyle I_n = \int_0^{\pi} x \sin^n x\,dx$, prove that $\displaystyle I_n = \frac{n-1}{n} I_{n-2}$

   and hence deduce that

   $$\int_0^{\pi} x \sin^4 x\,dx = \frac{3\pi^2}{16}$$

3. If $\displaystyle I_n = \int_0^{\pi/2} x^n \cos x\,dx$, prove that $I_n = (\pi/2)^n - n(n-1)I_{n-2}$

   and hence evaluate $\displaystyle\int_0^{\pi/2} x^4 \cos x\,dx$

4. (i) If $\displaystyle I_n = \int \frac{x^n\,dx}{\sqrt{ax^2 + 2bx + c}}$ (a, b and c are constants),

obtain the reduction formula

$$I_n = \frac{x^{n-1}\sqrt{ax^2 + 2bx + c}}{na} - \frac{(2n-1)b}{na} I_{n-1} - \frac{c(n-1)}{na} I_{n-2}$$

(ii) Evaluate

(a) $\displaystyle\int \frac{dx}{\sqrt{x^2 + 2x + 2}}$ (b) $\displaystyle\int \frac{x\,dx}{\sqrt{x^2 + 2x + 2}}$

and use the above formula to find

$$\int \frac{x^2\,dx}{\sqrt{x^2 + 2x + 2}}$$

5. Use standard formulae to evaluate:

(i) $\displaystyle\int_0^{\pi/2} \sin^6\theta\,d\theta$ (ii) $\displaystyle\int_0^{\pi/2} \sin^2\theta\cos^4\theta\,d\theta$

6. If $I_n = \displaystyle\int \cos^n x\,dx$, prove that $nI_n = (n-1)I_{n-2} + \cos^{n-1}x\sin x$

Hence find $\displaystyle\int_0^{\pi/4} \cos^3 x\,dx$

7. If $I_n = \displaystyle\int \frac{x^n}{\sqrt{a^2 + x^2}}\,dx$, show that $I_n = \dfrac{x^{n-1}}{n}\sqrt{a^2 + x^2} - \left(\dfrac{n-1}{n}\right)a^2 I_{n-2}$

where $n > 2$.

Evaluate $\displaystyle\int_0^2 \frac{x^5}{\sqrt{5 + x^2}}\,dx$

8. Find a reduction formula for $\displaystyle\int \tan^n\theta\,d\theta$ and use it to evaluate the integral where $n = 6$.

9. If $I_n = \displaystyle\int_0^\theta \frac{\sin(2n-1)\theta}{\sin\theta}\,d\theta$, show by considering $I_n - I_{n-1}$ that

$$I_n = I_{n-1} + \frac{\sin(2n-2)\theta}{n-1}$$

and deduce that $\displaystyle\int_0^{\pi/2} \frac{\sin(2n-1)\theta}{\sin\theta} = \frac{\pi}{2}$, when n is a positive integer.

If $J_n = \displaystyle\int_0^{\theta} \frac{\sin^2 n\theta}{\sin^2\theta}\,d\theta$, show that $J_n - J_{n-1} = I_n$ and hence evaluate

$$\int_0^{\pi/2} \frac{\sin^2 n\theta}{\sin^2\theta}\,d\theta$$

where n is a positive integer. (L.U.)

10. Differentiate $\sec^{n-2} x \tan x$ with respect to x. Hence, or otherwise, show that if

$$I_n = \int \sec^n x \, dx$$

where $n > 1$, then

$$(n-1)I_n = \sec^{n-2}x \tan x + (n-2)I_{n-2}$$

Prove that

$$8\int_0^{\pi/4} \sec^5 x \, dx = 7\sqrt{2} + 3\log(\sqrt{2}+1)$$

(L.U.)

SUMMARY

You should now be familiar with the principles behind integration as a means of reversing differentiation and as a powerful tool in the solution of problems in geometry and mechanics. The relationship between definite and indefinite integration must be understood. The analytical 'tricks of the trade' need practice in order that they become part of your stock of skills. You ought to be able to use Simpson's rule, preferably in conjunction with a digital computer, and should appreciate the possible errors involved. The applications of integration are important in many branches of engineering and you will have to acquire the necessary expertise.

Chapter Nine

First Order Differential Equations

9.0 INTRODUCTION

Many mathematical models are based on differential equations. In this chapter we first carry out a case study which will illustrate the ideas inherent in the solution of problems involving differential equations; we then set down some general definitions and ideas. The rest of the chapter is devoted to the formal investigation of the solution of first-order differential equations.

9.1 CASE STUDY – NEWTON'S LAW OF COOLING

We have already discussed the problem of the cooling of a hot liquid in Chapter 1 where we showed the nature of the graph of the experimental results. We now proceed to obtain a solution to this problem via a mathematical model.

Referring back to Section 1.1, we stated that the mathematical model equivalent to Newton's Law of Cooling was

$$\frac{d\theta}{dt} = -k(\theta - \theta_s), \qquad \text{with } \theta = \theta_0 \text{ at time } t = 0$$

In this model equation $\theta(t)$ is the temperature of the liquid at time t, θ_s the temperature of the surrounding air, θ_0 the initial temperature and k a constant of proportionality whose value depends on the particular liquid involved.

The equation is our mathematical model. It is one example of a **differential equation** which, in simplest terms, can be defined as an equation involving derivatives or rates of change.

We shall examine three main methods of obtaining information about the nature of the relationship between θ and t.

(a) The Analogue Computer Approach

The analogue computer is a device which can be made to produce an output voltage which varies with time in the same way as θ in our problem. This voltage can be displayed on a pen-recording trace or on a cathode-ray screen so that we can study the *qualitative* nature of the relationship between θ and t.

Furthermore, using the analogue computer, we can vary the parameters θ_s, θ_0 and k independently, so that we can study the role which each of these plays in the solution.

The circuit of an analogue computer contains high-gain direct current amplifiers (used as summers and integrators) and potentiometers operating on voltages varying with time. The diagrammatic representation and function of these is as follows.

(i) *Summers*

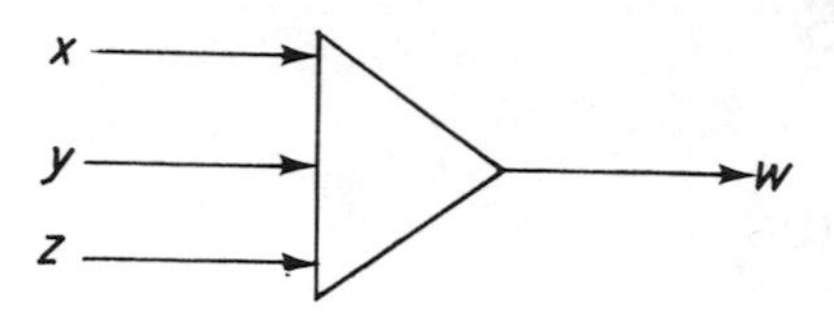

The input voltages to the amplifier are x, y and z. The output voltage produced is w and is such that

$$w = -(x + y + z)$$

Notice the reversal of sign.
In fact, if the only input is x, then the output is given by $w = -x$ so that the summer can be used to multiply an input by -1.

(ii) *Integrators*

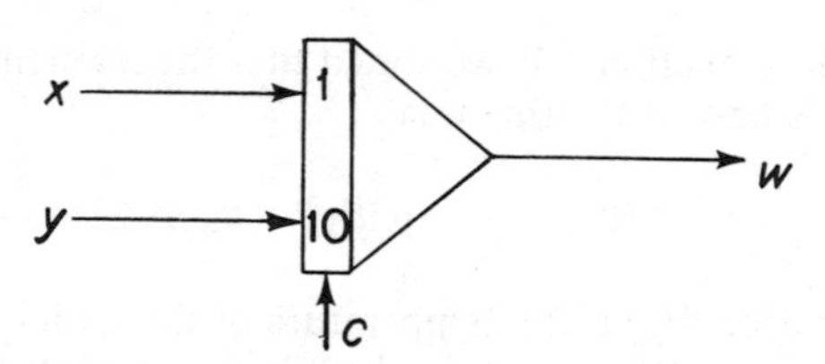

Here the input voltages are x and y. The integrator has the facility to multiply these voltages by the numbers in the box (here they are shown as 1 and 10), add them together, and then integrate with respect to time. It produces an output voltage w, where

$$w = -\int_0^t (x + 10y)\,\mathrm{d}t - c$$

t is time and c is the *initial condition* so referred to because when $t = 0$, $w = -c$. Notice again the sign reversal.

(iii) *Coefficient Multipliers*

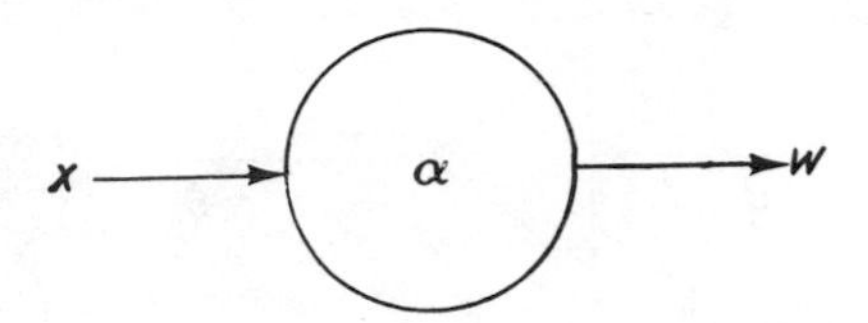

The coefficient multipliers are potentiometers which can be used to reduce a voltage. Here we have shown an input voltage x and output voltage w. The relation between these is of the form

$$w = \alpha x \quad (0 \leqslant \alpha \leqslant 1)$$

The above discussion, then, summarises the operations which can be performed by the analogue computer and which can be used to simulate the solution of a differential equation.

We return to the differential equation we were considering, that is,

$$\frac{d\theta}{dt} = -k(\theta - \theta_s), \qquad \text{with } \theta = \theta_0 \text{ when } t = 0$$

It can be written in integrated form as

$$\theta = -k \int_0^t (\theta - \theta_s)dt + \theta_0$$

The R.H.S. can be broken down as follows: $\theta - \theta_s$ is multiplied by k, the result is then integrated with an initial condition θ_0. So we have a scheme as shown in Figure 9.1.

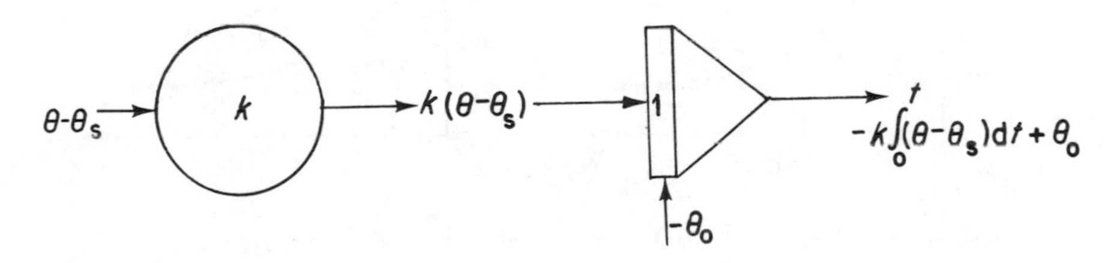

Figure 9.1

To produce $\theta - \theta_s$, which is $-(\theta_s - \theta)$, (remembering the sign reversal) we need a summer

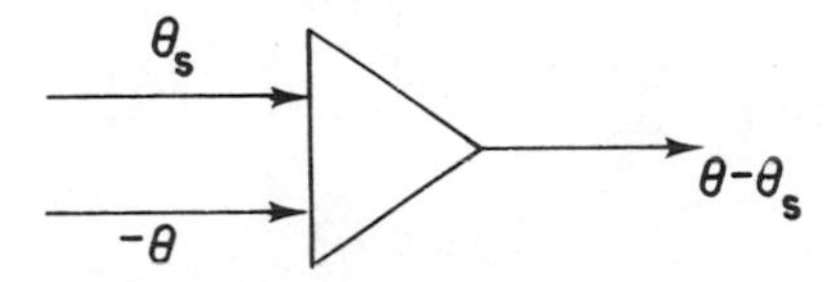

To produce $-\theta$ we can use another summer

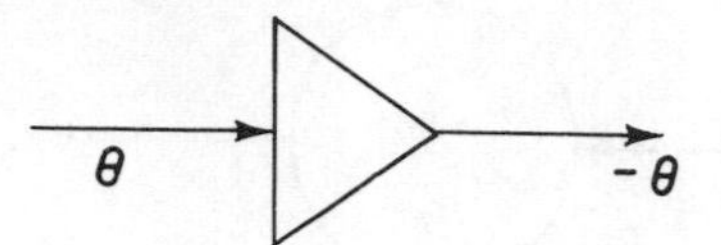

The full circuit is shown in Figure 9.2

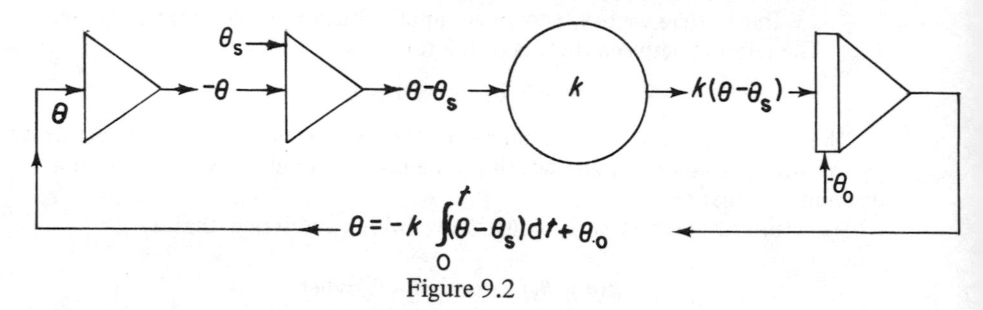

Figure 9.2

Note that we can vary the three parameters, θ_s, k and θ_0 at different points of the circuit.

A typical set of traces from the analogue computer set up in this way is sketched in Figure 9.3. In these diagrams, the axes have been inserted for clarity although we would not see these on the oscilloscope.

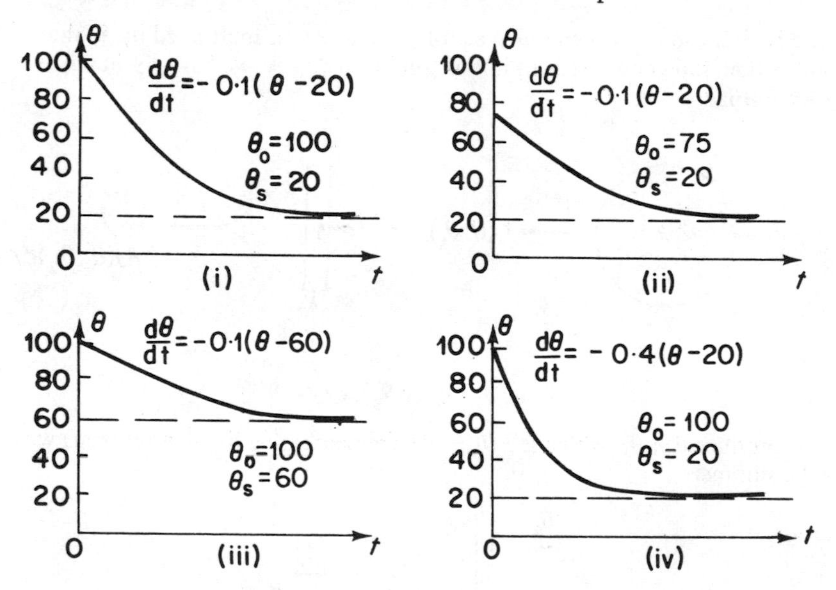

Figure 9.3

Notice that, in each case, the general shape of the curves is the same and this gives rise to the idea of a **general solution** of the differential equation from which a **particular solution** is found by selecting particular values of θ_0, θ_s and k. It is seen from the traces that

(i) decreasing the initial temperature θ_0 lessens the time taken to reach a specified temperature,
(ii) increasing θ_s decreases the time taken to cool to a specified temperature,
(iii) increasing k steepens the rate of decrease of temperature.

But, whatever the set of values taken for the three parameters, θ_0, θ_s and k, the general form of the solution curve is an initial rapid decay of temperature followed by a more gradual approach to the temperature of the surroundings and this agrees with the experimental curve in Figure 1.4

The same shape of curve is seen in many other physical systems. For example, when a beam of light passes through an absorbing medium, the model equation is $\mathrm{d}I/\mathrm{d}x = -\mu I$ where x is the distance from the light source, I is the intensity of the beam and μ is the absorption coefficient for the medium.

Similarly the rate of decay of a radioactive substance is governed by $\mathrm{d}m/\mathrm{d}t = -\lambda m$ where m is the amount of the substance remaining at time t. It is therefore clearly important to be able to get some more quantitative information about this kind of equation and to this end we now describe a numerical approach to the problem.

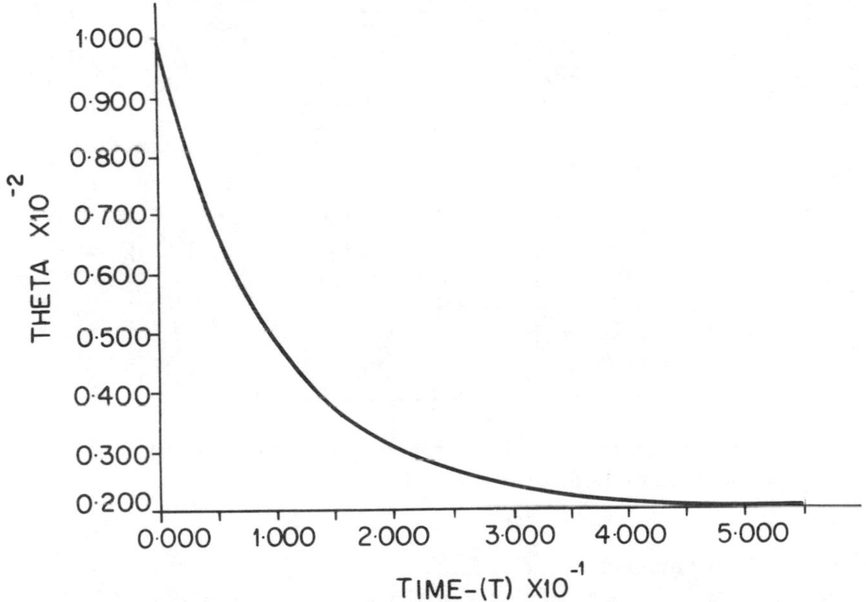

Graph plot from the digital computer of the numerical solution of $\mathrm{d}\theta/\mathrm{d}t = -0.1(\theta - 20)$

Figure 9.4

Table 9.1

t mins	θ degrees
0.0	100.0000
2.0	85.4985
4.0	73.6256
6.0	63.9049
8.0	55.9463
10.0	49.4304
12.0	44.0955
14.0	39.7278
16.0	36.1517
18.0	33.2239
20.0	30.8268
22.0	28.8643
24.0	27.2574
26.0	25.9419
28.0	24.8648
30.0	23.9830
32.0	23.2610
34.0	22.6699
36.0	22.1859
38.0	21.7897
40.0	21.4653
42.0	21.1996
44.0	20.9822
46.0	20.8041
48.0	20.6584
50.0	20.5390
52.0	20.4413
54.0	20.3613
56.0	20.2958
58.0	20.2422
60.0	20.1983

Printed output from the digital computer of the numerical solution of $d\theta/dt = -0.1(\theta - 20)$

(b) Step-by-step Method

We seek here to produce a table of the values of θ at definite values of t. There are several methods available **but** the *general principle* is that, given a starting value of θ_0, the temperature, θ, of the liquid after a small interval of time is estimated to be θ_1; this estimated value is then used to estimate θ_2, the temperature after the next time interval, and so on. The values of θ_0, θ_s and k would be read as data accompanying a program for a digital computer and the results could be output as in Table 9.1 or via a graph plotter in the form shown in Figure 9.4.

To obtain the solution under different conditions we can alter a data card, or have several separate data cards in the same program run.

It is seen that the graph thus produced has the same form as before. However, since we are using approximate methods, we must expect an error in the results. We shall examine later in this chapter a number of such methods and their accuracy.

(c) Analytical Method

We can also use calculus to obtain a formula for θ in terms of t. This is known, of course, as an *analytical* method of solving a differential equation and our solution is achieved as follows.

If we rearrange the equation

$$\frac{d\theta}{dt} = -k(\theta - \theta_s)$$

as

$$\frac{1}{\theta - \theta_s}\frac{d\theta}{dt} = -k$$

we can integrate both sides with respect to t to give

$$\log(\theta - \theta_s) = -kt + C$$

where C is an arbitrary constant of integration.

To find the value of this arbitrary constant we have the information that $\theta = \theta_0$ when $t = 0$. Substituting these values into the equation gives

$$\log(\theta_0 - \theta_s) = C$$

Therefore

$$\log(\theta - \theta_s) = -kt + \log(\theta_0 - \theta_s)$$

or

$$\log\left(\frac{\theta - \theta_s}{\theta_0 - \theta_s}\right) = -kt$$

and on taking antilogarithms of both sides we obtain

$$\frac{\theta - \theta_s}{\theta_0 - \theta_s} = e^{-kt}$$

Rearranging,

$$\theta = \theta_s + (\theta_0 - \theta_s)e^{-kt} \tag{9.1}$$

Then the value of θ at any time t can be found by substitution in (9.1); sketching this form of solution we get Figure 9.5.

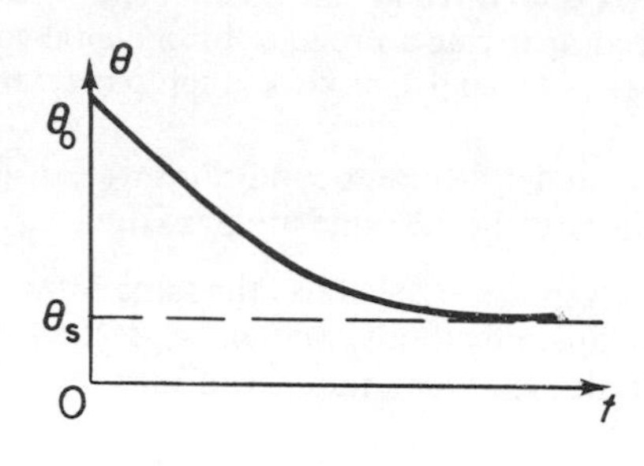

Figure 9.5

which is similar to the traces obtained from both the digital and analogue computers. We can, of course, obtain numerical values of θ by substituting particular values for k, θ_0 and θ_s into the formula (9.1).

Comparison of Methods

Of the above methods of solving differential equations, it would be fair to say that the analytical solution is the most desirable since it gives an exact† form of solution for any starting temperature θ_0 and any surrounding temperature θ_s. The analogue solution gives an idea of the solution enabling us to see how θ varies with time, at least in a qualitative way, for different values of θ_0 and θ_s. However, it is more difficult to obtain accurate quantitative answers. On the other hand, the digital computer will give us a table of values but requires to be told values for both the starting temperature θ_0 and the surrounding temperature θ_s. It gives us a *particular* solution rather than a *general* solution and the values for θ are not necessarily exact to the **last figure** quoted.

Hence if we were confronted with a differential equation we would aim as far as possible to get analytical expressions as our solution but in many practical cases we might not be able to find an analytical solution; furthermore, in some cases where we can find an analytical solution, this solution is of such a form that it cannot be used for finding quantitative and qualitative results. Let us consider an example of this latter case.

Example

A particle of unit mass is moving in a straight line and is being attracted towards a fixed point O by a force μx, where x is its distance from O. There is a frictional force opposing the motion equal to kv where v is the velocity of the particle. We can show that the equation governing this motion when the particle is moving away from O is

$$v\frac{\mathrm{d}v}{\mathrm{d}x} = -kv - \mu x \tag{9.2}$$

---

† see Section 1.2

Although it is outside our scope to solve this at the present time, the analytical expression giving v in terms of x can be found and is

$$\log(v^2 + kvx + \mu x^2) - (\frac{k}{\omega})\tan^{-1}(\frac{v}{\omega x} + \frac{k}{2\omega}) = \text{constant}$$

where $\omega^2 = \mu - k^2/4$, assumed positive.

The expression given above is much too complicated to be used for finding either qualitatively or quantitatively the explicit relationship between v and x without resorting to difficult numerical computation. Therefore, in this case it would be preferable to solve the equation in a non-analytical way from the start using analogue and/or digital computers.

9.2 ORDINARY DIFFERENTIAL EQUATIONS

Since we have said that the ideal situation is one where we can obtain an analytical solution to a differential equation, we shall spend some time on the different types of differential equations which are amenable to the analytical approach. These types are very important since they give us general types of solution and are of great use when developing theories in engineering subjects. However we recognise that we do not live in an ideal world and so we shall have to examine numerical and other techniques of solution which can be used as an alternative to,or as a substitute for, the analytical process.

First of all we must divide the differential equations we shall meet into different classes. We need some definitions for this task.

Definitions

1. An **ordinary differential equation** is an equation which contains only total derivatives and no partial derivatives. That is, it may contain dy/dx, d^2y/dx^2, etc. but not $\partial y/\partial x$, $\partial^2 y/\partial z^2$, etc. In other words, there is only one independent variable in the problem of which the differential equation is the mathematical model.
2. $y = y(x)$ is a **solution** of the equation if it satisfies this equation.
3. A **first order equation** is one which contains dy/dx, y and x only (as distinct from a second order equation which contains d^2y/dx^2 and could contain dy/dx, y and x as well). In general, the **order** of a differential equation is the order of the highest derivative in the equation.
4. The **degree** of a differential equation is the power to which the highest derivative in that equation is raised, for example,

$$\left(\frac{d^3y}{dx^3}\right)^2 + 2\frac{d^2y}{dx^2} + y\left(\frac{dy}{dx}\right)^4 = y + x$$

is a *third* order equation of *second* degree.

Problems

1. What are the order and degree of the following differential equations?

(a) $\dfrac{dy}{dx} = \dfrac{\sqrt{1+x}}{\sqrt{1+y}}$ (b) $y\dfrac{d^2y}{dx^2} + \left(\dfrac{dy}{dx}\right)^2 = 0$

(c) $EI\dfrac{d^2y}{dx^2} = \frac{1}{2}wx^2 - \frac{1}{2}wlx$ (d) $\dfrac{d^2y}{dx^2} - \left\{1 + \left(\dfrac{dy}{dx}\right)^2\right\}^{3/2} = 0$

(e) $2\dfrac{dy}{dx}\dfrac{d^3y}{dx^3} - 3\left(\dfrac{d^2y}{dx^2}\right)^2 = 0$ (f) $\left(\dfrac{dy}{dx}\right)^3 - y = x$

9.3 ARBITRARY CONSTANTS, GENERAL SOLUTIONS AND PARTICULAR SOLUTIONS

Consider the equation $y = x + A$, where A is arbitrary. We know that this equation can be represented by a straight line graph for any given value of A as follows (Figure 9.6)

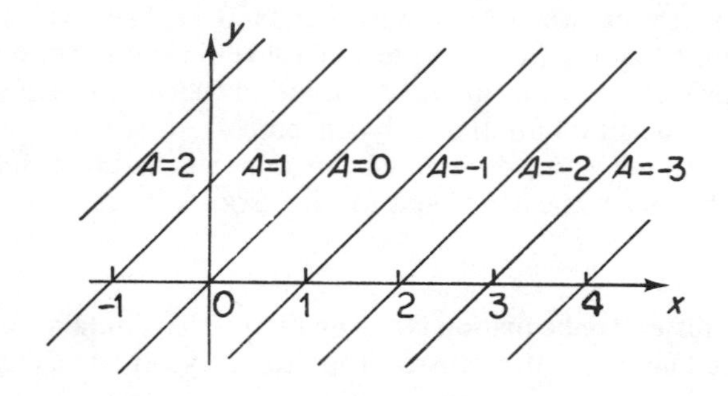

Figure 9.6

The straight lines obtained for different values of A are all parallel and, indeed, if we differentiate the given equation with respect to x, the arbitrary constant disappears and we get $dy/dx = 1$, showing that the slope always has value 1. Hence $dy/dx = 1$ is the differential equation of all these straight lines. Reversing the process, we integrate $dy/dx = 1$ w.r.t. x to obtain

$$y = x + \text{constant}$$

We say that this is the **solution** of the differential equation and we say it is a **general** solution because it represents any of the straight-line graphs which are such that $dy/dx = 1$. A **particular** solution will be obtained by taking a particular value for the constant of integration. This could be determined by specifying some point through which the straight line must pass. For example, we might require y to be 4 when $x = 1$.

There is only one of the lines that passes through the point (1, 4) and making this point satisfy our solution we get that the value of the constant = 4 − 1 = 3. The particular solution is then $y = x + 3$. (See Figure 9.7).

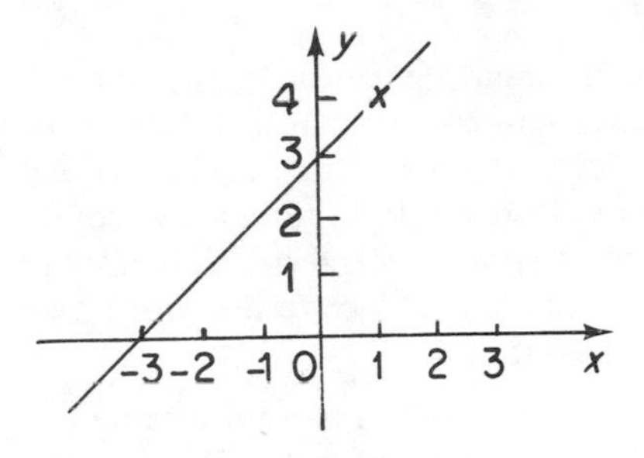

Figure 9.7

You will notice that to eliminate one arbitrary constant we needed to differentiate once. This is always the case.

Example 1

We start with the **family** of curves,

$$y = A\mathrm{e}^{-x}, \quad A \text{ arbitrary.}$$

Differentiating with respect to x, we obtain $\mathrm{d}y/\mathrm{d}x = -A\mathrm{e}^{-x}$. We can eliminate A by substituting from the equation of the family to obtain

$$\frac{\mathrm{d}y}{\mathrm{d}x} = -\left(\frac{y}{\mathrm{e}^{-x}}\right)\mathrm{e}^{-x}$$

or

$$\frac{\mathrm{d}y}{\mathrm{d}x} = -y$$

The whole family of curves $y = A\mathrm{e}^{-x}$ has an associated differential equation $\mathrm{d}y/\mathrm{d}x = -y$ and we say the **general solution** of the differential equation $\mathrm{d}y/\mathrm{d}x = -y$ is $y = A\mathrm{e}^{-x}$ where A is arbitrary. Again, a **particular solution** will be obtained by specifying some information which yields a particular value for A.

Example 2

Suppose now we have two arbitrary constants. Consider

$$y = A\mathrm{e}^{-x} + B\mathrm{e}^{2x}$$

Differentiation yields

$$\frac{\mathrm{d}y}{\mathrm{d}x} = -A\mathrm{e}^{-x} + 2B\mathrm{e}^{2x}$$

If we differentiate again w.r.t. x we get

$$\frac{\mathrm{d}^2y}{\mathrm{d}x^2} = A\mathrm{e}^{-x} + 4B\mathrm{e}^{2x}$$

We can now eliminate A and B from these two differential equations and you can verify that we obtain

$$\frac{d^2y}{dx^2} - \frac{dy}{dx} - 2y = 0$$

This is the second-order differential equation of the whole family of curves $y = Ae^{-x} + Be^{2x}$. We could not eliminate both arbitrary constants without differentiating w.r.t. x twice, so producing an equation involving d^2y/dx^2 and therefore an equation of second order. In reverse we would expect a second-order differential equation to give a general solution having *two* arbitrary constants since it must be integrated twice to produce y, each integration yielding one arbitrary constant.

In the same way, the general solution of a third-order differential equation should contain 3 arbitrary constants and so on. We therefore postulate the following theorem which will not be proved here:

There are n arbitrary constants in the general solution of an nth order linear ordinary differential equation.

(A **linear** differential equation is of the form

$$a_0 \frac{d^ny}{dx^n} + a_1 \frac{d^{n-1}y}{dx^{n-1}} + a_2 \frac{d^{n-2}y}{dx^{n-2}} + \ldots\ldots + a_{n-1} \frac{dy}{dx} + a_n y = f(x)$$

that is, one where all derivatives and y are raised to the first power and where such terms as

$$y\frac{dy}{dx}, \quad \left(\frac{dy}{dx}\right)\left(\frac{d^3y}{dx^3}\right)$$

do not occur.)

This theorem is very important, and can be shown to be true even if the equation is not linear. In particular, it enables us to recognise that the solution we have obtained is the most general solution by seeing that it contains the requisite number of arbitrary constants.

In books on more advanced mathematics, theorems are proved known as **existence theorems** which give the precise conditions under which the equation has a solution and **uniqueness theorems** are proved which explore the validity of considering any general solution of a differential equation as being the only solution of the equation.

Problems

1. Eliminate the arbitrary constants from the following equations to obtain the associated differential equations

   (i) $y = Ax + B$ (ii) $y^2 = Ax$

   (iii) $y = A\cos 2x + B\sin 2x$

   (iv) $y = (A + Bx)e^{2x}$ (v) $x = ½\log(A + Bt)$

2. Consider the equation $dy/dx = 2x$. Is there a solution curve which passes through (0, 1) and (2, 5)? Is there a solution curve which passes through (1, 3) and (4, 17)?

3. The general solution of

$$\frac{d^2y}{dx^2} + 3\frac{dy}{dx} + 2y = 0$$

is

$$y = Ae^{-2x} + Be^{-x}$$

A and B constants. Explain why $y(0) = 1$ is not sufficient information to determine a unique solution. [Hint: consider the Maclaurin expansion of y.]
Would the sole condition $y(1) = 2$ change your argument?

4. Examine

$$y = x\frac{dy}{dx} + \left(\frac{dy}{dx}\right)^2$$

Show that a *general* solution to the equation is

$$y = Ax + A^2, \quad A \text{ constant}$$

To find the envelope of this family of curves we find the partial derivatives of both sides with respect to the parameter A. This gives $0 = x + 2A$. Hence find the envelope as $y = -\frac{1}{4}x^2$. Show that this is a **singular** solution of the equation, that is, it cannot be obtained from the general solution. Sketch some of the family of curves and their envelope. If the boundary condition $y(3) = -2\frac{1}{4}$ were added would this give a unique solution?

5. In many branches of science the concept of an **orthogonal trajectory** is important. Remember from page 414 that this is concerned with finding a family of curves which is orthogonal to a given family of curves. That is to say, at each point where a member of the given family is intersected by a member of the new family, the intersection is at *right angles*. If the equation of the given family is $dy/dx = f(x, y)$ then the equation of the new family is $dy/dx = -1/[f(x, y)]$.
Show that for the family of circles $x^2 + y^2 = a^2$ the governing differential equation is $dy/dx = -x/y$. Show that the orthogonal family's differential equation is satisfied by $y = Kx$, K constant. Interpret geometrically, and give two examples from fluid flow of the applicability of the result.

6. (i) Show that the trajectory orthogonal to the family $xy = K$ is the family $x^2 - y^2 = B$. (K and B constants).

   (ii) Show that the trajectory orthogonal to the family $x^2 + 9y^2 = K$ is the family $y = Ax^9$. (K and A constants).

9.4 GRAPHICAL SOLUTIONS – ISOCLINES

Sometimes we can obtain a rough idea of the shape of solution curves by indicating their slopes locally. We illustrate the general principles by means of examples.

Example 1

$$\frac{dy}{dx} = x$$

First we sketch the curves on which dy/dx is constant. Now $dy/dx = c \Rightarrow x = c$ for this particular equation so the **isoclines** (lines of equal slope) are the straight lines x = constant, that is, parallel to the y-axis. In Figure 9.8 (i) some of the lines are shown, together with the appropriate values of c. The short thick lines indicate the local direction of a solution curve which crosses that isocline at the relevant point. Note that the value of the slope at the point is the value of c associated with the isocline and so, for example, on the isocline $x = 2$, the slope at all points is 2.

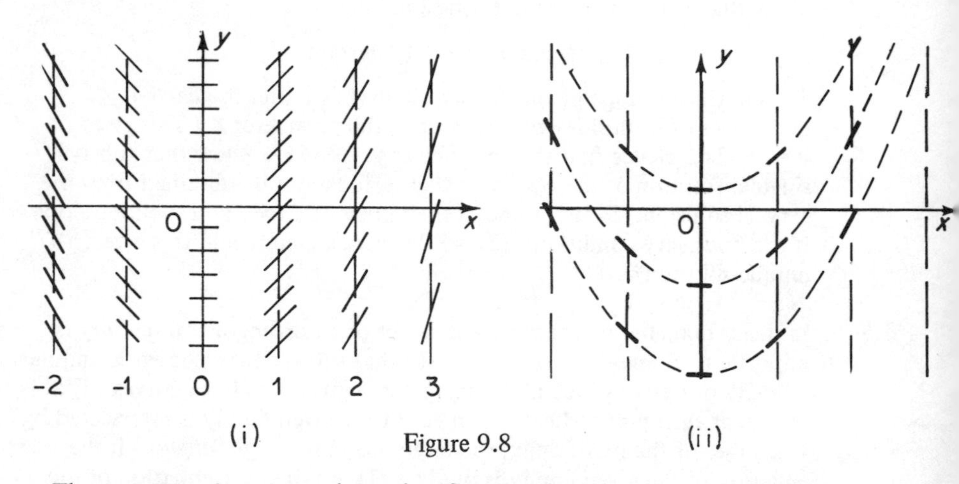

Figure 9.8

The next stage is to trace the paths of some solution curves and we usually start at a point on $x = 0$ and move off in either direction. In Figure 9.8 (ii) some solution curves are sketched.

The analytical solution of the differential equation is

$$y = \tfrac{1}{2}x^2 + A, \quad A \text{ constant}$$

You should be able to see the relationship between isoclines, local slopes and solution curves.

Example 2

$$\frac{dy}{dx} = y$$

dy/dx = constant $\Rightarrow y$ = constant and the isoclines are the lines $y = c$. These isoclines are drawn for some values of c in Figure 9.9(i). Again we start at a point on $x = 0$ from which we branch out in each direction. Several solution curves are shown in Figure 9.9(ii). The analytical solution is $y = Ae^x$ and it can be seen from the figure that the solution curves are of exponential form.

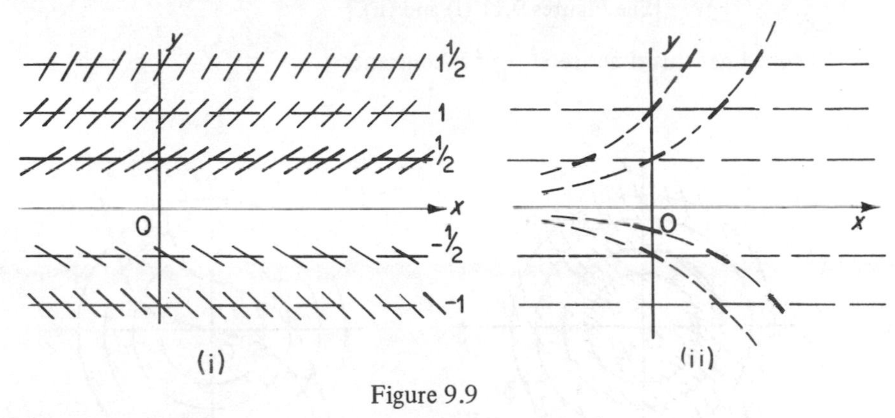

Figure 9.9

For this example we would require more isoclines close to the x-axis for greater precision in the construction of the solution curves. It should already be clear that this graphical method relies on skill and experience.

A point to remember is that $c = 0$ gives the locus of the stationary points on the solution curves, or as in this second example, where there are *no* such points in reality, it gives the *asymptote* to these curves.

There are many more complicated curves but we shall not consider them here, and, in any event, the method requires a certain amount of draughtsmanship. We provide two further examples for study.

Example 3

$\dfrac{dy}{dx} = x + y$ [Refer to Figure 9.10(i) and (ii)]

The isoclines here are $x + y$ = constant.

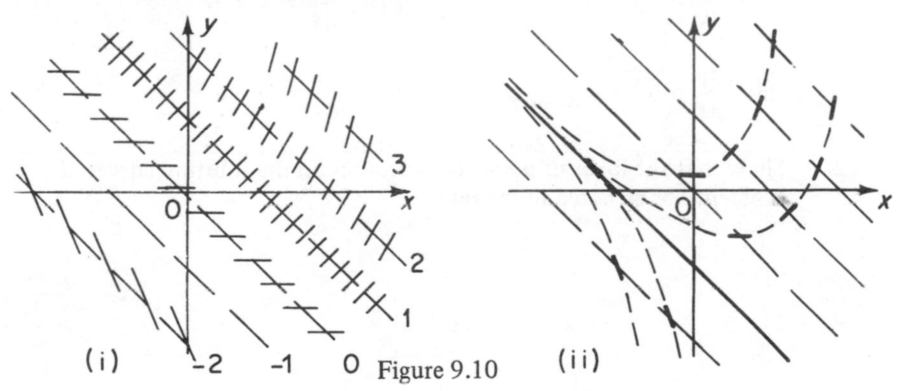

Figure 9.10

$y = -x - 1$ is an asymptote for large $-$ve x, as can be deduced from the analytical solution $y = -x - 1 + Ce^x$.

Example 4

$\dfrac{dy}{dx} = x^2 + y^2$ [See Figures 9.11 (i) and (ii).]

The isoclines this time are $x^2 + y^2 =$ constant.

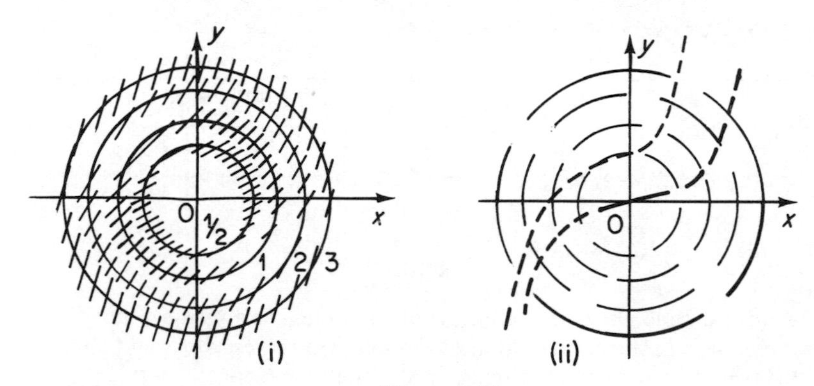

Figure 9.11

Here there is no simple analytical solution and the only place where $dy/dx = 0$ is at the origin.

Where the local slope is tangential to its isocline it can be shown that the solution curve which passes through the point has a point of inflection. These points of inflection move further away from the y-axis as $x^2 + y^2$ becomes larger.

Problems

1. Sketch the isoclines and the solution curve for

(i) $\dfrac{dy}{dx} = 2y - 1$ (ii) $\dfrac{dy}{dx} = y - x$

(iii) $\dfrac{dy}{dx} = x^2 + y$ (iv) $\dfrac{dy}{dx} = y - e^{-x}$

2. Show that the locus of points of inflection on the solution curves of Problem 1 is as stated in the text.

9.5 FIRST ORDER DIFFERENTIAL EQUATIONS OF THE FIRST DEGREE – VARIABLES SEPARABLE

In the rest of this chapter we shall be considering only those differential equations which are of first order and first degree. That is, we consider differential equations of the form

$$\frac{dy}{dx} = f(x, y) \tag{9.3}$$

We can recognise those equations of the form (9.3) for which the variables can be separated by the fact that $f(x, y)$ factorises into $F(x).G(y)$, that is, $f(x, y)$ is the product of two terms, one containing x only, the other containing y only.

For example,

$$\frac{dy}{dx} = x^2y + y$$

can be written

$$\frac{dy}{dx} = (x^2 + 1)y$$

and the equation

$$xy^2(1 + x^2)\frac{dy}{dx} - y^3 = 1$$

can be written

$$\frac{dy}{dx} = \frac{1 + y^3}{xy^2(1 + x^2)} = \frac{1}{x(1 + x^2)} \cdot \frac{(1 + y^3)}{y^2}$$

To solve differential equations of this form we note that if

$$\frac{dy}{dx} = F(x) \, . \, G(y)$$

then

$$\frac{1}{G(y)} \frac{dy}{dx} = F(x)$$

We then integrate both sides w.r.t. x, that is

$$\int \frac{1}{G(y)} dy = \int F(x) dx$$

Each integration could produce an arbitrary constant but, remembering that a first-order differential equation needs only one arbitrary constant for its general solution, we only need to put an arbitrary constant on one side of the equation as shown in the following example.

Example 1

$$\frac{dy}{dx} = x^2y + y$$

That is,
$$\frac{dy}{dx} = (x^2 + 1)y$$

or

$$\frac{1}{y}\frac{dy}{dx} = x^2 + 1$$

Integrating both sides w.r.t. x

$$\int \frac{1}{y}\,dy = \int (x^2 + 1)dx$$

This yields

$$\log y = \frac{x^3}{3} + x + A, \quad \text{where } A \text{ is arbitrary.}$$

Taking antilogarithms of both sides gives

$$y = e^{(x^3/3)+x+A} = e^A . e^{(x^3/3)+x}$$

Since A is arbitrary, e^A is also an arbitrary constant, and we can write $e^A = B$, giving finally the general solution

$$y = Be^{(x^3/3)+x}$$

Example 2

We now solve

$$xy^2(1 + x^2)\frac{dy}{dx} - y^3 = 1$$

which can be written

$$\frac{dy}{dx} = \frac{1}{x(1 + x^2)}\,\frac{(1 + y^3)}{y^2}$$

Separating the variables we get

$$\frac{y^2}{(1 + y^3)}\,\frac{dy}{dx} = \frac{1}{x(1 + x^2)}$$

Integrating both sides w.r.t. x produces

$$\int \frac{y^2}{1 + y^3}\,dy = \int \frac{1}{x(1 + x^2)}\,dx$$

This gives

$$\frac{1}{3}\log(1 + y^3) = \int \left(\frac{1}{x} - \frac{x}{1 + x^2}\right)dx$$

$$= \log x - \frac{1}{2}\log(1 + x^2) + A$$

In this case, where all the other terms are in logarithmic form it is best to write the arbitrary constant similarly, so we put $A = \log \alpha$ where α is arbitrary.

Then

$$\frac{1}{3}\log(1 + y^3) = \log x - \frac{1}{2}\log(1 + x^2) + \log \alpha$$

Taking antilogs, we obtain the general solution in the form

$$(1 + y^3)^{1/3} = \frac{\alpha x}{\sqrt{1 + x^2}}$$

Example 3

We now consider a practical example which can be modelled by an ordinary differential equation whose solution is obtained by the separation of variables method.

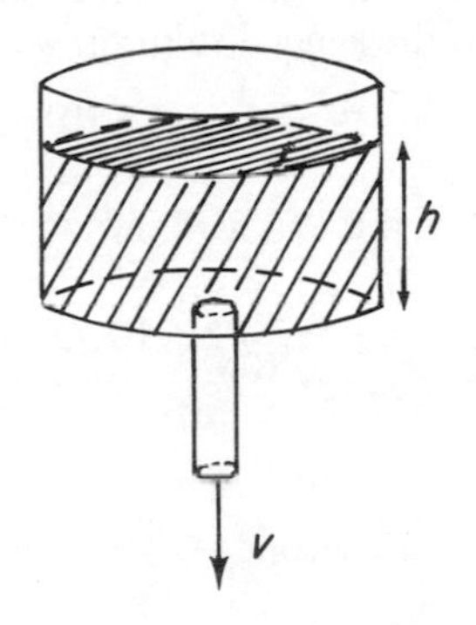

Figure 9.12

Liquid is contained in a cylindrical tank of cross-sectional area 1400 cm^2 and initially the depth of water is 50 cm. The tank is being emptied through an orifice of effective cross-sectional area 10 cm^2 (Figure 9.12). Find the time taken to empty the tank, *assuming* (i) that the effective area of the orifice remains constant, and (ii) that the velocity of the emerging liquid, v, is $\sqrt{2gh}$ cm/sec, when h cm is the depth of liquid in the tank. These are the assumptions for our physical model.

If h is the depth of liquid at time t sec, then the rate at which water is emerging through the orifice = $\sqrt{2gh}$. 10cm^3/sec. At this time the surface of the liquid must be descending at a rate of $\sqrt{2gh}$ 10/1400 cm/sec. [This is the rate of loss of liquid in the tank divided by the cross-sectional area.]

Hence

$$\frac{dh}{dt} = -\sqrt{2gh} \;.\; 10/1400$$

that is,

$$\frac{dh}{dt} = -\frac{1}{\sqrt{10}}\sqrt{h} \quad (g \text{ is taken as } 980 \text{ cm/sec}^2)$$

Separating the variables

$$\frac{1}{\sqrt{h}}\frac{dh}{dt} = -\frac{1}{\sqrt{10}} \tag{9.4}$$

and integrating both sides with respect to t, we get

$$\int \frac{1}{\sqrt{h}} dh = \int -\frac{1}{\sqrt{10}} dt$$

or

$$2\sqrt{h} = -\frac{1}{\sqrt{10}}t + A, \quad A \text{ any arbitrary constant.}$$

This is the general solution of equation (9.4), but we require a *particular* solution, since we were told that at time $t = 0$, the height $h = 50$ cm. Substituting these values into the general solution we find

$$2\sqrt{h} = -\frac{1}{\sqrt{10}}t + 10\sqrt{2}$$

When the tank is empty, then $h = 0$ and we find

$$0 = -\frac{1}{\sqrt{10}}t + 10\sqrt{2}$$

or

$$t = 10\sqrt{20} \text{ secs.}$$

Hence the tank empties in $10\sqrt{20}$ seconds $\simeq$ 45 seconds.

Example 4

A bacterial population of size P is known to have a rate of growth proportional to P itself. If between 9 a.m. and 10 a.m. the population doubles, at what time will P become 100 times what it was at 9 a.m.?

We are told $dP/dt \propto P$. Let

$$\frac{dP}{dt} = kP$$

therefore

$$\int \frac{dP}{P} = \int k dt$$

then

$$\log P = kt + \log \alpha$$

(putting the arbitrary constant = $\log \alpha$, for convenience)

Hence the general solution is

$$P = \alpha e^{kt}$$

For convenience we can assume that $P = 1$ at 9 a.m. (call this $t = 0$). Then $P = 2$ at $t = 60$ mins. (10.00 a.m.) and we want the time when $P = 100$.

Substituting this information into the general solution,

$$1 = \alpha e^{0} = \alpha$$

$$2 = \alpha e^{60k}$$

Hence, $\alpha = 1$, $e^{60k} = 2$, and so $60k = \log 2$, that is

$$k = \frac{\log 2}{60}$$

The law of growth is therefore

$$P = e^{(\log 2)\, t/60} = 2^{(t/60)}$$

When $P = 100$, $\quad 100 = e^{(\log 2)\, t/60} = 2^{(t/60)}$

therefore

$$t = \frac{60 \log 100}{\log 2} = 398.66 \text{ mins.} \;\hat{=}\; 6 \text{ hrs. } 39 \text{ mins.}$$

That is, the population is 100 times its 9 a.m. size at 15.39 (or 3.39 p.m.).

Problems

1. Solve the following differential equations

(i) $\sqrt{1+x^2}\,\dfrac{dy}{dx} = x e^{-y}$ given $y = 0$ when $x = 0$.

(ii) $\dfrac{dy}{dx} + \dfrac{1+y^{3/2}}{xy^{1/2}(1+x^2)} = 0$

(iii) $\dfrac{dy}{dx}\cos x = (\sin x + x \sec x)\cot y$

(iv) $x\dfrac{dy}{dx} = y \log x$

(v) $y\dfrac{dy}{dx} + x = xy^4$

(vi) $x\dfrac{dy}{dx} = y(1-y)$ given that $y = 2$ when $x = -4$

(vii) $\dfrac{dy}{dx} = (1 + 2x)\cos^2 y$

(viii) $xy\dfrac{dy}{dx} = 1 + x^2$

(ix) $xy(1 - x^2)\dfrac{dy}{dx} = 1 - y^2$

Show that the solution curves are symmetrical about each of the coordinate axes and that they all pass through four fixed points. (L.U.)

2. Solve the following differential equations

(i) $\dfrac{dy}{dx} = e^y \sin x,$ given $y = 0$ when $x = 0$

(ii) $\dfrac{dy}{dx} = \dfrac{xe^y}{\sqrt{1 + x^2}}$ given $y = 0$ when $x = 3/4$

(iii) $t\theta(1 + t^2)\dfrac{d\theta}{dt} - \theta^2 = 1,$ given $\theta = 0$ when $t = 1$

3. The rate at which a radio-active substance decays is proportional to n, the number of atoms present at time t. If the constant of proportionality is λ and initially there are N atoms present, express n as a function of t. Find the time taken for 1/4 of the initial amount to decay; find the time for the initial amount to be halved.

4. The radial compressive stress at a distance r from the axis of a thick cylinder is given by

$$r\frac{dp}{dr} = 2(A - p), \quad A \text{ constant}$$

Solve for p when $p = p_0$ at $r = r_0$ and $p = p_1$ at $r = r_1$.

5. Liquid fills a spherical vessel of radius a to a depth h_0. At time $t = 0$ the fluid is allowed to drain out of an orifice at the lowest point of the vessel at a volume rate $k\sqrt{h}$ where k is a constant. Derive the differential equation for the variation of depth h with time and find an expression for the time taken to empty the vessel. (C.E.I.)

6. A hemispherical water tank of radius R is full at $t = 0$. There is a hole of radius a at the bottom of the tank through which water escapes for time $t > 0$. Find the depth of water as a function of time.
(Assume that the velocity of emerging liquid is $\sqrt{2gh}$ where h is the depth of water.)

7. A particle moves along a straight line so that at time t sec its distance from a fixed point O is x cm and its velocity v cm/sec. If its acceleration is $(v^3 - v)$ cm/sec^2 directed away from O and $v = 3$ at $x = 0$, show that $x = \frac{1}{2} \log 2 - \coth^{-1} v$, proving any formula which you use for an inverse hyperbolic function. (L.U.)

9.6 INTRODUCTION TO NUMERICAL METHODS OF SOLUTION

Failure of the analytical methods

You must not think that all differential equations where the variables are separable can be solved by the above technique. We shall now consider a differential equation where these methods fail to work.

If we have the equation

$$y \frac{dy}{dx} = e^{x^2}$$

integrating both sides w.r.t. x produces

$$\int y dy = \int e^{x^2} dx \tag{9.5}$$

The integral on the right hand side of (9.5) cannot be evaluated analytically at this stage of our work. Therefore, although we have theoretically solved the equation, we cannot perform the integration and hence cannot get y explicitly in terms of x.

Given an initial pair of values of y and x, we could use an approximate method of integration to find an approximate value of y for some given value of x. For example, suppose we know that $y = 0$ when $x = 0$, then equation (9.5) can be written

$$\int_0^y y dy = \int_0^x e^{x^2} dx$$

By this notation we mean that we integrate both sides, putting $y = 0$ when $x = 0$ (bottom limits correspond) and obtain a particular solution for y in terms of x.

We get

$$\left[\frac{y^2}{2}\right]_0^y = \int_0^x e^{x^2} dx$$

that is,

$$\frac{y^2}{2} = \int_0^x e^{x^2} dx$$

For $x = 1$ we would get

$$\frac{y^2}{2} = \int_0^1 e^{x^2}\,dx$$

and using an approximate method of integration we find the approximate value of y when $x = 1$.

Similarly for $x = 2$, we get

$$\frac{y^2}{2} = \int_0^2 e^{x^2}\,dx$$

which produces an approximate value of y when $x = 2$.

Proceeding in the same way we can produce a whole table of values of y for corresponding values of x and we have found the solution. However the process is wasteful in time and it would be better to use a wholly numerical method in such a situation.

We now discuss a simple type of the step-by-step method mentioned when dealing with Newton's Law of Cooling in Section 9.1.

Euler's method

In the cooling problem we had to solve the equation

$$\frac{d\theta}{dt} = -0.1(\theta - 20) \text{ with } \theta = 100 \text{ when } t = 0 \tag{9.6}$$

We want to estimate θ at certain specified times and what we shall do is to start at $\theta = 100$, $t = 0$ and step forward a small interval of time to find an estimate of θ at this new time. From this point we step forward again and estimate θ after another short interval of time and so on. The estimate of θ at the new time is found by evaluating $d\theta/dt$ at the old time using the differential equation (9.6) and estimating the new value of θ by the formula

$$\theta_{\text{new}} = \theta_{\text{old}} + h\left(\frac{d\theta}{dt}\right)_{\text{old}} \tag{9.7}$$

where h is the step length. This is called **Euler's method.**

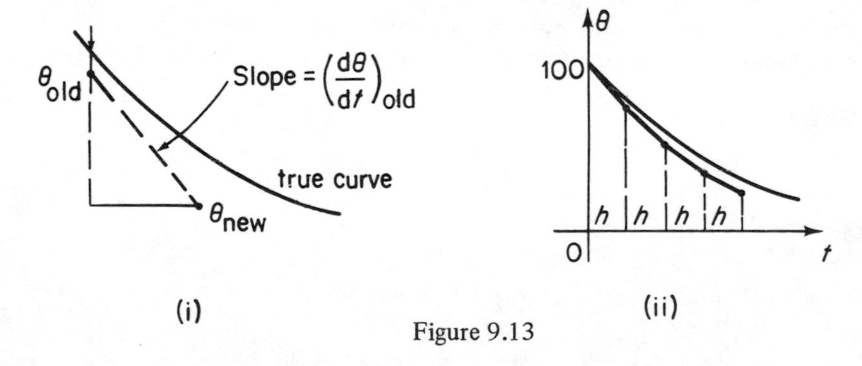

Figure 9.13

Essentially we are using a straight line approximation to the function in each interval as shown in Figure 9.13(i). This means that we shall obtain a graph for the estimated behaviour of θ consisting of straight-line segments as shown in Figure 9.13(ii).

Let us do a few steps of the process numerically. We shall assume that the constant k is such that we may use times in minutes.

We have $d\theta/dt = -0.1(\theta - 20)$ given that $\theta = 100$ when $t = 0$, i.e. $\theta(0) = 100$. Let us take $h = 1$. Then $d\theta/dt$ at $t = 0$ is $-0.1(100 - 20) = -8$.

Applying the Euler formula (9.7)

$$\theta(1) = \theta_1 = 100 + (1)\left(\frac{d\theta}{dt}\right)_{t=0} = 100 - 8 = 92$$

$$\theta(2) = \theta_2 = 92 + (1)\left(\frac{d\theta}{dt}\right)_{t=1} = 92 - 7.2 = 84.8$$

$$\theta(3) = \theta_3 = 84.8 + (1)\left(\frac{d\theta}{dt}\right)_{t=2} = 84.8 - 6.48 = 78.32$$

$$\theta(4) = \theta_4 = 78.32 - 5.832 = 72.488$$

This process could easily be used on a digital computer, the flow chart being as shown in Figure 9.14.

We know the general shape of the curve expected in this particular example (see Figure 9.3) and we expect that after a number of steps the change in θ will be very small. We could therefore modify the flow diagram so that instead of giving N, the number of steps to be taken, we could test whether at any stage the value of θ_{new} differs from θ_{old} by more than a specified amount and if not, stop the process.

You should now try to modify the flow chart to incorporate this sophistication and write the appropriate computer program. (Note that this sophistication is only possible here because we know the general form of the solution curve.)

Because this is an approximate method there is bound to be an error attached to the result quoted. Before making a formal estimation of the error expected we shall carry out some more arithmetic.

Take the same problem with a step length of 2 minutes. Then

$$\theta_1 = 100 - 16 = 84 = \theta(2)$$

$$\theta_2 = 84 - 12.8 = 71.2 = \theta(4)$$

(The notation θ_k means the kth value of θ estimated by Euler's method.)

Suppose we estimate $\theta(4)$ in one step of length 4, then

$$\theta(4) = 100 - 32 = 68$$

In order to try out the effect of other step sizes we programmed a digital computer to show the results of using steps of 1/2, 1/10, 1/20, 1/50, 1/100 minutes respectively.

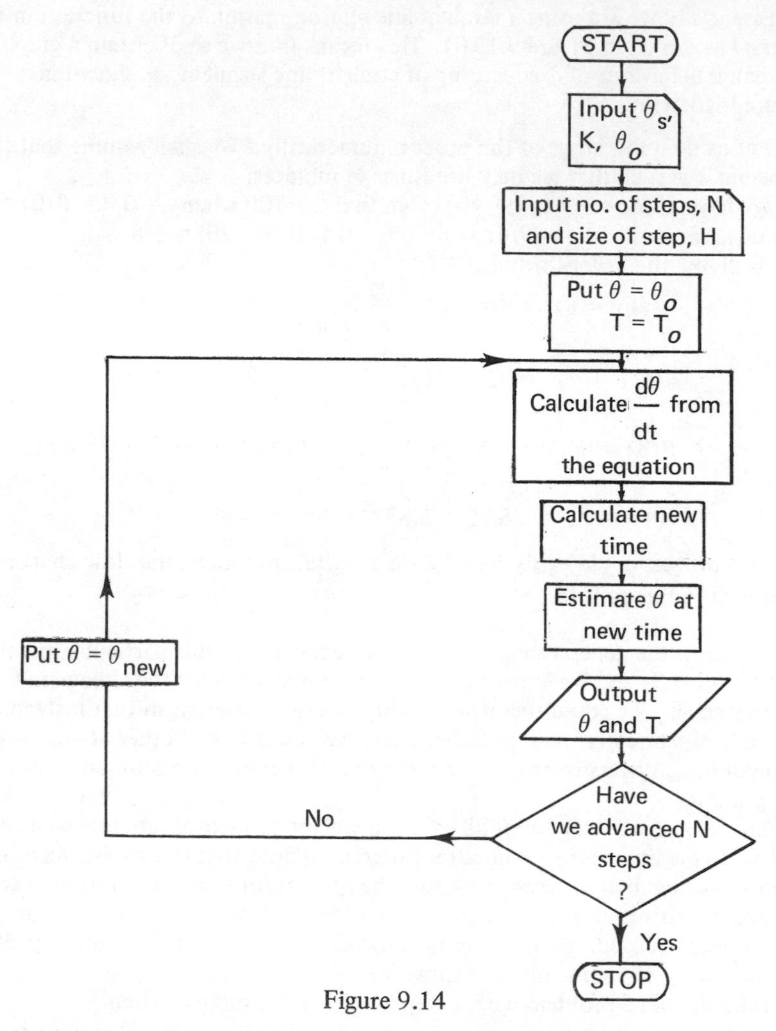

Figure 9.14

In Table 9.2 the results are compared with the value at 4 minutes obtained from the analytical solution to the problem which is

$$\theta = 20 + 80e^{-0.1t} \tag{9.8}$$

It would seem from these results that the smaller the step size the better our approximation, but this is not necessarily true because we have done calculations over the first four minutes only and we have no real idea what will happen later. In fact there is a **round-off error** at each step of the calculations so that the more steps we take then the more there is a danger of cumulative round-off error. Clearly what is needed is a balance between poor approximations and the number of steps.

Table 9.2

θ after 4 minutes

Step size h	Estimated θ (3 d.p.)	True θ (3 d.p.)	Error (3 d.p.)
4	68		−5.624
2	71.2		−2.424
1	72.488		−1.136
0.5	73.074		−0.550
0.1	73.518	73.624	−0.106
0.05	73.572		−0.052
0.02	73.604		−0.020
0.01	73.615		−0.009

Discussion of Round-off error

We cannot get much idea from Table 9.2 of the accumulation of error, since over the first four minutes the graph of θ against t does not differ much from a straight line. We need to look at the longer-term situation. If we ask at what time the temperature of the liquid will reach 22°, we can produce Table 9.3.

Table 9.3

t	Euler Estimates of θ with time steps 5.0 mins	1.0 mins	0.5 mins	Analytical value
	(3 d.p. where relevant)			3 d.p.
0	100	100	100	100
5	60	67.239	67.899	68.522
10	40	47.894	48.679	49.430
15	30	36.471	37.171	37.850
20	25	29.726	30.281	30.827
25	22.5	25.743	26.156	26.567
30	21.25	23.391	23.501	23.983
35	20.625	22.003	22.096	22.416
40	20.313	21.182	21.321	21.465

We see that for a step-size of 5 minutes, the temperature is estimated to reach 22° after a time between 25 and 30 minutes. For step size of 1 minutes, it is estimated to take just over 35 minutes; for a step of 0.5 minutes it takes a little longer but still just over 35 minutes. We can find an exact value from our analytical formula (9.8), as follows. We had

$$\theta = 20 + 80e^{-0.1t}$$

Now when $\theta = 22$, $22 = 20 + 80e^{-0.1t}$

so that

$$2 = 80e^{-0.1t}$$

and

$$e^{0.1t} = 40$$

Taking logarithms, $t = 10 \log 40 = 36.889$ minutes (3 d.p.).

Now, although it takes approximately 3700 evaluations with a step size of 0.01 minutes, we did program the digital computer to work with this step size and it gave a value of 22° at a time between 37.07 and 37.08 minutes. We can, therefore, assume that this discrepancy from the analytical value is almost entirely due to round-off error.

It can be seen that decreasing the step size provides better accuracy at each estimate examined. Although one might expect errors to accumulate, we find that for a fixed step size, the errors decrease with increasing time. There are clearly other sources of error besides round-off error and we examine them in turn.

(i) *Formula error*

The approximation during each step that the slope of the function stays constant will lead to a wrong estimate of θ_{new}. This could be alleviated by reducing the step size if the formula gave a consistently better approximation.

(ii) *Inherited error*

It has to be recognised that we are handicapped before we step forward each time because θ_{old}, the starting value for θ, is itself an estimation (except for the initial value). Under certain conditions this may cause an accumulation of errors but under other conditions this may lead to a better approximation.

9.7 GENERAL FORM OF EULER'S METHOD

Let us now put Euler's step by step process on a general footing for use with any first-order differential equation.

Take the first-order equation to be

$$\frac{dy}{dx} = f(x, y) \quad \text{with } y = y_0 \text{ at } x = x_0$$

The slope at x_0 is $f(x_0, y_0)$; call this f_0. Then, assuming the slope to be constant over a range of x, the value of y at $x_1 = x_0 + h$ is approximately $y_1 \doteq y_0 + f_0.h$.

We can now estimate the slope at x_1 as $f(x_1, y_1) = f_1$, say. Then the estimated value of y at $x_2 = x_1 + h = x_0 + 2h$ is

$$y_2 = y_1 + f_1 h$$

This procedure is repeated giving the general expression

$$y_{n+1} = y_n + f_n h$$

or

$$y_{n+1} = y_n + h\left(\frac{\mathrm{d}y}{\mathrm{d}x}\right)_{x=x_n} \tag{9.9}$$

Example 1

$\dfrac{\mathrm{d}y}{\mathrm{d}x} = 1$ and $y = 1$ when $x = 0$

We can easily obtain the analytical solution of this equation as $y = x + 1$, the graph of which is shown in Figure 9.15(i). Note we advance forward from $x = 0$.

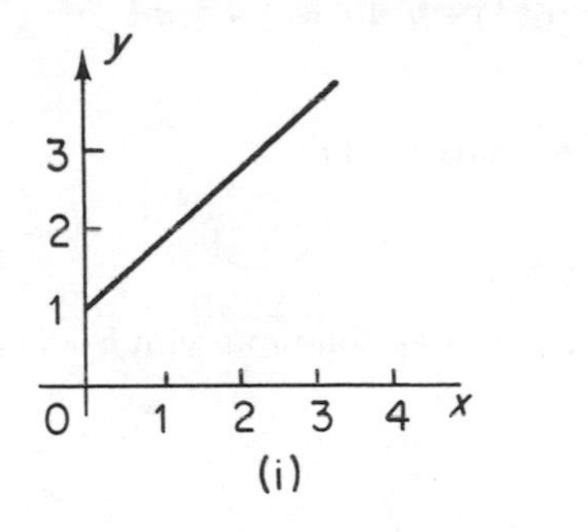

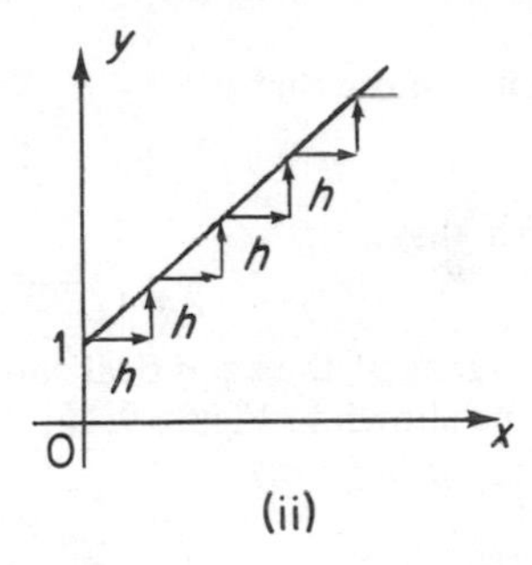

Figure 9.15

The step by step method is

$$y_{n+1} = y_n + h\left(\frac{\mathrm{d}y}{\mathrm{d}x}\right)_{x=x_n}$$

But $\mathrm{d}y/\mathrm{d}x = 1$ at all points, therefore

$$y_{n+1} = y_n + h$$

Starting at $y = 1, x = 0$, we shall get

$$\begin{aligned} y_1 &= 1 + h \\ y_2 &= (1 + h) + h = 1 + 2h \\ y_3 &= (1 + 2h) + h = 1 + 3h \\ y_n &= 1 + nh \\ y_{n+1} &= 1 + (n + 1)h \end{aligned}$$

The approximation is shown in Figure 9.15(ii). The approximation is, in this case, exact.

Example 2

$\dfrac{dy}{dx} = x$ with $y = 1$ when $x = 0$.

The analytical solution is

$$y = \frac{x^2}{2} + 1 \tag{9.10}$$

The step by step method gives

$$\begin{aligned} y_1 &= 1 + h.(0) = 1 \\ y_2 &= 1 + h.h = 1 + h^2 \\ y_3 &= (1 + h^2) + h.2h = 1 + 3h^2 \\ y_4 &= (1 + 3h^2) + h.3h = 1 + 6h^2 \\ y_5 &= (1 + 6h^2) + h.4h = 1 + 10h^2 \end{aligned}$$

You should show that

$$y_n = 1 + h^2.\tfrac{1}{2}n(n-1)$$

and hence that

$$y_{n+1} = y_n + hx_n$$

The graphs of the analytical and step by step solutions with $h = 1$ and $h = 0.5$ are shown in Figure 9.16.

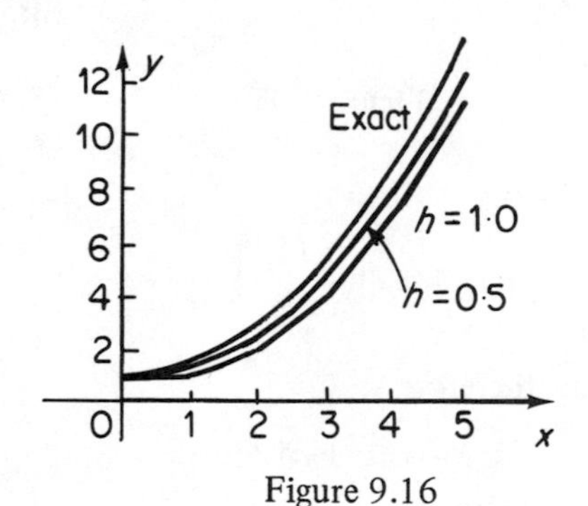

Figure 9.16

Note that in both these examples we started at $x = 0$; this was merely to simplify the arithmetic.

Problems I

1. In the following six problems estimate the dependent variable at suitable equally spaced values of the independent variable, using Euler's method. Compare your results with the analytical solution. (You should not advance the solution by more than 2 steps unless you program a digital computer.)

(i) $\dfrac{dy}{dx} = x^2 + 1; \quad y(0) = 1$ Estimate $y(1)$

(ii) $\dfrac{dy}{dx} = 2y - 1; \quad y(0) = 2$ Estimate $y(1)$

(iii) $y' + y = 0; \quad y(0) = 2, \ h = 0.2$ Estimate $y(1)$

(iv) $\dfrac{dy}{dx} = x^2 - y; \quad y(0) = 3$ Estimate $y(2)$

(v) $y^2 \dfrac{dy}{dx} = \dfrac{x+3}{y+1}; \quad y(0) = 1$ Estimate $y(1)$

(vi) The differential equation for the current i at time t in an L–R series circuit is

$$L\frac{di}{dt} + Ri = E$$

If $E = 240(1 - e^{-0.5t})$ and initially $i = 0$, find i at $t = 1$, given that $R = 100$ and $L = 100$.

2. A projectile is launched from the earth's surface with velocity V. Assuming no drag the equation of motion is

$$v\frac{dv}{dr} = -g\frac{R^2}{r^2}$$

where v is the velocity at a distance r from the centre of the earth of radius R. Take $g = 9.81$ m/sec^2, $R = 6.37 \times 10^6$ m and $V = 15000$ m/sec. [If you are *not* using a computer program, take $g = 10.0$ and $R = 6 \times 10^6$] Find the velocity when $r = 2R$.

Error estimates for Euler method

Returning to Example 2 on page 516, the true value of $y_n = y(nh)$ is $1 + \frac{1}{2}n^2h^2$ (see equation 9.10) and so the error accumulated by the nth stage is

$$\epsilon_n = 1 + \frac{h^2 n}{2}(n-1) - (1 + \tfrac{1}{2}n^2h^2) = -\frac{nh^2}{2}$$

Suppose we seek the formula error over one step. If we start from a correct value y_n, the Euler method gives

$$y_{n+1} = y_n + hx_n = y_n + nh^2$$

However, algebraically,

$$y_n = \frac{x_n^2}{2} + 1 = \frac{(nh)^2}{2} + 1 \quad \text{from (9.10)}$$

and

$$y_{n+1} = \frac{x_{n+1}^2}{2} + 1 = \frac{(n+1)^2h^2}{2} + 1$$

therefore

$$y_{n+1} = y_n + nh^2 + ½h^2$$

In proceeding from y_n to y_{n+1} the Euler method gives a result which is too small by an amount $½h^2$; hence, the total contribution from this source of error over n stages will be $-nh^2/2$. In other words, the formula error accounts for all the accumulated error (round-off not being considered).

Consider now the problem

$$\frac{dy}{dx} = y; \quad y(0) = 1$$

whose analytical solution is

$$y = e^x$$

The Euler formula is

$$y_{n+1} = y_n + hy_n = (1+h)y_n$$

The first two stages of application of Euler's method give

$$y_1 = (1+h)y_0 = (1+h)$$
$$y_2 = (1+h)y_1 = (1+h)^2 y_0 = (1+h)^2 = 1 + 2h + h^2$$

Analytically,

$$y(2h) = e^{2h} = 1 + 2h + 2h^2 + \frac{4}{3}h^3 + \ldots\ldots$$

The Euler estimate is short by an amount

$$e^{2h} - (1+h)^2 = h^2 + \frac{4}{3}h^3 + \ldots\ldots$$

Yet the formula error would be $e^h - (1+h)$ over the first stage and $e^{2h} - (1+h)e^h = e^h[e^h - (1+h)]$ over the second stage [use $y_2 = (1+h)y_1$ with $y_1 = e^h$]; this gives a total contribution of

$$\begin{aligned}[e^h - (1+h)](e^h + 1) &= e^{2h} - he^h - (1+h) \\ &= e^{2h} - h(1 + h + h^2/2! + \ldots\ldots) - (1+h) \\ &= e^{2h} - h(1+h) - (h^3/2! + \ldots\ldots) - (1+h) \\ &= e^{2h} - (1+h)^2 - (h^3/2! + \ldots\ldots)\end{aligned}$$

Thus the formula error does not account for all the error and clearly the inherited error is beginning to play its part. What allows us to distinguish when it does play a role and whether this role is to add to the formula error or not? If y_n is in error by an amount Δy_n and y_{n+1} by an amount Δy_{n+1}, we can define a **scale factor** $\Delta y_{n+1}/\Delta y_n$ which indicates the effect of inherited error. Should this factor be > 1 the effect is to make the next estimate worse and should it be < 1, its effect is to bring the next estimate nearer the true value. Of course, these effects have to be weighed against the formula error.

We approximate $\Delta y_{n+1}/\Delta y_n$ by dy_{n+1}/dy_n in order to estimate the scale factor. In our present example, $dy_{n+1}/dy_n = 1 + h$, and this indicates a growth in effect ($h > 0$).

For the cooling liquid problem it can be shown (see Problem 3 at the end of the section) that the scale factor < 1 which indicates that the successive estimates could get closer to the true values. This is in fact borne out by our results and shows that the inherited error is stronger than the formula error.

The overall error associated with an estimate is therefore seen to be a combination of factors and may be difficult to evaluate in a practical problem.

Table 9.4 shows the effect of accumulated error for the equation $dy/dx = y$ with $y(0) = 1$.

Table 9.4

Euler's method using a desk calculator which carries 22 significant figures

x	e^x 5 sig.figs.	h=1	h=0.5	h=0.1	h=0.01
1	2.7183	2	2.250	2.597	2.691
2	7.3891	4	5.062	6.730	7.245
3	20.085	8	11.39	17.46	19.50
4	54.598	16	25.62	45.29	52.47
5	148.41	32	57.68	117.5	141.3

Table 9.5 shows the effect of round-off by using four-figure tables instead of a desk calculator.

Table 9.5

Effect of round-off (4 figure tables)

x	h=0.5	h=0.01
1	2.250	2.705
2	5.063	7.316
3	11.39	19.79
4	25.63	53.52
5	57.67	144.8

Figure 9.17 gives a clear picture of the effects of decreasing the step size.

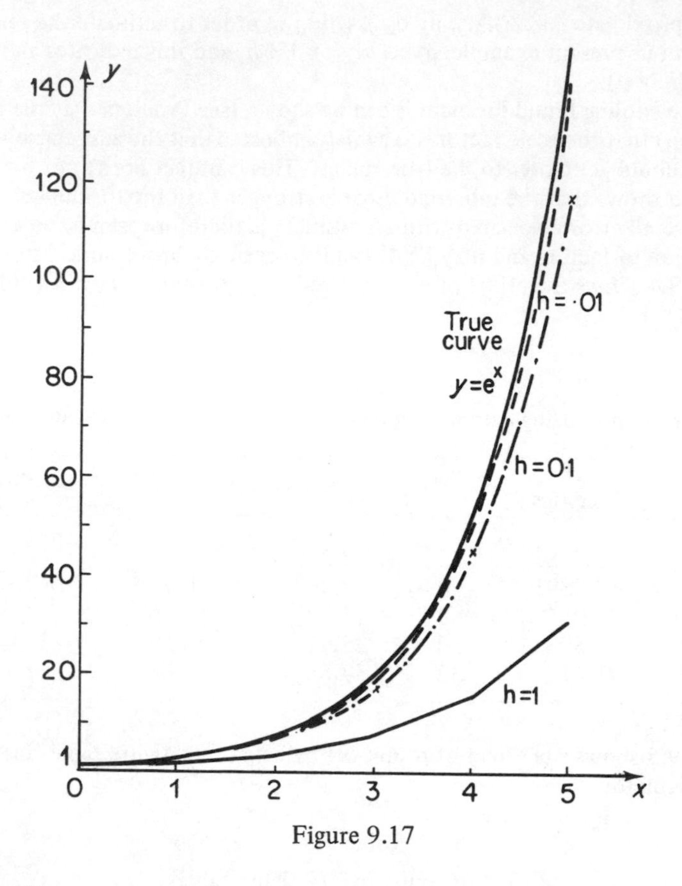

Figure 9.17

Problems II

1. Show that in Example 2 (page 516) $dy_{n+1}/dy_n = 1$; what conclusions about inherited error can be drawn?

2. Find the effect of inherited error for the problem $dy/dx = -y$ with $y = 1$ at $x = 0$.

3. For the cooling liquid problem show that the scale factor < 1.

4. For the problem $dy/dx = y$ with $y(x_0) = y_0$ show that the analytical solution is

$$y = y_0 e^{(x-x_0)}$$

Deduce that the Euler formula yields

$$y_n = (1+h)^n y_0$$

By replacing h by $(x_n - x_0)/n$ and taking the limit as $n \to \infty$ (keeping x_n fixed) prove that

$$y_n \to e^{(x_n - x_0)} . y_0$$

Interpret this result and verify your interpretation by estimating $y(2)$ given $x_0 = 1$ and step sizes of 1, 0.5, 0.1.

5. Given $d\theta/dt = -k(\theta - \theta_s)$ with $\theta(0) = \theta_0$, show that the Euler formula for $\theta_n \equiv \theta(nh)$ converges to the analytical solution as $h \to 0$.

   Note: $\lim_{n\to\infty} \left(1 + \frac{\alpha}{n}\right)^n = e^\alpha$

9.8 IMPROVED EULER METHOD

Graphically, we know that effectively Euler's method draws a tangent at the starting point P_0 (see Figure 9.18) and finds the point $P_1^{(0)}$ after a step h in the x-direction. The value of this ordinate is denoted by $y_1^{(0)}$. We have therefore assumed that the slope of the curve over the whole step takes the same value as that at P_0. Now the slope of the solution curve at $P_1^{(0)}$ will be approximately $f(x_1, y_1^{(0)})$ and this slope is indicated by the line $P_1^{(0)}T_1$.

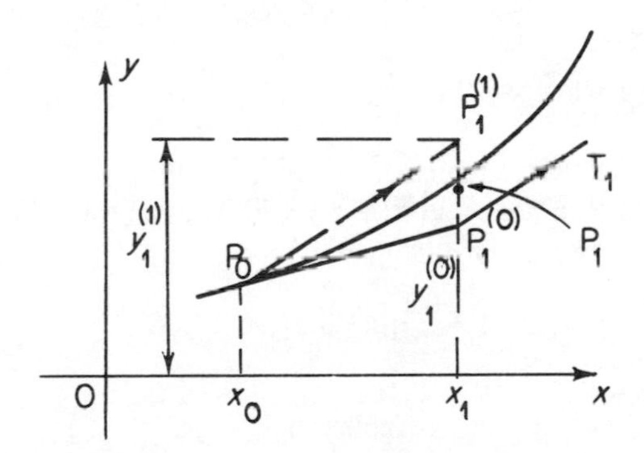

Figure 9.18

If we draw a parallel line through P_0 we get the line $P_0P_1^{(1)}$ and the ordinate of the point $P_1^{(1)}$ is given by

$$y_1^{(1)} = y_0 + hf(x_1, y_1^{(0)})$$

Here, $y_1^{(1)}$ denotes the first improved approximation to y_1.

We see that in this case $P_1^{(0)}$ and $P_1^{(1)}$ lie on either side of the exact solution so that half-way between these points should be a better approximation to the true value. This gives the point P_1 with ordinate

$$y_1 = \tfrac{1}{2}(y_1^{(0)} + y_1^{(1)})$$

which is our improved approximation to the value of y at $x = x_1$.

Now, since

$$y_1^{(0)} = y_0 + hf(x_0, y_0) \text{ ,}$$

$$y_1 = \tfrac{1}{2}[y_0 + hf(x_0, y_0) + y_0 + hf(x_1, y_1^{(0)})]$$

that is,

$$y_1 = y_0 + \frac{h}{2}[f(x_0, y_0) + f(x_1, y_1^{(0)})] \quad (9.11)$$

Formula (9.11) forms the basis for the **Improved Euler method.**

The next step will give

$$y_2 = y_1 + \frac{h}{2}[f(x_1, y_1) + f(x_2, y_2^{(0)})]$$

and, in general,

$$y_{n+1} = y_n + \frac{h}{2}[f(x_n, y_n) + f(x_{n+1}, y_{n+1}^{(0)})]$$

the value $y_{n+1}^{(0)}$ being obtained by Euler's method as

$$y_{n+1}^{(0)} = y_n + hf(x_n, y_n)$$

It can be shown that for $h < 1$ the error in using Euler's method is of the order of h^2 whereas that in the Improved Euler is of order h^3; thus the latter is a more accurate method.

Example

$$\frac{dy}{dx} = x \quad \text{and} \quad y = 1 \text{ when } x = 0$$

Let us use $h = 0.1$.

Here $x_0 = 0$, $y_0 = 1$ and hence $y_1^{(0)} = y_0 + hf(x_0, y_0) = 1 + (0.1).(0) = 1$. Thus, using the formula (9.11) with $x_1 = 0.1$,

$$y_1 = 1 + \frac{0.1}{2}[0 + 0.1] = 1.005$$

Table 9.6

x_n	y_n	$f(x_n, y_n)$	$y_{n+1}^{(0)}$	$f(x_{n+1}, y_{n+1}^{(0)})$	y_{n+1}
0.0	1.000	0.0	1.000	0.1	1.005
0.1	1.005	0.1	1.015	0.2	1.020
0.2	1.020	0.2	1.040	0.3	1.045
0.3	1.045	0.3	1.075	0.4	1.080
0.4	1.080	0.4	1.120	0.5	1.125
0.5	1.125	0.5	1.175	0.6	1.180
0.6	1.180	0.6	1.240	0.7	1.245
0.7	1.245				

The next step now gives

$$y_2^{(0)} = y_1 + hf(x_1, y_1) = 1.005 + (0.1)[0.1] = 1.015$$

and

$$y_2 = y_1 + \frac{h}{2}[f(x_1, y_1) + f(x_2, y_2^{(0)})] = 1.005 + \frac{0.1}{2}[0.1 + 0.2] = 1.020$$

Proceeding step by step we can construct Table 9.6.

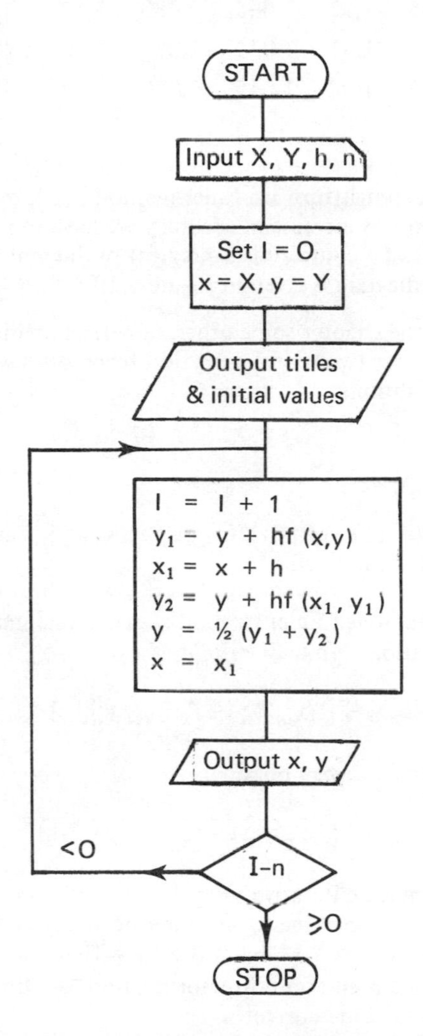

Here n is the number of estimations required and I is the counter which tells how many stages have been carried out to date.

Figure 9.19

In Table 9.7 we compare these results with the exact values obtained from the analytical solution $y = (x^2/2) + 1$ and with the values obtained using Euler's method.

Table 9.7

x	0.0	0.1	0.2	0.3	0.4	0.5	0.6	0.7
y_{exact}	1.0	1.005	1.02	1.045	1.080	1.125	1.180	1.245
y_{Euler}	1.0	1.0	1.01	1.03	1.06	1.10	1.15	1.21
$y_{\text{Improved Euler}}$	1.0	1.005	1.02	1.045	1.080	1.125	1.180	1.245

We can see that there is an increase in accuracy using the Improved Euler method as compared with straightforward Euler method but it must be remembered that to obtain this increase in accuracy we have to make twice as many function evaluations. Of course this is no great problem if we use a digital computer. A flow diagram is given in Figure 9.19.

We shall meet later in the chapter some other numerical methods but in the next two sections we consider two useful analytical techniques which are possible under certain conditions.

Problems

1. Try the problems 1 (i), (ii), (iii), (iv), (v), (vi) on page 517 using the Improved Euler technique.

2. We can regard the Improved Euler method as giving one improvement to the Euler approximation. We may write it as

$$y_{n+1}^{(1)} = y_n + \frac{h}{2}[f(x_n, y_n) + f(x_{n+1}, y_{n+1}^{(0)})]$$

If we repeat the process we can obtain

$$y_{n+1}^{(2)} = y_n + \frac{h}{2}[f(x_n, y_n) + f(x_{n+1}, y_{n+1}^{(1)})]$$

and so on, thus forming an iterative procedure.
Flow chart this method, working to a suitable accuracy and carry out two steps on the equation $\mathrm{d}y/\mathrm{d}x = x^2 + 1, y(0) = 1$ with $h = 0.1$.
This is an example of a **predictor-corrector** method, the first formula being the predictor and the second the corrector.

3. The **Modified Euler** method is, using customary notation,

$$y_{n+1} = y_n + hf[x_n + \tfrac{1}{2}h,\ y_n + \tfrac{1}{2}f(x_n, y_n)]$$

Estimate values of $y(0.1)$, $y(0.2)$ using this method on the equation $dy/dx = x^2 + 4$; $y(0) = 2$.

9.9 EXACT EQUATIONS

If we write a first-order differential equation in the form

$$P(x, y)\mathrm{d}x + Q(x, y)\mathrm{d}y = 0 \tag{9.12}$$

where $P(x, y)$ and $Q(x, y)$ are functions of x and y, then the equation is said to be **exact** if we can find a function $\phi(x, y)$ such that the differential $\mathrm{d}\phi$ is given by

$$\mathrm{d}\phi = P\mathrm{d}x + Q\mathrm{d}y$$

In such a case we have

$$\mathrm{d}\phi = 0$$

which integrates immediately to

$$\phi = \text{constant}$$

For example

$$(2x + 3x^2y)\mathrm{d}x + (x^3)\mathrm{d}y = 0$$

can be written

$$2x\mathrm{d}x + (3x^2y\mathrm{d}x + x^3\mathrm{d}y) = 0$$

i.e.

$$\mathrm{d}(x^2) + \mathrm{d}(x^3y) = 0$$

The second term has arisen from two terms, recognised as the derivative of a product.

Therefore we have

$$\mathrm{d}(x^2 + x^3y) = 0$$

and, integrating, we get

$$x^2 + x^3y = \text{constant}$$

This is all very well in a simple example, but can we recognise when it is possible to group the terms in such a form?

For example, the equation

$$y^3\,\mathrm{d}x + \frac{x}{y}\,\mathrm{d}y = 0$$

is not exact, but we might spend a long time discovering this if we first play around trying to find a suitable function $\phi(x, y)$.

The clue is to start in reverse: that is, if we assume we can find $\phi(x, y) = \text{constant}$, then we must have by differentiation

$$\mathrm{d}\phi(x, y) = 0$$

or

$$\frac{\partial \phi}{\partial x}\mathrm{d}x + \frac{\partial \phi}{\partial y}\mathrm{d}y = 0$$

Comparing this with $P\mathrm{d}x + Q\mathrm{d}y$, we see that

$$P = \frac{\partial \phi}{\partial x} \quad \text{and} \quad Q = \frac{\partial \phi}{\partial y}$$

Since, for most cases

$$\frac{\partial^2 \phi}{\partial x \partial y} = \frac{\partial^2 \phi}{\partial y \partial x}$$

then we must have

$$\frac{\partial P}{\partial y} = \frac{\partial^2 \phi}{\partial y \partial x} = \frac{\partial^2 \phi}{\partial x \partial y} = \frac{\partial Q}{\partial x}$$

So, if the equation (9.12) is exact, then

$$\frac{\partial P}{\partial y} = \frac{\partial Q}{\partial x}$$

Taking our first example,

$$\left.\begin{array}{ll} P = 2x + 3x^2 y & \dfrac{\partial P}{\partial y} = 3x^2 \\ Q = x^3 & \dfrac{\partial Q}{\partial x} = 3x^2 \end{array}\right\} \text{equal}$$

Hence the equation is **exact**.

The second example was such that

$$\left.\begin{array}{ll} P = y^3 & \dfrac{\partial P}{\partial y} = 3y^2 \\ Q = \dfrac{x}{y} & \dfrac{\partial Q}{\partial x} = \dfrac{1}{y} \end{array}\right\} \text{not equal}$$

The equation is **not exact**.

So we can recognise straight away whether or not an equation is exact.

Examples

(i) $(2x \log y)\mathrm{d}x + \left(\dfrac{x^2}{y} + 3y^2\right)\mathrm{d}y = 0$

$$P = 2x \log y \qquad Q = \frac{x^2}{y} + 3y^2$$

$$\frac{\partial P}{\partial y} = \frac{2x}{y} \qquad \frac{\partial Q}{\partial x} = \frac{2x}{y} \qquad \textit{Exact}$$

(ii) $(4x^2 - 2)\mathrm{d}x - \left(x^2 + \dfrac{3}{y}\right)\mathrm{d}y = 0$

$$P = (4x^2 - 2) \qquad Q = -\left(x^2 + \frac{3}{y}\right)$$

$$\frac{\partial P}{\partial y} = 0 \qquad \frac{\partial Q}{\partial x} = -2x$$

Not Exact

(iii) $(\mathrm{e}^x x \cos y + \mathrm{e}^x \cos y - 2x \sin y)\mathrm{d}x - (x\mathrm{e}^x \sin y + x^2 \cos y)\mathrm{d}y = 0$

$$P = x\mathrm{e}^x \cos y + \mathrm{e}^x \cos y - 2x \sin y$$

$$\frac{\partial P}{\partial y} = -x\mathrm{e}^x \sin y - \mathrm{e}^x \sin y - 2x \cos y$$

$$Q = -(x\mathrm{e}^x \sin y + x^2 \cos y)$$

$$\frac{\partial Q}{\partial x} = -(x\mathrm{e}^x \sin y + \mathrm{e}^x \sin y + 2x \cos y)$$

Exact

Integration of Exact Equations

We can now recognise when an equation is exact or not. Can we integrate it? In the case where $(P\mathrm{d}x + Q\mathrm{d}y)$ is exact, $P\mathrm{d}x + Q\mathrm{d}y = 0$ can be written

$$\frac{\partial \phi}{\partial x}\mathrm{d}x + \frac{\partial \phi}{\partial y}\mathrm{d}y = 0$$

Thus if we keep y constant in the function P and integrate with respect to x we have the opposite of partial differentiation and this will give us ϕ except for an unknown function of y (which would disappear on partially differentiating w.r.t. x). If we now differentiate this expression for ϕ partially with respect to y we should get Q.

Example 1

$(2x + 3x^2 y)\mathrm{d}x + x^3\,\mathrm{d}y = 0$

$$P = \frac{\partial \phi}{\partial x} = 2x + 3x^2 y$$

therefore $\phi = x^2 + x^3 y + f(y)$ where $f(y)$ is an unknown function of y.

We now differentiate with respect to y, therefore

$$\frac{\partial \phi}{\partial y} = x^3 + f'(y)$$

But

$$\frac{\partial \phi}{\partial y} = Q = x^3$$

therefore

$$f'(y) = 0$$

therefore

$$f(y) = A, \quad \text{an arbitrary constant}$$

Hence

$$\phi = x^2 + x^3 y + A = \text{constant}$$

or

$$x^2 + x^3 y = \text{constant}$$

Notice we do not need 2 arbitrary constants, so we could miss out A if we liked.

Example 2

$$(2x \log y)\mathrm{d}x + \left(\frac{x^2}{y} + 3y^2\right)\mathrm{d}y = 0$$

Since we have seen that this is exact

$$\frac{\partial \phi}{\partial x} = 2x \log y \tag{9.13a}$$

$$\frac{\partial \phi}{\partial y} = \frac{x^2}{y} + 3y^2 \tag{9.13b}$$

Integrating (9.13a)

$$\phi = x^2 \log y + f(y)$$

therefore

$$\frac{\partial \phi}{\partial y} = \frac{x^2}{y} + f'(y)$$

Comparing with (9.13b)

$$f'(y) = 3y^2$$

therefore

$$f(y) = y^3 + \text{constant}$$

therefore

$$\phi = x^2 \log y + y^3 + \text{constant}$$

The solution is thus

$$x^2 \log y + y^3 = \text{constant} = A, \text{ say.}$$

Example 3

Similarly, we consider

$$(x\mathrm{e}^x \cos y + \mathrm{e}^x \cos y - 2x \sin y)\mathrm{d}x - (x\mathrm{e}^x \sin y + x^2 \cos y)\mathrm{d}y = 0$$

which we showed to be exact.

$$\frac{\partial \phi}{\partial x} = x\mathrm{e}^x \cos y + \mathrm{e}^x \cos y - 2x \sin y$$

therefore

$$\phi = x\mathrm{e}^x \cos y - x^2 \sin y + f(y)$$

therefore

$$\frac{\partial \phi}{\partial y} = -xe^x \sin y - x^2 \cos y + f'(y)$$

Comparing with the given equation,

$$f'(y) = 0$$

therefore

$$f(y) = \text{constant}$$

therefore ϕ = constant gives

$$xe^x \cos y - x^2 \sin y = \text{constant}$$

[We could always perform the integration on $\partial\phi/\partial y$ to find $\phi = \ldots\ldots + f(x)$, then differentiate partially w.r.t. x and compare with the given equation.]

Sometimes an equation is not exact in the form it is given, but can be made exact by multiplying by a suitable factor called an **integrating factor**. It is beyond our scope to be able to find these factors in general for such equations but an example will explain what is meant.

Consider

$$(4x^2y^2 + 2\log y)\mathrm{d}x + \left(2x^3y + \frac{x}{y}\right)\mathrm{d}y = 0$$

$$P = 4x^2y^2 + 2\log y \qquad \frac{\partial P}{\partial y} = 8x^2y + \frac{2}{y}$$

Not Exact

$$Q = 2x^3y + \frac{x}{y} \qquad \frac{\partial Q}{\partial x} = 6x^2y + \frac{1}{y}$$

But to the experienced eye, it would be seen that if we multiply the given equation by x we do get an exact equation.

Thus

$$(4x^3y^2 + 2x\log y)\mathrm{d}x + \left(2x^4y + \frac{x^2}{y}\right)\mathrm{d}y = 0$$

P now is $4x^3y^2 + 2x\log y \qquad \dfrac{\partial P}{\partial y} = 8x^3y + \dfrac{2x}{y}$

Exact

Q now is $2x^4y + \dfrac{x^2}{y} \qquad \dfrac{\partial Q}{\partial x} = 8x^3y + \dfrac{2x}{y}$

Thus by multiplying by the **integrating factor** x we obtain an exact equation and we can proceed along the usual lines – you should complete the example yourself. Note that $2x$, $4x$, $-79.2x$ are also possible integrating factors (why?).

Problems

1. Test each of the following for exactness and solve the equation if it is exact.

   (i) $(3x^2y - x^3)\mathrm{d}x + (x^3 + y)\mathrm{d}y = 0$

(ii) $(x + 2y)\mathrm{d}x + 2(y + x)\mathrm{d}y = 0$

(iii) $(r + \sin\theta + \cos\theta)\mathrm{d}r + r(\cos\theta - \sin\theta)\mathrm{d}\theta = 0$

(iv) $\theta\,\mathrm{d}r = (\mathrm{e}^{\theta} + 2r\theta - 2r)\mathrm{d}\theta$

2. Show that the following equation is not exact but that $1/y^4$ is an integrating factor and hence solve the equation

$$(y^3 - 2xy)\mathrm{d}x + (3x^2 - xy^2)\mathrm{d}y = 0$$

3. Prove that x is an integrating factor of the equation $(x^2 + y^2 + x)\mathrm{d}x + xy\mathrm{d}y = 0$ and solve.

4. Prove the equation $(2x^3 + 3y)\mathrm{d}x + (3x + y - 1)\mathrm{d}y = 0$ is exact and solve.

5. Prove that x^n is an integrating factor of $x\mathrm{d}y = (y + \log x)\mathrm{d}x$ only for $n = -2$ and solve.

6. Solve the differential equation $(3x^2y - y^3)\mathrm{d}x + (x^3 - 3xy^2)\mathrm{d}y = 0$. (L.U.)

9.10 LINEAR EQUATIONS

Another type of equation which it is theoretically possible to solve analytically is the *linear, first-order differential equation.* The general form of such an equation is

$$\frac{\mathrm{d}y}{\mathrm{d}x} + py = q \tag{9.14}$$

where p and q are functions of x only.

For example

$$\frac{\mathrm{d}y}{\mathrm{d}x} + x^2y = (x - 3)$$

$$\frac{\mathrm{d}y}{\mathrm{d}x} - \frac{y}{(x+1)} = x^2\mathrm{e}^x$$

$$\frac{\mathrm{d}y}{\mathrm{d}x} + \tan x\,.\,y = \sec x$$

We have already met the idea of an integrating factor in the last section on exact equations. This is a factor by which we multiply the differential equation in order to make it into a suitable form for integration. This is just the technique we use here.

We take the l.h.s. of (9.14) and try to make it into the derivative of a product by multiplying it by some function of x.

That is, we take $(\mathrm{d}y/\mathrm{d}x) + py$ and multiply by a function of x, say $I(x) = I$ for short.

This gives

$$I\frac{\mathrm{d}y}{\mathrm{d}x} + pIy$$

The first term is the first part of

$$\frac{\mathrm{d}}{\mathrm{d}x}(Iy) = I\frac{\mathrm{d}y}{\mathrm{d}x} + \frac{\mathrm{d}I}{\mathrm{d}x}y$$

Comparing the second terms, we want

$$pIy = \frac{\mathrm{d}I}{\mathrm{d}x}y$$

or

$$\frac{\mathrm{d}I}{\mathrm{d}x} = pI$$

Separating the variables

$$\int \frac{\mathrm{d}I}{I} = \int p\mathrm{d}x$$

That is,

$$\log I = \int p\mathrm{d}x$$

or

$$I = \mathrm{e}^{\int p\mathrm{d}x}$$

Thus our integrating factor (I.F.) can, theoretically, be found.

Take

$$\frac{\mathrm{d}y}{\mathrm{d}x} + x^2y = (x - 3)$$

The function p is x^2, therefore

$$\text{I.F. is } \mathrm{e}^{\int x^2\,\mathrm{d}x} = \mathrm{e}^{x^3/3}$$

Similarly, for the example

$$\frac{\mathrm{d}y}{\mathrm{d}x} - \frac{y}{(x + 1)} = x^2\mathrm{e}^x$$

$$p \text{ is} - \frac{1}{(1 + x)}$$

therefore

$$\text{I.F. is } \mathrm{e}^{\int(-1/[1+x])\mathrm{d}x} = \mathrm{e}^{-\log(1+x)}$$

Thus, in the general equation $(dy/dx)+py = q$, if we can integrate p with respect to x, we can find the integrating factor and multiplying the equation by this factor reduces the l.h.s. to the form

$$\frac{d}{dx}(\text{I.F.} \times y)$$

The r.h.s. will have become $q \times$ I.F. Then the differential equation becomes

$$\frac{d}{dx}(\text{I.F.} \times y) = q \times \text{I.F.}$$

Integrating both sides w.r.t. x, we get

$$y \times \text{I.F.} = \int (q \times \text{I.F.})\,dx$$

Thus

$$y = \frac{1}{\text{I.F.}} \int (q \times \text{I.F.})\,dx \quad \text{is the solution.}$$

This solution will be general if it contains one arbitrary constant. This arbitrary constant will arise from the integral on the right-hand side.

Example 1

$$\frac{dy}{dx} + y = x$$

$$\text{I.F.} = e^{\int 1.dx} = e^x$$

Multiplying throughout by this we get

$$e^x \frac{dy}{dx} + ye^x = xe^x$$

or

$$\frac{d}{dx}(ye^x) = xe^x$$

Therefore

$$ye^x = \int xe^x\,dx$$

that is,

$$ye^x = xe^x - e^x + A$$

Hence

$$y = x - 1 + Ae^{-x}$$

(Notice that Ae^{-x} satisfies the equation $(dy/dx) + y = 0$.)

Example 2

$$\frac{dy}{dx} + \frac{y}{x} = 4$$

$$\text{I.F.} = e^{\int (1/x)dx} = e^{\log_e x}$$

[Whenever you see $e^{\log_e \alpha}$, you should be able to write this down without $\log_e$'s, for if

$$z = e^{\log_e \alpha}$$

$$\log_e z = \log_e(e^{\log_e \alpha}) = \log_e \alpha \,.\, \log_e e \quad (\text{using } \log_e a^x = x \log_e a)$$

But $\log_e e = 1$

therefore

$$\log_e z = \log_e \alpha$$

therefore

$$z = \alpha$$

We know then that

$$e^{\log_e \alpha} = \alpha$$

Similarly,

$$e^{-\log_e \alpha} = \frac{1}{e^{\log_e \alpha}} = \frac{1}{\alpha}$$

We have written $\log_e \alpha$ to emphasise the base of the logarithms.]

Thus, the

$$\text{I.F.} = e^{\log_e x} = x$$

Multiplying the equation by this factor gives

$$x\frac{dy}{dx} + y = 4x$$

which reduces to

$$\frac{d}{dx}(y\,.\,x) = 4x$$

therefore

$$y\,.\,x = \int 4x dx$$

that is,

$$y\,.\,x = 2x^2 + A$$

therefore

$$y = 2x + \frac{A}{x}$$

(Notice that A/x satisfies the equation $(dy/dx) + y/x = 0$)

Note that, once you are used to the method, there is no need to put in the step where you multiply the equation by the integrating factor. You can go straight to the following line

$$\frac{d}{dx}(y \times \text{I.F.}) = q \times \text{I.F.}$$

Example 3

$$x(x-1)\frac{dy}{dx} + y = (x+1)$$

We must first write this in the standard form $(dy/dx) + py = q$. In this case we have

$$\frac{dy}{dx} + \frac{1}{x(x-1)}y = \frac{x+1}{x(x-1)}$$

$$\text{I.F.} = e^{\int \frac{1}{x(x-1)}dx} = e^{\int\left(-\frac{1}{x}+\frac{1}{x-1}\right)dx} = e^{\log\left(\frac{x-1}{x}\right)} = \frac{x-1}{x}$$

The equation therefore reduces to

$$\frac{d}{dx}\left[y \cdot \left(\frac{x-1}{x}\right)\right] = \frac{x+1}{x(x-1)} \cdot \left(\frac{x-1}{x}\right) = \frac{x+1}{x^2}$$

Integrating both sides

$$y\frac{(x-1)}{x} = \int \frac{(x+1)}{x^2}\,dx$$

$$= \int \left(\frac{1}{x} + \frac{1}{x^2}\right) dx$$

Hence

$$y\frac{(x-1)}{x} = \log_e x - \frac{1}{x} + A$$

or

$$y = \frac{x}{x-1}\log_e x - \frac{1}{x-1} + \frac{Ax}{x-1}$$

Which equation do you think $Ax/(x-1)$ satisfies?

Example 4 (See Figure 9.20)

A circuit consisting of a resistance R ohms and an inductance L henrys is connected to an e.m.f. voltage of $E \cos \omega t$. Find the current i amps at a time t after the circuit is closed. We assume connecting wires are resistance-free.

The voltage $E \cos \omega t$ must be equal to the voltage drops Ri across the resistance and $L(di/dt)$ due to the inductance so that

$$L\frac{di}{dt} + Ri = E\cos\omega t$$

The integrating factor $= e^{\int (R/L)dt} = e^{(R/L)t}$

Hence

$$ie^{(R/L)t} = \int \frac{E}{L}\cos\omega t\, e^{(R/L)t}\,dt$$

then

$$ie^{(R/L)t} = \frac{e^{(Rt/L)}E}{(R^2 + L^2\omega^2)}(R\cos\omega t + L\omega\sin\omega t) + A$$

Since $i = 0$ when $t = 0$, then

$$0 = \frac{ER}{R^2 + L^2\omega^2} + A$$

giving

$$A = \frac{-ER}{R^2 + L^2\omega^2}$$

so that

$$ie^{(R/L)t} = \frac{e^{(Rt/L)}E}{R^2 + L^2\omega^2}(R\cos\omega t + L\omega\sin\omega t) - \frac{ER}{R^2 + L^2\omega^2}$$

therefore

$$i = \frac{E}{R^2 + L^2\omega^2}(R\cos\omega t + L\omega\sin\omega t - Re^{-(R/L)t})$$

Note that there is an oscillatory part to the solution and a decaying exponential or transient part. Where has this last part come from?

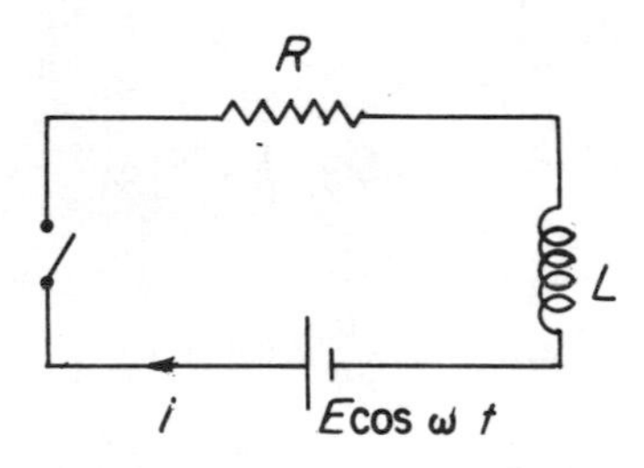

Figure 9.20

Example 5 (See Figure 9.21)

A body of mass m is projected with speed V on a rough table whose coefficient of friction is μ. If the air resistance is proportional to the (speed)^2, find the distance travelled before the body comes to rest.

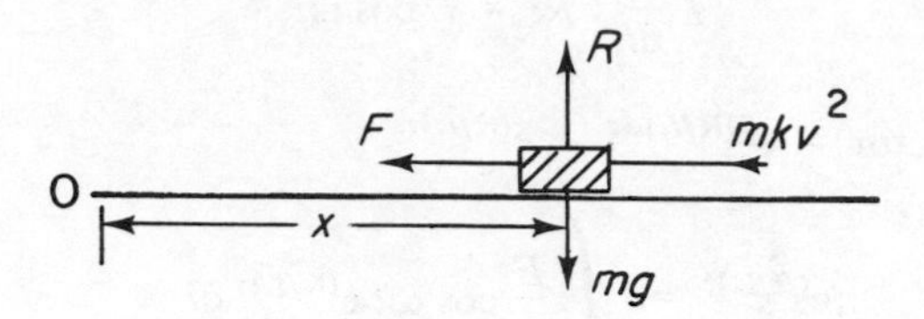

Figure 9.21

Take the resistance to be mkv^2. The laws of mechanics tell us that $R = mg$ and $F = \mu R = \mu mg$ so that the net resistance to motion is $F + mkv^2 = \mu mg + mkv^2$.

The equation of motion is now

$$m\frac{dv}{dt} = -m(\mu g + kv^2)$$

We can write dv/dt as

$$\frac{dv}{dx} \cdot \frac{dx}{dt} = \frac{dv}{dx} \cdot v = \frac{1}{2}\frac{d}{dx}(v^2)$$

Hence

$$m\frac{1}{2}\frac{dv^2}{dx} = -m(\mu g + kv^2)$$

which can be written

$$\frac{d(v^2)}{dx} + 2k(v^2) = -2\mu g$$

This equation is just like $(dy/dx) + 2ky = -2\mu g$ where $y = v^2$ and so is a linear first order differential equation.

$$\text{I.F.} = e^{\int 2k\,dx} = e^{2kx}$$

Hence

$$v^2 e^{2kx} = -2\mu g \int e^{2kx}\,dx$$

that is

$$v^2 e^{2kx} = -\frac{\mu g}{k}e^{2kx} + A$$

Now $v = V$ when $x = 0$, therefore

$$A = V^2 + \frac{\mu g}{k}$$

We now have

$$v^2 e^{2kx} = -\frac{\mu g}{k}e^{2kx} + \left(V^2 + \frac{\mu g}{k}\right)$$

or

$$v^2 = -\frac{\mu g}{k} + \left(V^2 + \frac{\mu g}{k}\right) e^{-2kx}$$

What significance has the term $(V^2 + \mu g/k)e^{-2kx}$?

The body will come to rest when $v = 0$. Then

$$0 = -\frac{\mu g}{k} + \left(V^2 + \frac{\mu g}{k}\right) e^{-2kx}$$

Rearranging,

$$e^{2kx} = \left(V^2 + \frac{\mu g}{k}\right) \bigg/ \frac{\mu g}{k} = \frac{kV^2}{\mu g} + 1$$

therefore

$$x = \frac{1}{2k} \log_e \left(1 + \frac{kV^2}{\mu g}\right)$$

This is the distance travelled before coming to rest.

Failure of the integrating factor method

Now this all appears very simple but many things can go wrong. Firstly, we may not be able to find the I.F. easily, if at all. For example,

$$\frac{dy}{dx} + \sqrt{1 + x^3}\, y = 1$$

$$\text{I.F.} = e^{\int \sqrt{1+x^3}\, dx} = ?$$

This is a difficult integral which would probably make a solution by this method unattainable.

Even if we can find the I.F. there will be many cases where we cannot integrate the resulting R.H.S. by analytical methods.

For example,

$$\frac{dy}{dx} + 2xy = (x - 2)$$

$$\text{I.F. is } e^{\int 2x dx} = e^{x^2}$$

The equation therefore reduces to

$$\frac{d}{dx}(y \cdot e^{x^2}) = (x - 2)e^{x^2}$$

or

$$ye^{x^2} = \int (x - 2)e^{x^2}\, dx$$

We can find

$$\int x e^{x^2} \, dx$$

but

$$\int -2e^{x^2} \, dx$$

is the stumbling block. We could evaluate the integral on the R.H.S. by numerical integration, but we must ask *should we?* Would it be better in these cases of failure to start from the beginning with a numerical technique? The answer is probably *Yes!*

Problems

1. For each of the following differential equations, indicate which can be solved by integrating factor and which by separation of variables. Write down the integrating factor where applicable.

(i) $\dfrac{dy}{dx} = y \cos x$ (ii) $y \dfrac{dy}{dx} = y + 3$

(iii) $\dfrac{du}{dt} = ue^{2t} - e^{t}$ (iv) $\dfrac{dz}{dt} = \sqrt{z(1+t)^2}$

(v) $\dfrac{d\theta}{dt} = \theta(\theta(t)) \times e^{t}$ (vi) $v \dfrac{dv}{dt} = v^2 e^{-2t}$

2. Solve the differential equations

(i) $x \dfrac{dy}{dx} + y = x^2$ (ii) $(x+1)\dfrac{dy}{dx} + 3y = x + 1$

(iii) $\dfrac{dy}{dx} + y \cot x = \operatorname{cosec} x$

(iv) $(1+t)\dfrac{d\theta}{dt} + (1+2t)\theta = (1+t)^2$, given $\theta = 0$ when $t = 1$

(v) $t \cos t \dfrac{d\theta}{dt} + (t \sin t - \cos t)\theta - t^2 = 0$, given $\theta = 0$ when $t = \dfrac{\pi}{4}$

3. Solve the differential equations

(i) $\dfrac{dy}{dx} + y \cot x = \sin x$, where $y = 1$ at $x = \dfrac{\pi}{2}$

(ii) $x(x^2 - 1)\dfrac{dy}{dx} + y = x^3$

(iii) $\dfrac{dy}{dx} + 2y = e^x$

(iv) $\sin x \dfrac{dy}{dx} + 2y \cos x = \cos x$, given that $y = -\tfrac{1}{2}$ when $x = \dfrac{\pi}{2}$

(v) $x \dfrac{dy}{dx} + 2y = x^2$, given that $y = \tfrac{1}{2}$ when $x = 1$

(vi) $\dfrac{d^2 z}{dt^2} + \dfrac{dz}{dt} = e^{2t}$

(vii) $(x+4)\dfrac{dy}{dx} + 3y = 3x$

(viii) $\dfrac{d^2 y}{dx^2} - \dfrac{dy}{dx} = x$, with the conditions $y = 2$, $\dfrac{dy}{dx} = 0$ at $x = 0$

4. If $\dfrac{dy}{dx} + 2y \tan x = \sin x$ and $y = 0$ when $x = \dfrac{\pi}{3}$ show that the maximum value of y is $\dfrac{1}{8}$. (C.E.I.)

5. Solve the differential equation

$$\frac{dy}{dx} + xy = \exp(-x^2/2)$$

(C.E.I.)

6. A rocket, when full of fuel, is of mass M lbs and it burns fuel at a constant rate, m lbs/sec. The relative backward velocity of the gases is constant and equal to u. The rocket is ignited and moves vertically upwards from rest against an air resistance kmV, where V is the velocity of the rocket. Using the equation of motion

$$(M - mt)\frac{dV}{dt} - mu = -(M - mt)g - kmV$$

show that

$$V = \frac{u}{k} + \frac{g(M - mt)}{m(1-k)} - \left(\frac{Mg}{m(1-k)} + \frac{u}{k}\right)\left(1 - \frac{mt}{M}\right)^k$$

7. A circular coil of n turns of area A whose inductance is L henrys and resistance R ohms is rotated with angular velocity ω about a diameter perpendicular to a field of strength H A m^{-1}. The current i induced in the coil is given by

$$L\frac{di}{dt} + Ri = n\omega\, HA \cos \omega t$$

Find the current at time t, assuming that initially it is zero.

8. A particle of mass m is attracted to a fixed point O by a force $m\mu x$ when it is at a distance x from O in a medium which offers a resistance mkv^2 to the motion (v being the velocity). Show that the equation of motion is

$$\frac{d(v^2)}{dx} + 2kv^2 + 2\mu x = 0$$

If the particle is initially at rest at a distance a from O show that it reaches O with velocity

$$\sqrt{\frac{\mu}{2k^2}\left\{1 + e^{2ka}[2ka - 1]\right\}}$$

9. A body is immersed in an atmosphere whose temperature in °C varies as $25(2 - t)°$. The temperature of the body is initially 60°C. Show that the governing equation for the temperature of the hot body, θ, is

$$\frac{d\theta}{dt} + k\theta = 25k(2 - t)$$

where k is some constant. If $k = 0.1$, obtain estimates of θ and compare with the analytical solution which you should show is

$$\theta = 300 - 25t - 240e^{-0.1t}$$

Try to sketch the solution and see whether it agrees with your formulation.

10. The differential equation for current i at time t in an $L - R$ series circuit is

$$L\frac{di}{dt} + Ri = E$$

If the applied voltage is $E = E_0(1 - e^{-\alpha t})$ and the initial current is i_0, obtain an expression for $i(t)$. (C.E.I.)

11. A reservoir has been contaminated by effluent from a factory. The capacity of the reservoir is 10^6 litres. The degree of contamination is 0.02% by weight. The (constant) average daily rate of consumption of water for non-drinking purposes is 2×10^4 litres and this is continuously replaced by pure water. How long will it be before the concentration of contaminant drops to the safe level of 10^{-5}% ?

12. A room of volume V m^3 is supplied with fresh air containing some CO_2. Let $c(t)$ be the concentration of CO_2 in the room at time $t > 0$, c_0 the initial concentration and c_f the concentration in the fresh air supply; all three are measured in parts per 10,000. Let Q m^3/sec be the rate of fresh air supplied and Q_p m^3/sec the volume of CO_2 produced by people inside the room.
Derive a differential equation for the concentration $c(t)$ and solve it.

Sketch on the same axes the solution in the case

$$\left(\frac{Q_{cf} + 10{,}000\, Q_p}{Q}\right) < c_0$$

and the solutions for the special cases :

(i) no people in the room and the fresh air free from CO_2,
(ii) the initial concentration of CO_2 in the room is zero.

A room of volume $170\,\text{m}^3$ receives a total change of air every 30 minutes and has a CO_2 content of 0.03% without people present. The concentration of CO_2 in the outside air is also 0.03% . If the production of CO_2 per person is $4.7 \times 10^{-6}\,\text{m}^3/\text{sec}$, what is the maximum number of people allowed in the room at any one time if the concentration of CO_2 in the room is not to exceed 0.1%? Sketch the solution curve in this case for $c(t)$.

[Hint : the long term solution only is important.]

9.11 ROCKET RE-ENTRY: A CASE STUDY

Before considering further numerical methods we should like to consider the following practical example which illustrates a failure of the analytical method.

A rocket re-entering the earth's atmosphere experiences a drag force proportional to the square of its velocity and proportional to the density of the air. This density, ρ, is approximated by the law $\rho = \rho_0 e^{-\lambda z}$ where z is the height above the earth; ρ_0 is the density at sea-level and λ is a positive constant.

Let A be the cross-sectional area of the rocket, C_D a constant called the drag coefficient and let us assume that re-entry is vertical and without lift. From Newton's second law, the equation of motion is

$$mv\,\frac{dv}{dz} = -mg + \frac{1}{2} C_D\, A\rho v^2$$

Note that we are measuring acceleration upwards. † (See Figure 9.22)

Now

$$\frac{d}{dz}(v^2) = 2v\,\frac{dv}{dz}$$

therefore, writing $V = v^2$ and $K = C_D A/m$ we obtain

$$\frac{d}{dz}(v^2) = -2g + \frac{C_D A}{m}\,\rho v^2$$

† It turns out to be more convenient this way. Imagine that we throw a stone upwards and then consider only its falling to the ground. In this example, the rocket should slow down as it approaches the earth and so the net force must be upwards.

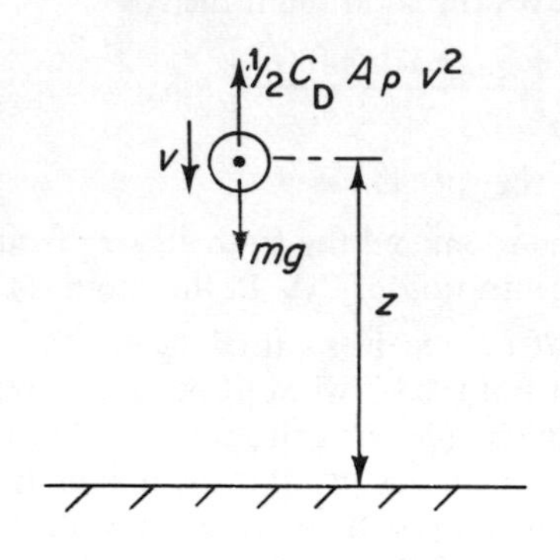

Figure 9.22

that is

$$\frac{dV}{dz} = -2g + K\rho_0 e^{-\lambda z} V$$

or

$$\frac{dV}{dz} - K\rho_0 e^{-\lambda z} V = -2g$$

The Integrating Factor is

$$e^{\int -K\rho_0 e^{-\lambda z} dz}$$

$$= e^{(K/\lambda)\rho_0 e^{-\lambda z}}$$

Multiplying the equation of motion by this factor produces

$$\frac{d}{dz}(Ve^{(K\rho_0/\lambda)e^{-\lambda z}}) = -2ge^{(K/\lambda)\rho_0 e^{-\lambda z}}$$

which, on integration, becomes

$$Ve^{(K\rho_0/\lambda)e^{-\lambda z}} = \int -2ge^{(K/\lambda)\rho_0 e^{-\lambda z}} dz$$

If we expand $e^{(K/\lambda)\rho_0 e^{-\lambda z}}$ as a Maclaurin's series and integrate term by term, we obtain

$$V = e^{(-K\rho_0/\lambda)e^{-\lambda z}} \left[-\frac{2g}{\lambda} \sum_{n=1}^{\infty} \frac{(\frac{\rho_0 K}{\lambda} e^{-\lambda z})^n}{n.n!} - 2gz + \text{constant} \right] \quad (9.15)$$

This formula would be very time-consuming to use as the basis for calculating V at several points on the rocket's descent. It is virtually impossible to use for qualitative information.

We shall return to this problem in Section 9.13.

9.12 TAYLOR SERIES METHOD

We start with the problem on Newton's Law of Cooling.

$$\frac{d\theta}{dt} = -0.1(\theta - 20); \quad \theta(0) = 100$$

We aim to form the Taylor's series for $\theta(t)$ centred at $t = 0$. To do this, we shall need the values of θ, $d\theta/dt$, $d^2\theta/dt^2$ ...... at $t = 0$. We know $\theta(0) = 100$ and from the differential equation itself we deduce

$$\left.\frac{d\theta}{dt}\right|_{t=0} \text{ is } -0.1(100 - 20) = -8$$

Differentiating the differential equation with respect to t yields

$$\frac{d^2\theta}{dt^2} = -0.1\frac{d\theta}{dt}$$

and so at $t = 0$,

$$\frac{d^2\theta}{dt^2} = 0.8$$

differentiating a second time produces

$$\frac{d^3\theta}{dt^3} = -0.1\frac{d^2\theta}{dt^2}$$

and at $t = 0$,

$$\frac{d^3\theta}{dt^3} = -0.08$$

In the same way we find

$$\left.\frac{d^4\theta}{dt^4}\right|_{t=0} = +0.008 \text{ etc.}$$

Hence

$$\theta(t) = \theta(0) + t.\theta'(0) + \frac{t^2}{2!}\theta''(0) + \frac{t^3}{3!}\theta'''(0) + \frac{t^4}{4!}\theta^{iv}(0) + \ldots\ldots$$

$$= 100 - 8t + \frac{0.8t^2}{2} - 0.08\frac{t^3}{3!} + 0.008\frac{t^4}{4!} - \ldots \tag{9.16}$$

In this example we can rearrange the right-hand side of (9.16) as

$$20 + 80\left(1 - 0.1t + 0.01\frac{t^2}{2!} - 0.001\frac{t^3}{3!} + 0.0001\frac{t^4}{4!} - \ldots\ldots\right)$$

that is,

$$20 + 80e^{-0.1t}$$

to recover the analytical solution. We can now see that the series on the right-hand side will converge for all values of t.

The next example shows the difficulty associated with finding the radius of convergence of the resulting series when the analytical solution is not known.

Example

$\frac{dy}{dx} = x - y^2$ with $y(0) = 0$

or

$$y' = x - y^2 \Rightarrow y'(0) = 0$$

Now

$$y'' = 1 - 2yy' \Rightarrow y''(0) = 1$$

$$y''' = 0 - 2(y')^2 - 2yy'' \Rightarrow y'''(0) = 0$$

$$y^{iv} = -4y'y'' - 2y'y'' - 2yy''' \Rightarrow y^{iv}(0) = 0$$

$$y^{v} = -6(y'')^2 - 6y'y''' - 2y'y''' - 2yy^{iv} \Rightarrow y^{v}(0) = -6$$

$$y^{vi} = -20y''y''' - 10y'y^{iv} - 2yy^{v} \Rightarrow y^{vi}(0) = 0$$

Therefore Taylor's series is

$$y = \frac{x^2}{2!} - \frac{6x^5}{5!} + \ldots\ldots = \frac{x^2}{2} - \frac{x^5}{20} + \ldots\ldots$$

Suppose we wish to estimate $y(0.1)$; the terms we have yield

$$y \simeq \frac{(0.1)^2}{2} - \frac{(0.1)^5}{20} = 0.005 - 0.0000005$$

$$= 0.0050 \quad (4 \text{ d.p.})$$

Now the first term to be ignored will at best contain x^7 and may be of the form $a_7(x^7/7!)$ where a_7 is a constant [in this case a_7 is zero and the first missing term is $a_8(x^8/8!)$]. This gives us a clue as to how big x has to be before the loss of the term is significant in so far as it affects the accuracy to which we are working. But it is *only a clue* and unless we calculate that next term we cannot be sure. There are two factors to consider in using this series method :

(i) for what range of x will the series converge?

(ii) for what range of x will the series converge fast enough for practical computation?

Since this example does not readily yield a formula for the nth term in the series the only real answer is to try to see. Of course, if we do not compute the next term we cannot make any safe predictions. In fact, the missing term is $252x^8/8!$, that is, $x^8/160$ and this will exceed 0.0001 if $x > 0.596$ (3 d.p.). We can thus theoretically use a truncated series for $x = 0.5$ but not for $x = 0.6$. We may use our approximation to estimate $y(0.2)$ as

$$\frac{(0.2)^2}{2} - \frac{(0.2)^5}{20} = 0.02 - 0.000016 = 0.0200 \quad (4 \text{ d.p.})$$

In order to avoid the danger of omitting a significant term, we may advance in steps of 0.1 (at the expense of round-off error) using a Taylor Series centred on $x = 0.1$. From the formulae for y', y'', etc. and using $y(0.1) = 0.0050$

$$y'(0.1) = 0.1 - (0.0050)^2 = 0.1000 \quad \text{(4 d.p.)}$$

$$y''(0.1) = 1 - 0.01 \times 0.1 = 0.9990 \quad \text{(4 d.p.)}$$

$$y'''(0.1) = -2(0.1)^2 - 2 \times 0.1 \times 0.9990 = -0.2198 \quad \text{(4 d.p.)}$$

therefore

$$y(x) = 0.0050 + (x - 0.1).(0.1) + \frac{(x-0.1)^2}{2!}(0.9990) - \frac{(x-0.1)^3}{3!}(0.2198) + \ldots\ldots$$

When $x = 0.2$

$$y(0.2) = 0.0050 + 0.01 + 0.0050 - \ldots\ldots$$

$$= 0.0200 \quad \text{(4 d.p.)}$$

which agrees with previous values.

The Taylor Series method is said to be **semi-analytical** and can only be used with a digital computer if the necessary calculus has first been performed.

Let us now put the method on a formal footing. We take the first-order differential equation

$$y' = f(x, y) \text{ with } y = y_0 \text{ at } x = x_0$$

Repeated differentiation gives

$$y'' = \frac{\partial f}{\partial x} + \frac{\partial f}{\partial y} \cdot y' = \frac{\partial f}{\partial x} + \frac{\partial f}{\partial y} \cdot f$$

$$y''' = \frac{\partial^2 f}{\partial x^2} + 2\frac{\partial^2 f}{\partial x \partial y} y' + \frac{\partial^2 f}{\partial y^2}(y')^2 + \frac{\partial f}{\partial y} y'' \text{ etc.}$$

We can find the values of y', y'' etc. near the point (x_0, y_0) and denote these $y_0', y_0'', \ldots\ldots$ We then form a Taylor expansion around this point as

$$y = y_0 + (x - x_0)y_0' + \frac{(x - x_0)^2}{2!} y_0'' + \ldots.$$

We then generate values of y for x near x_0 bearing in mind the requirements of convergence and practical convergence (i.e. speed of convergence)

Example

For the equation $\frac{dy}{dx} = x - y^2$; $y(0) = 0$ the Euler method would give

$$y_{n+1} = y_n + h(x_n - y_n^2)$$

and so

$$y(0.1) = y(0) + 0.1(0 - 0^2) = 0$$

$$y(0.2) = y(0.1) + 0.1(0.1 - 0^2) = 0.01$$

It is clear that the Euler method is of no use in this example and since it is a straight-line approximation it represents the first two terms only of the Taylor expansion. In this example both these terms vanish at the initial point and hence there is a total loss of information in estimating $y(0.1)$ which carries over into the estimate of $y(0.2)$.

The Taylor series method would be useful in such a case to obtain the first stage of a numerical solution as follows.

$$y' = x - y^2 \qquad y'(0) = 0$$

therefore

$$y'' = 1 - 2yy' \qquad y''(0) = 1$$

$$y''' = -2(y')^2 - 2yy'' \qquad y'''(0) = 0$$

$$y^{iv} = -6y'y'' - 2yy''' \qquad y^{iv}(0) = 0$$

$$y^{v} = -8y'y''' - 6(y'')^2 - 2yy^{iv} \qquad y^{v}(0) = -6$$

Hence

$$y = \frac{x^2}{2} - \frac{x^5}{20} + \ldots\ldots$$

$$\left.\begin{aligned} \text{at } x = 0.1,\ y &= \frac{0.01}{2} - \frac{0.00001}{20} + \ldots\ldots = 0.005 \\ \text{at } x = 0.2,\ y &= \frac{0.04}{2} - \frac{0.00032}{20} + \ldots\ldots = 0.020 \end{aligned}\right\} \quad [3 \text{ d.p.}]$$

Problems

1. For the equation $(dy/dx) + xy = 0$ with $y = 1$ at $x = 0$, show that

$$y = 1 - \frac{x^2}{2} + \frac{x^4}{8} - \frac{x^6}{48} + \frac{x^8}{384} - \ldots\ldots$$

and hence compute the solution at $x = 0.2, 0.4, 0.8$ correct to 4 d.p. Compare with the analytical solution.

2. Show, using Taylor's series, that the solution of $(dy/dx) = 2x - y$, $y(1) = 3$, near $x = 1$ is given numerically by

x	1.1	1.2	1.3
y	2.9145	2.8562	2.8225

Compare with the analytical solution.

3. How accurate is the approximation $y \triangleq (1/3)x^3$ to the solution of the differential equation

$$\frac{dy}{dx} = x^2 + y^2, \quad y(0) = 0 \quad ?$$

4. A Taylor expansion can be used to solve a differential equation numerically by a step-by-step process. Explain this statement with reference to the equation

$$y'(x) = [x + y(x)]^3, \quad y(0) = 1$$

with step length 0.1. (C.E.I.)

5. The differential equation

$$\frac{dx}{dt} + 2x + x^3 = 4$$

has the initial condition $x = 1$ when $t = 0$.
By using the first five non-zero terms of a Taylor's series, obtain values of x at $t = 0.10$ and $t = 0.15$, correct to 3 d.p. (L.U.)

9.13 RUNGE-KUTTA METHODS

We have just seen that, provided the series converges, the Taylor expansion provides an accurate estimation of the dependent variable, y. However it does require the evaluation of derivatives, which can be a tedious process. Methods have been developed which approximate the Taylor expansion by means of a combination of function evaluations. In this section we shall study one such method: the **Runge-Kutta fourth-order method.** This requires four evaluations of $f(x, y)$ and the particular combination of these chosen then provides an estimate of y which is equivalent to the Taylor expansion as far as the term in $(x - x_0)^4$.

We may proceed via a geometrical interpretation. We assume $y_0 = y(x_0)$ is known and we seek $y_1 = y(x_1) = y(x_0 + h)$ [see Figure 9.23 (i)]. At x_0, the slope of the curve is $y'(x_0) = f(x_0, y_0)$; this provides an estimate of y_1 as given by the Euler method. We calculate $k_1 = hf(x_0, y_0)$.

We now calculate the slope half-way along this line, that is, at the point A; this is $f(x_0 + h/2, y_0 + \tfrac{1}{2}k_1)$.† Through (x_0, y_0) we draw a line with this slope to produce a new estimate of y_1 and calculate $k_2 = hf(x_0 + h/2, y_0 + k_1/2)$ [see Figure 9.23 (ii)].

We calculate the slope half-way along this line, i.e. at B. It is $f(x_0 + h/2, y_0 + \tfrac{1}{2}k_2)$. Through (x_0, y_0) we draw a line with this slope to produce a third estimate of y, and calculate $k_3 = hf(x_0 + h/2, y_0 + \tfrac{1}{2}k_2)$ [see Figure 9.23 (iii)].

† Note that this is not the slope of the straight line.

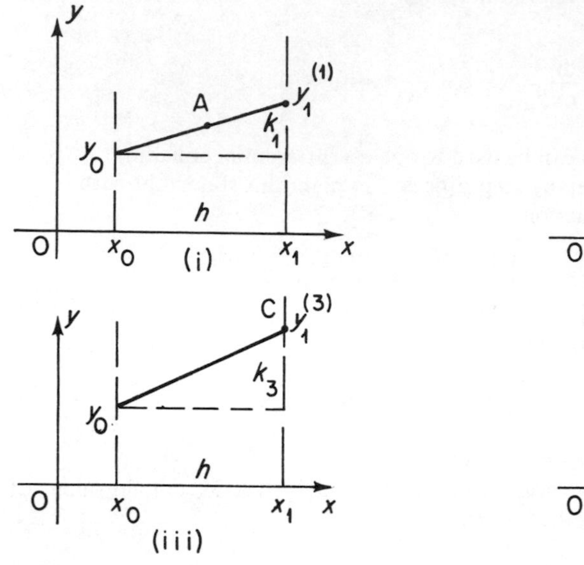

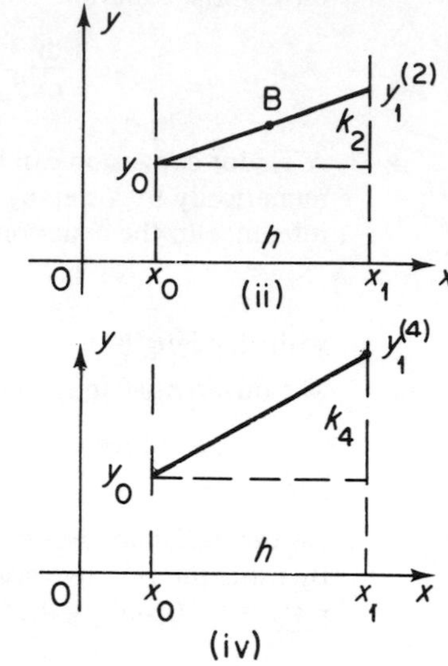

Figure 9.23

At this new estimated point, C, we calculate the slope $f(x_0 + h, y_0 + k_3)$ and draw through (x_0, y_0) a line with this slope to produce a fourth estimate of y, and calculate $k_4 = hf(x_0 + h, y_0 + k_3)$, [see Figure 9.23 (iv)].

The composite diagram for the estimates is shown in Figure 9.24. In general, the true value of y_1 will lie between the two extreme estimates. We take the *weighted average*

$$k = \frac{1}{6}(k_1 + 2k_2 + 2k_3 + k_4)$$

and estimate y_1 as $y_0 + k$.

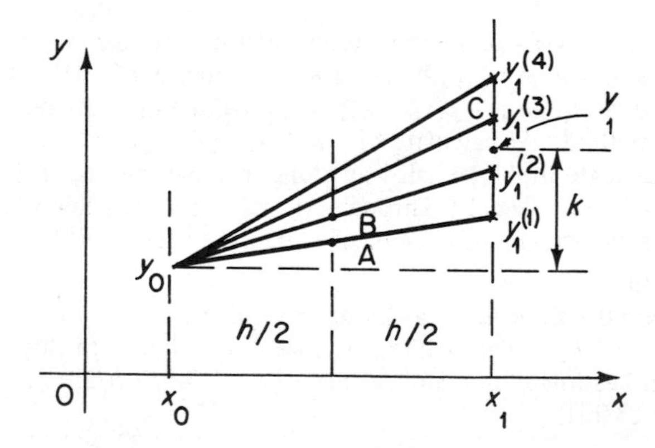

Figure 9.24

This seems a curious ritual, but our aim was to produce as estimates of y_1 as accurate as the Taylor expansion to the terms in $(x_1 - x_0)^4$ and analysis of the Runge-Kutta formula has shown this to be the case. [If $f(x, y)$ is independent of y then you should be able to show that the Runge-Kutta formula becomes Simpson's Rule.]

Example 1

First, the Newton's Law of Cooling example:

$$\frac{d\theta}{dt} = -0.1(\theta - 20); \quad \theta_0 = 100 \text{ and we seek to estimate } \theta(4)$$

$f(\theta, t) = f(\theta)$ only.

1. $\dot{\theta}_0 = -8$, therefore $k_1 = 4(-8) = -32$

2. At A, $\dot{\theta} = -0.1(100 - ½ . 32 - 20) = -6.4$,
   therefore $k_2 = 4(-6.4) = -25.6$

3. At B, $\dot{\theta} = -0.1(100 - ½ . 25.6 - 20) = -6.72$
   therefore $k_3 = 4(-6.72) = -26.88$

4. At C, $\dot{\theta} = -0.1(100 - 26.88 - 20) = -5.312$
   therefore $k_4 = 4 \times (-5.312) = -21.248$

$$k = -\frac{1}{6}(32 + 2 \times 25.6 + 2 \times 26.88 + 21.248) = -\frac{1}{6}(158.208)$$
$$= -26.368$$

therefore $\theta(4) = 100 - 26.368 = 73.63$ (2 d.p.)

This compares with the analytical value of 73.624 and an Euler estimate of 73.615 with a step size of 0.01 obtained using a digital computer.

Example 2

We next consider a worked example using a Runge-Kutta fourth order method to solve the ordinary differential equation

$$\frac{dy}{dx} = x^2 + y^2 \quad \text{with } y = 0 \text{ when } x = 0$$

For convenience in the argument to follow, let $x^2 + y^2 = f(x, y)$. The object is to start with the point defined in the boundary condition and then calculate the value of y at $x_1 = x_0 + h$ where h is a suitably chosen step size and x_0 the value of x at the boundary point. The process is then repeated to find the value of y at $x_2 = x_1 + h$.

The Runge-Kutta algorithm for making these calculations is:

(i) Suppose we have already found y_n at the point $x = x_n$, then calculate $k_1 = h \times f(x_n, y_n)$.

(ii) Use this value of k_1 to calculate k_2 from:
$k_2 = h \times f(x_n + \frac{1}{2}h, y_n + \frac{1}{2}k_1)$.

(iii) Use this value of k_2 to calculate k_3 from:
$k_3 = h \times f(x_n + \frac{1}{2}h, y_n + \frac{1}{2}k_2)$.

(iv) Use this value of k_3 to calculate k_4 from:
$k_4 = h \times f(x_n + h, y_n + k_3)$

(v) Calculate y_{n+1} from:

$$y_{n+1} = y_n + (k_1 + 2k_2 + 2k_3 + k_4)/6$$

(vi) Finally we have $x_{n+1} = x_n + h$.

Hence starting with the boundary values x_0, y_0, we can apply the above rules to find x_1, y_1, and then repeat to find x_2, y_2, and so on.

In the example, start at step (i). We have $x_0 = 0, y_0 = 0$ and let $h = 0.1$. Then $k_1 = 0.1 \times (x_0^2 + y_0^2) = 0.0$.

Step (ii) gives

$$k_2 = 0.1 \times [(x_0 + 0.05)^2 + (y_0 + \tfrac{1}{2} \times 0.0)^2] = 0.00025.$$

Step (iii) gives

$$k_3 = 0.1 \times [(x_0 + 0.05)^2 + (y_0 + \tfrac{1}{2} \times 0.00025)^2] = 0.25000156 \times 10^{-3}$$

and Step (iv) gives

$$k_4 = 0.1 \times [(x_0 + 0.1)^2 + (y_0 + 0.00025000156)^2] = 0.0010000063$$

From Step (v) we get

$$y_1 = y_0 + (0.0 + 2 \times 0.00025 + 2 \times 0.00025 + 0.001)/6 = 0.0003333349.$$

Now we repeat the process using $x_1 = 0.1$ and $y_1 = 0.0003333349$.

$$k_1 = 0.1 \times [x_1^2 + y_1^2] = 0.0010000111$$

$$k_2 = 0.1 \times [(x_1 + 0.05)^2 + (y_1 + \tfrac{1}{2} \times 0.0010000111)^2] = 0.0022500694$$

$$k_3 = 0.1 \times [(x_1 + 0.05)^2 + (y_1 + \tfrac{1}{2} \times 0.0022500694)^2] = 0.0022502127$$

$$k_4 = 0.1 \times [(x_1 + 0.1)^2 + (y_1 + 0.0022502127)^2] = 0.0040006675$$

and from these:

$$y_2 = y_1 + \frac{1}{6}(k_1 + 2k_2 + 2k_3 + k_4) = 0.0026668754, \text{ and}$$

$$x_2 = x_1 + h = 0.2.$$

In Table 9.8, the two sets of steps above are concisely written and the process is continued until $x = x_{10} = 1.0$.

Table 9.8

x	y	k_1	k_2	k_3	k_4
0.0	0.000000	0.000000	0.000250	0.0002500	0.001000
0.1	0.000333	0.001000	0.002250	0.0022502	0.004000
0.2	0.002667	0.0040007	0.006252	0.0062534	0.009008
0.3	0.009003	0.0090081	0.012268	0.0122729	0.016045
0.4	0.021359	0.0160456	0.020336	0.0203494	0.025174
0.5	0.041791	0.0251747	0.030545	0.0305756	0.036524
0.6	0.072448	0.0365249	0.043073	0.0431333	0.050336
0.7	0.115660	0.0503377	0.058233	0.0583460	0.067278
0.8	0.174081	0.0670304	0.076559	0.0767597	0.087292
0.9	0.250908	0.0872955	0.098926	0.0992723	0.112263
1.0	0.350233				

Case Study with Runge-Kutta

We return to the rocket re-entry problem of Section 9.11. This was programmed for a digital computer and the results are shown in Table 9.9. Two step sizes were used.

The dashes indicate that the numerical solution misbehaved by increasing the velocity of the rocket because no check was built-in to prevent such occurrence. Note that Runge-Kutta is almost independent of the two step sizes.

To see the effect of gravity the computations were repeated without it and the results are also shown in Table 9.9. A sketch of this is Figure 9.25. Since the Runge-Kutta is expected to be the most accurate method of the three, we

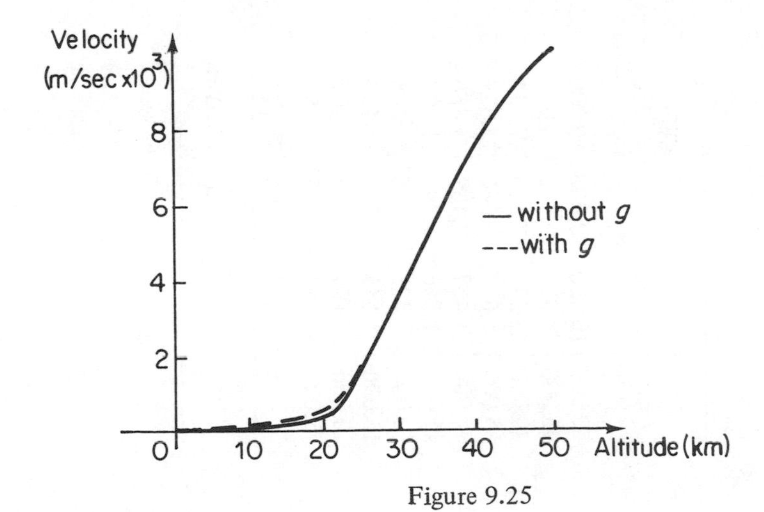

Figure 9.25

Table 9.9

z	Euler		Improved Euler		Runge-Kutta		Euler		Improved Euler		Runge-Kutta	
(m)	250	10^3	250	10^3	250	10^3	250	10^3	250	10^3	250	10^3
50,000	10,000	10,000	10,000	10,000	10,000	10,000	10,000	10,000	10,000	10,000	10,000	10,000
45,000	8,992	9,013	8,985	8,985	8,985	8,985	8,987	9,008	8,980	8,980	8,980	8,980
40,000	7,542	7,575	7,531	7,531	7,531	7,531	7,532	7,565	7,521	7,521	7,522	7,522
35,000	5,633	5,646	5,629	5,631	5,629	5,629	5,619	5,632	5,615	5,617	5,615	5,615
30,000	3,467	3,409	3,486	3,497	3,485	3,485	3,450	3,391	3,469	3,480	3,468	3,468
25,000	1,546	1,398	1,591	1,621	1,590	1,589	1,523	1,373	1,569	1,600	1,567	1,567
20,000	425	305	466	511	464	465	380	329	425	473	423	423
15,000	145	141	150	168	149	149	34	80	50	90	49	50
10,000	107	107	107	111	107	107	0	-	2	25	1	2
5,000	83	82	83	-	83	83	0	-	0	-	0	-
0	64	64	64	-	64	-	-	-	0	-	0	-
	With g						Without g					

consider the effects of including gravity on its calculation. Above 30,000 m the discrepancy is less than 0.5% but this quickly builds up as the altitude decreases. It has been shown that the exclusion of gravity is less important for higher re-entry velocities. If we return to the analytical solution then omission of gravity produces (see Equation 9.15):

$$V = \text{constant} \times e^{(-K\rho_0/\lambda)e^{-\lambda z}}$$

so that

$$v = \text{constant} \times e^{(-K\rho_0/2\lambda)e^{-\lambda z}}$$

(c.f. $v = \sqrt{V}$)

At altitudes of the order of 100 kilometres the exponential term as a whole ~ 1 if mg/C_DA is sufficiently large, and if v_e is the re-entry velocity at that speed

$$v = v_e e^{(-K\rho_0/2\lambda)e^{-\lambda z}}$$

This is one in a chain of approximations, but it illustrates the use of a simple method as a basis for the analysis of more complicated physical situations.

One question with which to leave this section is how reliable are the Runge-Kutta estimates. Write a computer program to check the tabulated estimates with gravity using Equation (9.15).

Problems

1. Solve any of the problems given earlier on Euler and Improved Euler using Runge-Kutta and compare the results.

2. Compute the solution of $dv/dt = 21.5 - 1.2v^2$ with initial condition $v = 0$ at $t = 0$, using the Runge-Kutta fourth-order method at $t = 0.001, 0.002$.

3. Solve the following:

   (i) $\dfrac{dy}{dx} = x - \sin x + y$ Initial condition: $x = 0,\ y = 1.0$

   (ii) $\dfrac{dy}{dx} = \dfrac{2yx}{y^2 + x^2}$ Initial condition: $x = 0,\ y = 3$

   (iii) $\dfrac{dv}{dx} = \dfrac{2x^2 + 3x}{4v}$ Initial condition: $x = 2,\ v = 1$

4. Consider $dy/dx = x + y$ with $y(0) = 2$; find $y(1)$.

5. The response y (> 0) of a certain hydraulic valve subject to sinusoidal input variation is given by

$$\frac{dy}{dt} = \sqrt{2\left(1 - \frac{y^2}{\sin^2 t}\right)} \quad \text{with } y_0 = 0,\ t_0 = 0$$

Why is the Taylor series method unsuitable as a solution?
Show that

$$\left(\frac{dy}{dt}\right)_0 = \sqrt{\frac{2}{3}}$$

and hence use Runge-Kutta (fourth-order) to obtain a solution at $t = 0.2$.

(C.E.I.)

6. Why (apart from round-off error) should the Runge-Kutta method give exact results for the equation $dy/dx = 2(x + 1)$; $y(0) = 1$? Show that an analytical solution of this equation and of $dy/dx = 2y/(x + 1)$ is $y = (x + 1)^2$. Compute $y(1), y(2)$ by Runge-Kutta for each equation and compare the results. What can you conclude?

7. The current i in a LR circuit at any time t after a switch is thrown at $t = 0$ can be expressed by the equation

$$\frac{di}{dt} = (E \sin \omega t - Ri)/L$$

where $E = 50$ volts, $L = 1$ henry, $\omega = 300$, $R = 50$ ohms and the initial condition is that at $t = 0, i = 0$.
Solve the differential equation numerically using the Runge-Kutta method and compare your answers with the analytical solution

$$i = \frac{E}{Z^2}(R \sin \omega t - \omega L \cos \omega t + \omega L e^{-Rt/L})$$

where

$$Z = \sqrt{R^2 + \omega^2 L^2}$$

8. A quantity of 10 kg of material is dumped into a vessel containing 60 kg of water. The concentration of the solution, c, in percentage at any time, t, is expressed as

$$(60 - 1.212c)dc/dt = (k/3)(200 - 14c)(100 - 4c)$$

where k, the *mass transfer coefficient*, is equal to 0.0589. The initial condition is that at $t = 0, c = 0$.
Find the $c-t$ relation by the Runge-Kutta and Euler methods and compare the result with the analytical solution.

9.14 COMPARISON OF METHODS

Now we compare results for the cooling liquid experiment. The step size for the numerical methods was 5 minutes. Values are quoted to 3 d.p. (Table 9.10)

Table 9.10

	Analytical	Euler	Imp. Euler	Runge-Kutta
0	100	100	100	100
5	68.52	70	100	68.511
10	49.432	51.25	100	49.430
15	37.848	39.532	100	37.846
20	30.824	32.208	100	30.827
25	26.568	27.630	100	26.567
30	23.984	24.768	100	23.983
35	22.416	22.881	100	22.417
40	21.464	21.801	100	21.465

Newton's Law of Cooling

You see that the Runge-Kutta method is the most accurate when compared with the "true" analytical solution. Try to produce a flow chart and write a computer program to obtain the above results.

Finally, we now give a summary comparison of the main methods used with a simple differential equation.

We seek the solution of $dy/dx = x + y$ when $y = 1$ at $x = 0$ for $x = 0(0.1)0.2$ (i.e. $h = 0.1$).

Analytical Solution

$y = 2e^x - x - 1$

x	e^x	y
0.1	1.10517	1.1103
0.2	1.22140	1.2428
0.3	1.34986	1.3997

Try to arrive at the analytical formula via the Taylor series approach.

Euler Method

$$y_1 = y_0 + h(x_0 + y_0)$$

$$y_1 = 1 + 0.1(0 + 1) = 1.1$$

$$y_2 = 1.1 + 0.1(0.1 + 1.1) = 1.22$$

$$y_3 = 1.22 + 0.1(0.2 + 1.22) = 1.362$$

Improved Euler Method

Let $f_0 = x_0 + y_0$ and $f_1 = x_1 + y_1$ where y_1 = previous estimate of y at $x_1 + x_0 + h$. Then a better estimate is

$$y_1 = y_0 + h(f_0 + f_1)/2 \tag{9.17}$$

First, $y_1 = y_0 + hf_0$.

x_0	y_0	x_1	y_1	f_0	f_1	y_1
0	1.0			1.0		1.1
		0.1	1.1		1.2	1.11
			1.11		1.21	1.1105
			1.1105		1.2105	1.1105
0.1	1.1105			1.2105		1.2316
		0.2	1.2316		1.4316	1.2426
			1.2426		1.4426	1.2432
			1.2432		1.4432	1.2432
0.2	1.2432			1.4432		1.3875
		0.3	1.3875		1.6875	1.3997
			1.3997		1.6997	1.4003
			1.4003		1.7003	1.4004
			1.4004		1.7004	1.4004

Note we have iterated by applying (9.17) successively until convergence of y_1 is obtained.

Runge-Kutta

$$
\begin{aligned}
k_1 &= h(x_0 + y_0) \\
k_2 &= h(x_0 + \tfrac{1}{2}h + y_0 + \tfrac{1}{2}k_1) \\
k_3 &= h(x_0 + \tfrac{1}{2}h + y_0 + \tfrac{1}{2}k_2) \\
k_4 &= h(x_0 + h + y_0 + k_3) \\
y_1 &= y_0 + (k_1 + 2k_2 + 2k_3 + k_4)/6
\end{aligned}
$$

x_0	y_0	k_1	k_2	k_3	k_4	y_1
0	1.0	0.1	0.11	0.1105	0.12105	1.1103
0.1	1.1103	0.12103	0.13208	0.13263	0.14429	1.2428
0.2	1.2428	0.14428	0.15649	0.15710	0.16999	1.3997

SUMMARY

The widespread use of differential equations in mathematical models makes it essential that you are in complete command of the material in this chapter. You should appreciate the significance of the relationship between a particular solution, the general solution and boundary conditions.

We have deliberately spent time on Euler's method since it illustrates the main ideas involved in the numerical solution of a differential equation; you should especially bear in mind the kinds of error involved in such a solution. In practice, the Runge-Kutta method is most widely used but again it is necessary to understand the principles on which it is based and the error involved, particularly in comparison with the Taylor series method.

The analytical techniques studied are of use in solving standard differential equations which arise in several branches of engineering, it is important to know the methods of separation of variables and Integrating Factor.

We hope you have read section 9.1 carefully, for here we see the interweaving of numerical and analytical approaches most clearly. The approach outlined here is the main theme in this book.

Chapter Ten

Second and Higher order Differential Equations

10.0 INTRODUCTION – VIBRATIONS

Vibrations are induced in many situations, for example when lorries cross a bridge, when a ship sets sail, when machines are set in motion. As in many other situations it is important to be able to predict the nature of these vibrations; failure to do so accurately may result in the snapping of a shaft, the breaking away of an engine from its mountings or, on a more spectacular scale, the collapse of a suspension bridge (as happened at Tacoma Narrows in 1940). Many of these vibrations can be modelled reasonably accurately by linear, second-order differential equations with constant coefficients (especially if the amplitude of the vibrations is small). In any event, such equations provide a starting point for the study of more complicated vibrations.

10.1 CASE STUDY: A CAR ON A BUMPY ROAD

The purpose of automobile suspension springing is two-fold: it must maintain the wheels in contact with the road to provide the degree of adhesion necessary to allow safe accelerating, cornering and braking; also it must provide a large measure of isolation for the vehicle and passengers from the road irregularities and from other possible causes of discomfort. With the advent of the pneumatic tyre even a simplified physical model of a car becomes complicated. In Figure 10.1 the body of the car is the sprung mass M and its connections to the tyres of masses m_1 and m_2 are represented by a spring and a dashpot to signify damping of oscillations (this damping is called shock absorption). Each tyre is assumed to have an elasticity which can be represented as a spring connection with the road. We have taken only a simple two-dimensional model but even so it comprises three separate bodies; each of these needs six coordinates to have its position in space specified completely (three to fix the centre of gravity and three to specify its orientation about the c.g.) and so our model would require 18 coordinates – we say the system possesses 18 **degrees of freedom**. As well as translational motion in three directions, each rigid body is capable of three rotational motions: *pitch, roll* and *yaw*. Most of the motions are unimportant, but even if we confine our attention to a purely vertical translation with the axles parallel we have still 2 degrees of freedom.

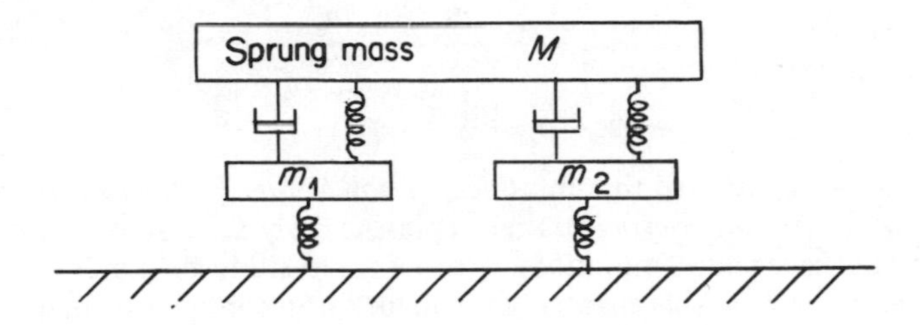

Figure 10.1

To make a start we consider just one wheel in vertical motion where the tyre is replaced by a solid wheel so that any irregularities in the road are transmitted without loss of form to the vehicle (c.f. railways where the track is relatively smooth and the danger of the wheel leaving the track is small). In effect, we have the situation depicted in Figure 10.2. Assume that the vehicle moves with a constant horizontal velocity, v. Let the profile of the road be described by $y(x)$; then the vertical velocity of the vehicle is

$$\frac{dy}{dt} = \frac{dy}{dx} \cdot \frac{dx}{dt} = \frac{dy}{dx} \cdot v$$

and its vertical acceleration is

$$\frac{d^2y}{dt^2} = \frac{d}{dt}\left(\frac{dy}{dt}\right) = \frac{d}{dx}\left(\frac{dy}{dx} \cdot v\right)v = v^2 \frac{d^2y}{dx^2}$$

Therefore we can treat the vertical acceleration of the vehicle, and hence the vertical force on the vehicle due to the road, as a function of t or of x.

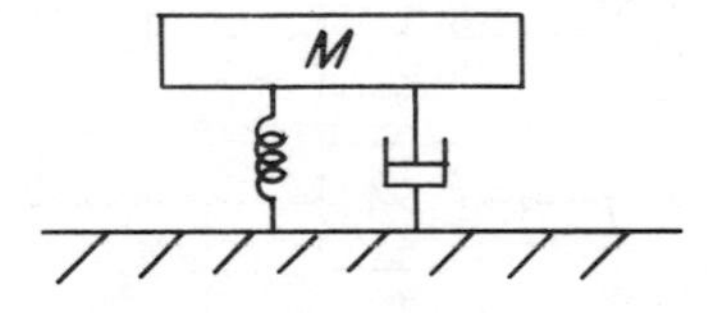

Figure 10.2

If we assume that the elastic effect of the suspension can be represented as a linear spring (i.e. it exerts a restoring force mn^2y *directly proportional* to displacement), and **if** we assume the damping effect of the suspension is viscous [i.e. the force it exerts $(2km\dot{y})$ is *proportional* to the vertical velocity], then the equation obtained by equating force to (mass × acceleration) is

$$m\ddot{y} = -mn^2y - 2km\dot{y} + g(t)$$

where n^2 is the spring stiffness of the suspension and $g(t)$ is the force due to the uneven road. This equation can be rearranged to give

$$\ddot{y} + \frac{2km}{m}\dot{y} + \frac{mn^2}{m}y = \frac{1}{m}g(t)$$

or, relabelling coefficients,

$$\ddot{y} + b\dot{y} + cy = f(t) \tag{10.1}$$

It should be emphasised that this *linear, second-order, ordinary differential equation with constant coefficients* can represent fairly accurately many forms of electrical or mechanical vibrations. Our task is to study the kinds of solution it can possess. Later in this chapter we shall look at techniques to handle vibrations which are governed by other equations, including those which are non-linear, as is sometimes the case when amplitudes are large.

10.2 THE ANALOGUE COMPUTER APPROACH

As we mentioned in Chapter 9, we can obtain a qualitative picture of the solution to an ordinary differential equation using an analogue computer. On this occasion we intuitively expect two initial (boundary) conditions since the equation is second-order. Figure 10.3 shows a block diagram of the circuit for the analogue computer simulation of Equation (10.1) with suitable values for b, c, $y(0)$ and $\dot{y}(0)$ and a suitable function $f(t)$. (Refer to Section 9.1 for details of the components of the circuit.)

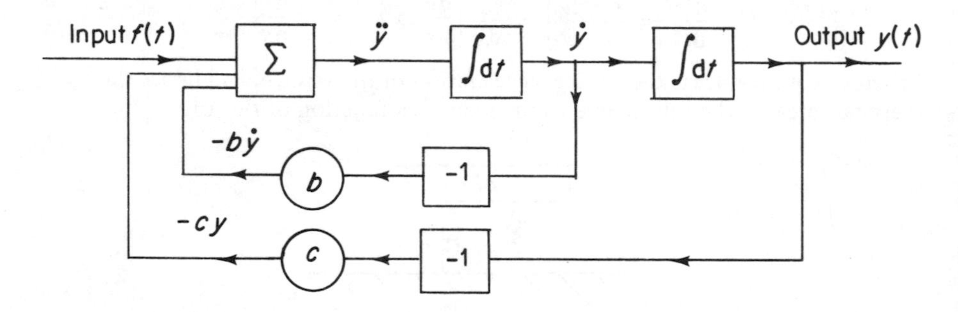

Figure 10.3

Figure 10.4(i) shows a typical trace from the computer; Figure 10.4(ii) shows the trace with the function $f(t)$ removed and Figure 10.4(iii) shows the trace of the $f(t)$ originally used.

We can see from Figure 10.4(iii) that the form of $f(t)$ is sinusoidal – an idealised bumpy road. Figure 10.4(i) seems to follow this shape, but superimposed on it is an oscillation which decays with time. Physically, this suggests that the *shock* of the bumpy road causes the car to oscillate rapidly and then the damping effect of the suspension causes these oscillations to die out rapidly and the car to follow the profile of the road in its motion. The *linearity* of the

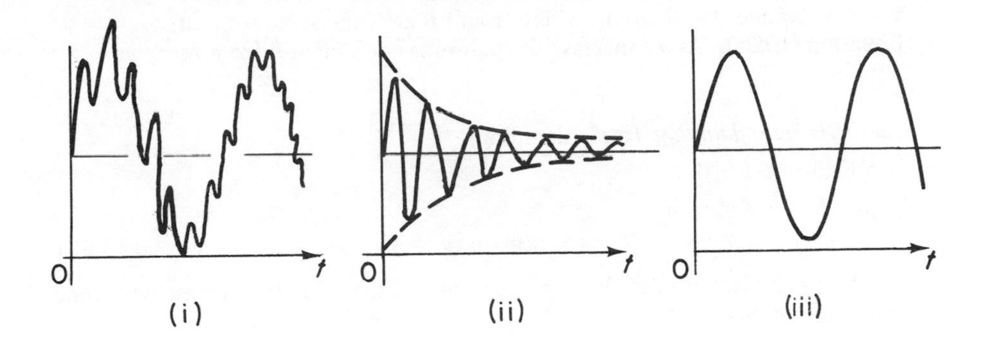

Figure 10.4

equation suggests a superimposing of solutions and since we know that part of the total trace is due directly to $f(t)$ we wonder from what the remainder arises. In fact, it is the so-called **natural vibration** or **free vibration** of the system: the motion which results when there is *no* imposed force and the system is slightly disturbed from an equilibrium position. When there is an external force acting, the motion is said to be **forced.**

If we recall from Chapter 1 the example of the oscillating spring, we remember that the amplitude of the oscillations was damped out with time: the motion of the spring corresponded to Figure 10.4 (ii).

We shall begin our study by examining the case of **free** oscillations, but first we shall rewrite Equation (10.1) as

$$\ddot{y} + 2k\dot{y} + n^2 y = f(t) \tag{10.2}$$

this is merely for algebraic convenience, as we shall see.

Case (i): no damping **(simple harmonic motion)**

Here

$$\ddot{y} + n^2 y = 0 \tag{10.3a}$$

The analogue trace Figure 10.5(i) shows two waveforms: the upper one represents the displacement, y, and the lower one represents the velocity, $\dot{y}$.

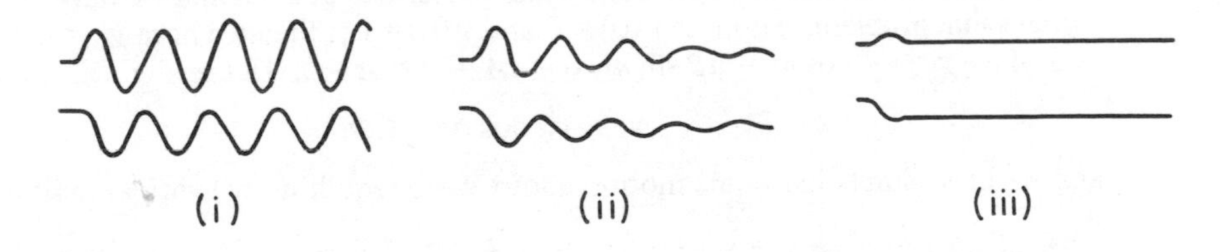

Figure 10.5

This suggests a sine or cosine solution and it is not too hard to see that $y = A \sin nt$ and $y = B \cos nt$, where A and B are constants, both satisfy Equation (10.3a). As we increase the damping coefficient k from zero, we obtain

Case (ii): light damping **(underdamping)**
[see Figure 10.5 (ii)]
Here

$$\ddot{y} + 2k\dot{y} + n^2 y = 0 \tag{10.3b}$$

The amplitudes of both the displacement and the velocity decrease with time in an exponential form.

Case (iii): heavy damping **(overdamping)**
For large values of k the waveforms cease to show any oscillatory behaviour [Figure 10.5 (iii)] and we have overdamping. The disturbance fails to cause the system to pass through equilibrium, merely approaching it asymptotically as $t \to \infty$. This condition is met with in a ballistic galvanometer. (Why might it not be desirable for automobile suspensions?) The larger the value of k, the longer the system takes to reach any particular position near equilibrium (why have we excluded equilibrium itself?).
There are therefore three distinct cases to consider and we take each in turn.

10.3 FREE OSCILLATIONS – ANALYTICAL APPROACH

The linearity of Equation (10.3a) means that the sum of any two solutions is also a solution. Hence we have a solution

$$y = A \sin nt + B \cos nt \tag{10.4}$$

But from Chapter 9, we know that the general solution of the equation contains two arbitrary constants. In fact, $A \sin nt$ and $B \cos nt$ are **linearly independent** solutions of (10.3a). We can, therefore, take the general solution to be (10.4) and the particular solution we want will be found by prescribing two conditions. If these are both values of y for two (different) values of t, they are called **boundary conditions** and we have a **boundary-value problem.** For the moment, we consider the situation when y and $\dot{y}$ are prescribed at $t = 0$ and we have an **initial-value problem.** Suppose $\dot{y}(0) = 0$ and $y(0) = y_0$, then we have $y_0 = 0 + B$ and, since $\dot{y} = nA \cos nt - nB \sin nt$, $0 = nA - 0 \Rightarrow A = 0$. Hence

$$y = y_0 \cos nt$$

and we have simple harmonic motion about $y = 0$ (equilibrium) with amplitude y_0.

Now we turn to the case of damping which is described by (10.3b). We note that the functions which reproduce themselves (save for a multiplicative constant) on differentiation are exponential (sine and cosine can each be defined as a linear combination of exponentials). We therefore substitute $y = e^{mt}$ into (10.3b) and obtain

$$m^2 e^{mt} + 2kme^{mt} + n^2 e^{mt} = 0$$

Since e^{mt} is never zero we may divide the equation by this factor to obtain the **auxiliary equation**

$$m^2 + 2km + n^2 = 0 \tag{10.5}$$

The general solution of (10.5) is

$$m = -k \pm \sqrt{k^2 - n^2}$$

so there are three cases to consider.

Case (a) $k < n$

Let $k^2 - n^2 = -p^2$, then $m = -k \pm ip$ and we have two possibilities: $y = e^{(-k+ip)t}$ and $y = e^{(-k-ip)t}$ to give a general solution

$$y = A' e^{(-k+ip)t} + B' e^{(-k-ip)t}$$

where A' and B' are constants. The values of A' and B' will be determined by the initial conditions. The solution is more readily interpreted if we note that the general solution can be written as

$$y = e^{-kt}(A' e^{ipt} + B' e^{-ipt})$$

and by using the identity

$$e^{i\theta} \equiv \cos\theta + i\sin\theta$$

we have

$$y = e^{-kt}[(A' + B')\cos pt + i(A' - B')\sin pt]$$

which we can write finally as

$$y = e^{-kt}(A\sin pt + B\cos pt) \tag{10.6a}$$

Example

$\ddot{y} + 8\dot{y} + 25y = 0$, with $y(0) = y_0$ $\quad \dot{y}(0) = 0$

The auxiliary equation is

$$m^2 + 8m + 25 = 0$$

Solving this we get

$$m = -4 \pm \sqrt{4^2 - 5^2} = -4 \pm 3i$$

Thus, in the above notation, we have $k = 4, p = 3$ and the solution is

$$y = e^{-4t}(A\sin 3t + B\cos 3t)$$

Now

$$\dot{y} = e^{-4t}[(-4A - 3B)\sin 3t + (-4B + 3A)\cos 3t]$$

Applying the initial conditions:

$$y_0 = 1(0 + B) \Rightarrow B = y_0$$
$$0 = 1[(-4A - 3y_0) . 0 + (-4y_0 + 3A) . 1]$$

so that

$$A = 4y_0/3$$

to give the solution

$$y = y_0 e^{-4t} \left\{ \frac{4}{3} \sin 3t + \cos 3t \right\}$$

whose graph resembles Figure 10.5(ii) – a case of light damping.

Case (b): $k > n$

Let $k^2 - n^2 = q^2$ so then $m = -k \pm q$ to give a general solution

$$y = Ce^{(-k+q)t} + De^{(-k-q)t} \tag{10.6b}$$

where C and D are constants.

Example

$\ddot{y} + 4\dot{y} + 3y = 0$ with $y(0) = y_0$, $\dot{y}(0) = u$.

Here

$$q^2 = 4 - 3 = 1$$

The auxiliary equation

$$m^2 + 4m + 3 = 0$$

leads to

$$y = Ce^{-t} + De^{-3t}$$

Now

$$\dot{y} = -Ce^{-t} - 3De^{-3t}$$

and you can check that the initial conditions lead to the solution

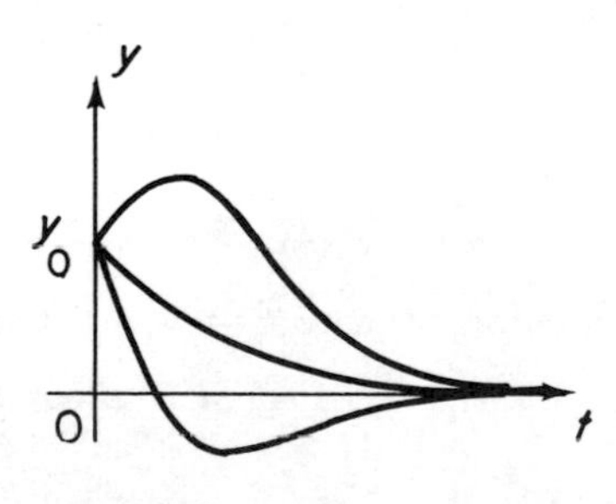

Figure 10.6

$$y = \frac{1}{2}\left\{(3y_0 + u)e^{-t} - (u + y_0)e^{-3t}\right\}$$

The possible motions are shown in Figure 10.6; the particular motion will depend on the size and sign of y_0 and u.

Case (c): $k = n$

$m = -k$, and we obtain $y = Ee^{-kt}$; this contains only one arbitrary constant, E. To attempt to find another linearly independent solution we use the technique of **Variation of Parameters.**

In this method we replace the constant E by a general function $f(t)$ so that

$$y = f(t)e^{-kt}$$

which yields

$$\dot{y} = [-kf(t) + f'(t)]\, e^{-kt}$$

and

$$\ddot{y} = [k^2 f(t) - 2kf'(t) + f''(t)]\, e^{-kt}$$

Substitution in (10.3b) provides

$$e^{-kt}[k^2 f(t) - 2kf'(t) + f''(t) - 2k^2 f(t) + 2kf'(t) + k^2 f(t)] \equiv 0$$

that is,

$$e^{-kt}[f''(t)] \equiv 0$$

which implies that

$$f''(t) \equiv 0$$

Hence

$$f(t) = Et + F$$

where E and F are constants so that

$$y = (Et + F)e^{-kt} \qquad (10.6c)$$

Example

Suppose we have the conditions

$$y(0) = y_0 \;(>0), \;\; \dot{y}(0) = u$$

Then

$$\dot{y} = (E - kEt - kF)e^{-kt}$$

to give

$$y = [(u + ky_0)t + y_0]\, e^{-kt}$$

We examine the effect of different values of u. If $u = 0$ then we obtain a graph similar to the middle one of Figure 10.6; if $u > 0$ there is an initial motion further away from equilibrium (the top graph), whilst if $u < 0$ (and $u + ky_0 < 0$) the graph at the bottom depicts a slight overshooting.

This case is called **critical damping** and the motion is said to be **dead beat.** There is a similarity between critical and heavy damping and the distinction is that the *critical* value of k is the *least* value to prevent one complete oscillation.

We now summarise the results obtained. The auxiliary equation of the differential equation

$$\frac{d^2y}{dt^2} + 2k\frac{dy}{dt} + n^2y = 0$$

is

$$m^2 + 2km + n^2 = 0$$

which can be solved to give

$$m = -k \pm \sqrt{k^2 - n^2}$$

with the three cases

(i) $k < n \Rightarrow m = -k \pm ip \Rightarrow$ solution $y = e^{-kt}[A \sin pt + B \cos pt]$

(ii) $k > n \Rightarrow m = -k \pm q \Rightarrow$ solution $y = Ce^{(-k+q)t} + De^{(-k-q)t}$

(iii) $k = n \Rightarrow m = -k \Rightarrow$ solution $y = (Et + F)e^{-kt}$

Problems

1. Write down and solve the auxiliary equation for each of the following differential equations

   (i) $\ddot{y} + 5\dot{y} - 6y = 0$ (ii) $\ddot{y} - 16\dot{y} + 64y = 0$

   (iii) $\ddot{y} + 6\dot{y} + 10y = 0$ (iv) $4\ddot{y} + 4\dot{y} + y = 0$

   (v) $\dddot{y} - \ddot{y} - \dot{y} - 2y = 0$ (Hint: you should get a cubic equation)

   (vi) $\ddot{y} + \dot{y} + 7y = 0$ (vii) $5\ddot{y} + 4\dot{y} - \frac{13}{4}y = 0$

   Hence write down the general solutions in the form of either (10.6a), (10.6b) or (10.6c). In the case of (v), try to guess what form the solution will take.

2. A uniform flexible cable hangs under its own weight w/unit length. Show that the profile $y(x)$ of the cable satisfies the differential equation

$$\frac{d^2y}{dx^2} = \frac{w}{H}\sqrt{1 + \left(\frac{dy}{dx}\right)^2}$$

   where H is the horizontal tension in the cable.
   Write $p = dy/dx$ and by choosing suitable boundary conditions obtain a formula for the profile.

3. An object moves in a straight line (which we can take as the x-axis) with constant speed V. A missile, initially a distance D away, at right angles

to the x-axis homes in on its target with constant speed nV in a direction always toward the target. Show that the equation of the path of the missile can be written

$$\frac{d^2x}{dy^2} = \frac{1}{ny}\sqrt{1 + \left(\frac{dx}{dy}\right)^2}$$

and show that the target is hit after a time

$$\frac{n}{n^2 - 1}\frac{D}{V}$$

4. Solve the equation

$$\frac{d^3y}{dx^3} + 4\frac{d^2y}{dx^2} + 4\frac{dy}{dx} + y = 0$$

5. In a mechanical vibration problem we have the following parameters: mass of spring, spring stiffness, damping constant. What are the analogues of these quantities for
   (i) a torsional vibration problem
   (ii) an LCR circuit?

6. A mass m, constrained to move along the x-axis, is attracted towards the origin with a force proportional to its distance from the origin. Find the motion
   (i) if it starts from rest at $x = x_0$, and
   (ii) if it starts from the origin with velocity v_0. (C.E.I.)

7. Solve the differential equation

$$\frac{d^4y}{dx^4} = 16y$$

(C.E.I.)

10.4 FORCED VIBRATIONS: COMPLEMENTARY FUNCTION AND PARTICULAR INTEGRAL

We return to equation (10.2):

$$\ddot{y} + 2k\dot{y} + n^2y = f(t)$$

We shall utilise the result that, *if $y_1(t)$ is a particular solution (called a* **particular integral***) of (10.2) and $y_2(t)$ (called the* **complementary function***) is the general solution of the associated homogeneous equation*

$$\ddot{y} + 2k\dot{y} + n^2y = 0$$

then the general solution of (10.2) is

$$y^*(t) \equiv y_1(t) + y_2(t)$$

The proof of this result is in two parts: first we show that $y^*(t)$ is a solution of (10.2) and then we show that any other solution differs only in the arbitrary constants of the complementary function. We know that

$$\ddot{y}_1 + 2k\dot{y}_1 + n^2 y_1 = f(t)$$

and

$$\ddot{y}_2 + 2k\dot{y}_2 + n^2 y_2 = 0$$

Since

$$\dot{y}^* = \dot{y}_1 + \dot{y}_2 \quad \text{and} \quad \ddot{y}^* = \ddot{y}_1 + \ddot{y}_2$$

it follows that

$$\ddot{y}^* + 2k\dot{y}^* + n^2 y^* \equiv (\ddot{y}_1 + 2k\dot{y}_1 + n^2 y_1) + (\ddot{y}_2 + 2k\dot{y}_2 + n^2 y_2) = f(t) + 0$$
$$= f(t)$$

Further, let $y_s(t)$ be a second solution of (10.2) and consider

$$y_d(t) \equiv y^*(t) - y_s(t); \; \ddot{y}_d + 2k\dot{y}_d + n^2 y_d \equiv f(t) - f(t) = 0$$

and so y_d is a solution of the associated homogeneous equation and simply means a relabelling of the coefficients of $y_2(t)$. If

$$y_2(t) = Ae^{m_1 t} + Be^{m_2 t}$$

then any solution of (10.2) can be written as

$$y(t) = y_1(t) + Ae^{m_1 t} + Be^{m_2 t}$$

The particular solution we want (i.e. the choice of the values of A and B) is determined by the initial (or boundary) conditions.

The rules may be summarised:

(i) find the complementary function (C.F.)
(ii) spot *any* particular integral (P.I.)
(iii) add (i) and (ii) to form the general solution (G.S.)
(iv) fit the initial (or boundary) conditions to obtain the particular solution.

It is important to note that the initial (or boundary) conditions are applied only at the last stage.

Example

$\ddot{y} + 8\dot{y} + 25y = 4; \; y(0) = 2, \; \dot{y}(0) = 1$

(i) C.F. This has been found in an earlier example to be

$$y_{CF} = e^{-4t}(A \sin 3t + B \cos 3t)$$

(ii) P.I. We *spot* that y = constant is a possible particular integral. Substitution yields

$$y_{PI} = 4/25$$

(iii) G.S. Addition gives

$$y = e^{-4t}(A \sin 3t + B \cos 3t) + 4/25$$

(iv) $y(0) = 2 \Rightarrow 2 = 1(0 + B) + 4/25 \Rightarrow B = 46/25$

Since

$$\dot{y} = e^{-4t}[(-4A - 3B)\sin 3t + (-4B + 3A)\cos 3t],$$

$$\dot{y}(0) = 1 \Rightarrow 1 = 1[0 + (-4B + 3A)] \Rightarrow A = \frac{209}{75}$$

The solution we require is therefore

$$y = e^{-4t}\left(\frac{209}{75} \sin 3t + \frac{46}{25} \cos 3t\right) + \frac{4}{25}$$

– do not be surprised by such unwieldy coefficients.

As long as we remember the scheme (i) to (iv) we can turn our attention now to (ii), that is, finding the particular integral which clearly depends (unlike the complementary function) on the form of $f(t)$. We shall develop techniques for finding the particular integral in the next section and in Sections 10.8 and 10.11.

10.5 PARTICULAR INTEGRAL: TRIAL SOLUTIONS

This technique is applicable when each term of $f(t)$ has a finite number of linearly independent derivatives; the functions we shall consider are: a constant, a polynomial, sine and cosine, exponential. We cannot cope, for example, with $1/x^2$ since this has an infinite number of linearly independent derivatives (in the sense that none of them can be expressed as a linear combination of the others).

We have a general result that *if $f(t)$ can be expressed as $f_1(t) + f_2(t)$ then a particular integral (P.I.) of $f(t)$ is the sum of a P.I. of $f_1(t)$ and a P.I. of $f_2(t)$.*

We consider again the differential equation

$$\ddot{y} + 2k\dot{y} + n^2 y = f(t)$$

(i) $f(t) = c$, constant. We try $y = A$, and since $\dot{y} = 0 = \ddot{y}$ it should be clear that $y = c/n^2$ is the particular integral we seek.

(ii) $f(t) = a_n t^n + a_{n-1} t^{n-1} + \ldots\ldots a_1 t + a_0$. We try a polynomial for y of the same form, but all powers of t lower than t^n must be present.

Example

$\ddot{y} + 2\dot{y} + y = 3t^2 + 4$

We try $y = At^2 + Bt + C$ (Note the presence of the term in t.)

Then

$$\dot{y} = 2At + B \quad \text{and} \quad \ddot{y} = 2A$$

to give, on substitution in the differential equation,

$$2A + 4At + 2B + At^2 + Bt + C = 3t^2 + 4$$

Equating the coefficients of like powers of t we have

$$t^2: \; A = 3$$
$$t: \; 4A + B = 0 \Rightarrow B = -12$$
$$\text{constant}: \; 2A + 2B + C = 4 \Rightarrow C = 22$$

Hence the P.I. is

$$y = 3t^2 - 12t + 22$$

(iii) $f(t) = Ae^{\lambda t}$. We try $y = Ce^{\lambda t}$, C constant.

Example 1

$\ddot{y} + 4\dot{y} + 4y = 2e^{3t}$

We try $y = Ce^{3t}$ so that $\dot{y} = 3Ce^{3t}$ and $\ddot{y} = 9Ce^{3t}$

Substitution yields

$$(9C + 12C + 4C)e^{3t} = 2e^{3t}$$

Hence $C = 2/25$ to give a P.I. of

$$y = \frac{2}{25}e^{3t}$$

Example 2

$\ddot{y} + 4\dot{y} + 3y = e^{-t}$

We try $y = Ce^{-t}$ so that $\dot{y} = -Ce^{-t}$ and $\ddot{y} = Ce^{-t}$, therefore

$$(C - 4C + 3C)e^{-t} = e^{-t}$$

The bracket on the left is zero so the method has failed.

The trouble here is that e^{-t} is part of the complementary function. The method of variation of parameters will show that $y = (Ct)e^{-t}$ is the form to try. This gives

$$\dot{y} = (C - Ct)e^{-t} \quad \text{and} \quad \ddot{y} = (-2C + Ct)e^{-t}$$

which on substitution leads to

$$(-2C + Ct + 4C - 4Ct + 3Ct)e^{-t} = e^{-t}$$

that is, $C = ½$ and $y = ½te^{-t}$ is the P.I. required.

Example 3

$\ddot{y} + 4\dot{y} + 4y = e^{-2t}$

It would seem that since e^{-2t} is part of the C.F., that we should try $y = Cte^{-2t}$, but fate is not often kind to the foolhardy. You can check that such a substitution yields $0 = 1$. The trouble here is that -2 is a double

root of the auxiliary equation.
Variation of parameters or intuition suggests the trial of $y = Ct^2 e^{-2t}$ and you should check the wisdom of this choice.

(iv) $f(t) = A \sin pt$ or $B \cos pt$

This is an important case practically and we consider first the undamped forced oscillations, i.e. we seek to solve $\ddot{y} + n^2 y = F \cos pt$.
We know that the only linearly independent derivatives of $\sin pt$ are of the form $\alpha \cos pt$ and $\beta \sin pt$, where α and β are constants; consequently we try for a P.I. in the form

$$y = \alpha \cos pt + \beta \sin pt$$

Then

$$\dot{y} = -p\alpha \sin pt + p\beta \cos pt$$

and

$$\ddot{y} = -p^2 \alpha \cos pt - p^2 \beta \cos pt$$

Substitution into the equation of motion produces

$$(n^2 - p^2)(\alpha \cos pt + \beta \sin pt) = F \cos pt$$

hence $\beta = 0$ and $\alpha = F/(n^2 - p^2)$ so that the particular integral is

$$y = [F/(n^2 - p^2)] \cos pt$$

Now the complementary function is

$$y = B \cos nt + C \sin nt$$

and so the general solution becomes

$$y = \underbrace{B \cos nt + C \sin nt}_{\text{free oscillations}} + \underbrace{[F/(n^2 - p^2)] \cos pt}_{\text{forced oscillations}}$$

Let us suppose that the body undergoing these forced oscillations is released from the origin at rest, i.e. at $t = 0$, $y = 0 = \dot{y}$; then you can check that the particular solution is

$$y = \frac{F}{n^2 - p^2}\left[\cos pt - \cos nt\right]$$

We see that F, the amplitude of the forcing term, has been *magnified* by a factor $1/(n^2 - p^2)$ in the P.I. which represents the forced oscillation. It is clear that our solution is invalid when $p = n$, i.e. when the frequency of the forcing motion is equal to the frequency of natural oscillations of the body. (The graph of magnification factor against p is shown in Figure 10.7.) Before dealing with that case, we examine the behaviour for other values of p. For p near zero, the forcing frequency is very slow and the body movement will approximate to free oscillations subject to a *static* force F;

on the other hand, for very high frequencies, the body is simply unable to follow the forcing motion and the amplitude is very small. For values of p approaching n, the amplitude of the vibration becomes large since the forcing motion is almost in sympathy with the natural oscillation and can push the body almost at the right time in the right direction.

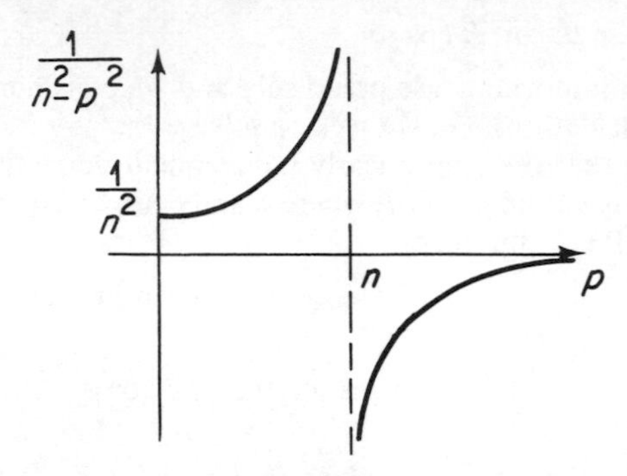

Figure 10.7

If the forcing term is $F \cos pt$ or $F \sin pt$, and if p is almost equal to n, then we have the phenomenon of **beats**. Let us take the case with a forcing term of $F \sin pt$. The solution is now

$$y = \frac{F}{n^2 - p^2}\left[\sin pt - \frac{p}{n}\sin nt\right]$$

Suppose that p is almost equal to n, then we have approximately

$$y \triangleq \frac{F}{n^2 - p^2}\left[\sin pt - \sin nt\right] = \frac{2F}{(n-p)(n+p)} \cos\left(\frac{p+n}{2}\right)t \sin\left(\frac{p-n}{2}\right)t$$

Denoting the small quantity $p - n$ by 2ϵ, so that

$$\frac{p+n}{2} = n + \epsilon \quad \text{and} \quad \frac{p-n}{2} = \epsilon$$

we can write

$$y \triangleq \frac{-F}{2n\epsilon} \cos nt \sin \epsilon t$$

Since ϵ is a small quantity, the period $2\pi/\epsilon$ of the term $\sin \epsilon t$ is large and we can regard y as a periodic function $\cos nt$ with slowly varying amplitude as shown in Figure 10.8 where the dashed curves are the curves to which the actual wave periodically rises. A similar analysis can, of course, be carried out for a forcing term of $F \cos pt$ and you should examine this for yourself.

Beats can be heard when an electric generator starts up or in a power house when the hum of the generator and that of the others on the line are almost of the same frequency.

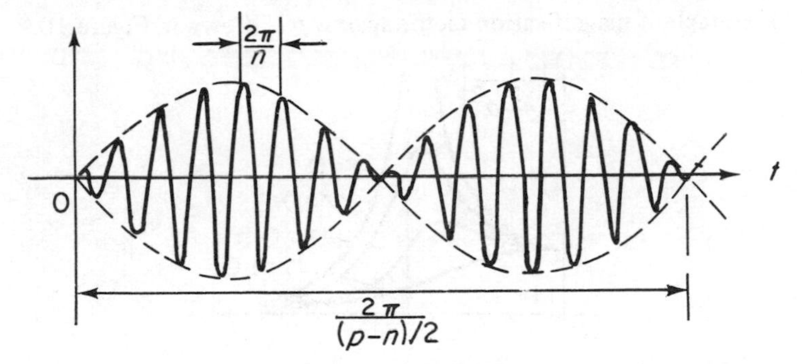

Figure 10.8

Now we return to the mathematical expression for y in the case of a forcing function $F \cos pt$. We had

$$y = \frac{F}{n^2 - p^2}\left[\cos pt - \cos nt\right]$$

which can be rearranged to

$$y = \frac{F}{n^2 - p^2}\, 2 \sin\left(\frac{p+n}{2}\right)t \,.\, \sin\left(\frac{n-p}{2}\right)t$$

Let $p = n + \Delta n$ so that

$$y \triangleq \frac{F}{n\Delta n} \sin(n + \tfrac{1}{2}\Delta n)t \,.\, \sin\frac{\Delta n}{2}t \tag{10.7}$$

Let t be fixed in (10.7) and $\Delta n \to 0$;

Now,

$$\left(\sin\frac{\Delta n}{2}t\right)\Big/\left(\frac{\Delta n}{2}\right) \to t$$

and

$$\sin(n + \tfrac{1}{2}\Delta n)t \to \sin nt$$

as $\Delta n \to 0$, therefore

$$y \to \frac{F}{2n}\, t \sin nt$$

(Why should we have supposed this from a trial solution technique for the case $p = n$?)

We see that the amplitude will grow indefinitely with time. This is the case of **resonance** and a well-known 'application' is the requirement for a group of marching soldiers to break step as they cross a bridge in case their 'p' should equal the 'n' of the bridge. Other examples of resonance will be found in the problems.

If we now introduce a little damping the danger is avoided; of course, the larger the damping, the less the maximum amplitude at $p = n$.

The graph of magnification factor against p is shown in Figure 10.9.

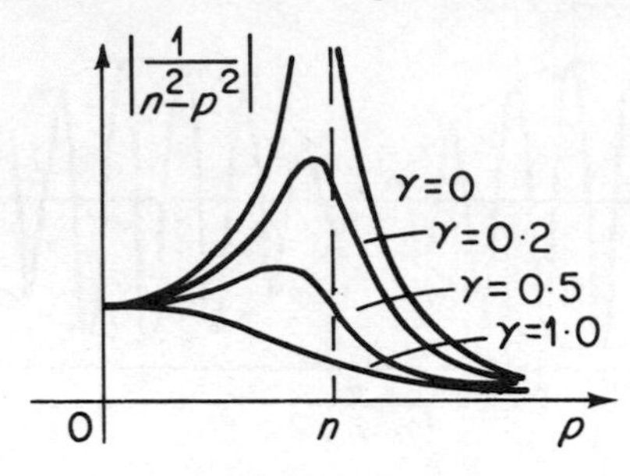

Figure 10.9

The ratio $\gamma = k/k_c$ measures the fraction of critical damping. We have shown the magnification factor to be positive in this case, so that a comparison is more readily seen.

Problems

1. Find the Particular Integral of the following differential equations

(i) $\dfrac{d^2y}{dx^2} + 4\dfrac{dy}{dx} + 5y = 2e^{-3x}$ (Try $y = ke^{-3x}$)

(ii) $4\dfrac{d^2y}{dx^2} - 4\dfrac{dy}{dx} + y = x^2 - x + 2$ (Try $y = \alpha x^2 + \beta x + \gamma$)

(iii) $\dfrac{d^2y}{dx^2} - 3\dfrac{dy}{dx} + 2y = \sin 2x$ (Try $y = \alpha \sin 2x + \beta \cos 2x$)

(iv) $\dfrac{d^2y}{dx^2} + \dfrac{dy}{dx} = 2e^{-2x} + 3x$ (Try $y = ke^{-2x} + \alpha x^2 + \beta x + \gamma$)

(v) $\dfrac{d^2y}{dx^2} + 4\dfrac{dy}{dx} + 4y = 2e^{-2x}$ (This time $y = ke^{-2x}$ will not work)

(vi) $\dfrac{d^2y}{dx^2} + 16y = 2\sin 4x$ (We shall have to try something different from $y = \alpha \sin 4x + \beta \cos 4x$)

2 Solve the differential equations

(i) $\dfrac{d^2y}{dx^2} + 4\dfrac{dy}{dx} + 5y = 8\cos x$ given that $y = 0$ when $x = 0$ and $dy/dx = 3$ when $x = 0$.

(ii) $\dfrac{d^2y}{dx^2} - \dfrac{dy}{dx} - 2y = 2e^{2x}$ given $y = 0$ and $dy/dx = 0$ when $x = 0$.

(iii) $\dfrac{d^2x}{dt^2} + 4x = 2e^t$ given $x = 0$ when $t = 0$ and $dx/dt = 3$ when $t = 0$.

(iv) $9\dfrac{d^2x}{dt^2} + x = \sin\dfrac{t}{3}$

(v) $\dfrac{d^2x}{dt^2} + 4\dfrac{dx}{dt} + x = t^3 - 2$

(vi) $\dfrac{d^2y}{dx^2} + 3\dfrac{dy}{dx} - 4y = e^{-x}\cos x$ with $y = 1$ when $x = 0$ and $y \to 0$ when $x \to \infty$. (C.E.I.)

3. Consider the equation

$$\frac{d^2y}{dx^2} + 4y = \sin 2x$$

Show that the complementary function is

$$y = C\cos 2x + D\sin 2x$$

Assume now that C and D are not constants and show that

$$\frac{dy}{dx} = -2C\sin 2x + 2D\cos 2x + C'\cos 2x + D'\sin 2x$$

where

$$C' \equiv \frac{dC}{dx}, \quad D' \equiv \frac{dD}{dx}$$

Since we have two arbitrary functions C and D to satisfy one equation, we can make the additional condition that

$$C'\cos 2x + D'\sin 2x = 0$$

Now find d^2y/dx^2 and substitute in the differential equation and you should obtain

$$-2C'\sin 2x + 2D'\cos 2x = \sin 2x$$

Solve the last two equations simultaneously and show that

$$C = -\frac{1}{4}x + \frac{1}{16}\sin 4x + A$$

$$D = -\frac{1}{16}\cos 4x + B$$

Hence show that the general solution is

$$y = A\cos 2x + F\sin 2x - \frac{1}{4}x\cos 2x$$

where

$$F = B + \frac{1}{16}$$

(Notice that we could have tried $y = kx \cos 2x$ for the P.I.)
This is another method of Variation of Parameters.
Repeat this approach for the equations

(a) $$\frac{d^2y}{dx^2} + 2\frac{dy}{dx} + y = 2e^{-x}$$

(b) $$\frac{d^2y}{dx^2} - 4\frac{dy}{dx} = \frac{1}{2}e^{4x}$$

4. A uniform beam of infinite length rests on an elastic foundation so that a principal axis of the beam is vertical. The reaction from the foundation at any point on the beam can be assumed proportional to the deformation there. If there is a uniformly distributed load of w/unit length acting over a length $2a$, it may be shown that the equation for the deflected profile is

$$EI\frac{d^4y}{dx^4} + ky = f(x)$$

where EI is the (constant) flexural rigidity of the beam,
k a constant of proportionality, and
$f(x)$ represents the load on the beam.

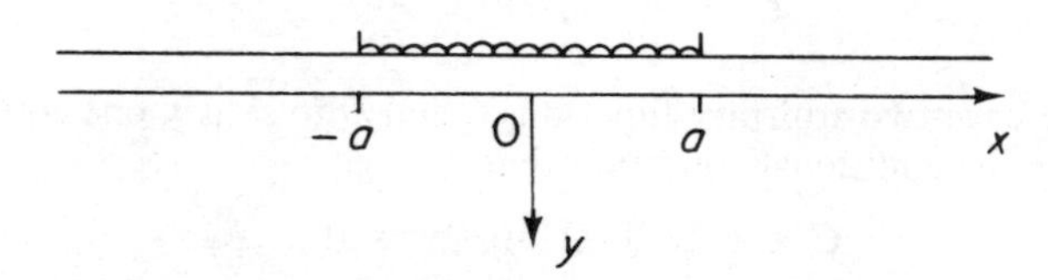

Using symmetry and splitting the profile into two parts, find the eight boundary conditions for the two resulting equations.
Hence determine the deflected profile.
(Hint: it helps to put $\beta^4 = k/4EI$)

5. Solve the equation

$$\frac{d^2y}{dx^2} + 4\frac{dy}{dx} - 5y = 2e^{-2x}$$

given that, at $x = 0$, $y = 0$ and $dy/dx = 1$. (C.E.I.)

6. Solve completely the differential equation

$$\frac{d^2s}{dt^2} + 2k\frac{ds}{dt} + 4s = \cos t$$

in each of the cases (a) $k = 2\frac{1}{2}$, (b) $k = 2$, (c) $k = 1$. (L.U.)

7. Obtain the general solutions of the differential equations

(i) $$\frac{d^2y}{dx^2} + 5\frac{dy}{dx} + 4y = e^{-2x} + \sin x$$

(ii) $$\frac{d^2y}{dx^2} - 6\frac{dy}{dx} + 10y = x + 1$$ (L.U.)

8. The current, I, in a circuit is given by the differential equation

$$\frac{d^2I}{dt^2} + 4\frac{dI}{dt} + 2504I = 250.4$$

If $I = 0$ and $dI/dt = 0$ when $t = 0$, find I in terms of t. Find also the maximum current in the circuit. (L.U.)

9. A body is attached by a spring to a point which is oscillating with sinusoidal motion. The displacement s of the body from a fixed reference point at any time t is given by the differential equation

$$\frac{d^2s}{dt^2} + 4s = \lambda \sin t$$

where λ is a constant. Given that $s = 0$ when $t = 0$, and $ds/dt = 2\lambda/3$ when $t = 0$, solve the equation for s in terms of t.
Show that the ratio of the maximum and minimum displacements is numerically equal to $5\sqrt{5} : 4$. (L.U.)

10. An electric cable has resistance R per unit length and the resistance per unit length of the insulation is n^2R, where n and R are constants. The voltage v and the current i at a distance x from one end satisfy the equations

$$\frac{dv}{dx} + Ri = 0 \quad \text{and} \quad \frac{di}{dx} + \frac{v}{n^2R} = 0$$

If current is supplied at one end A of a cable, of length l, at a voltage V and the cable is insulated at the other end, show that the current entering at A is

$$\frac{V}{nR}\tanh\frac{l}{n}$$ (L.U.)

10.6 NUMERICAL SOLUTIONS OF INITIAL-VALUE PROBLEMS

The techniques we have studied so far cope with linear equations with constant coefficients. The numerical method we shall describe can cope with most second-order differential equations; there are many numerical methods but we restrict ourselves to one.

First, we consider the equation

$$\ddot{y} + 2k\dot{y} + n^2y = f(t) ;$$

we write $\dot{y} = v$ and then cast the equation in the form

$$\dot{v} = -2kv - n^2y + f(t)$$

We have, therefore, two first-order differential equations to be solved simultaneously. In essence, we start at $t = 0$ knowing y and v; the equation $\dot{y} = v$ allows us to predict y at Δt by

$$y_{\text{new}} \simeq y_{\text{old}} + \Delta t \left(\frac{\mathrm{d}y}{\mathrm{d}t}\right)_{t=0}$$

(Euler's method) or by a more accurate formula. Likewise, the equation $\dot{v} = -2kv - n^2y + f(t)$ allows us to predict v at Δt by

$$v_{\text{new}} \simeq v_{\text{old}} + \Delta t \left(\frac{\mathrm{d}v}{\mathrm{d}t}\right)_{t=0}$$

or some more sophisticated formula. Thus at $t = \Delta t$, we know y and v and we are ready to step forward again. The choice of formula is subject to the same criteria as in Chapter 9 and the fourth-order Runge-Kutta method is popular.

As you can verify for yourself, this formula, with a suitably chosen step size, gives good agreement with the analytical solution.

First, we quote the formulae for a general second order equation and show the flow chart for the process in Figure 10.10.

Consider $\ddot{y} = f(t, y, v)$ with $\dot{y} = v$ so that $\dot{v} = f(t, y, v)$

Then

$$v_{n+1} = v_n + \frac{1}{6}(k_1 + 2k_2 + 2k_3 + k_4)$$

and

$$y_{n+1} = y_n + \frac{1}{6}(l_1 + 2l_2 + 2l_3 + l_4)$$

where

$$\left.\begin{aligned}
k_1 &= \Delta t \,.\, f(t_n, y_n, v_n) & l_1 &= \Delta t \,.\, v_n \\
k_2 &= \Delta t \,.\, f\left(t_n + \frac{\Delta t}{2}, y_n + \frac{l_1}{2}, v_n + \frac{k_1}{2}\right) & l_2 &= \Delta t \left(v_n + \frac{k_1}{2}\right) \\
k_3 &= \Delta t \,.\, f\left(t_n + \frac{\Delta t}{2}, y_n + \frac{l_2}{2}, v_n + \frac{k_2}{2}\right) & l_3 &= \Delta t \left(v_n + \frac{k_2}{2}\right) \\
k_4 &= \Delta t \,.\, f(t_n + \Delta t, y_n + l_3, v_n + k_3), & l_4 &= \Delta t (v_n + k_3)
\end{aligned}\right\} \quad (10.8)$$

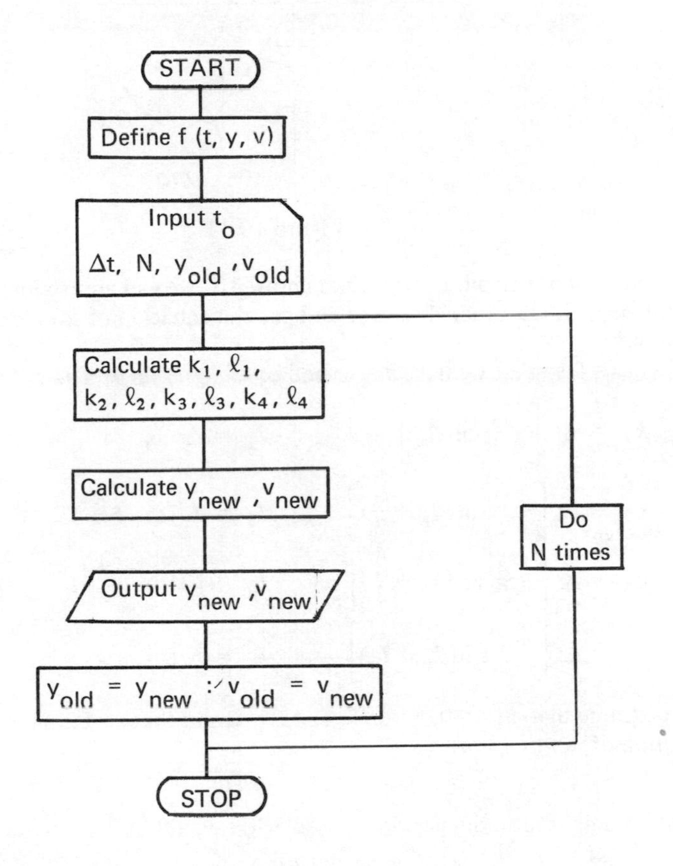

Figure 10.10

Example

We consider the application of the Runge-Kutta formula to the oscillations of a simple pendulum when the oscillations are sufficiently large in amplitude to prevent the usual assumption of a linear governing equation.

We now consider a simple pendulum consisting of a heavy bob at the end of a slender rod of negligible mass as shown in Figure 10.11.

The equation of motion neglecting air resistance etc. is

$$ml\ddot{\theta} = -mg \sin \theta$$

or

$$\ddot{\theta} = -\frac{g}{l} \sin \theta$$

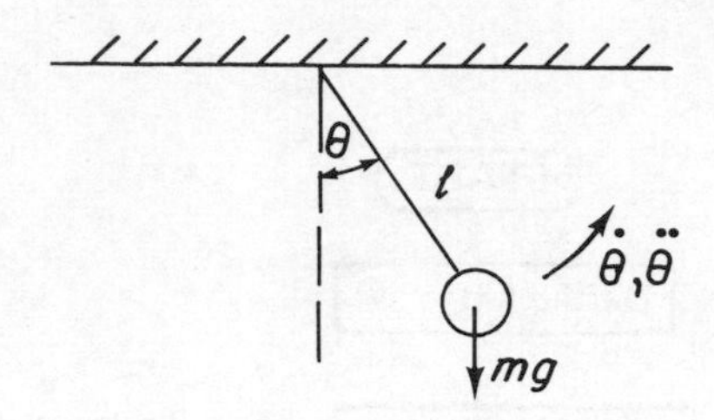

Figure 10.11

[Of course, if θ is small, i.e. less than about 10° we can approximate $\sin\theta$ by θ (radians) with less than 1% error and get the equation of simple harmonic motion.]

Let $\omega = \dot{\theta}$ so that we have $\dot{\theta} = \omega$ and $\dot{\omega} = -(g/l)\sin\theta$. Then

$$k_1 = \Delta t\left[-\frac{g}{l}\sin\theta_n\right] \qquad l_1 = \Delta t\,\omega_n$$

$$k_2 = \Delta t\left[-\frac{g}{l}\sin(\theta_n + \frac{l_1}{2})\right] \qquad l_2 = \Delta t(\omega_n + \frac{k_1}{2})$$

$$k_3 = \Delta t\left[-\frac{g}{l}\sin(\theta_n + \frac{l_2}{2})\right] \qquad l_3 = \Delta t(\omega_n + \frac{k_2}{2})$$

$$k_4 = \Delta t\left[-\frac{g}{l}\sin(\theta_n + l_3)\right] \qquad l_4 = \Delta t(\omega_n + k_3)$$

If we assume that, at $t = 0$, $\theta = \pi/3$, $\omega = \dot{\theta} = 0$ and that $l = 0.3$ metre, where the equation of motion is

$$\ddot{\theta} = -(g/l)\theta \quad ;$$

for small amplitudes, the period of oscillation would be

$$T = 2\pi\sqrt{(l/g)} \simeq 1 \text{ sec}$$

so we take a step size of 0.05 sec as a first attempt.
Then

$$k_1 = -0.05(32.7\,.\,\sin\frac{\pi}{3}) = -1.416 \qquad l_1 = 0.05(0) = 0$$

$$k_2 = -0.05\left[32.7\sin(\frac{\pi}{3} + 0)\right] = -1.416 \qquad l_2 = 0.05(0 - 0.708) = -0.035$$

$$k_3 = -0.05\left[32.7\sin(\frac{\pi}{3} - 0.0175)\right] = -1.401 \qquad l_3 = 0.05(0 - 0.708) = -0.035$$

$$k_4 = -0.05\left[32.7\sin(\frac{\pi}{3} - 0.035)\right] = -1.386 \qquad l_4 = 0.05(0 - 1.401) = -0.070$$

Hence

$$\theta_{\text{new}} = 1.012; \quad \omega_{\text{new}} = -1.406$$

The method was programmed for a digital computer; the departure from simple harmonic motion is depicted below.

t	Runge-Kutta soln for θ (3 d.p.)	SHM approximate soln for θ (3 d.p.)	Departure
0.00	1.047	1.047	0.000
0.05	1.012	1.005	0.007
0.10	0.908	0.881	0.027
0.15	0.739	0.685	0.054
0.20	0.516	0.434	0.082
0.25	0.253	0.147	0.106
0.30	−0.297	−0.151	0.146
0.35	−0.310	−0.437	0.127
0.40	−0.566	−0.688	0.122
0.45	−0.779	−0.883	0.104
0.50	−0.935	−1.006	0.061
0.55	−1.025	−1.047	0.022
0.60	−1.046	−1.004	0.042
0.65	−0.996	−0.879	0.117
0.70	−0.878	−0.682	0.186
0.75	−0.697	−0.430	0.267
0.80	−0.464	−0.144	0.320
0.85	−0.196	0.155	0.351
0.90	0.089	0.441	0.352
0.95	0.366	0.691	0.325
1.00	0.615	0.885	0.270
1.05	0.816	1.007	0.191
1.10	0.9	1.047	0.147

The analytical approach would be to write $\ddot{\theta}$ as $\omega(\mathrm{d}\omega/\mathrm{d}\theta)$ and obtain

$$\omega \frac{\mathrm{d}\omega}{\mathrm{d}\theta} = -\frac{g}{l}\sin\theta$$

that is,

$$\int \omega \mathrm{d}\omega = -\int \frac{g}{l}\sin\theta \, \mathrm{d}\theta$$

which gives

$$\tfrac{1}{2}\omega^2 = \frac{g}{l}\cos\theta + \text{constant}$$

When $t = 0$, $\omega = 0$, $\theta = \theta_0$, therefore

$$0 = +\frac{g}{l}\cos\theta_0 + \text{constant}$$

and so

$$\tfrac{1}{2}\omega^2 = \frac{g}{l}\cos\theta - \frac{g}{l}\cos\theta_0$$

that is,

$$\frac{\mathrm{d}\theta}{\mathrm{d}t} \equiv \omega = \sqrt{\frac{2g}{l}(\cos\theta - \cos\theta_0)}$$

and hence

$$t = \int_{\theta_0}^{\theta} \frac{\mathrm{d}\theta}{\sqrt{\frac{2g}{l}(\cos\theta - \cos\theta_0)}}$$

The integration then poses problems which we cannot tackle here; notice, however, that the integrand is infinite at $\theta = \theta_0$.

Problems (Unless you use a computer, advance *two* steps only)

1. If, in the above example, we assume air damping proportional to the square of the angular velocity of the pendulum, then the governing d.e. is

$$\frac{\mathrm{d}^2\theta}{\mathrm{d}t^2} + 2k\dot{\theta}^2 + \frac{g}{l}\sin\theta = 0$$

Evaluate numerically for $t = 0(0.05)2.5$ minutes, θ and $\dot{\theta}$.

2. In the circuit shown a coil is wound around an iron core; the core is in series with a resistance and an e.m.f. source. The self-inductance of the coil is L and the non-linearity of the magnetisation of the iron leads to the d.e.

$$\frac{\mathrm{d}\phi}{\mathrm{d}t} = E - 5\phi - 0.03\phi^3$$

where ϕ is the flux in the core and t is in milliseconds.
Take $E = 30$ and solve.

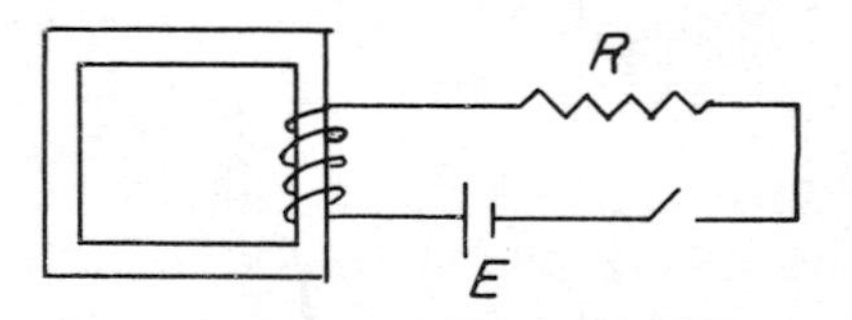

3. Van der Pol's equation, which arises in electronics, is

$$\ddot{x} + (1 - x^2)\dot{x} + x = 0$$

With initial conditions $x = 0.5$, $\dot{x} = 0$ at $t = 0$, solve for x, $\dot{x}$ and $\ddot{x}$ at $t = 0(0.1)4$.

4. A simple free vibrating system has a mass subjected to *Coulomb friction* so that the equation of motion is

$$m\ddot{x} + n^2 x = \begin{cases} -A, & \dot{x} > 0 \\ +A, & \dot{x} < 0 \end{cases} \quad \text{where } A \text{ is constant}$$

At $t = 0$, $\dot{x} = 0$, $x = 3$ m. Take $m = 1$ and $n = 0.8$ and $A = 2$ and solve for $t = 0(0.1)5$ sec.

5.

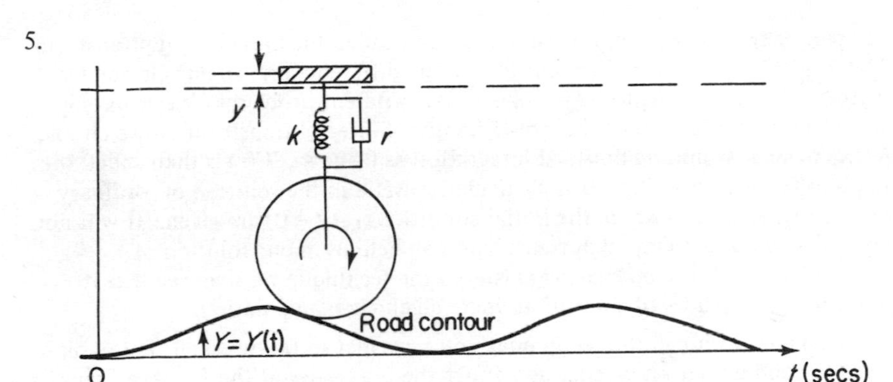

Figure 10.12

Figure 10.12 shows a wheel of a vehicle suspension which is connected to the body by a spring of stiffness coefficient k with a damping system (coefficient r) and it can be shown that the displacement of the body, when the wheel travels along a road of contour $Y = Y(t)$ is governed by

$$M\frac{d^2y}{dt^2} = -k(y - Y) - r\left(\frac{dy}{dt} - \frac{dY}{dt}\right)$$

where $y = 0$ and $dy/dt = 0$ when $t = 0$.

Take suitable values for M, k and r with $Y = \frac{1}{2}(1 - \cos 3t)$ and solve for $t = 0\,(0.1)\,10$ seconds.

10.7 COMPARISON BETWEEN ANALYTICAL AND NUMERICAL APPROACHES

Similar remarks to those made in Chapter 9 apply here. Wherever possible, the analytical approach should be used, but for some problems, linearisation of the model equation may give a false picture of the behaviour of the independent variable; alternatively, the solution of the full equation may be difficult to obtain. In such cases, the Runge-Kutta method is useful. To check the validity of its results we could halve the step size and repeat until sufficiently close agreement is reached. As before, the analogue solution of linear problems is a helpful guide to the behaviour to be analysed.

10.8 LAPLACE TRANSFORM METHOD

Integral Transforms have been used increasingly in recent times in the solution of differential equations, both ordinary and partial. The integral transform $F(s)$ of a function $f(t)$ of t in the range $[a, b]$ is defined as

$$F(s) = \int_a^b f(t).K(s, t)\,dt$$

where $K(s, t)$ is some given function of t and s called the **kernel** of the transform. This kernel is given many different forms, the different forms being chosen by mathematicians and engineers to solve many different problems. We shall only consider one special case of the whole family of integral transforms – we choose $K(s,t)$ to be e^{-st} and the limits of integration as 0 and ∞. $F(s)$ is then called the **Laplace Transform** of $f(t)$. It is particularly useful in the solution of ordinary differential equations where the initial conditions (at $t = 0$) are given. It will not be at all obvious yet why such transforms can help us in our solution of differential equations, but having mastered the technique we shall see that it reduces the procedure of solution to mere algebraic manipulation.

In common with all the techniques you have met so far, we shall first of all have to build up our knowledge and study the properties of the Laplace Transform before we can begin to apply it in the solution of differential equations.

The Laplace Transform

As indicated above we define the Laplace Transform $F(s)$ of a function $f(t)$ by the relation

$$F(s) = \int_0^\infty e^{-st}f(t)\,dt \tag{10.9}$$

it being assumed that the integral exists. (This places restrictions on s which we do not deal with here.)

At this stage we call your attention to the use of logarithms to aid multiplication. Suppose we want to multiply two numbers a and b. We can do this directly or via logarithms as shown schematically below.

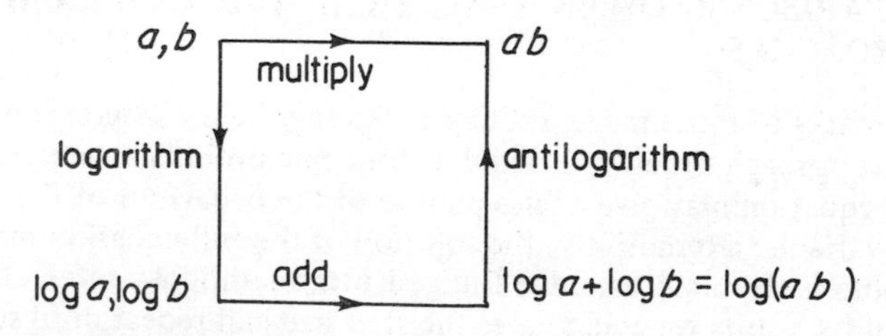

The method allows us to use the simple process of addition and is acceptable because the logarithmic function is one-one; taking antilogs (inverse of logs) is in effect applying a function because to each value $\log(x)$ only one value x is possible.

The idea behind Laplace Transforms is that we replace the process of integration by one of algebraic manipulation of linear equations. Just as the logarithmic function has an inverse which is a function, so Laplace Transforms have a unique inverse. Further, there are tables of Transforms just as there are tables of logarithms. But note that 'logarithm' is a *function* operating on *numbers* whereas 'Laplace Transform' is an *operator* applied to *functions.*

We shall now build up a table of transforms of given functions of t.

Example 1

$f(t) = 1$

$$F(s) = \int_0^\infty e^{-st} . 1 . dt = \left[-\frac{e^{-st}}{s}\right]_0^\infty = \frac{1}{s}$$

it being assumed that $s > 0$ in order that the integral exists.

Example 2

$f(t) = t$

$$F(s) = \int_0^\infty e^{-st} t \, dt = \left[-\frac{te^{-st}}{s}\right]_0^\infty + \frac{1}{s}\int_0^\infty e^{-st} dt \quad \text{(by parts)}$$

$$= 0 + \frac{1}{s} . \frac{1}{s} \quad \text{(using Example 1)}$$

$$= \frac{1}{s^2} \quad \text{(again } s > 0\text{)}$$

Example 3

$f(t) = t^2$

$$F(s) = \int_0^\infty e^{-st} t^2 \, dt$$

We could complete this integral by integration by parts, but consider the following technique. We know

$$\frac{1}{s^2} = \int_0^\infty e^{-st} t \, dt$$

If we differentiate both sides partially with respect to s we have

$$-\frac{2}{s^3} = \frac{\partial}{\partial s}\int_0^\infty e^{-st} t \, dt$$

We shall assume that

$$\frac{\partial}{\partial s}\int_0^\infty e^{-st}t\,dt = \int_0^\infty \frac{\partial}{\partial s}(e^{-st}t)\,dt$$

This gives

$$\int_0^\infty -te^{-st}t\,dt = -\int_0^\infty e^{-st}t^2\,dt$$

Hence

$$-\frac{2}{s^3} = -\int_0^\infty e^{-st}t^2\,dt$$

giving

$$F(s) = \int_0^\infty e^{-st}t^2\,dt = \frac{2}{s^3} \qquad (s > 0)$$

Example 4

$f(t) = t^n$

We differentiate the result of Example 1 n times with respect to the parameter s obtaining

$$\frac{\partial^n}{\partial s^n}\int_0^\infty e^{-st}\,dt = \frac{\partial^n}{\partial s^n}\left(\frac{1}{s}\right)$$

so that

$$\int_0^\infty (-t)^n e^{-st}\,dt = (-1)^n \frac{n!}{s^{n+1}}$$

Cancelling the factor $(-1)^n$, we obtain

$$F(s) = \int_0^\infty e^{-st}t^n\,dt = \frac{n!}{s^{n+1}} \quad (n \geqslant 0,\ s > 0)$$

Example 5

$f(t) = e^{at}$

$$F(t) = \int_0^\infty e^{-st}e^{at}\,dt = \int_0^\infty e^{-(s-a)t}\,dt = \left[-\frac{e^{-(s-a)t}}{(s-a)}\right]_0^\infty$$

$$= \frac{1}{s-a} \quad (\text{provided } s > a)$$

The linear property

If c_1 and c_2 are any constants and $f_1(t)$ and $f_2(t)$ are functions with Laplace transforms $F_1(s)$ and $F_2(s)$ respectively, then the Laplace transform of the function $c_1 f_1(t) + c_2 f_2(t)$ is

$$c_1 F_1(s) + c_2 F_2(s)$$

We use this result on the next function in our table.

Example 6

$f(t) = \cos kt$

Now

$$\cos kt = \frac{1}{2}(e^{ikt} + e^{-ikt})$$

Hence

$$\begin{aligned} F(s) &= \text{Laplace transform of } \tfrac{1}{2}(e^{ikt} + e^{-ikt}) \\ &= \text{Laplace transform of } (\tfrac{1}{2} e^{ikt}) + \text{Laplace transform of } (\tfrac{1}{2} e^{-ikt}) \\ &= \frac{1}{2}\frac{1}{(s - ik)} + \frac{1}{2}\frac{1}{(s + ik)} \qquad \text{(using Example 5)} \\ &= \frac{1}{2}\frac{2s}{s^2 + k^2} \\ &= \frac{s}{s^2 + k^2} \end{aligned}$$

Example 7

$f(t) = \sin kt$

Do this yourself to obtain

$$F(s) = \frac{k}{s^2 + k^2}$$

Example 8

$f(t) = \cosh kt$

$$\cosh kt = \tfrac{1}{2}(e^{kt} + e^{-kt})$$

Hence

$$F(s) = \frac{1}{2}\left(\frac{1}{s - k} + \frac{1}{s + k}\right) = \frac{s}{s^2 - k^2}$$

Notice we could have said

$$\cosh kt = \cos ikt$$

Hence using Example 6

$$F(s) = \frac{s}{s^2 + (ik)^2} = \frac{s}{s^2 - k^2}$$

Example 9

$f(t) = \sinh kt$

Use any method to obtain

$$F(s) = \frac{k}{s^2 - k^2}$$

We can now write out our table of Laplace transforms as far as we have gone

$f(t)$	$F(s)$
1	$\frac{1}{s}$
t^n	$\frac{n!}{s^{n+1}}$
e^{at}	$\frac{1}{s-a}$
$\cos kt$	$\frac{s}{s^2+k^2}$
$\sin kt$	$\frac{k}{s^2+k^2}$
$\cosh kt$	$\frac{s}{s^2-k^2}$
$\sinh kt$	$\frac{k}{s^2-k^2}$

Problems I

1. Find the transforms of the following functions

(i) $e^{-bt}\sinh at$ (ii) $e^{-bt}\cosh at$

(iii) $\sin(at - b)$ (iv) $(at - \sin at)/a^3$

(v) $(\cos at - \cos bt)/(b^2 - a^2)$ (vi) $t\cos at$

(vii) te^{-2t} (viii) $t^2 e^{-2t}$

We can now extend the table by employing two further theorems.

Translation or Shift Theorem

This states that *if $F(s)$ is the Laplace transform of a function $f(t)$, then the Laplace transform of $e^{+at}f(t)$ is $F(s-a)$. In other words, s is replaced by $s-a$ in the Laplace transform of $f(t)$.*

Before proving this theorem let us take a few examples.

Example 1

We have seen that the L.T. of 1 is $1/s$ and that of e^{at} is $1/(s-a)$

Example 2

Find the L.T. of $e^{-t}\cos 3t$

Considering

$$f(t) = \cos 3t$$

then

$$F(s) = \frac{s}{s^2+9}$$

Thus for $e^{-t}\cos 3t$ (with $a=-1$) the Laplace transform is

$$F(s+1) = \frac{s+1}{(s+1)^2+9}$$

$$= \frac{s+1}{s^2+2s+10}$$

Example 3

Find the L.T. of $t^3 e^{2t}$

Taking $f(t) = t^3$, then

$$F(s) = \frac{3!}{s^4}$$

Thus, using the theorem, the L.T. of $t^3 e^{2t}$

$$= F(s-2) = \frac{3!}{(s-2)^4}$$

Example 4

Find the L.T. of $e^{-2t}(3\cosh 4t - 2\sinh 4t)$

Taking $f(t) = 3\cosh 4t - 2\sinh 4t$, then

$$F(s) = \frac{3s}{s^2-16} - 2 \cdot \frac{4}{s^2-16} = \frac{3s-8}{s^2-16}$$

Thus, the L.T. of $e^{-2t}(3\cosh 4t - 2\sinh 4t)$ is

$$\frac{3(s+2)-8}{(s+2)^2-16} = \frac{3s-2}{s^2+4s-12}$$

Proof of the Theorem

By definition

$$F(s) = \int_0^\infty f(t)e^{-st}\,dt$$

The Laplace transform of $f(t)e^{at}$ is

$$\int_0^\infty f(t)e^{at}e^{-st}\,dt$$

$$= \int_0^\infty f(t)e^{-(s-a)t}\,dt$$

Now put $S = s - a$, then the L.T. =

$$\int_0^\infty f(t)e^{-St}\,dt = F(S) = F(s-a)$$

"Multiplication by t^n" Theorem

This states that *if $F(s)$ is the L.T. of $f(t)$, then the L.T. of $t^n f(t)$ is*

$$(-1)^n \frac{d^n F(s)}{ds^n}$$

To make the theorem clear, we take a few examples.

Example 1

Find the L.T. of $t^3 e^{2t}$

Taking $f(t) = e^{2t}$, then $F(s) = 1/(s-2)$.

Thus, using the theorem, the L.T. of $t^3 e^{2t}$

$$= (-1)^3 \frac{d^3}{ds^3}\left(\frac{1}{s-2}\right)$$

$$= -\frac{d^2}{ds^2}\left[\frac{-1}{(s-2)^2}\right]$$

$$= -\frac{d}{ds}\left[\frac{2}{(s-2)^3}\right]$$

$$= \frac{6}{(s-2)^4}$$

the same result as obtained earlier by another method.

Example 2

Find the L.T. of $t \sin 2t$

Taking $f(t) = \sin 2t$, then

$$F(s) = \frac{2}{s^2+4}$$

Thus, the L.T. of $t \sin 2t$ is

$$(-1)\frac{d}{ds}\left(\frac{2}{s^2+4}\right) = \frac{4s}{(s^2+4)^2}$$

Example 3

Find the L.T. of $te^{-2t}\cos 3t$

We need to use both theorems here.

Taking, first of all, $f(t) = \cos 3t$, then

$$F(s) = \frac{s}{s^2+9}$$

Thus the L.T. of $e^{-2t}\cos 3t$ is

$$\frac{s+2}{(s+2)^2+9} = \frac{s+2}{s^2+4s+13}$$

Now use the second theorem, and the L.T. of $te^{-2t}\cos 3t$

$$= (-1)\frac{d}{ds}\left(\frac{s+2}{s^2+4s+13}\right)$$

$$= -\frac{(s^2+4s+13)-(s+2)(2s+4)}{(s^2+4s+13)^2}$$

$$= \frac{s^2+4s-5}{(s^2+4s+13)^2}$$

Problems II

Find the transforms of the following functions:

(i) $(\sin at - at\cos at)/2a^3$ (ii) $e^{-at}t^{n-1}/(n-1)!$

(iii) $t\cos kt$ (iv) $t\sinh kt$

(v) $e^{-2t}\cos kt$ (vi) $(t\sin t)e^{-t}$

The full table of Laplace transforms is now as follows. Those that we have not proved you should obtain by any convenient method.

$f(t)$	$F(s)$
1	$\frac{1}{s}$
t^n	$n!/s^{n+1}$
e^{at}	$\frac{1}{s-a}$
$\cos kt$	$\frac{s}{s^2+k^2}$
$\sin kt$	$\frac{k}{s^2+k^2}$
$\cosh kt$	$\frac{s}{s^2-k^2}$
$\sinh kt$	$\frac{k}{s^2-k^2}$
$e^{at}\cos kt$	$\frac{(s-a)}{(s-a)^2+k^2}$
$e^{at}\sin kt$	$\frac{k}{(s-a)^2+k^2}$
$t\cos kt$	$\frac{(s^2-k^2)}{(s^2+k^2)^2}$
$t\sin kt$	$\frac{2ks}{(s^2+k^2)^2}$
te^{-at}	$\frac{1}{(s+a)^2}$
$e^{at}f(t)$	$F(s-a)$
$t^n f(t)$	$(-1)^n \frac{d^n F(s)}{ds^n}$

Laplace transform of derivatives

If we use Laplace transforms to solve differential equations we shall need to know the Laplace transforms of derivatives of $f(t)$. These are obtained as follows.

The L.T. of $\mathrm{d}f/\mathrm{d}t$ is

$$\int_0^\infty \mathrm{e}^{-st}\frac{\mathrm{d}f}{\mathrm{d}t}\,\mathrm{d}t = \Big[\mathrm{e}^{-st}f\Big]_0^\infty + s\int_0^\infty \mathrm{e}^{-st}f(t)\mathrm{d}t = -f(0) + sF(s) \qquad (10.10\text{a})$$

where $f(0)$ as usual denotes the value of $f(t)$ when $t = 0$ and we have assumed that

$$\lim_{t\to\infty}(\mathrm{e}^{-st}f) = 0$$

This is usually the case for physical problems for suitable values of s.

Similarly the L.T. of $\mathrm{d}^2f/\mathrm{d}t^2$ is

$$\int_0^\infty \mathrm{e}^{-st}\frac{\mathrm{d}^2f}{\mathrm{d}t^2}\,\mathrm{d}t = \left[\mathrm{e}^{-st}\frac{\mathrm{d}f}{\mathrm{d}t}\right]_0^\infty + s\int_0^\infty \mathrm{e}^{-st}\frac{\mathrm{d}f}{\mathrm{d}t}\,\mathrm{d}t$$

$$= -f'(0) + s[-f(0) + sF(s)]$$

$$= -f'(0) - sf(0) + s^2F(s) \qquad (10.10\text{b})$$

$f'(0)$ denoting the value of $\mathrm{d}f/\mathrm{d}t$ at $t = 0$. We have assumed that

$$\lim_{t\to\infty}\left(\mathrm{e}^{-st}\frac{\mathrm{d}f}{\mathrm{d}t}\right) = 0$$

We find generally, assuming

$$\lim_{t\to\infty}\left(\mathrm{e}^{-st}\frac{\mathrm{d}^{r-1}f}{\mathrm{d}t^{r-1}}\right) = 0 \quad \text{for } r = 1, 2, \ldots\ldots, n,$$

that the L.T. of $\mathrm{d}^nf/\mathrm{d}t^n$ is

$$-f^{n-1}(0) - sf^{n-2}(0) - \ldots\ldots - s^{n-1}f(0) + s^nF(s) \qquad (10.10\text{c})$$

The solution of ordinary differential equations by the Laplace transform

The method of solution will become clear if we consider an example:

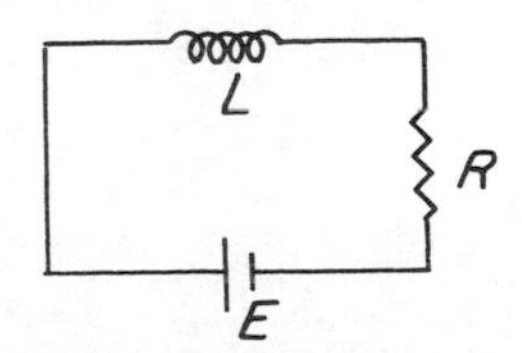

Figure 10.13

We study the circuit of Figure 10.13 with inductance L, resistance R and a constant electromotive force E, with zero current when $t = 0$.

The equation governing the current i as a function of time t is

$$L\frac{di}{dt} + Ri = E, \quad \text{given } i = 0 \text{ when } t = 0$$

Writing down the Laplace transform of each term we get

$$L[-i(0) + sI(s)] + RI(s) = \frac{E}{s}$$

where $I(s)$ is the L.T. of $i(t)$.
We are given $i(0) = 0$ and so

$$I(s)[Ls + R] = \frac{E}{s}$$

Hence

$$I(s) = \frac{E}{s(Ls + R)}$$

This is the L.T. of $i(t)$ and we should like to find its inverse to find the current at any time. We look at our table and ask – what function gives this L.T.? There is none there just like this but suppose we put the R.H.S. into partial fractions.

That is,

$$I(s) = \frac{\frac{E}{R}}{s} - \frac{\frac{LE}{R}}{Ls + R}$$

We can see in the table that these are L.T.'s of the form $1/s$ and $1/(s-a)$ so we write

$$I(s) = \frac{E}{R}\cdot\frac{1}{s} - \frac{E}{R}\cdot\frac{1}{s + \frac{R}{L}}$$

$1/s$ is the transform of unity and $1/[s + (R/L)]$ is the transform of $e^{-(R/L)t}$. Hence the inverse is

$$i(t) = \frac{E}{R} - \frac{E}{R}e^{-(R/L)t} = \frac{E}{R}\left[1 - e^{-(R/L)t}\right]$$

Thus we have solved the differential equation. The method has not involved integration (this was taken care of by writing down the Laplace transforms since in effect we multiplied by e^{-st} and integrated from 0 to ∞) and the technique reduced to one of algebraic manipulation. Many of the steps can be eliminated with a little practice and in the following examples we shall gradually leave out some of the detail.

Example 1

Solve

$$\frac{d^2y}{dt^2} + 4y = 12t$$

given that $y = 0$ and $dy/dt = 9$ when $t = 0$.

Writing the Laplace transform of each term we get

$$-y'(0) - sy(0) + s^2 Y(s) + 4Y(s) = 12 . \frac{1}{s^2}$$

where $Y(s)$ is the L.T. of $y(t)$.

Now $y'(0) = 9$ and $y(0) = 0$, so we get

$$(s^2 + 4) Y(s) = \frac{12}{s^2} + 9$$

or

$$Y(s) = \frac{9s^2 + 12}{s^2(s^2 + 4)} = \frac{3}{s^2} + \frac{6}{s^2 + 4}$$

Looking in the table we can find the inverse by writing

$$Y(s) = 3 . \frac{1}{s^2} + 3 . \frac{2}{s^2 + 2^2}$$

Hence

$$y(t) = 3t + 3 \sin 2t$$

Example 2

Solve

$$\frac{d^2y}{dt^2} - 4\frac{dy}{dt} + 5y = e^{2t}$$

given that $y = 0$ and $dy/dt = 0$ when $t = 0$.

Writing down Laplace transforms we get

$$-y'(0) - sy(0) + s^2 Y(s) - 4[-y(0) + sY(s)] + 5Y(s) = \frac{1}{s - 2}$$

so that

$$(s^2 - 4s + 5) Y(s) = \frac{1}{s - 2}$$

Hence

$$Y(s) = \frac{1}{(s - 2)(s^2 - 4s + 5)} = \frac{1}{s - 2} - \frac{s - 2}{s^2 - 4s + 5}$$

Looking at the term

$$\frac{s - 2}{s^2 - 4s + 5}$$

we complete the square in the denominator to get

$$\frac{s - 2}{(s - 2)^2 + 1}$$

which is of the form $S/(S^2 + 1^2)$ with $S = s - 2$.

Now you will remember that the shift theorem replaces s by $(s - 2)$ when the function is multiplied by e^{2t}. The inverse transform of $S/(S^2 + 1^2)$ is $\cos t$, so that the inverse of $(s - 2)/[(s - 2)^2 + 1^2]$ is $e^{2t} \cos t$. Thus we get

$$y(t) = e^{2t} - e^{2t} \cos t$$

Example 3

Solve

$$\frac{d^3x}{dt^3} - \frac{d^2x}{dt^2} + 2x = e^{-t}$$

given that $x = 1$, $dx/dt = 0$ and $d^2x/dt^2 = -2$ when $t = 0$.

Taking the Laplace transform of each term gives

$$-x''(0) - sx'(0) - s^2x(0) + s^3X(s) - [-x'(0) - sx(0) + s^2X(s)] + 2X(s) = \frac{1}{s+1}$$

which reduces to

$$(s^3 - s^2 + 2)X(s) = \frac{1}{s+1} + s^2 - s - 2 = \frac{s^3 - 3s - 1}{s+1}$$

Hence

$$X(s) = \frac{s^3 - 3s - 1}{(s+1)(s^3 - s^2 + 2)} = \frac{s^3 - 3s - 1}{(s+1)^2(s^2 - 2s + 2)}$$

$$= \frac{4}{25}\frac{1}{(s+1)} + \frac{1}{5}\frac{1}{(s+1)^2} + \frac{1}{25}\frac{(21s - 43)}{s^2 - 2s + 2}$$

(partial fractions)

$$= \frac{4}{25}\frac{1}{(s+1)} + \frac{1}{5}\frac{1}{(s+1)^2} + \frac{1}{25}\frac{[21(s-1) - 22]}{(s-1)^2 + 1}$$

$$= \frac{4}{25}\frac{1}{s+1} + \frac{1}{5}\frac{1}{(s+1)^2} + \frac{21}{25}\frac{(s-1)}{(s-1)^2 + 1^2} - \frac{22}{25}\frac{1}{(s-1)^2 + 1^2}$$

The inverse of each term then gives

$$x(t) = \frac{4}{25}e^{-t} + \frac{1}{5}te^{-t} + \frac{21}{25}e^t \cos t - \frac{22}{25}e^t \sin t$$

You will have noticed that in finding the inverses we are looking in the table of transforms and trying to find something similar to that which we have got. Then by completing the square in the denominator and other manipulation we suitably modify our terms. Take the last example: we have a term

$$\frac{1}{25}\frac{(21s - 43)}{(s^2 - 2s + 2)}$$

which by completing the square gave

$$\frac{1}{25}\,\frac{(21s - 43)}{(s-1)^2 + 1}$$

The nearest we have to this are the forms

$$\frac{(s-a)}{(s-a)^2 + k^2} \quad \text{and} \quad \frac{k}{(s-a)^2 + k^2}$$

which appear in our table as transforms of $e^{at} \cos kt$ and $e^{at} \sin kt$ respectively. Hence we split up the numerator to get such forms as

$$\frac{1}{25}\,\frac{21(s-1) - 22}{(s-1)^2 + 1^2} = \frac{21}{25}\,\frac{s-1}{(s-1)^2 + 1^2} - \frac{22}{25}\,\frac{1}{(s-1)^2 + 1^2}$$

You should now try the following examples. When trying to find the inverse in the first few of these, if you are finding difficulty, then look up the answer and find its Laplace transform. Then try to follow the steps from your form to the correct form of inverse.

One further point for the whole process of solution of differential equations using Laplace transform – always make a check that your solutions satisfy the boundary conditions.

Problems III

1. Solve the following equations with the conditions given at $t = 0$ in each case

(i) $\dfrac{di}{dt} + 2i = 100 \cos t,$ given that $i = 0$

(ii) $\dfrac{d^2r}{dt^2} + 4\dfrac{dr}{dt} + 3r = 0$ given that $r = 2,\ dr/dt = 3$

(iii) $\dfrac{d^2x}{dt^2} + 6\dfrac{dx}{dt} + 9x = 0$ given that $x = 2,\ \dot{x} = 4$

(iv) $\dfrac{d^2x}{dt^2} + 9x = t$ given that $x = 1,\ \dot{x} = 2$

(v) $\dfrac{d^2x}{dt^2} + n^2x = \cos nt$ given that $x = x_0,\ \dot{x} = x_1$

(vi) $\dfrac{d^2x}{dt^2} + 4x = 5 \cos 2t$ given that $x = 1,\ \dot{x} = 3$

2. Obtain, in each case, the Laplace transform of the dependent variable and so find the solution

(i) $2\dfrac{d^2y}{dt^2} - 5\dfrac{dy}{dt} - 3y = e^{3t}$ when $y = 1$ and $\dfrac{dy}{dt} = 0$ at $t = 0$

(ii) $\dfrac{d^2y}{dt^2} + 4\dfrac{dy}{dt} + 8y = 1$ if $y = 0$ and $\dfrac{dy}{dt} = 0$ at $t = 0$

(iii) $\dfrac{d^2\theta}{dt^2} + 2\dfrac{d\theta}{dt} + \theta = \sin 2t$ if $\theta = \dfrac{d\theta}{dt} = 0$ at $t = 0$

(iv) $\dfrac{d^2x}{dt^2} + 9x = \cos 3t$ if $x = 2$ and $\dfrac{dx}{dt} = -5$ when $t = 0$

(v) $\dfrac{dy}{dt} + y = t^2e^{-t}$ if $y = 10$ when $t = 0$

(vi) $\dfrac{d^2u}{dt^2} + 2\dfrac{du}{dt} + 5u = e^{-t}$ if $u = \dfrac{du}{dt} = 0$ at $t = 0$

(vii) $\dfrac{d^4y}{dt^4} - k^4y = 0$ if $y = \dfrac{dy}{dt} = \dfrac{d^2y}{dt^2} = \dfrac{d^3y}{dt^3} = 0$ at $t = 0$

10.9 SIMULTANEOUS ORDINARY DIFFERENTIAL EQUATIONS

In various problems in science and engineering more than one differential equation is required to describe the system. For example, consider the coupled electrical circuits shown in Figure 10.14 where E represents the e.m.f. in a circuit, R the resistance and L the inductance. M is the mutual inductance of the coils and i_1, i_2 represent the currents.

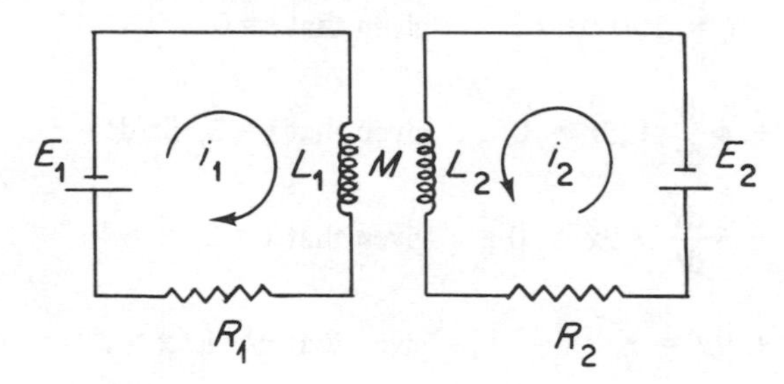

Figure 10.14

We know from Kirchhoff's laws that

$$R_1i_1 = E_1 - L_1\frac{di_1}{dt} - M\frac{di_2}{dt}$$

and

$$R_2i_2 = E_2 - L_2\frac{di_2}{dt} - M\frac{di_1}{dt}$$

Thus we have a pair of simultaneous differential equations in the two variables, i_1 and i_2 each dependent on time.

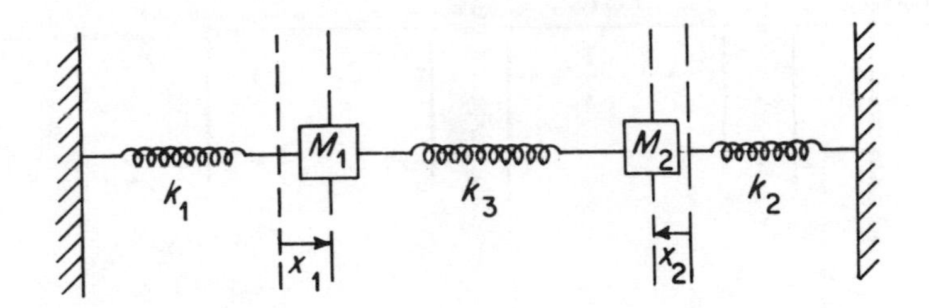

Figure 10.15

Similarly in a mechanical vibration system such as the one shown in Figure 10.15, where k_1, k_2, k_3 are the elastic constants of the springs, we obtain the two simultaneous differential equations

$$M_1 \frac{d^2x_1}{dt^2} = k_1x_1 - k_3(x_1 + x_2)$$

$$M_2 \frac{d^2x_2}{dt^2} = -k_2x_2 - k_3(x_1 + x_2)$$

to describe the vibrational motion of the masses M_1 and M_2. (The springs at the outside are stretched, the central one is compressed.) The Laplace transform method is particularly useful for solving such problems, if the initial values of the variables are given. The extension of the method to cover such problems will be clear if we take an example.

Example

Solve

$$\left.\begin{aligned} \frac{d^2x}{dt^2} - \frac{dy}{dt} + 2x &= 1 \\ \frac{dx}{dt} + \frac{d^2y}{dt^2} + 2y &= 0 \end{aligned}\right\} \quad \text{given that } x = 3,\ y = 0,\ dx/dt = 0,\ dy/dt = 0 \text{ when } t = 0.$$

Taking the L.T. of each term we get

$$[-x'(0) - sx(0) + s^2X(s)] - [-y(0) + sY(s)] + 2X(s) = 1/s$$

$$[-x(0) + sX(s)] + [-y'(0) - sy(0) + s^2Y(s)] + 2Y(s) = 0$$

giving

$$(s^2 + 2)X(s) - sY(s) = \frac{1}{s} + 3s$$

$$sX(s) + (s^2 + 2)Y(s) = 3$$

Solving these equations algebraically for the transforms $X(s)$ and $Y(s)$, we get

$$\frac{X(s)}{\begin{vmatrix} \frac{1}{s}+3s & -s \\ 3 & s^2+2 \end{vmatrix}} = \frac{Y(s)}{\begin{vmatrix} s^2+2 & \frac{1}{s}+3s \\ s & 3 \end{vmatrix}} = \frac{1}{\begin{vmatrix} s^2+2 & -s \\ s & s^2+2 \end{vmatrix}}$$

That is,

$$X(s) = \frac{3s^4+10s^2+2}{s(s^2+1)(s^2+4)}, \quad Y(s) = \frac{5}{(s^2+1)(s^2+4)}$$

The problem is now to invert these expressions to get $x(t)$ and $y(t)$. We start by taking partial fractions, that is

$$X(s) = \frac{(1/2)}{s} + \frac{(5/3)s}{(s^2+1)} + \frac{(5/6)s}{s^2+4}, \quad Y(s) = \frac{5/3}{s^2+1} - \frac{5/3}{s^2+4}$$

Writing these as

$$X(s) = \frac{1}{2}\cdot\frac{1}{s} + \frac{5}{3}\cdot\frac{s}{s^2+1^2} + \frac{5}{6}\cdot\frac{s}{s^2+2^2}$$

$$Y(s) = \frac{5}{3}\cdot\frac{1}{s^2+1^2} - \frac{5}{6}\cdot\frac{2}{s^2+2^2}$$

we find the inverses are

$$x(t) = \frac{1}{2} + \frac{5}{3}\cos t + \frac{5}{6}\cos 2t$$

$$y(t) = \frac{5}{3}\sin t - \frac{5}{6}\sin 2t$$

To emphasise the method used here: looking back through the example we see that the steps are

(i) take the L.T. of each term in the original d.e.'s
(ii) rearrange the equations as algebraic simultaneous equations
(iii) solve the equations algebraically
(iv) put the solutions in a suitable form for finding the inverse L.T.'s
(v) write down the inverses.

Problems

Find x and y, each as a function of t, in the following cases

1. $$\left.\begin{aligned}\frac{dx}{dt} + 2x + y &= 0\\ \frac{dy}{dt} + x + 2y &= 0\end{aligned}\right\}\quad \text{if } x = 1 \text{ and } y = 0 \text{ at } t = 0$$

2. $$\left.\begin{aligned}\frac{dx}{dt} + 5x - 2y &= t\\ \frac{dy}{dt} + 2x + y &= 0\end{aligned}\right\}\quad \text{if } x = y = 0 \text{ at } t = 0$$

3. $$\left.\begin{aligned}\frac{d^2x}{dt^2} - \frac{dy}{dt} + 2x &= 1\\ \frac{dx}{dt} + \frac{d^2y}{dt^2} + 2y &= 0\end{aligned}\right\}\quad \text{if } x = 3,\ y = \frac{dx}{dt} = \frac{dy}{dt} = 0 \text{ when } t = 0$$

4. In the coupled electrical circuit of Figure 10.14, with $E_1 = E, E_2 = 0$, let $i_1 = i_2 = 0$ at $t = 0$; show that

$$i_1 = \frac{E}{R}(1 - e^{-kt}\cosh mt)$$

$$i_2 = -\frac{E}{R}e^{-kt}\sinh mt$$

where $m = RM/(L^2 - M^2)$, $k = RL/(L^2 - M^2)$

5. In the mechanical vibrational system of Figure 10.15 suppose $\dot{x}_1(0) = \dot{x}_2(0) = 0$, $x_1(0) = x_1^0$ and $x_2(0) = x_2^0$, where x_1^0, x_2^0 are unknown fixed values. Take $k_1 = k_2 = k_3 = k$ and $M_1 = M_2 = 1$ and solve for x_1 and x_2.

6. In Problem 1 substitute for y in the second equation and obtain a second order differential equation in x only. Solve this and then obtain y. (Be careful which equation you use to obtain y.)
Repeat for Problem 2.

7. Solve the system

$$3\frac{dx}{dt} - 2x + 4y = 7e^{t} + 9e^{-t}$$

$$2x + 3\frac{dy}{dt} - 4y = 5e^{t} - 9e^{-t}$$

with the condition $x = 0$ and $y = 1$ at $t = 0$. (L.U.)

8. Solve the simultaneous differential equations

$$\frac{dx}{dt} - 2\frac{dy}{dt} + 2x - 4y = 16 \cos 2t$$

$$2\frac{dy}{dt} - 2y - x = 0$$

given that $x = 4$ when $t = 0$ and $y = 0$ when $t = 0$. (L.U.)

9. Find x in terms of t given that

$$\frac{d^2x}{dt^2} = \omega^2(y - 2x)$$

$$\frac{d^2y}{dt^2} = \omega^2(x - 2y)$$

where $x = b,\ y = 0,\ dx/dt = 0 = dy/dt$ at $t = 0$. (L.U.)

10. Find x in terms of t, given that

$$\frac{dx}{dt} + x - 3y = 0, \quad \frac{dy}{dt} - 2x + 2y = 0$$

and that $y = 0$, and $dx/dt = 20$ when $t = 0$. (L.U.)

11. In a radioactive series consisting of four different nuclides, starting with the parent substance N_1 and ending with the stable end product N_4, the amounts present at time t are given by

$$\frac{dN_1}{dt} = -\lambda_1 N_1 \qquad \frac{dN_2}{dt} = \lambda_1 N_1 - \lambda_2 N_2$$

$$\frac{dN_3}{dt} = \lambda_2 N_2 - \lambda_3 N_3 \qquad \frac{dN_4}{dt} = \lambda_3 N_3$$

$\lambda_1, \lambda_2, \lambda_3$ are different decay constants.

By means of Laplace transform method find $N_4(t)$, assuming that $N_1 = N_0$ (a constant) and $N_2 = N_3 = N_4 = 0$ at $t = 0$. (C.E.I.)

10.10 BOUNDARY VALUE PROBLEMS

We intend merely to introduce this class of problem by an example. In Figure 10.16 a straight uniform column of length l is hinged at both ends and an axial load P is applied.

The problem is to find the deflected profile $y(x)$ which is satisfied approximately by the d.e.

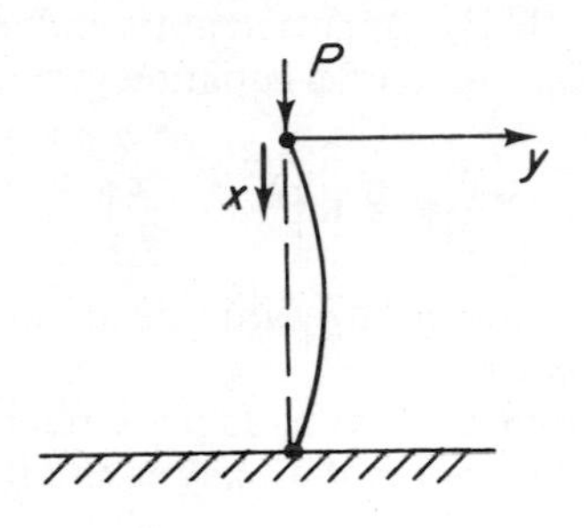

Figure 10.16

$$\frac{d^2y}{dx^2} + \frac{Py}{EI} = 0 \qquad (10.11)$$

where EI is constant for a given column.

The solutions must satisfy the boundary conditions: $y = 0$ at $x = 0$ and at $x = l$.

The general solution of equation (10.11) is

$$y = A \sin\sqrt{\frac{P}{EI}}\, x + B \cos\sqrt{\frac{P}{EI}}\, x$$

Applying $y = 0$ at $x = 0$ we find $B = 0$, and from the other boundary condition

$$A \sin\sqrt{\frac{P}{EI}}\, l = 0$$

Since $A = 0$ gives $y(x) \equiv 0$ we seek other solutions; we know $\sin n\pi = 0$ and so we conclude

$$\sqrt{\frac{P}{EI}}\, l = n\pi$$

and so we may find for a fixed beam several possible loads P_n which satisfy the d.e. and boundary conditions. The solutions are given by

$$P_n = \frac{n^2\pi^2}{l^2} EI, \qquad n = 1, 2, 3, \ldots\ldots$$

Such loads are called **critical loads** and in practice the first such load is the one which causes the column to buckle (according to the theory which produced equation 10.11).

Such boundary value problems do not easily allow a solution via Laplace transforms. Furthermore the numerical solution of such problems is usually much more complicated than the solution for the corresponding initial value problem. We shall not consider numerical solutions at this stage, but give two boundary value problems for you to try by analytical methods.

Problems

1. A beam of length l is freely supported at its ends and weighs w kg/unit length. The differential equation governing the deflection of the beam is

$$EI\frac{d^2y}{dx^2} = w\left(\frac{x^2}{2} - \frac{lx}{2}\right)$$

Solve this equation by straightforward integration given that $y = 0$ when $x = 0$ and $y = 0$ when $x = l$.
Now try to solve the equation using Laplace transforms – you will have to assume $y'(0) = Y =$ constant in the analysis and find Y eventually by using $y = 0$ when $x = l$.

2. Figure 10.17 shows a very long cooling fin of thickness t and length l attached to a hot wall which is maintained at a temperature of 100°C.

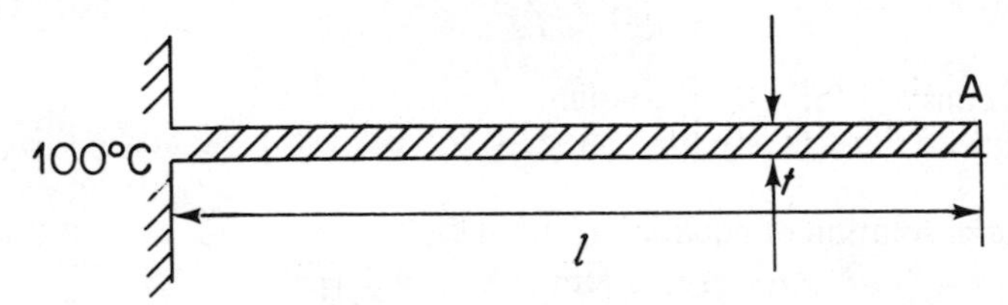

Figure 10.17

Heat is conducted steadily along the fin and is lost from the sides by convection to the surrounding air (which is at temperature 70°C). The fin temperature θ, assumed to depend only on the distance x along the fin obeys the differential equation

$$kt\frac{d^2\theta}{dx^2} = 2h(\theta - \theta_{air}) \quad \text{where } k \text{ and } h \text{ are constants}$$

We assume that the fin is long enough that the end A is at the same temperature as the surrounding air, so that we have the boundary conditions

$$\theta = 100 \quad \text{when } x = 0$$
$$\theta = 70 \quad \text{when } x \to \infty$$

Solve the equation, given that

$$h = 2\ \text{W/m}^2\text{K}$$
$$k = 20\ \text{W/mK}$$
$$t = 0.75\ \text{cm}$$

(Beware of the boundary condition $\theta = 70$ when $x \to \infty$.)

10.11 D-OPERATOR METHOD

We have already learned two ways of finding the solution to ordinary differential equations with constant coefficients – namely the trial method and the Laplace transform method. Here we shall discuss briefly an alternative method of solution called the **D-operator method.**

Given a differential equation of the form

$$A\frac{d^2y}{dx^2} + B\frac{dy}{dx} + Cy = F(x)$$

A, B, C being constants, we know that the general solution will consist of two parts

$$y = \text{C.F.} + \text{P.I.}$$

You will remember that the C.F. is the solution of the L.H.S. = 0, that is

$$A\frac{d^2y}{dx^2} + B\frac{dy}{dx} + Cy = 0$$

and the C.F. will contain two arbitrary constants in the case of a second order d.e.

The P.I. is some solution of the whole equation and contains no arbitrary constants. We found P.I.s by trial methods in Section 10.5.

The D-operator method is an alternative technique to the trial method for determining the P.I.

Definition

The symbol D is used to denote differentiation with respect to the independent variable. That is, if $y = f(x)$, then $Dy = dy/dx$; if $y = f(t)$ then $Dy = dy/dt$.

When $y = f(x)$, since $Dy = dy/dx = [d/dx](y)$, we can see that $D \equiv d/dx$ and it is said to *operate* on y.

We can extend this to say $D(Dy)$ means

$$\frac{d}{dx}\left(\frac{d}{dx}y\right) = \frac{d^2y}{dx^2}$$

Thus

$$D^2y = \frac{d^2y}{dx^2}$$

Similarly

$$D^3y = \frac{d^3y}{dx^3}$$

and, generally,

$$D^ny = \frac{d^ny}{dx^n}$$

Laws obeyed by the Operator D

(i) If u and v are functions of x, then $D(u+v) = Du + Dv$. We can see immediately that this result is true, for

$$\frac{d}{dx}(u+v) = \frac{du}{dx} + \frac{dv}{dx}$$

(ii) $D^p(D^q y) = D^q(D^p y) = D^{p+q}y$

Again it is easy to see that this is so, since

$$\frac{d^p}{dx^p}\left(\frac{d^q y}{dx^q}\right) = \frac{d^q}{dx^q}\left(\frac{d^p y}{dx^p}\right) = \frac{d^{p+q}y}{dx^{p+q}}$$

(iii) We can write a differential equation

$$A_0\frac{d^n y}{dx^n} + A_1\frac{d^{n-1}y}{dx^{n-1}} + \ldots\ldots + A_n y = F(x)$$

as $A_0 D^n y + A_1 D^{n-1}y + \ldots\ldots + A_n y = F(x)$

and further write this as

$$[A_0 D^n + A_1 D^{n-1} + \ldots\ldots + A_n]y = F(x)$$

Examples

The equation

$$\frac{dy}{dx} + 3y = 2x$$

can be written

$$(D+3)y = 2x$$

The equation

$$\frac{d^2y}{dx^2} + 3\frac{dy}{dx} + 2y = e^{2x}$$

can be written as

$$(D^2 + 3D + 2)y = e^{2x}$$

The expression $[A_0 D^n + A_1 D^{n-1} + \ldots\ldots + A_n]$ is itself an operator and operates on y to give the L.H.S. of the differential equation. We write for short

$$f(D) \equiv [A_0 D^n + A_1 D^{n-1} + \ldots\ldots + A_n]$$

(iv) We can factorise $f(D)$ and, for constant coefficients, the order of the factors is immaterial.
As an illustration consider the expression $(D+a)(D+b)y$. On multiplying out the operator we get

$$[D^2 + (a+b)D + ab]\,y$$

Now $(D+a)(D+b)$ represents

$$\left(\frac{d}{dx}+a\right)\left(\frac{d}{dx}+b\right)y$$

$$= \left(\frac{d}{dx}+a\right)\left(\frac{dy}{dx}+by\right)$$

$$= \frac{d^2y}{dx^2} + \frac{d}{dx}by + a\frac{dy}{dx} + aby$$

$$= \frac{d^2y}{dx^2} + a\frac{dy}{dx} + b\frac{dy}{dx} + aby$$

$$= [D^2 + (a+b)D + ab]y$$

Thus the operator in its factorised form produces the same expression as the non-factorised form.

Example

Another example is the expression $(d^4y/dx^4) - y$

In operator form this is

$$(D^4 - 1)y$$
$$= (D^2-1)(D^2+1)y$$
$$= (D-1)(D+1)(D-i)(D+i)y$$

(v) $f(D)e^{ax} = f(a)e^{ax}$

This is valid because, working systematically,

$$De^{ax} = \frac{d}{dx}e^{ax} = ae^{ax}$$

$$D^2e^{ax} = \frac{d^2}{dx^2}e^{ax} = a^2e^{ax}$$

$$\vdots \qquad \vdots$$

$$D^ne^{ax} \qquad = a^ne^{ax}$$

Thus

$$f(D)e^{ax} = [A_0D^n + A_1D^{n-1} + \ldots + A_n]e^{ax} = [A_0a^n + A_1a^{n-1} + \ldots + A_n]e^{ax}$$

$$= f(a)e^{ax}$$

Example 1

$(D+2)e^{-x} = (-1+2)e^{-x} = e^{-x}$

$$(D^2 + 2D + 5)e^{2x} = (2^2 + 2.2 + 5)e^{2x} = 13e^{2x}$$

$$(D^4 - 1)e^{3x} = (3^4 - 1)e^{3x} = 80e^{3x}$$

Note that the quantity 'a' could be a complex number.

Example 2

$(D^2 - 1)e^{2ix} = -5e^{2ix}$

so that, picking out real and imaginary parts

$$(D^2 - 1)\cos 2x = -5 \cos 2x$$

$$(D^2 - 1)\sin 2x = -5 \sin 2x$$

You should verify these results by working directly from

$$\left(\frac{d^2}{dx^2} - 1\right)\cos 2x \quad \text{and} \quad \left(\frac{d^2}{dx^2} - 1\right)\sin 2x$$

(vi) $f(D)e^{ax}V = e^{ax}f(D + a)V$ (V is a function of x)

This is called the **Shift theorem** and in effect we think of e^{ax} going through the operator, the operator being replaced by $f(D + a)$.

The verification of this rule follows by using the product rule, for

$$\begin{aligned} D(e^{ax}V) &= VD(e^{ax}) + e^{ax}DV \\ &= ae^{ax}V + e^{ax}DV \\ &= e^{ax}(D + a)V \end{aligned}$$

$$\begin{aligned} D^2(e^{ax}V) &= D[D(e^{ax}V)] \\ &= D[e^{ax}(D + a)V] \quad \text{(by the theorem)} \\ &= e^{ax}[D(D + a)V] + (D + a)V \,.\, ae^{ax} \quad \text{(product rule)} \\ &= e^{ax}[D(D + a)V + a(D + a)V] \\ &= e^{ax}(D + a)^2 V \end{aligned}$$

Repetition of this gives

$$D^n(e^{ax}V) = e^{ax}(D + a)^n V$$

and substitution in the polynomial expression $f(D)$ yields the result whereby D is replaced everywhere by $D + a$.

Examples

(i) $(D^2 + 2D + 1)e^{-x}x^2 = e^{-x}[(D - 1)^2 + 2(D - 1) + 1]x^2$

$$= e^{-x}[D^2]x^2$$

$$= 2e^{-x}$$

(ii) $(D^2 + 3D + 2)e^{-2x}\sin x = e^{-2x}[(D-2)^2 + 3(D-2) + 2]\sin x$

$$= e^{-2x}[D^2 - D]\sin x$$

$$= e^{-2x}[-\sin x - \cos x]$$

In particular

(iii) $(D^2 + 2D + 3)e^{2x} . 1 = e^{2x}[(D+2)^2 + 2(D+2) + 3] . 1$

$$= e^{2x}[D^2 + 6D + 11] . 1$$

$$= 11e^{2x}$$

We could have used rule (v), of course, to get

$$[D^2 + 2D + 3]e^{2x} = [2^2 + 2 . 2 + 3]e^{2x} = 11e^{2x}$$

However, we shall find it useful later on to sometimes treat e^{ax} as $e^{ax} . 1$ and then use the Shift theorem.

The inverse and finding the Particular Integral

The inverse of D, written as D^{-1}, must be such that $D^{-1}D(y) = y$ or $D(D^{-1}y) = y$. This means

$$\frac{d}{dx}(D^{-1}y) = y$$

Integrating both sides w.r.t. x then gives

$$D^{-1}y = \int y\,dx$$

Thus D^{-1} represents integration with respect to x as we might have expected (integration being the inverse of differentiation).

Examples

$$D^{-1}x^2 = \int x^2\,dx = \frac{x^3}{3}$$

$$(D^{-1} + D)e^{2x} = \frac{1}{2}e^{2x} + 2e^{2x} = \frac{5}{2}e^{2x}$$

Notice we do not need to include a constant of integration here.

Consider the general differential equation

$$A_0D^n + A_1D^{n-1} + \ldots\ldots + A_ny = F(x)$$

which we write as

$$[f(D)]y = F(x)$$

(Let the inverse of $f(D)$ be written as $1/f(D)$, so that

$$f(D) \cdot \frac{1}{f(D)} Q = Q$$

for any function Q.) Then operating on both sides with $1/f(D)$ gives

$$\frac{1}{f(D)} \cdot [f(D)]y = \frac{1}{f(D)} \cdot F(x)$$

That is,

$$y = \frac{1}{f(D)} f(x)$$

The inverse operator $1/f(D)$ does not yet mean anything to us except that it will undo the operation of $f(D)$ on y. If it does this we shall obtain y and **this will be a particular integral.**

Although difficult to prove, we shall assume that we can treat $1/f(D)$ as an algebraic quantity and use all the normal rules of algebra in its manipulation.

We shall consider various forms for the function $f(x)$ on the R.H.S. of the equation.

(a) *$F(x)$ is a polynomial expression*

Example

Consider

$$\frac{d^2y}{dx^2} - 3\frac{dy}{dx} + 2y = x^2 + 4x - 3$$

That is

$$[D^2 - 3D + 2]y = x^2 + 4x - 3$$

Thus

$$y = \frac{1}{[D^2 - 3D + 2]}[x^2 + 4x - 3]$$

$$= \frac{1}{2\left[1 - \frac{3}{2}D + \frac{D^2}{2}\right]}[x^2 + 4x - 3]$$

$$= \frac{1}{2}\left[1 - \frac{3}{2}D + \frac{D^2}{2}\right]^{-1}[x^2 + 4x - 3]$$

We shall expand the operator, using the Binomial theorem, and reject powers above D^2 since differentiating $[x^2 + 4x - 3]$ more than twice gives zero.
Thus

$$y = \frac{1}{2}\left[1 - \left(-\frac{3}{2}D + \frac{D^2}{2}\right) + \left(-\frac{3}{2}D + \frac{D^2}{2}\right)^2 - \ldots\ldots\right]\left[x^2 + 4x - 3\right]$$

$$= \frac{1}{2}\left[1 + \frac{3}{2}D - \frac{D^2}{2} + \frac{9D^2}{4} + \ldots\ldots\right]\left[x^2 + 4x - 3\right]$$

$$= \frac{1}{2}\left[1 + \frac{3}{2}D + \frac{7D^2}{4}\right]\left[x^2 + 4x - 3\right]$$

$$= \frac{1}{2}\left[x^2 + 4x - 3 + \frac{3}{2}(2x + 4) + \frac{7}{4} \cdot 2\right]$$

$$= \frac{1}{2}\left[x^2 + 7x + \frac{13}{2}\right]$$

This is the required P.I.

(b) $F(x) = ke^{ax}$

You will remember that we showed that $f(D)\,e^{ax} = f(a)e^{ax}$
This suggests that

$$\frac{1}{f(D)}\,e^{ax} = \frac{1}{f(a)}\,e^{ax}$$

provided $f(a) \neq 0$.

We shall not prove this here but will use it for our purposes. The case where $f(a) = 0$ will be discussed later.

Example

Consider the equation

$$\frac{d^2y}{dx^2} - \frac{dy}{dx} + 2y = 5e^{2x}$$

That is,

$$(D^2 - D + 2)y = 5e^{2x}$$

$$\text{The P.I.} = \frac{1}{(D^2 - D + 2)}\,[5e^{2x}] = \frac{5e^{2x}}{(2^2 - 2 + 2)} = \frac{5}{4}\,e^{2x}$$

(c) $F(x) = \sin \alpha x$ *or* $\cos \alpha x$

Here we consider $e^{i\alpha x} = \cos \alpha x + i \sin \alpha x$ and pick out either the real or the imaginary part as appropriate.

Example

$$\frac{d^2y}{dx^2} - \frac{dy}{dx} + 2y = 3 \cos 2x$$

$$\text{The P.I.} = \frac{1}{(D^2 - D + 2)}\,[3 \cos 2x]$$

Consider

$$\frac{1}{(D^2 - D + 2)}[3e^{i2x}]$$

This equals

$$\frac{3e^{i2x}}{[(2i)^2 - 2i + 2]} = \frac{-3e^{i2x}}{2(1 + i)}$$

$3 \cos 2x$ is the real part of $3e^{i2x}$. Hence we find the real part of the last expression , which can be written

$$-\frac{3}{2}e^{i2x}\frac{(1 - i)}{2}$$

$$\text{i.e.} \quad -\frac{3}{4}(1 - i)(\cos 2x + i \sin 2x)$$

Hence

$$\text{P.I.} = -\frac{3}{4}(\cos 2x + \sin 2x)$$

Again there will be a special case if the denominator becomes zero. This is one of the cases of failure which we shall discuss later.

(d) $F(x) = e^{ax}. V$ where V is some function of x.

We had the rule $f(D)e^{ax} V = e^{ax} f(D + a)V$ called the Shift theorem. It suggests that $[1/f(D)]\, e^{ax} V$ can be treated in a similar way. Thus we expect

$$\frac{1}{f(D)}e^{ax} V = e^{ax}\frac{1}{f(D + a)}V$$

This is found to be true – in effect we bring e^{ax} through and replace D by $D + a$.

Example

$$\frac{d^2y}{dx^2} + 3\frac{dy}{dx} + 2y = e^{-x}(x^2 + 1)$$

$$\text{The P.I.} = \frac{1}{(D^2 + 3D + 2)}[e^{-x}(x^2 + 1)]$$

$$= e^{-x} \cdot \frac{1}{[(D - 1)^2 + 3(D - 1) + 2]}[x^2 + 1]$$

$$= e^{-x}\frac{1}{(D^2 + D)}[x^2 + 1]$$

$$= e^{-x}\frac{1}{D(1 + D)}[x^2 + 1]$$

$$= e^{-x} . \frac{1}{D}(1 - D + D^2 - \ldots\ldots)[x^2 + 1]$$

(expanding by the Binomial theorem)

$$= e^{-x}\frac{1}{D}[x^2 + 1 - 2x + 2]$$

$$= e^{-x}\frac{1}{D}[x^2 - 2x + 3]$$

$$= e^{-x}\left(\frac{x^3}{3} - x^2 + 3x\right) \quad \text{(remember } 1/D \text{ means } \textstyle\int dx)$$

Problems I

1. Repeat Problem 1 (i), (ii), (iii), (iv) on page 574 using the D-operator method.

2. Find the particular integral for the differential equations given in Problem 2 (i), (iii), (v) on pages 574 and 575.

3. Find the general solutions of the following differential equations

(i) $\dfrac{d^2y}{dx^2} - \dfrac{dy}{dx} + 9y = 2e^{2x}$

(ii) $\dfrac{d^2y}{dx^2} - 16y = (x + 1)^2$

(iii) $\dfrac{d^2y}{dx^2} + \dfrac{dy}{dx} + 13y = 2\cos 2x$

Cases of failure

As pointed out earlier in Cases (b), (c) and (d), we can run into difficulties if the denominator becomes zero in our manipulations. These difficulties are called cases of failure and they occur for the most part when the R.H.S. of the differential equation is a constituent part of the C.F. For example, we expect that

$$\frac{d^2y}{dx^2} + 3\frac{dy}{dx} + 2y = e^{-x}$$

will produce a case of failure because the C.F. is

$$y = Ae^{-x} + Be^{-2x}$$

Similarly,

$$\frac{d^2y}{dx^2} + 4\frac{dy}{dx} + 13y = e^{-2x}\sin 3x$$

will produce a case of failure because the C.F. is

$$y = e^{-2x}(A\cos 3x + B\sin 3x)$$

In each case the R.H.S. is a constituent part of the C.F. As a counter – example, the equation

$$\frac{d^2y}{dx^2} + 4\frac{dy}{dx} + 13y = e^{-2x}$$

does not give a case of failure since the C.F. is

$$y = e^{-2x}(A \cos 3x + B \sin 3x)$$

and the R.H.S. is not one of the terms occurring in the C.F. (albeit a part of each of them).

All the cases of failure will occur in (b), (c), and in (d) when $F(x)$ contains an exponential term of the form $e^{\alpha x}$ (α can be complex) which makes $f(D + \alpha) = 0$.

We overcome the difficulty by writing $e^{\alpha x}$ as $e^{\alpha x}.1$ and using the Shift theorem.

Example 1

Take the equation

$$\frac{d^2y}{dx^2} + 3\frac{dy}{dx} + 2y = e^{-x}$$

$$\text{The P.I.} = \frac{1}{D^2 + 3D + 2} e^{-x}$$

$$= \frac{1}{(D^2 + 3D + 2)} [e^{-x} . 1]$$

$$= e^{-x} . \frac{1}{(D - 1)^2 + 3(D - 1) + 2} [1]$$

$$= e^{-x} \frac{1}{D^2 + D} [1]$$

$$= e^{-x} \frac{1}{D(1 + D)} [1]$$

$$= e^{-x} . \frac{1}{D}(1 - D + D^2 - ......) [1]$$

$$= e^{-x} . \frac{1}{D} . [1]$$

$$= xe^{-x}$$

Example 2

Again for the equation

$$\frac{d^2y}{dx^2} + 4\frac{dy}{dx} + 13y = e^{-2x} \sin 3x$$

we have

$$\text{the P.I.} = \frac{1}{D^2 + 4D + 13} [e^{-2x} \sin 3x]$$

$$= e^{-2x} \cdot \frac{1}{(D-2)^2 + 4(D-2) + 13} [\sin 3x]$$

$$= e^{-2x} \cdot \frac{1}{D^2 + 9} [\sin 3x]$$

Now

$$\frac{1}{[D^2+9]} [\sin 3x] = \mathscr{I}\left\{\frac{1}{D^2+9} [e^{i3x}]\right\} = \mathscr{I}\left\{\frac{1}{D^2+9} [e^{i3x} . 1]\right\}$$

$$= \mathscr{I}\left\{e^{i3x} \frac{1}{(D+3i)^2 + 9} [1]\right\}$$

$$= \mathscr{I}\left\{e^{i3x} \frac{1}{D^2 + 6iD} [1]\right\}$$

$$= \mathscr{I}\left\{e^{i3x} \frac{1}{6iD(1 + \frac{D}{6i})} [1]\right\}$$

$$= \mathscr{I}\left\{\frac{e^{i3x}}{6i} \cdot \frac{1}{D}(1 - \frac{D}{6i} + \ldots\ldots)[1]\right\}$$

$$= \mathscr{I}\left\{\frac{e^{i3x}}{6i} x\right\}$$

$$= \mathscr{I}\left\{-\frac{i}{6} \cdot e^{i3x} . x\right\}$$

$$= -\frac{x \cos 3x}{6}$$

Thus

$$\text{the P.I.} = \frac{-e^{-2x} x \cos 3x}{6}$$

Cases of failure are not produced in (d) when V is a polynomial. For example, consider

$$\frac{d^2y}{dx^2} + 4\frac{dy}{dx} + 4y = xe^{-2x}$$

The C.F. is $y = (Ax + B)e^{-2x}$ and the R.H.S. is a constituent part of this. However, we get

$$\text{P.I.} = \frac{1}{(D^2 + 4D + 4)} [xe^{-2x}]$$

$$= e^{-2x} \cdot \frac{1}{(D-2)^2 + 4(D-2) + 4}[x]$$

$$= e^{-2x} \cdot \frac{1}{D^2}[x]$$

$$= e^{-2x} \cdot \frac{x^3}{6} \qquad \left(\frac{1}{D^2} \text{ means integrate w.r.t. } x \text{ twice}\right)$$

Problems II (using D-operators)

1. Repeat Problems 1 (v), (vi); 2(ii), (iv) of pages 574 and 575.

2. Repeat Problems 2(i), 2(iv) and 2(v) on pages 597 and 598.

3. Solve $\frac{d^2y}{dx^2} - 4\frac{dy}{dx} + 5y = 2e^{2x}\cos x$

4. A body of unit mass moves along the x–axis under the action of an attractive force of magnitude $\omega^2 x$, directed towards the origin 0. It is also subject to a frictional resistance $2k\dot{x}$ and to a driving force $e^{-kt}\cos pt$. The equation of motion is

$$\frac{d^2x}{dt^2} + 2k\frac{dx}{dt} + \omega^2 x = e^{-kt}\cos pt$$

If $p^2 = \omega^2 - k^2$ and the body starts from rest at the origin show that

$$x = \frac{t}{2p}e^{-kt}\sin pt$$

gives the position x at any time t.

5. If D is defined as d/dx and V is a function of x, show that, in the notation of the D-operator method

$$\frac{1}{F(D)}[e^{kx}V] = e^{kx}\frac{1}{F(D+k)}V$$

Hence, or otherwise, solve the differential equation

$$\frac{d^2y}{dx^2} + 6\frac{dy}{dx} + 9y = xe^{-2x}$$

with $y = 0$ when $x = 0$ and when $x = 2$. (C.E.I.)

6. Solve the equation $(D^2 + 1)y = \sin 2x$ subject to the conditions $y = 0$, $Dy = 1$ at $x = 0$.

10.12 COMPARISON OF METHODS OF SOLUTION

Consider the following problem which we shall solve by each method in turn: a mass m rests on a horizontal table and is attached to one end of a light spring which, when stretched, exerts a tension mw^2 times the extension (w is constant) – refer to Figure 10.18. The other end of the spring is now moved with constant velocity u along the table in a direction away from the mass. The table offers a frictional resistance on the mass of amount mk times its speed (k is constant). Show that if $k = 2w$ then the extension of the spring after time t is

$$x = \frac{u}{w}[2 - (2 + wt)e^{-wt}]$$

Position of the mass after time t is $(l + x) - ut$
Speed after time t is $\dot{x} - u$
Acceleration after time t is $\ddot{x}$

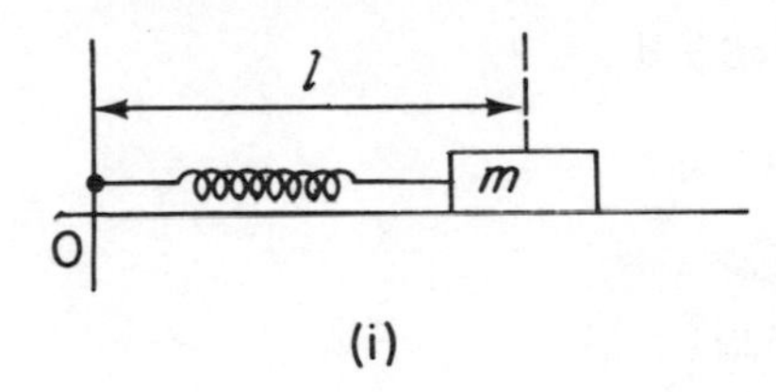

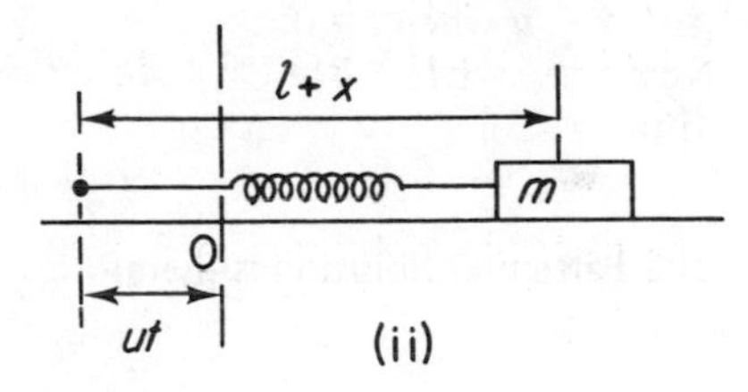

Figure 10.18

Hence

$$m\ddot{x} = -(\text{Resistance} + \text{Tension}) = -[mk(\dot{x} - u) + mw^2x]$$

therefore

$$\ddot{x} + k\dot{x} + w^2x = ku$$

Also we know $k = 2w$, so that

$$\ddot{x} + 2w\dot{x} + w^2x = 2wu$$

and $x = 0$ when $t = 0$, $\dot{x} = u$ when $t = 0$ since speed $= 0$ when $t = 0$.

Method I: Trial Method

$\ddot{x} + 2w\dot{x} + w^2x = 2wu$

For the C.F. $(m^2 + 2wm + w^2) = 0$
that is,

$$(m + w)^2 = 0$$

C.F. is

$$x = (At + B)e^{-wt}$$

For the P.I. we try $x = C$ giving

$$0 + 0 + w^2C = 2wu$$

Hence

$$C = \frac{2u}{w}$$

therefore

$$\text{P.I.} = \frac{2u}{w}$$

Thus the solution is

$$x = (At + B)e^{-wt} + \frac{2u}{w}$$

We know that $x = 0$ when $t = 0$ which implies that $0 = B + 2u/w$, hence

$$B = -\frac{2u}{w}$$

Also $\dot{x} = u$ when $t = 0$.
Now $\dot{x} = -w(At + B)e^{-wt} + Ae^{-wt} \Rightarrow u = -wB + A$.
Thus

$$A = u + wB = -u$$

The Particular Solution is therefore

$$x = \frac{2u}{w} - \left(ut + \frac{2u}{w}\right)e^{-wt}$$

Method II: Laplace Transform Method

We write down the L.T. of each term giving

$$[-x'(0) - sx(0) + s^2X(s)] + 2w[-x(0) + sX(s)] + w^2X(s) = \frac{2wu}{s}$$

Hence

$$(s^2 + 2ws + w^2)X(s) = \frac{2wu}{s} + u = \frac{u(2w + s)}{s}$$

Thus

$$X(s) = \frac{u\,(2w + s)}{s\,(s + w)^2} = \frac{2u/w}{s} - \frac{u}{(s + w)^2} - \frac{2u/w}{(s + w)}$$

The inverse is

$$x = \frac{2u}{w} - ute^{-wt} - \frac{2u}{w}e^{-wt}$$

as before.

Method III: D-Operator

Again we find the C.F. as

$$x = (At + B)e^{-wt}$$

The P.I. is

$$\frac{1}{(D+w)^2}[2wu] = \frac{1}{w^2\left(1+\frac{D}{w}\right)^2}[2wu]$$

$$= \frac{1}{w^2}\left(1+\frac{D}{w}\right)^{-2}[2wu]$$

$$= \frac{1}{w^2}\cdot\left(1-\frac{2D}{w}+\ldots\ldots\right)[2wu]$$

$$= \frac{1}{w^2}\cdot 2wu$$

$$= \frac{2u}{w} \quad \text{as before.}$$

The solution then proceeds as in Method I to find A and B.

Discussion

There is very little difference in the length and difficulty of the solutions but we observe the following

1. The P.I. is found much easier by the Trial Method than by the D-operator. This is because the R.H.S. of the differential equation is itself a very simple term.
2. The Laplace Transform Method eliminates the need to find the constants A and B. However it does involve us in finding the partial fractions which is not such an easy task. Nevertheless it is probably the best method to try if we are given *initial conditions* as was the case here.

Since this was an initial value problem we could of course have solved the equation by the method given in Section 10.6. As stated earlier, the analytical approach should be used wherever possible unless we have to so simplify the model of the particular physical situation that the results produced give a false picture.

Let us consider one further example. A body is dropped from a great height above the earth and falls freely. It is a fair approximation to say that the air resistance is proportional to the square of the velocity. Hence the equation of the motion is

mass × acceleration = mass × acceleration due to gravity − mass × k × (velocity)^2

↑ net force — ↑ gravitational force — ↑ air resistance (k constant)

Hence measuring y downwards from starting position we get

$$\frac{d^2y}{dt^2} = g - k\left(\frac{dy}{dt}\right)^2$$

That is,

$$\frac{d^2y}{dt^2} + k\left(\frac{dy}{dt}\right)^2 = g$$

where

$$y = 0$$

when $t = 0$, and

$$\frac{dy}{dt} = 0$$

when $t = 0$.

This is a non-linear second order differential equation because of the presence of the term $(dy/dt)^2$. Hence we cannot use the D-operator or the Laplace transform methods of solution.

Furthermore the trial method does not easily yield a solution. We would be led therefore to use the Runge-Kutta method or some other suitable numerical method. However an analytical solution is possible – we can write the equation in the form

$$\frac{d}{dt}\left(\frac{dy}{dt}\right) + k\left(\frac{dy}{dt}\right)^2 = g$$

which is

$$\frac{dv}{dt} + kv^2 = g \qquad \left[v = \frac{dy}{dt}\right]$$

and it is possible to integrate this equation by separation of variables giving

$$\log\left(\frac{1+\sqrt{\frac{k}{g}}\,v}{1-\sqrt{\frac{k}{g}}\,v}\right) = 2\sqrt{gk}\;t$$

We can rearrange this to get

$$v = \sqrt{\frac{g}{k}}\left(\frac{e^{2\sqrt{gk}\,t}-1}{e^{2\sqrt{gk}\,t}+1}\right)$$

and we can integrate both sides with respect to t to find y in terms of t. However this is by no means easy.

An alternative is to write d^2y/dt^2 as

$$\frac{d}{dy}\left(\frac{dy}{dt}\right) \cdot \frac{dy}{dt} = v\frac{dv}{dy}$$

The equation then becomes

$$v\frac{dv}{dy} + kv^2 = g$$

or

$$\frac{d}{dy}\left(\frac{v^2}{2}\right) + kv^2 = g$$

giving, by use of the Integrating Factor method

$$v^2 = \frac{g}{k}(1 - e^{-2ky})$$

Again it is possible but by no means easy to find y in terms of t from this equation.

It would seem, therefore, that for any second-order equation which is non-linear, the most straightforward method is to use a numerical approach.

Problem

Try completing the last problem by the analytical methods outlined above and compare with the Runge-Kutta solution for the first 10 seconds of the motion.

SUMMARY

This chapter concerns itself mainly with those differential equations which arise in the study of vibrations.

You should be conversant with the relationship between the complementary function and particular integral and you ought to master at least one method of finding the particular integral. You may be guided as to which method to choose, but it is unlikely that you will need to use all three. If you learn either the trial solution method or the D-operator method you must beware of the cases of failure; in the latter approach you will need to know standard results whereas, in the former you must become experienced in the kinds of trial function to employ. The relative merits of these methods and the Laplace Transform technique need to be known. If you use either of the first two methods, please remember that the boundary conditions can be fitted only after the general solution has been obtained. Most engineers will need to use Laplace Transforms and you should remember that this method does *not* find the particular integral directly; and in addition it incorporates the initial conditions in its particular solution. A comprehensive table of Laplace Transforms is provided in the appendix.

Chapter Eleven

Introduction to Statistical Methods

11.1 ROLE OF STATISTICS

We have seen already, in Section 1.4, that the repeated taking of a measurement produced a set of values which could be grouped and presented as a histogram. The presentation of the data thus collected and the drawing of conclusions as to the actual value of the quantity being measured fall in the realm of statistics, and in this section we look briefly at the application of the statistical method which will be pursued in the last three chapters of this book.

The first stage is the **collection of data.** Data is collected on a vast scale in every sphere of human activity: the national census, local and national surveys, university and college records – everywhere information is collected and analysed. The information can be of two kinds: *qualitative* (e.g. colour of people's eyes) or *quantitative* (e.g. heights of people) with a 'grey area' of information which is semi-quantitative (e.g. television programme ratings). It is necessary to take care in collecting the information. If a market survey is carried out, the organisers must ensure that the people asked are representative of the population as a whole in respect of the questions asked, and the choice and wording of the questions itself needs careful scrutiny. In this context, the statistician uses the term **population** to mean the *set of all possible measurements.* This may mean all the people who might be concerned in some issue, for example, the electorate or the house-owners in an urban area, or it might mean the diameters of all bolts manufactured by a given process or the lengths of life of all light bulbs made by a particular factory. If we wish to make some estimate of the behaviour of a population it may be necessary or desirable (on grounds of expense, or difficulty otherwise) to take measurements from a **sample**, that is, a *subset of the population.* Opinion polls take samples of the electorate since it would be pointless, and costly, to interview the whole population; inspectors in factories cannot always test non-destructively their product and they test a sample of the output. The choice of size of a sample is discussed later in this section and the choice of the members of the sample is taken up in Section 11.6.

Having collected the data, the next stage is to present the information in such a way that patterns and trends are discernible. this constitutes **descriptive statistics**. The information can be summarised pictorially (see the next section) and this is sometimes sufficient by itself; more usually, graphs and charts serve as adjuncts to numerical summaries. From quantitative data, certain numbers can be calculated which help describe various aspects of the data; such numbers are known as **sample statistics**. An average may be calculated to give an idea of the *middle* value of a set of readings (see Section 11.4) and a measure of the spread of values about this average is often useful (see Section 11.5); for example, the controller of a process designed to produce 1 kg bags of some commodity will hope that the average weight in each bag will be over 1 kg but will not want the weights to vary too much since this will produce some seriously underweight bags (bad for sales) and some seriously overweight ones (bad for profits).

The final, and most important, aspect of statistical work is the drawing of inferences from the data: **inferential statistics.** It must be borne in mind that *one only uses statistics in this aspect if there is insufficient data to draw firm conclusions.* Consequently, with *every* conclusion that is stated there must be a *risk* quoted: the risk that the conclusion is wrong. Sometimes the conclusion may call for further investigation; sometimes the evidence of the data is so strong that action can be taken on that evidence alone; but one must emphasise that *no amount of evidence is absolutely conclusive* unless the sample taken consists of the whole population. Let us consider an example of this sampling procedure to see how it works.

Items are produced by a manufacturing process at the rate of 200 per hour. In order to monitor the product, a sample is to be taken at specified times and some statistic measured from the sample which will give a clue as to the behaviour of the population (of items manufactured). How often should the sample be taken? How big should the sample be? The question of timing revolves around how continuous a monitoring is desired. That of sample size is somewhat easier to handle. Clearly the larger the sample, the more representative it is likely to be; a sample of 1 could be quite atypical, but the collective information from a sample of 20 is likely to be near to the population behaviour.

In order to make any statement about the quality of the product we need to compare the sample statistic with an expected value of the corresponding **population parameter** which is calculated from a *model* (in the case of observational errors, this is the **Gaussian**, or **normal** model). There will, in general, be a discrepancy between the observed and expected values of this parameter and it is the role of statistics to aid us in deciding how much of this discrepancy can be attributed to the fluctuations between samples (i.e. we may possibly have got a sample containing the highest values – this *can* happen however good our sample technique) and how much can be attributed to the underlying difference in the values. The basis of the method is to assume there is **no** discrepancy and calculate the chance of the observed discrepancy occurring. It is then up to the control inspector to decide whether the chance is acceptable in the light of his experience. Should he decide that the chance is too small to be acceptable (and in advance he sets a *critical level* below which he decides the chance is unacceptably low) then he could be rejecting when he should be accepting, if the sample was atypical of the population; in this case, he can improve the situation by choosing a low critical level. If he does this, he increases the danger of accepting that which should be rejected. We see here that the statistical method is not a clear-cut *yes* or *no* approach. We can merely state risks of conclusions being wrong, and leave the decision to experienced workers in the appropriate field.

On the other hand, we can be told in advance the acceptable risk, and work within that.

In this chapter, we look at descriptive statistics and the idea of a random sample, in Chapter 12 we examine models of chance and in Chapter 13 we study sampling and its applications.

11.2 GRAPHICAL REPRESENTATION OF DATA

We now look at some methods of displaying data which can be classified in categories rather than grouping according to ranges of values. The first method is the **pie chart** which shows the relationship of each constituent to the total: if a category takes a third of the whole then the area of the corresponding sector is one third of that of the circle or 'pie'. In Figure 11.1 is shown the distribution of seats in the House of Commons, according to party, after the General Election of 1970.

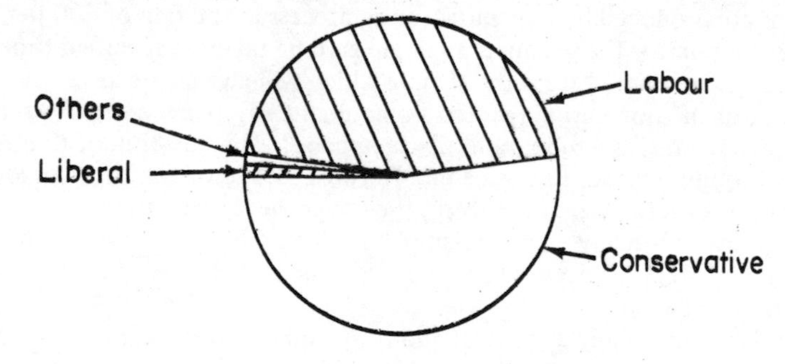

Figure 11.1

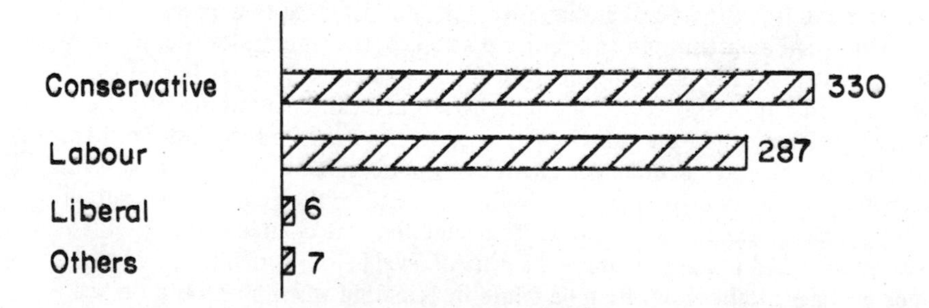

Figure 11.2

A second method is by means of the **bar chart** where the bars may be vertical or horizontal (if there are few categories and a wide range of values to be displayed). In Figure 11.2 the same data as for the pie chart is displayed, the widths of each bar are equal and the length is proportional to frequency.

Sometimes the bar chart may be modified to accommodate subdivisions. If the MPs in each party are split according to sex we obtain Figure 11.3.

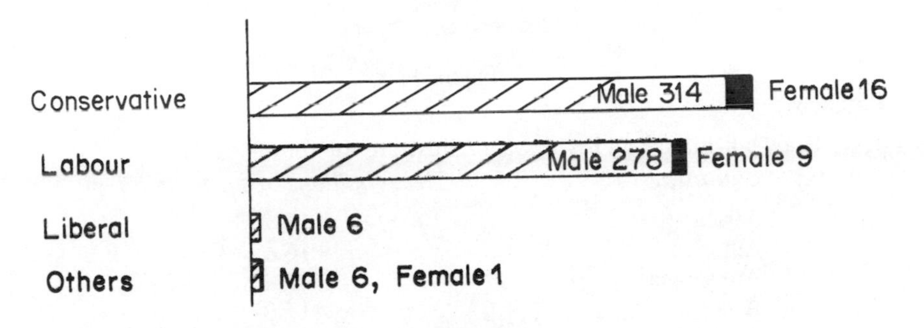

Figure 11.3

Next we mention the **pictogram** or **ideogram**. Here, to achieve greater visual impact, units in each category are depicted in a way which reflects the subject matter. In Figure 11.4 we show an example where the number of cars produced by a certain company is depicted.

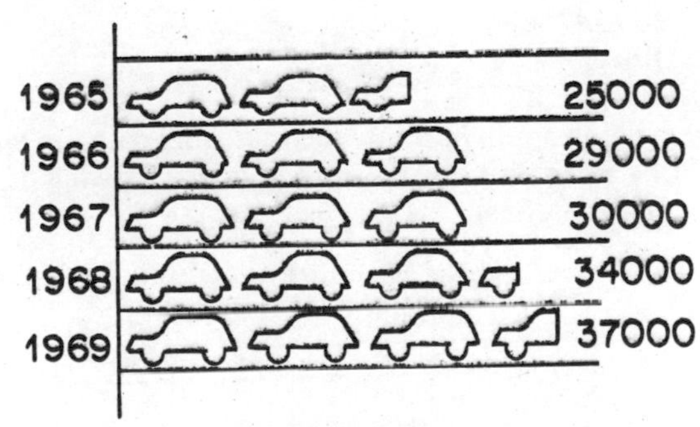

Figure 11.4

Finally, we can depict the changing situation of some quantity over a period of time by a **time series** diagram. An example is shown in Figure 11.5. In this way we can easily see any seasonal trends.

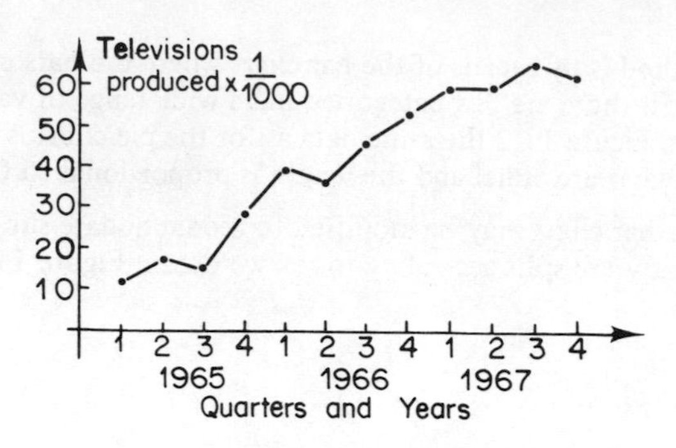

Figure 11.5

Problems

1. Draw pie charts and bar charts to illustrate the following data:

(a)

Continent	*Area (millions of square miles)*
Africa	11.7
America	16.3
Antarctica	5.3
Asia	17.0
Australia	3.0
Europe	3.9

(b)

Religion	*Followers (millions)*
Buddhism	269
Confucianism	280
Eastern Orthodox	220
Hinduism	300
Muslim	400
Protestant	210
Roman Catholic	420
Others	160

2. Draw a pictogram to illustrate the following data for the weight of fish caught by a certain fishing boat. (Take one drawing of a fish to represent 5000 tons).

Year	*Weight of fish caught (tons)*
1965	20,000
1966	25,700
1967	25,500
1968	30,000
1969	35,000
1970	33,000
1971	30,000
1972	40,000

3. Draw a time series graph for the data of Problem 2.

4. Draw a time series graph for the following data obtained from a weather station.

Year		1965	1966	1967	1968	1969	1970
	1	40	36	42	38	43	47
Rainfall	2	20	26	21	19	14	13
for quarters	3	15	14	23	15	17	12
(in inches)	4	30	40	62	31	37	33

11.3 FREQUENCY DISTRIBUTIONS

We now turn to situations where the data can be classified according to numerical values. The first task is to sort the data into the different values which can be assumed and calculate the **frequencies** of each value (the number of times it has occurred). First we give two examples of **discrete** variables (where the variable assumes only certain specified values). We have summarised the information in a frequency table in each case.

Example 1

A year's daily testing for faults in the sheath of an insulated cable revealed different numbers of faults on different days. The number of faults is a discrete variable.

Faults/day	0	1	2	3	4	5	6	7
Number of days	24	51	87	82	70	31	16	4

Example 2

The number of α-particles emitted in each second over a period of one hour was counted.

Number in a second	0	1	2	3	4	5	6	7	8	9 or more
Frequency	53	207	385	523	538	402	279	133	41	37

We can display each set of data by a line diagram as illustrated in Figures 11.6 (i) and (ii) on the next page.

In certain cases the variable is **continuous**; for example, the time taken to perform a given task; since the variable can only be measured to a given accuracy however, this gives it the appearance of being discrete. Since it would be senseless to record every observation even to the accuracy possible, it is customary to **group** the data and count the frequency of observations falling in each group: these are sometimes loosely called the frequencies of each group.

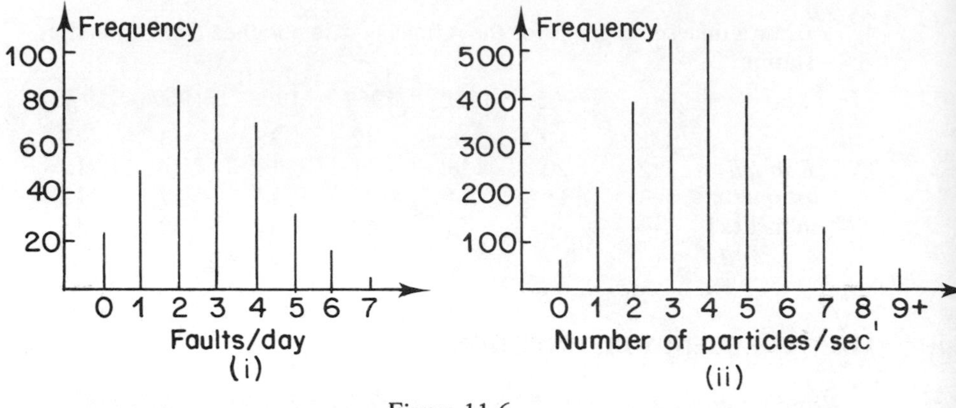

Figure 11.6

For both discrete and continuous variables the definitions of each group and their associated frequencies form the **frequency distribution.**

Example 3

A sample of 80 ball bearings was taken from a machine's output and the diameters measured to 0.001 mm. The results are shown in Table 11.1

Table 11.1

4.350	4.366	4.374	4.381	4.375	4.382	4.358	4.366	4.396	4.351
4.366	4.374	4.375	4.360	4.367	4.376	4.381	4.396	4.351	4.366
4.383	4.361	4.369	4.377	4.374	4.382	4.354	4.366	4.383	4.377
4.369	4.361	4.374	4.382	4.354	4.366	4.369	4.361	4.378	4.383
4.384	4.378	4.370	4.362	4.362	4.365	4.373	4.388	4.365	4.373
4.380	4.390	4.363	4.372	4.373	4.387	4.384	4.362	4.372	4.378
4.379	4.372	4.379	4.364	4.384	4.389	4.373	4.362	4.385	4.364
4.379	4.373	4.372	4.379	4.384	4.385	4.385	4.365	4.372	4.379

In Table 11.2 we first group the data and extend the end points of the groups to make them match at the end points even though the extensions contain no observations.

We can represent the data graphically by a **histogram** in Figure 11.7 – where the centre of each bar is the mid-point of each group; notice that the variable concerned is continuous, even though the measurements quoted give an impression of its being discrete.

Table 11.2

GROUP			RELATIVE FREQUENCY
end points included	*extended end points*		*2 d.p.*
4.350 – 4.354	4.3495 – 4.3545	5	.06
4.355 – 4.359	4.3545 – 4.3595	1	.01
4.360 – 4.364	4.3595 – 4.3645	11	.14
4.365 – 4.369	4.3645 – 4.3695	13	.16
4.370 – 4.374	4.3695 – 4.3745	15	.19
4.375 – 4.379	4.3745 – 4.3795	13	.16
4.380 – 4.384	4.3795 – 4.3845	13	.16
4.385 – 4.389	4.3845 – 4.3895	6	.07(5)
4.390 – 4.394	4.3895 – 4.3945	1	.01
4.395 – 4.399	4.3945 – 4.3995	2	.02(5)
		80	

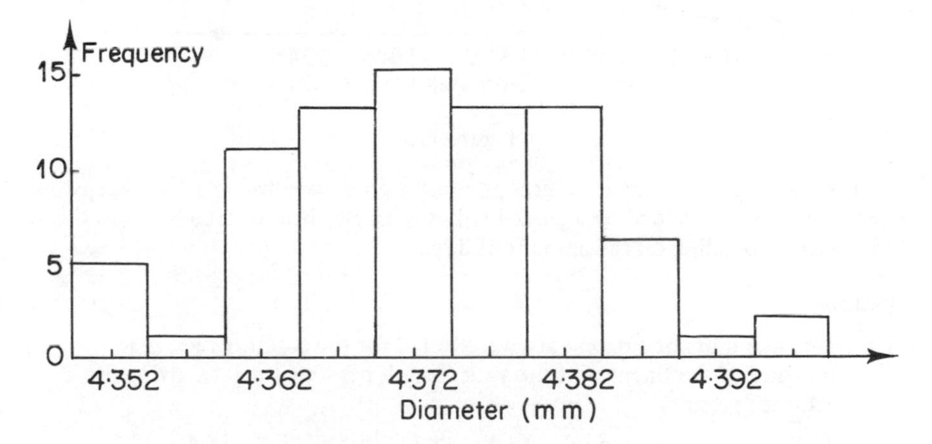

Figure 11.7

You should notice that in grouping the data, some detail has been sacrificed in order to achieve some clear picture of the behaviour of the data.

Two other terms can usefully be introduced here. First, the concept of **relative frequency**: this is simply the frequency of a certain variable or of a group divided by the overall total (see Table 11.2). Notice that the total relative frequency (which should be 1) is 0.99, due to rounding error. It is customary to denote the value taken by a discrete variable by x_i (this may refer to the group as we shall see in the next section) and its associated frequency by f_i. If there are N observations in all, the relative frequencies are f_i/N. Since

$$\sum_{i=1}^{n} f_i \equiv f_1 + f_2 + f_3 + \ldots\ldots f_n = N, \quad \sum_{i=1}^{n} \frac{f_i}{N} = \frac{N}{N} = 1$$

The second concept is that of **cumulative frequency distribution.** For this we accumulate frequencies so that we find the frequency of all values up to and including the one under consideration. In Example 3 the cumulative frequencies are 5, 6, 17, 30, 45, 58, 71, 77, 78, 80. These can be plotted on a **cumulative frequency diagram** or **ogive** and this is shown in Figure 11.8.

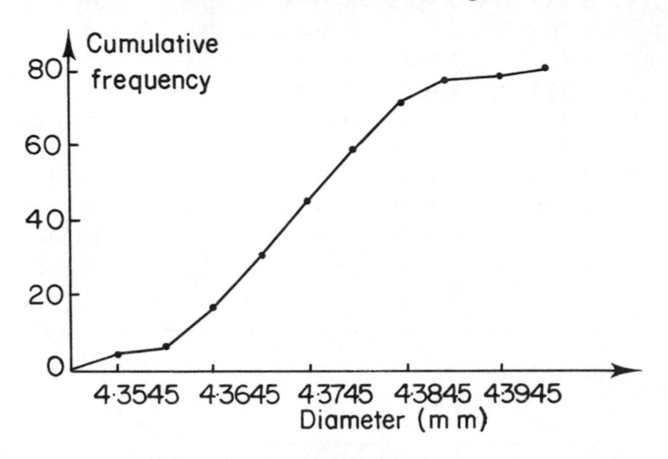

Figure 11.8

The choice of the number of groups and the group widths is partly a subjective matter, but can be helped by a guide-line; it is usually best to take between 6 and 15 groups, depending on the amount of data.

Problems

1. Separate units of cotton yarn were tested for the count, which is a measure of the fineness of the yarn (length per unit weight), giving the following results:

35.0	39.6	38.1	37.4	37.4	38.4	36.5	38.4	38.4	37.2
36.6	35.1	39.6	38.2	38.2	37.8	37.3	36.2	38.9	37.9
37.4	36.6	35.1	35.4	35.4	37.0	38.0	37.2	37.3	38.4
38.1	37.4	36.6	36.6	36.6	36.2	39.0	37.8	36.2	38.5
37.5	37.5	38.3	38.3	36.9	36.2	36.3	37.9	38.5	38.6
38.2	36.0	36.1	37.7	36.1	36.5	37.2	37.2	36.4	36.5
35.8	36.7	36.9	36.9	37.8	37.3	37.3	37.9	37.9	37.2
36.6	37.6	37.7	36.1	38.3	38.8	38.7	36.4	37.3	37.9

Draw up a frequency table taking groups 35 – 35.4, 35.5 – 35.9, ... 39.5 – 39.9

Draw a histogram, using the true end points 34.95 – 35.45, 35.45 – 35.95, ..., 39.45 – 39.95

Repeat for the grouping 35.0 to 35.9, 36.0 to 36.9, etc. Has this grouping lost any information or altered the histogram? Comment on your results.

2. For each of the tables of data given below
   (i) determine the greatest and least observation
   (ii) decide on a suitable number of groups
   (iii) form a frequency table
   (iv) draw a histogram and a cumulative frequency diagram.

   (a) Lengths of screws (to nearest 1/1000 cm)

3.239	2.671	3.114	3.078	2.830	2.700	3.314	3.055
3.060	3.155	3.211	2.965	2.956	2.611	2.933	2.942
3.130	2.792	3.353	3.149	2.996	2.964	2.765	2.800
2.713	3.125	2.654	2.773	3.142	3.070	2.931	3.202
2.921	3.198	3.054	3.197	3.153	3.201	3.474	3.394
3.111	3.084	2.755	2.516	3.194	2.907	2.905	3.061
3.186	3.228	3.026	3.427	2.821	2.865	2.852	3.018
2.850	3.166	3.439	2.819	2.716	2.861	2.754	3.231
2.751	3.221	2.631	3.020	3.258	3.333	3.111	3.371
2.800	2.836	2.723	3.051	3.242	3.245	3.063	3.211

   (b) Weights of bags of sugar (kilogrammes)

1.48	1.54	1.48	1.51	1.44	1.77	1.68	1.53	1.33	1.51
1.43	1.59	1.55	1.51	1.63	1.36	1.52	1.21	1.75	1.68
1.52	1.66	1.25	1.49	1.55	1.40	1.58	1.66	1.32	1.35
1.50	1.34	1.61	1.42	1.64	1.67	1.50	1.67	1.52	1.30
1.31	1.45	1.62	1.35	1.61	1.63	1.43	1.45	1.51	1.62

11.4 MEASURES OF CENTRAL TENDENCY

Given a frequency table we wish to provide an indication of the 'middle' or 'average' value. In this section we examine three possible measures.

Arithmetic Mean

This is the 'layman's' average. The sum of all the values is divided by the total number of values. Suppose we had five observations x_1, x_2, x_3, x_4 and x_5; the **arithmetic mean** (often shortened to *mean*) is

$$\frac{1}{5}(x_1 + x_2 + x_3 + x_4 + x_5) \quad \text{or} \quad \frac{1}{5}\sum_{i=1}^{5} x_i$$

In general, the mean of n observations $x_1, x_2, \ldots\ldots, x_n$ is

$$\frac{1}{n}\sum_{i=1}^{n} x_i$$

and is written $\bar{x}$ if the observations are regarded as a sample or μ if regarded as a population. The sample mean $\bar{x}$ can be used as an estimate of the population mean μ. For this chapter we shall use $\bar{x}$. We have so far said nothing about the observations being distinct. Suppose the five observations were x_1, x_1, x_1, x_2 and x_2 $(x_1 \neq x_2)$; then the above principles would apply, but we can use the fact that if the frequency of x_1 is f_1 (= 3 in this case) and that of x_2 is f_2 (= 2), then

$$x_1 + x_1 + x_1 + x_2 + x_2 = f_1x_1 + f_2x_2$$

and we have

$$\bar{x} = \frac{1}{5}\sum_{r=1}^{2} f_rx_r$$

We say $\bar{x}$ is the **weighted mean** of x_1 and x_2 with weights $f_1/5$ and $f_2/5$.

In general, if n observations fall into distinct values $x_1, x_2, \ldots\ldots, x_s$ with frequencies $f_1, f_2, \ldots\ldots, f_s$ then, since

$$\sum_{r=1}^{s} f_r = n$$

we have

$$\bar{x} = \frac{1}{n}\sum_{r=1}^{s} f_rx_r = \sum_{r=1}^{s} f_rx_r \Big/ \sum_{r=1}^{s} f_r \tag{11.1}$$

Note that

$$\sum_{i=1}^{n}(x_i - \bar{x}) = \sum_{i=1}^{n} x_i - \sum_{i=1}^{n} \bar{x} = n\bar{x} - \bar{x}\,.\,n = 0$$

and this property, that the sum of discrepancies from the mean is zero, may be used as a check in calculations of the mean. (c.f. the physical analogue of centre of mass.) The mean is best used to describe, for example, the typical tensile strength of a material, the typical diameter of washers produced by a machine or the typical life of a component.

Mode

The **mode** is that value which occurs most frequently. If the data is grouped we speak of the **modal group.**

In the three examples of Section 11.3 the modes of the first two are respectively 2 faults/day and 4 α-particles a second, and the modal group of the third is 4.370 – 4.374.

If a distribution has one mode it is called **unimodal**; if two, it is **bimodal**. In considering the donations to a retirement present, or the wages of members in a firm it may be more relevant to ask which value occurs most frequently rather than a mean value. In the latter case the use of the mean would take directors' salaries into account and one such salary might be equivalent to the wages of twenty factory hands. The mode is also useful for non-quantitative variables.

Median

First, the values of the variable are arranged in increasing or decreasing order; if the number of values is odd, the median is the middle value, whereas if the number of values is even it is the mean of the two middle values. For example, the median of [3,3,4,12,21] is 4 and of [3,3,4,12,21,25] is 8 (which is ½[4 + 12]).

The median divides the values in the population into two equal, or almost equal, parts. Half the values are less than or equal to the median. Note that the value 25 above could have been altered to 135 without changing the median, whereas the mean would have altered. In the example on ball-bearings there are 80 observations, both the 40th and 41st occurring in the group 4.370 – 4.374 and this is the **median group**.

Choice of the appropriate measure

Where the choice of appropriate measure of central tendency is not obvious the following comments may help. The mean is most readily understood and lends itself more readily to further statistical analysis; further, it takes all the values into account. It has the disadvantage of being affected by extreme values and where these occur the median may be a more useful measure. If these measures cannot readily be applied the mode can come into play.

Relationships between the measures

If a distribution is symmetrical then the mean, median and mode coincide; if the distribution is not symmetrical, but not too heavily *skewed*, then we have the approximate relationship (mean–mode) $\hat{=}$ 3 (mean–median) – see Figure 11.9.

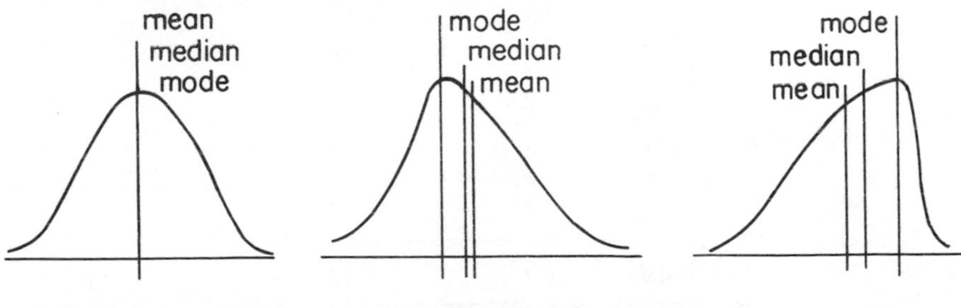

Figure 11.9

Calculation of the mean

We return to the ball-bearing example and calculate the mean of the values which are grouped.

We could tabulate f_i and $f_i x_i$, compute their totals and then calculate $\bar{x}$ as

$$\sum_{i=1}^{n} f_i x_i \Big/ \sum_{i=1}^{n} f_i$$

As a check we could compute

$$\sum_{i=1}^{n} f_i(x_i - \bar{x})$$

and this should be zero; in practice it may not be,because of the grouping of the values.

However, with grouped data, it is always advisable to **code** or transform the variable x_i into a new variable u_i. This transformation has the effect of simplifying the arithmetic and comprises two stages: First we guess or assume a mean somewhere in the middle of the values of x_i; in the following calculations resulting in Table 11.3 we guess the mean of the ball-bearing diameters to be 4.372, one of the group central values. In the second stage we divide $(x_i - 4.372)$ by the group width 0.005 to obtain the new variable

$$u_i = \frac{(x_i - 4.372)}{0.005}$$

Note that some of the values of u_i are negative.

Table 11.3

x_i	f_i	u_i	$f_i u_i$
4.352	5	−4	−20
4.357	1	−3	−3
4.362	11	−2	−22
4.367	13	−1	−13
4.372	15	0	0
4.377	13	1	13
4.382	13	2	26
4.387	6	3	18
4.392	1	4	4
4.397	2	5	10
Total	80		13

The values of u_i are also integral, for example if $x_i = 4.352$ then

$$u_i = \frac{4.352 - 4.372}{0.005} = -\frac{0.020}{0.005} = -4$$

We have chosen a new origin for the measurements and scaled their values. The calculations now proceed as in Table 11.3.

Now

$$\bar{u} = \frac{1}{n}\sum f_i u_i = 13/80 = 0.1625$$

But

$$\bar{u} = \frac{1}{n}\left(\sum f_i (x_i - 4.372)/0.005\right)$$

$$= \frac{1}{0.005}\,\frac{1}{n}\left(\sum f_i x_i - n\,.\,4.372\right)$$

$$= \frac{1}{0.005}\,[\bar{x} - 4.372]$$

therefore

$$\bar{x} = 0.005\bar{u} + 4.372 = 4.3728 \quad \text{(4 d.p.)}$$

The same result has been obtained with much simpler arithmetic and this second approach would often give a more accurate result. In general if the **coded variable** $u = (x - a)/b$ then

$$\bar{u} = (\bar{x} - a)/b \quad \text{or} \quad \bar{x} = b\bar{u} + a \qquad (11.2)$$

Problems

1. (i) The number of members and means of 3 distributions are as follows:

Number of members	280	350	170
Mean	45	54	50

Find the mean of the combined distribution.

(ii) A candidate's examination marks in three subjects were 86%, 47% and 63%; if these subjects were weighted in the ratio 1 : 3 : 2 find his weighted mean percentage mark.

2. The numbers of rejects from articles produced by a machine in 36 consecutive periods were:

3, 1, 2, 4, 0, 1, 2, 2, 3, 2, 1, 2, 5, 0, 1, 2, 1, 3, 2, 3, 4, 0, 1, 1, 2, 2, 3, 0, 3, 1, 2, 0, 1, 4, 3,2.

Draw up a frequency table showing the numbers of periods with 0, 1, 2, 3, 4 and 5 rejects. Draw up a cumulative frequency table.

Find (i) the median

(ii) the mode
and (iii) the arithmetic mean.

3. For each of the problems on Section 11.3 find the mean, mode and median.

4. The following table gives the distribution for the bursting strength of 225 samples of vinyl coated nylon.

Bursting strength (lb)	*Number of samples which burst in stated range*
61 – 70	2
71 – 80	8
81 – 90	17
91 – 100	28
101 – 110	36
111 – 120	40
121 – 130	34
131 – 140	29
141 – 150	19
151 – 160	10
161 – 170	2

Draw a histogram to represent the distribution.
Taking 115.5 as a new origin and the group width as a unit, calculate the arithmetic mean. Construct a cumulative frequency table.

5. The mean of 13 numbers is 11 and the mean of 23 other numbers is 7. Find the mean of the 36 numbers taken together.

11.5 MEASURES OF VARIABILITY

A manufacturer of bolts would hope for two things with regard to their diameter: a mean close to the desired value and not too wide a spread of values about that mean. In this section we look at measures of spread about the mean.

Range

This is simply the quantity (largest value – smallest value). For Example 1 of Section 11.3 the range is $(7 - 0) = 7$. For Example 3 it is $(4.396 - 4.350) = 0.046$.

Inter-Quartile Range

We can find three values which divide the ordered values of the variable into four groups each having the same frequency. The central value is then the median and the remaining two are the **first** and **third quartiles**. The **inter-quartile range** is (3rd quartile − 1st quartile). It has the advantage over the range that it measures a more typical spread since it does not rely on extreme values only. It should be noted that the inter-quartile range contains *half* the total frequency.

Mean Deviation

To take all the values into account we measure

$$\frac{1}{n}\sum_{i=1}^{n} |x_i - \bar{x}| \quad \text{or} \quad \frac{1}{n}\sum_{r=1}^{s} f_r\, |x_r - \bar{x}|$$

The modulus sign causes difficulty in manipulation, however.

Variance

To overcome that difficulty, we use the variance which is the mean squared deviation about the mean.

Variance is written

$$\text{var}(x) \quad \text{or} \quad V(x)$$

For a population of size n

$$V(x) = \frac{1}{n}\sum_{r=1}^{s} f_r(x_r - \mu)^2$$

This has the disadvantage that extremes are emphasised. For example, if we take the numbers 4, +1, 0, −1, the mean is 1, and

$$V(x) = \frac{1}{4}[(4-1)^2 + (1-1)^2 + (0-1)^2 + (-1-1)^2]$$

$$= \frac{1}{4}[9 + 0 + 1 + 4]$$

$$= 3.5$$

Now if the reading 4 had been in error it should have been ignored, yet it completely dominates the variance. We must emphasise that the variance is a particularly useful measure of variability.

Warning. There is a distinction to be drawn here between **population variance** σ^2 and **sample variance** s^2.

When we are dealing with n observations regarded as a sample we calculate

$$s^2 = \frac{1}{n-1}\sum_{i=1}^{n} f_i(x_i - \bar{x})^2$$

The reason for division by $(n - 1)$ is connected with the estimation of the population variance σ^2 by the sample variance s^2 (see Section 13.5)

For large samples, division by $(n-1)$ may be approximated by division by n. You should, however, make it quite clear whether you are treating the numbers in an example as a population or as a sample and find the appropriate variance.

Standard Deviation

In order to achieve a parity between the units of spread and the units of the values we take standard deviation = s.d. (or σ or s) as $\sqrt{\text{variance}}$. If the numbers of the previous example were lengths in cm we would have

$$\sigma = \sqrt{3.5} = 1.87\text{cm} \quad (3 \text{ s.f.})$$

It should be clear that the bigger σ, the larger the spread of values. Its physical analogy is radius of gyration.

We deduce a useful equivalent form for $V(x)$ which saves many calculations and sometimes yields a more accurate result.

We deal first with a population.

Now

$$nV(x) = \sum_{i=1}^{n} (x_i - \mu)^2$$

$$= \sum_{i=1}^{n} (x_i^2 - 2\mu x_i + \mu^2)$$

$$= \sum_{i=1}^{n} x_i^2 - \sum_{i=1}^{n} 2\mu x_i + \sum_{i=1}^{n} \mu^2$$

$$= \sum_{i=1}^{n} x_i^2 - 2\mu \sum_{i=1}^{n} x_i + n\mu^2$$

$$= \sum_{i=1}^{n} x_i^2 - 2\mu \,.\, n\mu + n\mu^2$$

$$= \sum_{i=1}^{n} x_i^2 - n\mu^2$$

therefore

$$V(x) = \left(\frac{1}{n} \sum_{i=1}^{n} x_i^2\right) - \mu^2 \qquad (11.3)$$

(Mean of the square − square of the mean)

The result also holds for grouped data.

For a sample, the result is

$$V(x) = \frac{1}{n-1}\sum_{i=1}^{n} x_i^2 - \frac{n}{n-1}\bar{x}^2 \tag{11.3a}$$

Example

We return to Example 3 of Section 11.3 We shall treat the observations as a sample and use the formula for variance given at the bottom of page 637. Then we have

$$V(x) = \frac{1}{79} \cdot 8.225 \times 10^{-3} = 1.041 \times 10^{-4}$$

and so

$$\sigma = 1.020 \times 10^{-2} = 0.010 \quad (3 \text{ d.p.})$$

(Notice we quote σ to as many d.p. as the original observations.)

If we use (11.3a) we have

$$V(x) = \frac{1}{79}\left(1529.727\right) - \left(\frac{349.825}{80}\right)^2 \cdot \frac{80}{79}$$

Working with a desk calculator with 8 digit capacity gave

$$V(x) = 19.363632 - 19.363532 = 0.000100$$

The rearrangement

$$\frac{1529.727 - (349.825)^2/80}{79}$$

gave

$$V(x) = 0.0001041 \quad (\text{more precision})$$

which yielded

$$\sigma = 0.0102 \quad (4 \text{ d.p.})$$

It is always advisable to use coded variables. It can be shown that for

$$u = \left(\frac{x-a}{b}\right)$$

the population variance

$$V(u) = \frac{1}{n}\sum_{i=1}^{n}(u_i - \bar{u})^2 = \frac{1}{n}\sum_{i=1}^{n}\left(\frac{x_i - a}{b} - \frac{\bar{x} - a}{b}\right)^2$$

$$= \frac{1}{n}\sum_{i=1}^{n}\left(\frac{x_i - \bar{x}}{b}\right)^2$$

$$= \frac{1}{b^2}\frac{1}{n}\sum(x_i - \bar{x})^2$$

$$= \frac{1}{b^2}V(x)$$

The same result holds for sample variance [despite division by $(n - 1)$ instead of n] and we summarise to say that if $u = (x - a)/b$

$$V(u) = \frac{1}{b^2}V(x) \tag{11.4}$$

We shall find $V(u)$ as

$$\frac{1}{n-1}\sum f_i u_i^2 - \bar{u}^2\left(\frac{n}{n-1}\right)$$

We tabulate the necessary information in Table 11.4; the coding is as before, ($a = 4.372$, $b = 0.005$).

Table 11.4

Group central value x_i	f_i	u_i	$f_i u_i$	$f_i u_i^2$
4.352	5	−4	−20	80
4.357	1	−3	−3	9
4.362	11	−2	−22	44
4.367	13	−1	−13	13
4.372	15	0	0	0
4.377	13	1	13	13
4.382	13	2	26	52
4.387	6	3	18	54
4.392	1	4	4	16
4.397	2	5	10	50
Totals	80		13	331

Then

$$V(u) = \frac{1}{79} \times 331 - \left(\frac{13}{80}\right)^2 \cdot \frac{80}{79}$$

$$= \frac{331 - 169/80}{79}$$

$$= 4.1631 \quad \text{(4 d.p.)}$$

Hence

$$V(x) = b^2 V(u) = (0.005)^2 \times 4.1631$$

and so

$$s = 0.005 \times \sqrt{4.1631} = 0.0102 \quad (4 \text{ d.p.})$$

We can see that we have retained accuracy by a modified approach; relationship (11.3) can be used if a calculating aid with good precision is available, but in general coding the variables is preferable.

Problems

1. The following table shows 80 measurements of the iron solution index of tin plate samples, used to estimate the corrosion resistance of tin plated steel.

Interval	0-.20	.21-.40	.41-.60	.61-.80	.81-1.00	1.01-1.30	1.31-1.60	1.61-2.00
Frequency	2	2	12	26	15	14	7	2

(a) Construct a suitable frequency diagram to illustrate the data.
(b) Draw a cumulative frequency diagram and estimate the percentage of the indices which exceed 1.25.
(c) Estimate the median and the inter-quartile range for the index.

2. During a piece of work study the normalised time for a certain operation was recorded on 46 occasions and these values were subgrouped as follows:

t (secs)	9.0	9.5	10.0	10.5	11.0	11.5	12.0	12.5	13.0
f	2	5	9	11	8	6	3	1	1

By introducing a suitable coded variable for the time, calculate the mean value, the mean deviation and the standard deviation.

3. The breaking load in N of 200 specimens of a certain material was measured and the results arranged in the following grouped frequency distribution.

Breaking load in N *(mid-interval values)*	*Number of Specimens*
982.5	3
987.5	10
992.5	27
997.5	62
1002.5	56
1007.5	24
1012.5	12
1017.5	5
1022.5	1

Determine the mean and standard deviation of this distribution. State the modal group and the median group. Find also the mean deviation and inter-quartile range.

4. The breaking strengths in N/m^2 of 100 metal specimens are given in the following table, with a class interval of 0.2 N/m^2.

Centre of interval N/m^2	32.0	32.2	32.4	32.6	32.8	33.0	33.2	33.4	33.6	33.8
No. of specimens	1	4	11	13	19	26	12	9	3	2

Using the method of taking an assumed mean, calculate the mean and standard deviation.
Define the median of a frequency distribution and using a graphical method (or otherwise) calculate it for the above distribution.

5. A set of 25 observations gave a mean of 56 and standard deviation 2. It was subsequently discovered that one observation recorded as 64 was very inaccurate. Find the mean and standard deviation of the remaining 24 observations when the faulty one is discarded.

6. If $\bar{x}$ is the arithmetic mean of n values of x, show that

$$\sum (x - \bar{x})^2 = \sum x^2 - \frac{(\Sigma x)^2}{n}$$

In a traffic count, the number of vehicles passing a check point in a two-minute interval was recorded for 63 such intervals giving the following results:

No. of vehicles passing in 2 min. interval	0	1	2	3	4	5	6	7	8
No. of intervals	1	11	12	11	10	9	4	2	3

Find the mean and standard deviation of this distribution.

7. 100 readings recorded in a certain experiment had a mean value of 0.913 and standard deviation of 0.156. If the following further 10 readings are then obtained, what would be the mean and standard deviation of the complete set of results?

0.875 0.898 0.912 0.913 0.917 0.920 0.923 0.924 0.928 0.937

8. The following frequency table records the diameter of 200 grains of moulding sand (in appropriate units)

Diameter	8	10	12	14	16	18	20
Frequency	25	40	67	35	23	7	3

State the modal group and median group and draw a histogram for the distribution. Choose a suitable assumed mean and hence calculate the mean and standard deviation, σ, of the distribution. The standard deviation calculated from the original measurements (before grouping) was 2.88. Why is there a discrepancy between this and your value for σ?

(C.E.I.)

9. The times taken by 35 people to perform a particular calculation were as follows (given in minutes and seconds)

2.48	2.00	2.34	1.44	1.58	2.27	2.25
2.24	1.57	1.41	2.32	3.13	3.39	1.41
1.11	2.00	2.50	3.27	1.04	1.25	1.55
4.24	1.26	2.32	2.43	3.11	2.14	3.52
1.36	1.47	3.21	3.04	2.30	1.43	1.05

Group the data into a frequency table, draw a histogram and calculate the sample mean, sample standard deviation, median group, modal group, mean deviation and inter-quartile range. Justify your choice of grouping and of coding the variable.

10. A quantity sometimes calculated is the **coefficient of variation** which is defined to be the ratio of the standard deviation to the mean. Find the coefficient of variation for Problems 2, 3, 4 and 6.

11. Grouping data leads to inaccuracies in the calculation of mean and standard deviation. **Sheppard's correction** partially compensates for the error in the standard deviation. It may be stated:

$$\text{corrected variance} = \text{calculated variance} - \frac{1}{12}(\text{group width})^2$$

Apply this to the data of Problem 2(b), page 631, and check its effect by calculating the variance of the original 50 observations and the variance of the data when grouped.

11.6 RANDOM SAMPLES

We have already remarked that a sample is a subset of a population and that to make some statement about the population it is necessary to infer from the behaviour of a sample. We hope that the estimate from a sample is not too far from the truth and that the average of the estimate from all possible samples is the value we seek. It is little use choosing a sample of the electorate in a particular constituency by choosing names from a telephone directory since this sample would reflect the better-paid members of the electorate.

We seek a technique of sampling which will give each member of the population an equal chance of selection. Put another way, if we fix on a sample size N from a given population there are many such possible samples; we aim to ensure that each possible sample has an equal chance of being chosen. We call a sample so chosen a **simple random sample.** If we wished to select 16 people from a population of 160, we could write each of the 160 names on a separate slip of paper, mix the slips in a container and draw 16 of them. This is a somewhat tedious process.

Alternatively we can make use of a table of random numbers. Random numbers are the digits 0, 1, 2, ..., 9 arranged, with repetitions, in no particular order. These digits can be produced by a computer in such a way that the sequence repeats only after several millions. A sample of such digits is shown in Table 11.5.

We now give an illustration of how to use the table.

Example

Explain how to choose a random sample of 4 from populations of 10, 7, 25, 155 respectively.

The first task is always to label the members of the population serially.

(i) For a population of size 10 we use single digits 0, 1, ..., 9 (0 corresponds to 10) and take a sequence of digits from the table starting anywhere and continue until we have met four different ones. Suppose we take the block 17512, we ignore the second 1 and obtain 1, 7, 5, 2; the items thus labelled will be the ones chosen.

(ii) For a population of size 7, the easiest way is to ignore the digits 0, 8 and 9 and repeat the above. If the sequence is 73570, we have only obtained 7, 3 and 5 and we take the next block 86860 and pick up 6 to complete the sample.

(iii) For a population of size 25 we could take digits in pairs and ignore all combinations 00 and above 26 (there are 100 combinations: 00 to 99); this would waste a possible three-quarters of the digits and, if we wanted a larger sample, could prove time-consuming. A useful alternative is to divide 25 into 100 and note we have an exact result of 4. Then, in taking pairs of digits, divide by 25 and take the remainder: the result is the member of the population chosen. This second approach can be illustrated by the block 91 30 76 42 20; remainders after division by 25 are: 16, 5, 1, 17, 20 and we take the members 16, 5, 1 and 17.

(iv) For a population of size 155 we need triples of digits. The second method above now becomes: divide 1000 by 155 to get 6+; now $6 \times 155 = 930$ and so the combinations 000 and above 930 are ignored. The triples of random numbers are divided by 155 and the remainders taken. Thus the blocks 235 954 329 410 479 give 80, ignore, 19, 100 and 14 as the sample.

Finally we mention that it may be desirable to use a different form of sampling. If it is wished to divide a population according to strata, depending on some feature, e.g. geographical area or income group, we take simple random samples of appropriate size from each stratum to give an overall **stratified random sample.** On the other hand, practical sampling may be too difficult, in which case, possibly with the aid of a digital computer, samples may be theoretically chosen by a technique known as **simulated sampling.**

Problems

1. Choose a random sample of size 8 from the data (a) of Problem 2 on page 631. Repeat for sample sizes 6, 14.

2. Repeat Problem 1 for the data (b) of Problem 2, page 631.

3. Arrange the numbers 1 to 8 in random order.

4. Arrange the numbers 1 to 16 in random order.

Table 11.5

A thousand random numbers

17512	73570	86860	91307	64220	23595	43294	10479
76841	09058	01305	60495	13421	71688	04120	80918
11052	32848	14058	88001	94641	70167	40104	35255
34311	42935	36458	04201	71573	37722	58698	46115
54641	26072	04705	27077	34834	14491	53407	22248
45749	23937	57052	53045	02583	30298	59306	50144
63243	59906	74883	31145	20350	47412	35309	02287
21051	96604	33444	52746	11929	77340	95053	84498
95766	17077	96760	96507	57473	22620	30675	76773
38800	29448	56232	61173	91526	86160	97255	79578
68736	08852	78657	91294	84045	76828	49909	80634
13607	36975	76285	95314	19047	07958	77110	95166
07671	34747	67528	90777	61004	04959	83438	57088
40736	06846	73412	59487	02897	09274	46440	13225
51282	73638	10025	54990	29162	38279	13792	09391
26501	43588	81906	69802	68634	16651	11125	77249
01557	13374	29465	27171	15987	59264	37949	03338
32264	32023	37468	04735	13468	21383	36507	77813
46169	29950	93968	48856	37585	86315	05745	76432
69995	63424	18966	12377	17669	07622	74232	23604
62212	31191	77839	72307	55308	54250	39561	25338
16413	40686	34371	52671	28770	36396	06696	28522
81999	34374	23814	38043	22069	95938	70678	64351
76141	69928	81966	03285	65663	10845	68747	29214
59349	34499	61480	12054	61602	09961	09195	28880

SUMMARY

You should now appreciate the scope and nature of the statistical method. The ideas of distribution, mean and variance must be understood thoroughly. The choice of the appropriate measures of central tendency and variability should be within your competence. Finally, the concept of sampling on a random basis is worth knowing.

Chapter Twelve

Discrete Probability Models

12.1 PROBABILITY MODELS

We first conduct a **statistical experiment** which will have one or more possible **outcomes** which can be specified in advance.

Some examples are:

(i)	tossing a coin with outcomes	head, tail
(ii)	throwing a die with outcomes	1, 2, 3, 4, 5, 6
(iii)	aiming at a target with outcomes	success, failure
(iv)	testing an electrical component	defective, non-defective

In each case, before the experiment is carried out, we are uncertain as to the outcome.

The set S of all possible outcomes of an experiment is called a **sample space** (or, sometimes, **outcome space**). The members of the set are called **sample points.** In the throwing of a die, the sample space is $S = \{1, 2, 3, 4, 5, 6\}$ and typical sample points are 2 and 5. Likewise in tossing a coin the sample space is $S = \{\text{head, tail}\}$. We take two much-used examples to consolidate ideas.

Example 1

A pair of dice are thrown and the total score observed. There are 36 possible outcomes as shown in Table 12.1.

Table 12.1

Score on 1st die	Score on 2nd die					
	1	2	3	4	5	6
1	(1,1)	(1,2)	(1,3)	(1,4)	(1,5)	(1,6)
2	(2,1)	(2,2)	(2,3)	(2,4)	(2,5)	(2,6)
3	(3,1)	(3,2)	(3,3)	(3,4)	(3,5)	(3,6)
4	(4,1)	(4,2)	(4,3)	(4,4)	(4,5)	(4,6)
5	(5,1)	(5,2)	(5,3)	(5,4)	(5,5)	(5,6)
6	(6,1)	(6,2)	(6,3)	(6,4)	(6,5)	(6,6)

Typical sample points are (4,3), (2,5), (6,6).

Example 2

Three coins are tossed once each. The sample space of the experiment is $S = \{HHH, HHT, HTH, HTT, THH, THT, TTH, TTT\}$ where H and T represent Heads and Tails respectively. If we are interested in those outcomes where there are more heads than tails we see that there are four possible outcomes: *HHH, HHT, HTH, THH.* These comprise a subset of the sample space. Any subset of a sample space is called an **event**. The subset $\{TTT\}$ is the event that exactly three tails occur and the subset $\{HHT, HTH, THH\}$ is the event that exactly one tail occurs.

Combination of events

We may now revert to Example 1 and note that the event that the total score is 8 is the subset of outcomes $A = \{(2,6), (3,5), (4,4), (5,3), (6,2)\}$; the event that the second die scores a six is the subset $B = \{(1,6), (2,6), (3,6), (4,6), (5,6), (6,6)\}$. The event that the total score is 8 **or** the second die scores 6 is the subset $A \cup B = \{(2,6), (3,5), (4,4), (5,3), (6,2), (1,6), (3,6), (4,6), (5,6), (6,6)\}$; the event that the total score is 8 **and** the second die score 6 is the subset $A \cap B = \{(2,6)\}$.

Relevant to the sample space S of Example 1, the **complement** of the event that there are more heads than tails is found by removing the outcomes of the first event from the sample space. This yields the complementary event $\{HTT, THT, TTH, TTT\}$.

Further ideas

Suppose we toss a coin many times, then this experiment consists of several simple experiments. If we decide to toss the coin 10 times and count the number of heads then we should find that if this experiment were carried out on different occasions, the number of heads would vary. The number of heads is a **discrete random variable**; it is discrete because it can only take *specific* values and it is random because we cannot predict its value in advance. What we can hope to do is to estimate the chances of a particular value occurring and to do this we shall need to form a mathematical model. We loosely define probability of an event as a measure of the likelihood of its occurrence and define it more rigorously in the next section. The model is called a **probability model.** The sequence $\{H, T, H, H, H, T, T, \ldots\ldots\}$ is called a **random sequence.**

An example of a **continuous random variable** would be the angle made with a fixed direction by a spun pointer mounted on a support. The angle could take any value in the range $[0, 360^\circ)$ and cannot be predicted in advance; again, we must point out that the inability to measure with infinite precision does not destroy the theoretical continuous aspect.

In a more realistic situation, suppose we are trying to build a model of how an epidemic would behave if it broke out in a town. We might be interested, for example, in how long it would last or whether it would affect more than half the population. To do this we must make assumptions as to what are the chances of a person being affected in all circumstances. Perhaps we could run several computer simulations and test the model. In any case we need an understanding of probability and this we develop in the next section.

Problems

1. Represent by points on a graph the 36 possible results of throwing two dice. Identify the sets of points for which the sum of scores of the two dice is
   (a) equal to 9
   (b) greater than or equal to 7.

2. A card is drawn from a pack. The sample space will be the set of all possible outcomes. Identify on a diagram the subsets that give the events
   (a) the drawn card is either a club or a queen, or both
   (b) the drawn card is a card lower than the 4 (ace is high)
   (c) the drawn card is a card higher than the 3.

   Identify the complements of the subsets in (a), (b) and (c).

12.2 PROBABILITY OF SIMPLE AND COMPOUND EVENTS

There are two ways of looking at probability: as a long-range relative frequency or as a pre-assigned entity. In the example of the tossing of a coin we may assume that the probability of a head occurring on any one throw is ½ and then say, for example, in 10 trials we expect 5 heads, in 1000 we expect 500, etc. attributing departures from these numbers to chance fluctuations. Alternatively, we may observe the relative frequency of heads as we toss the coin 1 000 000 times and note that, after early fluctuations, this relative frequency settles down and if after 1000000 trials we observe 501211 heads then we could say that the probability of a head on each throw is 0.501211 and accept that this is a better guide than the value ½. We shall adopt the former approach for ease of model building.

Each sample point is assigned a weight or measure of the chance of it occurring – the **probability**. This is a number p where $0 \leqslant p \leqslant 1$ such that the sum of the probabilities of all points in the sample space is 1. An outcome which has assigned to it a probability of 1 is a **certainty** and one which has assigned to it a probability of 0 is an **impossibility**. Hence the smaller the probability of an outcome, the less likely it is to occur.

The probability of an event A, written $p(A)$, is the sum of the probabilities of all sample points in A.

We shall restrict ourselves to discrete finite sample spaces for our illustrations in this section. In many situations we assume that each outcome is equally likely and assign to them equal probabilities. Then $p(A)$ = (number of outcomes in A)/(total number of outcomes). We shall obtain several results in this section which do not rely on this assumption unless so stated.

Example 1

If four 'fair coins' are tossed once each, what is the probability of obtaining exactly two tails?

By a 'fair' coin we mean that on each throw $p(\text{head}) = \frac{1}{2} = p(\text{tail})$. The sample space consists of 16 equally likely outcomes (check this) of which the six favourable ones are *HHTT, HTHT, HTTH, THHT, THTH, TTHH* and so the probability we seek is

$$\frac{6}{16} = \frac{3}{8} \quad (\text{or is } \frac{1}{16} + \frac{1}{16} + \frac{1}{16} + \frac{1}{16} + \frac{1}{16} + \frac{1}{16})$$

Example 2

If two fair dice are rolled, what is the probability of a total score of 4?

The sample space has 36 equally likely outcomes and the probability of each is 1/36. The subset A = total score of 4 = $\{(1, 3), (2, 2), (3, 1)\}$ and so $p(A) = 3/36 = 1/12$.

We now quote some important rules. Here, $\emptyset$ is the null set and S is the sample space.

(i) $p(S) = 1$
(ii) $p(\emptyset) = 0$
(iii) $p(A') = 1 - p(A)$

(If there are 4 defective fuses in a box of 100 then the probability of a fuse selected at random being defective is 4/100 = 1/25 and hence the probability that it is not defective is 24/25.)

We now look at the probability of events compounded from simpler ones.

General rule of addition

We illustrate by example. A pack of 52 playing cards is shuffled and a card drawn out; what is the probability that this card is either an ace or a red card?

Now the sample space comprises 52 equally likely outcomes and the event A = *either an ace or a red card* is the union of two other events B = *card is an ace* and C = *card is red.* There are 4 sample points in B and 26 in C, but only 28 in $B \cup C$ since we do not count the Ace of Hearts and Ace of Diamonds twice, i.e. number of sample points in A = number of aces + number of red cards − number of red aces

$$= 4 + 26 - 2$$

Then

$$p(A) = \frac{4}{52} + \frac{26}{52} - \frac{2}{52} = \frac{28}{52}$$

The event 'card is a red ace' is $B \cap C$.

In general, if B and C are events in S,

$$p(B \cup C) = p(B) + p(C) - p(B \cap C) \tag{12.1}$$

Example

The probability that a man watches TV in any one evening is 0.6; the probability that he listens to the radio is 0.3 and the probability that he does both is 0.15. What is the probability that he does neither?

Let V be the event *he watches TV*, R be the event *he listens to radio,* then

$$\begin{aligned} p(V \cup R) &= p(V) + p(R) - p(V \cap R) \\ &= 0.6 + 0.3 - 0.15 \\ &= 0.75 \end{aligned}$$

therefore

$$p(\text{he does neither}) = 1 - p(V \cup R) = 0.25$$

We show the results on the Venn Diagram (Figure 12.1). Note that the probability that the man listens to radio but doesn't watch TV is 0.15 and the probability that he watches TV but doesn't listen to the radio is 0.45.

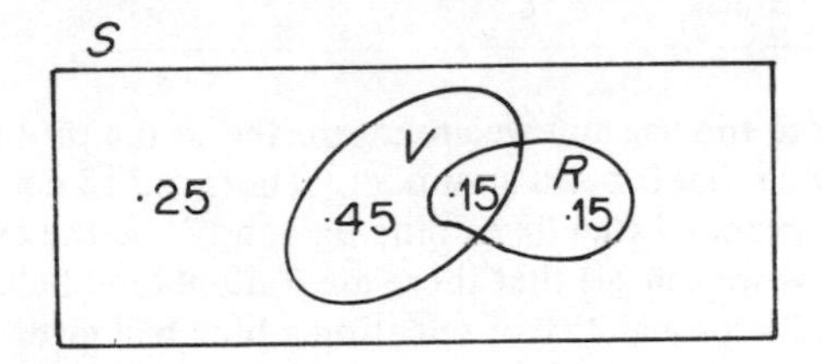

Figure 12.1

Addition rule for exclusive events

If the two events B and C are **mutually exclusive**, i.e. $B \cap C = \emptyset$ then $p(B \cap C) = 0$ and so, if B and C cannot occur simultaneously

$$p(B \cup C) = p(B) + p(C) \tag{12.2}$$

Example

If a fair die is rolled, what is the probability of obtaining a 1 or a 6? The events B = *obtaining 1* and C = *obtaining 6* are mutually exclusive and each has probability 1/6; by applying (12.2) we have

$$p(B \cup C) = \frac{1}{6} + \frac{1}{6} = \frac{1}{3}$$

Both addition rules can be extended to more than two events. For example, for three events

$$\begin{aligned} p(A \cup B \cup C) = {} & p(A) + p(B) + p(C) - p(A \cap B) - p(B \cap C) \\ & - p(C \cap A) + p(A \cap B \cap C) \end{aligned}$$

You can check this for yourselves by drawing a suitable Venn Diagram.

Conditional probability and the multiplication rule

If two events follow each other then the outcome of the second may depend on the outcome of the first. For example, if we draw a card from a pack and then draw a second card, the probability of the second being a heart is clearly affected by whether the first card was a heart or not. We make plausible a definition by a non-practical example, chosen for its simplicity.

Suppose we have a box containing red balls and white balls. Some balls are numbered and the composition is as shown below.

	numbered	blank	Totals
Red	3	5	8
Blue	4	1	5
Totals	7	6	13

We draw a ball from the bag and we are interested in the probability of the ball being blue knowing that it bears a number. There are 13 equally likely outcomes. Let B be the event of selecting a blue ball and N be the event of selecting a numbered ball. Now we can see that there are 7 numbered balls of which 4 are blue and we say that the probability of selecting a blue ball given that it is numbered is 4/7. We write this $p(B/N)$ and read it as *probability of B given N*. Now the probability $p(B)$ given no advance information is 5/13 and it does make a difference in this case if we know whether event N has occurred.

Further $p(N) = 7/13$ and $p(B \cap N) = 4/13$, since only four balls are both blue and numbered. In this example we can see that

$$\frac{p(B \cap N)}{p(N)} = \frac{4}{7}$$

and we might ask whether it is coincidence that this value is also $p(B/N)$. When we are told N has occurred we can cut down our sample space from the original 13 points each of which has probability 1/13 to a sample space of 7 members which are still equally likely and must have probability 1/7; we effectively have to reweight them by scaling the weights by a factor 13/7 and this is done by dividing by $p(N)$. The members we require are in $B \cap N$ and this leads us to identify $p(B \cap N)/p(N)$ with $p(B/N)$.

In general, we have the result that if two events A and B are such that $p(B) \neq 0$, then the **conditional probability** of A given B is

$$p(A/B) = \frac{p(A \cap B)}{p(B)} \tag{12.3}$$

Try to argue this equation from the Venn Diagram.

We consolidate these ideas by three examples.

Example 1

The probability $p(B)$ that it rains on July 15th in Loughborough is 0.6; the probability $p(A)$ that it rains there on July 15th and 16th is 0.35. If we know that it has rained there on July 15th, the probability of rain on the following day is

$$p(A/B) = \frac{0.35}{0.6} = \frac{7}{12}$$

Example 2

Two dice are rolled. What is the probability of the total score exceeding 8 given that the first die shows:

(a) 6 (b) 4 (c) 2?

(a) We have a sample space of 36 equally likely outcomes and we let A be the event *total score exceeds 8* and B the event *first die shows 6.*
Then

$$B = \{(6,1), (6,2), (6,3), (6,4), (6,5), (6,6)\} \quad \text{and} \quad p(B) = 6/36.$$

$$A = \{(3,6), (4,5), (4,6), (5,4), (5,5), (5,6), (6,3), (6,4), (6,5), (6,6)\}$$

and $p(A) = 10/36$.

Now $A \cap B = \{(6,3), (6,4), (6,5), (6,6)\}$ and $p(A \cap B) = 4/36$.
Then

$$p(A/B) = p(A \cap B)/p(B) = 4/36 \Big/ 6/36 = 4/6 = 2/3.$$

We can see directly that of the 6 outcomes in B, 4 give a total greater than 8. From (12.3) we can deduce that $p(A \cap B) = p(A/B) \, . \, p(B)$.
Equally well, we can say that

$$p(A \cap B) = p(B/A) \, . \, p(A).$$

In words, the first of these relationships can be stated that *the probability of both A and B occurring is the probability that B occurs multiplied by the probability that A occurs given that B has occurred.*

(b) In this case B is the event *first die shows 4.* Then

$$B = \{(4,1), (4,2), (4,3), (4,4), (4,5), (4,6)\}$$

and

$$p(B) = 6/36.$$

$A \cap B = \{(4,5), (4,6)\}$ and therefore

$$p(A \cap B) = 2/36$$

Then

$$p(A/B) = 2/36 \Big/ 6/36 = 1/3$$

(c) Here, $A \cap B = \emptyset$ and, therefore,

$$p(A/B) = 0$$

Example 3

Two cards are drawn without replacement from a pack. What is the probability that

(i) both cards are kings
(ii) one is a king and one is an ace?

(i) Let B be the event *the first card is a king*, A = *second card is a king.* Then

$$p(A \cap B) = p(B) \,.\, p(A/B) = \frac{4}{52} \cdot \frac{3}{51}$$

since if a king is drawn first we have 51 cards left, containing 3 kings. Hence

$$p(A \cap B) = \frac{4}{52\,.\,17}$$

(ii) Here we have two exclusive events: (ace, king) and (king, ace).

$$p(\text{king on } 1^{\text{st}} \text{ and ace on } 2^{\text{nd}}) = \frac{4}{52} \cdot \frac{4}{51}$$

$$= p(\text{ace on } 1^{\text{st}} \text{ and king on } 2^{\text{nd}})$$

Hence

$$p(\text{king and ace}) = \frac{16}{52\,.\,51} + \frac{16}{52\,.\,51} = \frac{32}{52\,.\,51}$$

Independent events

Suppose we consider two experiments: tossing a coin, and drawing a card from a pack; it would be difficult to imagine any interaction between them. Consequently, if we define events H = *tossing a head* and D = *drawing an ace* we would not expect $p(D)$ to be affected by whether H had occurred or not. In other words, in this situation

$$p(D/H) = p(D) = 4/52$$

We say that the two events are **statistically independent.**

If two events A and B are such that

$$p(A/B) = p(A) \quad \text{and} \quad p(B/A) = p(B)$$

then

$$p(A \cap B) = p(A) \,.\, p(B) \tag{12.4}$$

Notice that *physical independence implies statistical independence* and, conversely, *statistical dependence implies physical dependence.*

Example

A die is rolled twice. What is the probability of two successive sixes?

$$p(6 \cap 6) = p(6) \,.\, p(6) = \frac{1}{6} \cdot \frac{1}{6} = \frac{1}{36}$$

General Examples

All the rules can be extended to more than two events. We now consider some examples bringing in the rules but in an informal way. They serve to illustrate the basic ideas.

Example 1

A lot of 16 articles consists of 10 good ones, 4 with only minor defects and 2 with major defects.

(i) An article is drawn at random. Find the probability that

(a) it has no defects
(b) it has no major defects
(c) it is either good or has major defects.

(ii) Two articles are selected at random. Find the probability that

(a) both are good
(b) both have major defects
(c) at least one is good
(d) at most one is good
(e) exactly one is good
(f) neither has major defects
(g) neither is good.

(iii) Suppose a lot of 16 articles is to be accepted if three articles selected at random have no major defect. What is the probability that the lot described above is rejected?

(i) Let G = *article selected is good,* J = *article has major defect,* M = *article has minor defect.*

(a) We require

$$p(G) = \frac{10}{16} = \frac{5}{8}$$

(b) This is

$$p(G \cup M) = p(G) + p(M) = \frac{10}{16} + \frac{4}{16} = \frac{14}{16} = \frac{7}{8}$$

Alternatively

$$p(J') = 1 - p(J) = 1 - \frac{2}{16} = \frac{14}{16} \text{ as before}$$

(c) This is

$$p(G \cup J) = p(G) + p(J) = \frac{10}{16} + \frac{2}{16} = \frac{3}{4}$$

(ii) Let G_1 = *first article selected is good,* J_2 = *second article has a major defect,* and so on.

(a) $p(G_1 \cap G_2) = p(G_1) \,.\, p(G_2/G_1) = \frac{10}{16} \,.\, \frac{9}{15} = \frac{3}{8}$

(b) $p(J_1 \cap J_2) = p(J_1) \,.\, p(J_2/J_1) = \frac{2}{16} \,.\, \frac{1}{15} = \frac{1}{120}$

(c) If at least 1 is good, then we have p(1 is good) + p(2 are good).

$$p(\text{1 is good}) = p(G_1 \cap G_2' + G_1' \cap G_2)$$

$$= p(G_1 \cap G_2') + p(G_1' \cap G_2)$$

$$= \frac{10}{16} \,.\, \frac{6}{15} + \frac{6}{16} \,.\, \frac{10}{15}$$

$$= \frac{1}{4} + \frac{1}{4} \qquad = \frac{1}{2}$$

therefore

$$p(\text{at least 1 is good}) = \frac{1}{2} + \frac{3}{8} = \frac{7}{8}$$

Alternatively,

$$p(\text{at least 1 is good}) = 1 - p(\text{neither is good})$$

$$= 1 - \frac{6}{16} \,.\, \frac{5}{15} = 1 - \frac{2}{16} = \frac{7}{8}$$

(d) $p(\text{at most 1 is good}) = p(\text{1 is good or neither is good})$

$$= 1 - p(\text{both are good})$$

$$= 1 - \frac{3}{8} = \frac{5}{8}$$

(e) p(1 is good) = ½ [From (c)]

(f) $p(\text{neither has major defects}) = p(J_1' \cap J_2')$

$$= \frac{14}{16} \,.\, \frac{13}{15} = \frac{91}{120}$$

(g) $p(\text{neither is good}) = \frac{1}{8}$ [From (c)]

(iii) $p(J_1' \cap J_2' \cap J_3') = \frac{14}{16} \cdot \frac{13}{15} \cdot \frac{12}{14} = \frac{13}{20}$

therefore

$$p(\text{lot is rejected}) = 1 - \frac{13}{20} = \frac{7}{20}$$

Example 2

A circuit has three components in parallel; the probability of one of them failing is 0.05. If the circuit will work providing at least one of the components does not fail, what is the probability of the circuit working?

$$p(\text{all three components failing}) = (0.05)^3$$

therefore

$$p(\text{circuit working}) = 1 - (0.05)^3 = 1 - 0.000125 = 0.999875$$

Example 3

In a bag there are 40 balls, 3 of which are red. 5 balls are drawn without replacement. What is

(a) $p(\text{no red ball is drawn})$
(b) $p(\text{the first four are not red})$
(c) $p(\text{red ball appears on the fifth drawing only})$?

With the notation R_i = *red ball on draw i,*

(a) $p(R_1' \cap R_2' \cap R_3' \cap R_4' \cap R_5') = \frac{37 \,.\, 36 \,.\, 35 \,.\, 34 \,.\, 33}{40 \,.\, 39 \,.\, 38 \,.\, 37 \,.\, 36} = 0.662$ (3 d.p.)

(b) $p(R_1' \cap R_2' \cap R_3' \cap R_4') = \frac{37}{40} \cdot \frac{36}{39} \cdot \frac{35}{38} \cdot \frac{34}{37} = 0.723$ (3 d.p.)

(c) $p(R_1' \cap R_2' \cap R_3' \cap R_4' \cap R_5) = \frac{37}{40} \cdot \frac{36}{39} \cdot \frac{35}{38} \cdot \frac{34}{37} \cdot \frac{3}{36} = 0.060$ (3 d.p.)

Example 4

A circuit consists of three independent components, A, B, C in series. Let A = *component* A *is defective*, etc., then

$$p(A) = 0.03, \quad p(B) = 0.15, \quad p(C) = 0.01.$$

The circuit fails if any component is defective.

What is (a) $p(\text{circuit fails})$
(b) $p(\text{B alone fails})$?

(a) $p(\text{circuit fails}) = 1 - p(\text{all components work})$

$$= 1 - p(A' \cap B' \cap C')$$

$$= 1 - p(A') \,.\, p(B') \,.\, p(C')$$

$$= 1 - (0.97) \,.\, (0.85) \,.\, (0.99)$$

$$= 1 - 0.816$$

$$= 0.184 \quad (3 \text{ d.p.})$$

(b) $p(A' \cap B \cap C') = p(A') \,.\, p(B) \,.\, p(C')$

$$= (0.97) \,.\, (0.15) \,.\, (0.99)$$

$$= 0.144 \quad (3 \text{ d.p.})$$

Tree diagrams

A useful way of enumerating probabilities when there are several outcomes arising from a compounding of events is by means of a so-called **tree diagram.** We illustrate by an example.

Three dice are thrown and a success occurs with each die if the score exceeds 2. Then the probability of success is $p = 2/3$ and that of failure is $q = 1 - p = 1/3$ (see Figure 12.2).

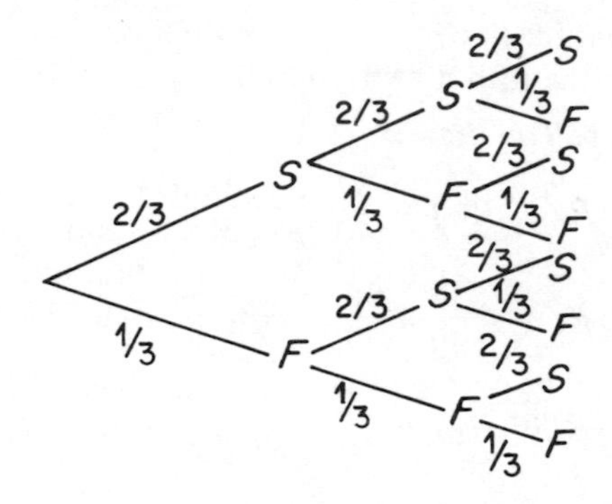

Figure 12.2

To evaluate the probability of an outcome we seek, such as *SFS*, we proceed along the relevant branches of the tree and then multiply the probabilities encountered (assuming the three dice give independent results) to obtain

$$\left(\frac{2}{3} \cdot \frac{1}{3} \cdot \frac{2}{3}\right) = \frac{4}{27}$$

If we do this for all outcomes we obtain Figure 12.3.

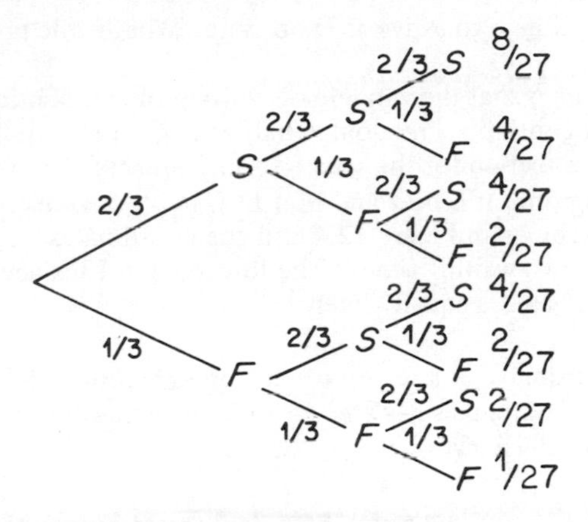

Figure 12.3

We check that these probabilities sum to 1. If we require the probability of at least two successes, we can see that the relevant mutually exclusive outcomes are *SSS, SSF, SFS, FSS* and their probabilities sum to

$$\left(\frac{8}{27} + \frac{4}{27} + \frac{4}{27} + \frac{4}{27}\right) = \frac{20}{27}$$

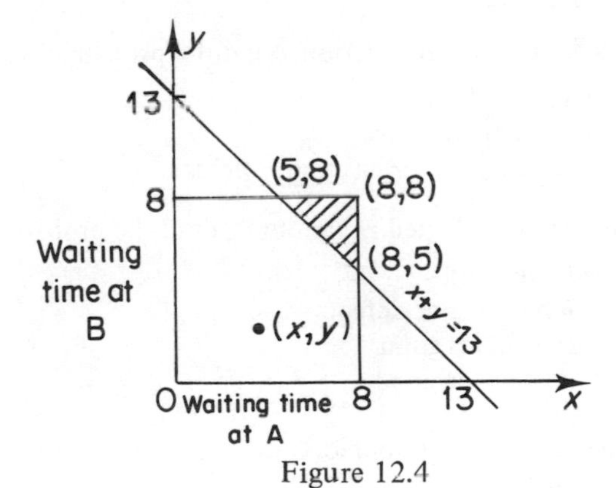

Figure 12.4

Diagram Solution

Finally, we mention one other technique for solving probability problems which we include for its simplicity despite its limited application. Again an example suffices to explain the method.

A traveller has to travel from A to D changing buses at stops B and C. He knows that he has a wait at B and one at C of at most 8 minutes each, but any time of waiting up to 8 minutes is equally likely. He can afford up to 13 minutes total waiting time if he is to arrive at D on time. What is the probability he will arrive late at D?

The trouble here is that the variable of waiting time is continuous and although we treat continuous random variables in Chapter 13, we are including the problem in this section for the sake of completeness.

Let the waiting time at B be x and that at C be y; we seek $p(x + y > 13)$. We draw a box as shown in Figure 12.4 and the coordinates (x, y) of any point in the box are the two waiting times. The line $x + y = 13$ is depicted and any point in the shaded area is equally likely.

We can say probability of a region of outcomes is proportional to its area and since the totality of outcomes has area 64 units we note that the shaded area is $\frac{1}{2}(3^2) = 9/2$ units, which gives

$$p(\text{late at D}) = \frac{9}{2} \Big/ 64 = \frac{9}{128}$$

Problems

1. Four articles are distributed at random among six containers. What is the probability that
   - (i) all objects are in the same container
   - (ii) no two objects are in the same container?

2. A lot consists of 12 good articles, 6 articles with only minor defects and 2 with major defects.
   - (i) One article is drawn at random. Find the probability that
     - (a) it has no defects
     - (b) it has no major defects
     - (c) it is either good or has major defects.
   - (ii) Two articles are selected at random. Find the probability that
     - (a) both are good
     - (b) both have major defects
     - (c) at least one is good
     - (d) at most one is good
     - (e) exactly one is good
     - (f) neither has major defects
     - (g) neither is good.
   - (iii) Suppose a lot of twenty articles is to be accepted if three articles selected at random have no major defect. What is the probability that the lot described in this question is
     - (a) accepted
     - (b) rejected?

3. In a bolt factory machines A, B, C manufacture 25, 35 and 40 per cent respectively of the total output. Of their outputs, 5, 4 and 2 per cent respectively are defective bolts. A bolt is drawn at random and found to be defective. What are the probabilities that it was manufactured by machines A, B or C respectively?

4. Two cards are drawn from a pack of 52 cards. What is the probability of drawing two aces if

   (i) the first card is returned to the pack before the second is drawn
   (ii) the first card is not returned to the pack?

5. A tester smokes cigarettes of each of three different brands, A, B and C. He then assigns the name A to one cigarette, B to another and C to the remaining one. If he were unable to discriminate between the brands (i.e. if the names were assigned at random) what would be the probability of just one brand being correctly identified?

6. A box contains 4 bad tubes and 6 good tubes.

   (i) Two are drawn together and one of them is tested and found to be good. What is the probability that the other is also good?

   (ii) If the tubes are checked by drawing a tube at random, testing it and repeating the process (not replacing the tube tested) until all four bad tubes are located, what is the probability that the fourth bad tube will be found

      (a) at the fifth test
      (b) at the tenth test?

7. A product is made up of three parts, A, B and C. The manufacturing process is such that the probability of A being defective is 0.2, of B being defective is 0.1, and of C being defective is 0.3. What is the probability that the assembled product will contain at least two defective parts?

8. An operator drilling for oil has options with six companies and must decide at the end of the year whether to take them up. The probability that he will strike oil by that time is 0.0125. If he does strike oil he will take up five of the options at random, and if he does not he will take up one at random. What is the probability that a particular company A will have its option taken up?

9. In a rifle shooting contest the target consists of a bull and outer, the contestants scoring 2 points for a bull, 1 point for an outer and zero otherwise.
   If the chance of a bull is 50%, and an outer 30%, what are the probabilities that with five shots a contestant will score

   (a) 10 points (b) 8 points (c) less than 8 points?

10. A product is made up of three parts, A, B and C. The manufacturing processes are such that the probability of A being defective is 0.07, of B being defective is 0.05 and of C being defective is 0.08. What is the probability that the assembled product will contain

    (a) no defectives
    (b) at least 2 defectives?

11. A box contains 3 white, 4 blue and 6 black balls. Two balls are drawn consecutively from the box without replacement. Find the probability that both are the same colour.

12. The odds for a particular event A are 5 to 1 against; for an independent event B the odds are 3 to 1 against, and for an independent event C the odds are 2 to 1 on. Find the probability that

    (a) all 3 events occur
    (b) only event A occurs
    (c) only one event occurs.

13. The probability that parents with a trace of a certain blue pigment in otherwise brown eyes will have a child with blue eyes is 1 in 4. If there are six children in a family what is the probability that at least half of them will have blue eyes? (C.E.I.)

14. (i) A die is thrown repeatedly until a 6 appears. What is the probability that a 6 does not appear until

    (a) the 4th throw (b) the 10th throw
    (c) the rth throw?

    (ii) Four players in turn draw (without replacement) a card from a pack to decide who will deal; the highest wins, aces count high. The first player draws the nine of hearts. What is the probability that he will deal?

15. (i) A, B and C take turns to throw a coin. Show that the probabilities of each of them throwing the first head are 4/7, 2/7 and 1/7 respectively.

    (ii) Two dice are thrown together repeatedly. Can you find the probability that a total score of 9 appears before a total score of 8?

16. Three bolts are selected at random from a box of six bolts and three nuts are selected at random from a box of six nuts. If each box contains 2 faulty items find

    (i) the probability that at least one of the six selected items is faulty
    (ii) the mean number of satisfactory nut and bolt pairs that can be obtained from the six selected items. (L.U.)

17. (i) Given that the probability that any one of 4 telephone lines is engaged at an instant is 1/3, calculate the probability
    (a) that two of the four lines are engaged and the other two are free,
    (b) that at least one of the four lines is engaged. (L.U.)

    (ii) A machine is powered by three similar storage batteries; it will function satisfactorily only if at least two of these batteries are serviceable. The probability of any one battery becoming unserviceable in less than 50 hours is 0.2, and of becoming unserviceable in less than 100 hours is 0.6. Find the probability that the machine will function satisfactorily for
    (a) at least 50 hours
    (b) between 50 and 100 hours. (L.U.)

18. A pond contains a large unknown number N of fish. A random sample of r fish is taken, marked and put back in the pond. A week later, a random sample of s fish is taken. Find the probability of the second sample containing n marked fish and estimate the value of N which maximises this probability. Hence estimate the likely value of N in the case where $r = 20$, $s = 25$ and $n = 2$.

12.3 PROBABILITY DISTRIBUTIONS

We shall again restrict our attention to discrete random variables. If an experiment has a given discrete finite sample space $S = \{x_1, x_2, x_3, \ldots\ldots, x_n\}$ and we assign probabilities $p_1, p_2, p_3, \ldots\ldots, p_n$ to these outcomes then we have defined for the experiment a **discrete probability distribution.** We call the function $p(X)$ which takes the values p_1, p_2, etc. when X takes the values x_1, x_2 etc. the **probability function** of X.

Example 1

For the tossing of a coin $S = \{H, T\}$ and $p(T) = ½$. It should be clear that

(i) $$0 \leqslant p(x_i) \leqslant 1$$

and, since $S = x_1 \cup x_2 \cup \ldots\ldots \cup x_n$, that

(ii) $$\sum_{i=1}^{n} p(x_i) = 1$$

If A is any event then

$$p(A) = \sum_{x_i \in A} p(x_i)$$

where we sum the probabilities of all outcomes in A.

Provided conditions (i) and (ii) are satisfied then any function will be a probability function; it is clear that we try to choose a probability function which closely matches observed frequency (see later section on Binomial and Poisson distributions). We can represent a discrete probability distribution by a line diagram.

Example 2

For the tossing of three coins we can specify the outcomes as *HHH, HTH, HTT*, etc. each of which has equal probability of occurring of 1/8 if the coins are fair. Alternatively, we may specify the outcomes as no heads, 1 head, 2 heads, 3 heads, which have probabilities respectively of 1/8, 3/8, 3/8, 1/8 and we represent these on a line diagram shown in Figure 12.5.

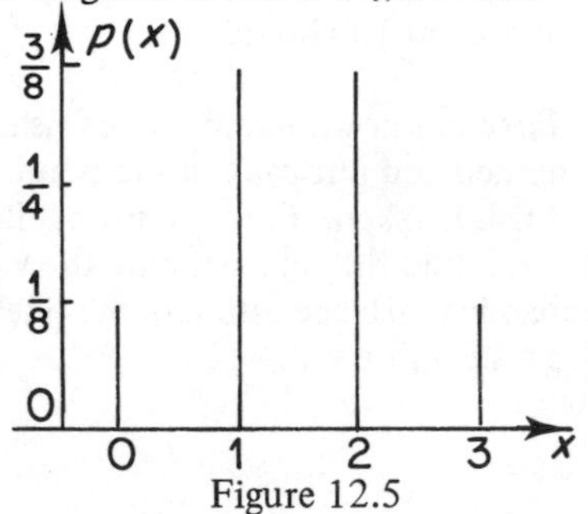

Figure 12.5

Example 3

The random variable X can take values 0, 1, 2, 3, 4, 5, 6 with probabilities 0.3, 0.14, 0.07, 0.08, 0.13, 0.18, 0.1. Verify that this is a description of a probability distribution, calculate $p(X \geqslant 3)$ and $p(X < 2)$ and display the distribution on a line diagram shown in Figure 12.6.

All probabilities are in the range $[0, 1]$ and their sum is 1.

$$p(X \geqslant 3) = 0.08 + 0.13 + 0.18 + 0.1 = 0.49$$

$$p(X < 2) = 0.3 + 0.14 = 0.44$$

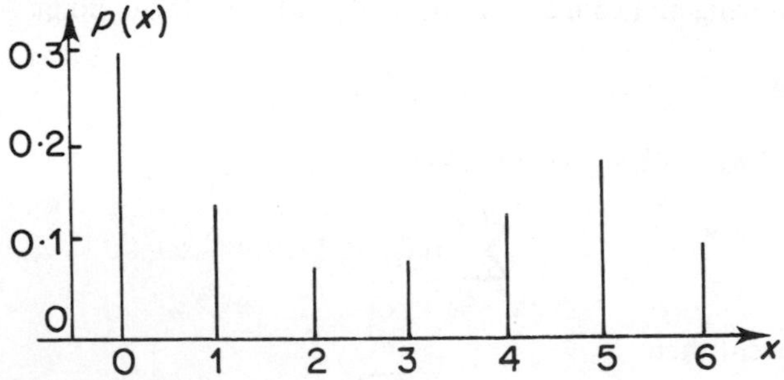

Figure 12.6

Problems

1. The sample space $X = \{0, 1, 2, 3\}$ is given. Which of the following can be probability functions? (Determine the value of C where necessary.) For those which are, find the probabilities of the events

   (a) x is odd (b) x is at least 1, (c) x is 1 or 2.

   For those which are not, state why not.

   (i) $\frac{1}{8}(1 + x)$ (ii) $\frac{1}{10}(1 + x)$

   (iii) $\frac{2}{x}$ (iv) $\frac{C}{x^2 + 1}$

   (v) $\frac{x^2 - 2}{4}$ (vi) $\frac{C}{2 + x}$

2. Find the probabilities associated with all the possible scores obtained when two dice (assumed fair) are thrown and draw a line diagram to represent the probability distribution.

3. A customer buys 20 transistors and tests a sample of 5 of them. If 4 of the transistors are faulty, how many of the possible samples of 5 will contain at least 1 faulty transistor? Find the probabilities of a sample containing 0, 1, 2, 3 or 4 faulty transistors. Draw a line diagram to represent the probability distribution.

4. If the sample space is $X = \{0, 1, 2, 3, 4\}$, verify that with a suitable choice of k, $p(X) = k/(X + 2)$ can be used as a probability function. Sketch the distribution.

5. The following is a simplified model representing the spread of measles within a family of three susceptible children, A, B and C. If A catches measles, there is a period during which B may get the disease from him and the chance that B does in fact succumb is θ; similarly for C. The probabilities for B and C are independent. Thus in the first stage none, one or two of the other children may catch the disease. If the first or last event occurs this completes the matter, but if B, say, gets the disease and not C, there is a further period during which C may, with probability θ, catch the disease from B.
   Show that, given that one child has measles, the chances that 0, 1, 2 of the other children get it are $(1 - \theta)^2$, $2\theta(1 - \theta)^2$, $\theta^2(3 - 2\theta)$. Obtain corresponding probabilities for families of four children.

12.4 MATHEMATICAL EXPECTATION

As usual in this chapter, we concentrate on discrete random variables. We have already seen that for a sample of n observations falling into s distinct values the arithmetic mean

$$\bar{x} = \sum_{r=1}^{s} \frac{f_r}{n} x_r \qquad (11.1)$$

Now we have already interpreted probability as long-range relative frequency and we might wonder what interpretation to put on the quantity

$$\sum_{r=1}^{s} p_r x_r$$

where the random variable X takes the values $x_1, x_2, \ldots\ldots, x_s$ with probabilities $p_1, p_2, \ldots\ldots, p_s$. We define the **expectation** or **expected value** of the discrete random variable X, written μ or $E(X)$,

$$\text{as} \sum_{r=1}^{s} p_s x_s \qquad (12.5)$$

For example, in the throwing of a die the probability of each outcome is 1/6 and so

$$E(X) = \frac{1}{6} \times 1 + \frac{1}{6} \times 2 + \frac{1}{6} \times 3 + \frac{1}{6} \times 4 + \frac{1}{6} \times 5 + \frac{1}{6} \times 6 = \frac{21}{6} = 3.5$$

What significance has this value? Before we answer this question let us consider a more simple experiment: the tossing of an unbiased coin; here we can code the outcomes Head and Tail as 0 and 1 respectively. Each outcome has probability ½ and the expected value is (½ × 0 + ½ × 1) = ½. But ½ is not a value assumed by either outcome; rather it represents the average score if the experiment were to be repeated many times. We will not get a score of ½ on one tossing, but the relative frequency of tails will approximate to ½ after several tossings and the approximation should get closer the more times the coin is tossed. In effect, the expectation ½ is the model prediction of the average score: for this reason, we use the alternative notation μ and speak of the **population mean.** The sample mean, $\bar{x}$, of n observations is then compared to the population mean since we are testing the sample observations against the model. (You might find it helpful to look back at Chapter 11.)

Returning to the throwing of a die, we can say that if we repeat the experiment many times we would expect the average score/throw to be 3½ and we may compare the observed average to 3½; if the discrepancy is large we may suspect that the die is not fair, since the model is built on the assumption of equiprobable outcomes, which leads to a population mean of 3½.

Example

A game consists of you drawing a ball from a bag containing 6 white balls and 4 blue balls. If the ball is white you win 40p, if it is blue you lose 80p. The ball is replaced, the bag shaken and another ball drawn. What are your expected winnings from this game?

Let $x_1 = +40\text{p}$, $x_2 = -80\text{p}$. Then $p(x_1) = 6/10$, $p(x_2) = 4/10$

$$E(X) = p(x_1)\,.\,x_1 + p(x_2)\,.\,x_2 = (6/10)\,.\,40 + (4/10)\,.\,-80 = -8\text{p}$$

On average you lose 8p per game, though you can either win 40p or lose 80p on each occasion. In n games you would expect to lose $8n$p and hence you would be unwise to play.

Properties of expectation

We state these properties without proof (they follow from the definition of expectation). X is a *discrete random variable* (d.r.v.)

(i) $E(aX + b) = aE(X) + b$ for any numerical constants a and b
(ii) $E(m) = m$ where m is a constant
(iii) If $E(X) = \mu$ then $E(X - \mu) = 0$
(iv) If Y is also a discrete random variable, $E(X + Y) = E(X) + E(Y)$
(v) If X and Y are independent, $E(XY) = E(X)\,.\,E(Y)$
(vi) If $f(X)$ is any function of X then

$$p[f(X)] = \sum_{i=1}^{n} p_i f(x_i)$$

for example,

$$E(X^2) = \sum_{i=1}^{n} p_i x_i^2$$

As a result of this last property we may define the variance of the d.r.v. X,

$$\text{var}(X) = E(X - \mu)^2 \qquad (12.6a)$$

Now

$$E[(X - \mu)^2] = \sum_{i=1}^{n} p_i (x_i - \mu)^2$$

$$= \sum_{i=1}^{n} p_i (x_i^2 - 2\mu x_i + \mu^2)$$

$$= \sum_{i=1}^{n} p_i x_i^2 - \sum_{i=1}^{n} 2\mu p_i x_i + \sum_{i=1}^{n} p_i \mu^2$$

$$= \sum_{i=1}^{n} p_i x_i^2 - 2\mu \sum_{i=1}^{n} p_i x_i + \mu^2 \sum_{i=1}^{n} p_i$$

$$= E(X^2) - 2\mu E(X) + \mu^2$$

$$= E(X^2) - 2[E(X)]^2 + [E(X)]^2$$

$$= E(X^2) - [E(X)]^2$$

Hence

$$\operatorname{var}(X) = E(X^2) - [E(X)]^2 \qquad (12.6b)$$

We usually denote population variance by σ^2.
It follows that

$$\operatorname{var}(aX + b) = a^2 \operatorname{var}(X) \qquad (12.6c)$$

where a and b are constants.

Further, if X and Y are independent, $\operatorname{var}(X + Y) = \operatorname{var}(X) + \operatorname{var}(Y)$.

Problems

1. A prize of £200 is given for naming the three winning dogs, in correct order, in a race for which 10 dogs were entered. What is the expectation of a person having no inside knowledge of the ability of the dogs?

2. If a single die is thrown (assumed fair) write down the probability distribution of the scores X and obtain $E(X)$ for the distribution. Also obtain var(X).

3. A man buys a sweepstake ticket. The first prize is £10000, the second prize £1000 and the third prize is £100. His probabilities of winning these prizes are 0.0001, 0.0008 and 0.005 respectively. What is a fair price to pay for the ticket if he can only win one prize?

4. Show that $E(X - \mu)^3 = E(X^3) - 3\mu E(X^2) + 2\mu^3$

5. The daily demand for a particular commodity at a grocery store has the following distribution

X	0	1	2	3	$\geqslant 4$
$p(X)$	0.36	0.35	0.19	0.09	0.01

Find the expected or average daily demand over a long period of time.

6. An event has a fixed probability p of success. Show that the expected number of failures before the first success occurs is

$$E = \sum_{r=0}^{\infty} rp(1-p)^r$$

Show that (i) $E = (1-p)/p$ and that (ii) the expected number of trials necessary for a success is $1/p$. [Hint for (i): find $E - (1-p)E$].

12.5 THE BINOMIAL DISTRIBUTION

Suppose a trial has two outcomes; we can label these success and failure in some sense. We denote the probability of success by p and that of failure by q $(= 1 - p)$. We are interested in experiments which consist of several such trials repeated. Some examples of such trials (sometimes called **Bernoulli trials**) are

(i) a marksman firing a shot at a target; a success is a bull's-eye
(ii) a die is thrown: success is a *six*; $p = 1/6$, $q = 5/6$
(iii) a large box of components contains 1% defectives – a component is selected: success is choosing a defective (if we are trying to test a theory which says there are defectives in the box); $p = 1/100$, $q = 99/100$.

If we repeat a **binomial** trial a **fixed** number of times, n, we can define a new discrete random variable which is the number of successes in the n trials.

Provided the probability of success is the same for each trial and the trials are independent of each other we define the experiment as a **binomial** experiment.

Note that four conditions must be satisfied:

(i) the trials must have only two outcomes,
(ii) the number of trials must be fixed,
(iii) the probability of success is the same for all trials,
(iv) the trials are independent.

These form the assumptions of a **binomial model.** We have to decide on the applicability of the model in any given situation. For example, if the marksman fires 5 rounds at the target we might reasonably assume his chance of a bull's-eye remains constant unless some sudden change in the light occurs, but if he fires 100 rounds this may not be the case. Again, if two components are selected from the box the probability of the second one being defective is not quite the same as that for the first, but if the box contains a large number of components, conditions (iii) and (iv) are nearly satisfied.

Let us now try to put some algebra into the problem. Let us find the probability of 0, 1, 2, 3, 4 successes in an experiment consisting of up to 4 repeated independent Bernoulli trials each with probability of success p. We enumerate the probabilities as follows:

Number of trials	1	2	3	4
Number of successes				
0	q	q^2	q^3	q^4
1	p	$2pq$	$3pq^2$	$4pq^3$
2	0	p^2	$3p^2q$	$6p^2q^2$
3	0	0	p^3	$4p^3q$
4	0	0	0	p^4

For example, for two successes in three trials the possible outcomes are *SSF, SFS, FSS.* Since the trials are independent the respective probabilities of these outcomes are *ppq, pqp, qpp,* that is, each is p^2q; since these outcomes are mutually exclusive, the probability of two successes in three trials is $3p^2q$. You can check any of the others by a tree diagram. We prove one other case: 2 successes in 4 trials. Each of the possible outcomes comprises 2 successes and 2 failures; each has probability p^2q^2. Now we can imagine a row of 4 pigeon-holes into which we must put a letter from a set consisting of 2 *S*'s and 2*F*'s. Once we decide where the *S*'s go, the *F*'s must go in the empty holes. These *S*'s can go in any 2 of the 4 holes and this can be accomplished in ${}^4C_2 = 6$ ways; hence the probability we seek is $6p^2q^2$.

It should be clear that we can summarise (and generalise) the above to n trials and say that the probabilities of the various outcomes are just the terms in the expansion of $(q + p)^n$. It follows from this that the sum of these probabilities is 1.

Example

A die is thrown with a success being a 'six'. We now calculate the probabilities of the various outcomes if the trial is repeated 5 times. The probabilities are the terms in the expansion of $(q + p)^5$, i.e. q^5, $5pq^4$, $10p^2q^3$, $10p^3q^2$, $5p^4q$, p^5 where $p = 1/6$, $q = 5/6$, and these become

$$\frac{3125}{7776}, \frac{3125}{7776}, \frac{1250}{7776}, \frac{250}{7776}, \frac{25}{7776}, \frac{1}{7776} \quad \text{(which, you can check, sum to 1)}$$

Binomial formula

An alternative generalisation is: if an experiment consists of n independent Bernoulli trials each with probability of success p and of failure q $(= 1 - p)$ the probability of the experiment yielding exactly r successes is

$$ {}^nC_r\, p^r\, q^{n-r}, \quad r = 0, 1, 2, \ldots\ldots, n \tag{12.7}$$

[which is sometimes written $b(r;\ n, p)$] – this is the **binomial formula.**

This expression is just a typical term in the expansion of $(q + p)^n$. We argue that r successes have probability p^r of occurring and the $(n - r)$ failures have probability q^{n-r}. The outcome r successes followed by $(n - r)$ failures has the probability $p^r q^{n-r}$; this will be the probability of any arrangement of r successes and $(n - r)$ failures and as there are ${}^nC_r$ ways of choosing the positions of the r successes we obtain the probability given by (12.7).

Example 1

A die is thrown 56 times; the probability of obtaining at least 3 sixes

$$= 1 - (\text{probability of } 0, 1, 2 \text{ sixes})$$

$$= 1 - [(5/6)^{56} + 56(5/6)^{55}\,.\,(1/6) + {}^{56}C_2(5/6)^{54}(1/6)^2]$$

Example 2

In a manufacturing process it is found that on average 0.5% of the articles produced are defectives. If the articles are packed in cartons of 100, what is

(i) p(the carton is free from defectives)
(ii) p(there are at least 2 defectives in a carton)?

We may take a binomial model of 100 trials with $p = 1/200$.

(i) $p(0) = \left(\frac{99.5}{100}\right)^{100}$. We can evaluate this most easily by taking logarithms.

$$\log p(0) = 100(\log 0.995) = 100 \times \bar{1}.99782 = \bar{1}.782$$

therefore

$$p(0) = 0.605 \quad (3 \text{ d.p.})$$

(ii) $p(1) = 100\left(\frac{99.5}{100}\right)^{99}\left(\frac{0.5}{100}\right)$; $\log \frac{p(1)}{0.5} = 99(\log 99.5 - \log 100)$

and hence

$$p(1) = 0.304 \quad (3 \text{ d.p.})$$

Hence

$$p(\geqslant 2) = 1 - p(0) - p(1) = 0.091 \quad (3 \text{ d.p.})$$

Example 3

A hurdler has a probability of 15/16 of clearing each hurdle; what is the probability that he knocks down less than 2 hurdles in a flight of 10?

Here the binomial model is not quite accurate since if a hurdler knocks one hurdle down he may be thrown out of his stride and the next hurdle may fall; the trials are not *strictly* independent. However, applying the model

$$p(0) = (15/16)^{10} = 0.524 \quad (3 \text{ d.p.})$$

and

$$p(1) = 10(15/16)^9(1/16) = 0.350 \quad (3 \text{ d.p.})$$

Then

$$p(\text{less than } 2) = p(0) + p(1) = 0.874 \quad (3 \text{ d.p.})$$

Expectation and Variance of the binomial distribution

$$\mu = E(X) = \sum_{r=0}^{n} {}^{n}C_r\, p^r(1-p)^{n-r} . r$$

We consider an example. Let us repeat the trial 3 times; then

$$\mu = q^3 . 0 + 3pq^2 . 1 + 3p^2q . 2 + p^3 . 3$$
$$= 3pq^2 + 6p^2q + 3p^3$$

$$= 3p(q^2 + 2pq + p^2)$$
$$= 3p(q + p)^2$$
$$= 3p$$

In general, since we would have in n trials with constant probability of success p an expected number of successes np we have

$$\mu = np$$

The number of successes is the sum of n random variables, each of which has two outcomes (0 for failure, 1 for success); the mean of one of these variables is $1 \times p + 0 \times q = p$ and by an extension of rule (iv) for expectations,

$$E(X) = np$$

For the variance of X, we have

$$\text{var}(X) = E[(X - \mu)^2] = E(X^2) - [E(X)]^2$$
$$= npq$$

Use of Probability Generating Function

We can obtain these last two results by means of a **probability generating function** (p.g.f.) for the binomial distribution. We consider the expansion

$(q + pt)^n$ where t is a dummy variable.

Note that the coefficient of t^r gives the probability of r successes in n trials (P_r). If we write $(q + pt)^n = P_0 + P_1 t + P_2 t^2 + \ldots\ldots + P_n t^n$ then differentiation with respect to t yields

$$np(q + pt)^{n-1} = P_1 + 2P_2 t + \ldots\ldots + nP_n t^{n-1} \tag{12.8a}$$

If we now multiply throughout by t we obtain

$$npt(q + pt)^{n-1} = P_1 t + 2P_2 t^2 + \ldots\ldots + nP_n t^n$$

and a further differentiation yields

$$np(q + pt)^{n-1} + npt(n - 1)p(q + pt)^{n-2} = P_1 + 2^2 P_2 t + \ldots\ldots + n^2 P_n t^{n-1} \tag{12.8b}$$

If we put $t = 1$ in (12.8a) we have

$$np(q + p)^{n-1} = 1 \,.\, P_1 + 2 \,.\, P_2 + \ldots\ldots + nP_n$$

or

$$np \,.\, 1 = 0 \,.\, P_0 + 1 \,.\, P_1 + 2 \,.\, P_2 + \ldots\ldots + nP_n = \sum_{r=0}^{n} rP_r = E(X)$$

If we put $t = 1$ in (12.8b) we obtain

$$np(q + p)^{n-1} + np(n - 1)p(q + p)^{n-2} = 1^2 P_1 + 2^2 P_2 + \ldots\ldots + n^2 P_n$$

that is,

$$np + np^2(n-1) = 0^2P_0 + 1^2P_1 + 2^2P_2 + \ldots\ldots + n^2P_n = \sum_{r=0}^{n} r^2P_r$$

$$= E(X^2)$$

that is,

$$E(X^2) = np + n^2p^2 - np^2 = n^2p^2 + np(1-p) = n^2p^2 + npq$$

so that

$$\text{var}(X) = E(X^2) - [E(X)]^2 = n^2p^2 + npq - (np)^2 = npq$$

Example

A fair coin ($p = ½$) is tossed 64 times. The Expected Number of Heads = np = 64 . ½ = 32. The variance is npq = 32 . ½ = 16, giving a standard deviation of 4 heads.

General Examples

We conclude this section with some examples.

Example 1

Of a large number of mass-produced articles, one-tenth are defective; find the probability that a random sample of 20 will contain

(a) exactly 3 defective articles,
(b) at least 3 defective articles.

The model used is Binomial with $n = 20$, $p = 0.1$, $q = 0.9$

(a) $p(3) = {}^{20}C_3(1/10)^3(9/10)^{17} = 0.190$ (3 d.p.)

(b) $$\begin{aligned} p(\text{at least } 3) &= 1 - p(0, 1, \text{or } 2) = 1 - p(0) - p(1) - p(2) \\ &= 1 - (9/10)^{20} - 20(9/10)^{19} . (1/10) \\ &\quad - [(20 . 19)/2]\,[(9/10)^{18} . (1/10)^2] \\ &= 0.323 \quad (3 \text{ d.p.}) \end{aligned}$$

Example 2 *(Activity sampling)*

A machine shop contains 4 operatives who use a particular machine. Random checks showed that each needed the machine for 22% of the time. How many machines would be needed for 8 operatives if the men were not to spend more than 4% of their working time waiting for a spare machine?

Let N be the number of machines. We can use a Binomial model, counting a success as wanting a machine with $p = 0.22$; clearly, at least one man will be waiting for a machine at a given time if more than N men want to use a machine at that time. The probability of this is

$$\sum_{r=N+1}^{8} {}^8C_r\, p^r q^{8-r} = 1 - \sum_{r=0}^{N} {}^8C_r\, p^r q^{8-r}$$

which is $\not> 0.04$.

We require that

$$\sum_{r=0}^{N} {}^8C_r\, p^r q^{8-r}$$

just exceeds 0.96 and look for the value of N which just tips the total over 0.96. We quote results to 3 d.p.

$$p(0) = (0.78)^8 = 0.137$$

$$p(1) = 8\,.\,(0.78)^7(0.22) = 0.309$$

therefore

$$p(0)+p(1) = 0.446$$

$$p(2) = \frac{8.7}{2}\cdot(0.78)^6(0.22)^2 = 0.305$$

therefore

$$p(0)+p(1)+p(2) = 0.751$$

Similarly,

$$\sum_{r=0}^{3} p(r) = 0.923, \quad \sum_{r=0}^{4} p(r) = 0.984$$

We see that with 3 machines the cumulative probability is less than 0.96 and with 4 machines the cumulative figure clears 0.96. Therefore we need 4 machines.

Example 3

Certain manufactured components are made containing 1% defective. They are made into assemblies which are classified as sub-standard if they contain one or more defective components.

(a) If not more than 10% of the assemblies are to be sub-standard what is the maximum number of components they can contain?

(b) If the assemblies consist of 40 components, what is the maximum percentage of defectives in the bulk of components if only 5% of the assemblies are to be sub-standard?

(c) A different kind of assembly is to be produced with the components in 40 pairs, a pair being defective only if both components which comprise it are defective. If the percentage of substandard assemblies is to be at most 5%, what maximum percentage of defective components is permissible?

(a) Let N be the number of components per assembly. At least 90% of assemblies must contain no defective components. The probability of no defectives in a random sample of N is

$$q^N = (1 - 0.01)^N$$
$$= (0.99)^N$$

We require that $(0.99)^N \geqslant 0.90$
that is,

$$N \log(0.99) \geqslant \log 0.90$$

that is,

$$N \times \bar{1}.9956 \geqslant \bar{1}.9542$$

$$N(-0.0044) \geqslant -0.0458$$

therefore

$$N \leqslant \frac{-0.0458}{-0.0044} \simeq 10.4$$

Hence at most 10 components are allowed.

(b) Let α be the proportion of defectives allowed; then $(1 - \alpha)^{40} \geqslant 0.95$, that is,

$$40 \log(1 - \alpha) \geqslant \log 0.95 = \bar{1}.9777$$
$$= \overline{40} + 39.9777$$

therefore

$$\log(1 - \alpha) \geqslant \frac{1}{40}(\overline{40} + 39.9777) = \bar{1}.9994$$

therefore

$$1 - \alpha \geqslant 0.9987$$

therefore

$$\alpha \leqslant 0.0013 = 0.13\% \quad (2 \text{ d.p.})$$

(c) If the components are paired, the probability of a defective pair is α^2. We then have

$$(1 - \alpha^2)^{40} \geqslant 0.95$$

This leads to

$$\alpha^2 \leqslant 0.0013$$

so that

$$\alpha \leqslant \sqrt{0.0013} \simeq 0.036 \quad (3 \text{ d.p.})$$

that is,

$$\alpha \leqslant 3.6\% \quad (2 \text{ s.f.})$$

(To get greater precision we should have had to use at least 5 figure logarithms.)

Problems

1. A box contains ten components, all apparently sound, but four of them are, in fact, substandard. If three components are removed from the box altogether, what is the probability that two are substandard?

2. A sampling inspection plan operates as follows. Take a random sample of size ten from a large batch. If none of the sample is defective, then accept the batch. If more than one are defective, then reject the batch. If exactly one is defective, take another sample of size ten, and accept the batch only if this second sample contains no defectives. If a batch which is 5% defective is tested by this plan, what is the probability that

   (i) it is accepted after the first sample,
   (ii) it is accepted?

3. In a series of trials, an anti-aircraft battery had, on average, 3 out of 5 successes in shooting down missiles, the trials taking place over a long period. What is the probability that if 8 came within range not more than 2 would get through?

4. Given a Binomial distribution, what is the value of p for the probability of 4 successes in 10 trials to be twice the probability of 2 successes in 10 trials?

5. On average, 3% of the articles produced by a certain manufacturer are defective. What is the probability that in a sample of 10 articles,

   (i) 2 are defective,
   (ii) at least 3 are defective?

6. A successful attack by an interceptor depends upon

   (a) the reliable operation of a computing system,
   (b) the transmission of correct directions, and
   (c) the proper function of the striking mechanism.

   When the probability of (a) is 0.7 and (b) is assured, the overall probability of success is 0.6. If the computing system is improved to 95% reliability, while (b) has only 0.8 probability and the probability of (c) is unchanged, what is the new overall probability of success?

7. Certain components are manufactured in quantity; the probability that any one taken at random will be faulty is q. Show that if we are to be 99% certain that any set of four components will contain at least two sound ones, then q must be given by the equation

$$300q^4 - 400q^3 + 1 = 0$$

8. In a survey of a day's production from 400 machines making similar components, 4 items selected at random from the output of each machine were inspected in detail. The number m of machines producing f faulty items was found to be

f	0	1	2	3	4
m	16	89	145	118	32

Represent this data by a binomial distribution and calculate the theoretical distribution of faulty components. (C.E.I.)

9. Of a large number of mass-produced articles, one in ten is defective. Find the probability that a random sample of 20 will contain

(i) exactly 2 defective articles
(ii) at least 2 defective articles.

10. A radio transistor has a probability of functioning for at least 800 hours of 0.25. Find the probability that in a test sample of 20 transistors

(i) none function for more than 800 hours
(ii) only one functions for more than 800 hours.

11. If two out of every ten vehicles turn right at a T junction, what is the probability that

(i) three consecutive vehicles turn right,
(ii) three out of five consecutive vehicles turn left?

12. For a binomial distribution with $n = 5$ and $p = 1/4$, tabulate the separate probabilities. Find the mean and variance of the distribution

(i) directly
(ii) from the general formulae $\mu = np$, $\sigma^2 = npq$.

13. In an experiment the probability of success is 5/7; the experiment is repeated 12 times. Show that the most probable number of successes is 9 with probability of occurrence of 0.248. Draw a line diagram for the distribution.

14. In an experiment the probability of success is 1/4. Find the most probable number of successes and the associated probability in 15 trials. Draw a line diagram.

15. In the manufacture of screws by a certain process it was found that 5% of the screws were rejected because they failed to satisfy tolerance requirements. What was the probability that a sample of 12 screws contained (a) exactly 2 (b) not more than 2 rejects? (L.U.)

16. A mass-produced article is packed in cartons each containing 40 articles. Seven hundred cartons were examined for defectives with the results given in the table.

Defective articles per carton	0	1	2	3	4	5	6	More than 6
Frequency	390	179	59	41	18	10	3	0

Obtain the proportion of defective articles p.
Assuming that the total number of articles examined is so large that p can

be taken to represent the proportion of defectives in the population, use the Binomial distribution to calculate the probability that a random sample of 5 of these articles will contain 2 defectives. (L.U.)

12.6 POISSON DISTRIBUTION

Consider the following situations:

(i) the number of accidents/year in a particular factory
(ii) the number of faults in a length of cable
(iii) the number of cars crossing a bridge per hour
(iv) the number of bacteria in a square metre of material.

(i) In any year there are almost endless opportunities for accidents to occur but several factors help reduce the actual number to a minimum. We can calculate the average number of accidents/year based on a long period of experience.

(ii) Here we must talk in terms of, say, an average number of faults/kilometre.

(iii) In the design of bridges we want to avoid congestion: our only sensible measure is an average number of cars/hour.

(iv) This is a two-dimensional extension of (ii).

Ideally we want a distribution which involves only an average rate λ. We say that for isolated events in space and time, that is, in those situations where the average number of events in a specified interval is λ, the probability of r such events occurring in that interval is given by

$$p(r) = \frac{\lambda^r e^{-\lambda}}{r!} \tag{12.9}$$

The resulting distribution is known as the **Poisson distribution.**

We illustrate its application by examples.

Example 1

The number of goals scored in 500 league games were distributed as follows:

Goals/match	0	1	2	3	4	5	6	7	8
Frequency	52	121	129	90	42	45	18	1	2

Compare the frequency distribution to a Poisson distribution.

We first compute the average number of goals/match, $\lambda = 1173/500 = 2.346$. Then we calculate the Poisson frequencies: $500p(0)$, $500p(1)$, ..., $500p(8)$ and compare.

$p(0) = e^{-2.346}$ therefore $500p(0) = 500 \times 0.09575 = 48$

$p(1) = e^{-2.346} \times 2.346,$ therefore $500p(1) = 500 \times 0.2246 = 112$

$$p(2) = e^{-2.346} \times 2.346^2/2, \text{ therefore } 500p(2) = 500 \times 0.2635 = 132$$

and so we obtain the following theoretical frequencies:

48	113	132	103	60	28	11	4	1

Example 2

The average rate of telephone calls received at an exchange of 8 lines is 6 per minute. Find the probability that a caller is unable to make a connection if this is defined to occur when all lines are engaged within a minute of the time of the call.

Provided the overall rate of calls is constant, we can use the Poisson formulation. Our time unit is 1 minute and so $\lambda = 6$; formula (12.9) becomes

$$p(r) = \frac{6^r . e^{-6}}{r!}$$

Now the probability of not being able to make a call is the probability of there being at least 9 calls in any one interval of a minute. This is

$$\sum_{r=9}^{\infty} \frac{6^r e^{-6}}{r!} = 1 - \sum_{r=0}^{8} \frac{6^r e^{-6}}{r!}$$

We speed the calculations by noting that

$$p(r+1) = \frac{\lambda}{r+1} p(r)$$

$p(0) = e^{-6} = 0.0025$; all results in Table 12.2 are quoted to 4 d.p.

Table 12.2

r	$\lambda/(r+1)$	$p(r)$
0	6	0.0025
1	3	0.0150
2	2	0.0450
3	1.5	0.0900
4	1.2	0.1350
5	1.0	0.1620
6	6/7	0.1620
7	0.75	0.1389
8		0.1042
	Total	0.8546

Hence the probability we seek is $1 - 0.8546 = 0.1454$ (4 d.p.).
Is this accuracy justifiable?

Example 3

On average, 240 vehicles per hour pass through a check-point and a queue forms if more than 3 vehicles attempt to pass in a minute; what is the probability that a queue will form in any given minute? If the check-point is closed for five minutes, what is the probability that a queue of at least two cars will form in the five minutes?

The unit we work with is a minute, and the average number of cars per minute is 4. Hence we use the Poisson formula with $\lambda = 4$. The probability of a queue forming is the probability of at least 4 vehicles arriving in one minute

$$= \sum_{r=4}^{\infty} p(r)$$

$$= 1 - p(0) - p(1) - p(2) - p(3)$$

Now $p(0) = e^{-4} = 0.0183$

Table 12.3

r	$\lambda/(r+1)$	$p(r)$
0	4	0.0183
1	2	0.0732
2	4/3	0.1464
3	1	0.1952
	Total	0.4331

From Table 12.3

$$p(\text{queue}) = 1 - 0.4331 = 0.5669 \quad (4 \text{ d.p.})$$

In a period of five minutes the expected number of cars arriving is 20 and the probability of 2 or more arriving is

$$\sum_{r=2}^{\infty} \frac{20^r e^{-20}}{r!} = 1 - e^{-20} - 20e^{-20} = 1.0000 \quad (4 \text{ d.p.})$$

Example 4

A shop sells on average 16 articles of product A per month; if the shop re-orders each week, to what number must the shop-keeper make up his stock if the risk of his running out of stock is not to exceed 5%, 10%?

The model is Poisson (why?) with $\lambda = 4$ articles/week. If we stock x articles then the stock will run out if the demand exceeds x, so for the risk of running out not to exceed 5%, we require

$$\sum_{r=x+1}^{\infty} \frac{4^r e^{-4}}{r!} \leqslant 0.05$$

that is,

$$1 - \sum_{r=0}^{x} \frac{4^r e^{-4}}{r!} \leqslant 0.05$$

or

$$\sum_{r=0}^{x} \frac{4^r e^{-4}}{r!} \geqslant 0.95$$

We can proceed as before to accumulate probabilities until the total just exceeds 0.9. You can check for yourselves that for 5% risk the number of articles required is 8, and repeating the calculations for 10% risk gives 7.

The real crux of applying the Poisson distribution is the decision as to the value of λ by selecting the appropriate interval.

Poisson approximation to the Binomial distribution

It is clear that the Binomial formula (12.7) becomes tedious to operate in situations where n is large and p (or q) is very small; in addition, accuracy may be lost in the course of the calculations. We would like a somewhat simpler formula which is a good approximation to (12.7). Suppose then, we take the situation to the extreme. We seek a distribution which will approximate the Binomial distribution when n is very large and p (or q) is very small.

We achieve this by taking the limit of the Binomial distribution as $n \to \infty$, $p \to 0$ and $np = \lambda$ stays fixed. (np is the mean of the Binomial distribution). The case q very small is achieved by reversing *success* and *failure.*

We give a derivation of the formula for the probability of r successes in our new distribution; this derivation may be omitted if desired.

For the Binomial distribution, the probability formula (12.7)

$${}^nC_r\, p^r q^{n-r}$$

may be written

$$\frac{n(n-1)(n-2)\ldots\ldots(n-r+1)}{r!} \; \frac{(1-p)^n p^r}{(1-p)^r}$$

$$= \frac{1 \cdot \left(1-\frac{1}{n}\right) \cdot \left(1-\frac{2}{n}\right) \ldots\ldots \left(1-\frac{r-1}{n}\right)}{r!} \cdot \left(1-\frac{\lambda}{n}\right)^n \frac{\lambda^r}{\left(1-\frac{\lambda}{n}\right)^r}$$

Now, as $n \to \infty$ and $\lambda = np$ stays fixed and finite , then p must $\to 0$,

$$^{n}C_r\, p^r q^{n-r} \to \frac{\lambda^r}{r!} e^{-\lambda}$$

since

$$\left(1 - \frac{1}{n}\right) \text{ etc.} \to 1, \quad \left(1 - \frac{\lambda}{n}\right)^r \to 1 \quad \text{and} \quad \left(1 - \frac{\lambda}{n}\right)^n \to e^{-\lambda}$$

This is, of course, the Poisson formula (12.9).

Example 1

Certain mass-produced articles, of which 0.5% are defective, are packed in cartons each containing 100. What proportion of cartons are free from defective articles and what proportion contain two or more defectives?

Using Binomial $n = 100,\ p = 0.005$

$$p(0) = {}^{100}C_0\,(0.995)^{100} = 0.6026 \quad \text{(or 0.60534 by 5 figure logs)}$$

$$p(1) = {}^{100}C_1\,(0.995)^{99}(0.005) = 0.3028$$

$$p(2) = {}^{100}C_2\,(0.995)^{98}(0.005)^2 = 0.0753$$

Using Poisson $\lambda = np = 0.5$

$$p(0) = e^{-0.5} = 0.6065$$

$$p(1) = (0.5)e^{-0.5} = 0.3033$$

$$p(2) = (1/8)e^{-0.5} = 0.0758$$

Using the Poisson approximation, the proportion free from defectives is 60.65% and the proportion containing 2 or more defectives is 9.02%.

If the number of items involved is not large enough, the approximation may be poor.

Example 2

The average number of defective articles produced by a machine is 1 in 20; a batch of 10 is tested: what is the probability that there will be at least three defectives in the batch?

For the Binomial model $p = 0.05$ with $\mu = np = 10 \,.\, (1/20) = 0.5$ and variance $= npq = 0.5 \times 0.95 = 0.495$ (for Poisson mean and variance are equal).

We use Poisson, since p is small, via the formula

$$p(r) = \frac{(0.5)^r \, e^{-0.5}}{r!}$$

$$p(0 \text{ defectives}) = e^{-0.5} = 0.6065$$
$$p(1) = 0.5\, e^{-0.5} = 0.3033$$
$$p(2) = [(0.5)^2/2]\, e^{-0.5} = 0.0758$$

therefore

$$p(>2) = 1 - p(0) - p(1) - p(2) = 0.0144 \quad (4 \text{ d.p.})$$

For comparison, the Binomial model gives

$$p(0) = (0.95)^{10} = 0.5984$$
$$p(1) = 10(0.95)^9 \,.\, 0.05 = 0.3150$$
$$p(2) = \frac{10.9}{2}(0.95)^8(0.05)^2 = 0.0709$$

therefore

$$p(>2) = 0.0157 \quad (4 \text{ d.p.})$$

The approximation is about 8% in error.

Problems

1. (i) In a factory, 2000 employees will be asked to use a barrier cream for which the probability of an initial reaction sufficient to cause absence from work is 0.001. Find the probability that more than two employees will be absent when the barrier cream is introduced.

   (ii) The number of days n in a 100-day period when x accidents occurred in a factory is shown.

No. of accidents (x)	0	1	2	3	4	5
No. of days (n)	42	35	14	6	2	1

   (a) Fit a Poisson distribution to this data to predict the probability of x accidents occurring.
   (b) Compare the variance of the given and calculated distributions.
   (C.E.I.)

2. Telephone calls are received at a switchboard at an average rate of 1.5 calls per minute. The switchboard is operated by two telephonists, and calls have an equal chance of being dealt with by either. What is the probability that a given telephonist does not have to deal with any calls in a particular two-minute interval? Do not attempt to evaluate the resulting expression.

3. A machine-shop storekeeper finds that, over a long period, the average demand per week for a certain machine tool is 3. His stocking policy is

to make up stock to 4 at the beginning of each week. Estimate the probability that he will fail to satisfy demand in a given week, and determine his stocking policy if the chance of running out is not to exceed 5%.

4. On average, telephone calls are received at an exchange at the rate of 3 per minute. The exchange has 4 lines. What is the probability of being unable to make a connection if this arises when all lines are engaged within a minute of the time of the call?

5. A factory uses a particular spare part at the average rate of 3 a week. If stocks are replenished weekly, use the Poisson distribution to determine the number which should be in stock at the beginning of each week so that not more than once every year on the average (i.e. probability 1/52) will the stock be insufficient.

6. If the chance that any one of 10 telephone lines is busy at an instant is 0.2, what is the chance that 5 of the lines are busy? What is the most probable number of busy lines?

7. Refer to Problem 6, page 642.
On the assumption that the number of vehicles passing in a two-minute interval follows a Poisson distribution with the same mean as this data, calculate the probability of less than 3 vehicles passing in a two-minute interval.

8. A process for manufacturing a device comprising s connected components can be taken to consist of s independent operations. The probability of r flaws in any component is Poisson with $\lambda = 1$. Show that the probability of a defective device (contains more than one flaw) is $1 - e^{-s}(1 + s)$.

9. Stoppages on a building site occur according to the Poisson law on the average once in three weeks. Obtain the probabilities of 0, 1, 2, 3, 4 stoppages occurring in five weeks and determine if the occurrence of 5 stoppages in five weeks is significant.

10. The probability of a fault in a component is 1/1000.
    (i) Find approximately the probability that there will be 4 faulty components in a daily output of 2000.
    (ii) Find the minimum daily output to ensure a probability of 95% of at least one faulty component.

11. If, on average, one in every thousand entries in a table contains an error, what is the probability that a set of 100 readings taken from the table are all correct?

12. Oranges are packed in crates of one gross (144). It is estimated that, on average, 1 in 96 oranges is bad. In a consignment of 1000 crates, how

many will contain 2 or more bad oranges? (Use the Poisson approximation to the Binomial distribution.)

13. Describe the Binomial and Poisson distributions stating the circumstances in which each might be used.
Random samples of fifty are examined from each batch of a large consignment of manufactured articles and in such samples the average number of defective articles was found to be 3.1. A batch is rejected if the sample taken from it contains three or more defective articles. Assuming a Poisson distribution, show that the probability of a batch being rejected is approximately 0.6. Calculate the probability that of six batches sampled three or more will be rejected. Find also the probability that of 100 batches sampled at least half will be accepted. (L.U.)

14. The probability that any one machine will become defective in the small interval between time t and $t + \delta t$ after maintenance is $\alpha e^{-\alpha t}\delta t$, where α is a constant.
A firm possessing 10 similar machines institutes a weekly maintenance of the machines, each of which operates for 44 hours per week, and $\alpha = 1/400$ when the unit of time is one hour. Find
(i) the probability that all ten machines will continue to function throughout the week,
(ii) the probability that more than two machines will become defective before the weekly maintenance is due. (L.U.)

SUMMARY

It should not be forgotten that the Binomial and Poisson distributions are mathematical models. The simple rules of addition and multiplication and conditional probability are important and you will recognise that even in apparently straight-forward problems much careful thought is involved.

The concept of expectation is very important and should be understood. You need to be able to decide when it is appropriate to use either the Binomial or the Poisson model and you should be able to calculate the required probability in such a situation by a combination of the appropriate formula and the rules of probability.

The two definitions of probability should be known.

Chapter Thirteen

The Normal Distribution and Significance Tests

13.1 CONTINUOUS PROBABILITY DISTRIBUTIONS

In the last chapter we concerned ourselves with discrete probability distributions. Consider the experiment of spinning a mounted pointer and measuring the final angle of deflection made with some fixed direction (easiest to visualise as an unmagnetised compass needle); again consider the time taken to perform a certain calculation. In both cases the distribution may seem to be discrete because of the limitations of measuring devices, but, in theory, the angle can be any of an infinite number of possible angles in the interval $[0, 360^\circ)$ and the time taken can be any value in the range $(0, \infty)$. With the spun pointer, if each outcome has an equal probability of occurrence this is infinitesimally small and cannot be distinguished from zero. In general, for continuous distributions it makes little sense to talk about the probability of a single outcome. Rather, we confine our attention to probabilities of an interval of outcomes; for example, we may say that the probability of the angle at which the pointer settles being in the range 0 to 180° is 1/2, in the range 90° to 180° is 1/4, in the range 150° to 180° is 1/12, and in the general range $[\theta_1, \theta_2]$ is $(\theta_2 - \theta_1)/360$. These ideas lead us to the idea of a **probability density function** (p.d.f.) which, in this example, is defined by

$$\lim_{\delta\theta \to 0} \left\{ \frac{1}{\delta\theta} p(\theta_0 \leqslant \theta \leqslant \theta_0 + \delta\theta) \right\} = \rho(\theta_0)$$

There is a clear analogy between the discrete distribution and a light rod bearing isolated point loads and between the continuous distribution and a heavy rod where the weight is distributed continuously along its length: hence the term 'density function'. The properties that the density function $\rho(x)$ must obey are

(i) $\rho(x) \geqslant 0$ for all x in the sample space

(ii) $\displaystyle\int_{-\infty}^{\infty} \rho(x)\,dx = 1$, or, if the interval of outcomes is $[a, b]$,

$$\int_a^b \rho(x)\,dx = 1$$

Notice how Σ for discrete distributions is replaced by $\int$ for continuous distributions.

The spun pointer experiment is an example of a **rectangular distribution** and the graph is shown in Figure 13.1. Note that the area under the graph is 1.

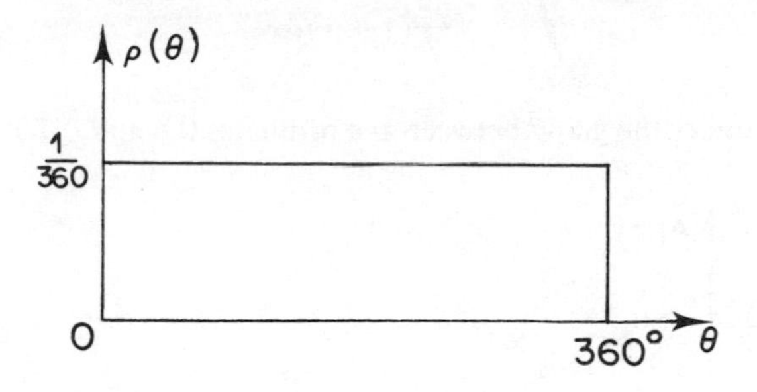

Figure 13.1

We can concoct any continuous distribution by defining a suitable p.d.f.

Example

Are any of the following functions p.d.f.'s?

(a) $\rho(x) = \frac{1}{2}x$, defined on $[1, 4]$;

(b) $\rho(x) = \frac{1}{4} - x$, defined on $[-2, 2]$;

(c) $\rho(x) = \lambda(1 - x)$, defined on $[\frac{1}{2}, \frac{3}{4}]$.

Sketch the graphs of those which are.

(a) $\rho(x) \geqslant 0$, but $\displaystyle\int_1^4 \tfrac{1}{2}x \, dx = [\tfrac{1}{4}x^2]_1^4 = 4 - \tfrac{1}{4} \neq 1$, therefore not p.d.f.

(b) $$\int_{-2}^{2} (\tfrac{1}{4} - x)dx = \left[\frac{1}{4}x - \frac{x^2}{2}\right]_{-2}^{2} = (\tfrac{1}{2} - 2) - (-\tfrac{1}{2} - 2) = 1,$$

but $\rho(x) < 0$ for x in $(\frac{1}{4}, 2]$, therefore not p.d.f.

(c) $\rho(x) \geqslant 0$ if $\lambda \geqslant 0$;
$$\int_{1/2}^{3/4} \lambda(1 - x)dx = \lambda\left[x - \frac{x^2}{2}\right]_{1/2}^{3/4}$$
$$= \lambda\left\{\left(\frac{3}{4} - \frac{9}{32}\right) - \left(\frac{1}{2} - \frac{1}{8}\right)\right\}$$
$$= \lambda\left\{\frac{15}{32} - \frac{3}{8}\right\}$$
$$= \lambda \cdot \frac{3}{32}$$

and thus $\lambda = 32/3$, if property (ii) is to be satisfied. Assuming this value for λ, the graph is shown in Figure 13.2. Again the area under the graph is 1. Notice that if we want $p(0.6 \leqslant x \leqslant 0.7)$ this is

$$\int_{0.6}^{0.7} \frac{32}{3}(1-x)\mathrm{d}x$$

(the area under the graph between the ordinates 0.6 and 0.7.)

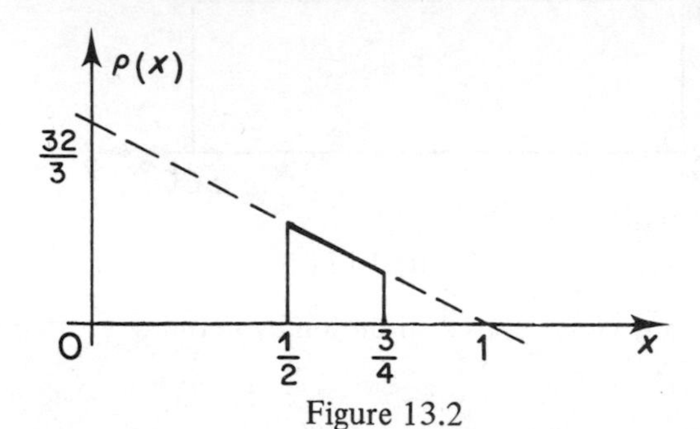

Figure 13.2

Cumulative distribution function

We find it useful to talk in terms of a **cumulative distribution function** (c.d.f.) $F(x)$ which gives the probability of all events up to and including a particular value, that is

$$F(x_0) = p(X \leqslant x_0) = \int_{-\infty}^{x_0} p(x)\mathrm{d}x \tag{13.1}$$

$$\text{for interval } [a, b],\ F(x_0) = \int_{a}^{x_0} p(x)\mathrm{d}x$$

For a discrete distribution we write

$$F(x_r) = \sum_{i=0}^{r} p(x_i) \tag{13.1a}$$

For the rectangular distribution of Figure 13.1,

$$F(\theta) = \int_0^{\theta} \frac{1}{360}\mathrm{d}\theta = \frac{\theta}{360}$$

and for the p.d.f. of Figure 13.2,

$$F(x) = \int_{1/2}^{x} \frac{32}{3}(1-x)\mathrm{d}x = \frac{32}{3}\left[x - \frac{x^2}{2}\right] - 4$$

Note that

$$\int_{-\infty}^{1/2} \rho(x)\,\mathrm{d}x$$

is not relevant since x is defined only on [½, ¾]. The graphs are shown in Figures 13.3(i) and (ii).

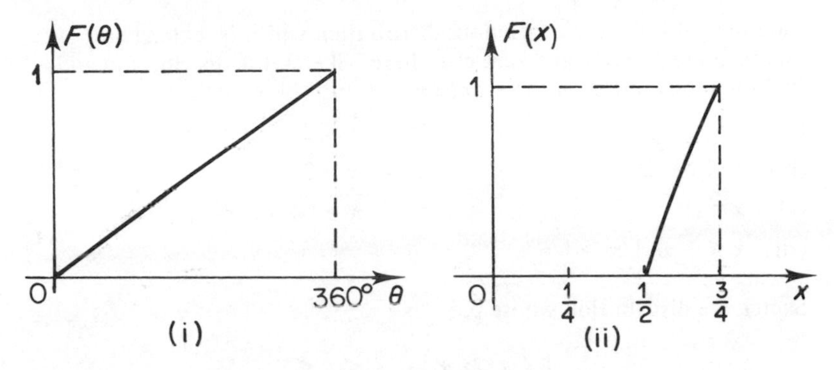

Figure 13.3

Measures of central tendency and spread

Other features which are analogous to those for discrete distributions are listed. X is assumed to be a continuous random variable defined on $(-\infty, \infty)$.

(i) $\mu = E(X) = \int_{-\infty}^{\infty} x\rho(x)\,\mathrm{d}x$

Properties (i) to (v) on page 667 hold here also.

(ii) For a unimodal distribution, the mode is found where $\dfrac{\mathrm{d}\rho(x)}{\mathrm{d}x} = 0.$

(iii) The median M is such that $\int_{-\infty}^{M} \rho(x)\,\mathrm{d}x = \dfrac{1}{2} = \int_{M}^{\infty} \rho(x)\,\mathrm{d}x$

(iv) The variance $= \int_{-\infty}^{\infty} (x-\mu)^2 \rho(x)\,\mathrm{d}x$

$$= \int_{-\infty}^{\infty} x^2\rho(x)\,\mathrm{d}x - 2\mu \int_{-\infty}^{\infty} x\rho(x)\,\mathrm{d}x + \mu^2 \int_{-\infty}^{\infty} \rho(x)\,\mathrm{d}x$$

$$= \int_{-\infty}^{\infty} x^2\rho(x)\,\mathrm{d}x - 2\mu\,.\,\mu + \mu^2\,.\,1$$

$$= \int_{-\infty}^{\infty} x^2 \rho(x)\mathrm{d}x - \mu^2$$

$$= E(X^2) - [E(X)]^2$$

Problems

1. Find the p.d.f. for the continuous distribution which is rectangular on the interval $[1, 4]$ and zero elsewhere. Sketch the distribution and find the probabilities of the events:

   (i) $x < 0$ (ii) $x < 2$

   (iii) $1 < x < 2$ (iv) $x > 2$

   (v) $x = 1\frac{1}{2}$ (vi) $x = 1\frac{1}{2}$ or $x = 2$

   (vii) $x < 2$ and $x > 1$

2. Sketch the distribution whose p.d.f. is

$$\rho(x) = \begin{cases} k(2x+1), & 0 < x < 2 \\ 0, & \text{elsewhere} \end{cases}$$

   Find the value of k. Find the probability of the events
   $A =$ '$x < 1$', $B =$ '$x > \frac{1}{2}$', $C =$ '$x \in (\frac{1}{2}, 1)$'

3. Sketch the distribution whose p.d.f. is

$$\rho(x) = \begin{cases} k\mathrm{e}^{-x}, & 0 < x < 1 \\ 0, & \text{elsewhere} \end{cases}$$

   Find the value of k and find the probabilities that $x < \frac{1}{2}$, $x > \frac{3}{4}$ and $x \in (\frac{1}{2}, \frac{3}{4})$.

4. Obtain and plot the cumulative distribution function for the variable which has probability density function

$$\rho(x) = \tfrac{1}{4}(4 - 2x) \quad 0 < x < 2$$

   Find the probabilities of '$x < 1$', '$x > 1\frac{1}{2}$', '$x > \frac{3}{4}$'.
   Also find x such that $p(X \leqslant x) = \frac{1}{2}$.

5. Assuming that the number of vehicles passing a checkpoint is a Poisson variable with a given average of vehicles per hour, then the length of time interval between two cars passing (headway) is a variable which follows the **negative exponential distribution.** If λ is the number of events/unit time then the probability density function for the N intervals is $\rho(t) = \lambda\mathrm{e}^{-\lambda t}$ where t is the number of unit time intervals (measured in seconds). Show that

$$\int_0^\infty \rho(t)\,\mathrm{d}t = 1$$

and graph both the distribution and the cumulative distribution function. Find the probabilities that the headway is $\geqslant 8$ seconds for vehicle flows of 100(100)1000 vehicles/hour; find the mean time headway and show that the mean headway for all intervals $\geqslant t$ is $t + (1/\lambda)$.

13.2 THE NORMAL DISTRIBUTION

We saw in Section 1.4 that a measurement subject to error had observational errors following a **normal** or **Gaussian distribution.** Many continuous distributions in practice, such as the heights or weights of a large group of people or the life of a class of electrical components follow approximately a normal distribution, a sketch of which is shown in Figure 13.4.

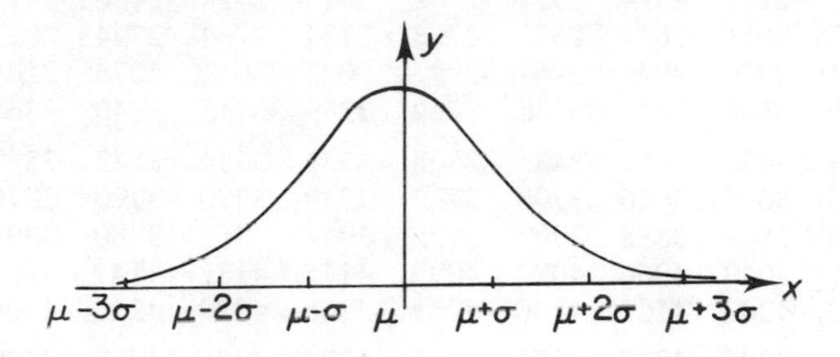

Figure 13.4

This distribution is defined entirely in terms of its mean μ and standard deviation σ. The defining formula is

$$y = \frac{1}{\sigma\sqrt{2\pi}}\,\mathrm{e}^{-\frac{1}{2}[(x-\mu)/\sigma]^2} \tag{13.2}$$

or we sometimes say *x is normally distributed with mean μ and standard deviation σ,* written

$$x \sim N(\mu, \sigma^2)\,\dagger \tag{13.2a}$$

Note the symmetry of the distribution about μ and the rapid decay on either side.

Although a whole host of such distributions exists, we make use of the fact that, if $x \sim N(\mu, \sigma^2)$ then the variable

$$z = \frac{x-\mu}{\sigma} \sim N(0, 1)$$

† Some authors use $N(\mu, \sigma)$; care is needed when reading such information.

Table 13.1

$$\frac{1}{\sigma\sqrt{2\pi}}\int_{\mu}^{x}\exp\left\{-\frac{(x-\mu)^2}{2\sigma^2}\right\}dx = \frac{1}{\sqrt{2\pi}}\int_{0}^{(x-\mu)/\sigma} e^{-z^2/2}\,dz$$

THE NORMAL PROBABILITY INTEGRAL

$\frac{x-\mu}{\sigma}$	0	1	2	3	4	5	6	7	8	9
0	0000	0040	0080	0120	0160	0199	0239	0279	0319	0359
.1	0398	0438	0478	0517	0557	0596	0636	0675	0714	0753
.2	0793	0832	0871	0909	0948	0987	1026	1064	1103	1141
.3	1179	1217	1255	1293	1331	1368	1406	1443	1480	1517
.4	1555	1591	1628	1664	1700	1736	1772	1808	1844	1879
.5	1915	1950	1985	2019	2054	2088	2123	2157	2190	2224
.6	2257	2291	2324	2357	2389	2422	2454	2486	2517	2549
.7	2580	2611	2642	2673	2703	2734	2764	2794	2822	2852
.8	2881	2910	2939	2967	2995	3023	3051	3078	3106	3133
.9	3159	3186	3212	3238	3264	3289	3315	3340	3365	3389
1.0	3413	3438	3461	3485	3508	3531	3554	3577	3599	3621
1.1	3643	3665	3686	3708	3729	3749	3770	3790	3810	3830
1.2	3849	3869	3888	3907	3925	3944	3962	3980	3997	4015
1.3	4032	4049	4066	4082	4099	4115	4131	4147	4162	4177
1.4	4192	4207	4222	4236	4251	4265	4279	4292	4306	4319
1.5	4332	4345	4357	4370	4382	4394	4406	4418	4429	4441
1.6	4452	4463	4474	4484	4495	4505	4515	4525	4535	4545
1.7	4554	4564	4573	4582	4591	4599	4608	4616	4625	4633
1.8	4641	4649	4656	4664	4671	4678	4686	4693	4699	4706
1.9	4713	4719	4726	4732	4738	4744	4750	4756	4761	4767
2.0	4772	4778	4783	4788	4793	4798	4803	4808	4812	4817
2.1	4821	4826	4830	4834	4838	4842	4846	4850	4854	4857
2.2	4861	4865	4868	4871	4875	4878	4881	4884	4887	4890
2.3	4893	4896	4898	4901	4904	4906	4909	4911	4913	4916
2.4	4918	4920	4922	4925	4927	4929	4931	4932	4934	4936
2.5	4938	4940	4941	4943	4946	4947	4948	4949	4951	4952
2.6	4953	4955	4956	4957	4959	4960	4961	4962	4963	4964
2.7	4965	4966	4967	4968	4969	4970	4971	4972	4973	4974
2.8	4974	4975	4976	4977	4977	4978	4979	4979	4980	4981
2.9	4981	4982	4982	4983	4984	4984	4985	4985	4986	4986
3.0	4987	4990	4993	4995	4997	4998	4998	4999	4999	4999

Note: the lowest line has entries at spacing 0.1.

$N(0, 1)$ is called the **standard normal distribution**; it is given by

$$y = \frac{1}{\sqrt{2\pi}} e^{-z^2/2}$$

and it has been tabulated in Table 13.1. It is of great significance, since we can easily convert statements about $N(\mu, \sigma^2)$ to ones about $N(0, 1)$, which can then be assessed.
The values quoted (to 4 d.p.) lie between 0 and 0.5 (why?).

The probability of x lying in the range $[\mu - \sigma, \mu + \sigma]$ is approximately 68%; in $[\mu - 2\sigma, \mu + 2\sigma]$ is approximately 95% and in $[\mu - 3\sigma, \mu + 3\sigma]$ is approximately 99.7%. Put another way, the probability of x lying outside $[\mu - \sigma, \mu + \sigma] \simeq 32\%$, outside $[\mu - 2\sigma, \mu + 2\sigma] \simeq 5\%$ and outside $[\mu - 3\sigma, \mu + 3\sigma]$ is about ¼%.

Notice that Table 13.1 gives the probability of x lying in a shaded area shown in Figure 13.5.

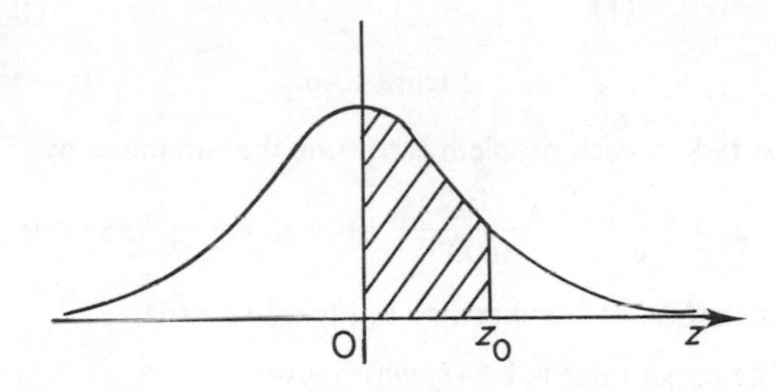

Figure 13.5

For example,

$$p(0 \leqslant z \leqslant 1.5) = 0.4332 \quad (= 43.32\%)$$
$$p(0 \leqslant z \leqslant 2.67) = 0.4962$$
$$p(0 \leqslant z \leqslant 3.6) = 0.4998$$

You should confirm these values from Table 13.1.

We now apply these ideas to an example which illustrates how to compute probabilities for various kinds of interval relative to the mean.

Example

A machine produces components of mean diameter 1.535 cm with standard deviation 0.005 cm. The diameters are assumed normally distributed.

(a) Find the probabilities

(i) that a component has diameter between 1.535 and 1.543 cm
(ii) that a component has diameter between 1.523 and 1.535 cm
(iii) that a component has diameter between 1.529 and 1.542 cm
(iv) that a component has diameter between 1.537 and 1.544 cm
(v) that a component has diameter between 1.527 and 1.533 cm.

(b) If all components with diameters outside the range 1.528 to 1.540 cm are rejected, what proportion of components are rejected? What is the probability that of 4 components selected at least 3 will be rejected?

(c) If 28% of components have diameters less than a given value, what is this value?

The distribution of the diameters is shown in Figure 13.6(i).

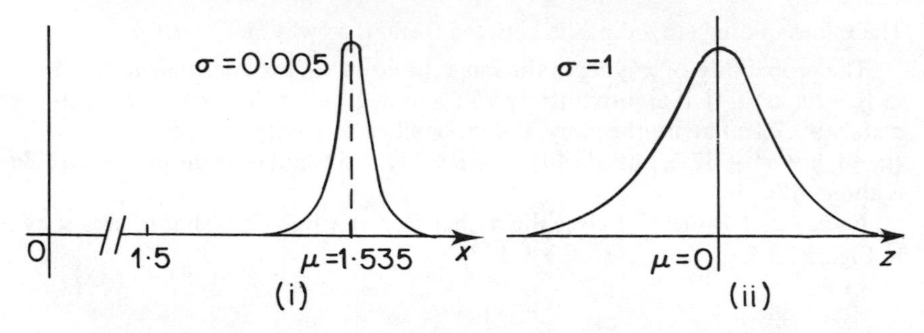

Figure 13.6

(a) The first task in each problem is to code the variable x by

$$z = \frac{x-\mu}{\sigma} = \frac{x-1.535}{0.005} \quad \text{so that, e.g. } 1.535 \equiv 0.$$

The distribution of z is sketched in Figure 13.6(ii).

(i) The upper limit is 1.543, which gives

$$z = \frac{1.543 - 1.535}{0.005} = 1.6$$

We want the value of the shaded area in Figure 13.7(i).
From the table we have that the area corresponding to $z = 1.6$ is 0.4452 and hence the probability we seek is 44.52% (2 d.p.).
Note we are saying that 1.543 is 1.6 standard deviations above the mean.

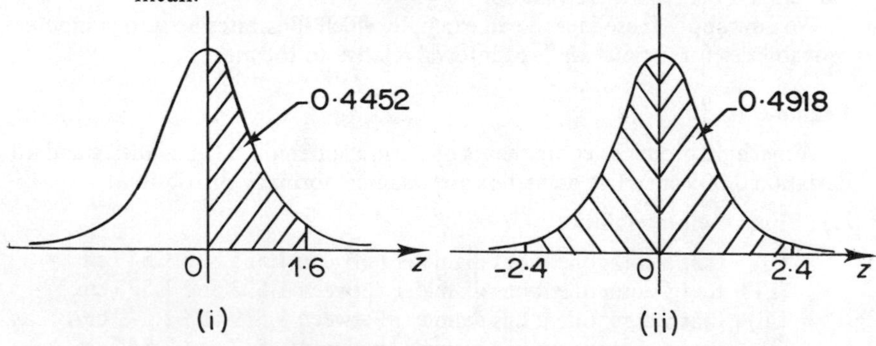

Figure 13.7

(ii) Here the lower limit is

$$z = \frac{1.523 - 1.535}{0.005} = -2.4$$

By the symmetry of the normal distribution, the shaded area we want is equal to the shaded area between 0 and + 2.4 as shown in Figure 13.7 (ii).
From Table 13.1 we pick up 0.4918 or 49.18% and this is the probability required.

(iii) The limits are

$$z_1 = \frac{1.529 - 1.535}{0.005} = -1.2$$

and

$$z_2 = \frac{1.542 - 1.535}{0.005} = 1.4$$

By reference to Figure 13.8 (i) we can see that the area we want is the sum of two areas. The one between 0 and z_2 has the associated probability 0.4192. Now, the shaded area between −1.2 and 0 is equal to one between 0 and +1.2 which is 0.3849; the sum of these areas is 0.8041 and hence we obtain a probability of 80.41% (2 d.p.)

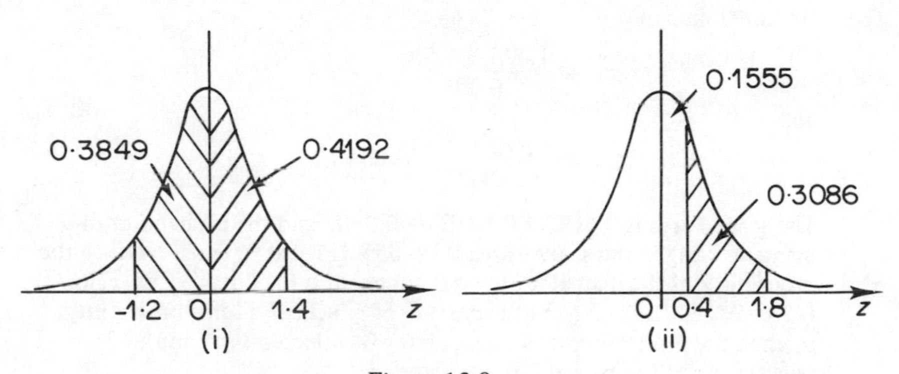

Figure 13.8

(iv) Here the limits are

$$z_1 = \frac{1.537 - 1.535}{0.005} = 0.4$$

and

$$z_2 = \frac{1.544 - 1.535}{0.005} = 1.8$$

The shaded area we want is the difference between the area from 0 to 1.8 and the area from 0 to 0.4 [see Figure 13.8 (ii)]. The first of these areas is 0.4641 and the second is 0.1555 and, therefore, we have the area difference of 0.3086 giving a probability of 30.86% (2 d.p.).

(v) In a like manner, these limits are

$$z_1 = \frac{1.527 - 1.535}{0.005} = -1.6$$

and

$$z_2 = \frac{1.533 - 1.535}{0.005} = -0.4$$

By symmetry, we want the area between 0.4 and 1.6; this is $0.4452 - 0.1555 = 0.2897$. Refer to Figure 13.9 (i).
Prob. = 28.97%

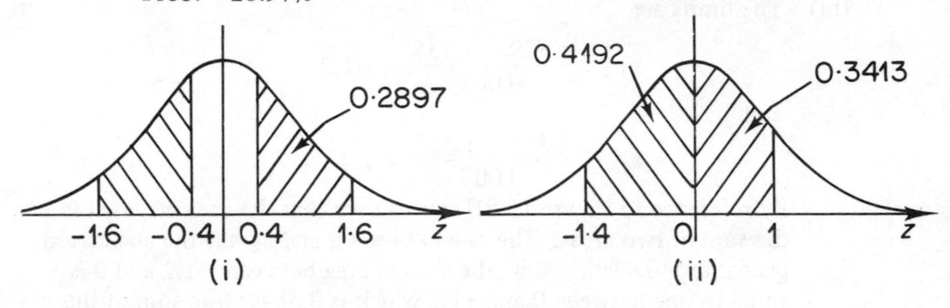

Figure 13.9

(b) We have limits of

$$z_1 = \frac{1.528 - 1.535}{0.005} = -1.4$$

and

$$z_2 = \frac{1.540 - 1.535}{0.005} = 1.0$$

The shaded area is $0.4192 + 0.3413 = 0.7605$ and the probability of a diameter lying in the shaded area is 76.05% [Figure 13.9 (ii)]. Hence the probability of the diameter of one component lying outside this region is 23.95%, which is also the proportion of rejects. Call this probability p; then the probability of 4 rejects out of 4 selected is p^4 and of 3 rejects is $4p^3$. The probability we want is

$$\begin{aligned} p^4 + 4p^3q &= (0.2395)^4 + 4(0.2395)^3(0.7605) \\ &= (0.2395)^3(3.2815) \\ &= \underline{4.58\%} \quad (2 \text{ d.p.}) \end{aligned}$$

(c) To find the line cutting off 28% of the area (see Figure 13.10) we note by symmetry we want the z-score corresponding to an area of 0.22 and this is between 0.58 and 0.59, from the table. We choose the nearer value 0.58 and hence have $z_1 = -0.58$; decoding, we obtain

$$\begin{aligned} x &= \mu - 0.58\sigma \\ &= 1.535 - 0.58 \times 0.005 \end{aligned}$$

$$= 1.535 - 0.0029$$

$$= \underline{1.532} \quad \text{(3 d.p.)}$$

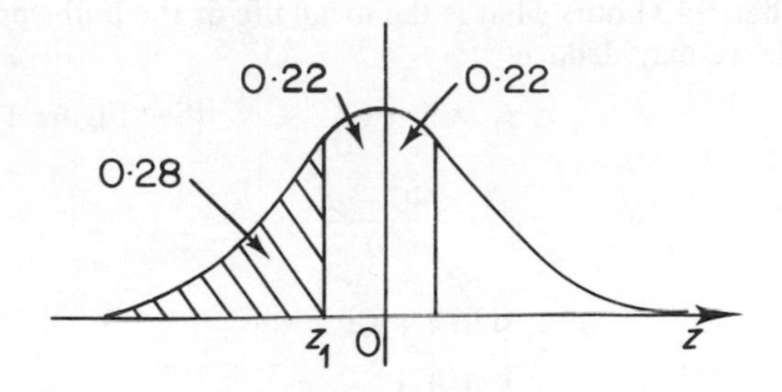

Figure 13.10

In the following three examples we illustrate other questions which might be asked.

Example 1

The weights of packets of a commodity have mean 18.2 gm and it is observed that 10% of the packets are above an acceptable level 18.7 gm. What is the standard deviation of the distribution, assuming it to be normal?

In this case we have

$$z_1 = \frac{x_1 - \mu}{\sigma} = \frac{18.7 - 18.2}{\sigma} = \frac{0.5}{\sigma}$$

But, from Table 13.1, $z_1 = 1.28$ (2 d.p.), therefore

$$\sigma = \frac{0.5}{1.28} = 0.39 \text{ gm} \quad \text{(2 d.p.) [See Figure 13.11 (i)]}$$

It is debatable whether the result should be quoted as 0.4 gm if the mean of 18.2 is accurate only to the last figure quoted; were it 18.20 then we should be justified in retaining 2 d.p. for σ.

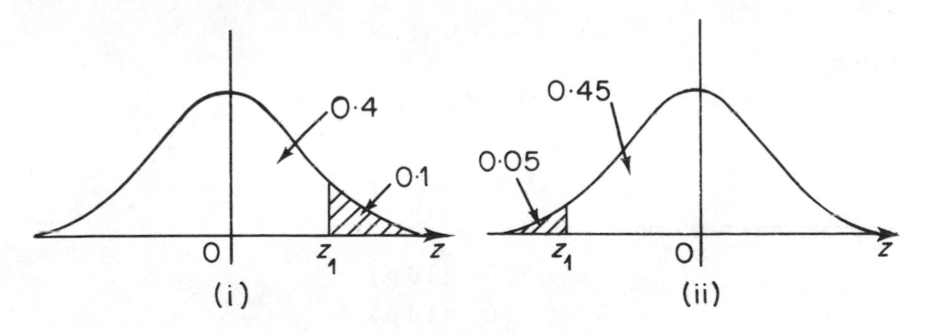

Figure 13.11

Example 2

The lengths of life of light bulbs produced in a particular factory have standard deviation 50 hours. If the lives are assumed normally distributed and 5% last less than 940 hours what is the mean life of the bulbs produced?

From the table we may deduce

$$z_1 = -1.64 \qquad \text{[See Figure 13.11 (ii)]}$$

But

$$z_1 = \frac{940 - \mu}{50}$$

therefore

$$\begin{aligned} \mu &= 940 + 1.64 \times 50 \\ &= 940 + 82 \\ &= 1022 \text{ hours} \end{aligned}$$

Example 3

Given that the lengths of bolts produced by a machine are normally distributed, that 12.8% have lengths greater than 12.60 cm and 15.1% less than 12.47 cm, what are the mean and standard deviation of the distribution? [See Figure 13.12]

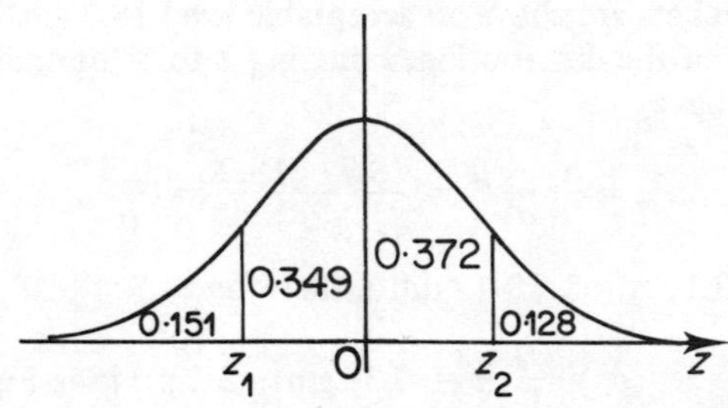

Figure 13.12

We obtain

$$z_2 = 1.14 \quad \text{(2 d.p.)}$$

and

$$z_1 = -1.03 \quad \text{(2 d.p.)}$$

Hence

$$\frac{12.60 - \mu}{\sigma} = 1.14$$

and

$$12.47 = -1.03\sigma + \mu$$

Hence we solve to obtain

$$\begin{aligned} \mu &= 12.53 \quad \text{(2 d.p.)} \\ \sigma &= 0.06 \quad \text{(2 d.p.)} \end{aligned}$$

Problems

1. Assuming that the variable z is distributed as $N(0, 1)$ use tables to find:
   (a) $P(z < 1.82)$ (b) $P(z > 0.58)$
   (c) $P(z > -2.03)$ (d) $P(z < -2.64)$
   (e) $P(1.20 < z < 1.35)$ (f) $P(-2.35 < z < 0)$
   (g) $P(-3 < z < -2)$ (h) $P(-0.05 < z < 0.25)$
   (i) $P(|z| > 1.82)$ (j) $P(|z| < 0.38)$
   (k) Find z' such that $P(z < z') = 0.75$
   (l) Find z' such that $P(0 < z < z') = 0.1$
   (m) Find z' such that $P(z < z') = 0.30$
   (n) Find z' such that $P(1 < z < z') = 0.08$
   (o) Find z' such that $P(-3 < z < z') = 4\%$
   (p) If x is $N(60, 16)$, find $P(50 < x < 70)$
   (q) If x is $N(0.36, 0.04)$, find $P(-0.1 < x < 0.5)$
   (r) If T is $N(2.5 \times 10^3, 10^4)$, find T' such that $T > T'$ in only 1% of all cases.
   (s) If T is $N(230, 120)$, find T' such that $T > T'$ in 99.9% of all cases.

2. Components are manufactured to be 18 in. in length, but they are acceptable inside the limits $17\frac{15}{16}$ in and $18\frac{1}{16}$ in. Observation indicates that about 2½% are rejected as too long and about 2½% as too short. Assuming that the lengths are normally distributed about the mean of 18 in. find the standard deviation of the distribution. Hence calculate the proportion of rejects if the tolerance limits are narrowed to $17\frac{61}{64}$ in and $18\frac{3}{64}$ in.

3. The average life of a 250 watt electric motor is 8 years, with a standard deviation of 2 years. The manufacturer replaces, free of charge, all motors that fail whilst under guarantee. Assuming that the motor lives are normally distributed, how long a guarantee should he provide if he is willing to replace no more than 2% of all the motors he sells? What proportion of motors will still be serviceable after 11 years?

4. Cylinders of diameter 1 cm are to fit into holes whose diameters are normally distributed with mean 1.002 cm and variance 3×10^{-6} cm? How many holes out of 500 would you expect to be too small?

5. The following table gives the electrical resistance of 138 carbon rods tested at the same temperature.

Resistance in ohms (mid-class value)	310	311	312	313	314	315	316	317	318	319	320
Frequency	1	2	6	21	25	32	24	18	5	3	1

Can the data reasonably be regarded as a sample from a normal distribution?

6. A machine is producing screws whose lengths are normally distributed with mean 1.002 cm and standard deviation 0.003 cm. Limit gauges are used to reject all screws greater than 1.010 cm in length and all screws less than 0.990 cm in length. What proportion of the screws will be rejected?

7. The breaking strength in N/m^2 of 100 mild steel specimens are arranged in the following frequency distribution with a class interval of 1000 N/m^2, the intervals being denoted by their central points.

Breaking strength	Frequency
65,500	1
66,500	0
67,500	4
68,500	10
69,500	14
70,500	22
71,500	18
72,500	14
73,500	8
74,500	5
75,500	3
76,500	1

Find the equation of the normal curve which has the same mean, standard deviation as this distribution and show that the frequencies deduced from its curve are: 0, 2, 4, 9, 14, 19, 19, 15, 10, 5, 2, 1.

8. In a normal distribution the mean is 80, the standard deviation 15 and the total number of items 1000.
   (i) Estimate the number of items between 65 and 95.
   (ii) Find the probability that an item chosen at random will be greater than 100.
   (iii) What is the least value that an item in the highest 30% can have?
   (iv) Between what values of the variable will the middle 60% of the frequencies lie?
   (v) The probability that a certain value of the variable will be exceeded is 5%. What is the value?
   (vi) What are the quartiles of the distribution?

9. The ground manoeuvring times of 84 aircraft landing at an airport on one day were

Time (min)	2	3	4	5	6	7
Frequency	8	13	27	23	10	3

Calculate the mean and standard deviation of this sample and derive the corresponding normal frequency distribution. (C.E.I.)

10. The lives in hours of electrical components is assumed to be distributed $N(1400, 300^2)$.
    (a) What is the probability that a component taken at random will have a life between 1400 and 1850 hours?
    (b) What is the percentage of components that will last longer than 2100 hours?
    (c) If the components are guaranteed for 1000 hours, what percentage of them are returnable?
    (d) What life should the manufacturer guarantee if he wants 95% of all components to satisfy that guarantee?

11. Analysis of past data has shown a manufacturer that the hub thickness of a particular type of gear is normally distributed about a mean thickness of 2.00 cm, with standard deviation 0.04 cm.
    Estimate, in a production run of 5000 such gears, how many
    (i) will have a thickness greater than 2.06 cm
    (ii) will have thicknesses between 1.97 and 2.03 cm
    (iii) will have thicknesses between 1.89 and 1.95 cm?

12. A machine is producing washers whose diameters have a standard deviation of 0.5 mm. If the tolerance allowed on diameter is ±0.8 mm, and diameter can be assumed to have a normal distribution, approximately what percentage of washers will be rejected?

13. Rods are made to a nominal length of 4 cm but in fact they form a normal distribution with mean 4.01 cm and standard deviation 0.03 cm. Each rod costs 6p to make and may be used immediately if its length lies between 3.98 and 4.02 cm. If its length is less than 3.98 cm, the rod cannot be used but has a scrap value of 1p. If its length exceeds 4.02 cm, it may be shortened and used at a further cost of 2p. Find the average cost per usable rod.

13.3 NORMAL APPROXIMATION TO BINOMIAL DISTRIBUTION

Although the normal distribution is continuous, it can be used under certain circumstances as a reasonable approximation to the Binomial distribution. It is not easy to give a hard and fast rule for when the approximation is reasonable, but, generally speaking, using the terminology for the Binomial distribution, if $p \triangleq \frac{1}{2}$, or if np and nq are both > 5 then the approximation can be taken to be a fair one. We show the approximation by examples.

Example 1

A die is thrown 18 times: what is the probability of 8 successful outcomes, if a successful outcome is defined to be a 'five' or 'six'?

The binomial model has $p = 1/3$, $q = 2/3$, $n = 18$ (note $np = 6$, $nq = 12$) which has mean $\mu = np = 6$ and standard deviation $\sigma = \sqrt{npq} = \sqrt{(1/3).(2/3).18} = 2$. We now approximate this model by a normal distribution with $\mu = 6$, $\sigma = 2$.

We approximate the probability of 8 successes by the area under this curve between 7.5 and 8.5. Now we cannot score 7.5 successes, but we are effectively splitting the difference between the last case we want to exclude – 7 successes – and the 8 we want to include; a similar consideration obtains at the 8.5 mark [Figure 13.13(i)]. We are effectively approximating a histogram by the normal curve and approximating the rectangle based on 8 by the area mentioned above. The normal curve and the histogram are superimposed in Figure 13.13 (ii)

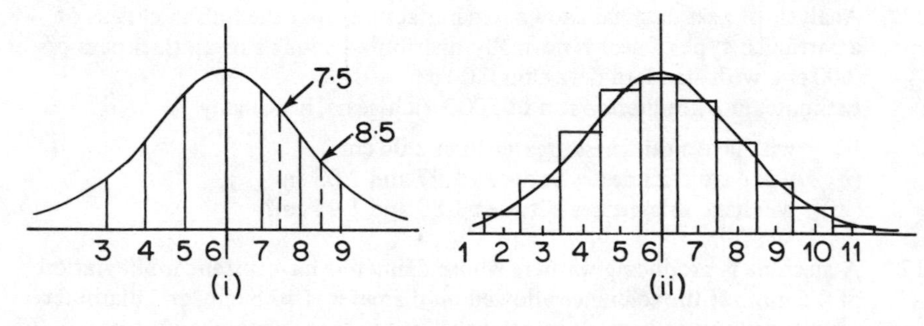

Figure 13.13

Table 13.2

r	0	1	2	3	4	5	6	7
B	0.0007	0.0051	0.0259	0.0690	0.1294	0.1818	0.1963	0.1682
N	0.0024	0.0092	0.0279	0.0655	0.1210	0.1747	0.1974	0.1747

r	8	9	10	11	12	13	$\geqslant 14$
B	0.1157	0.0643	0.0289	0.0105	0.0031	0.0007	0.0001
N	0.1210	0.0655	0.0279	0.0092	0.0024	0.0005	Negligible

Binomial probabilities shown sum to 1.0000
Normal probabilities shown sum to 0.9993

Now we find this area under the normal curve:

The limits are

$$z_1 = \frac{7.5 - 6}{2} = 0.75 \quad \text{and} \quad z_2 = \frac{8.5 - 6}{2} = 1.25$$

The required area is 0.3944 − 0.2734 = 0.1210 and the probability we obtain is therefore 12.10% (2 d.p.); the exact binomial result is

$$^{18}C_8\, p^8 q^{10} = {}^{18}C_8 (1/3)^8 (2/3)^{10} = 11.57\% \quad (2 \text{ d.p.})$$

A comparison of Binomial (B) and Normal (N) probabilities is shown in Table 13.2, where r is the number of successes.

Example 2

For the experiment above obtain the probabilities of

(i) at least 12 successes
(ii) between 8 and 10 successes (i.e. 8, 9 or 10).

These results can be obtained by adding entries from Table 13.2 or directly, as follows:

(i) p(at least 12 successes) $\hat{=}$ $p(11.5 \leqslant$ successes).
For 11.5 heads,

$$z = \frac{11.5 - 6}{2} = 2.75$$

The probability that z lies between 0 and 2.75 is 0.4970 and so the probability of at least 12 successes is at most 0.0030 or 0.30% (2 d.p.).
The corresponding Binomial result is 0.39% (2 d.p.)

(ii) Corresponding to 7.5 successes is

$$z_1 = \frac{7.5 - 6}{2} = 0.75$$

Corresponding to 10.5 successes is

$$z_2 = \frac{10.5 - 6}{2} = 2.25$$

The probability that z lies between 0.75 and 2.25 is

$$0.4878 - 0.2734 = 0.2144 = 21.44\% \quad (2 \text{ d.p.})$$

The corresponding Binomial result is 20.89% (2 d.p.)

Problems

1. Assume a gunner has a 50/50 chance of hitting a target each time he fires at one. Find the probability that after shooting at 12 targets he will have hit just 4 of them. Use the Binomial distribution and the Normal approximation and compare.

2. Use the Normal approximation to the Binomial to find the probabilities that in 100 spins of an unbiased coin, there will be

   (i) exactly 40 heads (ii) fewer than 42 tails
   (iii) more than 40 heads (iv) between 25 and 50 tails
   (v) exactly 20 heads (vi) fewer than 30 tails.

13.4 SAMPLING DISTRIBUTIONS AND STATISTICAL INFERENCE

We mentioned some of the reasons behind sampling in Section 11.1. Our aim is to use the sample as a basis for inferring certain things about the parent population. For the population we have certain **parameters**, for example, the mean, mode, standard deviation, etc. If similar quantities are computed from a sample, each is called a **sample statistic**. Whereas the value of a parameter is constant for a population, the corresponding statistic is variable from sample to sample. We can see this from an example: the sample $\{4, 3, 2, 6, 3\}$ is taken from a population with mean $\mu = 3.2$. The mean of the sample is

$$\bar{x} = (1/5)(4 + 3 + 2 + 6 + 3) = 3.6$$

yet, if we took only the first three items as our sample, the sample mean $\bar{x} = 3$. The sample mean $\bar{x}$ is calculated in the same way as the population mean μ. However, the sample standard deviation of a sample of size n, s, is calculated by the formula

$$s = \sqrt{\sum_{r=1}^{n} (x_r - \bar{x})^2 \Big/ (n-1)} \qquad (13.3)$$

Now, for large values of n, this gives almost the same result as the formula

$$\sqrt{\frac{1}{n} \sum_{r=1}^{n} (x_r - \bar{x})^2}$$

and this latter formula is sometimes used in that case. But for small samples there is a marked discrepancy; the reason for choosing (13.3) is discussed in the next section. Again, s will vary, in general, from sample to sample.

There are three main inferences to draw from a sample and these are dealt with in the next three sections.

In Section 13.5 we study the estimation of a parameter from a sample statistic. We shall mainly be concerned with the estimation of μ, the population mean, from the observed sample mean, $\bar{x}$. Of course, we cannot do more than quote a range of possible values for μ, around $\bar{x}$, outside which the chances of μ taking such a value are small. Since statistical inference is *not exact* we must attach a *risk* to our interval: the risk of a wrong statement.

Another aspect of sampling discussed in Section 13.6 is the testing of hypotheses about, for example, the population mean. If the sample mean is markedly different from the expected, or claimed, population mean does this indicate that μ is not what it should be? Of course, we cannot be certain, but we can state a conclusion and attach the risk of it being wrong. This is a branch of **statistical decision-making.**

In Section 13.7 we compare sample means and other statistics.

We turn for the rest of this section to the study of the behaviour of certain sampling statistics. On page 710 we deal with sums and differences.

Behaviour of sample means

A statistical population consists of the following six numbers, each printed on a ball: 1, 4, 4, 6, 6, 9. The balls are placed in a bag. The mean of this population $\mu = 5$ and the standard deviation $\sigma = \sqrt{6}$. Now consider the possible samples of size two we can draw from this population. There are ${}^6C_2$, that is, 15 of them. For each sample we calculate the sample mean, $\bar{x}$. The possible samples and their means are shown below in Table 13.3 Notice we have to allow for the fact there are two fours and two sixes, and although there are only six distinct pairings, some occur twice as often as others.

Table 13.3

Sample	Mean	Sample	Mean	Sample	Mean
(1, 4)	2.5	(4, 4)	4	(4, 6)	5
(1, 4)	2.5	(4, 6)	5	(4, 9)	6.5
(1, 6)	3.5	(4, 6)	5	(6, 6)	6
(1, 6)	3.5	(4, 9)	6.5	(6, 9)	7.5
(1, 9)	5	(4, 6)	5	(6, 9)	7.5

We have several distinct sample means; how well does any one of them represent the true population mean of 5?

We see that the samples (1, 9) and (4, 6) give exactly the population mean, yet the samples (1, 4) and (6, 9) give means differing by 2.5 from that of the population. The other samples give means 1 or 1.5 above or below the true mean. Had we presented the bag of balls to a third person and asked him to select a sample of two and find the sample mean, he would have no clue as to how accurate a prediction of the population mean this would be. The sample means constitute a random variable, the behaviour of which we now study.

First we compute its arithmetic mean, that is, the mean of the sample means. Since we have deliberately written down duplicates, we can simply average the entries in the above table, therefore

$$\begin{aligned} E(\bar{x}) &= \frac{1}{15}(2.5 + 2.5 + 3.5 + 3.5 + 5 + 4 + 5 + 5 + 6.5 \\ &\quad + 5 + 5 + 6.5 + 6 + 7.5 + 7.5) \\ &= \frac{1}{15}(5 + 7 + 4 + 25 + 6 + 13 + 15) \\ &= \frac{75}{15} \\ &= 5 \end{aligned}$$

We see that the expectation of the sample means is equal to the population mean.

This is quite generally true: however the individual sample means may be above or below the population mean, their average is always equal to the population mean (assuming all sample means are used in the computation of the average). Now you can argue that in this example we could easily enumerate the possible samples, but suppose we took a sample of ten components produced by a machine and found the sample mean diameter. How could we be sure that the above result applies since we would have to measure all the components! You will have to accept this result, because although there is a large number of possible samples, we would get on average the population mean.

Let us now consider samples of four taken from the population of six numbered balls; it should be easy to see that there are again fifteen of them (why?) and they are shown in Table 13.4, together with their means.

Table 13.4

Sample	Mean	Sample	Mean	Sample	Mean
(4, 6, 6, 9)	6.25	(1, 6, 6, 9)	5.5	(1, 4, 6, 9)	5
(4, 6, 6, 9)	6.25	(1, 4, 6, 9)	5	(1, 4, 6, 6)	4.25
(4, 4, 6, 9)	5.75	(1, 4, 6, 9)	5	(1, 4, 4, 9)	4.5
(4, 4, 6, 9)	5.75	(1, 4, 6, 6)	4.25	(1, 4, 4, 6)	3.75
(4, 4, 6, 6)	5	(1, 4, 6, 9)	5	(1, 4, 4, 6)	3.75

Note this time there are again five sample means agreeing with the population mean, but the sample means are bunched more closely about that mean – the furthest being 1.25 away.

If we compute $E(\bar{x})$ we find it is

$$\frac{1}{15}(2 \times 6.25 + 2 \times 5.75 + 5.5 + 5 \times 5 + 2 \times 4.25 + 4.5 + 2 \times 3.75) = 5$$

We now compute the standard deviation of the sample means for samples of size two. This is denoted by $\sigma_{\bar{x}}$ and can be calculated to be

$$\sqrt{\frac{36}{15}} = \frac{6}{\sqrt{15}}$$

Here we divide by 15 and not 14 since we are treating the sample as a set of numbers at the moment.

If we compute $\sigma_{\bar{x}}$ for samples of size 4 we obtain

$$\sqrt{\frac{9}{15}} = \frac{3}{\sqrt{15}}$$

and we see that doubling the sample size halves $\sigma_{\bar{x}}$.

It is clear that the larger sample size gives us less chance of being too far from the population mean, but to choose, for example, a sample size two-thirds that of the population size would in practice be ridiculous.

We have so far considered sampling *without* replacement. The case where the population is very large compared with the sample size approximates sampling *with* replacement and this is the case we now pursue in the rest of the chapter.

From the general results for variances on page 640 we have

$$\begin{aligned}
\text{var}(\bar{x}) &= \text{var}\left(\frac{x_1 + x_2 + \dots\dots + x_n}{n}\right) \\
&= \frac{1}{n^2}[\text{var}(x_1) + \text{var}(x_2) + \dots\dots + \text{var}(x_n)] \\
&= \frac{1}{n^2} \cdot n\sigma^2 \\
&= \frac{\sigma^2}{n}
\end{aligned}$$

Then the **standard error of the mean**

$$\sigma_{\bar{x}} = \frac{\sigma}{\sqrt{n}}$$

and generalises σ for a sample of size 1.

The relationship between the distributions of x and $\bar{x}$ is shown schematically in Figure 13.14.

We state that for such cases, with a sample of size n from which sample means $\bar{x}$ are calculated,

$$E(\bar{x}) = \mu, \quad \sigma_{\bar{x}} = \frac{\sigma}{\sqrt{n}} \tag{13.4}$$

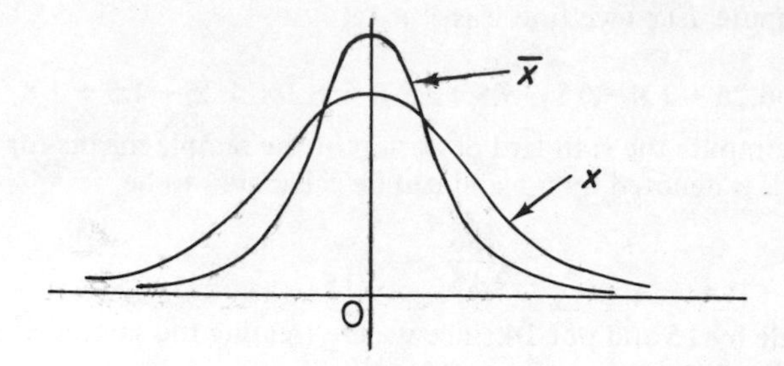

Figure 13.14

As the sample size increases, so the sample means become more closely bunched about the mean and hence any one of them is a better guide to the population mean than would be the case for a smaller sample size.

We now make use of the important **central limit theorem.** This states that the sampling distribution of $\bar{x}$ is approximately normal if the sample size n is large, the approximation improving with increasing n. This applies irrespective of the parent population, provided σ is finite.

Now this is a quite remarkable result and one which has allowed sampling theory to proceed. It says that *no matter what* the shape of the original population sample means are *approximately normally distributed.* How large n should be before the result holds is not clear-cut. A safe rule is that if $n \geqslant 30$ the theorem can be applied with confidence; for $n < 30$ we should need some assurance that the parent population is approximately normal.

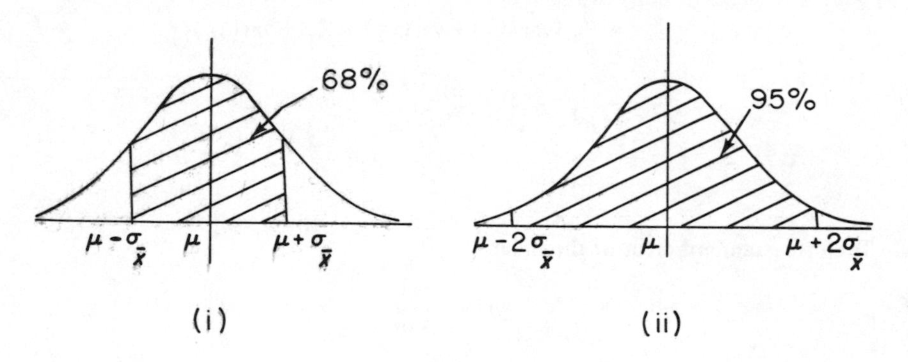

Figure 13.15

Example 1

From the student registration forms for a university an experiment is carried out. 1000 samples of 64 forms are taken (with replacement) and the ages of the students noted for each sample: the average age of the population is known to be 19.8 with $\sigma = 1.6$ years. There are 1000 sample means; the first mean may be 21, the second 19.5 and so on. But the 1000 means will be approximately normally distributed ($n \geqslant 30$) and the average of these will be 19.8 years with a standard error of $1.6/\sqrt{64} = 0.2$ years.

By our results for the normal distribution, we know that approximately 68% of these sample means will lie in the range $\mu \pm \sigma_{\bar{x}} = 19.8 \pm 0.2$, that is, between 19.6 and 20.0 years. [See Figure 13.15 (i)]

Likewise approximately 95% of sample means will lie in the range $19.8 \pm (1.96 \times 0.2)$ years. [See Figure 13.15 (ii)]

(Note there is a *double approximation*: the figures 68% and 95% are not exact and they would only be achieved approximately due to sampling variability.)

Example 2

It is known that the lives of certain machine components are normally distributed with mean $\mu = 5000$ hours and standard deviation $\sigma = 50$ hours. What is the probability

(i) that a single component will last longer than 5030 hours?
(ii) that the mean life from a sample of 16 is greater than 5030 hours?
(iii) that the mean life from a sample of 64 is greater than 5030 hours?

(i) We form

$$z = \frac{5030 - 5000}{50} = 0.6$$

and from the tables, the probability of a z-score greater than 0.6 is $0.5 - 0.2257 = 0.2743$, giving a probability of 27.43% (2 d.p.)

(ii) In this case, the standard deviation σ is replaced by the standard error

$$\frac{\sigma}{\sqrt{n}} = \frac{50}{\sqrt{16}} = 12.5$$

The z-score now formed is

$$\frac{5030 - 5000}{12.5} = 2.4$$

and the probability of a larger z-score is

$$0.5 - 0.4918 = 0.0082 \quad \text{or} \quad 0.82\% \quad \text{(2 d.p.)}$$

(iii) The standard error is $50/\sqrt{64} = 6.25$ giving a z-score of

$$\frac{5030 - 5000}{6.25} = 4.8$$

and so the probability of the mean life from a sample of size 64 exceeding 5030 is negligibly small. Note that all we have done is to replace the standard deviation σ by the more general standard error $\sigma/\sqrt{n}$.

Distributions of sums and differences

Given two independent random variables X and Y with means μ_1 and μ_2 and variances ${\sigma_1}^2$ and ${\sigma_2}^2$ respectively, then the random variable $X + Y$ has mean $\mu_1 + \mu_2$ and variance ${\sigma_1}^2 + {\sigma_2}^2$ and the random variable $X - Y$ has mean $\mu_1 - \mu_2$ and variance ${\sigma_1}^2 + {\sigma_2}^2$. *These results apply only for independent variables.*

Notice that we add variances in both cases. We have met this idea in the study of errors in Chapter 1.

Example 1

Two dice are thrown several times; the scores on each are recorded as X and Y. Find the mean and standard deviation of the difference in the scores.

For each die the mean score is 3.5 and the variance is $2\frac{11}{12}$. For the difference in the scores, the mean is $3.5 - 3.5 = 0$, and the variance is $2\frac{11}{12} + 2\frac{11}{12} = 5\frac{5}{6}$, which gives a standard deviation of 2.416 (3 d.p.).

Example 2

Batteries of type I have voltages which are distributed $N[6.0, (0.15)^2]$ whilst those of type II are distributed $N[12.0, (0.2)^2]$. A battery from each type is taken and the two connected in series, what is the probability that the combined voltage exceeds 17.4?

Let the voltages of the batteries of types I and II be x and y respectively. Then $(x + y)$ is distributed $N[18.0, (0.0225 + 0.04)]$, i.e. $N(18.0, 0.0625)$ which is $N[18.0, (0.25)^2]$. Then the critical z-score is

$$z = \frac{17.4 - 18.0}{0.25} = -2.4$$

The probability that $z > -2.4$

$$= 1 - (\text{probability that } z < -2.4)$$

$$= 1 - 0.0082$$

$$= 0.9918 \quad (4 \text{ d.p.})$$

Differences between sample means

A most important distribution is the difference between the means of two independent samples from two different populations. Given a sample of size n_A from population A with mean μ_A and standard deviation σ_A and a second sample of size n_B from population B with parameters μ_B and σ_B, the difference in means is distributed with mean $\mu_A - \mu_B$ (provided the difference is taken the corresponding way) and variance $(\sigma^2_A/n_A) + (\sigma^2_B/n_B)$.

Example

A bolt of diameter y cm must fit a nut of internal diameter x cm, both selected at random from cartons. If x and y are distributed $N[1.0, (0.0009)^2]$ and $N[0.9965, (0.0012)^2]$ respectively, what is the probability that the nut and bolt would fail to fit, given that the clearance for fitting must be at least 0.0008 cm?

Let $w = x - y$, then we require $w > 0.0008$ for a fit. The distribution of w is $N(0.0035, 0.00000081 + 0.00000144)$, that is, $N[0.0035, (0.0015)^2]$. We require for failure that

$$w < \frac{0.0008 - 0.0035}{0.0015} = -1.8$$

The required probability is

$$0.5 - 0.4641 = 0.0359$$

Problems

1. A population consists of 6 numbers, 1, 2, 3, 6, 7, 8. Samples of size 2 are taken from this population (without replacement).

   (a) Calculate the population mean and standard deviation;
   (b) calculate the mean of each possible sample;
   (c) calculate the mean of these means of samples;
   (d) find the standard error of the means.

   Comment.

2. A random sample of size 50 is drawn from a Normal population; the population mean is 52 and standard deviation is 24. Find

   (a) the probability that the sample mean will lie between 50 and 55;
   (b) the probability that the sample mean will lie betwen 40 and 60.

3. It is known that of a large group of students, 25% live in lodgings. If a random sample of 100 is chosen from the group, what is the probability that

   (a) more than 30 will live in lodgings,
   (b) less than 20 will live in lodgings?

4. Large numbers of plates of two types are produced. One type has thicknesses x cm following $N(0.24, 1.3 \times 10^{-5})$; the other has thicknesses y cm following $N(0.13, 0.8 \times 10^{-5})$. One plate of each type is taken at random and these are bolted together to give a plate of overall thickness $(x + y)$ cm. Between what limits can we expect 90% of all pairs to lie?

5. In a manufacturing process, a piston of circular cross-section has to fit into a similarly shaped cylinder. The distributions of diameters of pistons and cylinders are normal with parameters:

   pistons: mean diameter 10.42 cm, standard deviation 0.03 cm
   cylinders: mean diameter 10.52 cm, standard deviation 0.04 cm.

   If the pistons and cylinders are selected at random for assembly

   (a) what proportion of the pistons will not fit into cylinders?

(b) what is the chance that, in 100 pairs selected at random, all the pistons will fit? (C.E.I.)

6. A type of sensor has a mean life of 15000 hours with standard deviation of 1000 hours. Three of these sensors are connected in series in a fire-warning device so that when one fails another takes over. Assuming that the life times are normally distributed, what is the probability that the device

   (a) will function for 40000 hours
   (b) will fail before 39000 hours. (C.E.I.)

7. A manufactured product consists of a major part and three minor parts. The weight of the major part is normally distributed with mean weight 15 grams and standard deviation 1 gram. The weight of each minor part is normally distributed with mean weight 2 grams and standard deviation 0.2 grams. The product is rejected as unsuitable if its total weight is less than 19.5 grams or more than 22.5 grams. Find the proportion of the product which is unsuitable.

8. The weights of bags of sand have a distribution with mean 114 lbs wt and standard deviation 8 lbs wt. If a delivery consists of 16 such bags, what is the probability that the average weight per bag is less than 112 lbs wt?

9. Wire cables are formed from 10 separate wires, the strength of each wire being normally distributed with standard deviation of 24.5 lb about a mean of 645 lb. Assuming that the strength of each cable is the combined strength of the separate wires, what proportion of the cables have a breaking strain of less than 6350 lb? Assuming the variability to remain unchanged, to what must the mean strength of the individual wires be increased if only 1 in 1000 cables is to have a breaking strain of less than 6350 lb?

13.5 ESTIMATION FROM A SAMPLE

Point estimation

We first examine point estimation when a sample statistic is used to give an estimate of a population parameter; for example, $\bar{x}$ may be used to estimate μ and s^2 may be used to estimate σ^2. Often in statistical work we denote an estimated value by placing a *hat* on it: thus $\hat{\alpha}$ is an estimated value of α.

There are three properties of estimators worthy of attention: to define these we assume λ is a population parameter and l is a sample statistic used to estimate λ; l has a distribution with mean $E(l)$ and standard error σ.

(i) *Unbiasedness*

l is an **unbiased estimator** of λ, if $E(l) = \lambda$. If $E(l) \neq \lambda$, l is said to be a **biased estimator** of λ. Clearly we would prefer an estimator to be unbiased, for if an estimator is biased it will, on average, give too high or too low a value and the chances of obtaining a close estimate are decreased. $\bar{x}$ is an unbiased estimator of μ since $E(\bar{x}) = \mu$. For example,

$$\frac{1}{n} \sum_{r=1}^{n} (x_r - \bar{x})^2$$

is a biased estimator of σ^2 since

$$E\left(\frac{1}{n} \sum_{r=1}^{n} (x_r - \bar{x})^2\right) = \frac{\sigma^2(n-1)}{n}$$

however,

$$\frac{1}{n-1} \sum_{r=1}^{n} (x_r - \bar{x})^2$$

is an unbiased estimator of σ^2.

(ii) *Efficiency*

Given two unbiased estimators which is to be preferred? Since any one value of the estimator may be too high or too low, we would prefer that value not to be too far from the true value of the population parameter. We therefore choose the estimator whose distribution has the less variance. In general, in a set of unbiased estimators, the one with least variance is called the *most efficient* estimator.
It may be shown that of all possible unbiased estimators of μ, x is the one with least variance and is therefore the most efficient estimator of μ.

(iii) *Consistency*

Since the ideal of all values of the estimator coinciding with the parameter to be estimated cannot be realised in pactice, we look for an estimator which improves as the sample size n increases. If l is an unbiased estimator of λ and $\text{var}(l) \to 0$ as $n \to \infty$. then l is a **consistent estimator** of λ.
For example,

$$\text{var}(\bar{x}) = \sigma^2/n \to 0 \text{ as } n \to \infty$$

so $\bar{x}$ is a consistent estimator of μ.

Sometimes we wish to pool information from two independent samples. Let the suffix 1 refer to the one sample and 2 refer to the other, then if both samples come from a population of mean μ and variance σ^2, the most useful unbiased estimators of mean and variance are

$$\hat{\mu} = \frac{n_1\bar{x}_1 + n_2\bar{x}_2}{n_1 + n_2} \quad \text{and} \quad \hat{\sigma}^2 = \frac{(n_1 - 1){s_1}^2 + (n_2 - 1){s_2}^2}{n_1 + n_2 - 2}$$

You should show that these estimators are unbiased and note the obvious extension to more than two samples.

Interval estimation

Although we know that $\bar{x}$ is an unbiased, efficient and consistent estimator of μ, we need to do more than quote $\bar{x}$ as an approximate value of μ. Since we know that the sample means are normally distributed we may say that 95% of these sample means are expected to lie in the interval $[\mu - 1.96\sigma_{\bar{x}}, \mu + 1.96\sigma_{\bar{x}}]$. Alternatively, we may say that a single sample mean has a 95% probability of lying in that range or that it has a 5% chance of falling outside this interval. Yet another way of looking at the situation is that if for each sample we find $[\bar{x} - 1.96\sigma_{\bar{x}}, \bar{x} + 1.96\sigma_{\bar{x}}]$ then we would expect 95% of these intervals to contain μ.

We call the interval $[\bar{x} - 1.96\sigma_{\bar{x}}, \bar{x} + 1.96\sigma_{\bar{x}}]$ the 95% **confidence interval** for μ.

Sometimes we do not know σ^2 for the population and have to estimate it by the sample statistic s^2.

Example 1

A random sample of 100 bolts produced by a certain machine had a mean diameter of 12.5 mm with a standard deviation of 0.1 mm. Find

(i) the 95% confidence interval for the mean diameter of bolts produced by the machine;

(ii) the 99% confidence interval for the mean diameter of bolts produced by the machine;

(i) The 95% (or 0.95) confidence interval for μ is

$$\bar{x} \pm 1.96\sigma_{\bar{x}}$$

Now $\sigma_{\bar{x}} = \sigma/\sqrt{100}$ but we must approximate this by $s/\sqrt{100} = 0.01$ mm. Then our confidence interval is

$$(12.50 \pm 0.0196) \text{ mm}$$

or

$$(12.50 \pm 0.02) \text{ mm} \quad (2 \text{ d.p.})$$

(ii) The 99% confidence interval for μ is $\bar{x} \pm 2.58\sigma_{\bar{x}}$. That is,

$$(12.50 \pm 0.0258) \text{ mm}$$

or

$$(12.50 \pm 0.03) \text{ mm} \quad (2 \text{ d.p.})$$

Choice of sample size

The question which must always be asked in these kinds of problem is how large the sample should be? Now if we are given a maximum error which can be accepted, we can choose a large enough sample to try to cut down the error in our estimate, but we can never be quite sure that we shall exceed that maximum error.

For example, suppose we are asked to estimate the mean weight of bags of cement produced by a machine by estimation from a sample and the estimate is to be in error by no more than 50 gm. Clearly the larger the sample the more reliable the estimate, but whatever the size of the sample, we can never guarantee that the estimate will be within 50 gm of the correct value. All we can do is to try to ensure that the probability of the error exceeding 50 gm is less than, say, 5%.

We choose the sample size then so that $1.96\sigma_{\bar{x}}$ is just equal to 50 gm, that is, $1.96\sigma/\sqrt{n} = 50$, or $n = (1.96\sigma/50)^2$.

Note that if we wished to ensure that the error in our estimate did not exceed the stated maximum on more than 1% of occasions, we should have to take a larger value of n, viz. $(2.58\sigma/50)^2$.

Of course, if we have two populations with standard deviations σ_1, σ_2 where $\sigma_1 > \sigma_2$ then the sample sizes (n_1 and n_2) to achieve the same accuracy with the same confidence will be such that $n_1 > n_2$; this is intuitively obvious.

Example 2

Past records indicate that the lengths of rods produced by a machine are such that $\mu = 500$ cm and $\sigma = 5$ cm. Find the sample size needed if there should be a 99% confidence of the error in the sample estimate not exceeding 0.5 cm.

We have

$$\frac{2.58\sigma}{\sqrt{n}} = 0.5$$

that is,

$$\frac{2.58 \times 5}{\sqrt{n}} = 0.5$$

or

$$n = (25.8)^2 = 666, \text{ at least.}$$

Problems

1. Strength tests on steel parts gave a distribution with mean stress 27.56 and standard deviation 1.10 N/m^2. Find the standard error of the mean of 10 tests and find the probability that the mean of 10 tests will be less than 27.24 N/m^2.

2. A process produces items with a mean of 5.7 kg and standard deviation 2 kg. What are the 95% confidence limits for the mean of a sample of 50 items?

3. Strength tests conducted over a period of time have established that the breaking loads (in units of kN) of certain manufactured spars are distributed in a manner which is approximately normal, having a mean of 32.6 and a variance of 3.24.
   (i) What is the probability that a single spar selected at random will have a strength in excess of 34×10^3 N?
   (ii) Find the probability that the mean strength of a sample of 9 spars selected at random will be less than 34×10^3 N.

4. Measurement of the diameters of a random sample of ball bearings made by a certain machine during one week showed a mean of 0.824 cm, and a standard deviation of 0.042 cm. Find the 99% confidence limits for the mean diameter of all ball bearings made by this machine, assuming the normal distribution.

5. The standard deviation of the breaking load of certain cables is taken to be 150 N. A random sample of 5 cables is tested and has a mean breaking load of 2436 N. What are the 98% confidence limits for the mean breaking load of all such cables?

13.6 TESTING STATISTICAL HYPOTHESES ABOUT THE POPULATION MEAN

Having drawn a random sample and found its mean $\bar{x}$, we can state with a given confidence, limits within which we believe the population mean μ to lie. But what if we have been told what μ is supposed to be and this supposed value lies outside our confidence interval; do we believe the supposed value of μ or not? For example, if we throw a die 6000 times and we obtain 1250 6's we might suspect that the die were 'loaded'. If previous tests on a drug had shown it to be 80% effective and after modification of the drug sampling indicates it is 88% effective are we safe in concluding that the modified drug is more effective? (Notice we say 'safe' and not 'certain'.)

We now examine a specific example in detail.

A manufacturer claims that electrical components of a particular type last on average 1000 hours.

A simple random sample of 100 components is taken and found to give an average life of 988 hours; what can we conclude about the manufacturer's claim?

In statistical terms the claim is that if all the components made, and to be made, (forming an infinite population) were tested and the mean found, it would be 1000 hours.

Setting up of a statistical hypothesis

The first step is to set up a **statistical hypothesis**. This is usually of the form that the proportion of successes is unchanged, that the die is fair, that the components have the claimed average life (and, consequently, that any variability is due solely to sampling variations); this is called the **null hypothesis,** H_0. In this example we state $H_0 : \mu = 1000$.

We must state an **alternative hypothesis,** H_1, which in this case is $H_1 : \mu < 1000$. Now you may ask why we did not have $H_1 : \mu \neq 1000$. The reason is that the customer is unlikely to object if the component lasts more than 1000 hours: he is concerned only with getting at least 1000 hours from the component: his money's worth.

Now if the sample findings indicate a marked departure from what we had expected we would say that the departure is significant and we would tend to reject the null hypothesis, H_0; if the departure is not too large, we would tend either to accept H_0 or require further testing.

Risk of wrong judgment

In either eventuality we make a **statistical decision.** This decision must carry a **risk** of being wrong and a statement of such a risk should always be given alongside the statement of the decision.

In this example, we set a risk of wrong judgment of 5%. Assuming that a suitable sample size has been chosen we decide what statistic to use. In this section we are concerned with decisions concerning the population mean, μ, and we use as test statistic the sample mean $\bar{x}$. Having assumed that the null hypothesis H_0 is true, we then set limits on the values that $\bar{x}$ can take if we are to accept H_0. With a 5% risk, these limits are shown in Figure 13.16(i).

If $\bar{x}$ lies inside the **acceptance region** then we find no evidence to reject H_0; on the other hand, if $\bar{x}$ lies outside the acceptance region we reject H_0.

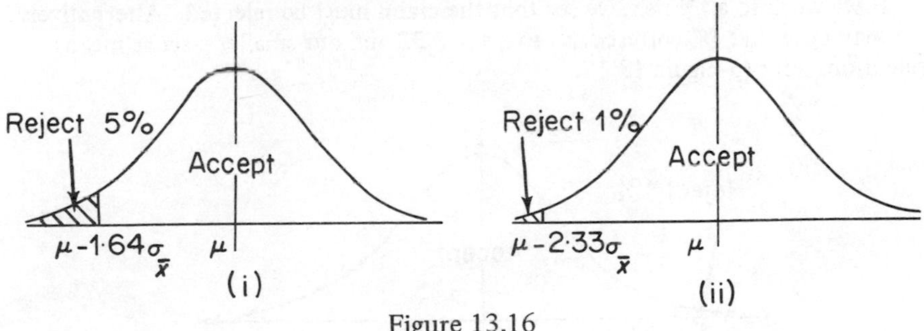

Figure 13.16

Notice here that we have a departure from the picture associated with confidence intervals in that the rejection region is on one side only; for this reason we say we are carrying out a **one-tailed test.**

It cannot be emphasised too strongly that it is possible that we *reject* a claim which should be *accepted,* (the sample gives misleading evidence), and to offset this we may reduce the risk of a wrong decision to 1%. However, this has the effect of increasing the chances of not rejecting a claim when it should be rejected since the region of acceptance is now widened: see Figure 13.16 (ii).

The risk of rejecting H_0 when it should be accepted is called an α-risk and the error of judgment is called a **Type I error.**

The error in accepting H_0 when it should be rejected is called a **Type II error**; the risk of acceptance is called a β-risk.

To decrease the chances of a Type I error we widen the acceptance region and cut down the rejection region; this increases the chances of a Type II error. However, this may be what is required, if we want to give a manufacturer the benefit of the doubt. On the other hand, it may be that we want to be stringent and cut down the chances of passing a wrong claim; in this eventuality we need to reduce Type II error by narrowing the acceptance region and increase the risk of rejecting H_0.

The safest way of reducing Type II errors is to increase the sample size.

Let us work through these steps with our example of the manufacturer's claim.

Example

We are concerned with the distribution of sample means which we know has mean μ and standard deviation $\sigma/\sqrt{n}$. Let us complete the data by saying that the population standard deviation is 50 hours. Then we have

$$\sigma_{\bar{x}} = \frac{50}{\sqrt{100}} = 5 \text{ hours}$$

We form a z-score

$$z = \frac{\bar{x} - \mu}{\sigma_{\bar{x}}} = \frac{988 - 1000}{5} = \frac{-12}{5} = -2.4$$

From the table we find that the probability of a z-score of -2.4 or less, that is, of a sample mean of 988 or less is $(0.5 - 0.4918) = 0.82\%$.

If we work to a 1% risk, we see that the claim must be rejected. Alternatively, we may note that 1% corresponds to $z = -2.33$ and our smaller z-score means rejection; refer to Figure 13.17.

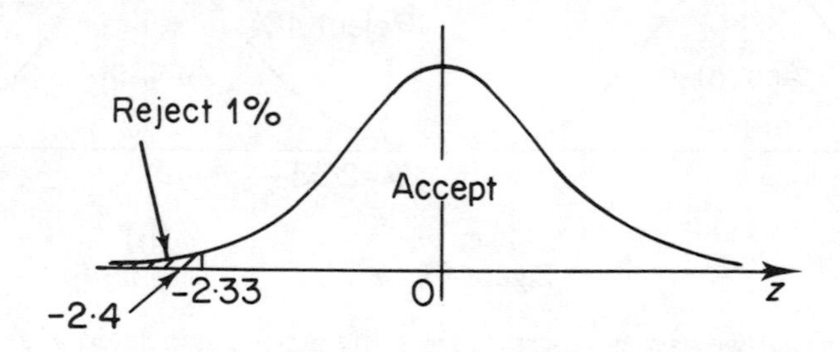

Figure 13.17

Suppose now that the manufacturer's claim is now modified to stating that the mean life of his components is 997 hours; how does this affect our decision?

The z-score is now

$$z = \frac{988 - 997}{5} = \frac{-9}{5} = -1.8$$

The corresponding probability is $(0.5 - 0.4641) = 3.59\%$ (2 d.p.).

If we work to a 5% risk, we shall still reject the null hypothesis, H_0; if we work to a 1% risk, we cannot reject H_0 on the evidence of the sample.

Example

Suppose a manufacturer claims his components last on average 1000 hours and that a sample of 100 had a mean life of 992 hours with a standard deviation of 40 hours. What can we conclude?

Here we have no information about the population standard deviation σ and we must use the sample standard deviation, s, in order to estimate $\sigma_{\bar{x}}$; we then have

$$\hat{\sigma}_{\bar{x}} = \frac{s}{\sqrt{n}} = \frac{40}{10} = 4$$

We calculate the probability that the sample mean of 100 components will be 992 hours or less.

We form

$$z = \frac{\bar{x} - \mu}{\hat{\sigma}_{\bar{x}}} = \frac{992 - 1000}{4} = -2.0$$

The probability that $\bar{x} \leqslant 992$ is $(0.5 - 0.4772) = 2.28\%$ (2 d.p.)

If we reject the claim, there is a risk of 2.28% that we are making a mistake and rejecting a valid claim.

Perhaps a better term than *accepting the null hypothesis* is *reserving judgment.* Also, another way of stating that we have a risk of 5% and the null hypothesis is rejected is to say that the null hypothesis is rejected at the 5% **level of significance.**

Two-tailed tests

So far we have examined cases where the alternative hypothesis H_1 is of the form $\mu < \mu_0$; it is clear that similar considerations obtain for the cases where H_1 is of the form $\mu > \mu_0$.

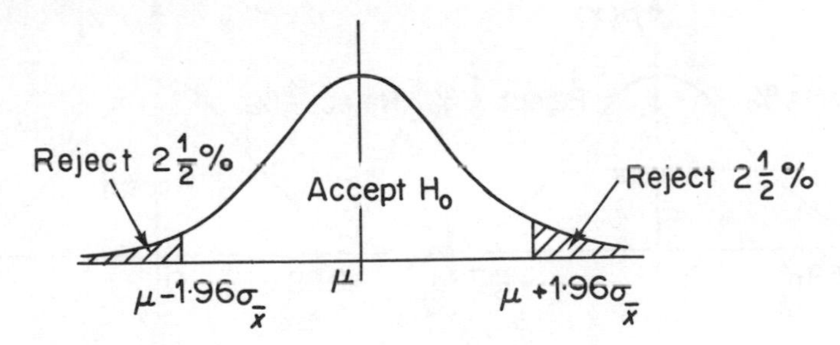

Figure 13.18

We now turn our attention to situations in which the null hypothesis $H_0 : \mu = \mu_0$ is accompanied by an alternative hypothesis $H_1 : \mu \neq \mu_0$. For example, suppose we are endeavouring to manufacture nuts with a diameter of 4 cm; it would be almost equally unfortunate whether the diameters were too small or too large. Clearly there will be two rejection regions, one on each side of the acceptance region.

But if we work to a 5% level of significance, we are now saying that the total probability of falling in the rejection region is 5%, and since the acceptance region is symmetrical about the mean μ this will yield a 2½% probability of landing on one side of the acceptance region; see Figure 13.18. We say we are carrying out a **two-tailed test.**

Example 1

A machine is supposed to pour on average 500 gm of a chemical into a jar. A sample of 64 jars is taken from a day's output of 5000 and the amount of chemical in a jar is found to have a mean of 504 gm with a standard deviation of 12.8 gm. Working to a 1% risk, is the machine correctly set?

Here the sample of 64 is sufficiently small with respect to the 5000 output that we can estimate the standard error of the mean as

$$\hat{\sigma}_{\bar{x}} = \frac{s}{\sqrt{n}} = \frac{12.8}{\sqrt{64}} = 1.6$$

We have the null hypothesis $H_0 : \mu = 500$, and the alternative hypothesis $H_1 : \mu \neq 500$.

We form

$$z = \frac{\bar{x} - \mu}{\hat{\sigma}_{\bar{x}}} = \frac{504 - 500}{1.6} = 2.5$$

Now the acceptance region is shown in Figures 13.19 (i) and (ii).

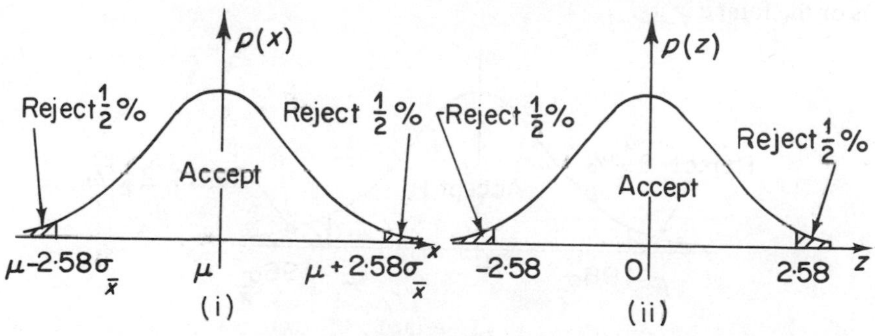

Figure 13.19

Clearly we are just inside the acceptance region and on the evidence of the sample, we reserve judgment.

It is important to notice that we have considered large-size samples here.

Example 2

A sample of 400 is taken from a population which is normally distributed with mean 6 and standard deviation 1.6. Is the sample truly random if its mean is 5.6?

The standard error of the mean

$$\sigma_{\bar{x}} = \frac{1.6}{20} = 0.08$$

The discrepancy $\bar{x} - \mu$ is 5 standard errors

$$\left(z = \frac{\bar{x} - \mu}{\sigma_{\bar{x}}} = \frac{5.6 - 6}{0.08} = -5 \right)$$

Therefore we must conclude that the sample is not random.

Problems

1. The distribution of hourly output of a certain type of machine has mean 700 units and standard deviation 50 units. With a new method of operation a sample of 10 single hour test runs gave a mean of 750 units. Show that the new method has made a statistically significant change in the hourly rate.

2. The lengths in mm of 50 components were recorded and the following summary values noted:

$$\sum_{i=1}^{50} x_i = 75.6, \sum_{i=1}^{50} x_i^2 = 120.80$$

where x_i = length − 100.

(a) Estimate the mean of the normal distribution which is assumed to describe the variation in length, and give an approximate 95% confidence interval for this mean.

(b) If 5 of the components are laid in line, their ends just touching, provide an estimate of the mean of the overall length ; give a 95% confidence interval.

Now assume the lengths are distributed as a $N[101.5, (0.4)^2]$ random variable. If the components are discarded when they are under 101 mm in length, what proportion will be discarded? With this variance, to what value should the mean length be increased if the proportion of discarded components is to be at most 0.01?

3. Routine strength tests have established that certain spars have a breaking stress normally distributed about 27.56 N/m^2 with a standard deviation of 1.01 N/m^2.

(a) What is the probability that the average breaking stress from a sample of 10 spars will be less than 27.24 N/m^2?

(b) Certain modifications are made in the construction of the spars and in order to test whether these modifications have changed the mean strength, a sample of 10 modified spars is tested. They gave a sample mean of 28.43 N/m^2. Does this suggest that the modifications have had a real effect?

13.7 OTHER TESTS OF SIGNIFICANCE

In this section we consider **three** other tests of significance.

The difference between sample means

Suppose we wish to compare mean scores from two samples. We have seen at the end of Section 13.4 that the sampling distribution for the difference in means $\bar{x}_A$ and $\bar{x}_B$ from two samples A and B containing n_A and n_B items respectively has expectation $\mu_A - \mu_B$, written μ_{A-B} and variance $({\sigma_A}^2/n_A + {\sigma_B}^2/n_B)$, where ${\sigma_A}^2$ and ${\sigma_B}^2$ are the variances of the two populations (which may be equal if the samples are drawn from the same population or they may be estimated by sample variances if necessary). Let us consider an example.

Example

Two brands of component were sampled for the following results:

	Brand A	*Brand B*
Number in sample	50	40
Sample mean	10.7 cm	10.2 cm
Sample s.d.	1.5 cm	1.7 cm

At the 5% level of significance what can we say about the difference in sample means?

Notice here that we shall have to estimate population variances by sample variances.

We set

$$H_0 : \mu_A = \mu_B \quad \text{and} \quad H_1 : \mu_A \neq \mu_B$$

Now $(\bar{x}_A - \bar{x}_B)$ is observed to be $(10.7 - 10.2) = 0.5$ cm, but under H_0 should, on average, be 0.

The standard error, s.e. $= \sigma_{\bar{x}_A - \bar{x}_B}$ is such that

$$(\text{s.e.})^2 = {\sigma_A}^2/n_A + {\sigma_B}^2/n_B$$

(the samples are large enough for a normal distribution approximation).
Then

$$(\text{s.e.})^2 = \frac{(1.5)^2}{50} + \frac{(1.7)^2}{40} = \frac{2.25}{50} + \frac{2.89}{40} = 0.045 + 0.07225$$

$$= 0.1173 \quad (4 \text{ d.p.})$$

therefore

$$\text{s.e.} = 0.3425 \quad (4 \text{ d.p.})$$

Now

$$z = \frac{(\bar{x}_A - \bar{x}_B) - (\mu_A - \mu_B)}{\text{s.e.}} = \frac{0.5 - 0}{0.3425} = 1.46 \quad (2 \text{ d.p.})$$

The probability associated with this z-score is

$$(0.5 - 0.4279) = 7.21\%$$

We require to show that the means are significantly different and we therefore double this figure to 14.42%.

At the 5% level of significance, we must accept H_0 and say that there is no significant difference in the sample means.

A sample proportion

We again assume a normal distribution applies.

The tests for a sample proportion follow the same path as those for a sample mean. All that is changed is the standard error, σ_p, which is now

$$\sigma_p = \sqrt{\frac{\pi(1-\pi)}{n}}$$

where π is the 'favourable' proportion and n the number in the sample.

The ideas of one-sided (or one-tailed) and two-sided testing hold here too and we provide one example to enable a comparison to be made.

Example

A medicine is claimed to be at least 85% effective in treatment. If, from a sample of 200 patients treated with the medicine, 160 showed improvement, do you consider the claim to be reasonable?

We set $H_0 : \pi = 0.85$ and $H_1 : \pi < 0.85$ (a higher effectiveness than 85% is acceptable), and work to a risk of 1%. The test is one-tailed.

The standard error is

$$\sigma_p = \sqrt{\frac{\pi(1-\pi)}{n}} = \sqrt{\frac{0.85 \times 0.15}{200}}$$

$$= \sqrt{6.375 \times 10^{-4}}$$

$$= 0.0252 \quad (4 \text{ d.p.})$$

$$z = \frac{0.80 - 0.85}{0.02525} = \frac{-0.05}{0.02525} = -1.98 \quad (3 \text{ s.f.})$$

Now the probability under H_0 of the sample proportion being 0.80 or less is

$$(0.5 - 0.4761) = 0.0239 = 2.39\% \quad (2 \text{ d.p.})$$

At the 1% level of significance we must therefore reserve judgment (the critical figure being $z = 2.33$).

Differences between sample proportions

In this case the standard error is

$$\sqrt{\frac{\pi_A(1-\pi_A)}{n_A} + \frac{\pi_B(1-\pi_B)}{n_B}}$$

We proceed straight to an example.

Example

In the previous example suppose the drug is tested after some modifications and a further sample of 200 patients treated with the new version showed 180 were cured. With a risk of 1%, can we conclude anything about the effectiveness of the new version?

We set

$$H_0 : \pi_A = \pi_B \quad \text{and} \quad H_1 : \pi_A \neq \pi_B$$

The standard error is

$$\sqrt{\frac{0.90(0.10)}{200} + \frac{0.80(0.20)}{200}} = \sqrt{4.5 \times 10^{-4} + 8.00 \times 10^{-4}}$$

$$= \sqrt{12.5 \times 10^{-4}}$$

$$= 0.0354 \quad (4\text{ d.p.})$$

$$z = \frac{(0.90 - 0.80) - 0}{0.0354}$$

$$= \frac{0.10}{0.0354}$$

$$= 2.82$$

Now for a two-tailed test the acceptable range for z is $[-2.58, 2.58]$ and we are outside this region so we must conclude that there is a difference in effectiveness of the two versions.

Problems

1. Population A has mean 13.7 cm and variance 2.5 cm^2, and population B has mean 12.5 cm and variance 3.7 cm^2. Find the probability that the average of 6 values taken from A is less than the average of 5 from B.

2. A machine produces items with a mean of 105 and variance 4. A sample of 400 items is found to have a mean of 104.45. Can this be regarded as a truly random sample? Justify your answer.

3. Two different processes are used to produce what should be identical items. Samples are taken from each to give the following measures (x) of some characteristic.

Process	A	B
Number in sample	144	96
Σx	2246	1354
$\bar{x}$	15.6	14.1
$\Sigma(x - \bar{x})^2$	548.2	426.4

Is there a significant difference in the means?

4. The lengths of life of two types of electric light bulb were recorded giving the following results:

	Type 1	*Type 2*
Number in sample	$N_x = 60$	$N_y = 80$
Mean of sample	$\bar{x} = 1070$	$\bar{y} = 1060$
	$\Sigma(x - \bar{x})^2 = 5480$	$\Sigma(y - \bar{y})^2 = 5800$

Are the mean lives of the two types of bulb significantly different at the 1% level of significance?

5. Tests on the moisture content of a product produced by two processes gave the following results (expressed as percentages).

A 100 samples had mean 8.1 and standard deviation 3.0
B 60 samples had mean 6.6 and standard deviation 2.6

Find 95% confidence limits for the difference of the means of the two populations.

6. The television tubes of manufacturer A have a mean lifetime of 6.5 years and a standard deviation of 0.6 years, while those of manufacturer B have a mean lifetime of 6.0 years and a standard deviation of 0.8 years. What is the probability that a random sample of 20 tubes from manufacturer A will have a mean lifetime that is at least 1 year more than the mean lifetime of a sample of 16 tubes from manufacturer B?

7. The following table shows the number of employees and of serious accidents in one year in each of two factories.

Factory	*No. of employees*	*No. of accidents*
1	15000	7
2	8000	16

How strong is the evidence that the chance of an individual being involved in a serious accident is greater in the second factory than in the first?

8. In a random sample poll of 200 individuals in constituency A the number favouring the party X was 96. Give 95% confidence limits for the proportion favouring this party in the constituency as a whole.
   In a neighbouring constituency B a random poll of 300 was taken and the number favouring party X was found to be 168. Does this indicate that a higher proportion favours party X in the second constituency than in the first. (L.U.)

9. X is a random variable representing the number of successes in n independent trials, such that the probability of success is a constant p. Show that the distribution of the proportion of successes $\hat{p} = x/n$ has a mean p and variance $p(1 - p)/n$.
   The results of a one hundred per cent inspection on a batch of 1000 components show that 190 were faulty. Do you consider these results are consistent with an expected proportion p of faulty components equal to 16%? (L.U.)

10. The following data relates to the speeds on a stretch of road before and after the introduction of a speed limit. What has been the effect of the limit?

Before	$\bar{x} = 30$ mph	$s = 10$	$N = 200$
After	$\bar{x} = 30$ mph	$s = 8$	$N = 400$

SUMMARY

Now you should be able to handle the normal distribution, both in its own right and in relation to sampling theory.

In applications of the normal distribution you must first decide what is the variable which is normally distributed and find the appropriate mean and standard error. It is important to know how to convert the problem into one concerning the standard normal distribution. You need to be able to set up the null and alternative hypotheses and to interpret your results in terms of these, and the risk factor which is attached.

If you are estimating from a sample you should know how to place confidence limits on your estimate.

Although this chapter uses only simple arithmetic, you need to think out your strategy very carefully and you must know how to interpret any results you obtain. A very important point to realise is that inferential statistics operates on incomplete data and the predictions you make are subject to a risk of being wrong. You can only estimate this risk and you can never be certain that you will have covered all contingencies even then.

Appendix

GRAPHS OF TRIGONOMETRIC FUNCTIONS

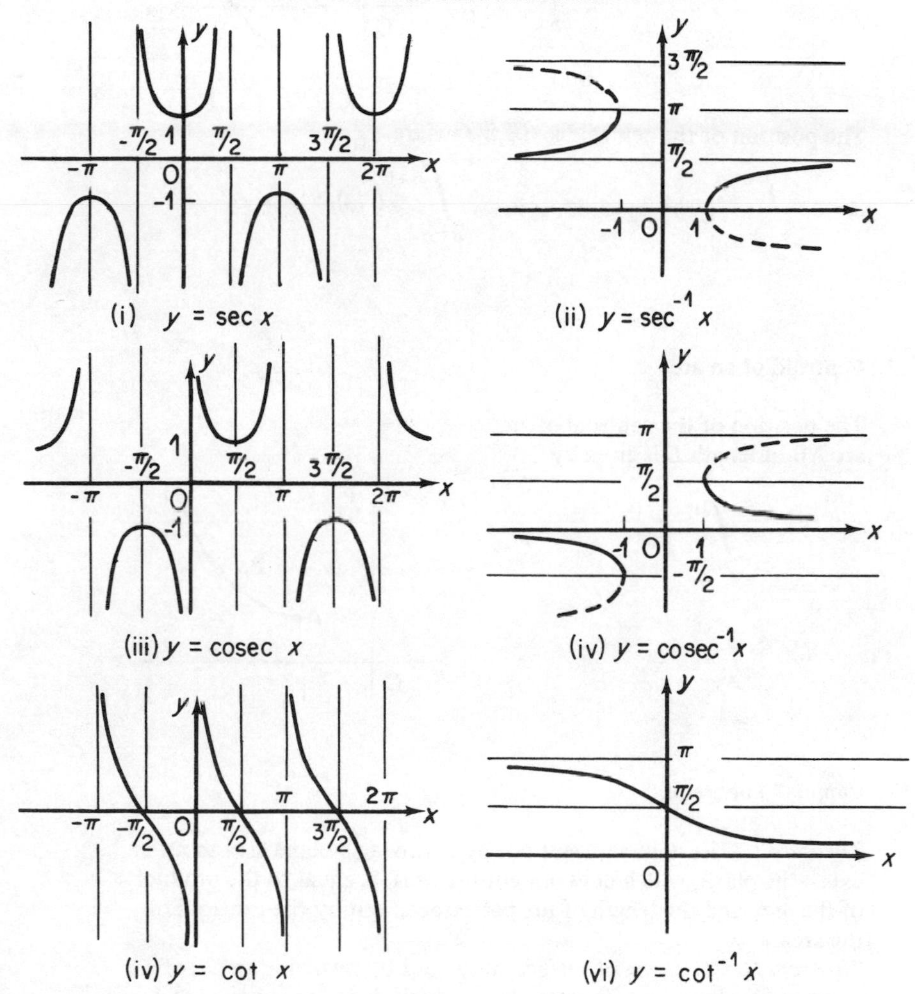

APPLICATIONS OF DEFINITE INTEGRALS – FURTHER FORMULAE

1. Area and Centroid of Area in polar coordinates

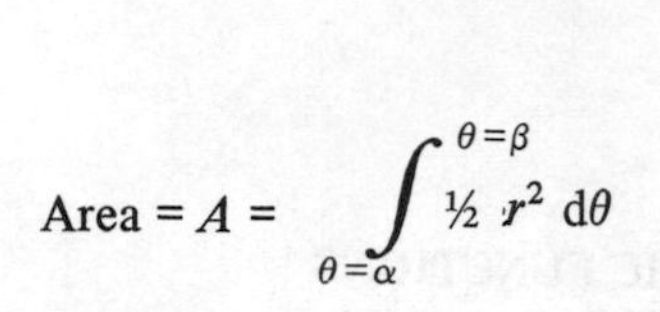

$$\text{Area} = A = \int_{\theta=\alpha}^{\theta=\beta} \tfrac{1}{2}\, r^2 \, d\theta$$

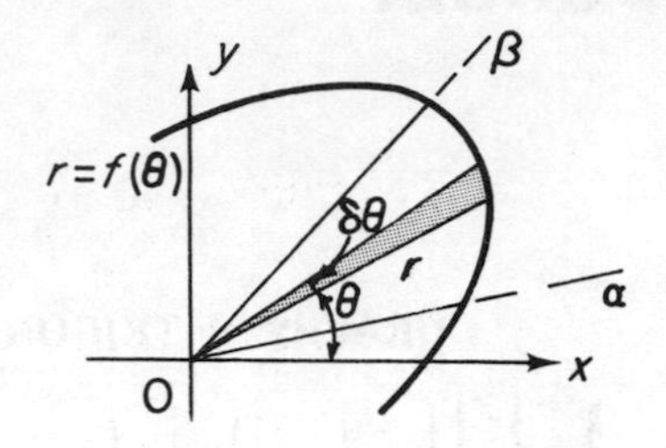

The position of the centroid is $(\bar{x}, \bar{y})$ where

$$A\bar{x} = \int_{\theta=\alpha}^{\theta=\beta} (1/3) r^3 \cos\theta \, d\theta, \quad A\bar{y} = \int_{\theta=\alpha}^{\theta=\beta} (1/3) r^3 \sin\theta \, d\theta$$

2. Centroid of an arc

The position of the centroid of an arc AB of length L is given by

$$L\bar{x} = \int_{A}^{B} x\,ds$$

$$L\bar{y} = \int_{A}^{B} y\,ds$$

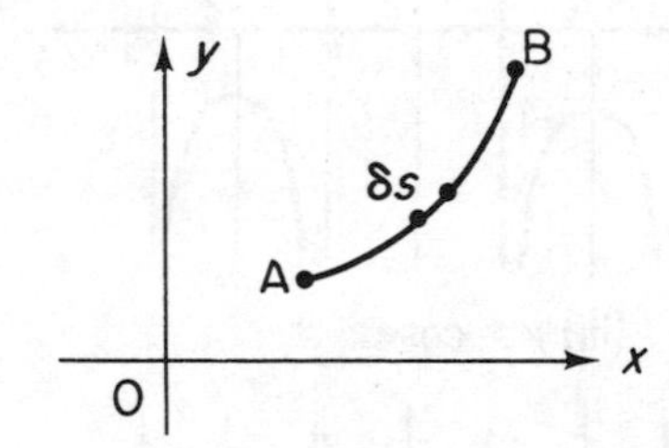

3. Pappus' Theorems

Theorem I. The volume swept out by revolving a plane area about an axis in its plane, which does not cut the area, is equal to the product of the area and the length of the path traced out by the centroid of the area.

Theorem II. The area of surface swept out by revolving an arc of a plane curve about an axis in its plane, which does not cut the arc, is equal to the product of the length of arc and the length of the path traced out by the centroid of the arc.

4. Moment of Inertia, Second Moment of area and Radius of gyration

For the plane area under a plane curve $y = f(x)$ between $x = a$ and $x = b$, the *second moments of area* about Ox and Oy are given by

$$S_x = \int_a^b (1/3) y^3 \, dx, \quad S_y = \int_a^b x^2 y \, dx \text{ respectively}$$

If we regard the area as a lamina having a mass distribution then the *moments of inertia* of this lamina about Ox and Oy are given by

$$I_x = \int_a^b (1/3) y^3 m \, dx, \quad I_y = \int_a^b x^2 y m \, dx \text{ respectively,}$$

where m is the mass/unit area.

The *radius of gyration, k,* is defined by

$$k^2 = \frac{\text{second moment of area}}{\text{total area}} = \frac{\text{moment of inertia}}{\text{total mass}}$$

For the volume of revolution generated by revolving the plane area through four right angles about the axis of x, the second moments about Ox and Oy are given by

$$S_x = \int_a^b \tfrac{1}{2} \pi y^4 \, dx, \quad S_y = \int_a^b \pi y^2 \left(\tfrac{1}{4} y^2 + x^2\right) dx.$$

5. Theorems on Moments of Inertia

I Perpendicular Axes Theorem for Areas
Let Ox and Oy be two perpendicular axes in the plane of an area and Oz a line through O perpendicular to the plane. If k_x, k_y and k_z are the radii of gyration of the area about Ox, Oy and Oz respectively then

$$k_z^2 = k_x^2 + k_y^2$$

II Parallel Axes Theorem
The moment of inertia of an area or volume with respect to any axis is equal to the moment of inertia with respect to a parallel axis through the centroid plus the product of the area or volume and the square of the distance between the parallel axes.

6. Mean Value

The *mean value* of a function $f(x)$ over an interval $x = a$ to $x = b$ is defined by

$$\text{mean value} = \frac{1}{b-a}\int_a^b f(x)\,\mathrm{d}x$$

7. Root Mean Square

The *root mean square* or R.M.S. value of a function over a given interval is the square root of the mean value of its square over that interval, i.e.

$$\text{R.M.S. value} = \sqrt{\left(\frac{1}{b-a}\int_a^b [f(x)]^2\,\mathrm{d}x\right)}$$

TABLE OF LAPLACE TRANSFORMS

This table is adapted from that used in the Part 2 Examination of the Council of Engineering Institutions.

$\mathcal{L}\{f(t)\}$ is defined by $\int_0^\infty f(t)\exp(-st)\,\mathrm{d}t$ and is written as $F(s)$.

Initial conditions are those just prior to $t = 0$.

	Time Function	Laplace Transform
Sum	$af_1(t) + bf_2(t)$	$aF_1(s) + bF_2(s)$
First derivative	$\dfrac{\mathrm{d}}{\mathrm{d}t} f(t)$	$sF(s) - f(0-)$
n^{th} derivative	$\dfrac{\mathrm{d}^n}{\mathrm{d}t^n} f(t)$	$s^n F(s) - s^{n-1} f(0-) - s^{n-2} f^{(1)}(0-) \ldots - f^{(n-1)}(0-)$
Definite integral	$\int_{0-}^{t} f(t)\,\mathrm{d}t$	$\dfrac{1}{s} F(s)$
Exponential multiplier	$\exp(-\alpha t) f(t)$	$F(s + \alpha)$
Shift	$f(t - T)$	$\exp(-sT) F(s)$
	$t^n f(t)$	$(-1)^n \dfrac{\mathrm{d}^n}{\mathrm{d}s^n} F(s)$
1. 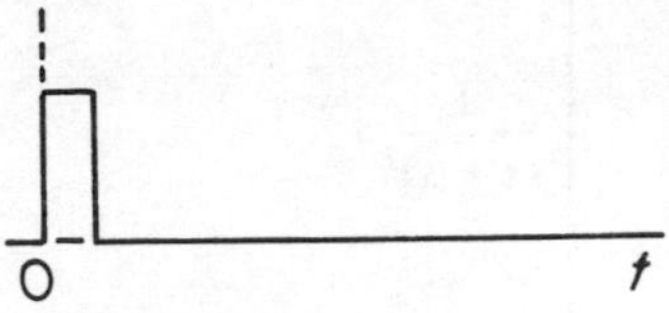 Unit impulse	$\delta(t)$	1
2. 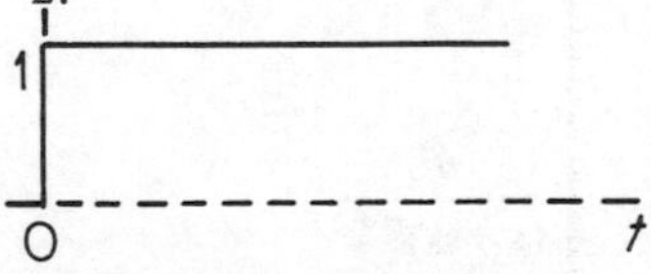 Unit step	$u(t)$	$\dfrac{1}{s}$

	Time Function	Laplace Transform
3. Ramp	t	$\frac{1}{s^2}$
4. n^{th} order ramp	t^n	$\frac{n!}{s^{n+1}}$
5. Exponential decay	$\exp(-\alpha t)$	$\frac{1}{s+\alpha}$
6.	$1 - \exp(-\alpha t)$	$\frac{\alpha}{s(s+\alpha)}$
7.	$t \exp(-\alpha t)$	$\frac{1}{(s+\alpha)^2}$
8.	$t^n \exp(-\alpha t)$	$\frac{n!}{(s+\alpha)^{n+1}}$

	Time Function	Laplace Transform
9.	$\exp(-\alpha t) - \exp(-\beta t)$	$\dfrac{\beta - \alpha}{(s+\alpha)(s+\beta)}$
10. 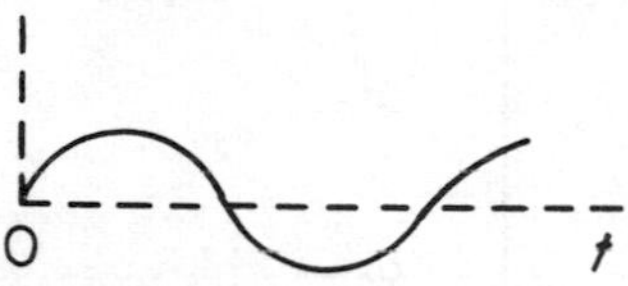 Sine wave	$\sin \omega t$	$\dfrac{\omega}{s^2 + \omega^2}$
11. 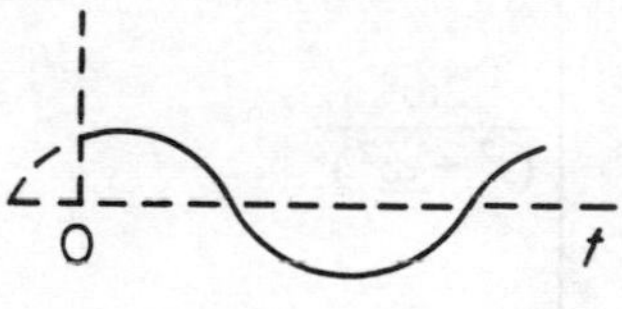	$\sin(\omega t + \phi)$	$\dfrac{\omega \cos \phi + s \sin \phi}{s^2 + \omega^2}$
12. 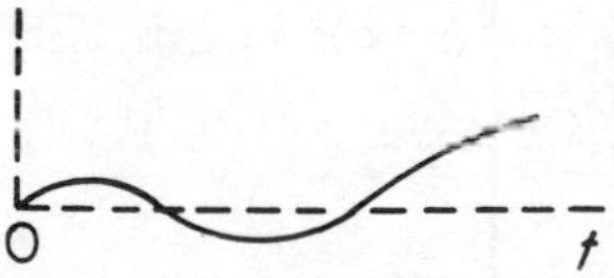	$t \sin \omega t$	$\dfrac{2\omega s}{(s^2 + \omega^2)^2}$
13. 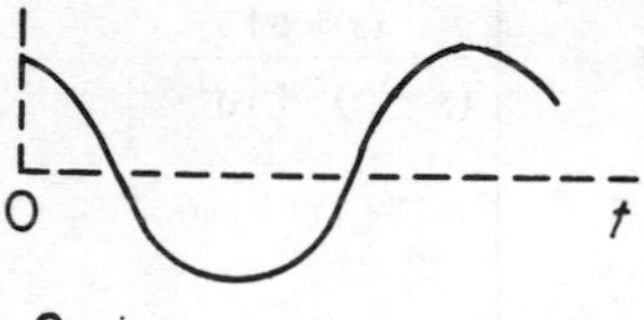 Cosine wave	$\cos \omega t$	$\dfrac{s}{s^2 + \omega^2}$
14. 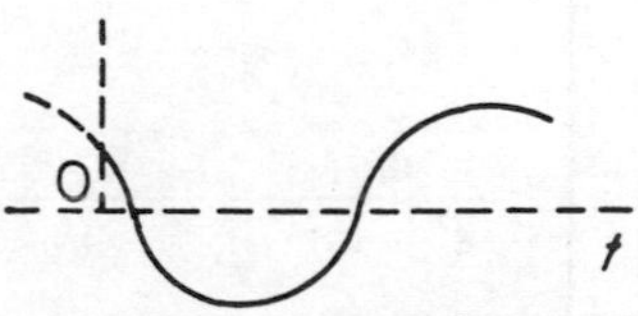	$\cos(\omega t + \phi)$	$\dfrac{s \cos \phi - \omega \sin \phi}{s^2 + \omega^2}$

	Time Function	Laplace Transform
15. 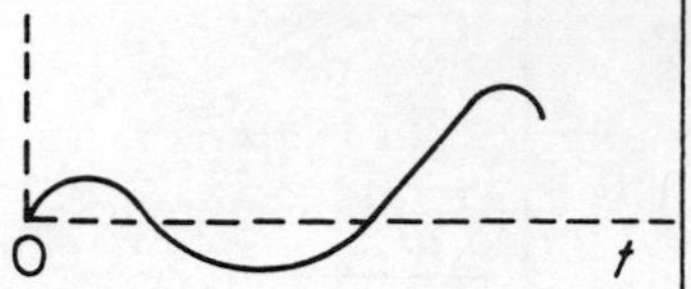	$t \cos \omega t$	$\dfrac{s^2 - \omega^2}{(s^2 + \omega^2)^2}$
16. 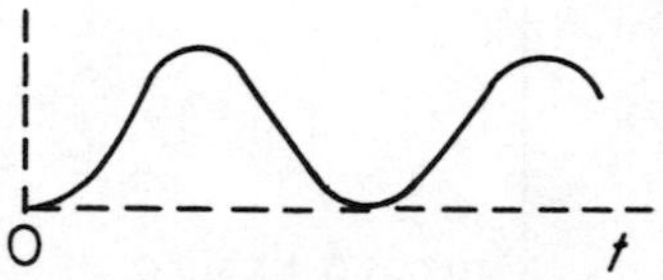	$1 - \cos \omega t$	$\dfrac{\omega^2}{s(s^2 + \omega^2)}$
17.	$\sin \omega t - \omega t \cos \omega t$	$\dfrac{2\omega^3}{(s^2 + \omega^2)^2}$
18. 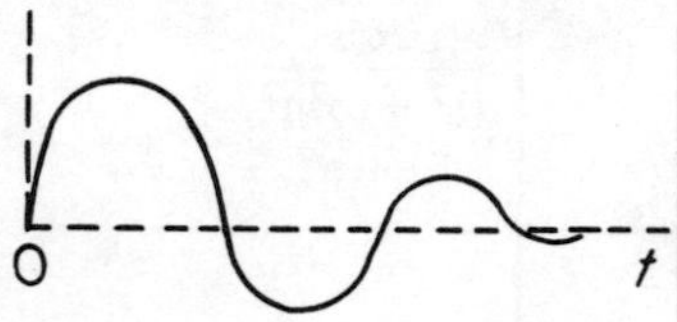	$\exp(-\alpha t) \sin \omega t$	$\dfrac{\omega}{(s + \alpha)^2 + \omega^2}$
19. 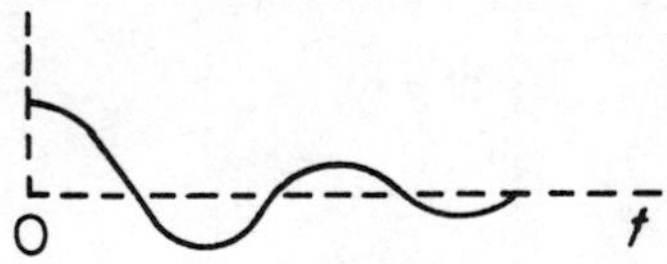	$\exp(-\alpha t) \cos \omega t$	$\dfrac{(s + \alpha)}{(s + \alpha)^2 + \omega^2}$
20.	$\exp(-\alpha t)(\sin \omega t - \omega t \cos \omega t)$	$\dfrac{2\omega^3}{\{(s + \alpha)^2 + \omega^2\}^2}$
21.	$\sinh \omega t$	$\dfrac{\omega}{s^2 - \omega^2}$

	Time Function	Laplace Transform
22.	$\cosh \omega t$	$\dfrac{s}{s^2 - \omega^2}$
23. 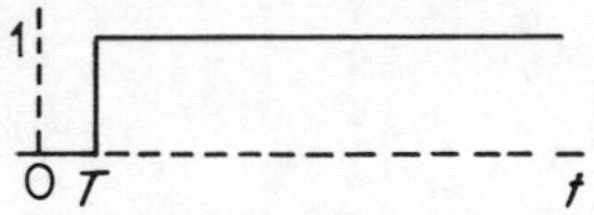 Delayed step	$u(t - T)$	$\dfrac{\exp(-sT)}{s}$
24. 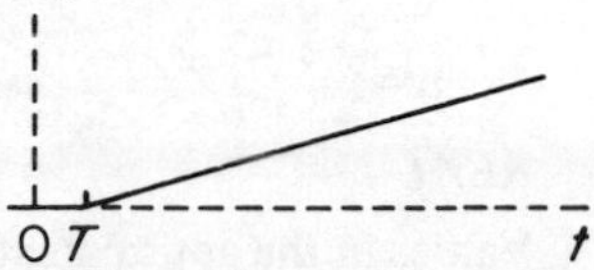 Delayed ramp	$(t - T)\,u(t - T)$	$\dfrac{\exp(-sT)}{s^2}$

A SIMPLIFIED SUMMARY OF FORTRAN IV INSTRUCTIONS

1. SYMBOLS

The following characters are used in FORTRAN

Alphabetic	A, B, ... , Z (Capitals only)
Numeric	0, 1, ... , 9
Operations	+ – * / **
Special characters	= ( ) , .

In general, blank spaces are ignored in FORTRAN statements. In a few special cases when a blank is important it is denoted by ^ in this Appendix.

2. ARITHMETIC STATEMENTS

	INTEGER	*REAL*
CONSTANTS	Whole number values in the range –8388607 to +8388607, + sign optional. No decimal point, comma or exponent must appear e.g. 29 –1760 999999 +144	Values in the approximate range $-5.6 * 10^{76}$ to $+5.6 * 10^{76}$ Recorded in floating point form to a precision of approximately 11 significant digits. Contains decimal point, decimal exponent or both. e.g. 298 + 37.6 –.33334 1234E7 + 98.4E – 16
VARIABLES	One to thirty-two alphabetic or numeric characters of which the first must be I, J, K, L, M or N. e.g. I, M2, JØBNØ9	One to thirty-two alphabetic or numeric characters of which the first is alphabetic but *not* I, J, K, L, M or N. e.g. A, B7, DELTA9

FUNCTIONS There are several functions available to the system. Some of these are given below: these require an argument A, which may be a real variable or a real expression enclosed in brackets.

ALØG(A) – natural logarithm; $\log_e A$
EXP(A) – exponential; e^A
CØS(A) – cosine of an angle given in radians; cos A
SIN(A) – sine of an angle given in radians; sin A
ATAN(A) – arc tangent of a number, result in radians; $\tan^{-1} A$

ASIN(A) – arc sine of a number
ACOS(A) – arc cosine of a number
SQRT(A) – square root, $\sqrt{A}$
ABS(A) – absolute value; $|A|$

e.g. Y = ALØG(X + B)

ARITHMETIC EXPRESSIONS

An expression is any sequence of constants, variables and functions separated by operation symbols and brackets.

The operation symbols used are

+	addition
–	subtraction
*	multiplication
/	division
**	exponentiation

These are written in reverse order of precedence; brackets take precedence over all else, otherwise control works from left to right. The expression is integer if *all* the variables and constants involved are integer.

The expression is *real* if *all* the variables and constants are real (exception: A**I is acceptable).

e.g. X * 3.5 * Y – Z ** 3.976 + SQRT(X * Y * Z)
I + J * 3 – K * L + L37 * M

An expression must be real or integer and it is not advisable for the expression to contain both integer and real quantities (see exception above).

A zero base must not be raised to a zero exponent *nor may a negative base be used.*

ARITHMETIC STATEMENT

General form V = E

where V is a variable and E an expression

e.g. (a) T = X * Y – 3.9 * V * W/Z
(b) X = X + H
(c) Q = K + 4
(d) N = I*J*K – 3*L
(e) M = X + 19.5

It is possible to have an integer variable on the left hand side and a real expression on the right hand side – see (e) above, and vice versa (c) above.
A statement may not consist of more than 330 characters (5 lines).

3. CONTROL STATEMENTS

STATEMENT NUMBER Any instruction may be labelled with an unsigned integer from 1 – 99999.

UNCONDITIONAL BRANCH GØ TØ n where n represents a statement number

e.g. GØ TØ 429

ARITHMETIC IF STATEMENT IF (E) n_-, n_0, n_+

where E is an expression and n_-, n_0, n_+ are statement numbers.

Control is transferred to statement number n_-, n_0, n_+ depending on whether the value of E is less than, equal to, or greater than zero respectively.

e.g. IF (A – B) 10, 5, 7 → statement number 10 if A < B
5 if A = B
7 if A > B

LOGICAL IF STATEMENT IF (EXPR) STA

where EXPR is an expression which can take the form e_1.L.e_2

and .L. could be one of

.LT.	less than
.LE.	less than or equal to
.EQ.	equal to
.NE.	not equal to
.GT.	greater than
.GE.	greater than or equal to

e_1 and e_2 are arithmetic expressions

STA is any executable statement except DØ or another IF. If EXPR is .TRUE. then STA is executed. If EXPR is .FALSE. then STA is ignored and control is transferred to the next statement after the IF.

e.g. IF (B*B.LT. 4.0*A*C) GØ$_\wedge$TØ$_\wedge$100 means that control is transferred to statement 100 if $B^2 < 4AC$, otherwise the next statement after the IF is executed.

DØ STATEMENTS DØ K I=m_1, m_2, m_3 causes the instructions up to and including the statement number K to be executed with the INTEGER control variable I having an initial value m_1 and increasing in steps of m_3 to a final value of m_2.

m_1, m_2, m_3 must be positive INTEGER constants, or INTEGER variables or INTEGER expressions taking positive values.

e.g. DØ 27 I= 1, K, 2

If m_3 and the preceding comma are omitted then steps of unity are assumed.

CØNTINUE A dummy statement resulting in no operation to which a statement number may be attached.

```
e.g. DØ ∧ 12 ∧ I = 2,80
        S = S + 2.0
    12∧CØNTINUE
```

STØP This causes the computer to delete the program

MASTER General form. MASTER∧name
where name is up to 32 alphanumeric characters the first of which is a letter.
This *must* be the first statement of a program

END END indicates the end of a segment and normally follows STØP

FINISH FINISH indicates that the program is complete and always follows an END

4. INPUT/OUTPUT STATEMENTS

READ General form:

READ(1,n)v_1,v_2, . . ,v_m

where n is the label of a FØRMAT statement and v_1,v_2,. . , v_m are variables. 1 refers to a card reader.

The effect is that m numbers are read off data cards according to the FØRMAT.

Variable v_1 is set equal to the first number, v_2 to the second, etc.

Example: READ(1,100)X1,X2,X3
100^FØRMAT (3F10.2)

WRITE General form:

WRITE (2,n)$v_1, v_2, \ldots, v_m$

where n is the label of a FØRMAT statement and 2 refers to a line printer.
$v_1, v_2, \ldots, v_m$ are variables.

The effect is that the values of $v_1, v_2, \ldots, v_m$ are written on to the printer according to the FØRMAT

Example: WRITE (2,200)X1,X2,X3
200^FØRMAT (1H^, 3F10.4)

Note that the numbers 1 and 2 may vary from one computer system to another.

5. FØRMAT STATEMENTS

FORMAT General form:

$$n_\wedge \text{FØRMAT}\ (S_1, S_2, \ldots\ldots, S_n)$$

where n is a statement number and $S_1, \ldots\ldots, S_n$ are format specifications. The FØRMAT specifications must be separated by commas or /. Output format should start with $1H_\wedge$, for each line of output.

NUMERIC FIELDS

Three types of specification for numerical data are:

(1) Ew.d e.g. E13.6 might produce −0.123456 E−02
w must be $\geqslant d + 7$

(2) Fw.d e.g. F7.3 might produce +12.123
w must be $\geqslant d + 2$

(3) Iw e.g. I5 might produce −1234

where w specifies the total number of characters to be input or output and d the number of decimal places.

Specifications for successive fields are separated by commas.

e.g. 3^FØRMAT(F10.3, E15.8, I5)

An oblique stroke, /, produces a new line. This must be followed by a carriage control symbol.

Repetition of Field Format

An integer ≤ 99 may be placed before E, F, I specifications

e.g. 2$_\wedge$FØRMAT(I2, 3 E 12.4)

Repetition of Groups

e.g. 1$_\wedge$FØRMAT(2(F10.6, E12.2),I4)
is equivalent to (F10.6, E12.2, F10.6, E12.2, I4)

Alphameric Fields

The specification wH is followed by w alphameric characters

e.g. 15H$_\wedge$CLASS$_\wedge$$_\wedge$EXAMPLE (note that blanks are included in the count)

(1) *Input* w characters are extracted from the input record and replace the w characters included in the specification.

(2) *Output* The w characters following the specification are written as output records.

e.g. WRITE(2,1)X
1$_\wedge$FØRMAT(1H$_\wedge$,5HXY$_\wedge$=$_\wedge$,F8.3)

could produce XY = –123.456

(3) *Blank Fields* The specification wX where 0 ≤ w ≤ 87 allows w characters to be skipped from an input record or w blank characters to be inserted in an output record.

(4) *Carriage Control* At the beginning of a set of FØRMAT specifications used for output, and after /, there must be a carriage control symbol. These symbols are

1H$_\wedge$ causes carriage to move to start of next line

1H0 causes carriage to move to start of next line but one

1H1 causes carriage to move to start of new page

1H+ prints on the same line

e.g. 1$_\wedge$FØRMAT(1H$_\wedge$,2HX=,E16.5/1H0,2HY=,F9.2)

6. COMMON PROGRAM ERRORS

Whilst a program is being compiled (translated into machine language) it is checked for errors in the Fortran text. Some errors are found during the translation of a line of Fortran and in these cases an error message is printed immediately following that line.

Certain types of error are not found until the end of a segment is reached and in these cases also an error message is printed at the end of the segment.

Example

```
        MASTER TEST
        READ(1,100)X1
        Ƶ=X1*X1
        WRITE (2,200)Ƶ
200     FØRMAT(1X,F10.3)
        END
```

In this case Label 100 is used in a read/write statement but is not a format label in this segment.

Warnings are also given. A warning may, for example, indicate that a character missing from a statement has been assumed.

Compilation errors will cause the run to fail but it will continue if only warnings are given.

7. COMMON EXECUTION ERRORS

A prohibited character found on an input card

'E' code in Format not associated with REAL variable

'I' code not associated with INTEGER variable

Input format record not long enough

Overflow set in simple arithmetic or attempt to divide by zero

Non-positive argument for ALOG

Reached END usually without meeting STOP

A SELECTION OF FORTRAN IV PROGRAMS

The programs which follow have been written for, and tested on an ICL 1904A computer. The input medium was punched cards and you should note that statement numbers are punched in columns 1 to 5 and the main punching takes place in columns 7 to 72.

Since the supporting statements which make a complete program deck vary from installation to installation, only the MASTER segment has been given in each case. The programs are as follows; the page numbers in parentheses refer to the appropriate part of the text.

1. GAUSS-SEIDEL ITERATION

(This program is for three equations in three unknowns)

```
    MASTER   GAUSS-SEIDEL
    DIMENSION A(3,3), B(3), X(3), OLDX(3)
    READ (1,2) TOL, ITMAX
  2 FORMAT (F10.6, I2)
    DO   1   I = 1,3
  1 READ (1,3) (A(I,J), J = 1,3), B(I)
  3 FORMAT (4F10.4)
    ITCOUN = 0
    READ (1,4) (OLDX(I), I = 1,3)
  4 FORMAT (3F10.4)
  8 X(1) = (B(1)–A(1,3)*OLDX(3)–A(1,2)*OLDX(2))/A(1,1)
    X(2) = (B(2)–A(2,3)*OLDX(3)–A(2,1)*X(1))/A(2,2)
    X(3) = (B(3)–A(3,2)*X(2)–A(3,1)*X(1))/A(3,3)
    ITCOUN = ITCOUN + 1
    GAP1 = ABS(X(1)–OLDX(1))
    GAP2 = ABS(X(2)–OLDX(2))
    GAP3 = ABS(X(3)–OLDX(3))
    IF(GAP1.GE.TOL) GO   TO   5
    IF(GAP2.GE.TOL) GO   TO   5
    IF(GAP3.LT.TOL)GO   TO   6
  5 IF(ITCOUN.EQ.ITMAX)GO   TO   7
    OLDX(1) = X(1)
    OLDX(2) = X(2)
    OLDX(3) = X(3)
    GO   TO   8
  6 WRITE (2,9)(X(I), I = 1,3), ITCOUN
  9 FORMAT (1H   , 3(F10.6, 5X), I2)
    GO   TO   10
  7 WRITE (2,11) ITCOUN
 11 FORMAT (1H   , 19H NO   SOLUTIONS   AFTER   , I2, 11H   ITERATIONS)
 10 STOP
    END
```

2. SQUARE ROOTS OF A NUMBER

```
      MASTER   ROOT
      READ (1,1) XOLD, A, TOL
    1 FORMAT (2F10.4, F10.6)
    2 XNEW = 0.5 * (XOLD + A/XOLD)
      DIFF = ABS (XNEW – XOLD)
      IF (DIFF.LT.TOL)GO   TO   3
      XOLD = XNEW
      GO   TO   2
    3 WRITE (2,4) XNEW
    4 FORMAT (1H   , F10.6)
      STOP
      END
```

3. SUCCESSIVE BISECTION

(The equation used here is $x^3 - 3x^2 + 2x + 1 = 0$).

```
      MASTER   BISECT
      F(X) = X**3 – 3.0*X*X + 2.0*X + 1.0
      READ (1,1) XL, XR, ACC
    1 FORMAT (2F10.4, F10.8)
    4 XM = 0.5*(XL + XR)
      FL = F(XL)
      FR = F(XR)
      FM = F(XM)
      D = FL*FM
      IF(D.LT.0.0) GO   TO   2
      XL = XM
      GO   TO   3
    2 XR = XM
    3 GAP = XR – XL
      IF (GAP.GE.ACC) GO   TO   4
      WRITE (2,5) XL, XM, XR, FL, FM, FR
    5 FORMAT (1H   , 6(E14.8,5X))
      STOP
      END
```

4. NEWTON-RAPHSON METHOD

(The equation used here is $x^3 - 3x^2 + 2x + 1 = 0$).

```
      MASTER   NEWRAP
      F(X) = X**3 – 3.0*X*X +2.0*X + 1.0
      FD(X) = 3.0*X*X – 6.0*X+2.0
      READ(1,1)XNEW, ACC, ITMAX
    1 FORMAT (F10.4, F10.8, I2)
      I = 0
    3 XOLD = XNEW
      XNEW = XOLD – F(XOLD)/FD(XOLD)
      I = I + 1
      GAP = ABS(XNEW – XOLD)
      IF (GAP.LT.ACC)GO   TO   2
      IF (I.LT.ITMAX)GO   TO   3
    2 WRITE (2,4)I, XOLD, XNEW
    4 FORMAT (1H   , I2, 2(5X, E14.8))
      STOP
      END
```

5. MACLAURIN'S SERIES FOR EXP(X)

```
      MASTER   SERIES
      READ(1,1) X, ACC, NMAX
    1 FORMAT (F10.5, F10.8, I2)
      TERM = X
      SUM = 1.0 + X
      N = 1
    3 N= N+1
      TERM = TERM*X/N
      IF (ABS(TERM).LT.ACC)GO   TO   2
      IF (N.EQ.NMAX)GO   TO   4
      SUM = SUM + TERM
      GO   TO   3
    2 WRITE (2,5) N, SUM
    5 FORMAT (1H   , I2, 5X, E14.8)
      STOP
    4 WRITE (2,6) N, TERM, SUM
    6 FORMAT (1H   , I2, 2(5X, E14.8))
      STOP
      END
```

6. GRID SEARCH

(The equation used here is $x^3 - 3x^2 + 2x + 1 = 0$).

```
     MASTER   GRID
     F(X) = X**3 – 3.0*X*X + 2.0*X + 1.0
     READ (1,1)H, XOLD, TOL
1    FORMAT (2F10.4, F10.8)
5    A = F(XOLD)
3    XNEW = XOLD + H
     B = F(XNEW)
     IF(A.LE.B)GO   TO  2
     XOLD = XNEW
     GO   TO  3
2    IF(H.LT.TOL)GO   TO  4
     XOLD = XNEW – 2.0*H
     H = H/10.0
     GO   TO  5
4    WRITE(2,6) XOLD, A, H
6    FORMAT (1H    , 2(E14.8, 5X), F10.8)
     STOP
     END
```

7. LEAST SQUARES STRAIGHT LINE FIT

```
     MASTER   LEASTSQ
     SX, SY, SXX, SXY = 0.0
     READ (1,1)N
1    FORMAT(I2)
     DO  2  I = 1,N
     READ (1,3)X(I), Y(I)
3    FORMAT (2F10.4)
     SX = SX + X(I)
     SY = SY + Y(I)
     SXX = SXX + X(I)*X(I)
     SXY = SXY + X(I)*Y(I)
2    CONTINUE
     AN = N
     B = (SX*SY – AN*SXY)/(SX*SX – SXX*AN)
     A = (SY – B*SX)/AN
     WRITE (2,4) A, B
4    FORMAT(1H    , 36H THE  EQUATION  OF  THE  STRAIGHT  LINE  IS/
1    1H   , 4HY  =   , E14.8, 5H    +     , E14.8, 3H     X)
     STOP
     END
```

8. EULER'S METHOD

(Applied here to Newton's Law of Cooling)

```
      MASTER   EULER
      READ (1,1) THETAS, THETAO, AK, TO
    1 FORMAT(4F10.4)
      WRITE (2,4)
    4 FORMAT (1H  , 7H        TEMP, 11X, 5H  TIME)
      READ (1,2)N, H
    2 FORMAT (I2, F10.4)
      THETA = THETAO
      T = TO
      I = 0
    5 DTHETA = -1.0*AK*(THETA - THETAS)
      T = T + H
      I = I + 1
      THETAN = THETA + DTHETA*H
      WRITE(2,3) THETAN, T
    3 FORMAT (1H  , F10.6, 5X, F10.4)
      IF (I.EQ.N) STOP
      THETA = THETAN
      GO TO 5
      END
```

9. RUNGE-KUTTA METHOD

(Applied here to $dy/dx = 1 - x - 4y$)

```
      MASTER   RUNKUT
      F(X, Y) = 1.0 - X - 4.0*Y
      READ (1,1) X0, Y0, H, N
    1 FORMAT (3F10.4, I2)
      WRITE (2,4)
    4 FORMAT (1H  , 7X, 1HX, 18X, 1HY)
      DO 2 I = 1,N
      AK1 = F(X0, Y0) *H
      AK2 = F(X0 + 0.5*H, Y0 + 0.5 *AK1) *H
      AK3 = F(X0 + 0.5*H, Y0 + 0.5 * AK2) *H
      AK4 = F(X0 + H, Y0 + AK3) *H
      X0 = X0 + H
      Y0 = Y0 + (AK1 + 2.0*AK2 + 2.0*AK3 + AK4)/6.0
      WRITE (2,3) X0, Y0
    3 FORMAT (1H  , 2 (E14.8, 5X))
    2 CONTINUE
      STOP
      END
```

10. SIMPSON INTEGRATION

(The integral is $\int_A^B \frac{1}{3x^3 + 2}\,dx$)

```
      MASTER  SIMPIN
      F(X) = 1.0/(3.0*X**3 + 2.0)
      READ (1,1)N, A, B
    1 FORMAT (I2, 2F10.4)
      AN = N
      H = (B - A)/AN
      SINT = 0.0
      X = A
      F1 = F(X)
      DO  2  I = 2, N, 2
      X = X + H
      F2 = F(X)
      X = X + H
      F3 = F(X)
      SINT = SINT + F1 + 4.0*F2 + F3
      F1 = F3
    2 CONTINUE
      SINT = SINT*H/3.0
      WRITE (2,3)SINT
    3 FORMAT (1H  , 29H THE  SIMPSON  APPROXIMATION  IS  , E14.8)
      STOP
      END
```

11. GAUSS-JORDAN ELIMINATION WITH PARTIAL PIVOTING

This program can cope with up to 10 equations in 10 unknowns. N is the number of equations (and unknowns and the coefficients are read into the array A).

The program can be modified to find the inverse of a matrix; refer to pages 248 and 249.

```
      MASTER  GJPP
      DIMENSION  A(10,11), X(10), P(10)
      READ(1,1) N
    1 FORMAT (I2)
      NA = N + 1
      READ(1,2)((A(I, J), J = 1, NA), I = 1, N)
    2 FORMAT (8F10.5)
      DO  3  L = 1, N
      AMAX = ABS (A(L, L))
      LN = I
      LA = L + 1
      DO  4  I = LA, N
      IF(AMAX.GE.ABS(A(I, L)))GO  TO  4
      AMAX = A(I, L)
      LN = I
    4 CONTINUE
      IF(LN.EQ.L)GO  TO  5
      DO  6  J = L, NA
      W = A(L, J)
      A(L, J) = A(LN, J)
    6 A(LN, J) = W
    5 DO  7  I = LA, N
    7 P(I) = –1.0*(A(I,L)/A(L, L))
      DO  8  I = LA, N
      DO  8  J = L, NA
    8 A(I, J) = A(I, J)+P(I)*A(L, J)
    3 CONTINUE
      X(N) = A(N, N + 1)/A(N, N)
      NB = N – 1
      DO  9  I = 1, NB
      NI = N – I
      X(NI) = A(NI, N + 1)
      NI1 = NI + 1
      DO  10  J = NI1, N
   10 X(NI) = X(NI)–A(NI, J) *X(J)
    9 X(NI) = X(NI)/A(NI, NI)
      DO  11  I = 1, N
   11 WRITE(2,12)I, X(I)
   12 FORMAT (1H  , I2, 5X, F10.4)
      STOP
      END
```

Bibliography

The following is a selected list of books related to the subject matter of this text.

1. BAJPAI, A.C., CALUS, I.M. and FAIRLEY, J.A. *Mathematics for Engineers and Scientists, Vol I.* Wiley, 1973.

2. McCALLA, T.R., *Introduction to Numerical Methods and FORTRAN programming.* Wiley, 1967.

3. McCRACKEN, D.D. and DORN, W.S., *Numerical Methods and FORTRAN programming.* Wiley, 1967.

4. COURT, R., *Fortran for Beginners.* Holmes McDougall, 1971.

5. BAJPAI, A.C., PAKES, H.W., CLARKE, R.J., DOUBLEDAY, J.M. and STEVENS, T.J., *Fortran and Algol.* Wiley, 1972.

6. LAMBE, C.G. and TRANTER, C.J., *Differential Equations for Engineers and Scientists.* E.U.P., 1961.

7. LEONARD, J.M., *Statistics, The arithmetic of Decision-making.* E.U.P., 1971.

8. KAYE, D., *Boolean Systems.* Longmans, 1968.

Answers

Chapter 1

Page 15

1. Write each number in the form $x = a \times 10^b$ where a is integral and use $x_1 x_2 = a_1 a_2 \times 10^{b_1 + b_2}$

2. (i) Place the first card face up.
   (ii) Put second card down face up – to the left if lower – to the right if higher.
   (iii) Repeat (ii) for all the cards, testing against the cards already down, working from the left to the right.

Page 20

3. Direct form has 5 multiplications and 3 additions. Nested form has 3 multiplications and 3 additions. Nested form is quicker and has less round-off.

6. The angle between two lines.

Page 28

1. 1/3500, 1/175

2. 0, 1/200, 1/19000, 17/1900

4. 2.6163×10^{-2}, 2.2437×10^{-2}, 4.52709×10^{-5}, 1.30435×10

Page 32

1. 1.83, largest = 1.89, smallest = 1.77, relative error $\leqslant$ 0.0336

2. Addition 6.5, 6.6, 6.4, absolute error $\leqslant$ 0.1; subtraction, 1.9, 2.0, 1.8, absolute error $\leqslant$ 0.1; multiplication, 9.66, 9.99, 9.34, relative error $\leqslant$ 0.034

3. 4.996, 0.042, 0.84%

Page 33

1. 0, 0.01, 0.012; 0.012, 0.0121, 0.01212

2. –48, 0.00; –0.0208

Page 35

2. Identical; $x = 0, y = -2; x = 0, y = -2$; no solution; $x = 2, y = 0$

3. $x = -1999, y = 1002$; no solution (199.8 changed to 200); $x = -1995, y = 1000$ (299.6 changed to 300)

Page 36

1. 34 (31.71); 0.7 (0.756)

2. –1.5 (–1.381); 15 (14.99); –0.013 (–0.018); 1.5 (1.4772)

4. Sum = 0.91, product = 32.7

Chapter 2

Page 44

1. (a) {3,4,5,6}; (b) {2,3,4}; (c) {(1,4), (2,3), (2,4), (3,2), (3,3), (3,4), (4,3), (4,2), (4,1), and negative combinations} (d) {1,3}

2. (a) $A \cup B = \{1,2,3,4,5,6,7,8,9,11\}$, $A \cap B = \{1,3,5,7\}$;
(b) $A \cup B = \{0, \pm1, \pm2, \pm3, \pm4, \pm5, \pm6, 7, 8, 9\}$, $A \cap B = \{4,5\}$;
(c) $A \cup B = \{0, \pm1, \pm2, \pm3, \pm4, 5, \ldots, 14, \pi, \sqrt{2}\}$, $A \cap B = \{0,5,6,8,9,10\}$

Page 48

1. (a) $(X \cup Y) \cap Z$; (b) $(X \cap Y) \cup (Z \cap W)$; (c) $(X \cap Y) \cup (Z \cap W)$;
(d) $(Z \cup X) \cap (Z \cup Y) \cap (W \cup X) \cap (W \cup Y)$;
(e) $(X \cup Y \cup Z) \cap (X \cup Y \cup W)$

5. 1(d) and 1(e)

Page 51

1. (a) $(X \cap Y') \cup (X' \cap Y)$
(b) $(X \cap Y \cap Z') \cup (X \cap Y' \cap Z) \cup (X' \cap Y' \cap Z) \cup (X \cap Y \cap Z)$;
(c) $(X \cap Y' \cap Z) \cup (X' \cap Y \cap Z) \cup (X \cap Y \cap Z)$

2. (b) $(X \cap Y) \cup Z$; (c) $Z \cap (X \cup Y)$

Page 56

2. (a) $a < 1$; (b) $b \geqslant -3/4$; (c) $p > 21/10$; (d) $x < ½$ or $x > 4$;
(e) $-2 < y < 5$; (f) $x = 0$

3. (a) $-4 < x < 4$, $(-4,4)$; (b) $1 < x < 5$, $(1,5)$;
(c) $-3\frac{2}{3} < x < 2\frac{1}{3}$, $(-3\frac{2}{3}, 2\frac{1}{3})$

4. (a) $|x-2|<5$; (b) $|x-6|<3$

5. 999

7. (a) $0.281<x<0.686$ and $-1.781>x>-2.186$;
(b) $-0.5275<x<2.5275$; (c) $x\geqslant 1, x\leqslant -1$; (d) $x>4, x<0$;
(e) $1/100>x>1/200$

Page 60

2. (a) $\{1,2,3,4,5\}$, $\{1,2,5,6,20\}$; (b) $\{1,2,3\}$, $\{-1,8,6,-2\}$;
(c) $\{1.5, 1.6, 1.7, 1.8, 1.9\}$, $\{2,3,5,7,9\}$
(d) $\{Z\}$, $\{Z\}$; (e) $\{Z\}$, $\{Z\}$

3. (a) one – many; (b) many – many; (c) one – one;
(d) many – many; (e) many – many

Page 66

1. (a) does; (b) does not; (c) does not; (d) does not

2. (a) $-\infty\leqslant x\leqslant\infty$; (b) $x\geqslant 1$; (c) $-1\leqslant x\leqslant 1$;
(d) $x\neq 0$; (e) $x\geqslant 1, x\leqslant -1$

3. (a) one – one; (b) one – one; (c) many – one;
(d) one – one; (e) many – one

5. (a), (b) and (d)

Page 70

1. one – one, $f^{-1}=x$; one – one, $f^{-1}=(1/x)-2$; one – one, $f^{-1}=\sin^{-1}x$;
many – one

2. (b) and (c)

3. (b) $\sqrt[3]{x}+3$; (c) x

Page 72

1. $f+g=x^2+x-4$, all x; $f-g=x-x^2-2$, all x; $f.\,g=(x^2-1)(x-3)$,
all x; $f\div g=(x-3)/(x^2-1), x\neq\pm 1$; $g\div f=(x^2-1)/(x-3), x\neq 3$

2. $f+g=x^2-3x$, all x; $f-g=x^2-3x+2$, all x; $f.\,g=(x-1)^3$, all x;
$f\div g=x-1, x\neq 1$; $g\div f=1/(x-1), x\neq 1$

3. $f(g)=x^2-4$; $g(f)=x^2-6x+8$; $f(g)=(x-2)^2$; $g(f)=x^2-2x$

4. $f(x)$ must lie in the domain; all x; $x=0,1$; $x=1$; $x=1,3$

5. $f(x)=x$, all x; $g(x)=x$, all x

Page 78

1. (a) $2/3(x-1)+5/3(2x+1)$; (b) $x+2+2/3(x-1)+5/3(2x+1)$;
(c) $1+1/(2-x)+1/(2+x)$; (d) $1/2(x-1)-3/(x+2)+11/2(x+3)$

2. (a) $1/10(x+1) - (x-1)/(x^2+9)$;
(b) $2/5(x-1) + 1/5(x-1)^2 - (2x+9)/5(x^2+3x+1)$;
(c) $4/x - 1/x^2 - 4/(2x+1)$; (d) $(x+1)/3(x^2-x+1) - 1/3(x+1)$

Page 92

1. 0.7078, –0.1001, 0.7493, 2.5607, 1.2764, –1.5040

7. 3.7622, 0.5211, –0.2913, 1.1948, 1.7627, 0.2554, –0.2554, 0.8671, –1.4436

8. 41/9, –40/41

Page 103

1. $\cos 11\,\pi/12 = 0$ by linear interpolation, should be –0.9659;
$\cos 160^\circ = -0.9107$, $\cos 170^\circ = -0.9553$, not accurate, interval too large

2. 0.5505 by mean differences, 0.5504 by linear interpolation

3. Useful as far as third differences

6. Third differences

7. $f(0.6)$ should be 24.48 (or 24.5)

8. 74.75 (Better, 75.00 due to inaccuracy of tables)

9. 0.33, 0.7

10. $\rho\,(5{,}500) = 0.6982$, $\rho\,(20{,}000)$ obtained by assuming third differences are constant and extending table

Chapter 3

Page 110.

1. Equality when $a = b$ 2. $-\sqrt{2}$ is also a root

3. $x_{n+1} = \frac{1}{3}\left(2x_n + \frac{A}{x_n^2}\right)$, $x_{n+1} = \frac{1}{3}\left(x_n + 2\sqrt{\frac{A}{x_n}}\right)$, $\sqrt[3]{10} = 2.154435$ (6d.p.), formula could give $\sqrt[3]{8}$ but it diverges

Page 116.

1. (i) $\{2, 12, 30, 56, \ldots\}$; (ii) $\{0, 2, 6, 16, 42, 110, \ldots\}$

2. 2, 10, 26, 82

3. (i) convergent, limit = 3; (ii) convergent, limit = 2;
(iii) divergent; (iv) convergent, limit = 2

4. (i) divergent; (ii) convergent, limit = 0;
(iii) convergent, limit = 1; (iv) convergent, limit = 10^{-4}

(v) convergent, limit = 2; (vi) divergent;
(vii) divergent

5. $(-\infty, 1 - 2^9/9) \cup (1 + 2^{10}/10, \infty)$, $(-\infty, 1 - 2^{999}/999) \cup (1 + 2^{1000}/1000, \infty)$

6. $\{2\frac{1}{2}, 2\frac{1}{3}, 2\frac{1}{4}, \ldots\}$; limits are 1, 1, 2; limits are 1,1; $\{2\frac{1}{2}, 4\frac{1}{3}, 6\frac{1}{4}, \ldots\}$, $\lim\{u_n\} = 1$, others do not exist; $\{4\frac{1}{2}, 3\frac{2}{3}, 3\frac{9}{20} \ldots\}$, 1, 2, 3, 2, ½

Page 120

1. Approx. gear ratio with $C_4 = \frac{1}{4}$, required ratio = 0.2504 (4 d.p.)

2. $\frac{1}{5+}\,\frac{1}{2+}\,\frac{1}{1+}\,\frac{1}{8}$; $\frac{1}{2+}\,\frac{1}{2+}\,\frac{1}{1+}\,\frac{1}{1+}\,\frac{1}{1+}\,\frac{1}{19}$; $\frac{1}{16+}\,\frac{1}{4+}\,\frac{1}{5+}\,\frac{1}{1+}\,\frac{1}{5}$;
$1 + \frac{1}{2+}\,\frac{1}{1+}\,\frac{1}{2+}\,\frac{1}{5}$

3. $\frac{1}{5}, \frac{2}{11}, \frac{3}{16}, \frac{26}{139}$; $\frac{1}{2}, \frac{2}{5}, \frac{3}{7}, \frac{5}{12}, \frac{8}{19}, \frac{157}{373}$; $\frac{1}{16}, \frac{4}{65}, \frac{21}{341}, \frac{25}{406}, \frac{146}{2371}$;
$1\frac{1}{2}, 1\frac{1}{3}, 1\frac{3}{8}, 1\frac{16}{43}$

4. Possible answers $\frac{70}{30} \times \frac{70}{120}$, $\frac{20 \times 95}{20 \times 70}$; $\frac{30}{50} \times \frac{30}{70}$

5. All except $\alpha = \beta = 0$

Page 128

1. 6, 2, ∞ **2.** $x = 0$; $x = 1$; none; none; $x = 1$

4. Split into $(x^2)(\sin x) + (2 \cos x) = f.g + h$ and use rules

7. $\to 0$, $\to 2 \cos x$ **8.** $\to \infty$, $\to 0$

Page 135

1. 33.4×10^{-6}, 33.4×10^{-6}, 33.5×10^{-6}

2. -0.1135×10^{-3}, -0.0865×10^{-3}, -0.0537×10^{-3}

3. 0.09, 0.49, 2.26 **4.** 0.3, 0.5625, 1.06, 1.18

Page 139

1. [1.6, 2.4], 0.507; [1.8, 2.2], 0.502; [1.9, 2.1], 0.500; [2.6, 3.4], 0.335
[2.8, 3.2], 0.334; [2.9, 3.1], 0.3335; [4.6, 5.4], 0.200; [4.8, 5.2], 0.200;
[4.9, 5.1], 0.200; derived function = $1/x$

2. Derived function = $-1/x^2$ odd function **3.** $f'(0) = 0$, no advantage in straddling,
4. 2.3524, 2.1293

5. $\sqrt[3]{9} = 2.0801$ **6.** $\sqrt{17} = 4.12$ (2 d.p.)

7. $f'(x) = 3ax^2 + 2bx + c$ 8. $-\sin x, 2\cos 2x, -3\sin 3x$

11. Estimates are 0.5650, 0.5575, 0.5400, 0.5225, 0.5150; then −0.625, −0.875, −0.625; then −12.5, 0, 12.5; finally, 625; accuracy rapidly gets worse

12. $g'(x_2) = [f(x_3) - f(x_1)]/2h$, more accurate since the curve is approximated by a parabola over three points

Chapter 4

Page 149

1. $2\sqrt{5}, \sqrt{26}, 2\sqrt{2}$

2. (i) $9y = 4x + 1$; (ii) $y = 13 - 2x$; (iii) $3x - 2y + 6 = 0$

3. (i) $\sqrt{2}$; (ii) $11/\sqrt{13}$; (iii) 0

4. (i) (2,1), 78°42′; (ii) same line
   (iii) (7/10, −1/10), 90°; (iv) parallel

5. $2y = 11x + 2, 22y + 4x = 67$ 6. (5, −2), (6,5) or (−8, 7), (−9, 0); 50

7. $\frac{1}{2}[x_1(y_2 - y_3) + x_2(y_3 - y_1) + x_3(y_1 - y_2)]$

8. (i) $x^2 + y^2 + 2x - 12y + 21 = 0$; (ii) $x^2 + y^2 - 6x - 8y + 16 = 0$

9. (i) not a circle; (ii) $(-\frac{1}{2}, 3/2), \sqrt{42}/2$; (iii) (0,1), 1

10. $3x + 5y + 15 = 0, y - y_1 = \dfrac{(x_1 - g)}{(y_1 - f)}(x - x_1)$

11. $\sqrt{x_1^2 + y_1^2 - 2gx_1 - 2fy_1 + c}; \sqrt{3}, \sqrt{19}, 0, \sqrt{-1}$; first two points outside circle, third point on the circle, fourth point inside circle

12. $2gg_1 + 2ff_1 = c_1 + c_2$; (i) yes; (ii) no; (iii) no

Page 152

1. $(y - y_1)/(y_1/b^2) = (x - x_1)/(x_1/a^2)$; $144y = -25x + 169$, $25y = 144x - 119$; $144y = 25x + 169$, $25y = -144x - 119$

2. $9y + 8x = 25, 8y = 9x - 10$; $9y - 8x = 25, 8y + 9x + 10 = 0$; foci $(\pm 5\sqrt{5}/6, 0)$, $x = \pm 3\sqrt{5}/2$; 5, 10/3

3. $y = -(1/m_1)x + (X + m_1 Y)/m_1, x^2 + y^2 = a^2 + b^2$

4. $x^2 + y^2 = a^2$ 5. $(\pm 15/13, \pm 60/13), 30\sqrt{17}/13$

6. $y = (3x/8) + 5/4, x = 2; y = 2 - \sqrt{3}x/2$; none

Page 154

2. $2y = x + 8, y + 2x = 24; y = \pm(x + 2), y = \pm(x - 6)$

3. 1/12, 1/3, 3/4

Page 156

2. $xy_1 + x_1y = 2c^2$ or $y + mx = \pm 2c\sqrt{m}$

3. $a^2(x - x_1)/x_1 + b^2(y - y_1)/y_1 = 0, x/x_1 + y/y_1 = 2, xx_1 - yy_1 = x_1^2 - y_1^2$

4. $x/3 - y/2 = \pm 1, 2y + 3x = \pm 13; y = \pm(2/3)x$

5. Latus rectum $= 2b^2/a = 2a(e^2 - 1)$; $2a(1 - e^2)$ for ellipse

6. $l = 2\sqrt{5} - 1, m = 2, n = \sqrt{5} - 1$

Page 157

1. $(y - 2)^2 = 12x$

2. $(Y + q)^2 = 4a(X + p), (X + p)^2/a^2 + (Y + q)^2/b^2 = 1, (X + p)^2/a^2 - (Y + q)^2/b^2 = 1$; (i) $h = 0, a = b$; (ii) $h = a = 0$ or $h = b = 0$ (iii) $h = 0, a \neq b, ab > 0$; (iv) $h = 0, a \neq b, ab < 0$

Page 159

2. $x + 2y < 6$ has no effect; $x + 2y < 3$ will cut down the region

3. (0,0), (2,1), (3,0), (0,2); no solution if new condition added

4. (i) $x \geqslant 0, y \leqslant 0$; (ii) $x + 3y > 3$; (iii) $0 \leqslant y \leqslant 2, 2x + 3y \geqslant 6, 4x + 3y \leqslant 12$ 5. $c = 6, c = 2$

6. $c = 9, c = 6$

7. $26X, 16Y$; $20X, 20Y$

Page 165

1. Parabola 2. Hyperbola

3. Ellipse 4. Touches axis of x at origin

5. Touches x – axis at $x = 3$ 6. Asymptotes are $x = 2, x = 3, y = 0$

7. Asymptotes are $x = 1, x = 2, y = 0$ 8. $y = -1$ is asymptote

9. $y = 0$ is asymptote 10. $x = 0$ and $y = 1 - x$ are asymptotes

11. $x = 0$ and $y = 1 - x$ are asymptotes

Page 169

1. (i) $r^2 = \sec^2\theta(1 - \tan^2\theta)$; (ii) $r^2 = \tan^2\theta$; (iii) $r^4 = 4/(1 - \sin 2\theta)$

2. (i) $x^2 + y^2 = 3x + 4y$; (ii) $x^4 - 6x^2y^2 + y^4 = 2(x^2 + y^2)$;
(iii) $x^6 + x^4y^2 = (x^2 - y^2)^2$

3. Figure of 8; circle; circle; figure of 8; four leaf clover; four leaf clover; three leaf clover

6. $r = (9 \pm 3\sqrt{5})/2$, $\cos\theta = (7 \pm 3\sqrt{5})/2$; curves do not cross at the origin

8. $x' = (x + y\sqrt{3})/2, y' = (-x\sqrt{3} + y)/2; x' = y, y' = -x$

9. $2x'y' = 1$.

Page 173

2. $x = 2ct_1t_2/(t_1 + t_2), y = 2c/(t_1 + t_2), (t_1 + t_2 \neq 0)$

3. $(x - a)^2 + y^2 = a^2$, circle

4. $x = 50(\cos 7t + \cos 6t), y = 50(\sin 7t + \sin 6t)$

5. $x^{2/3} + y^{2/3} = 1$ 6. $x = x_1 + \mu(x_2 - x_1), y = y_1 + \mu(y_2 - y_1)$

Page 177

2. $3\angle 0°, 2\angle 90°, 1\angle\pi, 2\angle(-\pi/2), 13\angle\tan^{-1}(12/5), 2\angle(5\pi/6)$,
$10\angle -126°52, 2\angle(-\pi/4), 4\angle(\pi/3), \sqrt{13}\angle 146°19, \sqrt{2}\angle(-3\pi/4)$,
$2.236\angle(-63°26)$

3. $1 + i, -1.532 + i\,1.2856, -\sqrt{3} - i, 2.868 - 4.096i$

4. $5 - 5i$ 5. (i) $-2 \pm 3i$, $3.606\angle \pm 123°41$

6. (i) $u = 2$ or -3; (ii) $z = (1 + i)/2$; (iii) $x = -5, y = -2$

10. $10, 26 + 2i, (5 - 6i)/7, (24 + 7i)/25$

11. $2\angle(2\pi/3), 5\angle 53°8, 2\sqrt{5}\angle -26°34$

12. $\arg\bar{z} = -\arg z$

13. $z = (3 \pm \sqrt{23})(1 + i)/2$

14. (i) $z = -1 \pm \sqrt{3i - 4}$
(ii) (a) arguments differ by 0 or π
(b) real parts equal, imaginary parts equal and opposite

17. $22i, 6/25$

Page 182

1. (i) $2 + 2\sqrt{3}i$ (ii) $-108\sqrt{2}(1 + i)$

2. (i) $3\angle 90°$; (ii) $2\sqrt{2}\,\overline{\angle 105°}$; (iii) $4\angle 60°$

4. (i) $-(3 + 2\sqrt{3}) + i(2\sqrt{3} - 1), (2\sqrt{3} - 3) - i(2\sqrt{3} + 1)$

5. $(4, -4)$. 6. (i) Interior of circle (ii) area between two lines making angle $\pm 30^\circ$ with the x-axis; (iii) area to the left of $x = 3/2$; (iv) area outside circle on $x = 1$ and 2 as diameter, $y \geqslant 0$ and inside circle, $y \leqslant 0$

8. (b) $\sqrt{5}/2$; circle, centre $1 + \frac{1}{2}i$, radius $\sqrt{5}/2$

9. Area inside ellipse, foci at $(1,0)$ and $(0,1)$, major axis $= 4$

10. $z\bar{z} + z + \bar{z} - 4 = 0$, centre $(-1,0)$, radius $\sqrt{5}$

11. $(\cos 80^\circ + i \sin 80^\circ)/2$ or $(\cos 40^\circ + i \sin 40^\circ)/2$ 12. $y = 1$

Page 186

1. $\cos 5\theta = 16 \cos^5\theta - 20 \cos^3\theta + 5 \cos\theta$, $\sin 5\theta = 16 \sin^5\theta - 20 \sin^3\theta + 5 \sin\theta$; $\cos 18^\circ = \sqrt{(5 + \sqrt{5})/8}$, $\sin 36^\circ = \sqrt{(5 - \sqrt{5})/8}$

4. (i) $a = \sqrt{3}/2$, $b = -1/2$

5. $\pi/2 + 2k\pi + i \cosh^{-1} 2$ 6. $\theta = 0, \pi, 2\pi, \ldots$

7. Z moves on the same locus, W lies on the real line between $-2a$ and $+2a$

8. 1 9. $\log z = \log r + i(\theta + 2k\pi)$, multivalued

10. i 11. $2.032 + i\,3.052$

12. $A = (\cosh^2 s - \sin^2 s)^{-\frac{1}{2}}$, $\cot \alpha = \tan s \tanh s$

Page 190

1. (i) $1.241 - 0.219i$, $-0.431 + 1.184i$, $-0.810 - 0.965i$; (ii) $\pm(1.621 + 1.396i)$, $\pm(1.369 - 1.621i)$

2. (i) $1 + 3i$; (ii) $1 - i, -2, -2, a = 2, b = -2$

3. (i) $0, \pm(1/\sqrt{2})(1 + i), \pm(1/\sqrt{2})(1 - i)$; (ii) $1.587, 1.587 \angle(\pm 2\pi/3), -1.442, 1.442 \angle(\pm\pi/3)$

4. (i) $\sqrt[3]{2} \angle 10^\circ$, $\sqrt[3]{2} \angle 130^\circ$, $\sqrt[3]{2} \angle -110^\circ$ (ii) $\sqrt[4]{5} \angle(13^\circ 17 + k.90^\circ)$

7. $1 \angle(k.72^\circ)$; $1 + \delta + \delta^2 + \delta^3 + \delta^4 = 0$

Page 192

1. Circuit impedance $= (R + j\omega L)/[(1 - \omega^2 LC) + j\omega RC]$, $i_1 = E/(R + j\omega L)$, $i_2 = jE\omega C$

Page 206

1. (a) $(-7, -2, 2)$; (b) $\sqrt{158}/2$

2. (3/7, 6/7, −2/7) **3.** $(\sqrt{3}-1)/4$

4. (2, −6, 3), 7

5. $\sqrt{6}, 5\sqrt{6}/2, \sqrt{114}/2$ **6.** $m = -19/11, n = -53/22$

Page 212

1. $(\mathbf{i}+\mathbf{j}+\mathbf{k})/\sqrt{3}$ **2.** (5**i** + 3**j**) . (2**i** + 5**j**) = 25 ft.lbs

4. −11 **5.** **a** . **b** = **a** . **c** = 0

Page 215

1. −4**i** + 5**j** + 13**k** **2.** $\sqrt{107}/2$

5. **b** = (41/157, 95/157, 21/157)

6. 2**i** − **j** + **k**

Page 217

1. 18

Page 218

1. each side = 7**j** − 7**k**

3. 24**i** + 7**j** − 5**k**, 15**i** + 15**j** − 15**k**

4. $(\mu/56\pi)(-6\mathbf{i} - 9\mathbf{j} + 11\mathbf{k})$

Chapter 5

Page 225

1. (a) 1/2, 3/2, 1; (b) 0, 1/3, 0; (c) equation (iv) = (i) − (ii) − (iii), 18/16, 30/16, 21/16, −3/16, 27/16; (d) no solution; (e) 81/11, −4/5, −73/55; (f) 1/2, 1/3, 1/4

2. (a) i, 1 − i; (b) (289 + 119i)/221, −2 − i

Page 230

1. (a) 3.36, 1.10, 1.24; (b) 5.00, −5.07, 0.99; (c) 1.59, −0.31, −0.20; (d) 2.28, 0.21, −0.44;

2. (a) 3.3610, 1.1026, 1.2384; (b) 5.01, −5.08, 0.99

3. 1,1,1,1

Page 236.

1. Determinants are all zero

3. 4, 8, 10; 4.28, 8.08, 9.88

4. Both sets of values almost satisfy the equations. Equations ill-conditioned.

Page 241

1. (i) $[A + B] = \begin{bmatrix} 3 & -5 \\ 2 & 7 \\ -4 & -2 \end{bmatrix}$, $[A - B] = \begin{bmatrix} 5 & -5 \\ 2 & -7 \\ -8 & 8 \end{bmatrix}$;

(ii) $[AB] = \begin{bmatrix} 20 \\ 29 \end{bmatrix}$;

(iii) $[BA] = \begin{bmatrix} 3 & 8 \\ 8 & -12 \end{bmatrix}$, $[AB] = \begin{bmatrix} -13 & 6 & -5 \\ 21 & 2 & 11 \\ 6 & -4 & 2 \end{bmatrix}$;

(iv) $[BA] = \begin{bmatrix} -22 & 4 & 42 \\ -8 & 16 & 48 \\ -1 & -2 & -3 \\ 0 & 0 & 9 \end{bmatrix}$;

(v) $[A + B] = \begin{bmatrix} 2 & 2 & -4 \\ 2 & 1 & 8 \\ 2 & -1 & 5 \end{bmatrix}$, $[A - B] = \begin{bmatrix} 2 & -4 & 6 \\ -2 & 1 & -4 \\ 0 & 1 & -3 \end{bmatrix}$,

$[AB] = \begin{bmatrix} -1 & 5 & -12 \\ 4 & -2 & 14 \\ 1 & 2 & -1 \end{bmatrix}$, $[BA] = \begin{bmatrix} -5 & 3 & 1 \\ 10 & -2 & 8 \\ 6 & -2 & 3 \end{bmatrix}$,

$[A + B]\ [A - B] = \begin{bmatrix} 0 & -10 & 16 \\ 2 & 1 & -16 \\ 6 & -4 & 1 \end{bmatrix}$, $A^2 - B^2 = \begin{bmatrix} 4 & -8 & 3 \\ -4 & 1 & -10 \\ 1 & 0 & -3 \end{bmatrix}$

Page 246

3. (a) pre-multiply by $\begin{bmatrix} 1 & 0 & 3 \\ 0 & 1 & 0 \\ 0 & 0 & 1 \end{bmatrix}$; (b) post-multiply by $\begin{bmatrix} 1 & 0 & 1 \\ 0 & -\frac{1}{2} & 0 \\ 0 & 0 & 1 \end{bmatrix}$;

(c) pre-multiply by $\begin{bmatrix} 1 & 0 & 0 \\ 0 & 1 & -1 \\ 0 & 0 & 1 \end{bmatrix}$

Page 252

1. (i) $\begin{bmatrix} -4/11 & 3/11 \\ 1/11 & 2/11 \end{bmatrix}$; (ii) not possible; (iii) $1/14 \begin{bmatrix} 5 & 3 & 1 \\ 3 & 6 & -5 \\ 1 & -5 & 3 \end{bmatrix}$;

(iv) $\begin{bmatrix} 7 & -3 & -3 \\ -1 & 1 & 0 \\ -1 & 0 & 1 \end{bmatrix}$; (v) not possible;

(vi) $1/29 \begin{bmatrix} 1 & -2 & 9 & 4 \\ -21 & 13 & 14 & 3 \\ 13 & 3 & 1 & -6 \\ -19 & 9 & 3 & 11 \end{bmatrix}$

Page 259

2. (i) −35; (ii) 0.765; (iii) −30; (iv) −232

Page 261

1. (i) 17/11, 1/11; (ii) 5/2, 3/2

2. (i) 3/2, 5/2, −2; (ii) 0, ½, 0

3. (i) 1/2, 3/2, 1; (ii) 81/11, −4/5, −73/55; (iii) 1/4, 1/2, 1/3

Page 265

1. $1/78 \begin{bmatrix} 15 & -3 & 9 \\ -7 & 17 & 1 \\ -2 & 16 & -22 \end{bmatrix}$; ½, 3/2, 1

2. $\begin{bmatrix} 7 & -3 & -3 \\ -1 & 1 & 0 \\ -1 & 0 & 1 \end{bmatrix}$ **3.** 5/11, 7/11; 1, −2, 5

Page 272

1. (i) True solution is 1, 2, 3; (ii) True solution is 3, 2, 1

Page 274

1. (i) One root $-3 < x < -2$; (ii) one root $2 < x < 3$;
(iii) two roots approx. 0 and 100, (iv) two roots approx. 2 and 6;
(v) infinite number of roots, $0 < x < \pi/2$ and near $n\pi$, $n = 1, 2, 3, \ldots$;
(vi) one root near $\pi/2$

Page 282

2. Rule holds **3.** $x_0 = -2$ converges to -1, $x_0 = 0$ to -1, $x_0 = 1$ to 2, $x_0 = 3$ to 2 **4.** $x = 1, 2$ **5.** 1.2022, 0.8526, 1.7456
6. $(3/2A) > x_0 > (1/2A)$ **7.** 3.5782, 0, −3.5782 **8.** Roots are ± 2

Page 286

1. $x_{n+1} = \frac{1}{2}(x_n + A/x_n)$; $x_{n+1} = (1 - 1/r)x_n + A/rx_n^{r-1}$

2. 4.493 3. $x_{n+1} = (x_n{}^2 + 2)/(2x_n - 1)$ 4. 2.83

5. 1.23 6. 1.114 (3 d.p.) 7. Real roots approx 1, 4

9. (i) 0.6155; (ii) 0.9643 10. 2.36 12. 0.87

Page 294

1. $x^2 + 6x + 9$, remainder 25
2. $3.6x - 11.9$, remainder 49.2
3. $x^2 + 2x + 2$, remainder 15
4. $2x^2 - 6x + 7$, remainder $-19x + 35$
5. $8x + 27$, remainder $72x - 130$
6. (i) $-182 + 442i$, (ii) $8.06 - 0.24i$
7. (a) -1.17; (b) 3.00

Page 298

1. (a) one positive root;(b) one positive, two negative roots, (c) one positive, two negative roots (d) two positive, two negative roots
2. (i) $-1.171 \pm 0.765i$, $1.171 \pm 0.399i$;
   (ii) $0.55 \pm 0.59i$, $-0.55 \pm 1.12i$
3. $x^2 - 2.075x + 1.906$
4. $x^2 - 4.006x + 5.004$
5. First root = 0.4, quadratic factors $x^2 + 0.625\,x + 0.146$, $x^2 - 0.225\,x + 0.155$

Chapter 6

Page 304

1. (i) $3x^2 \sin x + x^3 \cos x$; (ii) $e^x(\cos x - \sin x)$; (iii) $x^3(4 \log x + 1)$;
   (iv) $(2x + x^2)/(1 + x)^2$; (v) $\sec x\,(x \tan x - 1)/x^2$; (vi) $-10(3 - 2x)^4$;
   (vii) $-x/\sqrt{1 - x^2}$; (viii) $1 - x \sin^{-1} x/\sqrt{1 - x^2}$; (ix) $3 \cos 3x$;
   (x) $4 \sin 8x$; (xi) $-3 \cos 3x \sin(\sin 3x)$; (xii) $3e^{3x+2}$;
   (xiii) $2x \tan^2 2x + 4x^2 \tan 2x \sec^2 2x$; (xiv) $(2x - 1)/(1 - x + x^2)$;
   (xv) $2/(1 - x^2)$; (xvi) $-1/(\sin x \cos x)$;
   (xvii) $e^{2x}[2 \log(1 + x^2) + 2x/(1 + x^2)]$; (xviii) $\sec^2 x$;
   (xix) $-\operatorname{cosec} x \cot x$; (xx) $2 \sinh 2x$
2. -34 cm/sec; -14 cm/sec^2
3. 0.362 (3 d.p.)
4. $-e^{-0.2t}(2.32 \sin 4t + 1.72 \cos 4t)$
6. $7500\,\pi$ cm^3/sec
7. $-24 \sin 6t/(a^2 - 16 \sin^2 3t)^{1/2}$
8. $40/9\pi h^2$
9. $dy/dx = 3\sqrt{x}\,(1 + 2x^3)/2\sqrt{1 + x^3}$

Page 308

1. (i) -1; (ii) $1/x[1 + (\log x)^2]$; (iii) $2(x + y) \cos[(x + y)^2]/\{1 - 2(x + y) \cos[(x + y)^2]\}$;

(iv) $(25 - x - 2x^2)/\sqrt{25 - x^2}$;
(v) $[x + (1/x)]^x\,[\log[x + (1/x)] + (x^2 - 1)/(x^2 + 1)]$;
(vi) $\cot(a/b + 1)t/2$; (vii) $-\cot\theta$; (viii) $(1 + y^2)/(1 + 2y^2)$;
(ix) $-(2x + 3)/2(1 + y)$; (x) $(x^2 - 7)/(1 + x)^{3/2}(3 - x)^{½}$;
(xi) $\sqrt{3}/(2 + \sin x)$; (xii) $2/\sqrt{1 - x^2}$

4. (i) $-6(2x - 3)^{-3/2}(2x + 3)^{-½}$; (ii) $e^x(1 + x)(9 - x - 2x^2)/2(1 - x)^{5/2}$

5. (i) $-(2x + 3y)/(3x + 2y)$; (ii) $3x^2/10y$;
(iii) $[\cos(x + y) - 3y]/[3x - \cos(x + y)]$;
(iv) $y\cos x/(1 + y\sin y)$; (v) $e^{(x-y)}(x - 1)/2x^2$ or $1 - 1/x$

6. (i) $\lambda\cos\lambda t/\cos t$; (ii) $-b\cot\theta/a$; (iii) $4/3t$; (iv) $\sec\theta/4a$

Page 309

1. (i) $4/(3\cos x + 5)$; (ii) $4x/(x^4 - 1)$; (iii) $4e^{-2x}(1 - x^2)$;
(iv) $2/(1 + x^2)$

2. (i) $-\pi e$; (ii) $2/5$; (iii) $26/5$

3. (i) $2e^{-2}$; (ii) ½; (iii) 0; (iv) 1

5. (ii) $(3 - x)^2(12 - 11x + 4x^2)/(2 - x)^3$

7. (i) (a) $\sin^{-1}(½)$, (b) $-4/\sqrt{3}$, (c) $1 - \log 2$; (ii) $x/(1 + x)$

8. $(y + b\sin b\theta)(\theta + a\sin ax)/(b\cos b\theta - x)(a\cos ay - \theta)$

Page 314

1. (i) [−3,3], 20, −1/4; [−3,0], 20,2; [1,2], 0, −1/4;
(ii) [0,4], 2¼, −10; [−3,3], 2¼, −10; [−3,0], 2, −10; [1,2], 2,0

2. (i) local max at $x = 6/5$, local min at $x = 4$;
(ii) local min at $x = 0$;
(iii) local min at $x = 3$;
(iv) local max at $x = 3.281$, min at $x = 1.219$;
(v) turning points when $\theta\cot\theta = h$

3. $r = e^{-1}$ 4. $c\sqrt{2}$ 5. $(gT_0/3w)^{½}$ 6. $(CL)^{-½}$ 7. $R/2$

8. $\log x = 1 + 1/x$; 3.591

10. 2,2, −2, −1 11. Symmetry about $0y$, x axis is asymptote

12. Symmetry about $0x$, $x = -1$ is asymptote, loop between $x = 0$ and 1

13. (i) $f'(x) = x/(1 - x)^2$; (ii) max at $x = 1$, $2e^{-2}$; min at $x = -2$, $-e^4$

14. Local maximum at $x = -2$, point of inflexion at $x = 1$

Page 321

2. (i) $10(x^2 + 3xy + y^2)/(3x + 2y)^3$; (ii) $3x(20y^2 - 3x^3)/100y^3$;

(iii) $-[9(x-y)^2 \sin(x+y) + 6\{\cos(x+y) - 3y\}\{3x - \cos(x+y)\}]/\{3x - \cos(x+y)\}^3$;
(iv) $\{(1/y + \sin y)^2 \sin x + (1/y^2 - \cos y)\cos^2 x\}/(1/y + \sin y)^3$;
(v) $1/x^2$;
(i) $(-\lambda^2 \sin \lambda t \cos t + \lambda \cos \lambda t \sin t)/\cos^3 t$;
(ii) $-b \operatorname{cosec}^3\theta/a^2$ (iii) $-4/9t^4$ (iv) $-\sec^3\theta/16a^2$

3. $(4p^3k^3/27V^2) - F$

4. $r^6 = 2A/B$ 8. (i) $-3t/2 - t^3/2, 3(1+t^2)^3/4t$

12. Stable at $x = 0$, unstable at $|x| = a$

13. (i) $-12/x^4$; (ii) 2e

Page 327

2. $c \cosh^2(x/c)$. 3. (a) $\sqrt{12}$; (b) $5\sqrt{10}/3, \infty$; (c) $-\sqrt{2}$; (d) $-87\sqrt{29}/80$ 4. $-\{1 + (2 - 8x_1)^2\}^{3/2}/8$

6. $-ka/(a^2 + k^2)^{3/2}$, $2a + k^2/a$, $-(a^2 + k^2)/k$

7. Min = 0, max $= -16.6^3/5^5$, $1/\sqrt{2}$

Page 336

1. 3.3244 2. 0.72654, 0.93906. 3. 0.6088 (4 d.p.)

5. 1.60379, 2.28962 6. Error in $y(0.6)$, should be -0.8432, $y(0.45) = -0.2919$

7. $f(5)$ should be 156, $f(2.5) = 25.375$

8. $f(1.5) = 7.125$, $f(7) = 399$

Page 340

1. (a) 1.73; (b) 1.80, (c) 1.76. 2. $f(1) = 4, f(2) = -2$

Page 343

2. $f'(0.5) = -0.4795$, estimated error = 0.0002

3. -0.6400

Page 347

1. 0.011 2. 0.01, 0.0099

Page 352

1. Quadratic approximations
(a) $1 + x + x^2/2$, $e^{x_0}[1 + (x - x_0) + \frac{1}{2}(x - x_0)^2]$;
(b) $x - x^2/2$, $\log(1 + x_0) + (x - x_0)/(1 + x_0) - (x - x_0)^2/2(1 + x_0)^2$

(c) $1 - x^2/2$, $\cos x_0 - (\sin x_0)(x - x_0) - (\cos x_0)(x - x_0)^2/2$;
(d) $-2x^2 + 7x - 5$, $(3x_0 - 2)x^2 + (7 - 3x_0^2)x + x_0^3 - 5$

2. Quadratic approximations

	$x = 0$			$x = x_0$		
	0.10	0.25	0.50	0.10	0.25	0.50
(a)	1.1050	1.2813	1.6250	1.1054	1.2841	1.6490
(b)	0.0950	0.2188	0.3750	0.0955	0.2232	0.4055
(c)	0.9950	0.9688	0.8750	0.9951	0.9689	0.8776
(d)	−4.3200	−3.3750	−2.0000	−4.1380	−3.3593	−1.8740

Actual

	0.10	0.25	0.50
(a)	1.1052	1.2840	1.6483
(b)	0.0953	0.2231	0.4053
(c)	0.9950	0.9691	0.8776
(d)	−4.1390	−3.3594	−1.8750

Page 357

1. (i) Converges to $e - 1$; (ii) ratio $\to 1/3$, diverges; (iii) ratio $\to \sqrt{3}$, converges; (iv) ratio $\to 2$, converges

2. (i) converges; (ii) converges; (iii) diverges; (iv) diverges

3. 251

Page 360

1. (i) (1,3); (ii) (−2,0); (iii) (−1,1); (iv) (−1,1)

3. all x

Page 367

4. $-x^2/2 - x^4/12$ **5.** $-x^2/6 - x^4/180, -x^2/2 - x^4/12$

6. $|y| \leqslant 1$ **7.** $cx^3/2 - acx^4/4 + (3a^2c - 4bc)x^5/16$

8. 1.00061

9. 2.3026

11. $\log 2 + a/8 - a^2/64$ **12.** (i) $x - x^3/6$

Page 371

1. (a) $e[1 + (x - 1) + (x - 1)^2/2! + ...]$, all x;
(b) $2/3 - (2/9)(x - 2) + (4/27)(x - 2)^2/2! + ...\ |x| < 1$
(c) $\frac{1}{2} + (\sqrt{3}/2)(x - \pi/6) - (x - \pi/6)^2/4 - \sqrt{3}(x - \pi/6)^3/12 + ...$, all x

2. (ii) $\sin 51° = 0.7771$ (4 d.p), estimate = 0.7773
3. $\sqrt{2}/10$, $\sqrt[3]{60}/10$
4. Estimated error = 0.0008, actual error = 0.0002
5. 0.9325

Page 376

1. (i) 0; (ii) 1;
   (iii) 1; (iv) ½ 2. –3% 3. $(5/6\pi) \pm 2\%$ gm/cm^3
4. 6% 5. (i) 2; (ii) $1/2a$; (iii) –2
6. $\tan\theta = \theta + \theta^3/3 + 2\theta^5/15$
7. Et/L

Page 381

1. Minimum = –5 at $x = 0$ 3. Minimum at $x = 3$

Chapter 7

Page 384

1. $\Sigma e_i = 0.3$, $\Sigma\ |e_i| = 0.7$, $\Sigma\ e_i^2 = 0.19$; reject (2, 4.4); $\Sigma e_i = -0.1$, $\Sigma\ |e_i| = 0.3$, $\Sigma e_i^2 = 0.03$
2. $a_0 = 3$, $a_1 = 1$; $S = 0$; exact fit
3. $a_0 = 3.03, a_1 = 1.1, S \neq 0$ 4. $y = a_0 + a_1 x$, $S = \sum_{i=1}^{n} (y_i - a_0 - a_1 x_i)^2$

Page 388

1. (i) and (ii) represent planes; (iii) represents a line; (iv) represents a cylinder; (v) represents a circle
2. (i) Hyperboloid, all x, y; (ii) Cone, vertex downward at (1,0,0), all x, y; (iii) Cone, vertex downward at (1,0,1), all x, y; (iv) sphere, centre (2,0,1), radius 2, $0 \leqslant x \leqslant 4$, $-2 \leqslant y \leqslant 2$, $1 \leqslant z \leqslant 3$
3. (i) 13,10,1; (ii) 2/5, 2/5, not defined (iii) log 6, log 3, 0; (iv) $7e^2$, $e^{-4} - 6e^{-1}$, 0
4. (i) $(2,3,z)$; (ii) $(0,y,0)$; (iii) $(1 - z, 1 + z, z)$
5. Intersect if not all three of $a/a', b/b', c/c'$ are equal, planes must not be parallel
6. Ellipsoid
7. (i) $z = (y + 1)^2 + 1$ parabola, (ii) $z = (y + 1)^2 + 4$, parabola
   (iii) $z = (y + 1)^2 + a^2$ parabola (iv) $z = x^2 + 4$;

(v) $z = x^2 + 9$; (vi) $z = x^2 + (b + 1)^2$;
(vii) $(x + ½)^2 + (y + 3/2)^2 = 5/2$ circle

8. $(d/a, d/b, d/c)$ **9.** $-5 \sin 4$

Page 394

1. (i) 1/7; (ii) (a) 0, (b) 0

2. (a) −1; (b) 1; (c) 0; (d) 3/5; (e) 1; (f) $\cos 2\theta$

3. (i) 0, (ii) ∞, (iii) ∞, (iv) $2/\sin 2\theta$

4. (i) continuous everywhere; (ii) everywhere except (0,0);
(iii) continuous except at $x = -y$; (iv) continuous except at $x = y$;
(v) continuous except at (0,0), $f(0,0) = 0$

5. (i) $2x, 3y^2, 4z$; (ii) $2xy^3/z, 3x^2y^2/z, -x^2y^3/z^2$;
(iii) $2xye^{3z}, x^2e^{3z}, 3(1 + x^2y)e^{3z}$; (iv) $z \cos(xz + y), \cos(xz + y)$,
$x \cos(xz + y)$ (v) $ye^{(xy + 2y^2)}, (x + 4y)e^{(xy + 2y^2)}$;
(vi) $½y \sin z, ½x \sin z, ½xy \cos z$;
(vii) $y/(x^2 + y^2), -x/(x^2 + y^2)$;
(viii) $(x^3 + 2y^3)/x^3y^2, -(2x^3 + y^3)/x^2y^3$

6.

	f_x	f_y
(i)	$3x^2y^4$	$4x^3y^3$
(ii)	$2x/y^3$	$-3x^2/y^4$
(iii)	$2 \cos y$	$-2x \sin y$
(iv)	$\cos(x + y)$	$\cos(x + y)$
(v)	0	2
(vi)	e^x	$-\sin y$
(vii)	$2xye^{3y}$	$(3 + 3x^2y + x^2)e^{3y}$
(viii)	$-3y/(9x^2 + y^2)$	$3x/(9x^2 + y^2)$
(ix)	$x(x^2 - y^2)^{-½}$	$-y(x^2 - y^2)^{-½}$

	$f_x(0,0)$	$f_y(0,0)$	$f_x(1,2)$	$f_y(1,2)$	$f_x(-1,0)$	$f_y(-1,0)$
(i)	0	0	48	32	0	0
(ii)	–	–	1/4	−3/16	–	–
(iii)	2	0	2 cos 2	−2 sin 2	2	0
(iv)	1	1	cos 3	cos 3	cos 1	cos 1
(v)	0	2	0	2	0	2
(vi)	1	0	e	−sin 2	1/e	0
(vii)	0	3	$4e^6$	$10e^6$	0	4
(viii)	–	–	−6/13	3/13	0	−1/3
(ix)	—	—	—	—	−1	0

7. (iii) $n = 2, -3$

10. (a) $0, a/V^2$

Page 398

1. $-6x + 11y + 14z = 2$

2. (i) (0,0) local minimum; (ii) (–3,2) local minimum;
(iii) (–5/3, 1) saddle point, (–1,1) local minimum;
(iv) (0,0) saddle point, (1, –1) local minimum, (–1,1) local minimum
(v) (2/5, 4/5) saddle point, (0,0) local minimum; (vi) (1,1), (–1, –1) saddle points; (vii) (0,0) local maximum; (viii) (0,0), (–¼, 1) (0, – ½) saddle points, (–1/12, –1/6) local maximum (ix) (0,0) local minimum

4. (i) (0,0) saddle, $(6a, 18a)$ minimum; (iii) maximum value = 1/e

5. [16/9, –32/9, 32/9] **6.** $\theta = 30^\circ, x = l/6$

Page 404

1. (i) $0 = \Sigma y - na - b\Sigma x^3 = \Sigma x^3 y - a\Sigma x^3 - b\Sigma x^6$
(ii) $0 = \Sigma y - na - b\Sigma x^2 - c\Sigma x^3 = \Sigma yx^2 - a\Sigma x^2 - b\Sigma x^4 - c\Sigma x^5 = \Sigma yx^3 - a\Sigma x^3 - b\Sigma x^5 - c\Sigma x^6$
(iii) $0 = \Sigma xy - \Sigma x - a\Sigma x^2$
(iv) $0 = \Sigma xye^{bx} - a\Sigma xe^{2bx} = \Sigma ye^{bx} - a\Sigma e^{2bx}$
(v) $0 = \Sigma y/(a + bx) - \Sigma \log(a + bx)/(a + bx) = \Sigma xy/(a + bx) - \Sigma x \log(a + bx)/(a + bx)$
(vi) $0 = \Sigma y/x - a\Sigma 1/x^2$

2. $y = 461.44 - 48.39x, y = 487.82 - 75.12x + 3.06x^2$

3. $D = 15 + 0.329V + 0.057V^2$

4. $y = 13.26 - 18.90x, y = 13.07 - 18.43x$

5. (i) $y = 2.05x +$ [illegible] $/x$; (ii) $y = 2x^2 - 7\sqrt{x}$
(iii) $y = 0.0224/x^2$

6. $y = 1.524e^{3.014/x}$

7. $y = 3.66 - 2.08x + 0.42x^2$

8. $S = 3.743 + 1.7T - 0.071T^2$; (ii) $S = 3.6 + 1.7T$;
(iii) $S = 3.743 + 1.417T - 0.071T^2 + 0.083T^3$

9. $s = 86.9\, t^{-1.407}$

Page 410

1. (a) –1% ; (b) ±2.5%

2. 0.054π in^3 **3.** –2%

4. (i) $[-2/(x^3y) + y]\, dx + [-1/(x^2y^2) + x]\, dy$;
(ii) $e^{-(x+y)}\, [\cos(x + y) - \sin(x + y)]\, (dx + dy)$;
(iii) $[(-y/\sqrt{x^2 + y^2}) + x \tan^{-1}(y/x)/\sqrt{x^2 + y^2}]\, dx + [(x/\sqrt{x^2 + y^2}) + y \tan^{-1}(y/x)/\sqrt{x^2 + y^2}]\, dy$;

(iv) $(3x^2y + 3y^4)\mathrm{d}x + (x^3 + 12xy^3)\mathrm{d}y$;
(v) $2e^{(x^2+y^2)}(x\mathrm{d}x + y\mathrm{d}y)$, (vi) $-\sin(x + 2y)(\mathrm{d}x + 2\mathrm{d}y)$

6. -90 watts; exact answer $-$ 85.3 watts (1 d.p.)

7. 1.8%

Page 412

1. $\pi/2\,[8 + 77\sqrt{41}]$ cm²/sec

2. -14 cm/sec 3. $k/\sqrt{kTBR}\,[BR\,\mathrm{d}T/\mathrm{d}t + TR\,\mathrm{d}B/\mathrm{d}t + TB\,\mathrm{d}R/\mathrm{d}t]$

4. $3C\sqrt{60 + 2\mu}\,(0.3 + 0.1\mu)$ 5. $32(V + T/10)/V^2$

Page 415

1. (i) $-x^2/y^2$; (ii) $-[\sin xy + y(x + y)\cos xy]/[\sin xy + x(x + y)\cos xy]$; (iii) $(y\sin x - \cos y)/(\cos x - x\sin y)$;
(iv) $[6y(x + y)/x^2 - e^{(y/x)}]/[e^{(y/x)} + 6(x + y)/x]$

2. (i) $y = x\sqrt{-x/3} + (-x/3)^{3/2}$;
(ii) $2x^2/(1 + x^2 - y^2) + 2y^2/(1 - x^2 + y^2) = 1$

3. Circle is $x^2 + y^2 = 16$

7. $(3y^2z - 4x)/3(z^2 - xy^2)$, $2xyz/(z^2 - xy^2)$

8. (i) 1; (ii) $2t(1 + 2t^2 + t^3)/(1 + t)^3$

9. (i) $x/z, 2y/z$; (ii) $z/(1 - z), z/(1 - z)$; (iii) $2x + 8y, 8x$;
(iv) $-(x + z)/(y + x), -z/(y + x)$

Chapter 8

Page 422

1. 4 strips give Over-Estimate 284 and Under-Estimate 220; 8 strips give 271 and 239 respectively

2. For 4 strips we obtain 1502 and 1246 as the estimates; for 8 strips we obtain $1453\frac{7}{8}$ and $1325\frac{7}{8}$. The error is $512/n$ and this suggests 5120 strips for 1 d.p. accuracy and 51200 strips for 2 d.p. accuracy

Page 426

1. Value $-$ 2 predicted. Formula does not allow for discontinuity at $x = 2$

3. $0 \leqslant I \leqslant \pi/2$; $3/8 \leqslant I \leqslant \frac{1}{2}$; $3/10 \leqslant I \leqslant 3$

4. $\frac{1}{2}$; $1/\sqrt{3}$; $4/\sqrt[3]{4}$ or $\sqrt[3]{16}$

Page 446

1. (i) 5.83 (ii) 6
2. $S(0.72) = 0.186$. One step from 0.68 to 0.72 gives area 0.028 and therefore entry of $0.158 + 0.028 = 0.186$
4. True value $= \log 3 = 1.0986$ (4 d.p.) $I_S = 1.1$, therefore actual error is 0.0014; estimated error is 0.0167. Take 10 strips
5. $I_S = 64$ = analytical result; $I_S = (b^4 - a^4)/4$ = analytical result
6. (i) 0.938 (3 dp); (ii) 6.122, 1.791, – 2.080, –3.177, –1.421
7. 0.7854 analytically, $I_S = 0.7854$, error = 0.0000 (4 dp)
8. 0.2927; error bound is 4.2×10^{-4}
9. 0.6565 and 0.6689
10. (i) 1.3507 (ii) 1.3507
11. (i) 0.7980 (ii) 0.7981
12. 4.109 (3 dp)
13. 0.403 (3 dp) either method
14. (i) 0.540; (ii) 0.579; (iii) 0.618; Values are 0.540, 0.576, 0.616
15. (i) 0.1083 (ii) 0.386 and 4.047 (iii) 9.817
16. 1.418 a
17. 0.393(5) either way
18. 0.0801

Page 455

2. (a) $5/6\,(1 + 2x^2)^{3/2} + C$ (b) $4 - 2\sqrt{2}$ (c) 1/6
 (d) log 2 (e) $\sqrt{2} - \tfrac{1}{2}\sqrt{5}$ (f) $\sin^{-1}\sqrt{x} - \sqrt{x(1-x)} + C$

Page 457

1. (i) $-x\cos x + \sin x + C$ (ii) $1 - 2/e$ (iii) $2/3\, x^{3/2}\log x - 4/9\, x^{3/2} + C$,
 (iv) $x\log x - x + C$, (v) $1/5\, e^x(\sin 2x - 2\cos 2x) + C$
 (vi) $\pi/2 - 1$
 (vii) $x\tan^{-1}x + \log\cos(\tan^{-1}x) + C$ or $x\tan^{-1}x - \log\sqrt{1+x^2} + C$

Page 460

1. (i) $\tfrac{1}{2}\log 3/2$ (ii) $\log[(x-2)^2/(x-1)] + C$;
 (iii) $\tfrac{1}{2}\log[(2x-3)^5/(4x-1)] + C$;
 (iv) $4/13\log[(x-3)^2/(x^2+4)] - 11/26\tan^{-1}x/2 + C$
 (v) $\log\{(x+1)/(x-2)\} - 2/(x-2) + C$
 (vi) $\tfrac{1}{4}\log\{(x-1)/(x+1)\} - \tfrac{1}{2}\tan^{-1}x + C$
 (vii) $(-1/\sqrt{41})\coth^{-1}\{(x+4)/\sqrt{41}\} + C$ or $\dfrac{1}{2\sqrt{41}}\log\left(\dfrac{x+4-\sqrt{41}}{x+4+\sqrt{41}}\right) + C$

(viii) $(\sqrt{1/26})\tanh^{-1}\{(x+1)\sqrt{2/13}\}+C$ or $(1/2\sqrt{26})\log[(\sqrt{13}+\sqrt{2}x+\sqrt{2}/(\sqrt{13}-\sqrt{2}x-\sqrt{2})]+C$

(ix) $1+\log 5/2-\pi/2+2\tan^{-1}2$

(x) $(2/13)\log x-(1/13)\log(x^2-4x+13)+56/39\tan^{-1}\{1/3(x-2)\}+C$

Page 462

1. (i) $\tfrac{1}{2}x\sqrt{x^2-2}-\cosh^{-1}(x/\sqrt{2})+C$
   (ii) $\cosh^{-1}\{(2x-3)/\sqrt{5}\}+C$ (iii) $2\sqrt{x^2+5x-6}+C$
   (iv) $1/6\sqrt{3}\,[-\cosh^{-1}(3x+2)+(3x+2)\sqrt{9x^2+12x+3}+C$
   (v) $\tfrac{19}{8}\cosh^{-1}\{(2x-1)/\sqrt{5}\}+\tfrac{1}{4}(2x+7)\sqrt{x^2-x-1}+C$

Page 466

(i) $\tfrac{1}{4}\log\{(2+\tan x/2)/(2-\tan x/2)\}+C$
(ii) $-\log[1+\cot x/2]+C$; (iii) $(1/\sqrt{2})\tan^{-1}(\sqrt{2}\tan t)+C$;
(iv) $(1/5)\tan^{-1}[(1/5)\tan t]+C$;
(v) $(1/5)x+(14/15)\log[(2\tan(x/2)-1)/(\tan(x/2)-2)]+C$

Page 467

1. $-\tfrac{1}{4}$ 2. $1/a^2$ 3. (i) $\pi/2$ (ii) 2 (iii) 1/5

Page 468

1. (i) $1/\sqrt{2}$ (ii) $\pi/3+\sqrt{3}/2$
2. $\tfrac{1}{2}e^x[x\cos x+(x-1)\sin x]+C$ and $\tfrac{1}{2}e^x[x\sin x+(1-x)\cos x]+C$
3. $\tfrac{1}{2}x\sqrt{x^2-4}-2\cosh^{-1}(x/2)+C$ and $(1/5)(x^2-4)^{5/2}+(4/3)(x^2-4)^{3/2}+C$
4. (i) $\tfrac{1}{2}(x^2+1)\tan^{-1}x-\tfrac{1}{2}x+C$ (ii) 1/6
5. (i) (a) $\tfrac{1}{2}\log\tfrac{3}{4}+\pi/3\sqrt{3}$ (b) $4/5\,[e^{\pi}+\tfrac{1}{2}]$ (c) $\log\{(2+e)/3\}$
   (ii) $-2a/VRT-\log(V-b)+b/(V-b)+C$
6. $a^2\sigma/\epsilon_0 x$ 7. $2/3\,x^{3/2}\log x-4/9\,x^{3/2}+C$
8. (i) (a) $\pi/4$ (b) 8/105 (ii) $C=\tfrac{1}{2}e^{-x}[x(\sin x-\cos x)+\sin x]+C$
   $S=\tfrac{1}{2}e^{-x}[-x(\sin x+\cos x)-\cos x]+C$
9. $(x^2+x+1)^{1/2}-\tfrac{1}{2}\sinh^{-1}\{(1/\sqrt{3})(2x+1)\}+C$ and $\tfrac{1}{2}\log(x^2+x+1)-(1/\sqrt{3})\tan^{-1}\{(1/\sqrt{3})(2x+1)\}+C$
10. (i) $(1/3)\log(25/8)$; (ii) $\sinh^{-1}2$
11. (i) $e^{-x}(-x^2-2x-2)+C$; (ii) $\sin^{-1}x+\sqrt{1-x^2}+C$
12. (i) $\tfrac{1}{2}\log\tfrac{3}{4}+\pi/2\sqrt{3}$; (ii) $\tfrac{1}{2}\tan x\sec x+\tfrac{1}{2}\sinh^{-1}(\tan x)+C$
13. $3\pi/32-\tfrac{1}{4}$ 14. $k\,[(2/\sqrt{5})-(1/\sqrt{2})]/a^2$
15. (i) $(e^{-\pi}+1)/10$ (ii) $(1/\sqrt{2})\cosh^{-1}\{(4x+3)/\sqrt{17}\}+C$
    (iii) $1/3\cosh^3 x-\cosh x+C$

Page 477

1. $\pi a^2 (5^{3/2} - 1)/6$ **2.** $3a/8$ from base **3.** $2\pi rh$

4. (i) 128/3 and 4

5. (i) 256/15 (ii) $\bar{x} = [1 - (5/2) \log (9/7)] / \log (9/7)$, $\bar{y} = 0$

6. (i) $\bar{x} = 0, \bar{y} = ½ + 3\sqrt{3}/8\pi$ (ii) $8a^2/15$

7. $8\pi (3\sqrt{3} - 1)/3$ **9.** $8a^2/15\sqrt{3}, 4a/\sqrt{3}$

10. 0.35

11. (i) $256a^2/15$; (ii) $19a/7$ **12.** 16/3 **13.** $3\pi a^2/2$

14. $\log (2 + \sqrt{3})$ **15.** $\pi^2/2, 2\pi^2$ **19.** $4/\sqrt{3}, 8/15\sqrt{3}$

20. (i) $\log (2 + \sqrt{3})$ **21.** $3\pi\sqrt{E_1{}^2 + E_3{}^2}/2\sqrt{2}\,(3E_1 + E_3)$

22. $1/20, 1/28, \pi/6$ **23.** $(1/8)(e^{\pi} - 1)$

25. $\bar{x} = 5/6, \bar{y} = 109/30$ **26.** $\sqrt{2}$ **27.** 113

28. $11\pi a^2$; centroid at polar position $(20a/11, \pi)$

29. $c\,[1 + \sinh 1 - \cosh 1]/\sinh 1$ **30.** 2/3

31. $4a^2/15$ (i) $\pi a^3/12$ (ii) $32\pi a^3/105$; centroid is $(4a/7, 5a/32)$

Page 484

1. 8/105 **3.** $(\pi^4/16) - 3\pi^2 + 24$

4. (ii) (a) $\sinh^{-1} (x + 1) + C$; (b) $\sqrt{x^2 + 2x + 2} \quad \sinh^{-1} (x + 1) + C$;
$½ (x - 3) \sqrt{x^2 + x + 2} + ½ \sinh^{-1} (x + 1) + C$

5. (i) $5\pi/32$; (ii) $\pi/32$ **6.** $5/6\sqrt{2}$ **7.** $33\tfrac{3}{5} - 40\sqrt{5}/3$

8. $1/5 \tan^5 \theta - 1/3 \tan^3 \theta + \tan \theta - \theta + C$

9. $n\pi/2$

Chapter 9

Page 496

1. (a) 1,1; (b) 2,1; (c) 2,1; (d) 2,1; (e) (3,1); (f) 1,3

Page 498

1. (i) $d^2y/dx^2 = 0$; (ii) $2x\,dy/dx = y$; (iii) $d^2y/dx^2 = -4y$
(iv) $d^2y/dx^2 + 4\,dy/dx + 4y = 0$; (v) $d^2x/dt^2 + 2\,(dx/dt)^2 = 0$

2. $y = x^2 + 1$, no **4.** $y = 9/4 - (3/2)x$

Page 507

1. (i) $y = \frac{1}{2}\log(1+x^2)$; (ii) $(1+y^{3/2})^{2/3} = A\sqrt{1+x^2}/x$;
(iii) $\cos y = Ae^{-x\tan x}$; (iv) $y = Ae^{\frac{1}{2}(\log x)^2}$;
(v) $(y^2-1)/(y^2+1) = Ae^{2x^2}$; (vi) $y/(1-y) = x/2$ or $y = x/(2+x)$;
(vii) $\tan y = x + x^2 + A$; (viii) $y^2 = x^2 + 2\log x + A$;
(ix) $1 - y^2 = A(1-x^2)/x^2$

2. (i) $y = \log\sec x$; (ii) $y = -\log(9/4 - \sqrt{1+x^2})$;
(iii) $\theta = \sqrt{t^2-1}/\sqrt{t^2+1}$

3. $n = Ne^{-\lambda t}$; $(1/\lambda)\log(4/3)$, $(1/\lambda)\log 2$

4. $p = p_0 + r_1{}^2(p_1 - p_0)(r^2 - r_0{}^2)/[r^2(r_1{}^2 - r_0{}^2)]$

5. $t = (\pi/15k)(20a - 6h_0)h_0{}^{3/2}$

6. $(4/3)h^{3/2}R - (2/5)h^{5/2} = (14R^{5/2}/15) - a^2\sqrt{2g}\ t$

Page 516

1. (i) $2\frac{1}{8}$ ($2\frac{1}{3}$ analytically); (ii) $6\frac{1}{2}$ (4.58 analytically);
(iii) 0.6554 (0.7358 analytically); (iv) 1.0 (2.1353 analytically)
(v) 1.9578 (1.744 analytically); (vi) 0.2654 (0.37 analytically)

2. 1.285×10^4 ($g = 10, R = 6 \times 10^6$)

Page 520

1. No effect 2. Improves the estimates

Page 524

1. (i) $2\frac{3}{8}$; (ii) $9\frac{7}{8}$; (iii) 0.7414; (iv) 3/2; (v) 1.7503; (vi) 0.3688

2. $y(0.1) = 1.1005$(1.1003 analytically), $y(0.2) = 1.203$ (1.203 analytically)

3. $y(0.1) = 2.400$, $y(0.2) = 2.803$

Page 529

1. (i) E; $x^3y - x^4/4 + y^2/2 = C$;
(ii) E, $x^2/2 + 2xy + y^2 = C$;
(iii) E, $r^2/2 + r(\sin\theta + \cos\theta) = C$;
(iv) Not exact

2. $x/y - x^2/y^3 = C$

3. $x^2y^2/2 + x^4/4 + x^3/3 = C$

4. $x^4/2 + 3xy + y^2/2 - y = C$

5. $(y + 1 + \log x)/x = C$

6. $x^3y - xy^3 = C$

Page 538

1. (i) S.V., I.F. = $e^{-\sin x}$; (ii) S.V.; (iii) I.F. = $e^{-(½)e^{2t}}$; (iv) S.V.; (v) Neither; (vi) S.V., I.F. = $e^{(e^{-2t})}$ or $e^{(½ e^{-2t})}$
2. (i) $y = ⅓ x^2 + C/x$; (ii) $y = (¼)(x+1) + C/(x+1)^3$; (iii) $y = (x+C)$ cosec x; (iv) $\theta = 0.5(1+t)[1 - e^{(2-2t)}]$; (v) $\theta = t(\sin t - \cos t)$
3. (i) $y = (x/2 + 1 - \pi/4)$ cosec $x - (½)\cos x$; (ii) $y = x + Cx(x^2 - 1)^{-½}$; (iii) $y = ⅓ e^x + Ce^{-2x}$; (iv) $y = ½ -$ cosec$^2 x$; (v) $y = (¼)(x^2 + 1/x^2)$; (vi) $z = e^{2t}/6 - Ce^{-t} + D$; (vii) $y = (3x-4)/4 + C(x+4)^{3}$; (viii) $y = e^x + 1 - x^2/2 - x$
5. $y = (x + C)e^{-x^2/2}$
7. $[n\omega HA/(R^2/L + \omega^2 L)]\,[(R/L)\cos \omega t + \omega \sin \omega t - (R/L)e^{-Rt/L}]$
10. $i = E_0\,[1/R - \{1/(R - L\alpha)\}e^{\alpha t}] + [I_0 + E_0 L\alpha/R(R - L\alpha)]\,e^{-Rt/L}$
11. 380 days 12. $c = c_0 e^{-Qt/V} + (Q_{c_f} + 10^4 Q_p)(1 - e^{-Qt/V})/Q$, $N = 14$

Page 546.

1. 0.9802, 0.9231, 0.7262 2. Analytical values, 2.9144, 2.8561, 2.8224
3. First term neglected is $80x^7/7! = x^7/63$
5. $x(0.10) = 1.078$, $x(0.15) = 1.104$

Page 553

2. 0.021499, 0.042999
3. (i) $y = -x - 1 + (½)(\cos x + \sin x) + (3/2)e^x$; (ii) $x^2 = y^2 - 3y$; (iii) $v^2 = ⅓ x^3 + ¾ x^2 - 14/3$
4. In one step $y(1) = 6.0417$ (Analytically 6.1549)
5. Difficulty in differentiation; 0.1628
6. $y(1) = 4$, $y(2) = 9$; With $h = 1$; 3.9444, 8.8784; With $h = 0.25$; 3.9994, 8.9984
8. $t = 2.521 \log(100 - 4c) - 3.624 \log(100 - 7c) + 4.934$

Chapter 10

Page 566

1. (i) $y = \alpha e^{-6t} + \beta e^t$; (ii) $y = (\alpha t + \beta)e^{8t}$; (iii) $y = e^{-3t}(\alpha \cos t + \beta \sin t)$; (iv) $y = (\alpha t + \beta)e^{-t/2}$;

(v) $y = \alpha e^{2t} + e^{-\frac{1}{2}t}\,[\beta \cos(\sqrt{3}\,t/2) + \gamma \sin(\sqrt{3}t/2)]$;
(vi) $y = e^{-t/2}\,[\alpha \cos(3\sqrt{3}\,t/2) + \beta \sin(3\sqrt{3}t/2)]$;
(vii) $y = \alpha e^{-13t/10} + \beta e^{t/2}$

2. $y = H \cosh(wx/H)/w$

4. $y = \alpha e^{-t} + e^{-3t/2}\,[\beta e^{\sqrt{5}\,t/2} + \gamma e^{-\sqrt{5}\,t/2}]$

5. (i) moment of inertia, elastic constant, damping constant; (ii) L, $1/C$, R

6. (i) $x = x_0 \cos kt$; (ii) $x = (v_0 \sin kt)/k$, k constant

7. $y = Ae^{2x} + Be^{-2x} + C\cos 2x + D \sin 2x$

Page 574

1. (i) $y = e^{-3x}$; (ii) $y = x^2 + 7x + 22$; (iii) $y = (3\cos 2x - \sin 2x)/20$;
(iv) $y = e^{-2x} + 3x^2/2 - 3x$; (v) $y = x^2 e^{-2x}$;
(vi) $y = -(x \cos 4x)/4$

2. (i) $y = -e^{-2x}\cos x + \cos x + \sin x$; (ii) $y = (2e^{-x} - 2e^{2x} + 6xe^{2x})/9$
(iii) $x = (13 \sin 2t - 4\cos 2t + 4e^{t})/10$;
(iv) $x = A \sin(t/3) + \beta \cos(t/3) - t\cos(t/3)/6$;
(v) $x = e^{-2t}[Ae^{\sqrt{3}t} + Be^{-\sqrt{3}t}] + t^3 - 12t^2 + 90t - 338$;
(vi) $y = [57e^{-4x} + e^{-x}(\sin x - 7\cos x)]/50$

3. (a) $y = (Ax + B + x^2)e^{-x}$; (b) $y = (x/8 + B)e^{4x} + A$

4. $y = (w/k)[1 - e^{-\beta a}(\cos\beta a \cosh\beta x \cos\beta x + \sin\beta a \sinh\beta x \sin\beta x)]$, $(0 \leqslant x \leqslant a)$
$y = (w/k)\,e^{-\beta x}(\sinh\beta a \cos\beta a \cos\beta x + \cosh\beta a \sin\beta a \sin\beta x)$, $(x \geqslant a)$

5. $y = (5e^{x} - e^{-5x} - 4e^{-2x})/18$

6. (a) $s = Ae^{-t} + Be^{-4t} + (3\cos t + 5 \sin t)/34$;
(b) $s = (A + Bt)e^{-2t} + (3\cos t + 4\sin t)/25$;
(c) $s = e^{-t}(A\cos\sqrt{3}t + B\sin\sqrt{3}t) + (3\cos t + 2\sin t)/13$

7. (i) $y = Ae^{-4x} + Be^{-x} - e^{-2x}/2 + (3\sin x - 5\cos x)/34$;
(ii) $y = e^{3x}(A\cos x + B\sin x) + x/10 + 4/25$

8. $I = 1/10 - e^{-2t}(25\cos 50t + \sin 50t)/250$, 0

9. $s = \lambda(\sin 2t + 2\sin t)/6$

Page 582

1. Selected values, $\theta(0.5) = -0.0962$, $\theta(1.0) = 0.0024$, $\theta(1.5) = 0.0342$, $\theta(2.0) = -0.0514$, $\theta(2.5) = 0.0585$

2. Selected values, $\phi(0.5) = 4.9988$, $\phi(1.0) = 5.1663$, $\phi(2.0) = 5.1706$, $\phi(3.0) = 5.1706$

3. Selected values, $x(1) = 0.3203$, $x(2) = 0.0543$, $x(3) = -0.0765$
$x(4) = -0.0802$; $\dot{x}(1) = -0.2850$, $\dot{x}(2) = -0.2106$, $\dot{x}(3) = -0.0557$, $\dot{x}(4) = 0.0338$; $\ddot{x}(1) = -0.0646$, $\ddot{x}(2) = 0.1556$, $\ddot{x}(3) = 0.1319$, $\ddot{x}(4) = 0.0466$

4. Selected values, $x(1) = 2.9986$, $x(2) = 2.9975$, $x(3) = 2.9964$, $x(4) = 2.9953$, $x(5) = 2.9940$

Page 588

1. (i) $a/[(s+b)^2 - a^2]$; (ii) $(s+b)/[(s+b)^2 - a^2]$;
   (iii) $(a \cos b - s \sin b)/(s^2 + a^2)$;
   (iv) $1/[s^2(s^2 + a^2)]$; (v) $s/[(s^2 + a^2)(s^2 + b^2)]$;
   (vi) $(s^2 - a^2)/(s^2 + a^2)^2$; (vii) $1/(s+2)^2$;
   (viii) $2/(s+2)^3$

Page 591

1. (i) $1/(s^2 + a^2)^2$; (ii) $1/(s+a)^n$; (iii) $(s^2 - k^2)/(s^2 + k^2)^2$;
   (iv) $2ks/(s^2 - k^2)^2$; (v) $(s+2)/[(s+2)^2 + k^2]$;
   (vi) $2(s+1)/[(s+1)^2 + 1]$

Page 597

1. (i) $i = 40 \cos t + 20 \sin t - 40e^{-2t}$, (ii) $r = (9e^{-t} - 5e^{-3t})/2$;
   (iii) $x = 2e^{-3t} + 10\,t\,e^{-3t}$; (iv) $x = \cos 3t + (17 \sin 3t + 3t)/27$;
   (v) $x = x_0 \cos nt + (x_1 \sin nt)/n + (t \sin nt)/2n$;
   (vi) $x = \cos 2t + (3 \sin 2t)/2 + (5t \sin 2t)/4$
2. (i) $y = (5e^{3t} + 44e^{-t/2} + 7te^{3t})/49$; (ii) $y = [1 - e^{-2t}(\cos 2t + \sin 2t)]/8$
   (iii) $\theta = [2(2+5t)e^{-t} - 3 \sin 2t - 4 \cos 2t]/25$.
   (iv) $x = (t \sin 3t + 12 \cos 3t - 10 \sin 3t)/6$.
   (v) $y = (t^3 + 30)e^{-t}/3$; (vi) $u = (e^{-t} \sin^2 t)/2$;
   (vii) $y \equiv 0$

Page 601

1. $x = (e^{-3t} + e^{-t})/2$, $y = (e^{-3t} - e^{-t})/2$
2. $x = (1 + 3t - e^{-3t} - 6te^{-3t})/27$, $y = (4 - 6t - 4e^{-3t} - 6te^{-3t})/27$
3. $x = (3 + 10 \cos t + 5 \cos 2t)/6$, $y = (10 \sin t - 5 \sin 2t)/6$
5. $x_1 = [(x_1{}^0 - x_2{}^0) \cos \sqrt{k}t + (x_1{}^0 + x_2{}^0) \cos \sqrt{3k}\,t]/2$,
   $x_2 = [-(x_1{}^0 - x_2{}^0) \cos \sqrt{k}\,t + (x_1{}^0 + x_2{}^0) \cos \sqrt{3k}\,t]/2$
7. $x = 3e^t - e^{-t} - 2$, $y = e^t + e^{-t} - 1$
8. $x = 2e^{2t} + 4 \sin 2t + 2 \cos 2t$, $y = e^{2t} - \cos 2t$
9. $x = b(\cos \sqrt{3}\,\omega t + \cos \omega t)/2$
10. $x = -8e^{-4t} - 12e^t$
11. $N_4 = N_0 + N_0\,[\lambda_2 \lambda_3 (\lambda_2 - \lambda_3) e^{-\lambda_1 t} + \lambda_3 \lambda_1 (\lambda_3 - \lambda_1) e^{-\lambda_2 t} + \lambda_1 \lambda_2 (\lambda_1 - \lambda_2) e^{-\lambda_3 t}]/(\lambda_1 - \lambda_2)(\lambda_2 - \lambda_3)(\lambda_3 - \lambda_1)$

Page 604

1. $y = w\,(x^4 - 2lx^3 + l^3x)/24EI$
2. $\theta = 30\,e^{-4\sqrt{15}\,x/3} + 70$

Page 613

2. (i) $\cos x + \sin x$; (iii) $2e^t/5$; (v) $t^3 - 12t^2 + 90t - 338$
3. (i) $y = e^{x/2}\,[A\cos(\sqrt{35}x/2) + B\sin(\sqrt{35}\,x/2)] + 2e^{2x}/11$
   (ii) $y = Ae^{4x} + Be^{-4x} - (8x^2 + 16x + 9)/128$
   (iii) $y = e^{-x/2}\,[A\cos(\sqrt{51}\,x/2) + B\sin(\sqrt{51}\,x/2)] +$
   $(4\sin 2x + 18\cos 2x)/85$

Page 616

3. $y = e^{2x}\,(A\cos x + B\sin x) + xe^{2x}\sin x$
5. $y = (x - 2)\,(e^{-2x} - e^{-3x})$ 6. $y = (5\sin x - \sin 2x)/3$

Page 621

Analytical solution $y = (\log\cosh\sqrt{gk}\,t)/k$;
Runge – Kutta ($k = 1, g = 9.81$), $y\,(2.0) = 5.5710$, $y\,(4.0) = 11.8352$,
$y\,(6.0) = 18.0994$, $y\,(8.0) = 24.3636$, $y\,(10.0) = 30.6278$

Chapter 11

Page 630

1.

Group	35.0–35.4	35.5–35.9	36.0–36.4	36.5–36.9	37.0–37.4
Frequency	5	1	11	13	15

Group	37.5–37.9	38.0–38.4	38.5–38.9	39.0–39.4	39.5–39.9
Frequency	13	13	6	1	2

Group	35.0–35.9	36.0–36.9	37.0–37.9	38.0–38.9	39.0–39.9
Frequency	6	24	28	19	3

Information is lost in making the groups too wide.

2. (a) (i) 3.474, 2.516; (ii) 10 is a possible number;
   (iii)

Group	2.500–2.599	2.600–2.699	2.700–2.799	2.800–2.899
Frequency	1	4	10	10

Group	2.900–2.999	3.000–3.099	3.100–3.199	3.200–3.299
Frequency	10	12	14	11

Group	3.300–3.399	3.400–3.499
Frequency	5	3

(b) (i) 1.77, 1.21; (ii) 12 is a possible number;

(iii)

Group	1.20–1.24	1.25–1.29	1.30–1.34	1.35–1.39	1.40–1.44
Frequency	1	1	5	3	5

Group	1.45–1.49	1.50–1.54	1.55–1.59	1.60–1.64	1.65–1.69
Frequency	5	11	4	7	6

Group	1.70–1.74	1.75–1.79
Frequency	0	2

Page 635

1. (i) 50; (ii) $58\frac{5}{6}$ %

2.

Rejects	0	1	2	3	4	5
Frequency	5	9	11	7	3	1
Cumulative frequency	5	14	25	32	35	36

(i) 2; (ii) 2; (iii) 1.92

3. 37.3, 37.0–37.4, 37.3; 3.018, 3.100–3.199, 3.055; 1.51, 1.50–1.54, 1.51

4. 116.1

Group	61–70	71–80	81–90	91–100	101–110	111–120
Cumulative frequency	2	10	27	55	91	131

Group	121–130	131–140	141–150	151–160	161–170
Cumulative frequency	165	194	213	223	225

5. $8\frac{4}{9}$

Page 641

1. (b) 14.2%; (c) 0.785, 0.63–1.06

2. 10.64, 0.709, 0.892

3. 1000.2, 7.09, modal group = 997.5, median group = 997.5, mean deviation = 5.60, inter-quartile range = 995.8–1004.2

4. 32.9, 0.369, median = 32.91

5. $55\frac{2}{3}$, 1.18 6. 3.41, 1.98 7. 0.913, 0.150

8. Modal group = 12, median group = 12, mean = 12.24, s.d. = 2.74, grouping has condensed the information.

9.

Group	1.00–1.49	1.50–1.99	2.00–2.49	2.50–2.99	3.00–3.49
Frequency	11	3	12	1	6

Group	3.50–3.99	4.00–4.49
Frequency	1	1

mean = 2.17, s.d. = 0.84, median group = 2.00–2.49 = modal group, mean deviation = 0.66, inter-quartile range = 1.44–2.48

10. (2) 0.084; (3) 0.0071 (4) 0.011 (6) 0.580

11. Variance (grouped) = 0.0167, Variance (raw) = 0.01691, Sheppard's correction does not improve on the calculated variance in this case.

Chapter 12

Page 660

1. 1/216, 5/18

2. (i) (a) 0.6; (b) 0.9; (c) 0.7
(ii) (a) 33/95; (b) 1/190; (c) 81/95; (d) 62/95;
(e) 48/95; (f) 153/190; (g) 14/95
(iii) (a) 68/95; (b) 27/95

3. 25/69, 28/69, 16/69 **4.** (i) 1/169; (ii) 1/221

5. ½ **6.** (i) 5/9; (ii) (a) 2/105; (b) 4/10

7. 0.098 **8.** 7/40 **9.** (a) 1/32; (b) 7/40; (c) 7/10

10. (a) 0.813 (b) 0.013 **11.** 4/13 **12.** (a) 1/36; (b) 1/24; (c) 19/36

13. 347/2048 **14.** (i) (a) 125/1296; (b) $5^9/6^{10}$; (c) $5^{r-1}/6^r$;
(ii) ${}^{28}C_3/{}^{51}C_3$

15. (ii) 4/9

16. (i) 21/25; (ii) 1.68

17. (i) (a) 8/27; (b) 16/81; (ii) (a) 0.896; (b) 0.544

18. $\dfrac{{}^{r}C_n\ {}^{N-r}C_{s-n}}{{}^{N}C_s} \simeq {}^{s}C_n\left(\dfrac{r}{N}\right)^n\left(\dfrac{N-r}{N}\right)^{s-n}$; $\hat{N} = \dfrac{rs}{n}$; 250

Page 665

1. (i) Not a p.f. $\Sigma p_i \neq 1$
(ii) (a) 6/10; (b) 9/10; (c) 5/10;
(iii) Not defined at $x = 0$
(iv) C = 5/9; (a) 1/3; (b) 4/9; (c) 7/18;
(v) $p_0, p_1 < 0$; (vi) C = 60/77; (a) 32/77; (b) 47/77; (c) 35/77

3. 11,136; 91/323, 455/969, 70/323, 10/323, 1/969

4. Take $k = 20/29$

5. $(1-\theta)^3$, $3\theta(1-\theta)^4$, $3\theta^2(1-\theta)^3(3-2\theta)$, $\theta^3(16-33\theta+24\theta^2-6\theta^3)$

Page 668

1. £5/18 **2.** $p(x) = 1/6$ for each x; 7/2, 35/12

3. £2.30 **5.** 1.04

Page 675

1. 0.3 **2.** (i) 0.599 (ii) 0.787

3. $(3/5)^6 . (169/25) \simeq 0.315$ **4.** 0.396

5. (i) 0.031 (ii) 0.018 **6.** 0.651

8. 18.2, 84.8, 148.3, 115.2, 33.5

9. (i) 0.285; (ii) 0.608 **10.** (i) $(0.75)^{20}$; (ii) $5(0.75)^{19}$

11. (i) 1/125; (ii) 128/625

12. 243/1024, 405/1024, 270/1024, 90/1024, 15/1024, 1/1024; 5/4; 15/16

14. $p(3) = p(4) = {}^{15}C_3\ (\frac{1}{4})^3\ (\frac{3}{4})^{12}$

15. (a) 0.09874 (b) 0.9798 **16.** 0.02; 0.00376

Page 683

1. (i) 0.3235; (ii) (a) Poisson [0.94]; (b) $s^2 = 1.15, \sigma^2 = 0.94$

2. $e^{-0.75} = 0.472$ **3.** 0.185, stock 6

4. 0.185 **5.** 7 **6.** 0.036; 1 or 2 **7.** 0.337

9. 0.189, 0.315, 0.262, 0.145, 0.061; Yes

10. (i) 0.0902; (ii) 300

11. $e^{-0.1} \simeq 0.905$ **12.** 443

13. 0.8208; 0.07 **14.** (i) 0.3327; (ii) 0.0779

Chapter 13

Page 690

1. $\rho(x) \equiv 1/3$; (i) 0; (ii) 1/3; (iii) 1/3; (iv) 2/3; (v) 0; (vi) 0; (vii) 1/3

2. 1/6; 1/3; 7/8, 5/24

3. 1.582; 0.623, 0.165, 0.212 **4.** $x - \frac{1}{4}x^2$; 3/4, 1/16, 27/64; 5 and 6

5. 0.80, 0.64, 0.51, 0.41, 0.33, 0.26, 0.21, 0.17, 0.14, 0.11; $1/\lambda$

Page 699

1. 0.966, 0.281, 0.979, 0.004, 0.027, 0.491, 0.021, 0.119, 0.069, 0.296, 0.675, 0.253, −0.524, 1.414, −1.735, 0.988, 0.747, 2733, $\simeq$ 264

2. 0.032, 0.1416 **3.** 3.95 (4 years), 6.68%

4. $\simeq$ 62 **5.** Yes, qualitatively

6. 0.39% **7.** mean = 71,100, s.d. = 2040

8. (i) 683; (ii) 0.091; (iii) 87.9; (iv) 67.4, 92.6; (v) 105; (vi) 69.9, 80, 90.1

9. 4.27, 1.24

Time	1	2	3	4	5	6	7
Frequency	1	5	16	26	22	11	3

10. 0.43, 1%, 9%, 907 hours

11. 334, 2734, 513 **12.** 11% **13.** 7.8p

Page 704

1. Binomial 0.121, Normal 0.119

2. (i) 0.01; (ii) 0.04; (iii) 0.97; (iv) 0.54; (v) 0.13×10^{-8}; (vi) 0.21×10^{-4}

Page 711

1. (a) 4.5, 2.63; (b) 1.5, 2, 3.5, 4, 4.5, 2.5, 4, 4.5, 5, 4.5, 5, 5.5, 6.5, 7, 7.5; (c) 4.5; (d) 1.66

2. (a) 0.534; (b) 0.997 **3.** (a) 0.104; (b) 0.104

4. 0.37 ± 0.0074 **5.** (a) 2%, (b) 0.1

6. (a) 0.998, (b) 0.0003 **7.** 16%

8. 0.16 **9.** 10%, 659 lb

Page 715

1. 0.179 **2.** 5.15 – 6.25 **3.** (i) 0.221; (ii) 0.990

4. 0.824 ± 0.108 **5.** 2280, 2592

Page 721

2. $101.512 \pm 0.720, 507.56 \pm 1.61$, 10.6%, 102.43

3. (a) 0.16; (b) yes (0.5% level)

Page 724

1. 0.145 **2.** No **3.** Yes **4.** Yes **5.** 1.50 ± 0.88

6. 0.02 **7.** Significant at 0.5% level

8. Confidence interval [0.41, 0.55]; yes, significantly higher at 5% level

9. No, there is a significant difference at the 1% level

10. Cars travel at a more uniform speed

Index